Handbook of Experimental Pharmacology

Volume 87/II

Pharmacology of the Skin II

Methods
Absorption, Metabolism and Toxicity
Drugs and Diseases

Contributors

D.J. Atherton, D.N. Bateman, D. Brigden, J.L. Burton, S.K. Chandrasekaran, K.J. Collins, M. Corbett, W.J. Cunliffe, R.P.R. Dawber, R. Dover, P.M. Dowd, P.H. Dugard, P.M. Farr, R.J. Flower, J.R. Gibson, P. Goldsmith, M.W. Greaves, R.J. Hay, A. Herxheimer, K.T. Holland, B. Johnson, H. Kappus, A. Kobza Black, C.M. Lapière, C.M. Lawrence, T. Lotti, B. Lynn, F. Marks, J. Marks, R. Marks, J.R. Marsden, J.D. Middleton, N.E. Miller, B. Nusgens, D.A.A. Owen, E. Panconesi, J.A. Parrish, S.H. Peers, G.H. Rée, H. Schaefer, R.C. Scott, J.E. Shaw, B. Shroot, S. Shuster, N.B. Simpson, R. Stadler, R.C.D. Staughton, R.S. Stern, D.S. Thomson, T.M. Twose, M.I. White, N.A. Wright

Editors

Malcolm W. Greaves and Sam Shuster

Springer-Verlag Berlin Heidelberg New York
London Paris Tokyo Hong Kong

Professor MALCOLM W. GREAVES, M.D., Ph.D., F.R.C.P.

The Institute of Dermatology
St. Thomas's Hospital
Lambeth Palace Road
London SE1 7EH
Great Britain

Professor SAM SHUSTER, M.D., Ph.D., F.R.C.P.

The University of Newcastle upon Tyne
Department of Dermatology
The Royal Victoria Infirmary
Newcastle upon Tyne NE1 4LP
Great Britain

With 55 Figures

ISBN-13: 978-3-642-74056-5 e-ISBN-13: 978-3-642-74054-1
DOI: 10.1007/978-3-642-74054-1

Softcover reprint of the hardcover 1st edition 1989

Typesetting, printing and bookbinding: Brühlsche Universitätsdruckerei, Giessen
2122/3130-543210 – Printed on acid-free paper

List of Contributors

D. J. ATHERTON, Department of Dermatology, The Hospital for Sick Children, Great Ormond Street, London WC1N 3JH, Great Britain

D. N. BATEMAN, The Wolfson Unit, Northern Regional Clinical Pharmacology Unit, Claremont Place, Newcastle upon Tyne NE1 4LP, Great Britain

D. BRIGDEN, Division of Medical Sciences, The Wellcome Research Laboratories, Langley Court, Beckenham, Kent BR3 3BS, Great Britain

J. L. BURTON, Bristol and Weston Health Authority, Bristol Royal Infirmary, Bristol BS2 8HW, Great Britain

S. K. CHANDRASEKARAN, Pilkington, P.O. Box 39600, Phoenix, AR 85069, USA

K. J. COLLINS, Department of Geriatric Medicine, University College London and the Middlesex Hospital Medical School, St. Pancras Hospital, St. Pancras Way, London NW1 OPE, Great Britain

M. CORBETT, 26 High Street, Whittlesford, Cambridge CB2 4LT, Great Britain

W. J. CUNLIFFE, Department of Dermatology, Leeds Western Health Authority, The General Infirmary at Leeds, Great George Street, Leeds LS1 3EX, Great Britain

R. P. R. DAWBER, The Department of Dermatology, The Slade Hospital, Headington, Oxford OX3 7JH, Great Britain

R. DOVER, Histopathology Unit, Imperial Cancer Research Fund, Lincoln's Inn Fields, London WC2 3PX, Great Britain

P. M. DOWD, Department of Surgery, University College and Middlesex School of Medicine, Surgical Studies, The Middlesex Hospital, Mortimer Street, London WIN 8AA, Great Britain

P. H. DUGARD, Biochemical Toxicology Section, Imperial Chemical Industries PLC, Central Toxicology Laboratory, Mereside, Alderley Park, Macclesfield, Cheshire SK10 4TJ, Great Britain

P. M. FARR, The University of Newcastle upon Tyne, Department of Dermatology, The Royal Victoria Infirmary, Newcastle upon Tyne NE1 4LP, Great Britain

R. J. FLOWER, School of Pharmacy and Pharmacology, University of Bath, Claverton Down, Bath BA2 7AY, Great Britain

J. R. Gibson, Section of Clinical Dermatology, The Clinical Research Division, Wellcome Research Laboratories, Langley Court, Beckenham, Kent BR3 3BS, Great Britain

P. Goldsmith, Department of Dermatology, Westminster Children's Hospital, Vincent Square, London SW1, Great Britain

M.W. Greaves, The Institute of Dermatology, St. Thomas's Hospital, Lambeth Palace Road, London SE1 7EH, Great Britain

R.J. Hay, The Institute of Dermatology, United Medical and Dental Schools of Guy's and St. Thomas's Hospitals, 5 Lisle Street, London WC2H 7BJ, Great Britain

A. Herxheimer, Departments of Medicine and Pharmacology, 12th Floor, Laboratory Block, Charing Cross Hospital Medical School, Fulham Palace Road, Hammersmith, London W6 8RF, Great Britain

K.T. Holland, Department of Microbiology, The University of Leeds, Leeds LS2 9JT, Great Britain

B. Johnson, Department of Dermatology, University of Dundee, Level 8, Polyclinic Area, Ninewells Hospital, Dundee DD1 9SY, Great Britain

H. Kappus, Department of Dermatology, Free University of Berlin, Rudolf Virchow Clinic, Augustenburger Platz 1, D-1000 Berlin 65

A. Kobza Black, Professorial Unit, The Institute of Dermatology, United Medical and Dental Schools of Guy's and St. Thomas's Hospitals, Lambeth Palace Road, London SE1 7EH, Great Britain

C.M. Lapière, Clinique Dermatologique, Centre Hospitalier Universitaire de Liège, B-4000 Sart Tilman Par Liège 1

C.M. Lawrence, The University of Newcastle upon Tyne, Department of Dermatology, The Royal Victoria Infirmary, Newcastle upon Tyne NE1 4LP, Great Britain

T. Lotti, Istituto di Clinica Dermosifilopatica, Università degli Studi di Firenze, 37 via degli Alfani, I-50121 Firenze

B. Lynn, Department of Physiology, University College London, Gower Street, London WC1E 6BT, Great Britain

F. Marks, Institute of Biochemistry, German Cancer Research Center, Im Neuenheimer Feld 280, D-6900 Heidelberg 1

J. Marks, University Department of Dermatology, Royal Victoria Infirmary, University of Newcastle upon Tyne, Newcastle upon Tyne NE1 4LP, Great Britain

R. Marks, Department of Dermatology, Department of Medicine, Welsh National School of Medicine, University of Wales, Heath Park, Cardiff CF4 4XN, Great Britain

J. R. Marsden, Department of Dermatology, The General Hospital, Steelhouse Lane, Birmingham B4 6NH, Great Britain

J. D. MIDDLETON, Quest International, Ashford, Kent TN24 OLT, Great Britain

N. E. MILLER, Section on Endocrinology and Metabolism, The Bowman Gray School of Medicine, Wake Forrest University, 300 South Hawthorne Rd Winston-Salem, NC 27103, USA

B. NUSGENS, Clinique Dermatologique, Université de Liège, Hôpital de Bavière, Blvd. de la Constitution, 66, B-4020 Liège

D.A.A. OWEN, Pharmacology, RID Compound and Technology Acquisitions Group, Smith Kline & French Research Ltd., The Frythe, Welwyn, Herts. AL6 9AR, Great Britain

E. PANCONESI, Istituto di Clinica Dermosifilopatica, Università degli Studi di Firenze, 37 via degli Alfani, I-50121 Firenze

J.A. PARRISH, Department of Dermatology, Harvard Medical School, Massachusetts General Hospital, Boston, MA 02114, USA

S.H. PEERS, School of Pharmacy and Pharmacology, University of Bath, Claverton Down, Bath BA2 7AY, Great Britain

G.H. RÉE, Department of Health, Division of Specialized Health Services, Chest Clinic, 63–79 George Street, Brisbane 4000, Queensland, Australia

H. SCHAEFER, Department of Biochemistry, Centre International de Recherches Dermatologiques (CIRD), Sophia Antipolis, F-06565 Valbonne Cedex

R.C. SCOTT, Biochemical Toxicology Section, Imperial Chemical Industries PLC, Central Toxicology Laboratory, Alderley Park, Macclesfield, Cheshire SK10 4TG, Great Britain

J.E. SHAW, ALZA Corporation, 950 Page Mill Road, P.O. Box 10950, Palo Alto, CA 94303-0802, USA

B. SHROOT, Department of Biochemistry, Centre International de Recherches Dermatologiques (CIRD), Sophia Antipolis, F-06565 Valbonne Cedex

S. SHUSTER, Department of Dermatology, The Royal Victoria Infirmary, The University of Newcastle upon Tyne, Newcastle upon Tyne NE1 4LP, Great Britain

N.B. SIMPSON, Department of Dermatology, Glasgow Royal Infirmary, Wards 20 and 21, Glasgow G4 OSF, Great Britain

R. STADLER, Freie Universität Berlin, Universitätsklinikum Steglitz, Hautklinik und Poliklinik, Abteilung für Dermatologie, Hindenburgdamm 30, D-1000 Berlin 45

R.C.D. STAUGHTON, Department of Dermatology, Westminster Children's Hospital, Vincent Square, Westminster, London SW1, Great Britain

R.S. STERN, Department of Dermatology, Harvard Medical School, Massachusetts General Hospital, Boston, MA 02114, USA

D.S. THOMSON, Bioscience II Department, Imperial Chemical Industries Ltd., Pharmaceutical Division, Mereside, Alderley Park, Macclesfield, Cheshire SK10 4TG, Great Britain

T.M. Twose, Twose, Anthony & Associates, The Acreage, Goostrey, Cheshire CW4 8JY, Great Britain

M.I. White, Department of Dermatology, Ward 29, The Royal Infirmary, Foresterhill, Aberdeen AB9 2ZB, Great Britain

N.A. Wright, Department of Histopathology, Royal Postgraduate Medical School, Hammersmith Hospital, Ducane Road, London W12 OHS, Great Britain

Preface

The recent interest in the pharmacology of the skin and the treatment of its diseases has come about for two reasons. The first is a realisation that many aspects of pharmacology can be studied as easily in human skin as in animal models, where they may be more relevant to human physiology and disease. Examples of this are the action of various vasoactive agents and the isolation of mediators of inflammation after UV irradiation and antigen-induced dermatitis. The second reason is the fortuitous realisation that a pharmacological approach to the treatment of skin disease need not always await the full elucidation of aetiology and mechanism. For example, whilst the argument continued unresolved as to whether the pilo-sebaceous infection which constitutes acne was due to a blocked duct or to a simple increase in sebum production, 13-cis retinoic acid, was found quite by chance totally to ablate the disease; again, whilst cyclosporin, fresh from its triumphs in organ transplantation, has been found able to suppress the rash of psoriasis, it has resuscitated the debate on aetiology.

We are therefore entering a new era in which the pharmacology and clinical pharmacology of skin are being studied as a fascinating new way of exploring questions of human physiology and pharmacology as well as for the development and study of new drugs, use of which will improve disease control and at the same time help to define pathological mechanisms.

It was because of this burgeoning interest in pharmacology of skin and its diseases, that this book came about. Indeed it is long overdue and was planned several years ago; we console ourselves with the thought that the delay may have served to help define certain principles which were then only just emerging.

The book is divided into two volumes which are independent but complementary; the first being an account of the general pharmacology of skin, the second volume being more concerned with disease and drugs. The first volume is divided into two parts, the one dealing with the pharmacology of skin systems and their control and the other with autocoids in normal and inflamed skin. This second volume has three parts; the first part deals with the methods of measurement which are becoming of increasing importance both in studying the pharmacological effects of drugs and the clinical pharmacology of skin; the second deals with toxicology in its widest sense – including metabolism and percutaneous absorption; and the third is an account both of specific drugs and the drugs used for specific diseases.

The aim has been to give an up-to-date review in which there was sufficient background detail for an understanding of the subject without having to refer beyond the two volumes – but, of course, with sufficient referencing to serve as a guide to deeper reading. Authors were encouraged to present both consensus and personal views so that possibilities are presented as well as what appear today to be probabilities. In this way it is hoped that the two extremes of dogma and fantasy have been avoided. In presenting this account we were well aware of the problems of a rapidly advancing field; but we hope that the chapters are so written that the newer knowledge which is now almost continuously becoming available, can be more easily understood and incorporated into a body of knowledge. Finally but inevitably we have had to make certain omissions – for example in the field of antiviral drugs, AIDS and one or two other areas – if only for reasons of size.

The volumes are written for pharmacologists, clinical and non-clinical, and will be pertinent to pharmacists, and to many with an interest in the skin now working in the pharmaceutical industry, as well as to physiologists. The work too will be of major interest to dermatologists. Although the book is directed primarily at a postgraduate audience, there is much in it that will be of helpful to Honours and Ph. D. students. We hope that by including as much background as is relevant to an understanding within the two volumes, we will have reduced the likelihood of the reader having to look elsewhere for primary explanations. This plan may have led to some re-duplication and over-simplification, but we hope readers will agree that this was justified by the overriding objective. Finally we would like to thank our secretaries – Miss Angela Dell and Mrs. Madeline Young – for their inexplicably angelic assistance.

MALCOLM W. GREAVES
SAM SHUSTER

Contents

Section A: Methods

CHAPTER 1

Methods for the Study of Proliferative Rates in Epidermis

CHAPTER 2

Tissue and Fluids: Sampling Techniques

CHAPTER 3

Measurement of Sweating and Sweat Gland Function

CHAPTER 4

Measurement of Human Sebaceous Gland Function

CHAPTER 5

Methods for Assessing the Effect of Drugs on Hair and Nails

Section B: Absorption, Metabolism and Toxicity

CHAPTER 14

Drug Metabolism in the Skin

CHAPTER 15

Skin Cancer (Excluding Melanomas)

CHAPTER 16

Toxicology of Cosmetics

CHAPTER 17

Drug Sensitisation

Section C: Drugs and Diseases

CHAPTER 20

Glucocorticoids and Lipocortin

CHAPTER 21

Cutaneous Vasodilators

CHAPTER 22

Fibrinolysis and Fibrinolytic Drugs

CHAPTER 26

Hypolipidaemic Agents in the Treatment of Xanthomata

CHAPTER 27

Drugs Acting on Dermal Connective Tissue

CHAPTER 29

Bacterial Infections

Contents of Companion Volume 87, Part I

Section A: Pharmacology of Skin Systems

Section A: Methods

CHAPTER 1

Methods for the Study of Proliferative Rates in Epidermis

R. Dover and N. A. Wright

A. Introduction

Disturbances of growth are very common in the human epidermis, and there is mounting evidence that epidermal growth processes can be modified by the action of certain pharmacological agents; consequently, in many instances, investigators may find themselves in the situation where a measurement of proliferative rate in epidermis, possibly of the human, is mandatory. The question then arises, which measurement to make, and how to interpret the results? It is the purpose of this chapter to offer advice on both of these facets.

It is rather easy to become discouraged at the outset; attempts to understand cell kinetic methodology have been known to drive strong men to madness (see for example Marks 1985; Marks and Dykes 1983); however, a little consideration before embarking on an investigation is well worth while.

We should first consider what we want to measure. The commonest question is "can we say that cells in this tissue are being produced faster than in that", or "does this drug modify the rate of production of new epidermal cells?" Such questions demand a viable technique which gives reproducible and unequivocal results which have discriminatory power in statistical tests and it should be appreciated that, despite valiant attempts, such an ideal measurement has not yet emerged. However, there is something that can be done.

There has been some confusion in the terminology of cell proliferation as applied to epidermis: the term "proliferative rate" is a generic one which usually encompasses all factors which relate to the speed of production of new cells. Perhaps the most important parameter is the *birth rate* or *cell production rate*, which is usually expressed in cells per cell per hour, or more conveniently, in cells per 1000 cells per hour, and, in any experimental investigation, if we could measure the cell production rate complete with its interval estimate, in both control and experimental situations, then we would have gone a long way towards solving our problem. However, very few of the measurements which are available relate even tenuously to the cell production rate. At a fundamental level, the two parameters which jointly determine the cell production rate are the *cell cycle time*, or interval between successive cell divisions, and the *growth fraction*: unfortunately, measurement of both of these parameters in both human and animal epidermis is a difficult, expensive and frequently unrewarding exercise, and employs techniques which are really unsuitable for the question we are asking, i.e. how do we detect differences in proliferative rates?

B. Impractical Methods

For the purposes of the present discussion, there are certain methods which we should eschew at the outset, even though these are ingrained in the dermatological literature.

I. The Fraction Labelled Mitoses Method

The fraction labelled mitoses method (cf. Weinstein and Frost 1968; Goodwin et al. 1974; Duffill et al. 1976) can give information about the cell cycle time in certain circumstances. However, this procedure has a time scale which makes it inappropriate to monitor short-term changes in proliferative rate, is difficult to apply to the human situation and generates data in which the conficence limits are difficult to calculate, and even where available are of doubtful value in statistical tests; the method is therefore inappropriate for our purposes.

II. The Continuous Labelling Method

In the continuous labelling method (Gelfant et al. 1983) tritiated thymidine ($[^3H]TdR$) is made available to the cell population over a period of time usually greater than one cell cycle time; this method is considered to give information about the growth fraction, although this is debatable (Wright 1983). However, the technique suffers from the drawbacks already mentioned – time dependence and difficulties in analysis of the data.

C. Methods Suitable for Short-term Study

This leaves us with a smaller range of methods, but at the very best, we must be sure that the methods we select have the attributes which we need. The techniques which we should look at are discussed in the following sections.

I. Incorporation of Tritiated Thymidine into DNA

This is a simple method, beloved of experimental dermatologists, which has been the basic method for the investigation of many pharmacological phenomena in epidermis. For example, the greater number of investigations into the role of cyclic nucleotides in epidermal proliferative homeostasis have used this technique (cf. Vorhees 1976). Unfortunately, there is little doubt that this method has little to recommend it. Insufficient attention has been paid to the numerous potential defects inherent in the technique, which have been repeatedly pointed out (Maurer and Laerum 1976; Maurer 1981; Al-Mukhtar et al. 1982), but have been consistently ignored by many workers; it is not the purpose of this chapter to reiterate these reservations, except to advise that the method be abandoned by all those with even pretensions towards critical work; the only positive factor in its favour is ease of performance, and it is probably this that has led to its perpetuation and popularity. The more rapidly this method is consigned to the minor role of an extremely crude screening test, the better. We would doubt the significance of any investigation where the assay of proliferation rests solely on this method.

II. Proliferative Indices

These measurements relate to the fraction or percentage of cells in any particular phase of the cell cycle, such as the G_1, S or mitotic phase. Hitherto only two of these phases were exploitable: the fraction in DNA synthesis or the S phase, and the fraction of cells in mitosis. Cells in DNA synthesis are usually recognised by their ability to incorporate [^{3}H]TdR and to show up on subsequent autoradiography; cells in mitosis are readily recognised by their morphological characteristics. These methods have singular advantages: the main one is that there is little doubt that the cells in question, once recognised, are proliferative, and that their number, if correctly compared with the whole epidermal population, or part of it, usually gives a satisfactory assessment of the proliferative rate. Of course, there are technical problems, especially with the use of [^{3}H]TdR and the recognition of labelled cells in autoradiographs. The exposure of the appropriate cells to [^{3}H]TdR can be a problem in human work: there are clinical constraints upon the amount of [^{3}H]TdR that can be injected intradermally, and although the tissue is usually sampled after only 1 h, there remain problems in obtaining uniform labelling in the basal cells, even in such a small sample as a 4-mm punch biopsy.

There are also the criteria for regarding a cell as labelled; the *grain count* over labelled cells is a measurement which is usually conveniently ignored in human epidermal work, even though it is central, not only to the recognition of a labelled cell, but also to the confirmation of differences in the proportion of labelled cells in two populations. The factors which control the availability of [^{3}H]TdR, viz. endogenous and exogenous TdR pool sizes, TdR kinase activity etc. (Maurer 1981) also affect the grain count; therefore it is essential:

1. To define the autoradiographic parameters as definitively as possible, and to correct stringently for any background labelling. The grain count threshold for regarding a cell as labelled is most important in comparative work, and an arbitrary choice (the most usual method) is to be deplored. There are simple ways of correcting for background (Aherne et al. 1977; Wright and Alison 1984), and one of these should be used;

2. When comparing the numbers of labelled cells in say, control and experimental groups, it is essential to ensure that [^{3}H]TdR availability problems have not modified the count. If, for example, a particular drug selectively inhibited TdR kinase activity, the incorporation of [^{3}H]TdR via the salvage pathway would be reduced, even though DNA synthesis via the de novo pathway and proliferative rate will be unchanged; in this instance the interphase grain count in the treated epidermis will fall, taking some fraction of cells below the threshold value for recognition. The number of labelled cells will then fall, even though the proliferative rate is unchanged. It is therefore essential that interphase grain counts be quoted when comparing sets of control and experimental labelling data. There is, of course, less difficulty in recognising mitoses, although prophase and late telophase may be a problem.

Having recognised the target cell, we must relate its numbers to the total population in some way. This is a considerable problem in epidermis. In *thin* animal epidermis, such as that found in the mouse ear, there is no difficulty, since proliferative cells are confined to the basal layer, and it is a simple matter to constrain the count to that layer, and express the *labelling* or *mitotic index* with the basal cell population as the denominator, i.e. labelled cells as a percentage of basal cells. However, in human epidermis, especially in psoriasis, in thick animal epidermis such as the mouse foot or tail, or in thin epidermis which has undergone hyperplastic transformation, proliferative cells can be found well above the basal layer, and are not confined to any anatomical compartment. The alternatives here are several:

1. One can confine the count to the basal layer, working on the assumption that this layer is representative of the rest of the proliferative epidermis, but the major defect of this manoeuvre is that it would miss an increase in proliferative rate mediated by an expansion of the proliferative compartment, as occurs for example in hyperplastic transformation of the mouse ear after treatment with retinoids;

2. One can include all viable or nucleated cells in the epidermis in the count; this is more reasonable, but does present problems in parakeratotic epidermis, and if calculation of any derived parameter from the labelling index is envisaged, means that non-proliferative cells are definitely included in the count;

3. One can express the number of labelled or mitotic cells per unit length of surface. This is unwise unless potential changes in basal density are looked for, and also if hyperplastic transformation occurs, when there will be expansion of the epidermis in terms of rete peg elongation with lengthening of the basement membrane and increase in proliferative cell number, direct comparison with controls, *in terms of the proliferative index*, is impossible;

4. One can attempt to recognise so-called germinative cells in histological section. WEINSTEIN and FROST (1968) have laid down certain criteria for this, but these are likely to be subjective and to lack reproducibility, not only between investigators, but also when comparing a control epidermis with one which has undergone hyperplastic transformation.

In view of these reservations, it is difficult to know what to advise as a general rule for all situations; as far as possible we must ensure that, for comparative purposes, controls are counted in the same way as the experimental subjects, and that the method of counting will not affect the results. It is likely that the exact configuration of the experiment will determine the actual methodology.

The advent of flow cytometric techniques, and their application to the epidermis has introduced some additional proliferative indices (BAUER and de GROOD 1975; BAUER et al. 1980). In this method, an epidermal cell suspension is stained with a dye such as propidium iodide or ethidium bromide, which reacts stoichiometrically with DNA, and the cells are passed through the instrument, where the amount of dye per nucleus is measured by laser excitation and appropriate detection apparatus. The resulting histogram of the numbers of nuclei with a 2C, intermediate or 4C DNA content can be analysed to give the proportions of cells in G_0/G_1, S and G_2+M; the singular advantage of this automatic method of analysis over the laborious approach to the construction of DNA histograms through Feulgen staining and integrating microdensitometry of epidermal sections or squashes (GROVE 1979), is that many thousands of cells can be analysed in a very short time.

It therefore becomes possible to give the fraction of epidermal cells in G_0/G_1, S, and G_2+M, and of course the advantages of the method are several; (a) no radioactive tracers are necessary for the measurement of the S-phase index, which makes it particularly appropriate for human work; (b) only a single specimen is required; and (c) multiparameter analysis, e.g. staining with acridine orange and exciting at appropriate wavelengths can give a three-dimensional histogram incorporating DNA *and* RNA content against cell number, and this technique has been used to dissect the G_0/G_1 peak in human epidermis and to identify non-proliferative cells in normal and psoriatic epidermis (GELFANT et al. 1983), so that an appreciable amount of data is extractable from a single biopsy. The disadvantages of the procedure include: (a) problems in obtaining a clean single-cell suspension, although several groups have now overcome this (BAUER et al. 1981; FRENTZ et al. 1980); (b) the analysis of the data is usually computer based and can be quite complex; and (c) the cost of the equipment is not a negligible factor. There is the added problem, common to all proliferative indices, that such measurements are mere state parameters, sensitive not only to the rate of transit of cells through the phase, but also to the duration of the phase; this shows up particularly in epidermis treated with carcinogens such as 20-methylcholanthrene (EVENSEN 1962) where the early rise in the mitotic index is due to a prolongation in the mitotic duration rather than an increased transit rate into mitosis. Consequently, especially in pharmacological experiments where the effect of active substances on epidermal proliferative rate is to be assessed, this artifact should be considered. However, it is difficult to correct for, unless a *rate parameter measurement*, such as the rate of entry into DNA synthesis or mitosis, is also envisaged.

The commonest error found in work involving DNA histograms produced by flow cytometry is the assumption that a population with a higher proportion of S-phase cells is cycling faster than a population with a lower S-phase value. Used alone, the DNA histogram produces only a static picture and can thus yield only little information. New techniques now allow *rate measurements* with the cytometer (see Sect. C. III).

III. The Measurement of Rate Parameters in Epidermis

In our search for an optimum measurement of the epidermal proliferative rate, these measurements would at first appear to be ideal, since they should reflect the rate of cell production, which is the parameter we really want to measure. Two such measurements are routinely available: *the rate of entry or flux into DNA synthesis*, and the *rate of entry or flux into mitosis*.

The rate of entry into the S phase is usually measured by an appropriate *double-labelling method*; the rational is to label a cohort of epidermal cells in S with a first application of labelled DNA precursor, and, after an appropriate time (usually 1 h, but see later in this section), a further application labels those cells which have entered S in the time elapsed since the first application; the difference between the two labelling indices gives the flux into S during the time between the two injections. The main problem therefore is to isolate those cells labelled by the first application from those labelled by the second: there are three main ways of doing this:

1. By far the best method is to use thymidine labelled with isotopes of differing energies. The usual procedure is to use [^{14}C]TdR and [^{3}H]TdR; ^{14}C has a higher energy then ^{3}H, and can be detected by its higher penetration power; the section is therefore coated with a thick layer of emulsion, or preferably with two layers of emulsion (Schultze et al. 1976). The lower energy ^{3}H particles are detected by silver grains in the lower emulsion, while the higher energy ^{14}C reaches the uppermost emulsion layer. This procedure allows the separation of ^{3}H- and ^{14}C-labelled cells, and also those labelled with both ^{3}H and ^{14}C, and both influxes and effluxes from the S phase can be calculated (Burholt et al. 1976);

2. It is important to exert careful control over the specific activities and doses of the isotopes (Schultze et al. 1976), and over the autoradiographic procedure; a second method of separating the labelled cohorts is to use two applications of [^{3}H]TdR of differing doses, or concentration, which should result in a highly labelled and a heavily labelled cohort, theoretically distinguishable by the grain count. It is important to ensure that the distributions of grain counts over the respective cohorts do not overlap;

3. A third procedure is to give two applications of the same isotope, separated in time, and merely to measure the two labelling indices to provide the difference. One problem here is that of course only a small difference in labelling index is being sought, which could lead to problems in interpretation (Ralfs et al. 1981). However, this difference can be augmented by increasing the interval between the applications: this can only be done if cells which pass through G_2 and M are prevented from dividing and artificially elevating the first labelling index. This can be achieved by giving a metaphase arrest agent, such as demecolcine or colchicine at the time of the first injection (Tvermyr 1972).

In experimental animals, the two isotopes can easily be applied by systemic injection; in human work, however, where the intradermal route is mandatory, the first application can be given intradermally, followed by the second application in vitro (Pullman et al. 1974), although some workers have used two injections both given intradermally (Ralfs et al. 1981).

This measurement, the flux into DNA synthesis, is valuable in that it should reflect the cell production rate, and moreover, coupled with a measurement of the flash labelling index, will also allow the calculation of the duration of DNA synthesis. However, to equate the flux into S with the cell production rate we need to assume that there is no effective cell loss between the beginning of the S phase and cell division itself, and this may not be so in normal and psoriatic human epidermis (Ralfs et al. 1981; Camplejohn 1983). Nevertheless, the flux into S is an important technique in the context of the measurement of proliferative rate in the epidermis. *The rate of entry into mitosis* should, from a theoretical point of view, represent our elusive ideal measurement, since for every cell which undergoes mitosis in non-neoplastic epidermis, there should be a net gain of one cell – our cell production rate. The technique depends upon the ability of certain compounds, colchicine, demecolcine and the vinca alkaloids vincristine and vinblastine to arrest cells in metaphase by disrupting or preventing the formation of the mitotic spindle. Practically speaking, the agent is administered systemically (in animals) or into multiple intradermal sites in humans, and serial measurements are made of the mitotic or metaphase index with time after injection. The resulting linear rate of metaphase accumulation is usually found by a least-squares fit, giving the cell production rate with its confidence interval.

The general advantages and drawbacks of the method have been discussed in detail by WRIGHT and APPLETON (1980), and with reference to human epidermis by RALFS (1983), but important prerequisites include: (a) the use of optimal dosage; (b) the proof of linearity of metaphase collection; (c) the avoidance of a delay period before metaphase arrest becomes complete, and also of metaphase degeneration; and (d) the choice of an appropriate line-fitting procedure.

This method is valuable, and may be the method of choice in the epidermis of experimental animals, and has also been used to advantages in psoriatic epidermis (DUFFILL et al. 1976; RALFS et al. 1981), but in normal human epidermis, for some reason, there is no effective is rise in the metaphase index over the experimental period, which is usually 3–4 h (CAMPLEJOHN et al. 1981); why this should be is obscure. It could be because of the very slow proliferative rate in normal human epidermis and the fact that the consequent small increase over the experimental period is inadequate for the detection limits of the method. FISHER (1968) obtained measurable rates on normal human epidermis, with demecolcine cream, and extending the arrest period up to 12 h; this would obviously invite errors due to metaphase degeneration, and this limitation of the method should be remembered.

A recent development has made rate parameter measurements possible by flow cytometry. This involves using a marker to label a "window" of cells and following their progression relative to the other G_1, S and G_2/M phase cells. Initially [^{3}H]TdR was used (GRAY et al. 1977), but the technique was difficult to apply. Now bromodeoxyuridine (BrUdR) is used. BrUdR is a structural analogue of TdR and is taken up and incorporated by cells by the same mechanism. The advantage of BrUdR is that it can be detected with a monoclonal antibody. In practice cells are exposed to BrUdR and samples taken at various times. The cells are subjected to immunocytochemistry to give a green fluorescence (fluorescein isothiocyanate, FITC) for BrUdR and a fluorochrome (usually propidium iodide) is used for DNA. These can be plotted on a three-dimensional graph to show the position of the BrUdR-labelled cells relative to cells in the G_1, S and G_2/M phases. Immediately after labelling they would be in the S phase, but with time they move through S, into G_2/M and into G_1 (DOLBEARE et al. 1983; YANAGISAWA et al. 1985). From this one can estimate both cell cycle and phase transit times. Once this technique is established it is rapid, but because of the cost of the equipment and the technical difficulties involved it is not an easy option.

BrUdR can also be used to obtain a labelling index measurement on histological sections by immunocytochemistry, which can be completed in hours, rather than the days necessary for autoradiography. Unfortunately, all the drawbacks of [^{3}H]TdR also apply to BrUdR. BrUdR is mutagenic and in animal experiments has caused hair loss (R. DOVER 1987, unpublished work) or even death (W.J. HUME 1987, personal communication). As the same biological pathways are used by the cell to handle BrUdR and TdR, exogenous agents could affect both. Anomalous [^{3}H]TdR/BrUdR incorporation has been reported following drug treatment of cells (COLAMONICI et al. 1985). BrUdR could also be used in association with [^{3}H]TdR and autoradiography to estimate S-phase durations. This would avoid the technically difficult double-labelling method already discussed. The majority of studies using BrUdR have been carried out in vitro, but the methodology could be used for epidermis in vivo. Further information on monoclonal anti-BrUdR antibodies can be found in a special issue of *Cytometry* (vol. 6, no. 6, 1985) which was dedicated to this topic.

D. Conclusion

If the points we have made are accepted, there is no doubt that those who would measure the proliferative rate in epidermis, particularly in pharmacological experiments, must give a good deal of thought to what they measure. There is little point in performing high quality biochemistry and pharmacology if the method used to assay the proliferative rate is not acceptable to mainstream thought on cell proliferation, however convenient and rapid the measurement might be; much of the turmoil and disrepute the epidermal chalone field finds itself in at the present time stems directly from hopelessly inadequate methology where proliferative measurements are concerned (see Laurence et al. 1979 for a useful critique). In this chapter we have selected the methods which, if sufficient care is taken, might reasonably be expected to give reproducible and reliable results. Nevertheless, in employing these, one must remember, the provisos mentioned here, and also that different experimental situations may demand different experimental techniques.

References

Aherne WA, Camplejohn RS, Wright NA (1977) An introduction to cell population kinetics. Arnold, London

Al-Mukhtar MYT, Polak J, Bloom SR, Wright NA (1982) The search for appropriate measurements of proliferative and morphological status in studies of intestinal adaptation. In: Robinson JWL, Dowling RH, Riecken EO (eds) Mechanisms of intestinal adaptation. MTP Press, Lancaster, pp 3–28

Bauer FW, de Grood RM (1975) Impulse cytophotometry in psoriasis. Br J Dermatol 93:225–228

Bauer FW, Crombag NHMN, de Grood RM, de Jongh GJ (1980) Flow cytometry as a tool for the study of cell kinetics in epidermis. Br J Dermatol 102:629–639

Bauer FW, Crombag NHCMN, Boezemann JBM, de Grood RM (1981) Flow cytometry as a tool for the study of cell kinetics in skin. 2. Cell kinetic data for psoriasis. Br J Dermatol 104:271–276

Burholt DR, Schultze B, Maurer W (1976) Mode of growth of the jejunal crypt cells of the rat; an autoradiographic study using double labelling with ^{3}H and ^{14}C-thymidine in lower and upper parts of the crypts. Cell Tissue Kinet 9:107–117

Camplejohn RS (1983) Kinetic measurements in normal human epidermis in vivo. In: Wright NA, Camplejohn RS (eds) Psoriasis: cell proliferation. Livingstone, Edinburgh, pp 163–172

Camplejohn RS, Gelfant S, Chalker D (1981) An attempt to use vincristine and colcemid to measure proliferative rates in normal human epidermis in vivo. Br J Dermatol 104:243–248

Colamonici OR, Trepel JB, Neckers LM (1985) Phorbol ester enhances deoxynucleoside incorporation while inhibiting proliferation of K562 cells. Cytometry 6:591–596

Dolbeare F, Gratzner H, Pallavicini MG, Gray JW (1983) Flow cytometric measurement of total DNA content and incorporated bromodeoxyuridine. Proc Natl Acad Sci USA 80:5573–5577

Duffill M, Wright NA, Shuster S (1976) The cell proliferation kinetics of psoriasis examined by three in vivo techniques. Br J Dermatol 94:355–362
Evensen A (1962) Significance of the mitotic duration in evaluating the kinetics of cellular proliferation. Nature 195:718–719
Fisher LB (1968) Determination of the normal rate and duration of mitosis in human epidermis. Br J Dermatol 80:24–28
Frentz G, Moller U, Christensen I (1980) DNA flow cytometry on human epidermis. I. Methodological studies on normal skin. J Invest Dermatol 74:119–121
Gelfant S, Drewinko B, Chalker D, Eisinger M (1983) Cycling and non-cycling human germinative epidermal cells – the continuous ^{3}H-thymidine labelling method in vivo and flow cytometric studies. In: Wright NA, Camplejohn RS (eds) Psoriasis: cell proliferation. Livingstone, Edinburgh, pp 209–217
Goodwin PG, Hamilton S, Fry L (1974) The cell cycle time in psoriasis. Br J Dermatol 90:517–524
Gray JW, Carver JH, George YS, Mendelsohn MC (1977) Rapid cell cycle analysis by measurement of the radioactivity per cell in a narrow window in S phase (RCS). Cell Tissue Kinet 10:97–109
Grove GL (1979) Epidermal cell kinetics in psoriasis. Int J Dermatol 18:111–122
Laurence EB, Spargo DJ, Thornley AL (1979) Cell proliferation kinetics of epidermis and sebaceous glands in relation to chalone action. Cell Tissue Kinet 12:615–633
Marks R (1975) Is there a relationship between clinical morphology and epidermal cell kinetics? Proc R Soc Med 68:161–167
Marks R, Dykes PJ (1983) Steroids, squamous epithelium and psoriasis. In: Wright NA, Camplejohn RS (eds) Psoriasis: cell proliferation. Livingstone, Edinburgh, pp 327–335
Maurer HR (1981) Potential pitfalls of [^{3}H] thymidine techniques to measure cell proliferation. Cell Tissue Kinet 14:111–120
Maurer HR, Laerum OD (1976) Granulocyte chalone testing: a critical review. In: Houck JC (ed) Chalones. North-Holland, Amsterdam, pp 331–350
Pullman H, Lennartz KJ, Steigleder GK (1974) In vitro estimation of cell proliferation in normal and psoriatic epidermis with special regard to diurnal variations. Arch Dermatol Forsch 250:177–184
Ralf IG (1983) The metaphase arrest method in human epidermis. In: Wright NA, Camplejohn RS (eds) Psoriasis: cell proliferation. Livingstone, Edinburgh, pp 80–92
Ralfs IG, Dawber R, Ryan T, Duffill M, Wright NA (1981) The kinetics of metaphase arrest in psoriatic human epidermis; an examination of optimal experimental conditions for determining the birth rate. Br J Dermatol 104:231–242
Schultze B, Maurer W, Hagenbusch H (1976) A two emulsion atuoradiographic technique and the discrimination of the three different types of labelling after double labelling with ^{3}H and ^{14}C thymidine. Cell Tissue Kinet 9:245–255
Tvermyr EMF (1972) Circadian rhythms in hairless mouse epidermal DNA-synthesis as measured by double labelling with ^{3}H thymidine (H^{3}TdR). Virchows Arch [B] 11:43–54
Vorhees JJ (1976) Cyclic AMP and cyclic GMP in epidermal physiology and pathophysiology. Curr Probl Dermatol 6:107–153
Weinstein GD, Frost P (1968) Abnormal cell proliferation in psoriasis. J Invest Dermatol 50:257–262
Wright NA (1983) Statistical and operational problems in cell kinetic methodology in human epidermis. In: Wright NA, Camplejohn RS (eds) Psoriasis: cell proliferation. Livingstone, Edinburgh, pp 117–123
Wright NA, Alison MR (1984) The biology of epithelial cell populations. Clarendon, Oxford
Wright NA, Appleton DR (1980) The metaphase arrest method – a critical review. Cell Tissue Kinet 13:643–663
Yanagisawa M, Dolbeare F, Todoroki T, Gray JW (1985) Cell cycle analysis using numerical simulations of bivariate DNA/bromodeoxyuridine distributions. Cytometry 6:550–562

CHAPTER 2

Tissue and Fluids: Sampling Techniques

R. MARKS

A. Tissue Sampling

I. Surgical

Full-thickness samples of skin are usually obtained surgically with a scalpel. Trephines or "punches" are often used for small samples of skin. They are made in different sizes (1–8 mm diameter). Distortion from rotational movement can be minimised by using a high speed punch which also allows critical back reference of constituents or activity to surface area (BLACK et al. 1970). If the tissue to be removed is localised and raised it may be removed by curettage and pendulous tissue may also be removed by diathermy. If small pieces of skin consisting mostly of epidermis with little upper dermis are required, e.g. for study of epidermal proliferation, these can be obtained by raising the skin with a hypodermic needle, clamping with fine forceps and slicing the tissue off with a scalpel. A variety of specially designed knives are available for obtaining broad sheets of skin consisting of epidermis and a thin layer of dermis. The electrokeratotomes all rely on the reciprocating action of a sharp surgical blade driven by an electric motor. With the Castroviejo electrokeratotome sample thickness can be set by adjusting the blade. Other instruments include the Davis and the Davol keratotomes. The latter has a disposable head, but the thickness of the specimen to be removed cannot be varied. Of the many manual devices the most useful for small samples (e.g. 2 cm^2) is Silver's knife which uses a safety razor blade, thickness of sample being set by adjusting the distance between the blade and the skin contact bar. All these techniques require experience, e.g. the amount of manual stretching necessary for removal of a sheet of skin of even thickness. After removal of the skin the donor site should be subjected to firm pressure to stop bleeding before a non-adhesive dressing is applied. Although removal of skin to a depth of 0.3–0.4 mm should not result in scar formation a dull red mark is usually visible for weeks or months after the procedure and some changes in melanin pigmentation, especially in black or brown skin, may occur.

Treatment of the skin samples will depend on the investigation in hand. For light or electron microscopy the specimen should be placed in the appropriate fixative; for histochemical or biochemical studies the samples should be used directly or rapidly frozen for storage before use when appropriate; for microbial or cell culture it should be transferred aseptically to the appropriate culture medium.

II. Epidermal Samples

There are two methods. The first uses a suction blister (KISTALA 1968) made with a vacuum line and a suction cup or chamber containing one or several holes at the skin interface. A negative pressure of the order of 200 mmHg is usually applied for 1–2 h before blistering occurs. Blister formation varies with site, age, temperature and disease and formation can be seen through the plastic suction cup. When the blisters have formed their roofs are snipped off to obtain pure epidermal sheets and the sites are dressed. The problems with this method are that it is time-consuming, only small areas of epidermis can be sampled and there is some damage to the epidermis.

The second method of obtaining sheets of epidermis involves splitting off the epidermis from the underlying dermis in vitro (see Sect. A. III). Large samples of human skin are obtained from surgical operations (e.g. radical mastectomy or amputation). Epidermis from small mammals can be scraped off skin that has been pinned out.

III. Separation of Epidermis from Dermis In Vitro

None of the techniques available is completely satisfactory as they all damage the epidermis to a greater or lesser extent.

1. Physical Separation

Stretching is a satisfactory method, but requires special equipment. Epidermis may also be heat-separated by placing it on a surface heated to 50 °C or plunging it into saline at the same temperature.

2. Enzymic Techniques

Dermo-epidermal separation can be effected by incubation with proteolytic enzymes such as trypsin, Dispase, or collagenase. Trypsin is used more often, with concentrations of 0.025%–0.25%, depending on the source of the trypsin, the length of incubation, the temperature of the incubation and the degree of damage to the epidermis that is acceptable, e.g. 0.25% trypsin in Eagle's minimal essential medium (MEM) for 1 h at 37 °C, after which the epidermis is gently dissected from the underlying dermis with fine forceps. The site of cleavage is dependent on the temperature (SKERROW and SKERROW 1983). Dispase, a neutral proteinase obtained form *Bacillus polymyxa* does not disaggregate the epidermis as does trypsin after further incubation. Collagenase has been used to produce a dermo-epidermal split (HENTZER and KOBAYASI 1978); 0.1%–0.2% clostridial collagenase in Eagle's MEM is employed at a temperature of 37 °C and a 3-h incubation period.

3. Miscellaneous Agents

Acetic acid (0.5%) causes dermo-epidermal separation and is suitable for biochemical study of lipid metabolism (e.g. COOPER et al. 1976). Potassium bromide

(2 *M*), calcium chloride (1 *M*) and sodium iodide (2 *M*) are used where the viability of the epidermis is not important. Sodium thiocyanate produces a split at the dermo-epidermal junction through the basement membrane after incubation for 3–5 h in a 2 *M* solution (DIAZ et al. 1977).

IV. Stratum Corneum Sampling

Obtaining large samples of coherent stratum corneum poses major difficulties as it is tighlty bound down to the epidermis beneath. FERGUSON (1977) obtained samples from cadaver skin by means of a brass template stuck to the skin surface. Whole-thickness skin was removed with a scalpel and the stratum corneum obtained subsequently by heat and ammonia vapour was left on the template. Sheets of human stratum corneum can also be obtained from cadaver skin or keratotomed sheets of skin by using trypsin to remove dermis, followed by gentle scraping off the epidermis with a gauze swab (NICHOLLS 1981).

Staphylococcal epidermolytic toxin can also be used to remove stratum corneum from epidermal sheets (ELIAS et al. 1974), the split occurring in the granular layer. Cantharidin in acetone (0.2%) applied for 8–10 h in vivo will produce an intra-epidermal blister. The blister roof which is snipped off is virtually pure stratum corneum. Aqueous ammonium hydroxide (FROSCH and KLIGMAN 1977) and formic acid (LEHMANN and KLIGMAN 1983) have also been used. All these techniques suffer from the disadvantage that the stratum corneum obtained is chemically altered to a greater or lesser degree.

Partial thickness stratum corneum is more easily obtained; the simplest technique is scraping the skin surface with a scalpel, or shaving heel callus, although heel callus is not identical to stratum corneum elsewhere. A two- or three-cell layer of stratum corneum can also be obtained by one of the rapidly bonding cyanoacrylate adhesives (MARKS and DAWBER 1971). A drop of adhesive is placed on a microscope slide which is then pressed onto the skin area to be sampled. The adhesive rapidly polymerises (about 15 s) and the slide, with adhesive and stratum corneum, is removed with a rolling action. The skin surface biopsy can be examined by transmitted light or scanning microscopy (Dawber et al. 1972).

V. Corneocyte Sampling

Scraping the skin surface with a scalpel blade yields many clumps of corneocytes. A more certain method is to agitate the skin surface in the presence of 0.1% Triton X-100 in phosphate buffer at pH 7.2 which disperses the corneocytes, leaving few clumps. The horn cells can be packed by centrifugation and smears are easily made on glass slides for examination by light microscopy and measurement of corneocyte area (PLEWIG and MARPLE 1970). This principle has been used to quantify the rate of desquamation by employing an apparatus which delivers a standard mild rotational stimulus at the skin surface (NICHOLLS and MARKS 1977; ROBERTS and MARKS 1980), the corneocytes removed being counted in a haemacytometer electronically.

B. Sampling of Secretions

I. Sebum

It is difficult to obtain samples of pure sebaceous secretion and the material usually collected for studies is skin surface lipid. It differs somewhat in composition from pure sebum because it is modified by bacterial action and is contaminated by lipid from the stratum corneum. Skin surface lipid is collected for two main purposes: the estimation of sebum secretion rate and the analysis of lipid constituents, and the methods are discussed in Chap. 4.

II. Eccrine Sweat

Eccrine sweat can be collected on filter papers in airtight adherent chambers, sweating being induced thermally or pharmacologically (see Chap. 3).

III. Apocrine Sweat

Apocrine sweat secretion is much more difficult to collect. The axilla appears to be the site most easily studied by intradermal injections of adrenaline (1 in 10000) and collecting drops of a viscid material at the follicular orifices with capillary tubes (see Chap. 3).

IV. Tissue Fluid

Tissue fluid is sometimes required for investigation of the mediators released in the course of an inflammatory process in the skin or to determine the presence of specific metabolites or cellular constituents. All of the methods used suffer from the disadvantage of causing some tissue disturbance themselves. One of the earliest techniques used was skin perfusion (see GREAVES and SONDERGAARD 1970; GREAVES et al. 1971) with a buffered sterile physiological salt solution and a large bore needle, the perfusate being collected by a needle at the other side of the site. This technique is semi-quantitative and more recently suction blisters have been used for similar purposes (PLUMMER et al. 1978; MIDDLETON 1980). The disadvantage of this method is that it is not known to what extent the "tissue fluid" is contaminated by intracellular material leaking from damaged cells.

References

Black MM, Bottoms E, Shuster S (1970) Skin collagen content and thickness in systemic sclerosis. Br J Dermatol 83:552–555
Cooper MF, McGrath H, Shuster S (1976) Sebaceous lipogenesis in human skin: variability with age and with severity of acne. Br J Dermatol 94:165–172
Dawber RPR, Marks R, Swift JA (1972) Scanning electron microscopy of the stratum corneum. Br J Dermatol 86:272–281
Diaz LA, Heaphy MR, Calvanico NJ, Tomasi TB, Jordon RE (1977) Separation of epidermis from dermis with sodium thiocyanate. J Invest Dermatol 68:36–38
Elias PM, Mittermayer H, Tappeiner E, Fritsch P, Wolff K (1974) Staphylococcal toxic epidermal necrolysis (TEN): the expanded mouse model. J Invest Dermatol 63:467–475

Ferguson J (1977) A method to facilitate the isolation and handling of stratum corneum. Br J Dermatol 96:21–23

Frosch PJ, Kligman AM (1977) Rapid blister formation in human skin with ammonium hydroxide. Br J Dermatol 96:461–473

Greaves MW, Sondergaard J (1970) Urticaria pigmentosa and factitious urticaria: direct evidence for release of histamine and other smooth muscle contracting agents in dermographism. Arch Dermatol 101:418

Greaves MW, Sondergaard J, McDonald-Gibson W (1971) Recovery of prostaglandins in human cutaneous inflammation. Br Med J 2:258

Hentzer B, Kobayasi T (1978) Enzymatic liberation of viable cells of human skin. Acta Derm Venereol (Stockh) 58:197–202

Kistala U (1968) Suction blister device for separation of viable epidermis from dermis. J Invest Dermatol 50:128–137

Lehmann P, Kligman AM (1983) In vivo removal of the horny layer with formic acid. Br J Dermatol 109:313–320

Marks R, Dawber RPR (1971) Skin surface biopsy: an improved technique for the examination of the horny layer. Br J Dermatol 84:117–123

Middleton MC (1980) Evaluation of cellular injury in skin utilising enzyme activities in suction blister fluid. J Invest Dermatol 74:219–223

Nicholls R, Marks R (1977) Novel techniques for the estimation of intracorneal cohesion in vivo. Br J Dermatol 96:595–602

Nicholls S (1981) Development of physical methods for investigation of stratum corneum structure and function. PhD thesis, University of Wales

Plewig G, Marple RR (1970) Regional variations of cell size in the human stratum corneum. J Invest Dermatol 54:13–18

Plummer NA, Hensby CN, Warin AP, Camp RD, Greaves MW (1978) Prostaglandins E_2, $F_{2\alpha}$ and arachidonic acid levels in irradiated and un-irradiated skin of psoriatic patients receiving PUVA treatment. Clin Exp Dermatol 3:367

Roberts D, Marks R (1980) The determination of regional and age variations in the rate of desquamation: a comparison of four techniques. J Invest Dermatol 74:13–16

Skerrow D, Skerrow CJ (1983) Tonofilament differentiation in human epidermis, isolation and polypeptide chain composition of keratinocyte subpopulations. Exp Cell Res 143:27–35

CHAPTER 3

Measurement of Sweating and Sweat Gland Function

K. J. Collins

A. Introduction

Investigations of the sweat response are usually made for one or more of three purposes. First, for the evaporative component of human thermoregulation and water and electrolyte balance in hot ambient conditions; second, in diagnostic and clinical investigations (including galvanic skin response); third, for fundamental investigations of the physiology and pharmacology of the exocrine glands. The method selected depends largely on the requirements of the investigation. In fluid balance studies, measurement of total body sweat and cutaneous electrolyte loss calls for whole-body studies. Diagnostic tests of sweating usually involve the induction and measurement of sweating in localised areas. For investigations of secretory function at the cellular level, micro-techniques have been developed which can be applied to the study of the isolated gland.

B. Total Body Sweat Loss

Total sensible water loss is usually measured gravimetrically with a balance (e.g. Spido man-balance) capable of recording to ± 5 g. The method is fundamental, originating from Sanctorius in 1614 and is still exploited in both short- and long-term investigations of sweat loss (e.g. Weiner and Lourie 1981). In order to calculate cutaneous losses, corrections have to be applied for weight gain from food and fluid ingested and weight loss from respiratory water and CO_2, and excretory losses. Insensible transcutaneous water diffusion can be measured locally by an evaporimeter (see Chap. 1) in conditions when the sweat glands are inactive. Its magnitude depends on skin perfusion, temperature and the water vapour pressure of the environment (Shuster and Johnson 1969; Kerslake 1972). Continuous recording balances, operating on a strain gauge principle, are also valuable for the investigation of transient changes in sweat rate. A specialised method for collecting total body sweat (excluding that from the head) has been described for use with thermoregulatory function test equipment (Fox et al. 1967; Weiner and Lourie 1981). Sweat is removed by vacuum pump from a sealed PVC suit which completely encloses the body up to the neck. A disadvantage of this method is the imposition of a saturated micro-climate which results in sweat suppression (hidromeiosis) due to hydration occlusion of the sweat duct (Sarkany et al. 1965). A similar criticism applies to the regional collection of sweat in an impermeable arm-bag (Collins and Weiner 1962).

C. Local and Regional Sweat Responses

I. Qualitative Methods

Colorimetric methods are used to visualise local sweat responses either by imprints of sweat droplets or by skin surface reactions, e.g. the method of MINOR (1927) in which a solution consisting of 15 g iodine, 100 ml castor oil and 900 ml diluted alcohol is painted uniformly on the skin; when perfectly dry the skin is powdered with starch and the combined mixture reacts on the surface to make active sweat glands discernible. Other indicator combinations used are rhodamine, quinizarine and bromphenol blue. A starch-containing bonded paper onto which sublimated iodine is lightly deposited can be used to obtain a sweat gland imprint for investigating sweat gland density and the onset of sweating.

Changes in electrical resistance of the palmar skin as the result of sweating induced by mental activity has long been used for investigating arousal reactions in psychological tests. This psycho-galvanic skin response can be used as an analogue of sweat gland activity (see ROTHMAN 1954).

A semi-quantitative method, which enumerates active sweat glands and also indicates secretory output of individual glands, is the plastic impression technique (THOMSON and SUTARMAN 1953). A film of plastic solution containing 5% polyvinyl formal in ethylene dichloride with 1% dibutyl phthalate is first applied to the skin. Aqueous sweat droplets are immiscible with the drying plastic which can then be examined after the plastic film is removed on transparent tape (WEINER and LOURIE 1981). Similar results are obtained with silicone rubber impressions (SARKANY et al. 1965).

II. Quantitative Methods

The quantitative estimation of local sweat responses usually requires some form of air-tight capsule attached to the skin surface, the sweat being absorbed by weighed filter paper discs (COLLINS 1962). Ventilated capsule systems are more reliable if less convenient. A circulating stream of dry air or nitrogen passes through the sealed chamber and completely evaporates the sweat which is measured by a hygrometer or infrared water vapour detector. Sweat solute is measured in washings of the dry solutes remaining on the skin surface inside the capsule (FOSTER 1971).

A newer method for the measurement of water loss through the skin which achieves minimum alteration of temperature and humidity in the immediate micro-climate of the skin is based on the estimation of the vapour pressure gradient adjacent to the skin surface (NILSSON 1977). It offers high accuracy and improved sensitivity, especially at low rates of cutaneous water loss: the instrument's sensor head measures temperature and partial pressure of water vapour at two positional levels close to the skin and evaporation in g m^{-2} h^{-1} can be continuously displayed.

D. Induction of Sweating

In all these methods sweating is induced thermally, pharmacologically or psychologically. Stress-induced sweating is mostly confined to palms and soles and is induced by a variety of stimuli, e.g. mental arithmetic or other tasks requiring mental effort. Pharmacologically induced sweating uses intradermal injections or iontophoresis of cholinergic drugs to which dose-response curves can be constructed (FOSTER 1971). Thermally induced sweating is mostly used for whole-body studies.

I. The Isolated Sweat Gland

In vivo studies of the single human eccrine gland of the palm (KUNO 1956) have been described in which capillary pipettes have been inserted into the duct orifice; micro-puncture has also been used to study sweat formation in individual sweat glands (SCHULZ 1969). A remarkable develoment has been the modification of micro-techniques originally designed for studying the isolated kidney tubule for investigation of isolated eccrine and/or apocrine glands (SATO 1973, 1980).

II. Apocrine Sweating

Human apocrine glands are mostly found in the axillae and perineum. They have been studied by using a pipette in the hair follicle into which they drain (SHELLEY and HURLEY 1953) and more recently by micro-techniques (SATO 1980). Excretion is enhanced by adrenergic drugs which are thought to act on myoepithelial cells. For a review of animal studies see ROBERTSHAW (1983).

E. Localisation of Abnormalities Within the Sweat Gland

Dissociation of defects in sweat coil and duct has been attempted by dissociation of sweat rate and composition (SHUSTER and JOHNSON 1969), by a stop-flow method similar to that used in the kidney (SHUSTER et al. 1965) and by studies of histology and surface replicas.

References

Collins KJ (1962) Composition of palmar and forearm sweat. J Appl Physiol 17:99–102

Collins KJ, Weiner JS (1962) Observations on arm-bag suppression of sweating and its relationship to thermal sweat-gland "fatigue". J Physiol (Lond) 161:538–556

Foster KG (1971) Factors affecting the quantitative response of human eccrine sweat glands to intradermal injections of acetylcholine and methacholine. J Physiol (Lond) 213:277–290

Fox RH, Crockford GW, Hampton IFG, MacGibbon R (1967) A thermoregulatory function test using controlled hyperthermia. J Appl Physiol 23:267–275

Kerslake D McK (1972) The stress of hot environments. Cambridge University Press, Cambridge, p 126

Kuno Y (1956) Human perspiration. Thomas, Springfield, pp 365–367

Minor V (1927) Ein neues Verfahren zu der klinischen Untersuchung der Schweißabsonderung. Dtsch Z Nervenheilkd 101:302

Nilsson GE (1977) Measurement of water exchange through skin. Med Biol Eng 15:209–218
Robertshaw D (1983) Apocrine sweat glands. In: Goldsmith LA (ed) Biochemistry and physiology of the skin. Oxford University Press, Oxford, pp 642–653
Rothman S (1954) Physiology and biochemistry of the skin. University of Chicago Press, Chicago
Sarkany I, Shuster S, Stammers M (1965) Occlusion of the sweat pore by hydration. Br J Dermatol 77:101–104
Sato K (1973) Sweat induction from an isolated eccrine sweat gland. Am J Physiol 225:1147–1152
Sato K (1980) Pharmacological responsiveness of the myoepithelium of the isolated human axillary apocrine sweat gland. Br J Dermatol 103:235–243
Schulz I (1969) Micropuncture studies of the sweat formation in cystic fibrosis patients. J Clin Invest 48:1470–1477
Shelley WB, Hurley HJ (1953) The physiology of the human axillary apocrine gland. J Invest Dermatol 20:285–295
Shuster S, Johnson C (1969) Abnormality of the sweat duct function in psoriasis. Br J Dermatol 81:846–850
Shuster S, McKendrick T, Stammers M (1965) The site of the sweat gland defect in fibrocystic disease. Br J Dermatol 77:105–109
Thomson ML, Sutarman (1953) The identification and enumeration of active sweat glands in man from plastic impressions of the skin. Trans R Soc Trop Med Hyg 47:412–417
Weiner JS, Lourie JA (1981) Practical human biology. Academic, New York

CHAPTER 4

Measurement of Human Sebaceous Gland Function

W. J. CUNLIFFE

A. Introduction

Acne severity is related to sebum excretion rate (SER) (CUNLIFFE and SHUSTER 1969a) and can be controlled by drugs reducing sebum production. Therefore clinical pharmacological developments in this field require accurate measurements of sebaceous gland function. This chapter is concerned with measurement in humans, for a review of earlier animal methods see SHUSTER and THODY (1974) and THODY and SHUSTER (1989). Changes in composition of sebum and surface lipid (COTTERILL et al. 1972), and changes in rates of lipogenesis of specific components (see Chap. 14 Volume I and 35 Volume II) may provide insights into the mode of action of drugs and hormones, whilst changes in the production of free fatty acids by bacterial lipases provide evidence of antibacterial effect.

I. Histological Methods

Because the sebaceous glands are holocrine, early techniques included planimetry of histological sections. These are not satisfactory for therapeutic studies and gland size correlates poorly with sebum production rate. With tritiated thymidine and autoradiography, kinetic studies of sebaceous cell populations and estimates of turn-over times have been made, but these have a limited use (see CUNLIFFE and COTTERILL 1975).

II. Functional Methods

Early methods were extraction of surface lipid with organic solvents, lipid being determined gravimetrically, by monomolecular spread or by an osmic acid photoelectric technique (see CUNLIFFE and COTTERILL 1975).

B. Measurement of Sebum Excretion Rate

I. Gravimetric Technique

The method (STRAUSS and POCHI 1961) is simple, but is reproducible only if used with great care (CUNLIFFE and SHUSTER 1969b) and is therefore given in detail.

1. Materials

Glassware and absorbent paper (available from General Paper & Box Company) are well washed in ether and dried before use.

2. Collection of Sebum

The skin sampled is usually the forehead and make-up is not used on the day of collection. Zinc oxide adhesive plaster is applied leaving a 3 × 3 cm square of uncovered skin on each side of the forehead. Five sheets of absorbent paper cut to the length of the forehead are applied and covered with gauze, then a broad rubber band (20–25 cm) is attached at the back of the head with nylon mesh. The papers are replaced every 10 min, three times to remove the variable amount of surface lipid; if the papers are held to the light, small dots are seen, each representing sebum emerging from a pilosebaceous follicle. Timed collections can now be made of sebum excreted through the follicles with five sheets of absorbent paper left in place for a timed period (approximately 3 h) checking from time to time that the papers have not slipped.

3. Extraction and Weighing Lipid

The papers are transferred to beakers, discarding the outer one; sebum lipid is extracted with 3 × 25 ml Analar diethyl ether, transferred into large tared flasks and the ether removed by rotary evaporation (300 mmH_2O at 30 °C), leaving a small volume which is transferred to pre-weighed 15-ml flasks which are evaporated to dryness and re-weighed. Since the weight of lipid is very small the flasks are handled with forceps and kept in a dessicator for at least 1 h before weighing in a micro-balance accurate to 0.01 mg. More recently the much simpler method of weighing the papers before and after collection of sebum without extraction of lipid has been used (LOOKINGBILL and CUNLIFFE 1986) and appears to be as accurate.

SER is usually expressed in $\mu g\ cm^{-2} min^{-1}$, but as results vary with details of technique (e.g. duration of collection, absorbency of papers) changes in SER are more reliable than absolute rates in comparing results from different laboratories.

C. Photometric Technique

Instead of using absorption onto paper, SCHAEFER and KUHN-BUSSIUS (1970) used the phenomenon that contact with lipid increases light transmission through opalescent glass and at 460 nm transmission varies with the quantity of lipid adherent to the glass. After removing surface lipid with a ground-glass applicator, lipid reaccumulated over a timed period is removed by several applications of the glass extractor and transmission measured photometrically (see CUNLIFFE et al. 1980). This technique has the advantage that it can be used on a large number of subjects at the same time, but lipometers are not yet commercially available.

D. Sebum Production Rate

This is measured from the rate of incorporation of [^{14}C]glucose or [^{14}C]acetate into a dermal lipid usually over a 3-h period of incubation in vitro (COOPER et al. 1976), and corresponds reasonably well to SER. A 4-mm high speed punch biopsy, usually of intrascapular skin, is processed to separate the epidermis from the dermis. Sebaceous glands are not separated as the non-sebaceous contribution of dermis to lipid synthesis is negligible. After homogenisation the lipids are extracted and analysed by thin layer chromatography and scintillation counting. Thus total and individual components of lipogenesis can be determined. Lipid synthesis can be measured in isolated sebaceous glands (SUMMERLY and WOODBURY 1971), but this technique has not been applied to pharmacological studies.

E. Factors Affecting the Measurement of Sebum Excretion Rate

SER is only indirectly related to sebum production (CUNLIFFE and SHUSTER 1969b). More importantly it is influenced by many factors (see SHUSTER and THODY 1974; THODY and SHUSTER 1989), some of which are discussedin Chap. 14 Volume I, but are mentioned here because they are relevant to its measurement in practice.

1. Age and sex endocrine status all affect SER. Oral contraceptives will decrease SER, particularly if they contain 50 μg or more of ethinyloestradiol (POCHI and STRAUSS 1973) and may increase SER if high in progesterone (see Chap. 14 Volume I). The effect on the menstrual cycle is probably of minor importance to clinical studies of SER (CUNLIFFE and COTTERILL 1975).
2. Duration of collection affects the apparent SER and variance of the measurements is most marked in the first and second hour (CUNLIFFE and SHUSTER 1969b).
3. A 1 °C change in skin temperature produces a 10% change in SER (CUNLIFFE et al. 1970).
4. Sweating reduces the amount of lipid collected by absorbent papers and a thermally neutral environment is necessary during measurement of SER.
5. SER is maximal in the late afternoon and minimal during the early hours of the morning (BURTON et al. 1970).
6. Sebaceous glands are predominantly on the scalp, face, back and chest and SER on the forehead is five times that on the upper back (COOPER et al. 1976).
7. SER correlates with acne severity and past history of the disease. Changes also occur with endocrine disease, parkinsonism and certain drugs (see Chap. 14 Volume I and THODY and SHUSTER 1989).

Thus, in measuring SER it is desirable to make collections at the same time of the day, in a room of constant ambient temperature, preferably at comparable stages of the menstrual cycle in females, to stop all relevant therapy 4–6 weeks before the measurement, and to perform two base-line measurements before assessing the effect of therapy. Since the response of the sebaceous gland is slow such assessments should be after at least 4 weeks, treatment.

F. Measurement of Surface Lipid Composition

Many investigators have measured the effect of drug therapy on the composition of surface lipid collected from the forehead where it is mostly sebaceous and to a small extent epidermal; changes in the squalene and wax ester fraction are mostly associated with and reflect SER, whereas the level of bacterial activity mostly affects free fatty acid (FFA) excretion. Many chromatographic techniques are available for analysing sebum (see DOWNING and STRAUSS 1974). The effect of antimicrobial drugs is often assessed by changes in FFA measured by titration with base (see CUNLIFFE and COTTERILL 1975).

References

Burton JL, Cunliffe WJ, Shuster S (1970) Circadian rhythm in sebum excretion. Br J Dermatol 82:497–501

Cooper MF, McGrath H, Shuster S (1976) Sebaceous lipogenesis in human skin. Br J Dermatol 94:165–192

Cotterill JA, Cunliffe WJ, Williamson B, Bulus VL (1972) Further observations on the pathogenesis of acne. Br Med J 3:444–446

Cunliffe WJ, Cotterill JA (1975) The acnes, 1st edn. Saunders, Philadelphia

Cunliffe WJ, Shuster S (1969a) The pathogenesis of acne. Lancet 1:685–687

Cunliffe WJ, Shuster S (1969b) The rate of sebum excretion in man. Br J Dermatol 81:697–704

Cunliffe WJ, Burton JL, Shuster S (1970) Effect of local temperature variations on the sebum excretion. Br J Dermatol 83:650–654

Cunliffe WJ, Kearney JN, Simpson NB (1980) A modified photometric technique for measuring sebum excretion rate. J Invest Dermatol 75:394–398

Downing DT, Strauss JS (1974) Synthesis and composition of surface lipids of human skin. J Invest Dermatol 62:228–244

Lookingbill D, Cunliffe WJ (1986) A direct gravimetric technique for measuring sebum excretion rate. Br J Dermatol 114:75–81

Pochi PE, Strauss JS (1973) Sebaceous gland suppression with estradiol and diethylstilbestrol. Arch Dermatol 108:210–215

Schaefer H, Kuhn-Bussius H (1970) Methodik für die quantitative Bestimmung der menschlichen Talgsekretion. Arch Klin Exp Dermatol 238:429–435

Shuster S, Thody AJ (1974) The control and measurement of sebum excretion. J Invest Dermatol 62:172–190

Strauss JS, Pochi PE (1961) The quantitative gravimetric determination of sebum production. J Invest Dermatol 36:293–298

Summerly R, Woodbury S (1971) The in vitro incorporation of ^{14}C acetate into the isolated sebaceous glands and appendage freed epidermis of human skin. A technique for the study of lipid synthesis in the isolated sebaceous gland. Br J Dermatol 85:424–431

Thody AJ, Shuster S (1989) Control and function of sebaceous glands. Physiol Rev 69:383–416

CHAPTER 5

Methods for Assessing the Effect of Drugs on Hair and Nails

R. P. R. DAWBER and N. B. SIMPSON

A. Hair Loss

The complaint of hair loss may be related to an increased shedding rate or a decrease in the number of hairs per unit area (ROOK and DAWBER 1982). Apparent thinning of hair may relate to a decrease in hair numbers, in hair diameter or reduction of pigment. Microscopy or measurements will detect objective changes before the patient or physician is aware of any abnormality.

B. In Vivo Assessment of Hair Growth

I. Hair Cycle Status

The "trichogram" is an index of follicular activity and is based on the morphological appearance of plucked hairs at various stages of the growth cycle. It is usually expressed as the anagen : telogen (A : T) ratio. The technique is important; a minimum of 50 hairs must be plucked and the hair must not be washed for 1 week prior to sampling. The A : T ratio varies with age and sex (BRAUN-FALCO 1966), the presence of androgenetic alopecia and body site (SAITOH et al. 1970).

II. Cell Kinetics

WEINSTEIN and MOONEY (1980) used flash labelling curves, labelling indices and growth fraction studies to examine cellular kinetics of the hair bulb and found that the matrix cell cycle was shorter than that of normal epidermal basal cells. Changes in hair diameter are an indirect measure of matrix dynamics (DAWBER and MORTIMER 1982).

III. Linear Growth

There are both sex and body site differences in hair growth rate (SAITOH et al. 1969). Linear growth can be measured by direct assessment of individual hairs in situ by means of a graduated capillary tube (SAITOH et al. 1969; KATZ et al. 1979). This method can give daily assessment of hair growth, but is susceptible to observer error. Shaving or cliping followed by weighing hair (BARMAN et al. 1964) or microscopic measurement of length (EBLING et al. 1977) are accurate, but depend on extreme care in returning to the same body site. Injection of [^{35}S]cystine followed by autoradiography of plucked hairs (MUNRO 1966; COMAISH 1969) is a very accurate method. Measurements of hair shaft diameter of plucked anagen hairs by ocular micrometer is useful in monitoring the effects of nutrition (SIMS and HALL 1968) and the progression of androgenetic alopecia JACKSON et al. 1972). This technique has been reviewed by PEEREBOOM-WYNIA (1981).

IV. Hirsutism and Scalp Hair Density

The scoring system described by FERRIMAN and GALLWAY (1961) which divides the body into eleven areas has been adopted as the standard assessment for severity of hirsutism and for monitoring the effects of treatment. BARMAN et al. (1964) measured hair density and growth with a binocular microscope adapted for skin surface examination. In assessing the response of androgenetic alopecia to treatment RUSHTON et al. (1983) used hair density, two-dimensional shaft diameter of plucked hairs and the A:T ratio. DAWBER et al. (1982) used serial telogen counts (vertex and occipital), hair diameter and daily hair shedding rates. Computer-based digitised photography (GIBBONS et al. 1986) is a simple measure of hair density in alopecia areata and androgenetic alopecia.

V. Elemental Analysis

Many inorganic elements are incoporated into growing hair shafts (BROWN and CROUNSE 1980). Thus hair can be used as an index of environmental pollution.

C. In Vitro Methods for Detecting the Effect of Drugs

Plucked anagen hair roots have been used for biochemical analysis (ADACHI and UNO 1968; SCHWEIKERT and WILSON 1974a, b; SCHWEIKERT et al. 1975) and as a source for cell culture (WETTERINGS et al. 1981; WELLS 1982). Explant cell cultures from plucked hair roots yield only keratinocytes from the outer root sheath (LIMAT and NOSER 1986) with close biochemical similarity to epidermal keratinocytes (MAUDELONDE et al. 1986). A method for growing germinative (matrix) cells in culture has not been described, but recovery of large numbers of intact rat hair follicles (GREEN et al. 1986) may permit selective growth of other cellular components in the future. Dermal papilla (DP) cells cultured from the rat vibrissa follicle (JAHODA and OLIVER 1981) are distinct from dermal fibroblasts and early passage cells retain their capacity to induce hair growth (HORNE et al. 1986). MESSENGER (1984) developed a similar method for human DP cells. Androgen and glucocorticoid metabolism has been demonstrated in DP cells from human scalp and pubic skin (MURAD et al. 1985, 1986). Co-culture systems involving DP cells and hair bulb matrix cells may provide valuable evidence of the controlling mechanisms and interactions; results are awaited.

D. Nail Growth

Nail growth changes are noticed by patients only when growth falls to about one-third of the normal rate. Sudden large reduction of nail growth without absolute cessation leads to Beau's lines – transverse depressed lines, typically on all nails, corresponding to temporary nail matrix inhibition. This may also follow bolus cytotoxic therapy or X-irradiation. Continuous cytostatic therapy may cause onychomadesis – arrest of nail growth for the duration of therapy. Measurement of nail growth has been critically reviewed by BARAN and DAWBER (1984). The methods available depend on inscribing a standardised mark on the nail plate, measuring its position in relation to a fixed point, e.g. posterior nail fold or lunular margin, and repeating the measurement after a known period of time. With a T-shaped mark on the nail, and the proximal nail fold as fixed reference point, methotrexate and azathioprine were found to suppress nail growth whilst systemic steroid therapy had no effect (DAWBER 1970).

Because nail matrix tissue is difficult to obtain in humans and biopsy may leave scars and nail dystrophy, very few cell kinetic studies have been carried out (NORTON 1971). Techniques which require serial biopsies (WRIGHT 1980) are therefore less applicable than the flash labelling technique (RALFS et al. 1982) which requires only a single biopsy.

E. Penetration of Topical Agents

Only a few studies have been done. STÜTTGEN and BAUER (1981) assessed the permeation and penetration of imidazole antifungal agents both in vivo and in vitro, with a ^{14}C label. Dimethylsulphoxide increases penetration and with some substances alcoholic vehicles produce greater concentrations through the nail than do ointments, presumably because of preferential partitioning.

In view of the importance of nail permeability to topical therapeutics WALTERS et al. (1981) developed a stainless steel diffusion cell which allows measurement of permeability coefficients over periods of up to 4 h. As expected of a diffusional process the rates of diffusion across the nail plate are inversely proportional to its thickness.

F. Radiation Penetration

In view of the success of photochemotherapy for conditions such as psoriasis, UV penetration studies have been undertaken in avulsed nail plates. PARKER and DIFFEY (1983) studied the transmission of optical radiation of wavelength 300–600 nm through human nails. Transmitted radiation at 15 different wavelengths was measured with an irradiation monochromator in conjunction with an integrated sphere and photodiode. The results have shown that the nails exhibit decreasing transmission as wavelength shortens, very little UV-B (290–320 nm) penetrating. The nail plate therefore acts as an efficient sun-screen.

X-irradiation penetrates the nail plate tissue (GREY and BOWLT 1978; GAMMELTOFT and WULF 1980) sufficiently to cause carcinoma. The minimum voltage capable of causing therapeutic effect or significant damage is not known.

G. Wood's Light

Wood's light is a source of UV radiation from which all visible rays are excluded by a nickel oxide filter. In the nail it can be used to show fluorescence at the site of drug incorporation, e.g. tetracyclines, and to assist recognition of pigmentation due to mepacrine.

References

Adachi K, Uno H (1968) Glucose metabolism of growing and resting human hair follicles. Am J Physiol 215:1234–1239
Baran R, Dawber RPR (1984) Diseases of the nails and their management, 1st edn. Blackwell, Oxford
Barman JM, Pecoraro V, Astore I (1964) Method, technique and computations in the study of the trophic state of human scalp hair. J Invest Dermatol 42:421–425
Braun-Falco O (1966) Dynamik des normalen und pathologischen Haarwachstums. Arch Klin Exp Dermatol 227:419–452
Brown AC, Crounse RG (1980) Hair, trace elements and human illness. Praeger, New York
Comaish S (1969) Autoradiographic studies of hair growth in various dermatoses: investigation of a possible circadian rhythm in human hair growth. Br J Dermatol 81:283–288
Dawber RPR (1970) The effect of methotrexate, corticosteroids and azathioprine on fingernail growth in psoriasis. Br J Dermatol 83:680–683
Dawber RPR, Mortimer PS (1982) Hair loss during lithium treatment. Br J Dermatol 197:124–125
Dawber RPR, Sonnex T, Ralfs I (1982) Antiandrogen treatment of common baldness in women. Br J Dermatol 107:247
Ebling FJ, Thomas AK, Cooke ID, Randall VA, Skinner J, Cawood M (1977) Effect of cyproterone acetate on hair growth, sebaceous secretion and endocrine parameters in a hirsute subject. Br J Dermatol 97:371–381
Ferriman D, Gallway JD (1961) Clinical assessment of body hair growth in women. J Clin Endocrinol Metab 21:1440–1447
Gammeltoft M, Wulf HC (1980) The transmission of 12 kV Grenz rays and 29 kV X-rays through normal and diseased nails. Acta Derm Verereol (Stockh) 60:431–432
Gibbons RD, Fiedler-Weiss VC, West DP, Lepin G (1986) Quartification of scalp hair – a computer-aided methodology. J Invest Dermatol 86:78–82
Green MR, Clay CS, Gibson WT, Hughes TC, Smith CG, Westgate GE, White M, Kealey T (1986) Rapid isolation in large numbers of intact, viable, individual hair follicles from skin: biochemical and ultrastructural characterisation. J Invest Dermatol 87:768–770
Grey LJ, Bowlt C (1978) An attempt to use thermally stimulated currents in human nail to estimate dose in case of accidental exposure to ionizing radiation. Phys Med Biol 23:759–760
Horne KA, Jahoda CAB, Oliver RF (1986) Whisker growth induced by implantation of cultured vibrissa dermal papilla cells in the adult rat. J Embryol Exp Morphol 97:111–124
Jackson D, Church R, Ebling F (1972) Hair diameter in female baldness. Br J Dermatol 87:361–367
Jahoda CAB, Oliver RF (1981) The growth of vibrissa dermal papilla cells in vitro. Br J Dermatol 105:623–627
Katz M, Wheeler KE, Radowsky M, Gordon W (1979) Assessment of rate of hair growth using a simple trichometer. Med Biol Eng Comput 17:333–336
Limat A, Noser FK (1986) Serial cultivation of single keratinocytes from the outer root sheath of human scalp hair follicles. J Invest Dermatol 87:485–488
Maudelonde T, Rosenfield RL, Shuter CF, Schwartz SA (1986) Studies of androgen metabolism and action in cultured hair and skin cells. J Steroid Biochem 24:1053–1060

Messenger AG (1984) The culture of dermal papilla cells from human hair follicles. Br J Dermatol 110:685–689

Munro DD (1966) Hair growth measurement using intradermal sulfur 35 cystine. Arch Dermatol Syphilol 93:119–122

Murad S, Hodgins MB, Simpson NB, Oliver RF, Jahoda C (1985) Androgen receptors and metabolism in cultured dermal papilla cells from human hair follicles (Abstr). Br J Dermatol 113:768

Murad S, Hodgins MB, Oliver RF, Jahoda C (1986) Comparative studies of androgen receptors and metabolism in dermal papilla cells cultured from human and rat hair follicles (Abstr). J Invest Dermatol 87:158

Norton LA (1971) Incorporation of thymidine methyl ^{3}H and glycine 2^3H in the nail matrix and bed of humans. J Invest Dermatol 56:61–68

Parker SG, Diffey BL (1983) The transmission of optical radiation through human nails. Br J Dermatol 108:11–16

Peereboom-Wynia JDR (1981) Comparative studies of the diameters of hair shafts in anagen and in telogen phases in male adults without alopecia and in male adults with androgenetic alopecia. In: Orfanos CE, Montagna W, Stüttgen G (eds) Hair research. Springer, Berlin Heidelberg New York

Ralfs I, Dawber RPR, Ryan TJ, Wright NA (1982) The epidermal cell kinetics of the erythrokeratodermas. Br J Dermatol 107:565–567

Rook A, Dawber R (1982) Diseases of the hair and scalp. Blackwell, London

Rushton J, James KC, Mortimer CH (1983) The unit area trichogram in the assessment of androgen-dependent alopecia. Br J Dermatol 109:429–438

Saitoh M, Uzaka M, Sakamoto M, Kobori M (1969) Rate of hair growth. In: Montagna W, Dobson RL (eds) Hair growth. Pergamon, Oxford, pp 183–201 (Advances in biology of skin, vol 9)

Saitoh M, Uzaka M, Sakamoto M (1970) Human hair cycle. J Invest Dermatol 54:65–81

Schweikert HU, Wilson JD (1974a) Regulation of human hair growth by steroid hormones. I. Testosterone metabolism in isolated hairs. J Clin Endocrinol Metab 38:811–819

Schweikert HU, Wilson JD (1974b) Regulation of human hair growth by steroid hormones. II. Androstenedione metabolism in isolated hairs. J Clin Endocrinol Metab 39:1012–1019

Schweikert HU, Milewich L, Wilson JD (1975) Aromatisation of androstenedione by isolated human hairs. J Clin Endocrinol Metab 40:313–417

Sims RT, Hall T (1968) X-ray emission microanalysis of the density of hair proteins in kwashiorkor. Br J Dermatol 80:335–338

Stüttgen G, Bauer E (1981) Bioavailability, skin and nail penetration of topically applied antimycotics. Mykosen 25:74–80

Walters KA, Flynn GL, Marvel JR (1981) Physiochemical properties of the human nail. J Invest Dermatol 76:76–79

Weinstein GD, Mooney K (1980) Cell proliferation kinetics in the human hair root. J Invest Dermatol 74:43–46

Wells J (1982) A simple technique for establishing cultures of epithelial cells. Br J Dermatol 107:481–482

Wetterings PJJM, Vermorken AJM, Bloemendal J (1981) A method for culturing human hair follicle cells. Br J Dermatol 104:1–5

Wright NA (1980) The kinetics of human epidermal cell populations in health and disease. In: Rook AJ, Savin JA (eds) Recent advances in dermatology 5. Livingstone, Edinburgh, pp 317–343

CHAPTER 6

Measurement of Drug Action in the Skin: Sensation

B. LYNN

A. Introduction

Skin sensations are of three main types. First, there are sensations caused by innocuous mechanical stimuli, such as touch, pressure, vibration and tickle. Second, there are sensations of warmth and cold due to small, innocuous, temperature changes. Third, unpleasant sensations of pain or itch are caused by the presence of certain chemical agents in the skin or, in the case of pain, by potentially or actually damaging levels of thermal or mechanical stimuli. In addition, more complex sensory blends, such as wetness, also occur.

Drugs may act to modify skin sensations caused by applied stimuli or by endogenous chemicals (e.g. analgesics). Drugs may also themselves generate sensations (e.g. bradykinin causes pain). Whether one is dealing with directly generated sensations or with changes in responsiveness, broadly similar methods of measurement of drug action are applicable. In this chapter aspects of the methods will be considered first, followed by an examination of the effects on cutaneous sensation of certain selected classes of drugs. No extensive review of base-line cutaneous sensation will be given since several such reviews are available (e.g. KENSHALO 1978; LYNN 1983; MOUNTCASTLE 1980; SINCLAIR 1981). Additional information about itch may be found in Chap. 7.

B. Methods for Studying Cutaneous Sensation

Several aspects of cutaneous sensory responses can be measured and these various types of measurement will be considered first in this section. Some widely used methods for applying mechanical, thermal and chemical stimuli, both noxious and innocuous, will then be described. Finally, the important problem of experimental design, both for human and animal studies, will be discussed.

I. Types of Measurement

1. Intensive

The most commonly used measures of cutaneous sensation concern the ability to detect and discriminate stimuli of different intensity. Frequently the parameter measured is the *sensory threshold* and the methods for such measurements are well established (e.g. KLING and RIGGS 1971; Chap. 2; DYCK et al. 1984). Thresholds for noxious stimuli pose special problems since these require a judgment about the quality of the sensation and not just an assessment of whether a stimulus is present or not. One way round this problem is to use the flexion reflex as a response measure (WILLER et al. 1979). In behavioural experiments in animals the assessment of thresholds for noxious stimuli frequently depends on observing aversive reactions such as tail withdrawal.

The measurement of sensory thresholds can be influenced by the subjects, willingness to respond as well as by the performance of the sensory apparatus. The methodology of signal detection theory (SDT) (GREEN and SWETS 1966) allows the analysis of subjects' responses, both positive and negative, and provides a measure both of the subjects' sensitivity and of their bias towards making a response. SDT has been applied extensively to the problem of measuring cutaneous thresholds for heat pain (CLARK 1969; FEATHER et al. 1972; CLARK and YANG 1974).

The classical measure of sensitivity of suprathreshold stimuli is the *difference threshold*, i.e.the increment in stimulus intensity that produces a just-noticeably different sensation or response. Difference thresholds have been determined for innocuous pressure (GATTI and DODGE 1929; see also WEBER 1978) for cold and warmth (DARIAN-SMITH and JOHNSON 1977) and for noxious heat (HARDY et al. 1952).

An alternative approach for assessing sensitivity to suprathreshold stimuli involves simply asking subjects to rate the intensity of sensations on arbitrary scales. Perhaps surprisingly this method of *magnitude estimation* yields quite reliable power function relations between stimulus intensity and subjective estimate (STEVENS 1957, 1970). However, the exact methods used (e.g. range of stimuli, range of allowed responses) do also influence the results (POULTON 1968). Magnitude estimation methods have been used in several studies with mechanical skin stimulation, including vibration (STEVENS 1959; FRANZEN 1969; VERRILLO and CAPRARO 1975) and step indentations (HARRINGTON and MERZENICH 1970). Similar methods have also been used for innocuous warmth and cold (STEVENS and STEVENS 1960) and for heat pain (LAMOTTE and CAMPBELL 1978). Magnitude estimates may also be made by marking a linear scale (visual analogue scale) or squeezing a hand-grip. For example, the visual analogue scale method has been used to estimate both itch (WESTERMAN et al. 1986) and pain (HUSKISSON 1974).

2. Time-Dependent and Spatial Measures

Measures of time-dependent properties of cutaneous sensation, such as rates of adaptation and responses to different vibration frequencies, have only been used in a few studies of drug action. Similarly, spatial measures such as two-point discrimination thresholds have hardly been utilised, despite their importance and the ease with which such tests can be made (e.g. ROBINSON and SHORT 1977).

II. Stimulation Techniques

Most skin areas are easily accessible and so it is a straightforward matter to carry out rapid, and at least semi-quantitative sensory testing with hand-held stimulators. Stimulation methods that allow fully quantitative studies can also often be relatively simple, although sophisticated methods have been developed for some specific tasks (DYCK et al. 1984). A brief review of several of the more useful methods for mechanical, thermal and chemical stimulation will be given in this section.

1. Mechanical Stimulation

Von Frey hairs, often constructed these days from monofilament nylon, still provide the most convenient method for investigating pressure sensitivity and are particularly useful for experiments where several widely separated areas are to be studied (WEINSTEIN 1977). The amplitude of stimuli from calibrated bristles is most readily expressed in terms of the force exerted although the sensation produced depends also on the diameter or cross-sectional area of the hair (VON FREY and KIESOW 1899). Very stiff von Frey hairs or hairs with a needle point stuck to them can be used to assess pain sensitivity.

For many studies control over the time course of stimuli is necessary, for example in studies of adaptation or of responses to sinusoidal vibration. For these purposes various electromagnetic devices have been used. For precise stimulation such stimulators should be controlled by negative feedback from a transducer that monitors the output. For studies of touch or vibration sense, the displacement of the stimulus is usually controlled (e.g. WERNER and MOUNTCASTLE 1965). A useful refinement is automatic compensation for small skin movements such as those that occur with each heart beat (WESTLING et al. 1976). It is possible to control the force applied (BYRNE 1975), and this may be preferable for high intensity, painful stimuli (LYNN 1977a). This is because the skin becomes rather stiff for large indentations (PETIT and GALIFRET 1978) and so it is easier to grade the force applied rather than the displacement.

Various relatively simple "pressure algesimeters" for applying strong, steady pressures to the skin have been devised (GREEN and YOUNG 1951; RANDALL and SELITTO 1957; MERSKEY and SPEAR 1964). However, as normally used, for example in the paw pressure test (GREEN and YOUNG 1951) these devices must excite both cutaneous and subcutaneous receptors. Devices for controlled pinching of small skin folds (e.g. LYNN and PERL 1977) may provide a more specific test of cutaneous sensitivity.

2. Thermal Stimulation

The skin temperature may be altered by conduction or by radiation. The simplest conduction methods involve immersion in water at a controlled temperature and this is a good method for the hands and feet. It is also used for the rat tail immersion test of analgesic potency (JANSSEN et al. 1963). Cooling or heating by conduction is, however, usually carried out with metal "thermodes", and several different sorts of thermode have been developed. Devices using circulating liquid from thermostatically controlled reservoirs can be used for rapid "step" stimuli (KENSHALO et al. 1960). If the time course is to be controlled, for example to generate linear "ramp" transitions, some form of electronic control is needed. One suitable device has a powerful heating element that operates against a continuous circulation of cold liquid (HILDER et al. 1974). Another approach is to use Peltier elements. The temperature gradient generated by these devices is proportional to the size and direction of current flow through them. Because of the significant thermal capacity of the elements, rates of temperature change are limited to 2 °C/s or less. Nevertheless, Peltier stimulators can provide very precise hot and cold stimuli (KENSHALO and BERGEN 1975; JAMAL et al. 1985). For controlled skin heating, without control of the rate of cooling, simple electrically heated thermodes can be used (e.g. GUILLEMIN et al. 1952; CARPENTER and LYNN 1981).

A novel method for the semi-quantitative assessment of cold sensitivity involves using a set of discs made from materials of differing thermal conductivity. Those with high conductivity feel colder than those with low conductivity, and the use of four different discs allows a useful assessment of the sensitivity of the skin to cold (DYCK et al. 1974).

Radiant thermal stimulation is normally only used for heating, although slow radiant cooling with a very cold source (solid CO_2) is possible (HARDY and OPPEL 1938). Some radiant heaters monitor the total heat supplied. This was true for the "dolorimeter" devised by HARDY et al. (1952) and is also the basis of rat tail flick tests involving radiant heating (D'AMOUR and SMITH 1941). However, the effectiveness of these methods depends upon maintaining a steady base-line skin temperature (HARDY et al. 1952). More reliable stimulation can be achieved by monitoring the skin temperature and using this signal to provide negative feedback control of the radiant heat source (e.g. BECK et al. 1974; FITZGERALD and LYNN 1977). These devices also allow the time course of the heating to be varied. For example, linear increases in skin temperature (ramps) may be generated and these provide a quick method for assessing warmth and heat pain thresholds (see LYNN 1980).

A problem with controlled radiant heaters is to measure skin temperature accurately. Usually a small polished thermocouple is used, but there is inevitably some direct absorption of energy by the thermocouple so that this method overestimates the skin temperature (STOLL and HARDY 1950). One solution is to use a radiant thermometer operating at an infrared bandwidth different from the lamp source (thus avoiding problems with reflected energy). The rather complex instrument developed by MEYER et al. (1976) adopts this approach.

Thermal stimuli are widely used for measuring cutaneous pain sensitivity, as well as sensitivity to innocuous temperatures. Although such stimuli are relatively easy to generate and control, absolute calibration in relation to the true skin temperature may be difficult. The problem with radiant methods has already been mentioned; there is also a difficulty with conduction thermodes since these may make only partial contact with the uneven skin surface. Differences in the method of skin temperature assessment probably account for some of the discrepancies between the heat pain thresholds reported by different laboratories.

3. Chemical Stimulation

Many chemicals produce pain, including several endogenous substances that may be involved in generating pain during inflammation. Studies have also been made of agents that specifically inhibit the actions of endogenously produced substances. However, there is a major problem in delivering chemical agents to the skin in a reliable and non-traumatic manner (KEELE and ARMSTRONG 1964, Chap. 3).

Topical application can be used (e.g. GAMMON and STARR 1941; WEINSTEIN 1977), but most agents penetrate the epidermis too slowly for this to be a generally useful method. Topical iontophoresis has been used to aid the penetration of acetylcholine and histamine (SKOUBY 1951; MELTON and SHELLEY 1950) and has also been used for local anaesthetics (TALBOT et al. 1968). Penetration can also be aided by abrading the skin (GREAVES and MCDONALS-GIBSON 1973). Direct injection into the skin by fine needles has been used frequently, but causes considerable discomfort and produces rather variable results. A major factor is probably the difficulty of making injections at a consistent depth within the skin (KEELE and ARMSTRONG 1964). This problem is particularly acute in many experimental animals with their relatively thin skin. An alternative is high pressure jet injection (LINDAHL 1961), but this method places most of the injected solution subcutaneously and so is not effective with substances that only diffuse slowly or that are rapidly inactivated. To apply chemicals for long periods it is possible to set up a subcutaneous perfusion with two needles placed several centimetres apart in the skin (FOX and HILTON 1958; FERREIRA 1972).

An alternative approach is to remove all or most of the epidermal barrier and to apply solutions directly to the exposed surface. This can be done by physically cutting away the skin surface or more commonly by creating a blister by application of cantharidin (ARMSTRONG et al. 1951) or suction (KIISTALA 1968). The blister base technique has been widely used (SHELLEY and ARTHUR 1957; KEELE and ARMSTRONG 1964). However, it does produce substantial damage and this inevitably makes it only usable with highly motivated subjects and limits the number of tests that can be carried out.

III. Experimental Design

Experiments to measure cutaneous sensation need to be planned with care. Adequate controls are important (D'AMATO 1970) as is the need for efficient statistical analysis (WINER 1971). Other aspects of experimental design differ depending on whether the studies are on experimental animals or on humans.

1. Design of Experiments on Cutaneous Sensation in Laboratory Animals

A common use of laboratory animals is for assessing the potency of analgesic drugs by examining "unconditioned" reactions to noxious cutaneous stimuli. An example is the tail flick test (D'AMOUR and SMITH 1941) where the rat's tail is reflexly removed from a hot light beam. Unconditioned reactions to innocuous stimuli can also be used as measures of cutaneous sensitivity. For example, placing reactions to paw pressure (RUCH and PATTON 1979, p. 264) or orientation responses to light touch (WALL 1970).

The use of unconditioned responses is restricted to a rather limited set of situations. Results are also strongly influenced by habituation and in addition largely measure reaction, rather than purely sensory, thresholds. Examination of a wider range of parameters, including difference thresholds and spatial measures, requires the use of specially trained animals. Although classical conditioning can provide a way of doing this, in practice most studies have used various instrumental learning procedures and particularly operant conditioning methods (BLOUGH and BLOUGH 1977). For example, in the monkey absolute and differential thresholds for cutaneous vibration (MOUNTCASTLE et al. 1972; LAMOTTE and MOUNTCASTLE 1975), tactile discriminations of stimulus size and direction of movement (VIERCK 1973, 1974) and innocuous thermal thresholds (CRAGG and DOWNER 1967) have all been studied. Cutaneous thresholds for von Frey hairs have also been determined in the cat by a conditioned avoidance training procedure (TAPPER 1970). It is also possible to design behavioural experiments that require a clear differential response to the noxious component of a cutaneous stimulus, rather than indicating perhaps only the occurrence of an unusual event (DUBNER and BEITEL 1976). In general, careful studies with trained animals have yielded important data on cutaneous sensory responses, especially after lesions of sensory pathways. However, such studies are expensive to carry out and have not been used for examining the effects of drugs.

There are important ethical considerations with all types of animal experimentation (HUME 1962) and the testing of analgesic drugs may, because of the necessity of using intense stimuli, raise particular problems. It is clearly important to ensure that no needless suffering occurs, for example by reducing numbers through efficient experimental design and by keeping stimulus levels below those where irreversible tissue damage occurs. Long-term monitoring of nociceptive reactions is also possible in decerebrate animals (WOOLF 1983).

2. Design of Experiments on Cutaneous Sensation in Humans

Most human experiments depend on verbal reports about stimuli, although some experiments on pain sensitivity have used the flexion reflex response (WILLER et al. 1979). Methods for psychophysical studies are well discussed in many psychology texts (e.g. D'AMATO 1970). A major problem when investigating drug effects is the placebo reaction and the classical description by BEECHER (1959) should be studied by anyone contemplating making such determinations.

C. Effects of Drugs on Cutaneous Sensation

I. Drugs that Produce Sensation

1. Pain-producing Agents

A variety of agents produce pain in the skin and several useful reviews of such agents are available (KEELE and ARMSTRONG 1964; KEELE 1970; CHAHL and KIRK 1975). When assessing reports of pain production it is important to consider whether adequate checks have been made for the presence of unsuspected pain-producing contaminants in the solutions tested. For example, hypotonic and hypertonic solutions are painful, as are solutions of low or high pH (KEELE and ARMSTRONG 1964; LINDAHL 1961). Tissue extracts may contain potassium, an ex-

Table 1. Cutaneous pain thresholds for four endogenous agents applied by three different methods

	Acetylcholine (g/ml)	Histamine (g/ml)	Serotonin (g/ml)	Bradykinin (g/ml)
Topical application to blister base[a]	10^{-5}	3×10^{-5}	3×10^{-6}	10^{-7}
Intradermal jet injection[b]	10^{-3}	10^{-4}	10^{-3}	
Intradermal needle injection	10^{-2}–10^{-3} [a]	3×10^{-5} [c]	10^{-4} [d]	10^{-5} [d,e]

[a] KEELE and ARMSTRONG (1964).
[b] LINDAHL (1961).
[c] BROADBENT (1955).
[d] GREAVES and SHUSTER (1967).
[e] CORMIA and DOUGHERTY (1960).

cellent pain-producing substance. That these are real problems may be illustrated by the example of substance P. Human and animal studies had revealed potent pain reactions to substance P extracts (e.g. LEMBECK 1957; KEELE and ARMSTRONG 1964). However, when the pure peptide finally became available in the 1970s it did not show such actions (STEWART et al. 1976). Subsequent investigation revealed that various extracts contained potassium or bradykinin and that these contaminants had been responsible for the previously reported pain responses (LEMBECK and GAMSE 1977).

Table 1 gives pain threshold concentrations for four endogenous substances that have been studied extensively because they are thought to play a role in generating the pain and hyperalgesia that accompany skin injury and inflammation (LYNN 1977b; see also Chap. 10 in Part 1 of this volume). Bradykinin is the most potent of these agents, especially on a molar basis, and injections of as little as 0.5 μg (0.5 nmol) will cause strong pain (CORMIA and DOUGHERTY 1960). In the intact skin histamine is almost as potent on a weight basis, but acetylcholine (ACh) and serotonin (5-HT) are much less effective. Bradykinin, ACh and 5-HT are markedly more effective on the blister base than in the intact skin. This difference may reflect the hyperalgesia present on blister bases produced by cantharidin, perhaps owing to the presence of subthreshold amounts of various endogenously produced mediators. Pain responses to mixtures of histamine and ACh, as found in nettle stings, are much greater than from either agent alone (EMMELIN and FELDBURG 1947). Similarly, responses to bradykinin and histamine are enhanced by the prostaglandin PGE_1 (FERREIRA 1972) and vascular pain from bradykinin is enhanced by 5-HT (SICUTERI et al. 1965).

Many other endogenous substances produce cutaneous pain, such as prostaglandin PGE, and related fatty acid hydroperoxides (FERREIRA 1972), leukotrienes LTC_4 and LTD_4 (BISGAARD et al. 1982) and adenosine, xanthosine and related compounds (MOULTON et al. 1957; KEELE 1970). Interestingly, prostaglandins and leukotrienes are much less potent as pain-producing agents than other inflammatory mediators such as bradykinin and histamine. However, PGE_1 is a potent sensitising agent for bradykinin and histamine (FERREIRA 1972) whilst LTD_4 has been reported to desensitise the skin to bradykinin (SCHWEIZER et al. 1984). Large numbers of exogenous toxins are also painful, as reviewed by CHAHL and KIRK (1975) and so are several pungent substances extracted from plants, such as capsaicin (from peppers) and mustard oil (JANCSO 1968). An extensive range of pain-producing agents is described in the monograph by KEELE and ARMSTRONG (1964).

2. Substances Producing Itch and Other Non-painful Sensations

Histamine will produce itch in concentrations 10–100 times smaller than those required for pain production. However, no unequivocal demonstration of similar effects of low concentrations of other pain-producing agents has been possible, so this appears to be a special feature of histamine itself. The itch response to histamine is easiest to elicit when it is introduced into the most superficial layers of skin (BROADBENT 1955; KEELE and ARMSTRONG 1964). As with pain production, the itch-inducing effects of histamine are enhanced by low concentrations of PGE_1 (GREAVES and MCDONALD-GIBSON 1973). Several other substances that produce itch do so by releasing endogenous histamine from mast cells (e.g. compound 48/80, KEELE and ARMSTRONG 1964; substance P, HAGERMARK et al. 1978). However, itching can be produced without histamine release by some proteolytic enzymes (e.g. papain) and this is part of the mechanism of action of itching powder (spicules of cowage, *Macuna pruriens*, ARTHUR and SHELLEY 1959).

By way of contrast with the irritant sensations described so far, mention must be made of menthol. When this is applied to skin it gives rise to a sensation of cold, just as it does when applied to mucous membranes (ZOTTERMAN 1959).

II. Drugs that Modify Sensation

1. Local Anaesthetics

Local anesthetic agents such as procaine and lignocaine block nerve conduction by interfering with sodium permeability changes (RITCHIE and GREENE 1980). However, not all types of nerve fibre are equally sensitive to local anaesthetics and so it is possible by using moderate concentrations (e.g. 0.25%–0.5% for lignocaine or procaine, MACKENZIE et al. 1975) to produce, at least transiently, a differential block. In general, the smaller fibres block before the larger fibres, although there is some overlap and some large myelinated axons will be affected before all the non-myelinated axons are blocked (HALLIN and TOREBJORK 1976). During the onset of a local anaesthetic block sensations are lost in a repeatable sequence that is clearly related to the fact that fibres carrying nociceptive information are mostly of small calibre (A–δ and C) whilst those responding to touch and pressure are predominantly large and myelinated (A–β) (see Chap. 10 in Part 1 of this volume). Thus, the usual pattern is that pain responses are reduced or abolished before responses to touch and pressure. Responses to warm stimuli are also affected during the early stages, with cold responses being blocked or diminished later, but still before those for touch. Responses to sharp, pricking pain can be felt when all the C fibres are blocked, although under these circumstances slow, dull pain responses are abolished (TOREBJORK and HALLIN 1973; MACKENZIE et al. 1975; HALLIN and TOREBJORK 1976). However, there are many reported variations in the sequence of sensory loss during local anaesthetic nerve blocks, as discussed by SINCLAIR (1981, p. 210).

Local anaesthetic applied topically to the skin, or by injection or iontophoresis, will also produce anaesthesia (e.g. WEINSTEIN 1977). However, sensations from high frequency (> 50 Hz) vibration are not affected, presumably because the receptors involved (Pacinian corpuscles) are mostly located subdermally (TALBOT et al. 1968). An additional action of intracutaneous local anaesthetic is to block the spread of secondary hyperalgesia around an area of injury (LEWIS 1935).

2. Opioid Analgesics

Morphine and related drugs can produce profound analgesia through their ability to combine with highly specific opiate receptors present in many parts of the central and peripheral nervous systems (JAFFE and MARTIN 1980). In humans, pain thresholds, for example to skin heating, are elevated whilst the reactions to suprathreshold pain are very markedly reduced (BEECHER 1959). In experimental animals parallel effects occur; e.g. tail flick responses are reduced (D'AMOUR and SMITH 1941). In general, direct opioid agonist drugs, such as morphine and meperidine, have a clear action on a wide range of human and animal pain responses. Mixed agonist-antagonist drugs, such as pentazocine, are also effective analgesics in humans, but may not produce clear effects on animal tests of cutaneous pain reactivity unless fairly mild noxious stimuli are used (GRAY et al. 1970). Opioid analgesics have a highly selective action on pain responses; sensations from tactile and pressure stimuli are unaffected (MULLIN and LUCKHARDT 1935; WIKLER et al. 1945; YAKSH and REDDY 1981).

The major actions of opioids are on the central nervous system where they activate inhibitory neurones concerned with the control of pain transmission in the brain stem and spinal cord, an action that mimics the effects of enkephalins and endorphins (FIELDS and BASBAUM 1978; YAKSH 1981; PERT 1982). Recently, evidence for a peripheral action of opioids that reduces responses of rats to painful stimuli applied to inflamed tissue has been presented (FERREIRA 1980).

In view of the existence of endogenous opioids, one might expect morphine antagonists such as naloxone to produce an enhancement of pain sensations. In fact no consistent effects on base-line cutaneous pain sensitivity have been found in humans (EL-SOBKY et al. 1976; DEHEN et al. 1978). However, increased pain sensitivity following administration of opiate antagonists has been found in individuals with unusually high base-line pain thresholds (BUCHSBAUM et al. 1977), in a patient with generalised insensitivity to pain (DEHEN et al. 1978), in patients suffering from post-operative pain (LEVINE et al. 1978) and in subjects stressed by strong electric shocks (WILLER et al. 1981). Similar increases have also been found in rodents with low pain sensitivity at night (FREDRICKSON et al. 1977) or following stressful electric shocks (MAIER et al. 1980).

It thus appears that in normal, unstressed mammals the endogenous opioid system is either quiescent or, possibly, acting via a receptor system not affected by naloxone. However, there are some individuals and some situations (e.g. pre-existing pain) where tonic activity does occur and where naloxone has a pain-enhancing effect.

3. Anti-inflammatory Agents

Many anti-inflammatory agents can reduce cutaneous sensitivity for itch and for pain. Glucocorticoid steroids such as cortisone may reduce cutaneous pain and itch when the skin is diseased or inflamed. However, they do not reduce itch caused by histamine injections or by application of itch powder (CORMIA and KUYKENDALL 1954; SHELLEY and ARTHUR 1957). Anti-histamines of the H_1 type reduce histamine-induced itch (CORMIA and KUYKENDALL 1954) and may (BROADBENT 1953) or may not (SHELLEY and ARTHUR 1957) reduce itch from cowage. H_2-receptor antagonists may also reduce histamine itch, but the results are very variable (ROBERTSON and GREAVES 1978).

Prostaglandin synthesis inhibitors such as aspirin have marked analgesic and anti-pruritic actions (e.g. see FERREIRA 1979). These drugs act partly via a peripheral anti-inflammatory action (e.g. RANDALL and SELITTO 1957), but also appear to act on the central nervous system, an action that must be particularly important for drugs such as paracetamol and phenacetin that have only weak anti-inflammatory effect (FERREIRA 1979; PARKHOUSE et al. 1979). The effects of aspirin on experimentally measured pain thresholds are small (e.g. HARDY et al. 1952; SMITH and BEECHER 1969) and aspirin does not affect heat pain responses such as the tail flick reaction in experimental animals (JANSSEN et al. 1963). It has been reported that histamine-induced itch is reduced by aspirin (CORMIA and KUYKENDALL 1954), but SHELLEY and ARTHUR (1957) found cowage-induced itch to be unaffected by sodium salicylate. Aspirin does not affect cutaneous touch thresholds (MULLIN and LUCKHARDT 1935).

4. Adrenaline, Acetylcholine, Capsaicin, Gonadal Hormones

Cutaneous sensation is little affected in normal subjects by *catecholamines* or catecholamine antagonists. Injection of noradrenaline has been found to produce a slight decrease in touch sensitivity (WALLIN et al. 1976), an effect that is consistent with the results of studies on receptor units (see Sect. G2 of Chap. 10 in Part 1 of this volume). A slight decrease in histamine itch after intracutaneous adrenaline injection has been reported (CORMIA and KUYKENDALL 1954), but no effect on cowage itch was found after intravenous adrenaline injection (SHELLEY and ARTHUR 1957). The alpha-blocker phenoxybenzamine also failed to alter itch sensation in response to cowage (SHELLEY and ARTHUR 1957). Heat pain thresholds are also unaffected by locally applied adrenaline (BEECHER 1959; CHAPMAN and JONES 1944).

In certain abnormal conditions in humans, however, catecholamines or sympathetic blocks can have striking effects. In causalgic skin areas, iontophoresis of noradrenaline, but not of adrenaline, produced a hyperalgesic region where previously innocuous tactile stimuli became painful (WALLIN et al. 1976). The ongoing pain that occurs following some peripheral nerve injuries is also often reduced or abolished by sympathectomy or by guanethidine block (LOH and NATHAN 1978). Chronic peripheral pain (e.g. phantom limb pain) may be abolished by beta-blockers (MARSLAND et al. 1982), but worthwhile effects only occur in a minority of patients with pain from traumatic injury (SCADDING et al. 1982). These findings may be related to the special sensitivity of injured peripheral nerve fibres to catecholamines (WALL and GUTNICK 1974) (see Sect. G 2 of Chap. 10 in Part 1 of this volume).

As described in Sect. C. I. 1, ACh produces pain when applied to the skin in fairly high concentrations. At lower concentration it produces hyperalgesia for chemical stimuli (SKOUBY 1951, 1953; EMMELIN and FELDBURG 1947). SKOUBY (1953) also reported a reduction in heat pain thresholds, although CHAPMAN and JONES (1944) found no such effect with subcutaneous acetyl-β-methylcholine. Atropine has been claimed to decrease heat pain thresholds in humans (SKOUBY 1953), but does not alter reactions to tail immersion in the rat (JANSSEN et al. 1963). Small decreases in heat pain threshold have also been found following application of anti-cholinesterases (SKOUBY 1951), although such agents apparently do not grossly affect tactile or pricking pain thresholds (HURLEY and KOELLE 1958). Finally, BING and SKOUBY (1950) found an enhancement of cutaneous cold sensitivity following local application of ACh or acetyl-β-methylcholine.

Capsaicin produces burning pain when applied to the skin, as described in Sect. C. I. 1. However, repeated application leads to a long-lasting desensitisation to further applications of capsaicin and some other irritants (JANCSO 1960). Such repeated applications also lead to an elevation of heat pain thresholds and a reduction in itch responses to histamine that lasts for several days (CARPENTER and LYNN 1981; TOTH-KASA et al. 1986), although immediately after the first few applications there is considerable hyperalgesia and a large reduction in heat pain threshold (SZOLCSANYI 1977; CARPENTER and LYNN 1981). Capsaicin can also cause long-lasting reductions in cutaneous heat pain sensitivity following its application to cutaneous nerve trunks (JANCSO et al. 1980), an effect that may be related to its ability to damage afferent C fibres (LYNN and PINI 1985; JANCSO et al. 1985; LYNN et al. 1987).

Finally, tactile two-point discrimination on the skin of the breast, including nipple and areola, has been found to be improved in a group of nulliparous females when taking a contraceptive pill containing synthetic *oestrogen* and *progestin* (ROBINSON and SHORT 1977).

5. **Anti-pruritic Agents** (see Chap. 7)

In the past most anti-pruritic agents were H_1-receptor antagonists and their mode of action was thought to be their anti-histaminic effect. It has since been shown that the anti-pruritic action of H_1 anti-histamines is related to their central sedative effect (KRAUSE and SHUSTER 1983) the more potent but less sedative anti-histamines such as terfenadine and astemizole having no anti-pruritic effect, whilst the less potent but more sedative anti-histamines such as trimeprazine are anti-pruritic. The role of sedation is not clear as the sedative benzodiazepines decrease itch whilst the comparably sedative barbiturates increase it (SHUSTER 1981). Naloxone has also been shown to be effective both in morphine- and histamine-induced itch (BERNSTEIN et al. 1982). Aspirin is without effect on the anti-histamine-resistant itch of eczema (DALY and SHUSTER 1986).

D. Final Comments

Detailed assessments of the ability of drugs that enhance or reduce pain to affect cutaneous sensitivity have been carried out in both humans and experimental animals and these studies have utilised much of the available methodology, although few studies with instrumental conditioning procedures have been made. There has also been quite a lot of work, some of it rather crude, on the effects of drugs on that most characteristic of cutaneous sensations, itch. For both itch and pain the main drugs studied have been opiates (e.g. morphine) and non-steroidal anti-inflammatory agents such as aspirin; little work has been done on other classes of drugs with known actions on the central nervous system. Our knowledge of the pharmacology of sensations produced by innocuous mechanical and thermal stimuli is still only fragmentary.

References

Armstrong D, Dry RML, Keele CA, Markham JW (1951) Method for studying chemical excitants of cutaneous pain in man. J Physiol (Lond) 115:59–61P
Arthur RP, Shelley WB (1959) The peripheral mechanism of itch in man. In: Wostenholme G, O'Connor M (eds) Ciba Foundation Study Group No 1. Livingstone, London, pp 84–95
Beck PW, Handwerker HO, Zimmermann M (1974) Nervous outflow from the cat's foot pad during noxious radiant heat stimulation. Brain Res 67:373–386
Beecher HK (1959) Measurement of subjective responses. Quantitative effects of drugs. Oxford University Press, New York
Bernstein JE, Swift RM, Soltani K, Lorincz AL (1982) Antipruritic effect of an opiate antagonist naloxone hydrochloride. J Invest Dermatol 78:82–83
Bing HI, Skouby AP (1950) Sensitization of cold receptors by substances with acetylcholine effect. Acta Physiol Scand 21:286–302
Bisgaard H, Kristensen J, Sondergaard J (1982) The effect of leucotriene C_4 and D_4 on cutaneous blood flow in humans. Prostaglandins 23:797–801
Blough D, Blough P (1977) Animal psychophysics. In: Konig WK, Staddon JER (eds) Handbook of operant behaviour. Prentice Hall, Englewood Cliffs, pp 514–539
Broadbent JL (1953) Observations on itching produced by cowhage and on the part played by histamine as a mediator of the itch sensation. Br J Pharmacol 8:263–270
Broadbent JL (1955) Observations on histamine-induced pruritis and pain. Br J Pharmacol 10:183–185
Buchsbaum MS, Davis GC, Bunney WE (1977) Naloxone alters pain perception and somatosensory evoked potentials in normal subjects. Nature 270:620–622
Byrne J (1975) A feedback controlled stimulator that delivers controlled displacements or forces to cutaneous mechanoreceptors. IEEE Trans Biomed Eng 22:66–69
Carpenter SE, Lynn B (1981) Vascular and sensory responses of human skin to mild injury after topical treatment with capsaicin. Br J Pharmacol 73:755–758
Chahl IA, Kirk EJ (1975) Toxins which produce pain. Pain 1: 3–49
Chapman WP, Jones CM (1944) Variations in cutaneous and visceral pain sensitivity in normal subjects. J Clin Invest 23:81–91

Clark WC (1969) Sensory-decision theory analysis of the placebo effect on the criterion for pain and thermal sensitivity (d'). J Abnormal Psychol 74:363–371
Clark WC, Yang JC (1974) Acupunctural analgesia? Evaluation by signal detection theory. Science 184:1096–1098
Cormia FE, Dougherty JW (1960) Proteolytic activity in development of pain and itching. Cutaneous reactions to bradykinin and kallikrein. J Invest Dermatol 35:21–26
Cormia FE, Kuykendall U (1954) Experimental histamine pruritis. III. Influence of drugs on itch threshold. Arch Dermatol 69:206–218
Cragg BG, Downer J de C (1967) Behavioural evidence for cortical involvement in manual temperature discrimination in the monkey. Exp Neurol 19:433–442
Daly BM, Shuster S (1986) Effect of aspirin on pruritus. Br Med J 293:907
D'Amato MR (1970) Experimental psychology. Methodology, psychophysics and learning. McGraw-Hill, New York
D'Amour FE, Smith DL (1941) A method for determining loss of pain sensation. J Pharmacol 72:74–79
Darian-Smith I, Johnson KO (1977) Temperature sense in the primate. Br Med Bull 33:143–148
Dehen H, Willer JC, Prier S, Boureau F, Cambier J (1978) Congenital insensitivity to pain and the "morphine-like" analgesic system. Pain 5:351–358
Dubner R, Beitel RE (1976) Neural correlates of escape behaviour in rhesus monkey to noxious heat applied to the face. In: Bonica JJ, Albe-Fessard D (eds) Advances in pain research and therapy, vol 1. Raven, New York, pp 155–160
Dyck PJ, Curtis DJ, Busher W, Offord K (1974) Description of "Minnesota thermal discs" and normal values of cutaneous thermal discrimination in man. Neurology (Minneap) 24:325–330
Dyck PJ, Karnes J, O'Brien PC, Zimmermann IR (1984) Detection thresholds of cutaneous sensation in humans. In: Dyck PJ, Thomas PK, Lambert EH, Bunge R (eds) Peripheral Neuropathy, 2nd edn. Saunders, Philadelphia, pp 1103–1138
El-Sobky A, Dostrovsky JO, Wall PD (1976) Lack of effect of naloxone on pain perception in humans. Nature 263:783–784
Emmelin N, Feldburg W (1947) The mechanism of the sting of the common nettle (*Urtica urens*). J Physiol (Lond) 106:440–455
Feather BW, Chapman CR, Fisher SB (1972) The effect of a placebo on the perception of painful radiant heat stimuli. Psychosom Med 34:290–294
Ferreira SH (1972) Prostaglandins, aspirin-like drugs and analgesia. Nature [New Biol] 240:200–203
Ferreira SH (1979) Prostaglandins. In: Houck JC (ed) Chemical messengers of the inflammatory process. Elsevier, Amsterdam, pp 113–151, Handbook of inflammation, vol 1
Ferreira SH (1980) Peripheral analgesia: mechanism of the analgesic action of aspirin like drugs and opiate antagonists. Br J Clin Pharmacol 10:237S–245S
Fields HL, Basbaum AI (1978) Brainstem control of spinal pain transmission. Annu Rev Physiol 40:217–248
Fitzgerald M, Lynn B (1977) The sensitization of high threshold mechanoreceptors with myelinated axons by repeated heating. J Physiol (Lond) 265:549–563
Fox RH, Hilton SM (1958) Bradykinin formation in human skin as a factor in heat vasodilatation. J Physiol (Lond) 142:219–232
Franzen O (1969) The dependence of vibrotactile threshold and magnitude functions on stimulation frequency and signal level. Scand J Psychol 10:289–298
Fredrickson RCA, Burgis V, Edwards JD (1977) Hyperalgesia induced by naloxone follows diurnal rhythm in responsivity to painful stimuli. Science 198:756–758
Gammon GD, Starr I (1941) Studies on the relief of pain by counterirritation. J Clin Invest 20:13–20
Gatti H, Dodge R (1929) Über die Unterschiedsempfindlichkeit bei der Reizung eines einzelnen isolierten Tastorgans. Arch Gesamte Psychol 69:405–426
Gray WD, Osterberg AC, Scuto TJ (1970) Measurement of the analgesic efficacy and potency of pentazocine by the D'Amour and Smith method. J Pharmacol Exp Ther 173:154–162

Greaves MW, McDonald-Gibson W (1973) Itch: role of prostaglandins. Br Med J 3:608–609
Greaves MW, Shuster S (1967) Responses of skin blood vessels to bradykinin, histamine and 5-hydroxytryptamine. J Physiol (Lond) 193:255–267
Green AF, Young PA (1951) A comparison of heat and pressure analgesiometric methods in rats. Br J Pharmacol 6:572–587
Green DM, Swets JA (1966) Signal detection theory and psychophysics. Wiley, New York
Guillemin V, Benjamin F, Cornbleet T, Grossman MI (1952) A method of quantitative heat application to small skin areas at controlled temperature. J Appl Physiol 4:920–924
Hagermark O, Hokfelt T, Pernow B (1978) Flare and itch induced by substance P in human skin. J Invest Dermatol 71:233–235
Hallin RG, Torebjork HE (1976) Studies on cutaneous A and C fibre afferents, skin nerve blocks and perception. In: Zotterman Y (ed) Sensory functions of the skin in primates. Pergamon, Oxford, pp 137–148
Hardy JD, Oppel TW (1938) Studies in temperature sensation. IV. The stimulation of cold sensation by radiation. J Clin Invest 17:771–778
Hardy JD, Wolff HG, Goodell H (1952) Pain sensations and reactions. Williams and Wilkins, Baltimore
Harrington T, Merzenich MM (1970) Neural coding in the sense of touch: human sensations of skin indentation compared with the responses of slowly adapting mechanoreceptive afferents innervating the hairy skin of monkeys. Exp Brain Res 10:251–264
Hilder R, Ramey E, Darian-Smith I, Johnson KO, Dally LJ (1974) A contact stimulator for the study of cutaneous thermal sensibility. J Appl Physiol 37:252–255
Hume CW (1962) Avoidance of pain in the laboratory. In: Keele CA, Smith R (eds) The assessment of pain in man and animals. UFAW, London, pp 309–314
Hurley HJ, Koelle GB (1958) The effect of inhibition of non-specific cholinesterase on perception of tactile sensation in the human volar skin. J Invest Dermatol 31:243–245
Huskisson EC (1974) Measurement of pain. Lancet 2:1127–1131
Jaffe JH, Martin WR (1980) Opioid analgesics and antagonists. In: Gilman AG, Goodman LG, Gilman A (eds) The pharmacological basis of therapeutics, 6th edn. McMillan, New York, pp 494–534
Jamal GA, Hansen S, Weir AI, Ballantyne JP (1985) An improved, automated method for the measurement of thermal thresholds. I. Normal subjects. J Neurol Neurosurg Psychiatry 48:354–360
Jancso G, Kiraly E, Jancso-Gabor A (1980) Direct evidence for an axonal site of action of capsaicin. Naunyn-Schmiedeberg's Arch Pharmacol 313:91–94
Jancso G, Ferencsik M, Such G, Kiraly E, Nagy A, Bujdoso M (1985) Morphological effects of capsaicin and its analogues in newborn and adult mammals. In: Hakanson R, Sundler F (eds) Tachykinin antagonists. Elsevier, Amsterdam, pp 35–44
Jancso N (1960) Role of the nerve terminals in the mechanism of inflammatory rections. Bull Millard Fillimore Hosp 7:53–77
Jancso N (1968) Desensitization with capsaicin and related acylamides as a tool for studying the function of pain receptors. In: Lim RKS, Armstrong D, Pardo EG (eds) Pharmacology of pain. Pergamon, Oxford, pp 35–55 (Proceedings of the international pharmacology meeting, vol 9)
Janssen PAJ, Niemigeers CJE, Dony JGH (1963) The inhibitory effect of fentanyl and other morphine-like analgesics on the warm water induced tail withdrawal reflex in rats. Arzneimittelforschung 13:502–507
Keele CA (1970) Chemical causes of itch and pain. Annu Rev Med 21:67–74
Keele CA, Armstrong D (1964) Substances producing itch and pain. Arnold, London
Kenshalo DR (1978) Biophysics and psychophysics of feeling. In: Carterette EC, Friedman MP (eds) Handbook of perception, vol 6B. Academic, New York, pp 29–74
Kenshalo DR, Bergen DC (1975) A device to measure cutaneous temperature sensitivity in humans and subhuman species. J Appl Physiol 39:1038–1040

Kenshalo DR, Nafe JP, Dawson WW (1960) A new method for the investigation of thermal sensitivity. J Psychol 49:29–41
Kiistala U (1968) Suction blister device for separation of viable epidermis from dermis. J Invest Dermatol 50:129–137
Kling JW, Riggs LA (1971) Experimental psychology. Holt Rinehart and Winston, New York
Krause LB, Shuster S (1983) The mechanism of action of antipruritic drugs. Br Med J 287:1199–1200
LaMotte RH, Campbell JN (1978) Comparison of responses of warm and nociceptive C-fiber afferents in monkey with human judgements of thermal pain. J Neurophysiol 41:509–528
LaMotte RH, Mountcastle VB (1975) Capacities of humans and monkeys to discriminate between vibratory stimuli of different frequency and amplitude: a correlation between neural events and psychophysical measurements. J Neurophysiol 38:539–559
Lembeck F (1957) Untersuchungen über die Auslösung afferenter Impulse. Arch Exp Pathol Pharmakol 230:1–9
Lembeck F, Gamse R (1977) Lack of effect of substance P on paravascular pain receptors. Naunyn-Schmiedeberg's Arch Pharmacol 299:295–303
Levine JD, Gordon ND, Fields HL (1978) The mechanism of placebo analgesia. Lancet 2:654–657
Lewis T (1935) Experiments relating to cutaneous hyperalgesia and its spread through somatic nerves. Clin Sci 2:373–423
Lindahl O (1961) Experimental skin pain induced by injection of water soluble substances in human. Acta Physiol Scand 51:Suppl 179
Loh L, Nathan PW (1978) Painful peripheral states and sympathetic blocks. J Neurol Neurosurg Psychiatry 41:664–671
Lynn B (1977a) A mechanical stimulator for applying graded forces to the skin. J Physiol (Lond) 269:7–8P
Lynn B (1977b) Cutaneous hyperalgesia. Br Med Bull 33:103–108
Lynn B (1980) Heat pain sensitivity of human skin after mild heat injury and its lack of dependence on the local blood flow. Pain 8:189–196
Lynn B (1983) Cutaneous sensation. In: Goldsmith L (ed) Biochemistry and physiology of the skin. Oxford University Press, New York, pp 654–684
Lynn B, Perl ER (1977) A comparison of four tests for assessing the pain sensitivity of different subjects and test areas. Pain 3:353–365
Lynn B, Pini A (1985) Long-term block of afferent C-fibres following capsaicin treatment in the rat. J Physiol (Lond) 362:115–124
Lynn B, Pini A, Baranowski R (1987) Injury of somatosensory afferents by capsaicin: selectivity and failure to regenerate. In: Pubols LM, Sessle BJ (eds) Effects of injury on trigeminal and spinal somatosensory systems. Liss, New York, pp 115–124
Mackenzie RA, Burke D, Skuse NF, Lethlean AK (1975) Fibre function and perception during cutaneous nerve block. J Neurol Neurosurg Psychiatry 38:865–873
Maier SF, Davies S, Grau JW, Jackson RL, Morrison DH, Moye T (1980) Opiate antagonists and long-term analgesic reaction induced by inescapable foot shock in rats. J Comp Physiol Psychol 94:1172–1183
Marsland AR, Weekes JWN, Atkinson RL, Leong MG (1982) Phantom limb pain: a case for beta blockers? Pain 12:295–297
Melton FM, Shelley WB (1950) The effect of topical antipruritic therapy on experimentally induced pruritis in man. J Invest Dermatol 15:325–332
Merksey H, Spear FG (1964) The reliability of the pressure algometer. Br J Soc Clin Psychol 3:130–136
Meyer RA, Walker RE, Mountcastle VB (1976) A laser stimulator for the study of cutaneous thermal and pain sensations. IEEE Trans Biomed Eng 23:54–60
Moulton R, Spector WG, Willoughby DA (1957) Histamine release and pain production by xanthosine and related compounds. Br J Pharmacol 12:365–370
Mountcastle VB (1980) Medical physiology. Mosby, St Louis

Mountcastle VB, LaMotte RH, Carli G (1972) Detection thresholds for vibratory stimuli in humans and monkeys: a comparison with threshold events in mechanoreceptive afferent nerve fibres innervating the monkey hand. J Neurophysiol 35:122–136
Mullin FJ, Luckhardt AB (1935) Effects of certain analgesic drugs on cutaneous, tactile and pain sensitivity. Am J Physiol 113:100–101
Parkhouse J, Pleuvry BJ, Rees JMH (1979) Analgesic drugs. Blackwell, Oxford
Pert A (1982) Mechanisms of opiate analgesia and the role of endorphins in pain suppression. Adv Neurol 33:107–122
Petit J, Galifret Y (1978) Sensory coupling function and the mechanical properties of the skin. In: Gordon G (ed) Active touch. Pergamon, Oxford, pp 19–27
Pini A (1983) Effects of capsaicin on conduction in a cutaneous nerve of the rat. J Physiol (Lond) 338:60–61P
Poulton EC (1968) The new psychophysics: six models for magnitude estimation. Psychol Bull 69:1–19
Randall LD, Selitto JJ (1957) A method for measurement of analgesic activity on inflamed tissue. Arch Int Pharmacodyn 61:409–419
Ritchie JM, Greene NM (1980) Local anesthetics. In: Gilman AG, Goodman LS, Gilman A (eds) The pharmacological basis of therapeutics, 6th edn. Macmillan, New York, pp 300–320
Robertson I, Greaves MW (1978) Responses of human skin blood vessels to synthetic histamine analogues. Br J Clin Pharmacol 5:319–322
Robinson JE, Short RV (1977) Changes in breast sensitivity at puberty, during the menstrual cycle, and at parturition. Br Med J 1:1188–1191
Ruch T, Patton HD (1979) The brain and neural function. Saunders, Philadelphia (Physiology and biophysics, vol 1)
Scadding JW, Wall PD, Wynn Parry CB, Brooks DM (1982) Clinical trial of propranolol in post-traumatic neuralgia. Pain 14:283–292
Schweizer A, Brom R, Glatt M, Bray MA (1984) Leukotrienes reduce nociceptive responses to bradykinin. Eur J Pharmacol 105:105–112
Shelley WB, Arthur RP (1957) The neurohistology and neurophysiology of the itch sensation in man. Arch Dermatol 76:296–323
Shuster S (1981) Reason and the rash. Proc R Inst GB 53:136–163
Sicuteri F, Franchi G, Fanciulacci M,del Bianco PL (1965) Serotonin bradykinin potentiation of pain receptors in man. Life Sci 4:309–316
Sinclair D (1981) Mechanisms of cutaneous sensation. Oxford University Press, Oxford
Skouby AP (1951) Sensitization of pain receptors by cholinergic substances. Acta Physiol Scand 24:174–191
Skouby AP (1953) The influence of acetylcholine, curarine and related substances on the threshold for chemical pain stimuli. Acta Physiol Scand 29:340–352
Smith GM, Beecher HK (1969) Experimental production of pain in man: sensitivity of a new method to 600 mg aspirin. Clin Pharmacol Ther 10:213–216
Stevens JC, Stevens SS (1960) Warmth and cold: dynamics of sensory intensity. J Exp Psychol 60:183–192
Stevens SS (1957) On the psychophysical law. Psychol Rev 64:153–181
Stevens SS (1959) Tactile vibration: dynamics of sensory intensity. J Exp Psychol 57:210–218
Stevens SS (1970) Neural events and the psychophysical law. Science 170:1043–1050
Stewart JM, Getto CJ, Neldner K, Reeve EB, Kirvoy WA, Zimmermann E (1976) Substance P and analgesia. Nature 262:784–785
Stoll AM, Hardy JD (1950) Study of thermocouples as skin thermometers. J Appl Physiol 2:531–543
Szolcsanyi J (1977) A pharmacological approach to elucidation of the role of different nerve fibres and receptor endings in mediation of pain. J Physiol (Paris) 73:251–259
Talbot WH, Darian-Smith I, Kornhuber HH, Mountcastle VB (1968) The sense of flutter vibration: comparison of the human capacity with response patterns of mechanoreceptive afferents from the monkey hand. J Neurophysiol 31:301–334
Tapper DW (1970) Behavioural evaluation of the tactile pad receptor system in hairy skin of the cat. Exp Neurol 26:447–459

Torebjork HE, Hallin RG (1973) Perceptual changes accompanying controlled preferential blocking of A and C fibre responses in intact human skin nerves. Exp Brain Res 16:321–332
Toth-Kasa I, Jancso G, Bognar A, Husz S, Obal F (1986) Capsaicin prevents histamine-induced itching. Int J Clin Pharmacol Res 6:163–169
Verillo RT, Capraro AJ (1975) Effect of stimulus frequency on subjective vibrotactile magnitude ratings. Percept Psychophys 17:91–96
Vierck CJ (1973) Alterations of spatio-tactile discrimination after lesions of primate spinal cord. Brain Res 58:69–79
Vierck CJ (1974) Tactile movement detection and discrimination following dorsal column lesions in monkeys. Exp Brain Res 20:331–346
Von Frey M, Kiesow F (1899) Ueber die Function der Tastkoerperchen. Z Psychol Physiol Sinnesorgane 20:126–163
Wall PD (1970) The sensory and motor role of impulses travelling in the dorsal columns towards cerebral cortex. Brain 93:505–524
Wall PD, Gutnick M (1974) Ongoing activity in peripheral nerves: the physiology and pharmacology of impulses originating from a neuroma. Exp Neurol 43:580–593
Wallin G, Torebjork E, Hallin R (1976) Preliminary observations on the pathophysiology of hyperalgesia in the causalgic pain syndrome. In: Zotterman Y (ed) Sensory functions of the skin in primates. Pergamon, Oxford, pp 489–499
Weber EH (1978) The sense of touch. Academic, London
Weinstein S (1977) Effects of local anaesthetics on tactile sensitivity thresholds for cutaneous and mucous membranes. J Invest Dermatol 69:136–145
Werner G, Mountcastle VB (1965) Neural activity in mechanoreceptive cutaneous afferents: stimulus-response relations, Weber functions and information transmission. J Neurophysiol 28:359–397
Westerman R, Magerl W, Handwerker H, Szolcsanyi J, Pratt A (1986) Itch sensation evoked by percutaneous histamine iontophoresis and cowhage spicules applied to human skin. Proc Int Union Physiol Sci 16:571
Westling G, Johansson R, Vallbo AB (1976) A method for mechanical stimulation of skin receptors. In: Zotterman Y (ed) Sensory functions of the skin in primates. Pergamon, Oxford, pp 151–158
Wikler A, Goodell H, Wolff HG (1945) Studies on pain. The effect of analgesic agents on sensations other than pain. J Pharmacol Exp Ther 83:294–299
Willer JC, Boureau F, Berny J (1979) Nociceptive flexion reflexes elicited by noxious laser radiant heat in man. Pain 7:15–20
Willer JC, Dehen H, Cambier J (1981) Stress-induced analgesia in humans: endogenous opioids and naloxone-reversible depression of pain reflexes. Science 212:689–691
Winer BJ (1971) Statistical principles in experimental design. McGraw-Hill, New York
Woolf CJ (1983) Evidence for a central component of post-injury pain hypersensitivity. Nature 306:686–688
Yaksh TL (1981) Spinal opiate analgesia: characteristics and principles of action. Pain 11:293–346
Yaksh TL, Reddy SVR (1981) Studies in the primate on the analgetic effects associated with intrathecal actions of opiates, alpha-adrenergic agonists and baclofen. Anesthesiology 54.451–467
Zotterman Y (1959) Thermal sensations. In: Field J, Magoun HW (eds) Neurophysiology, vol 1. American Physiological Society, Washington, pp 431–458 (Handbook of physiology, sect 1)

CHAPTER 7

The Measurement of Itch

S. SHUSTER

A. Introduction

Itch is a uniquely cutaneous sensation related to the superficial organisation of pain sensory nerves in the skin (see Chap. 6; see also Chap. 10 in Part 1 of this volume). It has been suggested that its anatomical localisation and relationship to pain and scratch are likely to have evolved as an integrated response to irritants such as parasites, successful removal of which by scratching away the superficial layers of the skin is signalled when itch is replaced by pain (SHUSTER 1981). Itch and skin appearance are the main symptoms of cutaneous disease and the treatment of itch is therefore a major concern. Although treatment of itch is primarily that of the causal disease, e.g. the itch of scabies or the rash of dermatitis herpetiformis, anti-pruritics are used during the early stages of response of several diseases and in those diseases where control is incomplete (e.g. see Chap. 32). Thus, the development of more potent and specific anti-pruritic drugs and the selection of the most satisfactory drugs from amongst those already available is desirable and to this end methods for the measurement of itch are required. Since by its very nature sensation cannot be measured directly, either the sensation is recorded *subjectively, but directly* by the individual experiencing it or else its consequences are measured *objectively, but indirectly*. The former integrates quality and magnitude of stimulus with its interpretation which includes cortical modification; the latter simply measures the consequence of a response which, depending on the type of measurement, may be subject to cortical modulation. The subjective methods are better for gauging relative changes in an individual than differences between individuals: they are far more susceptible to extraneous variables and are less susceptible to mathematical analysis than objective methods. Although objective methods give more precise results, their very indirectness makes it necessary to exclude changes due to other unrelated mechanisms and if possible to control them out of the experimental study.

B. Subjective Methods

These have mostly been of threshold and degree and there has been little work on characterisation of type of itch.

I. Threshold

This has mostly been studied by using various concentrations of a pruritogen such as histamine on abraded or blistered epidermis; e.g. the effect of prostaglandins on itch threshold (GREAVES and McDonald-Gibson 1973). Blind testing and considerations of statistical design can greatly improve the precision of the methods (LOVELL et al. 1976).

II. Degree

Three methods used are: (a) simple nominal grading, e.g. +, + +, + + + or 1–3; (b) simple graded response, e.g. putting on a recording lever as used by KEELE and ARMSTRONG (1964); and (c) marking severity on a 10-cm line (e.g. KRAUSE and SHUSTER 1983). There has been no direct comparison of these methods and until such studies are done the simple, reproducible and easily analysed linear analogue response seems to be the method of choice. The recently introduced portable computerised data recorder on which subjective severity can be noted at hourly intervals may have advantages for studying the effect of drugs during the day (WAHLGREN et al. 1988).

C. Objective Methods

The only methods developed so far are based on the measurement of itch as scratch. SAVIN et al. (1975) used needle electrodes to measure number and duration of nocturnal scratch bouts. The studies were done in a sleep laboratory and could be related to sleep patterns. The disadvantage of the method is that only one patient can be studied at a time and the equipment is distracting. More importantly, as the method only measures duration and frequency, but not amplitude of scratch movement, it is only semi-quantitative. Thus, using a double-blind placebo-controlled plan they were only able to show modest changes with trimeprazine and trimipramine (SAVIN et al. 1979) although much larger changes could be detected by measuring scratch as total limb movement in open studies of these and other drugs (FELIX and SHUSTER 1975a, b; MUSTON et al. 1979; SHUSTER 1981; KRAUSE and SHUSTER 1983).

Two methods which are more quantitative and do not disturb the patients are the measurement of itch-provoked scratch as total body movement from bed movement and from limb meters.

I. Nocturnal Bed Movement

By means of a transducer on one bed leg the symmetrical movements of scratch can easily be distinguished from the asymmetrical movements and change of base-line of restlessness. This permits the measurement of duration, amplitude and frequency of bouts of itch-provoked scratch, their relationship to levels of sleep and to restlessness, as well as defining the time of scratch and its relationship to other nocturnal events. Thus, the method can be used for detailed study of itch (FELIX and SHUSTER 1975a, b; SHUSTER 1981). The disadvantage of the method as published is that measurement of amplitude is semi-quantitative because of the positional response of the sensor.

II. Limb Meters

The limb meters which have been used are a specially adapted self-winding watch (Felix and Shuster 1975a, b) or a modification of it (Summerfield and Welch 1980). Because scratch is done mostly by the hands, and the legs mostly represent restlessness, by measurement of arm and leg movement an estimate of both scratch and restlessness can be obtained. Since, however, there is evidence that itch provokes restlessness (Felix and Shuster 1975a, b) it appears that, provided measurement of movement in all four limbs is made to show that movement is predominantly by the arms (usually $>90\%$), total body movement is the best measure of the response to itch. The method is easy to use and has been used by patients in a home study of uraemic pruritus (H. Muston, S. Shuster and D. Kerr 1979, unpublished work).

III. Relationship of Itch and Scratch

By these methods the patient's subjective retrospective assessment of itch on the morning corresponds well to the objective measurement of body movement and the response to drugs (Felix and Shuster 1975a, b; Savin 1980). The exception is barbiturate which visibly and measurably increases nocturnal itch-provoked scratch, but seems to impair its subjective recall (Shuster 1981). The relationship of itch and scratch differs in different diseases and to some degree in the same disease, dependent upon its extent. This is because the symptom of itch is more related to point severity whilst the response of scratch is also related to extent. Thus, comparisons should consider and match for disease type and severity. These problems can be minimised in drug studies by comparing responses sequentially in individual patients, e.g. two nights before, during and after administration of a drug (see Krause and Shuster 1983).

IV. Short-term Measurement of Itch as Scratch

The limb meters in use are too insensitive for short-term measurement. Bed movement detects minor movement, but conscious control of short-term bouts of scratch makes its use unsatisfactory other than during sleep.

References

Felix RH, Shuster S (1975a) A new method for the measurement of itch and the response to treatment. Br J Dermatol 93:303–312

Felix RH, Shuster S (1975b) Measurement of itch. Mechanisms of topical corticosteroid activity. Livingstone, London, pp 106–113

Greaves MW, McDonald-Gibson W (1973) Itch: role of prostaglandins. Br Med J 3:608–609

Keele CA, Armstrong D (1964) Substances producing itch and pain. Arnold, London

Krause LB, Shuster S (1983) The mechanism of antipruritic drugs. Br Med J 287:1199–1200
Lovell CR, Burton PA, Duncan EHL, Burton JL (1976) Prostaglandins and pruritus. Br J Dermatol 94:273–275
Muston H, Felix R, Shuster S (1979) Differential effect of hypnotics and anxiolytics on itch and scratch. J Invest Dermatol 72:283
Savin JA (1980) Do systemic antipruritic agents work? Br J Dermatol 103:113–118
Savin JA, Paterson WD, Oswald I, Adam K (1975) Further studies of scratching during sleep. Br J Dermatol 93:297–302
Savin JA, Paterson WD, Adam K, Oswald I (1979) Effects of trimeprazine and trimipramine on nocturnal scratching in patients with atopic eczema. Arch Dermatol 115:313–315
Shuster S (1981) Reason and the rash. Proc R Inst GB 53:136–163
Summerfield JA, Welch ME (1980) The measurement of itch with sensitive limb movement meters. Br J Dermatol 102:275–281
Wahlgren C-F, Hagermark O, Bergstrom R, Hedin B (1988) Evaluation of a new method of assessing pruritus and antipruritic drugs. Skin Pharmacol 1:3–13

CHAPTER 8

Measurement of Skin Thickness, Wealing, Irritant, Immune and Ultraviolet Inflammatory Response in Skin

P. M. FARR, C. M. LAWRENCE, and S. SHUSTER

A. Measurement of Skin Thickness

Measurement of skin and lesion thickness is one of the few objective ways of measuring the cutaneous response to inflammatory and therapeutic agents in humans. The three techniques described here measure epidermal and dermal thickness, although the normal epidermis contributes only about 5%. Variation in skin thickness occurs with site, age and sex and is due principally to differences in dermal collagen layer (SHUSTER et al. 1975).

I. The Harpenden Skin Fold Caliper

The Harpenden skin fold caliper is a hand-held instrument consisting of one fixed and one spring-loaded handle which open the rectangular jaws. It was designed to measure subcutaneous fat thickness (TANNER and WHITEHOUSE 1955), but, at sites where the skin can be picked up separate from the underlying fatty tissue, such as the dorsum of the hand or the flexor aspect of the forearm, the caliper can be used to measure a double thickness of skin without subcutaneous tissue. To minimise the compression achieved at the jaw faces the instrument is used with one spring removed which produces a pressure at the jaw face of 5 gm^{-2} and only a small degree of compression by fluid displacement (COOK and SHUSTER 1980). The calipers have a resolution of 0.1 mm and give reproducible measurements (LAWRENCE and SHUSTER 1985) with little observer bias (COOK and SHUSTER 1980). Harpenden calipers have been used to measure both normal and atrophic skin thickness (MCCONKEY et al. 1963), skin lesions (MARSDEN et al. 1983; KRAUSE and SHUSTER 1985) and their response to inflammation (MOSS et al. 1981; LAWRENCE and SHUSTER 1987). The advantages of the Harpenden caliper are that it is inexpensive, portable, easy to use and robust, and gives a reproducible skin thickness measurement in normal and inflamed skin. The disadvantages are that it can only be used on certain sites such as the flexor forearm where a fold of skin can be lifted separate from the underlying fat; and in some patients it is impossible to lift a forearm skin fold because of obesity, muscularity or tight skin (LAWRENCE et al. 1986).

II. X-ray

The radiographic method for measuring skin thickness (Meema et al. 1964) involves directing an X-ray beam through the skin and using a wooden block to flatten and immobilise the area to be measured; additional metal plates are used to facilitate focusing of the beam and to detect rotation of the block (Black et al. 1973). Skin thickness is measured directly from the radiographic image by a magnifying lens. The technique gives results comparable to Harpenden calipers (Dykes and Marks 1977) and ultrasound (Tan et al. 1981, 1982). It has been used to measure corticosteroid-induced dermal thinning (Black et al. 1973; Tan et al. 1981). The disadvantages are the need to use ionising radiation, the expensive, non-portable equipment and the fact that the method can only be used where the X-ray beam runs tangentially to the skin surface, effectively limiting the technique to limb skin.

III. Ultrasound

Skin thickness can be measured by using ultrasound (Alexander and Miller 1979; Kirsch et al. 1984). A transducer, offset from the skin by a water bath, detects the pulse transit time which is displayed on an oscilloscope, with peaks representing the water-skin interface, the dermis-fat interface and the fat-muscle interface. The distance between the first and second peaks represents the time taken for the ultrasound signal to travel through the skin. Skin thickness is calculated from this using a constant for the speed of waveform in human skin (Tan et al. 1981). The method has been used to measure normal and atrophic skin thickness and gives comparable results to the X-ray and caliper techniques (Tan et al. 1981, 1982), although the values are lower (Lawrence and Shuster 1985), and histological measurements show that ultrasound underestimates skin thickness (Tan et al. 1982). The similar poor correlation between X-ray and histological skin thickness (Dykes and Marks 1977; Tan et al. 1982) is in part due to fixation shrinkage. The advantage of ultrasound is that the equipment is portable and the method is quick and harmless and can be used at any skin site. The disadvantages are that it gives poor results in inflamed and lesional skin (Lawrence and Shuster 1985) partly because it fails to detect the lower limit of the dermis in that situation and partly because inflammation may extend beyond that interface.

IV. Use of Skin Thickness for Lesion Measurement and Response to Treatment

Plaque thickness on the forearm has been measured with Harpenden calipers as a method of monitoring the effect of treatment of psoriasis (Marsden et al. 1983; Farr et al. 1987) and the degree of inflammation of the plaque (Paramsothy and Lawrence 1987). There are no published studies of the use of the X-ray and ultrasound techniques, but our own unpublished evidence is that ultrasound gives variable and unsatisfactory recordings.

V. Corticosteroid Atrophy of Skin

Skin thickness measurements have been widely used to assess changes in skin connective tissue, particularly collagen loss after administration of corticosteroids since the first study of BLACK et al. (1973). However, there are serious reservations with this because changes in skin thickness are only an indirect measure of dermal collagen content (SHUSTER et al. 1975; BLACK et al. 1973). More importantly, most studies of the atrophic potency of corticosteroids are concerned with small changes in skin thickness which occur in the first weeks of treatment and there is no evidence that these changes are due to collagen loss. Thus, the measurement of skin thickness first introduced by BLACK et al. (1973) for the analysis of atrophic effects of corticosteroids has to be used critically in closely defined situations.

B. Measurement of Different Types of Inflammation and Their Response to Therapy

I. Measurement of Ultraviolet Erythema

The present account deals with ultraviolet erythema because this has been a major use of measurement of erythemal response. However, the methods discussed are applicable equally to other forms of erythema.

1. Minimal Erythema Dose

The erythemal response of the skin to ultraviolet radiation is usually inferred from the minimal erythema dose, determined by exposing adjacent areas of skin to increasing doses of radiation, usually a geometrical series, and recording the lowest dose of radiation to achieve erythema at a specified time, usually 24 h after irradiation.

The visual detection of erythema is subjective and is affected by unrelated factors such as viewing geometry, intensity and spectral composition of ambient illumination, colour of unexposed surrounding skin (CHAMBERLIN and CHAMBERLIN 1980) and the experience and visual acuity of the observer. The difficulty in judging a minimal erythema response accurately is reflected by the varying definitions which range from the dose required to initiate a faint, but easily discernible erythema (EPSTEIN 1962), to that dose which will just produce a uniform redness with sharp borders (WILLIS and KLIGMAN 1970). More importantly, the minimal erythema dose is a single point on the dose-response curve and an apparent threshold measurement; it cannot therefore characterise a biological event: it provides no information about the effect of higher doses and may consequently fail to reveal significant differences between responses.

2. Quantification of Erythema

The methods used to quantify erythema are: (a) visual grading; (b) colour comparison charts; (c) red-coloured optical filters; and (c) reflectance spectrophotometry.

a) Visual Grading

The ordinal scale, in which the observer allots a grade to a particular intensity of erythema (e.g. HAWK and PARRISH 1982), has several drawbacks. It remains subjective and, although under ideal viewing conditions it is possible to detect small differences in erythemal intensity between adjacent irradiated sites, the eye is poor at estimating the exact value of the difference. Also it is often assumed that the erythemal increments between grades are equal, whereas many ordinal scales in fact have a skewed distribution. Dose-response curves constructed in this way for different radiation wavelengths have been produced, but very few data points are possible, no indication of the measurement uncertainty can be given and the widely spaced dapta points encourage the siting of readings between allotted grades (see HAWK and PARRISH 1982).

b) Colour Comparison Charts

HAUSSER and VAHLE (1927) graded the intensity of erythema by comparison with dilutions of a red solution put on to matt skin-coloured paper. Although it remains subjective, this method has advantages over the visual grading of erythema as the eye performs well in a null system such as the matching of two colours. Dose-response curves for five wavelengths of radiation with erythema graded by colour comparison charts (HAUSSER and VAHLE 1927) show that erythema increases more rapidly with multiples of the minimal erythema dose for radiation around 300 nm (UV-B) than for 254 nm radiation (UV-C).

c) Red-Coloured Optical Filters

A series of photographic filters with a high transmittance for red light and decreasing transmittance for blue-green light has been used (BERGER et al. 1968) to quantify erythema by noting which of the filters just caused the erythema to disappear.

d) Reflectance Spectrophotometry

Reflectance spectrophotometry can be used to measure erythema because haemoglobin in the blood vessels of the superficial dermis is the main cutaneous chromophore of green light. In areas of erythema there is a reduction in reflectance of green light (around 550 nm) compared with normal non-erythematous skin. Simple photoelectric reflectance meters designed to quantify erythema by measuring the amount of green light re-emitted from the skin have been available for some years (see TRONNIER 1969), but these instruments were too cumbersome and imprecise for routine clinical use. Reflectance instruments using modern electro-optical technology (DIFFEY et al. 1984) can record the degree of cutaneous erythema easily and precisely and have been used to measure objectively the erythemal response of human skin to different wavelengths of ultraviolet radiation (Fig. 1) (FARR and DIFFEY 1985).

The development of computer-controlled spectrophotometers has allowed reliable measurement of the spectral reflectance of human skin in vivo (DAWSON et al. 1980; WAN et al. 1983); they provide full, diffuse spectra over the ultraviolet and visible regions, but the time required for scanning and recording at each site,

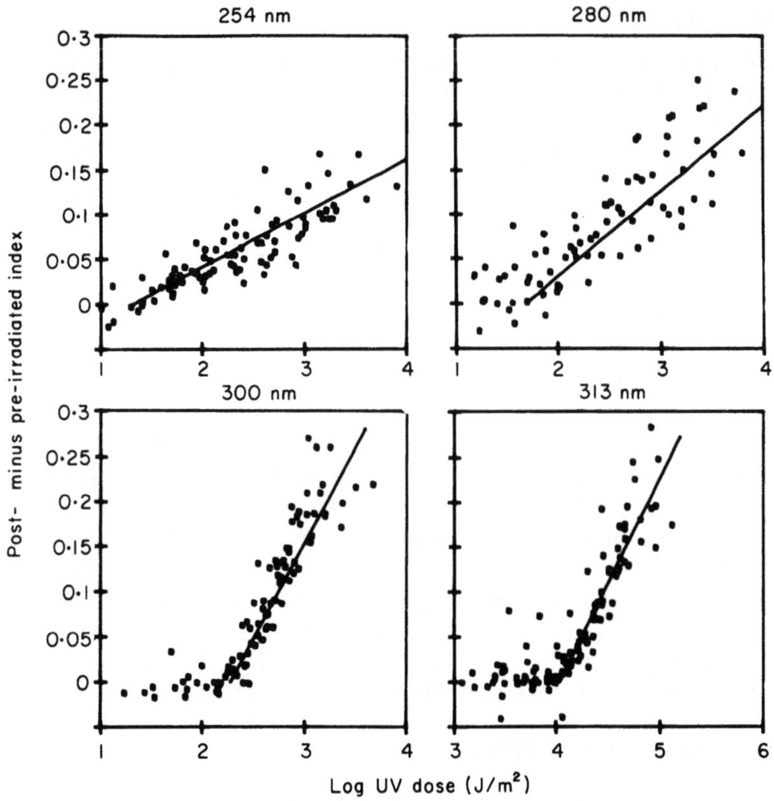

Wavelength (nm)	No. subjects	Pooled slope ± 1 s.e.	Threshold dose (J/m²) Geometric mean	95% Confidence limits
254	15	0·060 ± 0·001	21	3–167
280	10	0·095 ± 0·002	49	14–173
300	13	0·211 ± 0·006	190	63–576
313	17	0·233 ± 0·004	10,855	4100–28800

Fig. 1. Pooled dose-response curves for four wavelengths of ultraviolet radiation

and the bulkiness and expense of the equipment present practical limitations to their use.

II. Weal Reactions

The response to vasoactive agents such as intradermal histamine and antagonists can be measured as weal area, weal thickness and the calculated volume (COOK and SHUSTER 1980). Weal area can be measured at the time of maximum size and definition by marking the edge with ink, transferring the outline on cellophane-tape-stripped stratum corneum and computing the area on a digitising pad or planimeter (KRAUSE and SHUSTER 1984; HUMPHREYS and SHUSTER 1987). Weal thickness is measured by Harpenden calipers. Weal dose-response curves can be used to study the effect of disease and treatment. Skin thickness can also be used to measure the rate of formation and disappearance of weals; rate constant have been established for histamine wealing (COOK and SHUSTER 1980) and differences between various wealing agents, e.g. histamine, compound 48/80 and antigen weals (HUMPREYS et al. 1987).

1. Dermographic Wealing

A variety of spring-loaded stylus devices are used to give a reproducible force as the stylus is moved across the skin. Using weal diameter as a measure of response, weal force-response curves are constructed for a series of different forces. Site and skin frictional resistance are important and this may be the reason that the force-response curves (KRAUSE and SHUSTER 1985) are less linear than those for intradermal histamine injections. The response to H_1-receptor antagonists can be demonstrated by a shift in the force-response curve in dermographic urticaria (KRAUSE and SHUSTER 1984, 1985). The clinical response correlates well with the increase in dermographic threshold which in turn corresponds to subjective itch.

III. Irritant Inflammation

The term irritant has the connotation of tissue damage not due to an immune mechanism or to a primary pharmacological effect. The response to irritants is varied and it is simplistic to consider all such reactions as a homogeneous group. A variety of "irritants" are used experimentally, e.g. detergents and weak acids, and the response is measured as erythema, oedema and transepidermal water loss. The response to dithranol (anthralin) and the effect on it of drugs has been used as an example of an "irritant" non-immune reaction perhaps due to an effect of free radicals on cell membranes (FINNEN et al. 1984). By means of Harpenden calipers to measure the inflammatory oedema and reflectance spectrophotometry for erythema, the effect of various topical promoters and inhibitors of anthralin inflammation have been studied (LAWRENCE et al. 1985, 1987; LAWRENCE and SHUSTER 1987; KINGSTON and LOWE 1986).

1. Transepidermal Water Loss

This has been used to measure the response to detergent application and inflammatory reactions to various agents as well as the loss of barrier function in scaly disorders such as ichthyosis, eczema and psoriasis. The methodology used has been the measurement of water vapour leaving the skin surface into a stream of dry nitrogen;in this system water loss is a function of gas flow (JOHNSON and SHUSTER 1969), or more easily, if less precisely, by comparing relative humidity at two points at different distances from the skin surface (NILSSON 1977).

IV. Immune Reactions

1. The Immediate (Type I) Response

The type I response can be measured by the wealing reaction to antigen most commonly introduced by pricking. However, this is less satisfactory than intradermal injection for creating dose-response curves (HUMPHREYS et al. 1987). The delayed response to antigen such as house dust mite which occurs in patients with atopy remains semi-quantitative, and dose-response curves have not yet been established.

2. The Delayed (Type IV) Response

The type IV immune reaction of contact eczema is most easily examined with dinitrochlorobenzene (DNCB) (MOSS et al. 1981). The capacity of a particular population to become sensitised is measured by using different sensitising doses of DNCB and the degree of sensitisation is measured 3–4 weeks later by the magnitude of the response to increasing concentrations of DNCB. The dose-response curve is measured as thickness with Harpenden calipers: this is more discriminating than change in area. Some measurements of response have been made using intensity of erythema and transepidermal water loss, but these measures have not yet been validated in humans.

References

Alexander H, Miller DL (1979) Determining skin thickness with pulsed ultrasound. J Invest Dermatol 72:17–19

Berger D, Urbach F, Davies RE (1968) The action spectrum of erythema induced by ultraviolet radiation. Preliminary report. In: Jadassohn W, Schirren CG (eds) XIII Congressus Internationalis Dermatologiae. Springer, Berlin 2:1112–1217

Black MM (1969) A modified radiographic method of measuring skin thickness. Br J Dermatol 81:661–668

Black MM, Shuster S, Bottoms E (1973) Skin collagen and thickness in Cushing's syndrome. Arch Dermatol Forsch 246:365–368

Chamberlin GJ, Chamberlin DG (1980) Colour: its measurement, computation and application. Heyden, London, p 46

Cook J, Shuster S (1980) Histamine weal formation and absorption in man. Br J Pharmacol 69:579–585

Dawson JB, Barker DJ, Ellis DJ, Grassam E, Cotterill JA, Feather JW (1980) A theoretical and experimental study of light absorption and scattering by in vivo skin. Phys Med Biol 25:695–709

Diffey BL, Oliver RJ, Farr PM (1984) A portable instrument for quantifying erythema induced by ultraviolet radiation. Br J Dermatol 111:663–672

Dykes RJ, Marks R (1977) Measurement of skin thickness; a comparison of 2 in vivo techniques and conventional histometric methods. J Invest Dermatol 69:275–278

Epstein JH (1962) Polymorphous light eruptions. Wavelength dependency and energy studies. Arch Dermatol 85:82–88

Farr PM, Diffey BL (1985) The erythemal response of human skin to ultraviolet radiation. Br J Dermatol 113:65–76

Farr PM, Diffey BL, Marks JM (1987) Phototherapy and dithranol treatment of psoriasis: new lamps for old. Br Med J 294:205–207

Finnen MJ, Lawrence CM, Shuster S (1984) Inhibition of anthralin inflammation by free radical scavengers. Lancet 2:1129–1130

Hausser KW, Vahle W (1927) Sonnenbrand und Sonnenbräunung. Wissenschaftliche Veröffentlichungen des Siemens Konzerns 6:101–120
Hawk JLM, Parrish JA (1982) Responses of normal skin to ultraviolet radiation. In: Regan JD, Parrish JA (eds) The science of photomedicine. Plenum, New York
Humphreys F, Shuster S (1987) The effect of nedocromil on weal reactions in human skin. Br J Clin Pharmacol 24:405–408
Humphreys F, Krause LB, Shuster S (1987) The effects of astemizole and indomethacin on weal and flare reactions to histamine 48/80 and house dust mite antigen. Br J Dermatol 116:435
Johnson C, Shuster S (1969) The measurement of transepidermal water loss. Br J Dermatol [Suppl 4]81:40–45
Kingston T, Lowe NJ (1986) Factors influencing anthralin irritancy. Br J Dermatol [Suppl 31]115:80–81
Kirsch JM, Hanson ME, Gibson JR (1984) The determination of skin thickness using conventional diagnostic ultrasound equipment. Clin Exp Dermatol 9:280–285
Krause LB, Shuster S (1984) The effect of terfenadine on dermographic wealing. Br J Dermatol 110:73–80
Krause LB, Shuster S (1985) A comparison of astemizole and chlorpheniramine in dermographic urticaria. Br J Dermatol 112:447–453
Lawrence CM, Shuster S (1985) Comparison of ultrasound and caliper measurement of normal and inflamed skin thickness. Br J Dermatol 112:195–200
Lawrence CM, Shuster S (1987) Effect of arachidonic acid on anthralin inflammation. Br J Clin Pharmacol 24:125–131
Lawrence CM, Howel D, Shuster S (1986) Site variations in anthralin inflammation on forearm skin. Br J Dermatol 114:609–613
Lawrence CM, Shuster S, Collins M, Bruce JM (1987) Reduction of anthralin inflammation by potassium hydroxide and Teepol. Br J Dermatol 116:171–177
Marsden JR, Coburn PR, Marks JM, Shuster S (1983) Measurement of the response of psoriasis to short term application of anthralin. Br J Dermatol 109:209–218
McConkey B, Fraser GM, Bligh AS, Whiteley H (1963) Transparent skin and osteoporosis. Lancet 1:693–695
Meema HE, Sheppard RH, Rappoport A (1964) Roentgenographic visualisation and measurement of skin thickness and its diagnostic application in acromegaly. Radiology 82:411
Moss C, Friedmann PS, Shuster S (1981) Impaired contact hypersensitivity in untreated psoriasis and the effects of photochemotherapy and dithranol/UV-B. Br J Dermatol 105:503–508
Nilsson GE (1977) Measurement of water exchange through the skin. Med Biol Eng Comput 15:209–218
Paramsothy Y, Lawrence CM (1987) Time course and intensity of anthralin inflammation on involved and uninvolved psoriatic skin. Br J Dermatol 116:517–519
Shuster S, Black MM, MacVitie E (1975) The influence of age and sex on skin thickness, skin collagen and density. Br J Dermatol 93:639–643
Tan CY, Marks R, Payne P (1981) Comparison of xeroradiographic and ultrasound detection of corticosteroid induced dermal thinning. J Invest Dermatol 76:126–128
Tan CY, Statham B, Marks R, Payne PA (1982) Skin thickness measurement by pulsed ultrasound: its reproducibility, validation and variability. Br J Dermatol 106:657–667
Tanner JM, Whitehouse RH (1955) The Harpenden skinfold caliper. Am J Phys Anthropol 13:743–746
Tronnier H (1969) Evaluation and measurement of ultraviolet erythema. In: Urbach F (ed) The biologic effects of ultraviolet radiation with emphasis on the skin. Pergamon, Oxford, pp 255–256
Wan S, Jaenicke KF, Parrish JA (1983) Comparison of the erythemogenic effectiveness of ultraviolet-B (290–320 nm) and ultraviolet-A (320–400 nm) radiation by skin reflectance. Photochem Photobiol 37:547–552
Willis I, Kligman AM (1970) Aminobenzoic acid and its esters. Arch Dermatol 102:405–417

CHAPTER 9

Measurement of Drug Action in Skin: Dermal Connective Tissue

B. NUSGENS and C. M. LAPIÈRE

A. Introduction: Selection of Procedures and Biological Models

Investigating the activity of a pharmacological agent on the dermis requires some knowledge of its potential effect. As described earlier (Chap. 27) the connective support of the skin is a complex tissue resulting from the biosynthetic activity of differentiated cells producing a variety of components interacting to achieve defined supportive properties. A modification induced in any one of these products of biosynthesis may lead to an alteration in the function of the dermis. There exists a large array of procedures that can measure the amount, the quality and often the physicochemical organisation of the macromolecules composing the dermis. The development of specific cDNA probes for the macromolecules forming the dermal connective tissue allows quantitative and qualitative analysis at the transcriptional level. However, one needs to define what component is involved before investigating the mechanism by which it occurs.

To obtain meaningful measurements on skin the samples to be compared should correspond to a constant portion of the organ, i.e. a full-thickness biopsy of a constant surface of skin, as originally proposed by BOTTOMS and SHUSTER (1963) (see also SHUSTER et al. 1975) and not a unit weight of skin or the ratio to another constituent; or, in small experimental animals as an aliquot of the whole homogenised skin. Besides considering the dermis under physiological conditions, one might prefer to investigate the effect of a drug on granulation tissues induced by a foreign material (cotton pellet, Ivalon sponge, carrageenan granuloma, etc.). The total granuloma can be collected easily and reproducibly. Granulomas are temporally and spatially well-defined masses of connective tissue where cell-cell and cell-matrix interactions remain intact. They have the advantage that such connective tissues are newly formed, that their constituents are often easier to extract and that their synthesis triggered at a defined time can be monitored to occur under the influence of the drug to be tested. Granuloma studies may however provide information unrelated to the connective tissue proper, but to the mechanism inducing the granulomatous reaction (for further discussion see KULONEN 1976; HANSEN 1980). The healing wound is another type of induced proliferation of connective tissue in vivo that can also be used to assess the activity of drugs (ZEDERFELDT 1980).

To investigate the pharmacological activity of a compound we would consider the following strategy. The first step should aim at demonstrating the reality of the effect of a drug on the dermis on the basis of modified physical properties or morphological parameters and/or change in chemical composition. In terms of physical properties one can consider rheological properties and dermal thickness. Morphological analysis would include standard microscopy, histochemistry, immunohistochemistry and ultrastructural microscopy. Hybridisation in situ with specific cDNA probes provides information on the precise localisation of the cells involved in the synthesis of a defined molecule. Chemical procedures for measuring the variation of a specific component can then be applied to precisely defined samples.

The second step should provide information on the mechanism or mechanisms responsible for an effect of the drug. It would consist in estimating the biosynthesis and/or the degradation rates of a defined macromolecule by radiotracing in experimental animals in vivo or by short-term organ culture in a suitable experimental model. Both procedures can investigate the effect of systemic or topical administration of a pharmacological agent on the balance between synthesis and degradation. Pulse-chase measurement and evolution of this balance with time can provide information about the effect of disturbances, modified synthesis or degradation, or both. Short-term organ culture systems can be a convenient way of performing such studies. The pharmacological agent can be applied in vivo followed by measurement of its effect by radiotracing in vitro (applicable to human tissue). The drug can also be applied in vitro before and/or at the time of radiolabelling.

The third step represents a detailed analysis of the mechanism or mechanisms responsible for the activity of a drug at the translational and transcriptional level. It can be performed in fibroblast culture. In cell culture, the mechanisms involved in the biosynthetic activity are similar to those operating in vivo. However, their regulation may be completely different. This should always be kept in mind before extrapolating the information to an in vivo situation. The development of a model in which the fibroblasts, embedded in a three-dimensional fibrous collagen gel, display differentiated properties similar to those observed in the dermis has provided a suitable model for studying cell responses to pharmacological agents under more physiological conditions (COULOMB et al. 1984). The tissue and cell culture studies represent a potential screening procedure for defined families of drugs to compare the extent of their activity.

The rest of this chapter is a critical selection of procedures required for these steps. They are presented in order of increasing specificity and precision in terms of biochemical mechanisms. They can be applied to the various models cited in this introduction.

B. Physical Parameters

Estimation of skin thickness and measurement of rheological properties of the dermis can be performed in situ by various non-invasive methods (see Chap. 8). It should be noted that whereas skin thickness and collagen content correspond in general (see SHUSTER et al. 1975) the many divergencies both in health and disease limit the use of measurements of skin thickness.

C. Morphology

The examination of a skin biopsy by various techniques, including transmission and scanning electron microscopy, provides topographical information about the organisation of the dermal connective tissue components and the cells. Specific antisera recognising a variety of fibrous proteins, glycoproteins and proteoglycans, combined with morphological techniques, allow their immunolocalisation at the microscopic level (for review see TIMPL et al. 1980) or at the ultrastructural level (WICK et al. 1978). Hybridisation in situ with specific cDNA probes is the method of choice to visualise the cells actively involved in the synthesis of these components (HAYASHI et al. 1986).

D. Cell Kinetics

A large number of metabolic parameters are only significant if expressed as a function of the number of cells in the sample. The most accurate and rapid technique is the measurement of DNA by fluorimetry (LABARCA and PAIGEN 1980). The dynamic aspect of cell multiplication can be estimated by measuring the incorporation of [^{3}H]thymidine in DNA. Histoautoradiography performed after labelling allows one to locate and sometimes to identify the cells in proliferation. Extraction of the labelled DNA and measurement of the incorporated radioactivity provide an estimate of the rate of multiplication (RONNEMAA and DOHERTY 1977).

E. Quantitative Measurements

The concentration of each component of the connective tissue can be measured by specific chemical or immunological assays on fragments of tissue and in various biological fluids.

I. Collagen

Hydroxyproline is a specific imino acid present in large amounts in collagen molecules. It is observed in much smaller concentration in elastin, C1q and acetylcholinesterase. Besides chromatography, all techniques for hydroxyproline determination depend on the oxidation of hydroxyproline to pyrrole which is then reacted with Ehrlich's reagent to give a chromophore (for review see MITCHELL and TAYLOR 1970). Hydroxylysine is another specific amino acid of collagen (and C1q) that can be measured by a chemical technique (BLUMENKRANTZ and ASBOE-HANSEN 1973) or by column chromatography. The collagen concentration of normal human dermis, determined in 48 normal human skin samples, is 63% on a dry weight basis.

II. Elastin

The measurement of elastin remains a difficult problem since no specific marker is present in a concentration permitting its detection in the unfractionated tissue.

Elastin has first to be isolated by exhaustive autoclaving, resulting in the removal of most other proteins prior to measurement of the specific cross-links, desmosine and isodesmosine. These complex amino acids are quantitated by reaction with ninhydrin after purification by ion exchange chromatography (for review see PARTRIDGE 1970). Antisera directed towards elastin (CHRISTNER et al. 1976) and microfibrillar glycoprotein (KEWLEY et al. 1977) can be used to measure these proteins by immunoassay of their soluble forms in tissue culture. Cross-linked elastin can be determined by radioimmunoassay (STARCHER and MECHAM 1981) or enzyme-linked immunosorbent assay (ELISA) (GUNJA-SMITH 1985) using antiserum directed against desmosines.

III. Proteoglycans and Glycoproteins

Proteoglycans in the dermis are quantified by measurement of uronic acids after mild acid hydrolysis by a colorimetric reaction with a borate-carbazole reagent (BITTER and MUIR 1962). The simultaneous measurement of hexosamines (ELSON and MORGAN 1933) and the determination of the ratio of hexosamines to uronic acid also allows an indirect estimation of glycoproteins. These latter do not contain uronic acid while proteoglycans display a repeating structure made up of uronic acids and hexosamines in roughly equimolar amounts. More specific techniques are given in Sect. F. III.

F. Qualitative Measurements

Most of the macromolecules forming the connective tissues are families of closely related, but specific biosynthetic products. The type of product or the proportion of different products are often specific for a tissue, expressing the phenotypic differentiation of the cells that ensure its edification.

I. Collagen

The two main types of interstitial collagen found in the dermis are type I and type III. Type IV collagen is found in basement membranes while type V collagen has a pericellular location. Type VI collagen forms the beaded filaments and is present throughout the dermis. The fibrils anchoring the basement membrane to the papillary dermis are made up of type VII collagen (for details see Chap. 27).

Quantification of each type of collagen is difficult owing to the insolubility of this protein in most tissue. This problem can be overcome by exhaustive digestion of the tissue with pepsin and extraction in acid medium. The helical part of collagen in its native form below the denaturation temperature is resistant to this enzyme. Identification of the various types of collagen can be accomplished by differential salt fractionation (TRELSTAD et al. 1976), ion exchange chromatography (UITTO et al. 1980), sodium dodecyl sulphate-polyacrylamide gel electrophoresis (SDS-PAGE) in the absence and presence of reducing agents (LAEMMLI 1970), by late reduction (SYKES et al. 1976) or immunoprecipitation. Digestion of the whole tissue by cyanogen bromide will solubilise all the proteins and collagen. Typical peptides from the various types of collagen can then be identified by acrylamide electrophoresis or ion exchange chromatography. This technique has been used to determine the relative proportion of collagen types I and III in human skin during development (EPSTEIN 1974). Although not quantitative, the Western blot technique with specific antiserum allows one to identify the different types of collagen after SDS-PAGE (TOWBIN et al. 1979). A derived method, the dot or slot-blot technique, may give semi-quantitative information on the amount of a particular collagen type and is mainly used for comparison between samples.

Modifications induced by pharmacological agents in the type of collagen synthesised in cell or organ culture can be performed by radiolabelling and analysis of the secreted products by one- or two-dimensional slab gel electrophoresis followed by fluorography (BONNER and LASKEY 1974). The identification of collagen polypeptides can also be done by peptide mapping after limited proteolysis by protease from *Staphylococcus aureus* or mast cells (SAGE et al. 1981) or by cyanogen bromide cleavage of collagen chains within the polyacrylamide gel, followed by a second-dimension electrophoresis (BARSH et al. 1981). The quantification of each collagen type may be performed by ELISA (RENNARD et al. 1980) or radioimmunoassays (RISTELI and RISTELI 1986). Localisation of the different types of collagen in tissue or culture is performed by immunofluorescence or by immunoperoxidase with specific antisera (TIMPL et al. 1980). This technique does not allow quantification.

II. Elastin

The amino acid composition of insoluble elastin isolated from different tissues in a defined species is remarkably similar, suggesting a single genetic type of this protein in all tissues. A survey of sensitive techniques for examining the in vitro synthesis of elastin is provided by FOSTER et al. (1981).

III. Proteoglycans and Glycoproteins

Proteoglycans can be sequentially extracted in isotonic or hypertonic salt solution, taking great care to avoid proteolysis by the continuous presence of a "cocktail" of protease inhibitors. Two different techniques and a large variety of analytical procedures related to skin proteoglycans have been published (FUJI and NAGAI 1981; PEARSON and GIBSON 1982; MATSUNAGA and SHINKAI 1986).

The physical and chemical nature of the proteoglycans is determined by ultracentrifugation in cesium chloride gradients, by sieve chromatography on Sepharose, by ion exchange chromatography, by fractionation in cetylpyridinium chloride and analysis of glycosaminoglycans by one- or two-dimensional electrophoresis on cellulose acetate and after hydrolysis by specific glycanases. Physical and compositional analysis of the protein core of the proteoglycans is performed by standard techniques of protein chemistry or immunological characterisation.

Fibronectin is soluble in plasma at body temperature and in tissue culture fluid and precipitates at low temperature. Its extraction from tissues is facilitated by heparin (BRAY et al. 1981). The different forms of fibronectin (plasma, cellular or oncofoetal) arise from alternative splicings of a primary RNA transcript

(Kornblihtt et al. 1985). Antisera directed against specific domains can differentiate these various forms (Borsi et al. 1987). Extraction of the microfibrillar glycoprotein of the elastic fibre is performed by denaturation in 6 *M* guanidine hydrochloride in the presence of a reducing agent and protease inhibitors according to Sear et al. (1981).

In skin, besides plasma proteins extractable in isotonic saline, a residual fraction resisting extraction in the absence of denaturation contains the so-called non-collagen glycoproteins (NCP). They can be solubilised by denaturing solvents after removal of collagen (review and references in Anderson 1976, 1981). The analysis of these NCP is performed by electrophoresis and chromatography and for some of them by immunological procedures.

G. Estimation of Turnover Rate

I. Collagen

Radioactive proline is incorporated into newly synthesised collagen and roughly one-half of the residues are converted into hydroxyproline by prolyl hydroxylase provided adequate co-factors and co-substrates are present. Three techniques are available for measuring radioactive hydroxyproline or radiolabelled collagen.

Incubation of labelled proteins with a highly purified collagenase followed by a precipitation of the non-digested protein (Peterkosky and Diegelmann 1971) solubilises all the collagenase-susceptible material. This method is rapid, but requires the use of absolutely protease-free collagenase. It is the procedure of choice when a defect in hydroxylation of proline is suspected since the unhydroxylated collagen is a substrate for bacterial collagenase.

The determination of radiolabelled hydroxyproline is performed after hydrolysis and separation from labelled proline by ion exchange chromatography on Dowex-50 × 8 (Cutroneo et al. 1972) or strong anionic resin (Spackman et al. 1958). This procedure is accurate, but time-consuming. A third method for determining radioactive hydroxyproline is the technique of Juva and Prockop (1966) relying on the oxidation of hydroxyproline into pyrrole, its extraction in toluene and its separation from oxidation products of proline by chromatography of the toluene extract on a short column of silicic acid. Meaningful results can only be obtained if adequate controls are performed, i.e. estimation of the yield of conversion of hydroxyproline into pyrrole by running in parallel known amounts of [^{14}C]hydroxyproline and determination of contamination of the column eluate by labelled proline and impurities. This problem is discussed in detail by Bienkowski and Engels (1981).

Measurement of the metabolic activity of fibroblasts in culture as well as in vivo has to take into consideration the intracellular pool of the tracer. In the case of labelled proline it can be measured by the technique of Robins (1979). An appreciation of the biosynthetic activity of fibroblasts can be obtained from measurements of mRNA as described in Sect. H.I.

II. Elastin

In skin, the metabolism of elastin is difficult to study owing to a rapid insolubilisation of the newly synthesised protein, its low concentration and its slow rate of turnover except in rapidly growing animals. Studies on chick aorta and copper-deficient pig skin have demonstrated the synthesis of a soluble precursor of elastin (ROSENBLOOM et al. 1980). The preparation of specific antisera against elastin (CHRISTNER et al. 1976) allowing its quantification by a rocket immunoelectrophoretic system has been described by FOSTER et al. (1980). Measuring specific cross-links of elastin desmosine and isodesmosine by immunological techniques (STARCHER and MECHAM 1981; GUNJA-SMITH 1985), isolation and translation of mRNA (BURNETT and ROSENBLOOM 1979) and construction of a chick elastin cDNA clone (BURNETT et al. 1981) can be applied to skin elastin for studying regulation of its biosynthesis. The synthesis of the microfibrillar components of the elastic fibre can be measured in fibroblast culture derived from human skin (SEAR et al. 1977) and bovine ligamentum nuchae by immunoprecipitation of [^{3}H]fucose-labelled glycoproteins by a specific antiserum (SEAR et al. 1981).

III. Proteoglycans and Glycoproteins

Incorporation of $^{35}SO_4$ into proteoglycans or labelled glucosamine into glycoproteins and proteoglycans is a widely used and sensitive method for measuring synthesis in vivo or in cell and organ culture. Various methods have been used for isolation and purification of the labelled proteoglycans (WIEBKIN and MUIR 1973; DORFMAN and HO 1970; NAKAMURA et al. 1983).

H. Measurement of Defined Biochemical Parameters

I. Messenger RNA

Developments in molecular biology related to fibrous proteins and associated glycoproteins and proteoglycans have provided valuable tools for estimating biosynthetic activities in vivo (VUORIO et al. 1983) and in vitro by measurement of mRNA. The mRNAs are extracted in the presence of ribonuclease inhibitors, partially purified by chromatography or gradient centrifugation, dotted on nitrocellulose filters and tested with specific labelled cDNA probes. The amount of hybridised label is determined by autoradiography and scanning densitometry (BRESSER et al. 1983). The extraction and purification of mRNA directing collagen synthesis can also be applied to skin (KAUFMAN et al. 1980).

II. Intracellular Post-translational Enzymes

Assaying the activity of enzymes catalysing specific post-translational modifications of procollagen (also elastin and C1q) has been used for measuring collagen metabolism although increased enzyme activity is not always paralleled by enhanced collagen synthesis (HIRAMATSU et al. 1982).

Prolyl hydroxylase activity is measured in fibroblast culture and in tissues extracted with buffer containing dithiothreitol and Triton X-100. The technique relies on the use of 3,4-[^{3}H]proline-labelled unhydroxylated collagen polypeptide as substrate, prepared by incubating chick embryo calvaria in the presence of a chelator of Fe^{2+} or under anaerobic conditions. In the presence of adequate co-substrates (O_2,α-ketoglutarate) and co-factors (Fe^{2+}, ascorbic acid), the peptide-bound prolyl is hydroxylated in position 4 and the released ^3H_2O is collected either by distillation (HUTTON et al. 1966) or by passage on a Dowex-50 column (PETERKOFSKY and DIBLASIO 1975). A similar assay has been described for measuring lysyl hydroxylase activity using 4,5-[^{3}H]lysine to label unhydroxylated collagen. The activity of prolyl hydroxylase can also be determined by the formation of [^{14}C]hydroxyproline from [^{14}C]proline-labelled substrate by a specific assay (JUVA and PROCKOP 1966) or by using α-keto-1-[^{14}C]glutarate in the assay and measuring the $^{14}CO_2$ produced by its decarboxylation (RHOADS and UDENFRIEND 1968).

Prolyl hydroxylase is synthesised in the form of inactive precursor subunits (MCGEE et al. 1971) cross-reacting with the antibody to the purified active prolyl hydroxylase. Determination of total antigenic prolyl hydroxylase and of enzyme activity allows one to estimate biosynthesis and activation of the precursor. The amount of mRNA directing the synthesis of prolyl hydroxylase has also been estimated by specific immunoprecipitation of the translation product of mRNA extracted from dermal polysomes according to ROKOWSKI et al. (1981).

Activity measurement of the two enzymes catalysing the glycosylation of the hydroxylysyl residues is performed on purified skin gelatin as substrate using UDP-[^{14}C]galactose or UDP-[^{14}C]glucose as the source of sugar, in the presence of $MnCl_2$ according to the technique described by MYLLYLA et al. (1975) and modified by ANTTINEN and OIKARINEN (1977). The hydroxylysyl glycosides produced are released by alkaline hydrolysis, purified and quantified by ion exchange chromatography.

III. Extracellular Processing Enzymes

Lysyl oxidase is responsible for the formation of lysyl aldehydes, precursors of intramolecular cross-links in collagen and elastin and for the progressive insolubilisation of the proteins. Its activity can be estimated either directly on labelled substrate or indirectely by the concentration of the reaction products. Extractability of collagen under various non-denaturing conditions seems inversely related to the enzyme activity. Extraction has been performed on skin biopsy (FRANCIS et al. 1973) and in culture (DI FERRANTE et al. 1975). The ratio between polymers and monomers of collagen α-chains after denaturation, measured by scanning polyacrylamide gel electrophoretic patterns, also reflects the cross-linking process. Reducible cross-links, intermediate products before stabilisation, are measured and identified after reduction by [^{3}H]borohydride, alkaline or acid hydrolysis and separation by ion exchange chromatography (for review see TANZER 1973). Stable cross-links can also be measured (FUJIMOTO and MORIGUCHI 1978). Lysyl oxidase is a copper metalloenzyme that can be extracted from tissue or cell culture by a solution containing 4–6 *M* urea to stabilise the enzyme. Its assay depends on the release of ^3H_2O from a 6-[^{3}H]lysine-labelled collagen or elastin substrate prepared in cell or organ culture by incubation in the presence of lathyrogen, a specific inhibitor of the enzyme (for review see SIEGEL 1979).

Procollagen peptidases (PCP) are specific endopeptidases responsible for the excision of the amino (PCP-N) and carboxy terminal (PCP-C) precursor sequences of procollagen and are type specific, at least for procollagen type I (PCP-N-I) and III (PCP-N-III). In skin, PCP-N-I is readily extractable in 0.05 *M* Na cacodylate, 1 *M* KCl, 0.1% Triton X-100, 2.5 m*M* *N*-ethylmaleimide and 0.5 m*M* phenylmethanesulphonylfluoride. Enzyme activity is measured with [^{14}C]carboxymethylated p-I dermatosparactic collagen as substrate (NUSGENS and LAPIERE 1979) or a labelled collagen precursor from cell culture. PCP-C-I is purified from the medium of cultured tendon fibroblasts and can be measured with labelled procollagen in cell or organ culture (LEUNG et al. 1979). PCP-N-III is not detectable in skin, perhaps owing to the presence of inhibitors. It can be purified from normal tendon fibroblast culture medium and measured with [^{14}C]carboxymethylated p-N-III collagen as substrate (NUSGENS et al. 1980).

IV. Degradation, Enzymes and Products

Collagen degradation in vivo is mediated by collagenase, a specific neutral protease, that exists mostly in its latent form. Determination of collagenase activity can be performed in fibroblast or organ culture medium after activation of the latent enzyme by mild trypsin, sulphocyanate or organic mercury treatment (BAUER et al. 1975; SHINKAI et al. 1977). In the safest assay, introduced by LAPIERE and GROSS (1963) and developed by NAGAI et al. (1966), the substrate for testing collagenase activity is radiolabelled reconstituted native collagen fibrils. Several other techniques have been developed (for reviews see HARRIS and VATER 1980; JOHNSON-WINT 1980; MOOKHTIAR et al. 1986). Collagenase can also be measured by a direct technique on samples of tissue. The insoluble fibres upon which the enzyme is bound are incubated in vitro and the hydroxyproline released is measured by a chemical procedure (RYAN and WOESSNER 1971). The technique has been applied to skin in scleroderma by BRADY (1975). Antisera raised against purified collagenase (for review see WOOLLEY et al. 1980) provide another approach to measuring immunoreactive protein concentration (COOPER et al. 1982) and allow its immunolocalisation in various connective tissues. A cDNA clone for collagenase is useful for studying the mechanisms of regulation of this enzyme and their pharmacological modifications (BRINCKERHOFF et al. 1987).

A large variety of neutral and acid endopeptidases of lower specificity and various exopeptidases could also be considered, such as prolidase and prolinase (JACKSON et al. 1975). A specific end-product-degrading enzyme is hydroxyproline oxidase (EFRON et al. 1968).

Elastin is degraded by elastases. The elastolytic activity in organ or cell extracts is estimated by measuring the release of labelled fragments from radiolabelled insoluble elastin (TAKAHASHI et al. 1973) or the clearing of elastin entrapped in a gel medium (ROBERT et al. 1974) or by the lysis of a synthetic peptide according to BIETH et al. (1974). Serum contains potent inhibitors of elastase (α_1-antitrypsin, α_2-macroglobulin) that can interfere with the measurement of the enzyme activity, mainly in culture and in serum. The development of immunological identification and quantitation of elastase has proved most useful (KUCICH et al. 1980).

There is a vast number of enzymes involved in the complete degradation of proteoglycans and glycosaminoglycans. Proteoglycan-degrading activity can be measured by incubation in vitro of ^{35}S-labelled cartilage or from labelled proteoglycan / collagen-coated plates (VAES et al. 1981). Hyaluronidase is measured by the release of acetylglucosamine from hyaluronic acid (SILBERT et al. 1965). A large array of lysosomal glycanases can be assayed by using the appropriate substrate and the basic technique of VON FIGURA and WEBER (1978).

A defined proportion of the products of degradation of the various macromolecules circulate in the body fluid and are excreted in the urine. Measuring these products provides indications of metabolic activity. In humans, the overall catabolism of collagen is measured by determining the urinary excretion of either hydroxyproline, by the technique of PROCKOP and UDENFRIEND (1960) or, less frequently, the glycosides of hydroxylysine (ASKENASI 1974). The ratio of hydroxyproline to proline is also an index of collagen catabolism (NUSGENS and LAPIÈRE 1973). A radioimmunoassay measuring the circulating amino terminal precursor sequence of type III collagen has proved particularly useful for following the evolution of various fibrotic processes mainly in hepatic pathology (ROHDE et al. 1978).

The determination of urinary desmosines in humans, measured as [^{3}H]borohydride-reducible compounds, can be used as a biochemical marker of elastin degradation (GUNJA-SMITH and BOUCEK 1981). The catabolism of proteoglycans is estimated by measuring the urinary excretion of uronic acids after proteolysis and precipitation of glycosaminoglycans by quaternary ammonium salts. These glycosaminoglycans can be characterised by electrophoresis on strips of cellulose acetate (LINKER and HOVINGH 1972) and digestion by specific enzymes.

References

Anderson JC (1976) Glycoproteins of the connective tissue matrix. Int Rev Connect Tissue Res 7:251–322

Anderson JC (1981) Structural glycoproteins and other connective tissue proteins. Connect Tissue Res 7:63–71

Anttinen H, Oikarinen A (1977) Age-related changes in human skin collagen galactosyltransferase and collagen glucosyltransferase activities. Clin Chim Acta 76:95–101

Askenasi R (1974) Urinary hyroxylysine and hydroxylysyl glycoside excretion in normal and pathologic states. J Lab Clin Med 83:673–679

Barsh GS, Peterson KE, Byers PH (1981) Peptide mapping of collagen chains using CNBr cleavage of proteins within polyacrylamide gels. Coll Relat Res 1:543–548

Bauer EA, Stricklin GP, Jeffrey JJ, Eisen AZ (1975) Collagenase production by human skin fibroblasts. Biochem Biophys Res Commun 64:232–240

Bienkowski RS, Engels CJ (1981) Measurement of intracellular collagen degradation. Anal Biochem 116:414–424

Bieth J, Spiess E, Wermuth C (1974) The synthesis and analytical use of a highly sensitive and convenient substrate of elastase. Biochem Med 11:350–357

Bitter T, Muir H (1962) A modified uronic acid carbazole reaction. Anal Biochem 4:330–334

Blumenkrantz N, Asboe-Hansen G (1973) An improved method for the assay of hydroxylysine. Anal Biochem 56:10–15

Bonner WM, Laskey AB (1974) A film detection method for tritium labelled proteins and nucleic acids in polyacrylamide gels. Eur J Biochem 46:83–88
Borsi L, Carnemolla B, Castellani P, Rosellini C, Vecchio D, Allemanni G, Chang SE et al. (1987) Monoclonal antibodies in the analysis of fibronectin isoforms generated by alternative splicing of mRNA precursors in normal and transformed human cells. J Cell Biol 104:595–605
Bottoms E, Shuster S (1963) Effect of ultraviolet light on skin collagen. Nature 199:192–193
Brady AH (1975) Collagenase in scleroderma. J Clin Invest 56:1175–1180
Bray BA, Mandl I, Turino GM (1981) Heparin facilitates the extraction of tissue fibronectin. Science 214:793–794
Bresser J, Doering J, Gillespie D (1983) Quick-blot: selective m-RNA or DNA immobilization from whole cells. DNA 2:243–253
Brinckerhoff CE, Ruby PL, Austin SD, Fani ME, White HD (1987) Molecular cloning of human synovial cell collagenase and selection of a single gene from genomic DNA. J Clin Invest 79:542–549
Burnett W, Rosenbloom J (1979) Isolation and translation of elastin mRNA from chick aorta. Biochem Biophys Res Commun 86:478–484
Burnett W, Yoon K, Rosenbloom J (1981) Construction, identification and characterization of chick elastin cDNA clone. Biochem Biophys Res Commun 99:364–372
Christner P, Dixon M, Cywinski A, Rosenbloom J (1976) Radioimmunological identification of tropoelastin. Biochem J 157:525–528
Cooper TW, Bauer EA, Eisen AZ (1982) Enzyme-linked immunoadsorbent assay for human skin collagenase. Coll Relat Res 3:205–216
Coulomb B, Dubertret L, Bell E, Touraine R (1984) The contractibility of fibroblasts in a collagen lattice is reduced by corticosteroids. J Invest Dermatol 82:341–344
Cutroneo KR, Guzman NA, Liebelt AG (1972) Elevation of peptidylproline hydroxylase activity and collagen synthesis in spontaneous primary mammary cancers of inbred mice. Cancer Res 32:2828–2833
Di Ferrante N, Leachman RD, Angelini P, Donnelly PV, Francis G, Almazan A, Segni G et al. (1975) Ehlers-Danlos type V (X-linked form) a lysyl oxidase deficiency. Birth Defects 11:31–37
Dorfman A, Ho P (1970) Synthesis of acid mucopolysaccharides by glial tumor cells in tissue culture. Proc Natl Acad Sci USA 66:495–499
Efron ML, Bixby EM, Hockaday TDR, Smith LH, Meshorer E (1968) Hydroxyprolineamia. III. The origin of free hydroxyproline in hydroxyprolinemia. Collagen turnover. Evidence for a biosynthetic pathway in man. Biochim Biophys Acta 165:238–250
Elson LA, Morgan WTJ (1933) A colorimetric method for the determination of glucosamine and chondrosamine. Biochem J 27:1824–1828
Epstein EH (1974) Alpha 1 (III)3 human skin collagen. Release by pepsin digestion and preponderance in fetal life. J Biol Chem 249:3225–3231
Foster JA, Wy SC, Marzullo M, Rich C (1980) A sensitive assay for the quantitation of soluble elastin. Anal Biochem 101:310–315
Foster JA, Rich C, Karr S, Desa M (1981) A survey of sensitive techniques for examining the in vitro synthesis of elastin. Connect Tissue Res 8:259–262
Francis MJO, Smith R, Macmillan DC (1973) Polymeric collagen of skin in normal subjects and in patients with inherited connective tissue disorders. Clin Sci 44:429–438
Fujii N, Nagai Y (1981) Isolation and characterization of a proteodermatan sulfate from calf skin. J Biochem 90:1249–1258
Fujimoto D, Moriguchi T (1978) Pyridinoline, a non reducible cross-link of collagen. Quantitative determination, distribution, and isolation of a cross-linked peptide. J Biochem (Tokyo) 83:863–867
Gunja-Smith Z (1985) An enzyme-linked immunosorbent assay to quantitate the elastin cross-link desmosine in tissue and urine samples. Anal Biochem 147:258–264
Gunja-Smith Z, Boucek RJ (1981) Desmosines in human urine. Amounts in early developments and in Marfan's syndrome. Biochem J 193:915–198

Hansen TM (1980) Sponge induced granulation tissue as a pharmacological test system. In: Viidik A, Vuust J (eds) Biology of collagen. Academic, New York, pp 339–346
Harris E, Vater C (1980) Methodology of collagenase research: substrate preparation, enzyme activation and pruification. In: Woolley D, Evanson J (eds) Collagenase in normal and pathological connective tissues. Wiley, New York, pp 37–64
Hayashi M, Ninomiya Y, Parsons J, Hayashi K, Olsen BR, Trelstad RL (1986) Differential localization of mRNAs of collagen types I and II in chick fibroblasts, chondrocytes, and corneal cells by in situ hybridization using cDNA probes. J Cell Biol 102:2302–2311
Hiramatsu M, Kumegawa M, Hatakeyama K, Yajima T, Minami N, Kodama H (1982) Effect of epidermal growth factor on collagen synthesis in osteoblastic cells derived from newborn mouse calvaria. Endocrinology 111:1810–1816
Hutton JJ, Tappel AL, Udenfriend S (1966) A rapid assay for collagen proline hydroxylase. Anal Biochem 16:384–394
Jackson SH, Dennis AW, Greenberg M (1975) Iminodiptinuria: a genetic defect in recycling collagen; a method for determining prolidase in erythrocytes. Can Med Assoc J 113:759–763
Johnson-Wint B (1980) A quantitative collagen film collagenase assay for large number of samples. Anal Biochem 104:175–181
Juva K, Prockop DJ (1966) Modified procedure for the assay of ^{3}H or ^{14}C labeled hydroxyproline. Anal Biochem 15:77–83
Kaufmann R, Belayew A, Nusgens B, Lapière CM, Gielen JE (1980) Extraction and translation of collagen mRNA from fetal calf skin. Eur J Biochem 106:593–601
Kewley MA, Steven FS, Williams G (1977) Preparation of a specific antiserum towards the microfibrillar protein of elastic tissues. Immunology 32:483–489
Kornblihtt AR, Umezawa K, Vibe-Pedersen K, Baralle FE (1985) Primary structure of human fibronectin: differential splicing may generate at least 10 polypeptides from a single gene. EMBO J 4:1755–1759
Kucich U, Abrams WR, James HL (1980) Solid phase immunoassay of dog neutrophil elastase. Anal Biochem 109:403–409
Kulonen E (1976) Experimental granuloma as a tool in connective tissue research. In: Hall D (ed) The methodology of connective tissue research. Johnson-Bruvvers, Oxford, pp 29–36
Labarca C, Paigen K (1980) A simple rapid and sensitive DNA assay procedure. Anal Biochem 102:344–352
Laemmli UK (1970) Cleavage of structural proteins during the assembly of the head of bacteriophage T4. Nature 227:680–685
Lapière CM, Gross J (1963) Animal collagenase and collagen metabolism. In: Sognnaes R (ed) Mechanisms of hard tissue destruction. American Association for the Advancement of Science, Washington, pp 663–694
Leung MKK, Fessler LI, Greenberg DB, Fessler JH (1979) Separate amino and carboxyl procollagen peptidases in chick embryo tendon. J Biol Chem 254:224–232
Linker A, Hovingh P (1972) Heparinase and heparitinase from flavobacteria. Methods Enzymol 28:902–908
Matsunaga E, Shinkai H (1986) Two species of dermatan sulfate proteoglycans with different molecular sizes from newborn calf skin. J Invest Dermatol 87:221–226
McGee JO'D, Langness U, Udenfriend S (1971) Immunological evidence for an inactive precursor of collagen proline hydroxylase in cultured fibroblasts. Proc Natl Acad Sci USA 68:1585–1589
Mitchell A, Taylor I (1970) Spectrophotometric determination of hydroxyproline. Analytical investigation. Analyst 95:1003–1011
Mookhtiar KA, Mallya SK, van Wart HE (1986) Properties of radiolabeled type I, II and III collagens related to their use as substrates in collagenase assays. Anal Biochem 158:322–328
Myllyla R, Risteli L, Kivirikko KI (1975) Assay of collagen-galactosyltransferase and collagen-glucosyltransferase activities and preliminary characterization of enzymic reactions with transferases from chick embryo cartilage. Eur J Biochem 52:401–410

Nagai Y, Lapière CM, Gross J (1966) Tadpole collagenase. Preparation and purification. Biochemistry 5:2123–3130
Nakamura T, Matsunaga E, Shinkai H (1983) Isolation and some structural analyses of a proteodermatan sulphate from calf skin. Biochem J 213:289–296
Nusgens BV, Lapière CM (1973) The relationship between proline and hydroxyproline urinary excretion in human as an index of collagen catabolism. Clin Chim Acta 48:203–211
Nusgens BV, Lapière CM (1979) A simplified procedure for measuring aminoprocollagen peptidase type I. Anal Biochem 95:406–412
Nusgens BV, Goebels U, Shinkai H, Lapière CM (1980) Procollagen type III N-terminal endopeptidase in fibroblast culture. Biochem J 191:699–706
Partridge SM (1970) Isolation and characterization of elastin. In: Balazs E (ed) Chemistry and molecular biology of the intercellular matrix. Academic, London, pp 593–616
Pearson CH, Gibson GJ (1982) Proteoglycans of bovine periodontal ligament and skin. Occurrence of different hybrid sulphated galactosaminoglycans in distinct proteoglycans. Biochem J 201:27–37
Peterkofsky D, Diblasio R (1975) Modification of the tritium-release assays for prolyl and lysyl hydroxylase using Dowex-50 columns. Anal Biochem 66:279–286
Peterkofsky D, Diegelmann R (1971) Use of a mixture of proteinase-free collagenases for the specific assay of radioactive collagen in the presence of other proteins. Biochemistry 10:988–993
Prockop DJ, Udenfriend S (1960) A specific method for the analysis of hydroxyproline in tissues and urine. Anal Biochem 1:228–239
Rennard SI, Berg R, Martin GR, Foidart JM, Gehron-Robey P (1980) Enzyme linked immunoassay (ELISA) for connective tissue components. Anal Biochem 104:205
Rhoads R, Udenfriend S (1968) Decarboxylation of alpha-ketoglutarate coupled to collagen proline hydroxylase. Proc Natl Acad Sci USA 60:1473–1478
Risteli L, Risteli J (1986) Radioimmunoassays for monitoring connective tissue metabolism. Rheumatology 10:216–245
Robert B, Derouette JC, Robert L (1974) Mise en évidence de protéase à activité élastolytique dans les extraits d'aortes humaines et animals. C R Acad Sci (Paris) 278:3251–3254
Robins SP (1979) Metabolism of rabbit skin collagen. Differences in the apparent turnover rates of type I and type III collagen precursors determined by constant intravenous infusion of labelled aminoacids. Biochem J 181:75–82
Rohde H, Hahn E, Timpl R (1978) Radioimmunoassay for aminoterminal procollagen peptides in liver disease. Fresenius Z Anal Chem 290:151–152
Rokowski RJ, Sheehy J, Cutroneo KR (1981) Glucocorticoid mediated selective reduction of functioning collagen messenger ribonucleic acid. Arch Biochem Biophys 210:74–81
Ronnemaa T, Doherty NS (1977) Effect of serum and liver extracts from hypercholesterolemic rats on the snythesis of collagen by isolated aortas and cultured aortic smooth muscle cells. Atheroscloris 26:261–272
Rosenbloom J, Harsch M, Cywinski A (1980) Evidence that tropoelastin is the primary precursor in elastin biosynthesis. J Biol Chem 255:100–106
Ryan JN, Woessner JF Jr (1971) Mammalian collagenase: direct demonstration in homogenates of involuting rat uterus. Biochem Biophys Res Commun 44:144–149
Sage H, Pritzl P, Bornstein P (1981) A new mapping technique of collagen chains. Coll Relat Res 1:3–15
Sear C, Kewley MA, Grant ME, Steven FS, Jackson DS (1977) Biosynthesis of a structural glycoprotein component of elastic tissues by cultured human skin fibroblasts. Biochem Soc Trans 5:430–431
Sear C, Jones C, Knight K, Grant M (1981) Elastogenesis and microfibrillar glycoprotein by bovine ligamentum nuchae cells in culture. Connect Tissue Res 8:167–170
Shinkai H, Kawamoto T, Hori H, Nagai Y (1977) A complex of collagenase with low molecular weight inhibitors in the culture medium of embryonic chick skin explants. J Biochem (Tokyo) 81:261–263
Shuster S, Black MM, MacVitie E (1975) The influence of age and sex on skin thickness, skin collagen and density. Br J Dermatol 93:639–643

Siegel RC (1979) Lysyl oxidase. Int Rev Connect Tissue Res 8:73–118
Silbert JE, Nagai Y, Gross J (1965) Hyaluronidase from tadpole tissue. J Biol Chem 240:1509–1511
Spackman D, Stein W, Moore S (1958) Automatic recording apparatus for use in the chromatography of aminoacids. Anal Chem 30:1190–1206
Starcher BC, Mecham RP (1981) Desmosine radioimmunoassay as a means of studying elastogenesis in cell culture. Connect Tissue Res 8:255–258
Sykes B, Puddle B, Francis M, Smith R (1976) The estimation of two collagens from human dermis by interrupted gel electrophoresis. Biochem Biophys Res Commun 72:1472–1480
Takahashi S, Seifter S, Yang FC (1973) A new radioactive assay for enzymes with elastolytic activity using reduced tritiated elastin. The effect of sodium dodecyl sulfate on elastolysis. Biochim Biophys Acta 327:138–145
Tanzer ML (1973) Cross-linking of collagen. Science 180:561–566
Timpl R, von der Mark K, von der Mark H (1980) Immunochemistry and immunohistology of collagens. In: Viidik A, Vuust J (eds) Biology of collagen. Academic, London, pp 211–222
Towbin H, Staehlin T, Gordon J (1979) Eletrophoretic transfer of proteins from polyacrylamide gels to nitrocellulose sheets: procedure and some applications. Proc Natl Acad Sci USA 76:4350–4354
Trelstad RL, Catanese VM, Rubin DF (1976) Collagen fractionation: separation of native types I, II and III by differential precipitation. Anal Biochem 71:114–118
Uitto J, Booth BA, Polak KL (1980) Collagen biosynthesis by human skin fibroblasts. II. Isolation and further characterization of type I and type III procollagens synthesized in culture. Biochim Biophys Acta 624:545–561
Vaes G, Huybrechts-Godin G, Peeters-Joris C, Laub R (1981) Cell cooperation in collagen and proteoglycan degradation. In: Dingle J, Gordon E (eds) Cellular interactions. Biomedical, Elsevier Amsterdam, pp 241–251
Von Figura K, Weber E (1978) An alternative hypothesis of cellular transport of lysosomal enzymes in fibroblasts. Effect of inhibitors of lysosomal enzyme endocytosis on intra and extracellular lysosomal enzyme activities. Biochem J 176:943–950
Vuorio T, Vuorio E, Lehtinen P, Upholt WB, Dorfman A (1983) Measurement of type I collagen expression in human and rat tissues using c-DNA clones for chick type I collagen messenger RNA's. Coll Relat Res 3:69
Wick G, Olsen BR, Timpl R (1978) Immunohistologic analysis of fetal and dermatosparactic calf and sheep skin with antisera to procollagen and collagen type I. Lab Invest 39:151–156
Wiebkin O, Muir H (1973) The inhibition of sulphate incorporation in isolated adult chondrocytes by hyaluronic acid. FEBS Lett 37:42–46
Woolley D, Tetlow L, Evanson J (1980) Collagenase immunolocalization studies of rheumatoid and malignant tissues. In: Wolley D, Evanson J (eds) Collagenase in normal and pathological connective tissue. Wiley, New York, pp 105–126
Zederfeldt B (1980) Factors influencing wound healing. In: Viidik A, Vuust J (eds) Biology of collagen. Academic, New York, pp 247–362

CHAPTER 10

Microbiological Sampling Techniques

K. T. HOLLAND

A. Introduction

The purpose of microbiological sampling is to allow statements of density, types and locations of microorganism which reside on the skin. The problem is that different answers are given by different sampling techniques. In laboratory experiments microorganisms are grown in ideal homogeneous culture conditions and a single sample of the culture will reflect the entire culture. The human skin is a heterogeneous environment composed of many small homogeneous environments, often distributed in patterns which differ with site and individual. Since environment determines the microbial inhabitants (bacteria, fungi, viruses, protozoa and mites) their distribution on human skin varies from site to site and from person to person. In practice the investigator makes many compromises: simple techniques produce data of limited value whilst the complex techniques are restricted by the type of site that can be sampled and the number of samples that can be taken.

B. Surface Distribution

Various solid vehicles can be used to transpose the microorganisms from the skin for identification and viable and total cell counting. The main methods are solid nutrient medium (ULRICH 1964); velvet pad (RAAHAVE 1975); cardboard disc (SEEBERG et al. 1981) and cellophane tape (KOOYMAN and SIMONS 1965; MALCOLM 1980). These methods underestimate the total flora present at the site: there may be failure to replicate a part of the total area, either during sampling or in transferring the nutrient medium; they cannot detect intrafollicular microorganisms; in most cases because only one medium is used only a limited range of microorganisms can be detected; and in the absence of a dispersion procedure low numbers of specific types of microorganisms are likely to be missed. At best these methods give a rough picture of the surface distribution of the major microbial inhabitants. The cellophane tape method will detect intrafollicular microorganisms if repeated samples are taken from the same site, or by the method of MALCOLM (1980) in which the skin is stained in vivo after which cellphane tape strips are examined microscopically. The advantages of these methods are speed and simplicity; they can be used to sample diseased skin which may be sensitive.

C. Swabbing Methods

A defined area of skin is rubbed with a moistened swab which is transferred to a dispersion fluid which may be diluted and plated out for viable cell counts, or concentrated by centrifugation or filtration for viable cell counting, staining and microscopy. The technique samples a variable and low percentage of surface microorganisms, depending on swab fibre, time and pressure of rubbing. Variable numbers of microorganisms are removed from the swab to the dispersion fluid, depending on the agitation procedure and swab fibre. These methods are therefore useful for qualitative, but not quantitative work. Their advantages are simplicity, speed and use for pathologically affected skin and sites such as toe-webs.

D. Washing Methods

These techniques can be divided into two groups. The first is where part of the body, e.g. the hands, is washed and the fluid used for microbiological analysis (AYCLIFFE et al. 1975; MICHAUD et al. 1976). The method is simple and because a fluid sample is obtained many quantitative and qualitative bacteriological tests can be done, but it has great variability and only samples surface microorganisms. The data obviously give only an average microbiological picture and cannot be used to examine microflora over small areas of skin.

The second type of method is the most commonly used for quantitative work and is based on the method of WILLIAMSON and KLIGMAN (1965). First of all, 1 ml buffered non-ionic detergent, Triton X-100, is pipetted into a metal cup held in position on the skin site to be sampled. The skin is rubbed with a smooth polytetrafluoroethene abrader for 1 min and the fluid removed. The procedure is repeated, the two fractions pooled and analysed. BIBEL and LOVELL (1976) used a mechanised system with a defined 250-g load and STRINGER and MARPLES (1976) used ultrasonic energy delivered by a probe. These methods are superior to the previous methods since defined areas of skin can be sampled and greater recoveries of microorganisms obtained. However, they cannot be used for open lesions, some skin sites, e.g. toe-webs and nose, and they do not remove follicular microorganisms. Repeated sampling at the same site gives an exponential decrease in microorganisms with individual differences in the negative gradient presumably owing to differences in the stratum corneum (ALY et al. 1978).

E. Follicular Sampling

Apart from comedones and pustules (IZUMI et al. 1970), sampling of follicles is difficult. HOLLAND et al. (1974) used a cyanoacrylate glue to remove follicular contents. Material from single or pooled follicles could be processed and analysed. Although the method is simple the degree of removal of follicular material is variable. The most accurate method was developed by PUHVEL et al. (1975); a skin punch biopsy is processed to separate the epidermis and the intact follicles from the dermis. The follicles are then dissected from the epidermal disc under the stereomicroscope and investigated.

F. Miscellaneous Techniques

Fungal pathogens are usually sought by scraping with a blunt scalpel blade and examined by microscopy after clearing with KOH (as with hair and nail clippings); they may also be grown in culture. Sampling by cellophane tape or canoacrylate glue are alternatives. Yeasts are usually sampled by a saline-moistened swab. The sampling of microorganisms on hair, without removal, can be accomplished by a hairbrush (NOBLE and MIDGLEY 1978).

G. Comment

There is no universally applicable sampling technique and the investigator must choose a method most likely to answer a defined question. Thus, the swab technique cannot be used to study the effect of an agent on intrafollicular microflora. It should also be noted that if temporal changes in microflora are studied samples should not be taken repeatedly from the same site because its microbial population will be abnormal: it is preferable to use symmetrically opposite skin sites on many people.

References

Aly R, Maibach HI, Bloom E (1978) Quantification of anaerobic diphtheroids on the skin. Act Derm Venereol (Stockh) 58:501

Aycliffe GAJ, Babb JR, Bridges K, Lilly HA, Lowbury EJL, Varney J, Wilkins MD (1975) Comparison of two methods for assessing the removal of total organisms and pathogens from skin. J Hyg (Cambridge) 75:259

Bibel DJ, Lovell DJ (1976) Skin flora maps: a tool in the study of cutaneous ecology. J Invest Dermatol 67:265

Holland KT, Roberts CD, Cunliffe WJ, Williams M (1974) A technique for sampling microorganisms from the pilosebaceous ducts. J Appl Bacteriol 37:397

Izumi AK, Marples RR, Kligman AM (1970) Bacteriology of acne comedones. Arch Dermatol 102:397

Kooyman DJ, Simons RW (1965) "Sticky disc" sampling of skin microflora. Arch Dermatol 92:581

Malcolm SA (1980) An in vivo staining technique for the demonstration of microorganisms on the stratum corneum. Proc R Soc Edinburgh [Biol] 79:201

Michaud RN, McGrath MB, Goss WA (1976) Application of a gloved-hand model for multiparameter measurements of skin degerming activity. J Clin Microbiol 3:406

Noble WC, Midgley G (1978) Scalp carriage of *Pityrosporum* species: the effect of physiological maturity, sex and race. Sabouraudia 16:229

Puhvel SM, Reisner RM, Amirian D (1975) Quantification of bacteria in isolated pilosebaceous follicles in normal skin. J Invest Dermatol 65:525

Raahave D (1975) Experimental evolution of the velvet pad rinse technique as a microbiological sampling method. Acta Pathol Microbiol Scand [B]83–416

Seeberg S, Lindberg A, Bergman BR (1981) Preoperative showever bath with 4% chlorhexidine detergent solution: reduction of *Staphylococcus aureus* in skin carriers and practical application. In: Maibach H, Aly R (eds) Skin microbiology – relevance to clinical infection. Springer, Berlin Heidelberg New York, p 86

Stringer MF, Marples RR (1976) Ultrasonic methods for sampling human skin microorganisms. Br J Dermatol 94:551

Ulrich JA (1964) Techniques of skin sampling for microbial contaminants. Health Lab Sci 1:133

Williamson P, Kligman AM (1965) A new method for the quantitative investigation of cutaneous bacteria. J Invest Dermatol 45:498

CHAPTER 11

Clinical Trial Methods

M. CORBETT and A. HERXHEIMER

A. The Beginnings

The Medical Research Council investigation of 1948 of the effect of streptomycin treatment for acute pulmonary tuberculosis was the first controlled clinical trial which conforms to present-day principles. This trial and many that followed are useful models for controlled trials in other disciplines (see FOX 1971). In dermatology, clinical trials on any scale started with the topical corticosteroids in eczema (MORGAN 1955). As other topical corticosteroids were developed small trials multiplied, many of them funded by the drug companies concerned. Even large multi-centre trials (e.g. WILLIAMS et al. 1964) in which over 800 patients were collected at 7 different centres, were actually a collection of several brief pair-wise studies. Although many of these small trials were not of very good quality, they helped to disseminate ideas about controlled clinical trials. A good trial was the study by BOWERS et al. (1966) which compared the short-term value of ultraviolet radiation, tar baths and dithranol in psoriasis by means of a factorial design. Although the analysis was somewhat naive it contributed substantially to current knowledge.

B. Clinical Trial Principles Applied to Dermatology

A clinical trial is an experiment designed to answer a primary question, usually about the response of patients representative of a particular population to a new treatment or method of management. If the primary question is carefully defined and set down at the outset (FRIEDMAN et al. 1981) then the study can be designed to answer it. It is tempting to include subsidiary questions, but their consideration may seriously dilute the attention given to the primary question, and so may lead to an inconclusive study.

C. The Subjects

The patients or subjects should be as representative as possible of the population which may be able to benefit from the intervention to be tested. The criteria for admission to, or exclusion from, the trial define a "study population" which is a subset of the wider population. Selection of patients is usually not random because the investigator must use patients who present and are willing to take part. When the results of the study are reported, the investigator has to generalise, as far as possible, from the patients in the sample who were actually studied, to the study population, and then to all those who might be able to benefit from the intervention. Inclusion and exclusion criteria must therefore be clearly defined. Patients must give informed consent to their inclusion. They should be people who might benefit from the treatment and in whom the response is likely to be measurable. To minimise the number who drop out it is best to exclude patients who are at high risk of leaving the trial and avoid complex trial designs. Those in whom adverse effects might be specially serious should also be excluded.

D. Trial Design

Controlled trials can be classified as follows:

I. Concurrent comparisons
   1. Independent groups
   2. Related groups: (a) cross-over plans; (b) matched pairs; (c) repeated measurements, including right/left comparisons; and (d) factorial plans

II. Historical comparisons

I. Concurrent Comparisons

1. Independent Groups

Independent groups are specially suitable for investigations in which: (a) the disease is of limited duration or may be cured; (b) it is suspected that the effect of treatment in one period might carry over into the second period; and (c) the study is over a relatively long time so that seasonal influences might make different treatment periods non-comparable. Independent groups consist of separate individuals. Since measurements made repeatedly on one set of subjects must be analysed by a special method, it is best to avoid a mixed trial design. ROSARIO et al. (1979) compared skin histological changes after different numbers of treatments with UV-A, UV-B and UV-C and PUVA (UV-A plus oral methoxsalen). As the observations are not fully independent the results are difficult to interpret.

2. Related Groups

Related groups are not independent and the form and practical details of the analysis must take this into account when the trial design is selected. In a study of the effect of oral disodium cromoglycate in systemic mastocytosis (SOTER et al. 1979), because the cross-over courses of treatment were seriously unbalanced, the results could not be economically analysed taking full account of the clinical effort put into the trial. The authors ignored possible period effects, and their optimistic conclusions are ill-founded.

a) Cross-over Plans

Here each subject has first one treatment and then the other. Since subjects can thus serve as their own controls, the variability of the individual subjects will to some extent cancel out and the investigator hopes for economy in the use of patients and for greater precision. However, a threefold price is paid for this potential economy (MAINLAND 1963):

1. Confounding of time and treatment is a special hazard of the cross-over plan and can be countered by random allocation of patients to the two groups, the first given A then B, while the other receives B followed by A. If in both groups one of the treatments is associated with better responses, interpretation is straightforward. But when this is not the case it is harder to draw reasonable conclusions. If a clear period effect emerges, the investigator may decide to discard the second, cross-over, set of results and use only the responses to the first treatment. Such a decision converts the trial design from a cross-over to one with two independent groups and is rigorous but wasteful.
2. Interference of treatments with each other is a difficulty in cross-over trials and also in right/left comparisons (see Sect. D.I.2.c). An effect carried over from one treatment period to the next may be difficult to exclude unless the drug or its metabolites can be estimated in body fluids.
3. Duration of the experiment is prolonged because each patient has to take two courses of treatment to complete the study, and fewer patients will persevere.

Many investigators have ignored the point that a cross-over study is valid only if the earlier course of treatment does not alter the patient's response to a later course of treatment. The analysis of response measurements from a cross-over trial ought to examine the observations for such a carry-over effect. Unfortunately, available tests are weak, and it is never possible completely to exclude a carry-over effect. For this reason cross-over trial plans are not recommended (FRIEDMAN et al. 1981). ARMITAGE and HILLS (1982) discuss the two-period cross-over trial in detail.

b) Matched Pairs

Matched paris were used in an acne study (CUNLIFFE et al. 1979). The authors matched their patients for their overall grade of acne severity and the number of inflamed lesions. This form of stratification is of value only when one factor has prognostic importance, but wastes patients who are recruited and remain unmatched. For this reason matching done on more than one factor is likely to show sharply diminishing returns.

c) Repeated Measurements

Repeated measurements are taken at different times on the same individuals. In studies of treatments for polymorphic light eruption (CORBETT et al. 1982), for each patient up to 15 weekly measurements of response were related to exposure to sunlight over 3–4 months. The proportion of patients who reported the symptom in a given week appeared to be linearly related to the mean exposure for that week. Straight lines were fitted and treatments compared by analysis of covariance. Nevertheless, because the weekly summary figures refer to the same groups of patients they are not independent. This weakens the statistical foundation of the analysis and conclusions.

The right/left comparison method of repeated measurement (SIEMENS 1952) is popular as it is straightforward and patients act as their own controls. Translocation of active substance (MARPLES and KLIGMAN 1973) is not a problem with low potency substances. In some circumstances a right/left comparison plan is both appropriate and economical. MURRAY et al. (1980) studied bilaterally symmetrical persistent palmo-plantar pustulosis treated with PUVA on one randomly selected side, using a visual analogue scale (AITKEN 1969) and colour transparencies. This trial showed not only that PUVA is better than placebo, but also answered an important subsidiary question about the frequency and timing of relapses.

d) Factorial Plans

The factorial design, in which two treatments and their combination are studied, can be much more economical and informative than the designs discussed, but cannot be used with ordinally scaled variables (see Sect. J. I) and is therefore rarely applicable in dermatology.

II. Historical Comparisons

Historical controls save time and cost, but are particularly vulnerable to bias and to uncontrolled variation and thus difficult to interpret. It is probably best to keep the historical controlled study for special cases, where a concurrent control is impossible.

E. Power and Statistical Analysis

Is the study large enough to give a high probability of detecting an important therapeutic improvement? FREIMAN et al. (1978) discussed this question and noted the hazard that a potentially useful intervention might be dismissed as useless on the strenght of one or more small controlled trials with low power to detect improvement. FEINSTEIN (1977) discussed the power problem and likened the confidence with which the new hypothesis can be sustained to that of the specificity of a diagnostic test, and gave a worked example.

In a cross-over study in children with atopic eczema ATHERTON et al. (1982) concluded that reasonable doses of oral disodium cromoglycate given over 4 weeks had no more effect than did the placebo in the control period. The study was very detailed, but was done on only 30 children and the authors did not quantify their conclusions with any statement about the power of the study. The power can be roughly estimated by substituting in FEINSTEIN'S (1977) example the proportion of atopic children whose eczema was better at the end of the study after treatment with the drug or with placebo; it suggests that the study was capable of detecting clinically relevant improvement in response had it existed, and a statement to this effect in the paper would have been valuable.

The power of a significance test depends on the total sample size, the chosen significance level of the test and the size of the difference in responses which we want to be able to detect. Both in planning and retrospectively, it is sensible to use the size of the difference in responses which is considered both clinically interesting and realistic. The close relationship between a powerful test and the conclusions which are drawn is emphasised by our review of 26 published dermatological clinical trials which showed non-significant results. In only one of these studies, dating from 1979 to 1982, was there a power statement. This study (SWINYER et al. 1980) is remarkable because the required size of the trial was calculated in advance, given the power desired, the magnitude of the difference considered clinically interesting and the estimated standard deviation of acne lesion counts. ALTMAN (1982) in discussing the misuse of statistics in papers published in medical journals, suggests that instead of a power statement, the observed effect should be quoted with a confidence interval, which the general reader may understand more easily.

F. Allocation to Treatment

The study by BOWERS et al. (1966), mentioned in Sect. A, had eight treatment groups. How are we to allocate patients to these treatments? The patients are first recruited from the study population into the trial sample and then allocated to treatment. It is important that every patient who has been recruited into the study should be suitable for inclusion in any one of the treatment groups. If this is not the case, a form of sorting of patients will occur, so that some treatment groups have fewer and some have more of the most severely affected patients. Such systematic variation of severity or any other feature of prognostic significance is undesirable. To avoid bias each patient must be given an equal chance of allocation to each treatment group.

In large trials *randomised allocation* tends to produce comparable groups in which factors known and unknown which may affect the response to treatment will be evenly balanced. But a good balance will not necessarily be obtained in any one study and in small studies the balance is less likely to be completely satisfactory. Even so, randomisation is the best method of attempting to produce comparable groups for small and medium-sized clinical trials and it also ensures that statistical tests of significance are valid. *Restricted randomisation* is used in trials of all sizes to equalise the number of patients in each treatment group. Half the patients are randomly allocated to one treatment; the other half then receive the other treatment. This method is easily adapted to allocation to more than two treatments (FRIEDMAN et al. 1981). Random allocation of patients to treatment groups prevents the introduction of bias which may be in either direction and can easily invalidate the comparison between treatments (FRIEDMAN et al. 1981). The investigator should decide only that the patient, who has consented, is suitable for admission to the trial and may receive either treatment. Random allocation to treatment group then follows.

G. "Intention to Treat"

Particularly in studies with a rigid dose schedule, some patients may not tolerate the treatment. If many such patients drop out, the results may become seriously biased. Such an impasse can be avoided if patients are asked to continue in the trial to the end, without necessarily taking any treatment. The groups formed by randomisation at the "intention to treat" stage can then be compared, knowing the number who have had to stop the treatment (SCHWARTZ et al. 1980). It follows that the policy about withdrawals from treatment should be decided as part of the definition of treatment, along with exact details of the dose, timing and mode of application of treatments.

H. Blindness

So-called blindness guards against bias in the measurement of responses. In studies where double-blinding is unethical or impossible because of the nature of the treatments, a single-blind study, in which the investigator but not the patient knows which treatment each patient receives, is appropriate, if possible with extra precautions to reduce bias.

J. Measurements

I. Measurement Scales

A crucial part of a trial design is the choice of what is to be measured to assess each patient's response. It is often impossible to measure exactly what we want to know, but the observations should reflect them as accurately as possible, i.e. the measurements should be valid. For instance, the mechanical recording of scratching movements has been used as an indicator of itching (FELIX and SHUSTER 1975).

The most primitive form of measurement is the grouping of individual observations into qualitative classes. Such a nominal scale is used in all studies in which the patient's response is classified as "success" or "failure". For example, in that by PETERSEN et al. (1980) of ketoconazole as treatment for chronic mucocutaneous candidiasis in which patients whose mucosae were worse after 5 days or still showed infection after 14 days, and those whose cutaneous candidiasis had persisted or progressed after 4 weeks' treatment, were classed as failures. Several classes can be used, but they must be mutually exclusive and exhaustive so that each observation can be placed in one, and only one, class.

If the measurement classes can be arranged in order of magnitude, for instance the overall grade of acne severity described by BURTON et al. (1971), then we have an ordinal scale. In such a scale each observation can be placed only in one class and the classes can be arranged in order. Although ordinal measurements are often expressed as numbers, arithmetical manipulation may not truly summarise the response measured. For instance, the mean score may be a poor way of summarising responses measured on an improvement rating scale; the median, a non-parametric measure of location, is usually better.

Quantitative measurements of lesion area (MICHELL et al. 1980) thickness or some other property (see Chap. 8), duration or time to clearing (ROGERS et al. 1979) and biological measurements (see Cunliffe Chap. 4) are on the higher ratio scale where observations, and their mean (transformed if appropriate) give a faithful summary. Such measurements add greatly to the discriminatory power of a trial and an attempt should always be made to use them.

The investigator needs to consider the type of measurement chosen so that the numerical results which have been obtained can be interpreted correctly. Special non-parametric statistical techniques exist for analysis of ordinal scale measurements (CONOVER 1980). Whether the parametric methods used for interval and ratio scale measurements should ever be used for ordinal measurements is unsettled; HAYS (1963) argues convincingly for the choice of any technique, but points out that it is easier to make measurements and express them as numbers than to make justifiable statements about them.

While the trained eye is a highly sensitive measuring instrument and both convenient and clinically convincing, we are bad at visual quantitation and our memory of previous observations is poor (MARKS 1979). Thus, clinical observations are not ideal for comparisons, particularly when separated in time.

II. Types of Measurement

The types of measurement used in 84 controlled clinical trials in dermatology published in the period 1976–1982 were: nominal scale 4; rating scales only 43; measurements only 18; rating scales + measurements 19.

Most rating scales in dermatology are used only by their author who assumes, sometimes mistakenly, that the scale measurements correlate with the responses under investigation. It would be an important step forward in dermatological research to develop a consensus on the most meaningful rating scales, so that findings can be replicated and validated. The disparity between the ease of use of the trained eye and the difficulty of measurement has for decades deprived dermatology of the powerful advances in understanding which apposite measurements can provide and an attempt should always be made to incorporate such measurements into clinical trials.

Lesion counts are a well-established method for assessing response in acne. Several studies have used the area, diameter or thickness of lesions or the amount of skin involved expressed as an "intensity score" (VAN DER RHEE et al., 1980) or as a percentage of body area (VELLA BRIFFA et al. 1981). Assessment of infections and infestations is more easily measured, e.g. change in number of parasites in skin snips and qualitative changes in fungi, bacteria and viruses in various samples (Chaps. 2 and 10). Published studies using photographic assessment are surprisingly few (e.g. MURRAY et al. 1980) given the obvious dependence of dermatology on visual skills for assessment. This is because the accuracy of photographic reproduction has generally been found unsatisfactory, presumably because the many variables are not easily standardised. Position, camera distance, background and lighting, type of camera, lens, film and its development or projection are all critical. In photographic terms, low flare and good micro-contrast together make up the gradation capability needed for photographic assessment of therapeutic responses. Most inexpensive lens systems do not provide these qualities, and an expensive system designed for general purposes, or for a use very different from skin photography may also be unsatisfactory.

K. The Protocol

This is a comprehensive description of what has to be done and preferably also why. Check lists to help in writing the protocol have been published (e.g. CHAPUT DE SAINTONGE 1977; WARREN 1978). A precise definition of details of treatments and their supervision should be included, together with details of permitted and forbidden ancillary treatments. The use of topical corticosteroids in a study of etretinate as treatment for persistent palmo-plantar pustulosis (FREDERIKSSON and PETTERSON 1979) is an example of the haphazard use of an ancillary treatment.

L. The Reasons for Performing Therapeutic Trials

Clinical trials are done for many different reasons including the personal motivation of the trialist, all of which will affect how the trial is carried out and often indeed what conclusions may be drawn. In their simplest form they are a study of an individual remedy. In other instances they may be used as a test of a package of treatment modalities or policies. The two-centre PUVA trial (see ROGERS et al. 1979; VELLA BRIFFA et al. 1981) compared the merits of treating psoriasis by the Ingram dithranol regime with those of PUVA. It was a trial of treatment policy, and also assessed the costs of the two treatment packages, the possible long-term risks to the patients and the costs of those risks.

Many trials are supported by the manufacturer of the product under test because they are needed for submission to drug regulatory authorities as part of the application for a licence to market the product. In the United Kingdom the treatment needs only to be reasonably effective and safe. For yet another topical corticosteroid preparation this is of negligible scientific interest, though it may be important commercially. When the new treatment is cheaper the results of trials will also be important for medical practice. Yet the design of many trials appears to be chosen as often to put a new preparation in a favourable light as to illuminate its pharmacology. An example is the exceptionally poor multi-centre general practice study (BAYMAN et al. 1982) which reaches the remarkable and totally unjustified conclusion that a combination of 2% miconazole and 1% hydrocortisone made by the trial's sponsor may be used to treat a wide range of skin conditions at the initial presentation and without the need for a correct diagnosis. A less transparent example is a study of fluocinonide cream by eight investigators working independently (FISHER and KELLY 1979). Furthermore, if a trial is done for routine purposes it may well lack the enthusiasm necessary for a reliable study. However, there are many examples of trials which are of both scientific and commercial value.

M. Applying the Results of Trials in Practice

A controlled clinical trial is a complex and difficult enterprise and is a uniquely effective way of gaining information about the responses to the intervention as well as benefits and risks (DOLLERY 1981). One of the problems is the generalisation of results obtained from a sample of patients usually collected on a basis of convenience and availability rather than randomly, to the study population from which they were drawn, and thence to the general population. In the end generalising becomes a matter of judgment, more or less informed. All clinicians, not only those interested in clinical trials, should learn to make such judgments competently. They will then be able to identify those new developments that can continue to improve their therapeutic practices.

References

Aitken RCB (1969) A growing edge of measurement of feelings. Proc R Soc Med 62:989:993

Altman DG (1982) Statistics in practice. British Medical Association, London

Armitage P, Hills M (1982) The two-period crossover trial. Statistician 31:119–131

Atherton DJ, Soothill JF, Elvidge J (1982) A controlled trial of oral sodium cromoglycate in atopic eczema. Br J Dermatol 106:681–686

Bayman I, Wakelin J, Newby M (1982) The problem of accurate initial diagnosis of skin conditions in general practice and the case for routine treatment with Daktacort. Clin Res Rev 2:81–88

Bowers RE, Dalton D, Fursdon D, Knowelden J (1966) The treatment of psoriasis with UVR, dithranol paste and tar baths. Br J Dermatol 78:273–281

Burton JL, Cunliffe WJ, Stafford I, Shuster S (1971) The prevalence of acne vulgaris in adolescence. Br J Dermatol 85:119–126

Chaput de Saintonge DM (1977) Aide-memoire for preparing clinical trial protocols. Br Med J 1:1323–1324

Conover WJ (1980) Practical nonparametric statistics, 2nd edn. Wiley, New York

Corbett MF, Hawk JLM, Herxheimer A, Magnus IA (1982) Controlled therapeutic trials in polymorphic light eruption. Br J Dermatol 107:571–581

Cunliffe WJ, Burke B, Dodman B, Gould DJ (1979) A double-blind trial of a zinc sulphate / citrate complex and tetracycline in the treatment of acne vulgaris. Br J Dermatol 101:321–325

Dollery CJ (1981) Summary and conclusions. In: Cavalla JF (ed) Risk-benefit analysis in drug research. MTP Press, Lancaster

Feinstein AR (1977) Clinical biostatistics. Mosby, St Louis

Felix RH, Shuster S (1975) A new method for the measurement of itch and the response to treatment. Br J Dermatol 93:303–312

Fisher M, Kelly AP (1979) Multicenter trial of fluocinonide in an emollient cream base. Int J Dermatol 18:660–664

Fox W (1971) The scope of the controlled clinical trial, illustrated by studies of pulmonary tuberculosis. Bull WHO 45:559–572

Fredriksson T, Petterson U 81979) Oral treatment of pustulosis palmo-plantaris with a new retinoid, Ro 10-9359. Dermatologica 158:60–64

Freiman JA, Chalmers TC, Smith H, Kuebler RR (1978) The importance of beta, the type II error and sample size in the design and interpretation of the randomised control trial. N Engl J Med 299:690–694

Friedman LM, Furberg CD, DeMets DL (1981) Fundamentals of clinical trials. Wright, Boston

Hays WL (1963) Statistics. Holt, Rhinehart and Winston, New York

Mainland D (1963) Elementary medical statistics, 2nd edn. Saunders, Philadelphia

Marks R (1979) Design and conduct of dermatological trials. In: Marks R (ed) Investigative techniques in dermatology. Blackwell, Oxford

Marples RR, Kligman AM (1973) Limitations of paired comparisons of topical drugs. Br J Dermatol 88:61–67

Medical Research Council Streptomycin in Tuberculosis Trials Committee (1948) Streptomycin treatment of pulmonary tuberculosis. Br Med J 2:769–782

Michell P, Hawk JLM, Shafrir A, Corbett MF, Magnus IA (1980) Assessing the treatment of solar urticaria: the dose-response as a quantifying approach. Dermatologica 160:198–207
Morgan JK (1955) Observations on the use of hydrocortisone ointment. Br J Dermatol 67:180–188
Murray D, Corbett MF, Warin AP (1980) A controlled trial of photochemotherapy for persistent palmoplantar pustulosis. Br J Dermatol 102:659–663
Petersen EA, Alling DW, Kirkpatrick CH (1980) Treatment of chronic mucocutaneous candidiasis with ketoconazole. Ann Intern Med 93:791–795
Rogers S, Marks J, Shuster S, Vella Briffa D, Warin A, Greaves M (1979) Comparison of photochemotherapy and dithranol in the treatment of chronic plaque psoriasis. Lancet 1:455–458
Rosario R, Mark GJ, Parrish JA, Mihm MC (1979) Histological changes produced in skin by equally erythemogenic doses of UV-A, UV-B, UV-C and UV-A with psoralens. Br J Dermatol 101:299–308
Schwartz D, Flamant R, Lellouch J (1980) Clinical trials. Academic, New York
Siemens HW (1952) Die Technik der Rechts-Linksbehandlung für den Praktiker. Hautarzt 3:307–309
Soter NA, Austen KF, Wasserman SI (1979) Oral disodium cromoglycate in the treatment of systemic mastocytosis. N Engl J Med 301:465–469
Swinyer LJ, Swinyer TA, Britt MR (1980) Topical agents alone in acne: a blind assessment study. JAMA 243:1640–1643
Van der Rhee HJ, Tijssen JGP, Herrmann WA, Waterman AH, Polano MJ (1980) Combined treatment of psoriasis with a new aromatic retinoid (Tigason) in low dosage orally and triamcinolone acetonide cream topically: a double blind trial. Br J Dermatol 102:203–212
Vella Briffa D, Greaves MW, Warin AP, Rogers S, Marks J, Shuster S (1981) Relapse rate and long-term management of plaque psoriasis after treatment with photochemotherapy and dithranol. Br Med J 282:937–940
Warren MD (1978) Aide-memoire for preparing a protocol. Br Med J 1:1195–1196
Williams DI, Wilkinson DS, Overton J, Milne JA, McKenna WB, Lyell A, Church R (1964) Betamethasone 17-valerate: a new topical corticosteroid. Lancet 1:1177–1179

Section B:
Absorption, Metabolism and Toxicity

CHAPTER 12

The Properties of Skin as a Diffusion Barrier and Route for Absorption

R. C. SCOTT and P. H. DUGARD

A. The Location and Nature of the Percutaneous Absorption Barrier

Mammalian skin is a complex, integrated organ. It is the largest organ in mammals and in an adult human male covers an area of 18–22000 cm^2. Essentially, it is composed of the outermost epidermis, an inner dermis and associated hair follicles and sweat ducts. Our main interest in relation to the permeability properties is the epidermis which is responsible for the production of the outermost layer of the skin, the stratum corneum. The whole of the stratum corneum is believed to constitute the permeability barrier, with perhaps the exception of the outermost desquamating corneocytes (KLIGMAN 1964).

As the epidermal cells, keratinocytes, move through the layers outwards from the basal layer, they undergo an extensive differentiation before becoming mature corneocyte cells in the stratum corneum. This process of differentiation has been well studied. In the keratinocytes filaments form and increase in size (BRODY 1960; CHARLES and SMIDDY 1957). These filaments are formed of keratin which itself is composed of polypeptide chains held together by disulphide and Σ(-glutamyl) lysine bonds between the chains; adjacent keratin molecules are linked similarly. The mature keratin has been reported (ODLAND 1964) to be organised into bundles of 5–10 filaments producing structures of diameter 250 Å. A cationic protein, filaggrin, has been reported to act as an interfilamentous matrix. The mature corneocytes, therefore, contain a mixture of complex surface active proteins that form concentric hydrophilic interfaces about the closely packed fibrils.

The potential corneocyte membrane also undergoes a remarkable transformation. Organelles, known as membrane coating granules (MCG), are synthesised whilst the cell is in the stratum granulosum (MATOLTSY and PARAKKAL 1965; ODLAND 1964; OLSON et al. 1969). The MCGs extrude their lipid-rich contents into the intracellular spaces (MATOLTSY and PARAKKAL 1965; ELIAS and FRIEND 1975; SQUIER 1973) and their membranes fuse with those of the maturing corneocytes causing a change in thickness from 70 to 150–200 Å (MATOLTSY and PARAKKAL 1965; BRODY 1960). Most of this thickening occurs in the intercellular leaflet of the tripartite membrane of the cell (BONNEVILLE et al. 1968) which, thus, becomes essentially a proteinaceous coat. The extrusion of the MCG contents causes the intercellular space to increase from 1% of the total volume in the stratum granulosum to as much as 30% in the stratum corneum (ELIAS and LEVENTHAL 1979). The extruded contents are principally ceramides (40%), cholesterol, cholesteryl sulphate and free fatty acids which form multiple lipid bilayers to produce a water diffusion barrier. This role was first suggested by MIDDLETON (1968) who reported that the lipids organised into "shells" that accounted for water retention. These bilayers were believed to be present only in the lower layers of the stratum corneum (MENON et al. 1986) however, more recent studies indicate that they are found throughout this cell layer (MADDISON et al. 1987). The bilayers are composed of hydrocarbon chains which are straight and almost entirely saturated. The lack of structural perturbation among the chains allows the formation of tightly packed interiors.

Between the intercellular bilayer lamellae and the proteinaceous coat of the corneocyte envelope is a horny cell lipid membrane (SWARTZENDRUBER et al. 1987). This plasma membrane is composed of *o*-acylceramides (a sphingolipid based on 30-carbon hydroxyacid). It is bonded to the surface of the proteinaceous coat by ester linkages and acts as an interlamellar bridge holding together the bilayers. This important *o*-acylceramide is formed from linoleic acid which has been known for some time to be important in barrier function.

This complex arrangement of lipid and protein has led to the barrier being described as a two-compartment diffusion barrier (MICHAELS et al. 1975) with lipophilic and hydrophilic areas. The precise mechanism by which molecules diffuse across this barrier is not well understood. Perhaps, as our knowledge of the molecular structure of the barrier improves it may be possible to elucidate the mechanism further and to define the actual pathways by which molecules with different physicochemical properties are absorbed into the body. At present we must accept a model based on this intimate relationship of lipid and keratin which, if either is incompletely formed owing to a failure of the differentiation process, leads to a malformed stratum corneum and abnormal permeability properties.

B. Methods of Measuring Percutaneous Absorption

I. In Vivo Techniques

Many compounds are poorly absorbed through the skin and the small amounts are subsequently very difficult to detect once in the body. Various techniques have been devised to overcome this problem. The application of a test chemical to the skin surface and quantitation of its rate of "disappearance" from the surface can be used to calculate a (crude) rate of absorption. This technique, known as remainder analysis, is inaccurate. Only a small amount of chemical will be absorbed from a relatively large surface depot and, therefore, the resolution of the technique must be very high. It is also possible that decreases in surface concentration have been caused by sweating or transepidermal water loss, thus leading to an overestimation of absorption. Because of the limited usefulness of the technique it has not been used to study the absorption of many compounds (AINSWORTH 1960; WAHLBERG 1965).

The use of autoradiography to study percutaneous absorption has also been attempted. In this technique radiolabelled compound is applied to skin. Its presence is identified by placing skin slices, cut with a freeze-microtome in contact with a light-sensitive film or emulsion. The radioactivity is visualised by photographic development. This technique is only semi-in vivo as a punch biopsy of the application site must be made. The technique, at best, provides only a static picture of the location of the chemical at the time of processing the tissue. It cannot provide information relevant to the rates or routes of absorption. Interpretation is made difficult by local high concentrations of a substance (as is likely in the stratum corneum) causing shadows on the autoradiograph. However, this technique has been found useful (ZESCH and SCHAEFFER 1975) in the development of topical formulations where local high concentrations of drugs are desired.

Some penetrating chemicals induce a response at nerve ending, blood vessel walls or in sweat ducts. Such pharmacological responses have been used to indicate absorption. It must be recognised that not only the rate of absorption, but also the potency of the chemical are involved in such a reaction. A poorly absorbed, but extremely potent chemical might induce a similar response as a well-absorbed, but less active chemical. The absorption of sarin has been quantified by anti-cholinesterase activity (BLANK et al. 1958), nicotinic acid by erythema production (ALBERY and HADGRAFT 1979) and nitroglycerin by changes in blood pressure (FRANCIS and HAGEN 1977).

The topical steroids induce vasoconstriction which leads to "blanching" of the skin. This has been exploited to study the absorption of these molecules in the development of the "vasoconstrictor assay" (for a comprehensive review see BARRY 1983). The relative potencies of different steroids have been assessed by preparing formulations with the same thermodynamic activity (WOODFORD and BARRY 1982). By using the same steroid in different vehicles the relative merit of topical formulations has also been assessed (BENNET et al. 1985).

Absorbed molecules may be detected in body fluids or tissues. Most compounds are poorly absorbed through intact skin and, therefore extremely sensitive assay techniques are needed. Essential to the development of these studies has been the use of radiolabelled (principally ^{14}C) compounds, which has facilitated the analytical procedures. Usually, only the radiolabel is assayed and there is no attempt to discern whether the radioactivity is associated with parent compound or a metabolite. However, the degree of accuracy and specificity achieved is often adequate, particularly when the absorption of the same chemical from a range of formulations is studied.

The problems of using radiolabelled compounds in humans, together with metabolism data from animal studies, often lead to the development of sensitive assays for the unlabelled parent molecule and principal metabolites. Such assay procedures can then be used to study non-labelled parent compound absorption together with metabolism in humans.

The protocols used in animal studies can be broadly divided into two groups. In the first, the test penetrant is applied to a defined area of skin. The site is either occluded or left open to the atmosphere and the animal is placed in a metabolism cage or a restraining cage. Such cages separate and collect urine and faeces and can trap exhaled CO_2. After a specific time the animal is killed. Various organs, the site of application, blood, urine and faeces, and the remainder of the carcass are then analysed for radiolabel or parent molecule and metabolites. Variants of this protocol have been used by many workers (KNAAK et al. 1984; BLACK et al. 1975; BARTNIK et al. 1985; SHAH et al. 1981).

This type of protocol is limited. It provides detailed information about the amount of compound absorbed, routes of elimination and disposition at a specific time point. If it is necessary to study the time course of the absorption process, then groups of animals must be dosed and killed sequentially to build up data to generate an adequate profile.

An alternative method has been developed for use in humans (Feldmann and Maibach 1969) although it can be used in other species and has been used extensively in monkeys (van Lier 1985; Bronaugh et al. 1985). The procedure relies on the principle that when a chemical is absorbed through the skin it enters the bloodstream directly. Therefore, intravenous dosing has been used to mimic this direct access to the circulatory system via the skin. Following the intravenous dose, urine and faeces are collected for a defined period (usually 5 days) to determine the major route of elimination from the body (usually no allowance is made for biliary excretion). The percentage of the dose which accumulates in the urine and faeces is then determined. It is assumed that this percentage remains the same irrespective of the achieved plasma levels.

Following application to the skin, urine and faeces (though most often only urine) are collected and assayed for the applied chemical. The amount of applied chemical collected is corrected by means of the data after intravenous dosing and the total percutaneous absorption is thus calculated. This type of study can be used to provide data on the time course of absorption by collecting urine (or faeces) over measured time intervals. It can be used for the majority of chemicals, provided the chemical is eliminated from the body faster than it is absorbed through the skin. Alternatively, changes in plasma levels can be used to compare absorption of the same chemical from different formulations.

II. In Vitro Techniques

The development and use of in vitro methods has been pursued as the permeability properties of skin have been demonstrated to be maintained after excision (Burch and Winsor 1942; Harrison et al. 1984). This makes the in vitro study of percutaneous absorption possible. Experiments have shown that there is no active transport involved in the absorption process (Tregear 1966) and absorption is driven by simple diffusion through the stratum corneum and epidermis into the capillary network beneath the dermo-epidermal junction. For these reasons it is possible to measure the absorption of chemicals in vitro through the skin.

Skin membranes are mounted in glass diffusion cells of which there are many types and designs (Fig. 1 and 2) suitable for use in experiments in vitro (Dugard et al. 1984). Essentially, these cells consist of two chambers, the donor and receptor chambers, which sandwich the skin. The epidermal surface (stratum corneum) of the skin faces into the donor chamber and the lower surface (dermal) into the receptor chamber fluid. The receptor chamber contains a receptor fluid and by analysis of this fluid, absorption of the penetrating molecule is detected.

The donor chamber can be completely closed, holding a solution or neat liquid in contact with the skin (Behl and Barrett 1981) or open, thus exposing the applied chemical to the atmosphere (Franz 1975). In the closed donor type of cell the skin membrane becomes artificially hydrated (Idson 1978) and this can alter

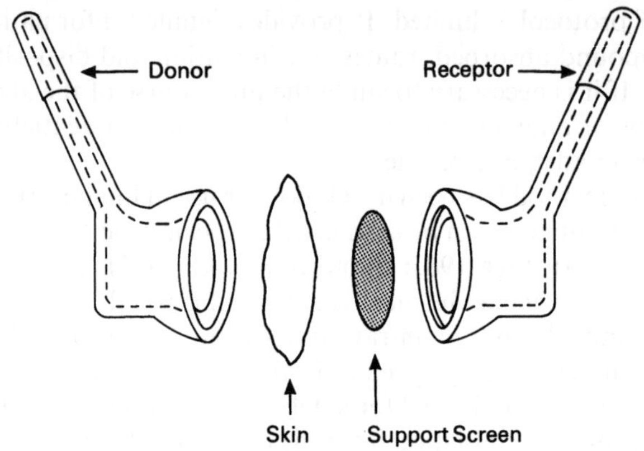

Fig. 1. A two-sided or vertical membrane glass diffusion cell

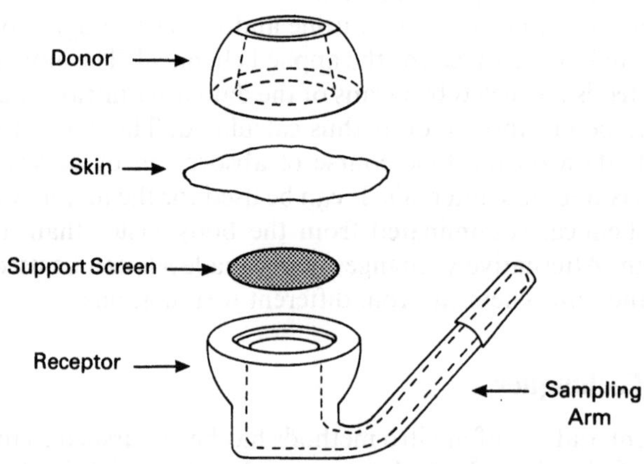

Fig. 2. An open-top or horizontal membrane glass diffusion cell

the permeability of the skin. However, this type of cell has proven very useful for comparative measurements under the described, specific conditions. In order to mimic dermal applications in vivo the open-top donor chamber should be used. This allows measurements to be made under more physiologically natural conditions and so the results are more relevant to many human exposure situations.

The skin membranes which are used may be whole skin membranes (i.e. epidermis and dermis) from which subcutaneous fat and any attached muscle have been removed, or epidermal membranes. Epidermal membranes are fairly simple to prepare from human whole skin (KLIGMAN and CHRISTOPHERS 1963), but are more difficult to produce from animal skins because of the denser distribution of hairs. A technique has been developed for rat skin (SCOTT et al. 1986 b) and skin slices approximately 300–400 μm have also been cut from whole skin (BRONAUGH and STEWART 1985). This slicing technique produces a membrane of epidermis and some upper dermis. For water-soluble molecules the use of full-thickness skin presents no problem. For these molecules, the diffusion barrier is located in the stratum corneum and the resistance to diffusion from the epidermis and dermis is negligible and cannot be determined experimentally. Thus, full-thickness skin and epidermal membranes have, at least within the resolution of the experimental technqiue, the same permeability (SCOTT et al. 1986b). However, for lipophilic molecules, the dermis can act as a very significant in vitro barrier (SCOTT et al. 1986b). This dermal barrier is an artificial barrier as molecules do not have to diffuse through the dermis in vivo. Consequently, rates of absorption for lipophilic molecules, measured through full-thickness skin in vitro are depressed.

It is essential that the penetrating molecule has a high solubility in the receptor fluid and that the receptor fluid does not alter the permeability properties of the skin. For water-soluble molecules saline can be used. For lipophilic molecules, saline has poor solubility properties, and if it is used, the penetrating molecule will not partition out of the skin, into the receptor phase. Various receptor fluids have been used to increase the solubility properties for lipophilic molecules. A 6% solution of Volpo N20 (a non-ionic surfactant) in saline has proved useful (BRONAUGH and STEWART 1985), though our preference is for a 50% aqueous solution of ethanol. We have found (SCOTT and RAMSEY 1986) that 50% aqueous ethanol leads to rates of absorption indistinguishable from those determined by in vivo techniques, whereas, the Volpo N20 solution caused an underestimation of the in vivo absorption in some instances.

C. In Vivo – In Vitro Comparisons

The early in vitro investigations of percutaneous absorption used two-sided glass diffusion cells (SCHEUPLEIN 1965). The penetrant molecule was placed in the donor chamber (either as a neat liquid or aqueous solution) and saline or distilled water in the receptor chamber. The tissue became hydrated in such a system and consequently this is a component of the permeability characteristics determined (SCHEUPLEIN and ROSS 1974). These data, whilst accurate, are relevant only to the conditions of stratum corneum hydration under which they are made. Differences between in vitro data measured under these conditions, and in vivo data obtained under conditions of non-occlusion must be anticipated. However, many in vivo studies are done under conditions of occlusion and such skin has altered permeability (IDSON 1978). Therefore, results obtained in vivo under occlusion might be predicted from this type of in vitro experimental protocol, although this has not been determined.

Open-top donor chambers were designed to mimic in vivo conditions more accurately (FRANZ 1978; SCHEUPLEIN and ROSS 1974). In vitro data for [^{14}C]cortisone (WURSTER and KRAMER 1961) have been compared with in vivo data (WAHLBERG 1965). Almost identical absorption rates were determined over 4 days of skin contact, which suggested that the conditions of the stratum corneum in such a "dry" in vitro diffusion cell are similar to the in vivo condition.

The percutaneous absorption of 21 organic compounds has been studied in vivo through human forearm skin (FELDMANN and MAIBACH 1970). The absorption of 12 of these was measured (Franz 1975) through full-thickness skin. These 12 compounds were selected from the 21 on a random basis, although highly water-insoluble compounds, like hexachlorophene, were excluded since it was believed that insolubility in the receptor fluid might limit absorption. Generally, the in vivo and in vitro data showed a good correlation in the rates of absorption, with the exception of thiourea. New in vivo data were obtained for thiourea the abdomen, rather than forearm, and a much better agreement determined with the in vitro data. This might have been due to the body site differences although there are question marks against the in vivo calculations. When the total amounts of compounds absorbed were compared there was a poorer agreement between the in vivo and in vitro data. This was believed due to differences in sites of application (forearm or abdomen) and lack of corneocyte desquamation in vitro. By comparing the amount absorbed in vitro after 2 days with the amount absorbed in vivo after 5 days, the two methods showed very similar total absorption, except for the poorly absorbed compounds (hippuric acid, nicotinic acid and thiourea). Further studies were done (Franz 1978) in vivo and in vitro, with abdominal skin which was washed after 24 h. This produced excellent agreement for these compounds between the in vivo and in vitro data for both total absorption and rate of absorption.

Excellent agreement between in vivo and in vitro data through rabbit ear skin has been shown for three molecules with different physicochemical properties (Creasey et al. 1978). The penetrant molecules were tritiated water (T_2O), [^{32}P]tri-*n*-propylphosphate (TPP) and T-dibenz [*b*,*f*] (1,4) oxazepine (CR). Good agreement was obtained through full-thickness skin for T_2O and TPP. However, in vitro accumulation of CR occurred which caused the in vitro rates of absorption to be lower than those measured in vivo. Removal of the adipose tissue led to in vitro data similar to the in vivo data. As already discussed, in vivo, chemicals do not have to penetrate through the dermis and the aqueous dermis in vitro markedly affects the absorption of CR by acting as an additional barrier to the stratum corneum.

Use of full-thickness skin in vitro has been a major reason for poor predictions of the in vivo absorption of lipophilic molecules. The problem is not encountered with water-soluble chemicals as they penetrate full-thickness skin at a rate which is indistinguishable from that through the stratum corneum (Scheuplein 1965; Scott et al. 1986 b). This has been a major reason for the reluctance to use in vitro techniques to develop formulations, however, with a better understanding of the potential technical problems associated with the technique, it will be more readily and widely used in future.

Good agreement between in vivo and in vitro data has been observed for rat skin (Bronaugh et al. 1982 b; Scott and Ramsey 1986, 1988). The test penetrants used (benzoic acid, urea, caffeine, acetylsalicylic acid, cypermethrin, carbaryl) had a range of physicochemical properties and were applied to the skin in a variety of vehicles. Comparable patterns and rates of absorption were detected in these studies. Similarly, good agreement was found for monkey skin with a series of well-absorbed nitroaromatic compounds [1-nitro-*p*-phenylenediamine (1NPD), 4-amino-2-nitrophenol (4A2N), nitrobenzene (NB), *p*-nitroaniline (*p*-N) and 2,4-dinitrochlorobenzene (DNCB). DNCB and NB were also measured in vitro through human skin and the data compared with previously published human data (BRONAUGH and MAIBACH 1985). Good agreement was obtained with DNCB, but lower values were measured in vivo for NB and this was believed due to loss by evaporation.

Comparisons of in vitro absorption through pig skin and previously reported in vivo human data have also been published (REIFENRATH et al. 1984). Good agreement was seen with test penetrants with log log (octanol: water partition coefficient) greater than 3, but above this value there was poor agreement. For the in vitro experiments, skin slices 1.9 mm thick were used and these would have contained some dermis. The dermis would have acted as an additional artificial barrier to the penetration of the more lipophilic chemicals, thus leading to poor prediction of the in vivo absorption. If experimental conditions are carefully matched and epidermal skin membranes are used, then excellent predictions of in vivo absorption can be obtained by the in vitro techniques.

D. Mathematical Derivation of Absorption Parameters

Our understanding of the mechanisms by which formulations and molecules interact to affect absorption has been enhanced by the use of in vivo, but mainly in vitro techniques. This absorption has been examined with the use of non-complex physical chemistry and mathematics.

In high dermal exposure situations, the chemical contacts the skin as a "bulk" solution. The precise definition of the volume which constitutes a "bulk" solution is difficult. The simplest, workable definition is that the volume is sufficiently large so that the penetrant content is not significantly reduced by the absorption process, i.e. the absorption process is not appreciably reduced, or altered in nature, during the time course of an experiment through a fall in penetrant concentration.

When such a solution contacts the skin, the absorption profile (Fig. 3) will start with a "lag phase" of increasing absorption rate (as a concentration gradient is established across the stratum corneum) which is followed by a steady state period. This steady state period is in reality a pseudo-steady state period, but practically will equate with the maximum rate (in the absence of progressive chemical alteration of stratum corneum) achievable from the formulation.

In such circumstances, the absorption rate will be proportional to the concentration in the vehicle. The relevant concentration of importance is that in true solution, not the "total" concentration applied (i.e. in a suspension). It can be shown that the steady state absorption rate J_s is proportional to the applied concentration C_v and this is a simple representation of Fick's Law.

If the absorption rate is known, together with the area of skin to which the solution was applied then the absorption rate can be presented in units of amount (μg), area (cm^2) and time (h). This allows exposure areas and times to be manipulated to predict absorption under other conditions.

If the absorption rate is divided by the applied concentration (in solution) a proportionality coefficient, the permeability coefficient, K_p (units cm h^{-1}) can be calculated.

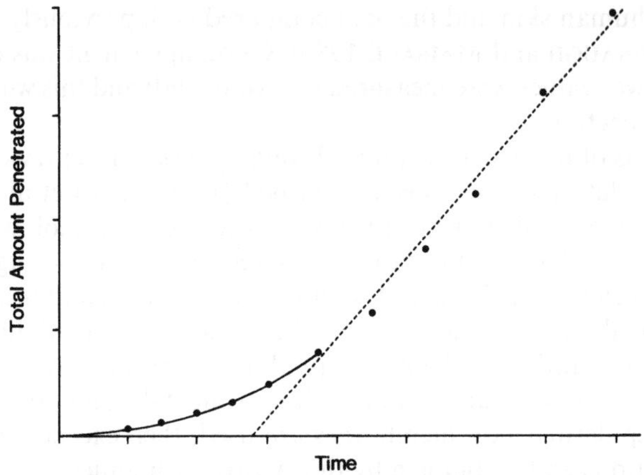

Fig. 3. Time course of absorption from an infinite or "bulk" application. The initial curved region represents the "lag phase" during which the concentration gradient across the skin membrane is being established and the pseudo-linear phase is the "steady state period" of absorption

$$K_p\ (\mathrm{cm\,h^{-1}}) = \frac{\text{Absorption rate } (\mu\mathrm{g\,cm^{-2}\,h^{-1}})}{\text{Applied concentration } (\mu\mathrm{g\,cm^{-3}})}$$

or

$$K_p = \frac{J_s}{C_v} \tag{1}$$

and

$$J_s = K_p C_v, \tag{2}$$

This permeability coefficient is ideally independent of concentration, and is only applicable to a single vehicle. The permeability coefficient allows subsequent rates of absorption to be predicted. The calculated values for human skin, of permeability constants through intact, normal skin, cover at least five orders of magnitude. Generally, under the same experimental conditions, values for animal skins are higher.

The permeability coefficient can be expanded to show the factors which control the absorption process (SCHEUPLEIN 1965):

$$K_p = \frac{K_m D_m}{\delta}, \tag{3}$$

where K_m is the stratum corneum: vehicle partition coefficient (distribution of the penetrant between the vehicle and outer layer of stratum corneum), D_m is the diffusion coefficient (mobility of the penetrant molecule within the stratum corneum) and δ is the thickness of the stratum corneum.

If Eqs. 2 and 3 are combined, the factors controlling the absorption process are displayed

$$J_s = \frac{C_v K_m D_m}{\delta}. \qquad (4)$$

Thus, the rate of absorption (steady state) is proportional to the concentration (in solution) in the vehicle, the partition coefficient and diffusion coefficient, but is inversely related to the thickness of the membrane.

Theoretically, for a particular chemical, D_m should remain constant and the value is a biological variable. Manipulation of C_v and K_m make it possible to alter the rates of absorption from various formulations. Linear extrapolation of the steady state period of a plot of total penetration versus time to the time axis defines the lag time, τ. Note this time is not the start of the steady state period, which is achieved after approximately 2.7 times the lag time (BARRY 1983). The lag time τ can be used to calculate the diffusion coefficient, thus

$$\tau = \frac{\delta^2}{6 D_m}. \qquad (5)$$

If the parameters K_m, D_m and δ are determined during the steady state period, then there are mathematical expressions (largely untested) to predict rate of absorption and total amount dissolved or within the stratum corneum for any specific time after dermal applications (DUGARD 1977).

Whether an applied dose acts as a "finite" (or small) amount, is determined by whether it is depleted as result of absorption. This situation is probably more relevant to many real dermal contact situations, as experienced with products formulated specifically for topical application. From a finite application to the skin (Fig. 4), the absorption rate increases in the first instance, during a period equivalent to the lag phase. Following a maximum rate of absorption, the rate then falls as depletion of the surface material occurs. This type of pattern is very difficult to detect, owing to the inevitable analytical difficulties of quantifying very low levels of a penetrant. It is probably impossible, technically, to measure by in vivo methods and only by an in vitro technique could the necessary degree of resolution be obtained.

Equations have been developed (SCHEUPLEIN and ROSS 1974) which can be used to predict the time at which the maximum rate T_{max} will occur.

$$T_{max} = \frac{\delta^2 - h^2}{6 D_m}. \qquad (6)$$

Usually, the thickness of the penetrant layer h is small relative to the stratum corneum thickness δ, therefore Eq. 6 reduces to

$$T_{max} = \frac{\delta^2}{6 D_m}.$$

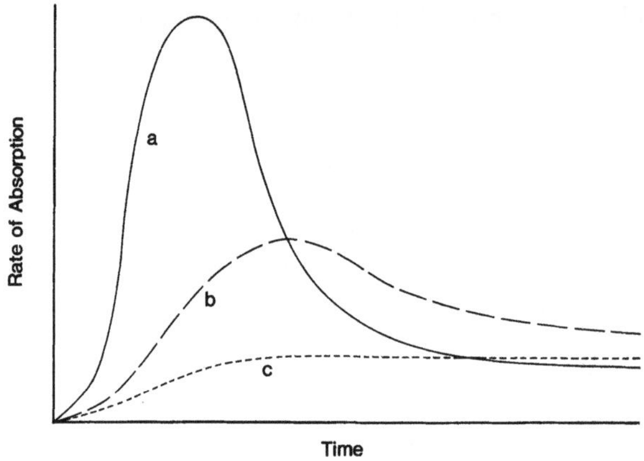

Fig. 4. Profiles of percutaneous absorption from "finite" or small skin depots. *Curve a*: fast rate of absorption leads to a short peak phase, but high absorption. *Curve b*: slower absorption produces a broader peak and lower absorption. *Curve c*: with very poorly absorbed penetrants the surface amount is not significantly reduced by absorption and a steady state is apparent

Peak absorption rate J_{max} is proportional to A, the amount of penetrant per unit area (termed the specific dose) as follows

$$J_{max} = \frac{1.89\, D_m A}{\delta^2}. \tag{7}$$

These equations are only suitable for situations where considerable depletion of the surface deposit occurs and cannot be applied to "bulk" infinite exposure.

E. Species Comparisons: Relevance of Animal Data to Humans

The development of topical therapeutic formulations, or assessment of absorption to determine the hazard of toxic chemicals contacting skin usually involves the use of animals as models for humans. The skins of humans and animals vary in biochemical composition and structure. The presence (in humans) or absence (on other species) of sweat ducts is an obvious difference. The numbers and size of hair follicles (TREGEAR 1966) and the thickness of the stratum corneum and viable skin layers (BRONAUGH et al. 1982b) are other, documented differences. These differences, in part, help to explain the observed relative permeability properties of different skin types.

The permeability properties of human and animal skin have been studied by in vivo methods. In a series of human volunteer studies, the percutaneous absorption of some pesticides, steroids and other organic chemicals has been investigated (FELDMANN and MAIBACH 1969, 1970, 1974). These studies form the basis for much of the subsequent series of in vivo studies in various animal models. In the human studies the chemicals were applied to the ventral forearm in acetone which was allowed to evaporate, depositing the test penetrant: the application site

Table 1. Comparison of the in vivo absorption of several chemicals through human, rat, rabbit, and mini pig skin

Penetrant	% Applied dose absorbed			
	Human	Pig	Rabbit	Rat
Haloprogrin	11.0	19.7 (1.8)[a]	113.0 (10.3)	95.8 (8.7)
Acetylcysteine	2.4	6.0 (2.5)	2.0 (0.8)	3.5 (1.5)
Cortisone	3.4	4.1 (1.2)	30.3 (8.9)	24.7 (7.3)
Caffeine	47.6	32.4 (0.7)	69.2 (1.5)	53.1 (1.1)
Butter yellow	21.6	41.9 (1.9)	100.0 (4.6)	48.2 (2.2)
Testosterone	13.2	29.4 (2.2)	69.6 (5.3)	47.4 (3.6)

[a] Values in parentheses indicate factor of difference between animal and human skin permeability.

was left unoccluded. Invariably, ^{14}C-labelled compounds were used and absorption was assessed by analysis of urine for ^{14}C. In order to correct for elimination by extrarenal pathways, a single parenteral bolus of the ^{14}C-labelled test penetrant was given and the proportion eliminated in the urine estimated. The amount accumulating in the urine after dermal application was thus corrected to calculate the total amount absorbed through the skin.

By a similar method (Bartek et al. 1972) the absorption of haloprogin, acetylcysteine, cortisone, caffeine, butter yellow and testosterone were measured through rat, rabbit, mini-pig and human skin: the results are summarised in Table 1. In this study it was necessary to shave the fur from the animal skin and this might have altered the permeability. The data showed that the relative permeability of the skins was compound dependent. However, the rat and rabbit skins were more permeable than pig and human skin which had comparable permeabilities to these test penetrants.

The in vivo absorption of some pesticides (DDT, lindane, parathion and malathion) has been measured through rabbit, pig and squirrel monkey skin (Bartek and LaBudde 1975) and the data compared with available human data (Feldmann and Maibach 1974). The results (summarised in Table 2) showed that the relative permeability of the different skins was compound dependent. Rabbit skin was most permeable and the monkey and pig skins were the most predictive of human skin permeability.

A more stringent comparison of the absorption of testosterone, hydrocortisone and benzoic acid through rhesus monkey skin (Wester and Maibach 1976) has shown that with these penetrants, monkey skin behaves very similarly to human skin. Not only were the amounts absorbed comparable, but similar increases in absorption were seen for each compound as the topical dose was increased.

Table 2. Relative in vivo absorption of some pesticides through human and animal skins

Pesticide	% Applied dose absorbed			
	Human	Pig	Monkey	Rabbit
DDT	10.4	43.4	1.5	46.3
		(4.2)[a]	(0.14)	(4.5)
Lindane	9.3	37.6	16.0	51.2
		(4.0)	(1.7)	(5.5)
Parathion	9.7	14.5	30.3	97.5
		(1.5)	(3.12)	(10.1)
Malathion	8.2	15.5	19.3	64.6
		(1.9)	(2.4)	(7.9)

[a] Values in parentheses indicate factor of difference between animal and human skin permeability.

Rat and cynomolgus monkey skin have been shown (WICKREMA SINHA et al. 1978) to be over 50 times more permeable than human skin to diflorasone diacetate (a corticosteroid). Skin from the Mexican hairless dog, however, was less permeable than human skin (HUNZIKER et al. 1978) to the model penetrants benzoic acid, progesterone and testosterone.

Strictly comparable data with which to relate in vivo human and animal results are not available. When differences are found, the influence of body site, age and pharmacokinetics cannot be satisfactorily dissociated from absolute permeability properties. The in vivo data, however, indicate that the skins of rat and rabbit are more permeable than human skin, whilst Mexican hairless dog skin is less permeable and so these species should be regarded as poor models. Any data derived from these species will probably not be relevant to humans. There is a growing data bank which indicates that the pig and monkey are better models for human percutaneous absorption.

A further series of studies have compared the in vitro permeability properties of human and animal skins (SCOTT et al. 1987; BRONAUGH and MAIBACH 1985) for a wide range of compounds and skin samples of various ages from different parts of the body. Table 3 shows the permeability ranking of various animal skins used in these studies. The general impression is that the animal skins are more permeable than human skin, with the exception of the pig and monkey. Further confirmation of the similarity of human and monkey skin in vitro has been demonstrated with a series of nitroaromatic compounds (BRONAUGH and MAIBACH 1985).

Rat and rabbit skin have also been shown (MCGREESH 1965) to be more permeable to naproxen than human skin. The hairless mouse, however, has been reported (STOUGHTON 1975) to have similar permeability to human skin for steroids (tolnaftate, 5-fluouracil and thiabendazole) and the *n*-alkanols (DURRHEIM et al. 1980). This contrast markedly with the data obtained through hairless mouse skin for paraquat (SCOTT et al. 1987). Hairless mouse skin was approximately 1000 times more permeable than human skin.

Table 3. In vitro absorption of several nitroaromatic compounds through human and monkey skin

Compound	% Applied dose absorbed	
	Human	Monkey
p-Nitroaniline	48.0	62.2 (1.3)[a]
4-Amino-2-nitrophenol	45.1	48.2 (1.1)
2, 4-Dinitrochlorobenzene	32.5	48.4 (1.5)
2-Nitro-*p*-phenylenediamine	21.7	29.6 (1.4)
Nitrobenzene	7.8	6.2 (0.8)

[a] Values in parentheses indicate factor of difference between human and monkey skin permeability.

Table 4. In vitro absorption of water and paraquat through human and some common laboratory animal skins

Species	Absorption rate (permeability constant; cm $h^{-1} \times 10^5$)	
	Water	Paraquat
Human	93	0.7
Rat	103 (1.1)[a]	27.2 (39)
Hairless rat	130 (1.4)	35.3 (50)
Nude rat	152 (1.6)	35.5 (51)
Mouse	163 (1.8)	97.2 (139)
Hairless mouse	254 (2.7)	1065.0 (1520)
Rabbit	253 (2.7)	92.9 (133)
Guinea pig	442 (4.8)	196.0 (280)

[a] Values in parentheses indicate factor of difference between human and animal skin.

In Table 4, factors of difference, relative to human skin, have been calculated for a number of compounds. This, together with Tables 1–3, supports the suggestion of the in vivo data, that monkey and pig skin have the closest permeability to human skin. Differences between the permeability of human skin and that of animal skins are real and we do not fully understand the reasons. At present, the best model for human skin remains human skin.

If we are restricted to a limited number of experimental measurements with human skin then an animal model must be used. It is probable (though more data are needed) that any factor of difference in the relative permeability of the chosen species and human skin will remain constant for a particular molecule from a range of applications (formulations). If this difference can be quantified, then sensible extrapolation can be made and the information gained can be made more quantitatively relevant to humans. However, it should be realised that the factor of difference is compound specific and may vary from chemical to chemical.

F. Metabolism

As the last section demonstrated, the skin has traditionally been regarded as an inert, dead layer in relation to the absorption process, primarily because of the evidence that diffusion through the stratum corneum dominates the absorption process. The majority of studies have used radiolabelled penetrants, mainly ^{14}C, and absorption has been quantitated by analysis for ^{14}C. Whether the diffusing molecule is the parent molecule or a metabolite has not been resolved in many experiments. It is now apparent that the skin should not be regarded as such a simple layer, but a complex structure which is metabolically active. The concept of a "metabolic barrier" present in the epidermis has been raised (Ando et al. 1977) and signifiant levels of enzymes are present which may play a role in metabolising environmental and medicinal compounds to which the skin is exposed. However, the significance of metabolism on the absorption process is not clearly defined and, as already discussed, investigations using ^{14}C-labelled compounds have shown good agreement between the absorption profiles in vivo and in vitro, at least for the ^{14}C marker. Because the significance of metabolism on the percutaneous absorption of chemicals is unknown we first illustrate this metabolic activity and then discuss briefly its possible role in the absorption process.

The epidermal metabolism of polycyclic aromatic hydrocarbons (PAH) has been studied as they can induce carcinomas. Following dermal application, benzo[*a*]pyrene is detoxified by human skin and reactive metabolites are formed. The enzyme aromatic hydrocarbon hydroxylase (AHH, a non-specific enzyme which contains cytochrome P-450) is involved and two phenols, three quinones and various hydrodiols are produced (Fox et al. 1975). The hydrodiols are different from those formed by liver microsomes, suggesting a different metabolic route. The enzyme epoxide hydratase is involved in subsequent metabolism of benzpyrene epoxide (Bentley et al. 1976) and hydroxylated metabolites are further metabolised to glucuronide-conjugated benzpyrenols (Harper and Calcutt 1960). Generally, relative to liver, AHH activity in the skin is considered low. However, such comparisons have used whole skin and when the epidermis alone is considered, then the activity can approach that of the liver (Noonan and Wester 1985). The metabolites of 3-methylcholanthrene (3-MC) have been reported to have tumorigenic activity. The enzyme epoxide hydratase metabolises the 11,12-oxide metabolite of 3-MC to the very potent dihydrodiol.

7,12-Dimethylbenz[*a*]anthracene (DMBA) is an example of a benzo[*a*]anthracene PAH which as a group are potent carcinogens. DMBA undergoes aromatic and aliphatic oxidation (DiGiovanni et al. 1977). The 7-methylbenz[*a*]anthracene (7-MBA) metabolite is further metabolised to several dihydrodiols (Vigny et al. 1977) of which the 3,4-dihydrodiol has the highest tumour-promoting activity (Chouroulinhov et al. 1977) in the mouse.

Mouse, rabbit and rat skin contain monoamine oxidase (MAO) which metabolises noradrenaline to dehydroxymandelic acid (DHMA) (Hakanson and Moller 1963). The activity of DOPA (3,4-dihydroxyphenylaniline), decarboxylase and dopamine-β-oxidase has been demonstrated. Also, mixed-function oxidase (MFO) which dialkylates 7-ethoxycoumarin (7-EC) (Pohl et al. 1976) and deaminase activity involved in metabolism of vidarabine (9-α-D-arabinofuranoladenine).

Detoxification of dithranol by the skin has been reported (Kammerau et al. 1975). This compound is readily absorbed via the skin yet toxicity has never been seen although the compound has a high systemic toxicity. The enzyme catechol-*o*-methyltransferase which controls the conversion of noradrenaline to normetanephrine has been reported (Bamshad 1969) as have glucuronidation reactions controlled by glucuronyl transferase (Stevenson and Du Ton 1960).

Steroid molecules are deliberately applied to skin for therapeutic reasons and their metabolism has been studied. Hydrocortisone is metabolised in the skin primarily to cortisone plus several other metabolites (Hsia et al. 1965). The esters of hydrocortisone are also extensively metabolised. Metabolism of hydrocortisone in vitro has been demonstrated to occur in both the epidermis and the dermis (Witten 1968). This metabolism is believed to occur before hydrocortisone enters the target cell which reduces its potency relative to fluorinated corticosteroids which are rapidly metabolised in guinea pig and rat skin, but are slowly degraded in human skin (Tauber and Toda 1976).

The absorption process does depend naturally on diffusion, but whether the diffusing molecule is the parent compound or a metabolic product has not been ascertained in many experiments. The previous sections have shown that the skin is capable of metabolising many chemicals, yet the majority of percutaneous absorption experiments have used ^{14}C-labelled penetrants and absorption has been quantified by analysis of ^{14}C. The ^{14}C has been equated with the parent molecule and no analysis for metabolic products has been attempted. The fact that, for many chemicals, similarities of absorption have been found by in vitro and in vivo methods suggests that for these chemicals (assuming that metabolism does not occur in vitro after periods of skin storage) metabolism of the parent molecule is not important for the absorption. However, it has been shown with testosterone, benz[*a*]pyrene (Kao et al. 1985), diisopropylfluorophosphate (Loden 1985), T-2 toxin (in a range of species) and aldrin (Kemppainen et al. 1987) that metabolism of the parent molecule can occur during the absorption process. This was particularly true for benz[*a*]pyrene which, apparently, had to be metabolised to less lipophilic metabolites before any absorption was detected. It is possible that highly lipophilic chemicals can enter the stratum corneum, but not partition into the more aqueous epidermis. This would prevent absorption unless the parent chemical was metabolised to more water-soluble compounds.

This is a relatively new and exciting area of interest in our knowledge and study of percutaneous absorption. As more data are generated, the significance of metabolism during percutaneous absorption will become clearer as will an understanding of which type of chemical is metabolised during diffusion through the epidermis into the systemic circulation. Indeed, this activity might be exploited to deliver pro-drugs into the skin.

G. Control and Prediction of Absorption

The published literature over the past 20 years supports the view that we now have a better knowledge of the factors which influence the percutaneous absorption of chemicals. However, we are still in the position of having to use experimental methods either in vivo or in vitro to assess absorption. Our ability to predict, with confidence, the absorption of chemicals from chemical structure or physicochemical properties is still in its infancy, but this is an area which is now receiving a great deal of attention.

This is not a new concept and some of the early studies on the absorption of non-electrolytes (SCHEUPLEIN 1965) from non-saturated aqueous solution investigated both the mechanism and how the physical characteristics of the penetrants influenced absorption. In these experiments a series of alcohols (methanol to octanol) was used as test penetrants. The molecular volume of these penetrants increased by only a factor of 4, yet the permeability coefficient K_p covered a 50-fold range. The permeability coefficients increased as the alcohols became more lipid soluble, demonstrating that there was an important relationship between the penetrant and its solubility in the membrane. As the same vehicle, water, was used with each alcohol the important factor which influenced the absorption rate was the membrane vehicle partition coefficient K_m. For these molecules, the K_m values covered approximately a 100-fold range with the K_m and K_p values changing together throughout the series.

Absorption studies with model steroids, much larger molecules than the diols, with a wide range of polarity, but with a narrow molecular weight band, have also been reported (SCHEUPLEIN et al. 1969) from aqueous solution through human skin. There was a difference of 1000 between the absorption rates of estrone (the slowest) and hydrocortisone (the fastest). With these steroids there was only a factor of 15 difference in the calculated K_m values: so clearly some other factor was determining the absorption. This was found to be due to changes in the diffusion coefficient D_m values, which alter within this series of steroids owing to molecular volume and the presence of an increasing number of polar groups. Consequently, with these molecules there was no relationship between K_p and K_m alone.

The absorption of a wide range of phenolic compounds has been reported (ROBERTS et al. 1977). Absorption was measured in vitro through human skin from an aqueous vehicle. The results showed that the compound with the lowest absorption rate (resorcinol) also had the lowest partition coefficient and those with the highest measured rates (chlorocresol, chloroxylenol, 2,4,6-trichlorophenol and 2,4-dichlorophenol) had the highest partition coefficients. These data again suggest an intimate relationship between absorption rate and partition coefficient. The partition coefficients used in this study were different from those used with the linear alcohols and steroids in that these were measured in an octanol: water system rather than calculated from the data obtained from a skin membrane. This is a departure from determining all parameters related to percutaneous absorption by experiment and introduces the possibility of predicting properties and thus rates rather than measuring.

A more complex approach to model percutaneous absorption, with the goal of predicting, has recently been presented (KASTING et al. 1987) combining terms for lipid solubility and a size-dependent diffusion coefficient. In all, 35 test penetrants were chosen with a range of physical, structural and pharmacological properties. The two most important parameters for controlling the absorption of these lipophilic molecules were melting point and size (molecular weight). Absorption rates which were measured covered a range of five orders of magnitude and the model was able to predict absorption rates to within one order of magnitude.

An attempt to predict absorption of a single penetrant (water) from different vehicles has also been made (DUGARD and SCOTT 1986). In this study the absorption rate of water was measured from 28 vehicles with the penetrant present at the same proportion of saturation in each vehicle: thus, present at the same level of thermodynamic activity in each vehicle. Thermodynamic activity, rather than simple concentration in a vehicle is important in regulating absorption. The 28 absorption rates measured were essentially the same (absorption rates from 23 vehicles were within a factor of 2 of predicted values). Absorption rate from a saturated solution is the maximum rate achievable and only chemical in solution can be absorbed. The practical reliability of this concept has been elegantly demonstrated in a series of in vivo human studies of topical steroids (BENNETT et al. 1986). This concept is often misunderstood or forgotten (e.g. GIBSON et al. 1987). Comparisons of vasoconstriction after changes in corticosteroid concentration in vehicles did not reflect the changes in concentration. However, the concentrations *in solution* in the vehicles were, unfortunately, not considered.

Further understanding of the important parameters which influence the absorption of a chemical, together with an increase in our ability to predict these parameters might lead to a reduction in the numbers of experiments. In this way the cost of developing new formulations will decrease. This will promote more efficacious preparations (controlled, enhanced absorption) and a decrease in toxicity (reduce absorption).

References

Ainsworth M (1960) Methods for measuring percutaneous absorption. J Soc Cosmet Chem 11:69–74

Albery WJ, Hadgraft J (1979) Percutaneous absorption: in vivo experiments. J Pharm Pharmacol 31:140–147

Ando HY, Ho NFH, Higuchi WI (1977) Skin as an active metabolising barrier. I. Theoretical analysis of topical bioavailability. J Pharm Sci 66:1525–1528

Bamshad J (1969) Catechol-*o*-methyltransferase in epidermis and whole skin. J Invest Dermatol 52:351–252

Barry BW (1983) Dermatological formulations: percutaneous absorption. Dekker, New York

Bartek MJ, LaBudde JA (1975) Percutaneous absorption in vitro. In: Maibach HI (ed) Animal models in dermatology. Churchill Livingstone, New York, pp 103–120

Bartek MJ, LaBudde JA, Maibach HI (1972) Skin permeability in vivo: comparison in rat, rabbit, pig and man. J Invest Dermatol 58:114–123
Bartnik FG, Gloxhuber C, Zimmermann V (1985) Percutaneous absorption of formaldehyde in rats. Toxicol Lett 25:167–172
Behl CR; Barrett M (1981) Hydration and percutaneous absorption. II. Influence of hydration on water and alkanol permeation through Swiss mouse skin: comparison with hairless mouse. J Pharm Sci 70:1212–1215
Bennet SL, Barry BW, Woodford R (1985) Optimisation of bioavailability of topical steroids: non-occluded penetration enhancers under thermodynamic control. J Pharm Pharmacol 37:298–304
Bentley P, Schussman H, Sims P, Oesch F (1976) Epoxides derived from various polycyclic hydrocarbons as substrates of homogeneous and microsome bound epoxide hydratase. Eur J Biochem 69:97–103
Black JG, Howes D, Rutherford T (1975) Percutaneous absorption and metabolism of Irgasan® DP300. Toxicology 3:33–47
Blank IH, Griesmer RD, Gould E (1958) A method for studying the rate of sarin penetration into the living rabbit. J Invest Dermatol 31:255–258
Bonneville MA, Weinstock M, Wilgram GF (1968) An electron microscope study of cell adhesion in psoriatic epidermis. J Ultrastruct Res 23:15–43
Bowden G, Slaga T, Shapas B, Boutwell R (1974) The role of aryl hydrocarbon hydroxylase in skin tumour initiation by 7,12-dimethylbenz[*a*]anthracene and 1,2,5,6-dibenzanthracene using DNA binding and thymidine-^{3}H incorporation into DNA as criteria. Cancer Res 34:2634–2642
Brody I (1960) The ultrastructure of the tonofibrils in the keratinisation process of normal human epidermis. J Ultrastruct Res 4:264–297
Bronaugh RL, Maibach HI (1985) Percutaneous absorption of nitroaromatic compounds: in vivo and in vitro studies in human and monkey. J Invest Dermatol 84:180–183
Bronaugh RL, Stewart RF (1985) Methods for in vitro percutaneous absorption studies. III. Hydrophobic compounds. J Pharm Sci 73:1255–1258
Bronaugh RL, Stewart RF, Congdon ER (1982a) Methods for in vitro percutaneous studies. II. Animal models for human skin. Toxicol Appl Pharmacol 62:481–488
Bronaugh RL, Stewart RF, Congdon ER, Giles AL (1982b) Methods for in vitro percutaneous absorption studies. 1. Comparison with in vivo studies. Toxicol Appl Pharmacol 62:474–480
Bronaugh RL, Stewart RF, Wester RC, Bucks D, Maibach HI (1985) Comparison of percutaneous absorption of fragrances by human and monkeys. Fd Chem Toxic 23:1, 111–114
Burch CE, Winsor T (1942) Diffusion of water through dead plantar, palmar and dorsal human skin and through toe nails. Arch Dermatol Syph 53:39–41
Charles A, Smiddy FC (1957) The tonofibrils of human epidermis. J Invest Dermatol 29:327–338
Chouroulinhov I, Gentil A, Tierney B, Grover P, Sims (1977) The metabolic activation of 7-methylbenz[*a*]anthracene in mouse skin: high tumour initiating activity of the 3,4-dihydrodiol. Cancer Lett 3:247–253
Creasey NH, Battersby J, Fetcher JA (1978) Factors affecting the permeability of skin. The relation between in vivo and in vitro observation. Curr Probl Dermatol 7:95–106
DiGiovanni J, Slaga T, Berry D, Juchan M (1977) Metabolism of 7,12-dimethylbenz[*a*]anthracene in mouse skin with high pressure liquid chromatography. Drug Metab Dispos 5:295–301
Dugard PH (1977) Skin permeability theory in relation to measurement of percutaneous absorption in toxicology. In: Marzulli FN, Maibach HI (eds) Dermatotoxicology and pharmacology. Hemisphere, Washington, pp 525–550 (Advances in modern toxicology, vol 4)
Dugard PH, Scott RC (1986) A method of predicting percutaneous absorption rates from vehicle to vehicle: an experimental assessment. Int J Pharm 28:219–227
Dugard PH, Walker M, Mawdsley SJ, Scott RC (1984) Absorption of some glycol ethers through human skin in vitro. Environ Health Perspect 57:193–197

Durrheim H, Flynn GL, Higuchi WI, Behl CR (1980) Permeation of hairless mouse skin. I. Experimental methods and comparison with human epidermal permeation by alkanols. J Pharm Sci 69:781–786

Elias PM, Friend DS (1975) The permeability barrier in mammalian epidermis. J Cell Biol 65:180–191

Elias PM, Leventhal ME (1979) Intercellular volume changes and cell surface expansion during cornification. Clin Res 27:525 a

Elias PM, Goerke J, Friend DS (1977) Mammalian epidermal barrier layer lipids: composition and influence on structure. J Invest Dermatol 69:535–546

Feldmann RJ, Maibach HI (1969) Percutaneous penetration of steroids in man. J Invest Dermatol 52:89–94

Feldmann RJ, Maibach HI (1970) Absorption of some organic compounds through skin in man. J Invest Dermatol 54:399–404

Feldmann RJ, Maibach HI (1974) Percutaneous penetration of some pesticides and herbicides in man. Toxicol Appl Pharmacol 28:399–404

Fox C, Selkirk J, Price F, Croy R, Sandford K, Fox M (1975) Metabolism of benzo-[*a*]pyrene by human epithelial cell in vitro. Cancer Res 35:3551–3557

Francis GS, Hagen AD (1977) Nitroglycerin ointment: a review. Angiology 28:873–875

Franz TJ (1975) Percutaneous absorption – on the relevance of in vitro data. J Invest Dermatol 64:190–195

Franz TJ (1978) The finite dose technique as a valid in vitro model for the study of percutaneous absorption. Curr Probl Dermatol 7:58–68

Gibson JR, Hough JE, Marks P, Webster A (1987) Effect of concentration on the clinical potency of corticosteroid ointment formulations. In: Shroot B, Schaefer H (eds) Skin pharmacokinetics. Karger, Basel, p 138 (Pharmacology and the skin, vol 1)

Hakanson R, Moller H (1963) On metabolism of noradrenaline in the skin. Activity of catechol-*o*-methyltransferase and monoamine oxidase. Acta Derm Venereol (Stockh) 43:552–553

Harper K, Calcutt G (1960) Conjugation of 3:4-benzpyrenols in mouse skin. Nature 186:80–81

Harrison SM, Barry BW, Dugard PH (1984) Effects of freezing on human skin permeability. J Pharm Pharmacol 36:261–262

Hsia S, Musallem J, Witten V (1965) Further metabolic studies of hydrocortisone-4-^{14}C in human skin. J Invest Dermatol 45:384–390

Hunziker N, Feldmann RJ, Maibach HI (1978) Animal models of percutaneous penetration: comparison in Mexican hairless dog and man. Dermatologica 156:79–88

Idson B (1978) Hydration and percutaneous absorption. Curr Probl Dermatol 7:132–141

Kammerau B, Zesch A, Schaefer H (1975) Absolute concentration of dithranol and triacetyldithranol in the skin layers after topical treatment – in vivo investigations with four different types of pharmaceutical vehicles. J Invest Dermatol 64:145–148

Kao J, Patterson FK, Hall J (1985) Skin penetration and metabolism of topically applied chemicals in six mammalian species, including man: an in vitro study with benzo-[*a*]pyrene and testosterone. Toxicol Appl Pharmacol 81:502–516

Kasting GB, Smith RL, Cooper ER (1987) Effect of lipid solubility and molecular size on percutaneous absorption formulations. In: Shroot B, Schaefer H (eds) Skin pharmacokinetics. Karger, Basel, p 214 (Pharmacology and the skin, vol 1)

Kemppainen BW, Riley RT, Joyave JL, Hoerr FJ (1987) In vitro percutaneous penetration and metabolism of [^{3}H] T-2 toxin: comparison of human, rabbit, guinea pig and rat. Toxicol 25:185–194

Kligman AM (1964) The biology of the stratum corneum. In: Montagna W, Lobitz WC (eds) The epidermis. Academic, New York, p 387

Kligman AM, Christophers E (1963) Preparation of isolated sheets of human skin. Arch Dermatol 88:702–705

Knaak JB, Yee K, Ackerman CR, Zweig G, Wilson BW (1984) Percutaneous absorption of Triadimefon in the adult and young male and female rat. Toxicol Appl Pharmacol 72:406–416

Loden M (1985) The in vitro hydrolysis of diisopropyl fluorophosphate during penetration through full-thickness skin and isolated epidermis. J Invest Dermatol 85:335–339

Maalson KC, Swartzendruber DC, Wertz PW, Downing DT (1987) Presence of intact intercellular lipid lamellae in the upper layers of the stratum corneum. J Invest Dermatol 88:714–718
Marzulli FN, Brown DWC, Maibach HI (1969) Techniques for studying skin penetration. Toxicol Appl Pharmacol 3(Suppl):79–83
Matoltsy AG, Parakkal PE (1965) Membrane-coating granules of keratinizing epithelia. J Cell Biol 24:297–307
McGreesh AH (1965) Percutaneous toxicity. Toxicol Appl Pharmacol 2(Suppl):20–26
Menon GK, Grayson S, Brown BE, Elias PM (1986) Lipokeratinocytes of the epidermis of a cetacean (*Phocaena phocaena*): histochemistry, ultrastructure and lipid composition. Cell Tissue Res 244:385–394
Michaels AS, Chandrasekaran SK, Shaw JE (1975) Drug permeation through human skin: theory and in vitro experimental evidence. Am Inst Chem Eng J 21:985–996
Middleton JD (1968) The mechanism of water binding in stratum corneum. Br J Dermatol 80:437–450
Noonan PK, Wester RC (1985) Cutaneous metabolism of xenobiotics. In: Bronaugh RL, Maibach HI (eds) Percutaneous absorption. Dekker, New York, pp 65–85 (Dermatology, vol 6)
Odland GF (1964) Tonofibrils and keratohyalin. In: Montagna W, Lobitz W (eds) The epidermis. Academic, New York, pp 237–249
Olson RL, Nordquist RE, Everett MA (1969) Small granules of the superficial epidermis. Arch Klin Exp Derm 234:15–24
Pohl R, Philpot, Fouts J (1976) Cytochrome P-450 content and mixed function oxidase activity in microsomes isolated from mouse skin. Drug Metab Dispos 4:442–450
Reifenrath WG, Chellquist EM, Shipwash EA, Jederberg WW (1984) Evaluation of animal models for predicting skin penetration in man. Fundam Appl Toxicol 4:334–346
Roberts MS, Anderson RA, Swarbrick J (1977) Permeability of human epidermis to phenolic compounds. J Pharm Pharmacol 29:677–683
Scheuplein RJ (1965) Mechanism of percutaneous absorption. I. Routes of penetration and the influence of solubility. J Invest Dermatol 45:334–346
Scheuplein RJ, Ross LA (1974) Mechanism of percutaneous absorption. V. Percutaneous absorption of solvent deposited solids. J Invest Dermatol 62:353–360
Scheuplein RJ, Blank IH, Brauner GJ, MacFarlane DJ (1969) Percutaneous absorption of steroids. J Invest Dermatol 52:63–70
Scott RC, Ramsey JD (1986) Percutaneous absorption: in vitro assessment. Fed Cosmet Toxicol 24:763–764
Scott RC, Ramsey JD (1988) The in vivo and in vitro percutaneous absorption of carbaryl (a water-soluble penetrant). In preparation
Scott RC, Dugard PH, Doss AW (1986a) Permeability of abnormal rat skin. J Invest Dermatol 86:201–207
Scott RC, Walker MW, Dugard PH (1986b) In vitro percutaneous absorption experiments: a technique for the production of intact epidermal membranes from rat skin. J Soc Cosmet Chem 37:35–41
Scott RC, Walker M, Dugard PH (1987) A comparison of the in vitro permeability properties of human and some laboratory animal skins. Int J Cosmet Sci 8:189–194
Shah PV, Monroe RJ, Guthrie FE (1981) Comparative rates of dermal penetration of insecticides in mice. Toxicol Appl Pharmacol 59:414–423
Squier CA (1973) The permeability of keratinised and nonkeratinised oral epithelium to lanthanum in vitro. J Ultrastruct Res 43:160–177
Stevenson I, Dutton G (1960) Mechanism of glucuronide synthesis in skin. Biochem J 77:19p
Stoughton RB (1975) Animal models for in vitro percutaneous absorption. In: Maibach HI (ed) Animal models in dermatology. Churchill Livingstone, New York, pp 121–132
Swartzendruber DC, Wertz PW, Madison KC, Downing DT (1987) Evidence that the corneocyte has a chemically bound lipid envelope. J Invest Dermatol 88:709–713
Tauber V, Toda T (1976) Biotransformation of diflucortolone valerate in the skin of rat, guinea pig and man. Arzneim-Forsch (Drug Res) 26:1484–1487

Tregear RT (1966) Physical functions of the skin. Academic, London
Van Lier RBL (1985) The use of monkey percutaneous absorption studies. In: Honeycutt RC, Zweig G, Ragsdale NC (eds) Dermal exposure related to pesticide use: discussion of risk assessment. American Chemical Society, Washington DC, pp 81–91 (Acs symposium series no 273)
Vigny P, Duquesne M, Coulomb H, Tierney B, Grover P, Sims P (1977) Fluorescence spectral studies on the metabolic activation of 3-methylcholanthrene and 7,12-dimethylbenz[*a*]anthracene in mouse skin. FEBS Lett 82:278–282
Wahlberg JE (1965) "Disappearance measurements" a method for studying percutaneous absorption of isotope-labelled compounds emitting gamma-rays. Acta Derm Venereol (Stockh) 45:397–414
Wester RC, Maibach HI (1976) Relationship of topical dose and percutaneous absorption in rhesus monkey an man. J Invest Dermatol 67:518–520
Wester RC, Maibach HI (1985) In vivo methods for percutaneous absorption measurements. In: Bronaugh RL, Maibach HI (eds) Percutaneous absorption. Dekker, New York, pp 245–266 (Dermatology, vol 6)
Wickrema Sinha A, Shaw SR, Weber DJ (1978) Percutaneous absorption and excretion of tritium-labelled diflorasone diacetate, a new topical corticosteroid in the rat, monkey and man. J Invest Dermatol 71:372
Witten VH (1968) Topical corticosteroid therapy. In: Jadassohn W, Schirren CG (eds) XIII Congressus Internationalis Dermatologiae, 31 July–5 Aug 1968. Springer, Berlin Heidelberg New York
Woodford R, Barry BW (1982) Optimisation of bioavailability of topical steroids: thermodynamic control. J Invest Dermatol 79:388–391
Wurster DE, Kramer SF (1961) Investigation of some factors influencing percutaneous absorption. J Pharm Sci 50:288–293
Zesch A, Schaefer H (1975) Penetration of radioactive hydrocortisone in human skin from various ointment bases. II. In vivo experiments. Arch Dermatol Forsch 252:245–256

CHAPTER 13

Skin as a Mode for Systemic Drug Administration

J. E. SHAW and S. K. CHANDRASEKARAN

A. Introduction

Skin is a more permeable organ than is commonly believed. Despite the effective barrier that it provides to protect the body against the invasion of microorganisms or the loss of water, it can allow permeation of drugs from topically applied creams and ointments (SCHEUPLEIN and BLANK 1971) in quantities sufficient to produce systemic actions. Until recently such actions have usually been a cause for concern; for example, topical steroids valuable in dermatology are capable of disturbing the pituitary adrenal axis (VAN SCOTT and YU 1980). More recently, however, the transdermal route for administration of drugs to the systemic circulation has become a part of therapeutics. That development is attributable in large part to the advent of dosage forms capable of overcoming what WEPIERRE and MARTY (1979) described as the chief obstacle to transdermal drug delivery: variability in drug absorption through skin, resulting in lack of precision regarding the real dose absorbed.

B. Therapeutic Objectives of Transdermal Delivery

Lack of precision in drug dosage has characterized the use of topically applied ointments and creams for systemic therapy. Topical ointments, creams, and gels are available for systemic administration of nitroglycerin (Physicians' Desk Reference 1987, pp 1177–1178, 1521, 1726, 2104), etofenamate (HEINDL et al. 1977), 17β-estradiol (LYRENÄS et al. 1981; Physicians' Desk Reference 1987, pp 1203–1206), and progesterone (Dictionnaire Vidal 1986).

Ointments containing nitroglycerin require multiple applications per day (Physicians' Desk Reference 1987, pp 1177–1178, 1521, 1726, 2104). The amount and duration of drug input are variable according to differences in the area of skin covered with the ointment, and the thickness of the ointment layer applied, either of which can result in varying efficacy with repetitive applications.

More recently, pharmaceutical companies have evaluated more convenient topical dosage forms for systemic drug therapy; this discussion will focus on examples of these new transdermal dosage forms that deliver drug to the bloodstream at a defined rate over prolonged periods of time. Their design emphasis has been directed toward having control of systemic drug input reside in the dosage form rather than in the skin, because of the known interindividual differences in skin permeability.

The first such dosage form to be commercially available (Transderm Scōp) releases scopolamine to the bloodstream at the rate necessary for prevention of motion sickness (SHAW and URQUHART 1980). Subsequently, three transdermal dosage forms for the prevention and treatment of angina have become commercially available. One of them – Transderm-Nitro, which is widely used – is known to provide rate-controlled systemic input of nitroglycerin (GÉRARDIN et al. 1985; PLACE 1981; SHAW 1981, 1984). Two additional transdermal systems – Catapres-TTS and Estraderm – have recently become available and provide rate-controlled delivery of clonidine and 17β-estradiol, respectively.

Expanding interest in transdermal drug delivery systems has led to identification of a number of perceived therapeutic advantages (Table 1) that vary from drug to drug. The principal aim in development of a rate-controlled transdermal dosage form for scopolamine – a drug with an extremely narrow therapeutic index – was to provide control over plasma concentrations, such that one could separate the required, highly potent antiemetic effects from the marked parasympatholytic effects that have hampered use of this drug in conventional dosage forms (SHAW and URQUHART 1980). Secondary aims were the achievement of a more convenient regimen than that provided by four times daily injections or tablets. For nitroglycerin, a drug with a wide therapeutic index, but an extremely short

Table 1. Therapeutic advantages of transdermal drug delivery[a]

Functional capability of transdermal dosage form	Therapeutic advantages
Multiday, continuous drug delivery	Reduced dosage frequency Use of short-half-life drugs Simplified patient compliance More consistent treatment of chronic disorders
Avoidance of gastrointestinal tract	Prevention of unpredictable absorption owing to gastrointestinal tract variables (motility, acidity, food content) Avoidance of unpredictable absorption owing to first pass through the liver Avoidance of gastric irritation Reduced total daily dosage without loss of therapeutic effect because of avoidance of drug degradation in liver or gastrointestinal tract Alternative route when gastrointestinal tract is unsuitable
Noninvasive parenteral route	More convenient than intravenous or intramuscular routes Less risk of complications
Precision of systemic drug input[b]	Blood concentrations of drug controllable within and between patients in a narrow range Enhanced selectivity of drug action (less side effects)
External dosage form	Safety

[a] Not all advantages will apply to all transdermally delivered drugs or to all transdermal dosage forms.
[b] Applies only to dosage forms providing rate-controlled transdermal delivery.

biological half-life (3–5 min), the therapeutic objective was a dosage form permitting convenient, prolonged use of this drug for chronic prophylaxis of anginal pain.

C. Design of a Rate-Controlled Scopolamine System

Prior to the development of a transdermal dosage form for scopolamine, clinical rate-ranging studies were conducted, and it was found that antiemetic prophylaxis required a scopolamine release rate to the systemic circulation of approximately 5 μg/h (SHAW and URQUHART 1980). In vitro studies showed that scopolamine flux through skin in the postauricular area was in the range of 20 $\mu g\ cm^{-2}\ h^{-1}$; about a fivefold difference existed among individuals with the most and least permeable skin (SHAW and URQUHART 1980). To negate the effect of this difference, drug release from the adhesive dosage form to the circulation was set at only 2 $\mu g\ cm^{-2}\ h^{-1}$. To deliver 5 μg/h scopolamine, therefore, the area of the system had to be 2.5 cm^2, a size that is suitable for postauricular application. (Total system area and drug delivery area are identical.) Studies have demonstrated that before steady state drug release can produce steady state input into the systemic circulation, binding sites for drug in skin must be saturated (SHAW and URQUHART 1980). For drugs that are significantly absorbed by skin, and for which skin has low permeability, saturation may be long delayed: scopolamine is such a drug. Therefore, to hasten saturation of binding sites, the system incorporates a priming dose of about 140 μg, delivered at an asymptotically declining rate over the 6 h immediately following application to the skin (SHAW and URQUHART 1980); drug input then stabilizes at 5 μg/h for the remainder of the product's functional lifetime. Steady state plasma levels of drug are established within approximately 6 h after placement of the system on the skin.

The system has the appearance of a small circular adhesive film. It has the following four functional layers (from the innermost outward): (1) an adhesive inner face (containing the priming dose of scopolamine), which is placed securely against postauricular skin; (2) a rate-controlling microporous polypropylene membrane; (3) a steady state drug reservoir containing scopolamine in mineral oil; and (4) an impermeable backing layer. (A peel strip is removed from the system prior to use.) After application to the skin, the drug diffuses through microscopic channels in the membrane in the direction of the concentration gradient. Drug release is governed by modifications of the Fick diffusion equation; the energy source for drug release is the difference in the drug's chemical potential between the reservoir and the system's exterior. Virtual constancy of rate, after delivery of the priming dose, is assured as long as drug is present in excess in the reservoir (i.e., a saturated solution in the vehicle).

Figure 1 demonstrates the consistent rate of scopolamine permeation (as reflected by rate of urinary excretion of unchanged scopolamine) in subjects wearing a transdermal scopolamine system for 72 h compared with subjects receiving intramuscular administration of the drug (200 μg scopolamine hydrobromide).

Uniform systemic absorption of scopolamine is achieved by placing the transdermal scopolamine system on the postauricular area – the site where its perme-

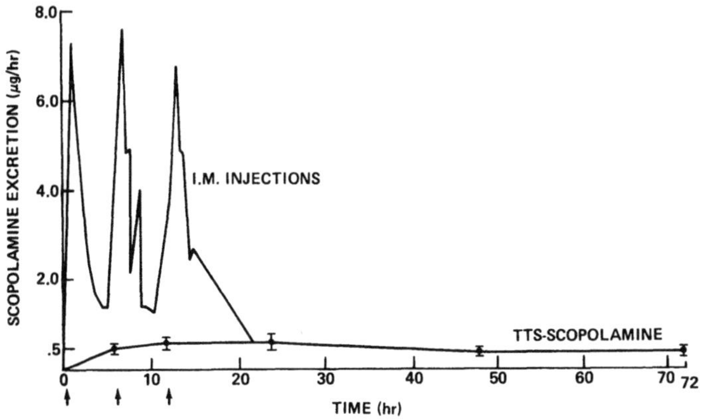

Fig. 1. Urinary excretion of free scopolamine by subjects wearing the Transderm Scōp for 72 h, compared with subjects receiving intramuscular injections of 200 μg scopolamine hydrobromide

ation rate is highest – and choosing a delivery rate that: (a) will maintain therapeutic blood concentrations; but (b) will be far below the rate at which even the least permeable postauricular skin can accept it. Thus, control over systemic drug input resides in the dosage form rather than in the skin.

The drug delivery pattern from Transderm Scōp has been demonstrated to be bioequivalent to a 72-h constant intravenous infusion of scopolamine at a rate of 5 μg/h (SHAW and URQUHART 1980; SCHMITT et al. 1981). Transderm Scōp prevents motion sickness in approximately 75% of susceptible patients (Physicians' Desk Reference 1987, pp 865–866; SCHMITT et al. 1981; MCCAULEY et al. 1979); partial and transient reduction of salivary secretion occurs, but other recognized parasympatholytic effects usually produced by conventional means of administering scopolamine (BRAND and WHITTINGHAM 1970; GRAYBIEL et al. 1976) are infrequent or rare (SHAW and URQUHART 1980).

D. Controlled Systemic Absorption of Nitroglycerin

Nitroglycerin's wide therapeutic index suggests that a transdermal dosage form that simply limits the maximum dose of drug delivered to the circulation (to prevent overdosage) is theoretically feasible for this drug. Studies (SHAW 1981) have shown, however, that lack of any rate-controlling component can theoretically allow variations of as much as sixfold in drug absorption between subjects, leading to the possibility of overdosage if the system is placed on highly permeable skin. A degree of rate control sufficient to prevent overdosage has therefore been an objective in the development of the Transderm-Nitro system.

The four functional components of this system are, like the transdermal scopolamine system, an impermeable backing layer, a steady state drug reservoir, a rate-controlling membrane, and an adhesive layer. The reservoir contains nitroglycerin on lactose in silicone medical fluid with colloidal silicon dioxide added

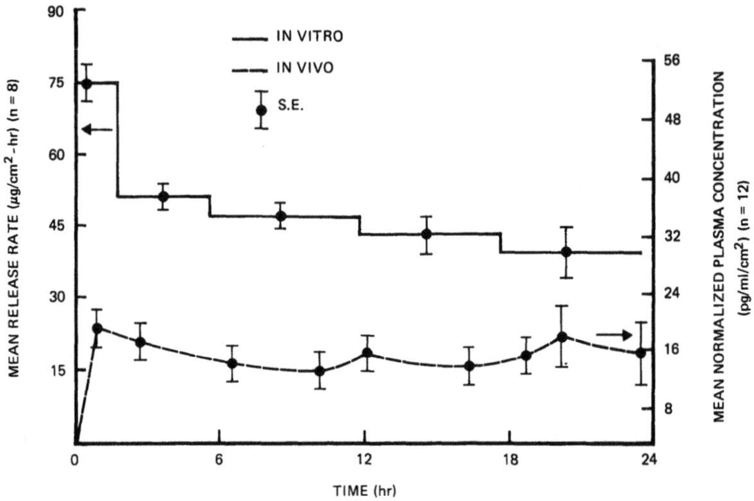

Fig. 2. In vitro release rate of nitroglycerin from the Transderm-Nitro and normalized in vivo plasma concentration on normal skin following application of the system

to increase viscosity. The rate-controlling membrane is an ethylene-vinyl acetate copolymer; in this case, drug dissolves in the membrane and diffuses toward the gradient between concentrated drug in the reservoir and the much lower concentration external to the system.

In vitro, this system releases nitroglycerin at a rate of 40–50 μg cm^{-2} h^{-1} (Fig. 2), which is the maximum rate of drug delivery to the systemic circulation that can be achieved, no matter if the permeability of skin is much greater than this value. For the average subject, when the system is placed on intact skin, clinical studies demonstrate that absorption of drug occurs at a rate of 25 μg cm^{-2} h^{-1}, which translates to approximately 500 μg/cm^2 in 24 h or to total daily doses of 5 or 10 mg nitroglycerin from 10- and 20-cm^2 systems, respectively. During the 24-h wearing of a system of size 10 cm^2, concentrations of nitroglycerin in plasma remain fairly uniform (110–160 pg/ml).

Controlled studies on angina patients, involving a variety of study designs, have shown reduction of the frequency of anginal attacks by more than 60%–70% in repeat-application studies (PLACE 1981; IMHOF 1985; TATTERSALL et al. 1985); no signs were evident of the development of nitroglycerin tolerance with regard to anti-anginal effect. Tolerance did develop, however, with respect to the headaches that are a typical side effect of nitrate therapy. Other than such headaches, few adverse systemic effects have been noted.

E. New Advances in Transdermal Drug Delivery

Transderm Scōp and Transderm-Nitro are the first in a growing family of rate-controlled transdermal delivery systems. Two additional systems, Catapres-TTS and Estraderm, have recently become available.

I. Catapres-TTS

Catapres-TTS, marketed by Boehringer Ingelheim, provides 7 days of treatment of hypertension with a single application. It is available in three sizes delivering 0.1–0.3 mg clonidine per day (SHAW 1984). Catapres-TTS incorporates the same technology as does Transderm Scōp.

Clinical tests have demonstrated that Catapres-TTS provides a hypotensive effect comparable to that of oral regimens of the drug, and with diminished occurrence of unwanted pharmacologic effects (MROCZEK et al. 1982; BOEKHORST and VAN TOL 1985; LAWSON 1985).

II. Estraderm

Estraderm provides continuous, controlled treatment of menopausal symptoms with only two transdermal applications per week. Estraderm is constructed similarly to Transderm-Nitro (GOOD et al. 1985) and delivers 0.025–0.1 mg/day of the natural female hormone 17β-estradiol. Estraderm recently received marketing approval from the United States Food and Drug Administration; it is marketed both in the United States and in Europe by Ciba-Geigy.

In clinical trials with Estraderm, postmenopausal women experienced serum levels of estradiol typical of the early follicular phase of premenopausal women and an estradiol:estrone ratio that approximated 1:1 (POWERS et al. 1985). Another study showed that Estraderm is as effective as conjugated equine estrogens for controlling postmenopausal symptoms, and is well tolerated (PLACE et al. 1985).

F. The Future

Transdermal therapy appears to be on the brink of rapid expansion for the rate-controlled administration of potent, nonirritating, nonallergenic agents – with suitable physicochemical properties – whose current method of administration causes any of the following problems:

1. Troublesome side effects or unreliable therapeutic action with repetitive dosing in conventional dosage forms
2. Patient compliance difficulties, e.g., in chronic treatment of an asymptomatic disease
3. Need for frequent dosing in conventional dosage forms because of the drug's short biologic half-life
4. Gastric irritation with oral therapy

Because of constraints arising from drug potency, skin permeability, and/or topical reactions, however, transdermal administration probably will not become the preferred dosage form for all agents. Nevertheless, a number of agents widely utilized for chronic therapy – such as anti-arthritics, analgesics, and hormones – appear to be candidates for this mode of administration.

For new drug research, rate-controlled transdermal delivery offers one means of circumventing certain problems. These problems include the inability of bolus dosage forms to maintain therapeutic concentrations of short-half-life drugs over reasonable dosage intervals, and the inability of conventional dosage forms to control the concentrations of drugs with narrow therapeutic indexes so as to prevent expression of toxic effects. Transdermal delivery is, of course, only one of several rate-controlled dosage forms now available for use: oral (THEEUWES 1981), rectal (ECKENHOFF and YUM 1981; BREIMER et al. 1985), and simplified intravenous systems (THEEUWES 1981; AUSMAN et al. 1982), usable with virtually any drug, are likely to find use in a large share of clinical studies aimed at understanding a new agent's pharmacodynamics. Once the agent's pharmacodynamics are understood, it is then possible to make a rational judgment about the optimal kinetic requirement for the drug dosage form needed for use of the drug in therapeutics.

References

Ausman RK, Caballero GA, Quebbeman E, Ausman DC (1982) Long-term ambulatory, continuous intravenous infusion of 5-fluorouracil for the treatment of metastatic adenocarcinoma in the liver. Wis Med J 81:25–28

Boekhorst JC, van Tol RGL (1985) Catapres transdermal therapeutic system (TTS) for long-term treatment of hypertension. In: Weber MA, Drayer JIM, Kollach R (eds) Low dose oral and transdermal therapy of hypertension. Steinkopff, Darmstadt, pp 106–110

Brand JJ, Whittingham P (1970) Intramuscular hyoscine in control of motion sickness. Lancet 2:232–235

Breimer DD, de Leede LGJ, de Boer AG (1985) Rate-controlled rectal drug delivery. In: Prescott LF, Nimmo WS (eds) Rate control in drug therapy. Churchill Livingstone, Edinburgh, pp 54–64

Dictionnaire Vidal, 62nd edn. (1986) OVP, Paris, p 1230

Eckenhoff B, Yum SI (1981) The osmotic pump: novel research tool for optimizing drug regimens. Biomaterials 2:89–97

Gérardin A, Gaudry D, Moppert J, Theobald W, Fankhauser P (1985) Glycerol trinitrate (nitroglycerin) plasma concentrations achieved after application of transdermal therapeutic systems to healthy volunteers. Arzneimittelforschung 35:530–532

Good WR, Powers MS, Campbell P, Schenkel L (1985) A new transdermal delivery system for estradiol. J Control Rel 2:89–97

Graybiel A, Knepton J, Shaw J (1976) Prevention of experimental motion sickness by scopolamine absorbed through the skin. Aviat Space Environ Med 47:1096–1100

Heindl I, Lorenz D, Siebers S, Blumberger W (1977) Klinische Prüfung des neuen perkutan wirksamen Antirheumatikums Etofenamat: Zusammenfassender Bericht. Arzneimittelforschung 27:1357–1363

Imhof PR (1985) Anti-anginal therapy with transdermal nitroglycerin. In: Prescott LF, Nimmo WS (eds) Rate control in drug therapy. Churchill Livingstone, Edinburgh, pp 201–214

Lawson AAH (1985) Clinical and pharmacological studies with transdermal clonidine. In: Prescott LF, Nimmo WS (eds) Rate control in drug therapy. Churchill Livingstone, Edinburgh, pp 215–219

Lyrenäs S, Carlström K, Backström T, Von Schoultz B (1981) A comparison of serum oestrogen levels after percutaneous and oral administration of oestradiol-17β. Br J Obstet Gynecol 88:181–187

McCauley ME, Royal JW, Shaw JE, Schmitt LG (1979) Effect of transdermally administered scopolamine in preventing motion sickness. Aviat Space Environ Med 50:1108–1111

Mroczek WJ, Ulrych M, Yoder S (1982) Weekly transdermal clonidine administration in hypertensive patients. Clin Pharmacol Ther 31:252

Physicians' Desk Reference, 41sh edn. (1987) Medical Economics Company, Oradell pp 1177–1178, 1521, 1726, 2104, 1203–1206, 865–866
Place VA (1981) Transdermal therapeutic system for nitroglycerin: bioavailability, hemodynamic and clinical data. Seminar on transdermal therapeutic system: a major advance in angina prophylaxis. Presented in conjunction with the 16th American Society of Hospital Pharmacists Midyear Clinical Meeting, 6 December 1981, New Orleans
Place VA, Powers M, Darley PE, Schenkel L, Good WR (1985) A double-blind comparative study of Estraderm and Premarin in the amelioration of postmenopausal women. Am J Obstet Gynecol 152:1092–1099
Powers MS, Schenkel L, Darley PE, Good WR, Balestra JC, Place VA (1985) Pharmacokinetics and pharmacodynamics of transdermal dosage forms of 17β-estradiol: comparison with conventional oral estrogens used for hormone replacement. Am J Obstet Gynecol 152:1099–1106
Scheuplein RJ, Blank IH (1971) Permeability of the skin. Physiol Rev 51:702–747
Schmitt LG, Shaw JE, Carpenter PF, Chandrasekaran SK (1981) Comparison of transdermal and intravenous administration of scopolamine. Clin Pharmacol Ther 29:282
Shaw JE (1981) Development of the transdermal therapeutic system. Seminar on transdermal therapeutic system: a major advance in angina prophylaxis. Presented in conjunction with the 16th American Society of Hospital Pharmacists Midyear Clinical Meeting, 6 December 1981, New Orleans
Shaw JE (1984) Pharmacokinetics of nitroglycerin and clonidine delivered by the transdermal route. Am Heart J 108:217–223
Shaw J, Urquhart J (1980) Programmed, systemic drug delivery by the transdermal route. Trends Pharmacol Sci 1:208–211
Tattersall AB, Bridgmen KM, Carr M (1985) Retrospective post-marketing surveillance of Transiderm-nitro 5 in general practice in the United Kingdom. J Int Med Res 13:222–228
Theeuwes F (1981) Drug delivery systems. Pharmacol Ther 13:149–191
Van Scott E, Yu RJ (1980) Drugs in dermatologic therapy. In: Modell W (ed) Drugs of choice: nineteen eighty to nineteen eighty-one. Mosby, St Louis, pp 681–694
Wepierre J, Marty JP (1979) Percutaneous absorption of drugs. Trends Pharmacol Sci 1:23–26

CHAPTER 14

Drug Metabolism in the Skin

H. KAPPUS

A. Introduction

Drug metabolism is a general phenomenon which is essential in order to remove foreign compounds from the body. Although the liver is the major organ which metabolizes systemically applied drugs, it has become evident that extrahepatic tissues contain drug-metabolizing enzymes. Various aspects of the metabolism of xenobiotics in the skin have been reviewed (WATTENBERG and LEONG 1971; WIEBEL and GELBOIN 1977; PANNATIER et al. 1978; BICKERS and KAPPAS 1980; NOONAN and WESTER 1983; BICKERS 1983; VERMORKEN et al. 1985; BICKERS et al. 1986; GOERZ and MERK 1987).

In general, the enzymes involved in drug metabolism detoxify foreign chemicals, but some endogenous compounds, e.g., steroid hormones, are metabolized by these enzymes as well. However, this subject will not be dealt with in this chapter. **Steroid metabolism** is described in Chap. 15, Vol. I, and Chap. 36, Vol. II.

Besides excretable drug metabolites, the enzymatic conversion of chemicals may lead to chemically reactive metabolites which can covalently bind to tissue constituents, resulting in toxicity, mutagenicity, and carcinogenicity (ANDERS 1985). In this respect the qualitative nature of drug metabolite formation by a specific organ may be much more relevant than the quantitative. This is of particular importance in the skin, because it is not only exposed to therapeutic drugs, but also to many environmental chemicals. Therefore, in this chapter the enzymatic activation of drugs, especially of chemical carcinogens, to reactive metabolites will also be discussed.

B. Drug Metabolism in General

Organic xenobiotics are metabolized by a number of enzymes which were first identified in the liver (for reviews see JAKOBY 1980; JAKOBY et al. 1982; CALDWELL and JAKOBY 1983), but are also present in many extrahepatic tissues (GRAM et al. 1980). Because most drugs and chemicals entering the body are lipophilic, they have to be transformed to hydrophilic compounds in order to be excreted in the urine.

Organic compounds which do not contain functional groups or side chains are oxidized in the body by the cytochrome P-450 system. It is comprised of a heme protein which in its reduced form shows a characteristic carbon monoxide spectrum with an absorption maximum near 450 nm. Reduction of cytochrome P-450

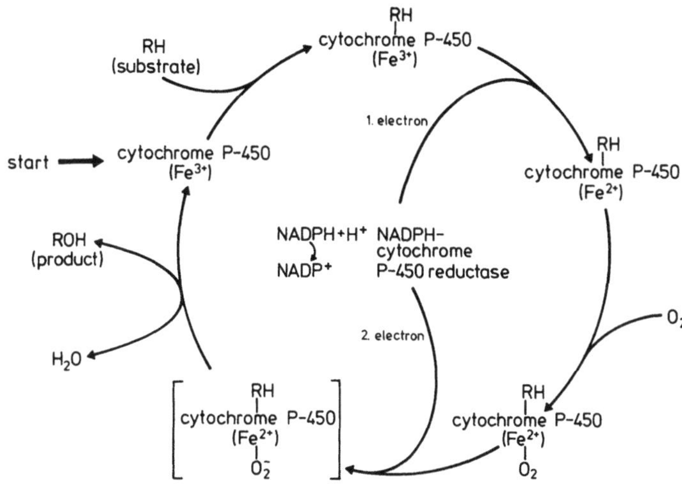

Fig. 1. Reaction scheme of the mixed function oxygenation of a drug (RH) to a hydroxylated metabolite (ROH) by the cytochrome P-450 enzyme system. (Fe^{3+}) is the oxidized heme moiety of cytochrome P-450, (Fe^{2+}) the reduced

is catalyzed by cytochrome P-450 reductase, an enzyme requiring NADPH as cofactor. Depending on the isoenzymes present (see later in this section), the absorption peak can be lower or higher than 450 nm, e.g., cytochrome P-448. In contrast to most enzymes of the intermediate metabolism, the substrate specificity of cytochrome P-450 is low. Therefore, it can oxidize a great variety of different organic compounds.

Mostly, hydroxylation at a carbon atom occurs. The underlying mechanism of hydroxylation by the cytochrome P-450 system is shown in Fig. 1. First of all, the substrate is bound to the protein moiety of oxidized cytochrome P-450. Then, the heme iron (Fe^{3+}) is reduced by one electron delivered by NADPH-cytochrome P-450 reductase. The reduced cytochrome P-450-substrate complex binds molecular oxygen and takes up another electron from NADPH-cytochrome P-450 reductase. During this step the substrate is attacked by the oxygen molecule, one oxygen atom being inserted in the substrate which is released from the enzyme as the hydroxylated product. The second oxygen atom yields water. Therefore, the whole enzyme reaction is named "mixed function oxidase" or "monooxygenase."

Figure 2 shows a number of oxidation reactions of drugs catalyzed by the cytochrome P-450 system. Besides hydroxylations at carbon atoms, cytochrome P-450 can also oxidize heteroatoms, e.g., nitrogen and sulfur. When aliphatic double bonds or aromatic hydrocarbons are oxidized by cytochrome P-450, epoxides occur, either as labile intermediates or as stable products. Epoxides can be hydrolyzed by water, leading to dihydroxy metabolites, "diols." This step can also be catalyzed by epoxide hydrolase. Many drugs already contain hydroxyl or other functional groups which are masked by esterification. Nonspecific esterases are able to split these compounds.

Free hydroxyl, carboxyl, or amino groups can be conjugated with glucuronic acid, resulting in highly hydrophilic metabolites. This conjugation is catalyzed by

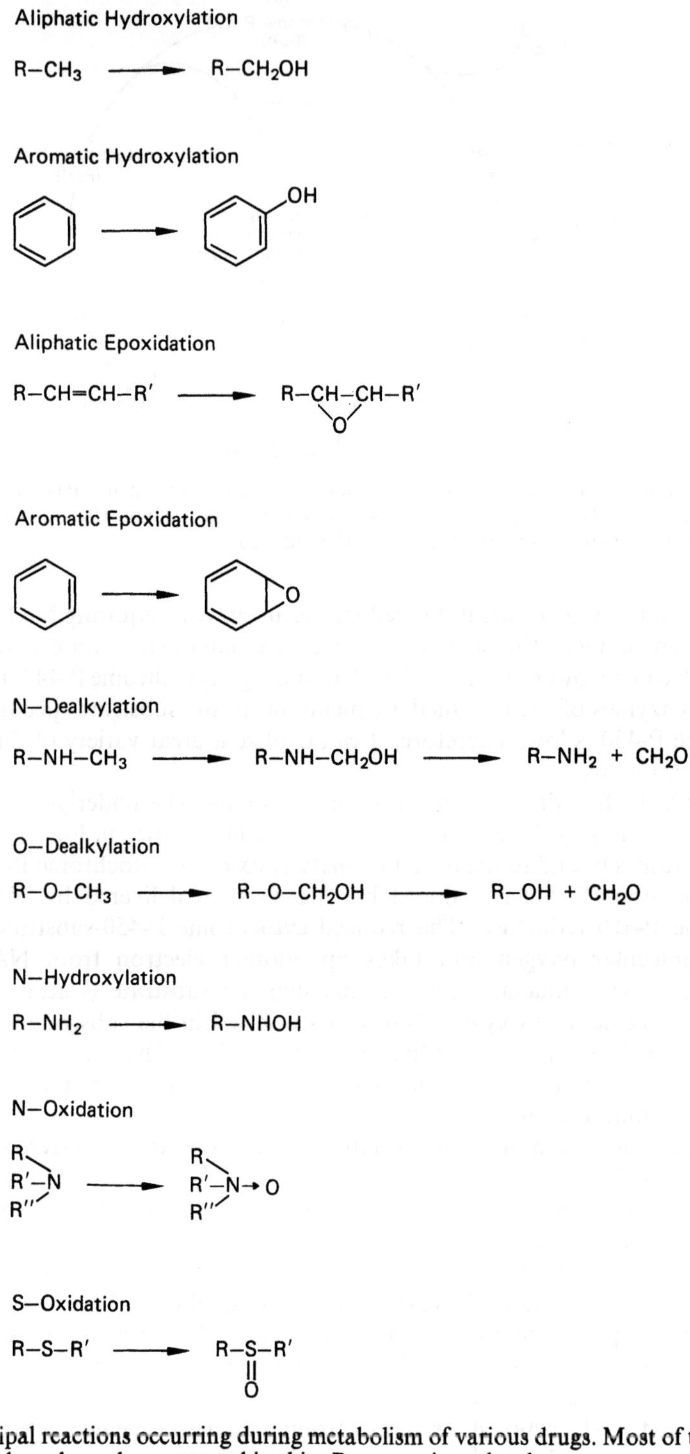

Fig. 2. Principal reactions occurring during metabolism of various drugs. Most of the reactions shown have been demonstrated in skin. R = organic molecule

Epoxide Hydrolysis

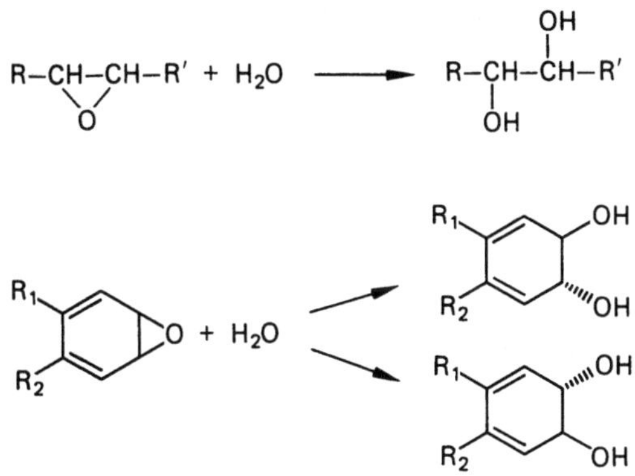

Ester Hydrolysis

$$R{-}C(=O){-}OR' + H_2O \longrightarrow R{-}COOH + R'{-}OH$$

$$R{-}C(=O){-}NH{-}R' + H_2O \longrightarrow R{-}COOH + R'{-}NH_2$$

Glucuronidation

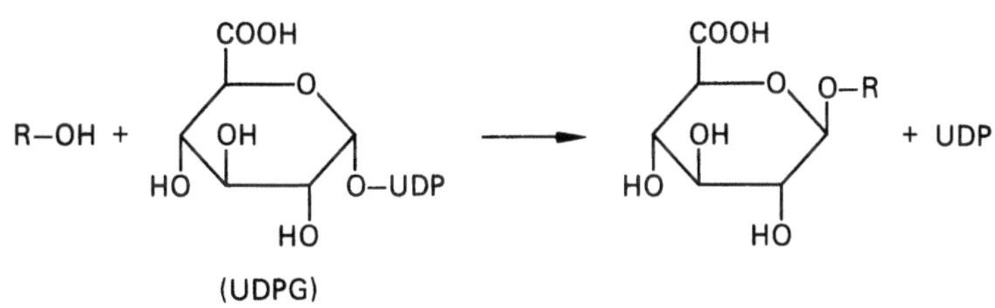

$$R{-}C(=O){-}OH + UDPG \longrightarrow R{-}C(=O){-}O{-}G + UDP$$

$$R{-}NH_2 + UDPG \longrightarrow R{-}NH{-}G + UDP$$

Sulfation

$$R{-}OH + \underset{(PAPS)}{PAP{-}O{-}SO_3H} \longrightarrow R{-}O{-}SO_3H + PAP$$

$$R{-}NH_2 + PAPS \longrightarrow R{-}NH{-}SO_3H + PAP$$

Fig. 2. (continued)

Glutathione Conjugation

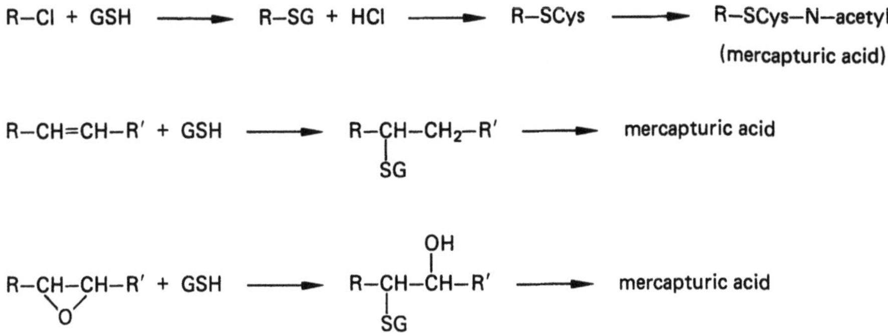

Fig. 2. (continued)

glucuronosyl transferase with uridine diphosphoglucuronic acid as cosubstrate. Conjugation with sulfate, which is catalyzed by sulfotransferase with the cosubstrate phosphoadenine phosphosulfate, also leads to hydrophilic metabolites (Fig. 2).

Not only compounds with free hydroxyl, carboxyl, or amino groups can be conjugated. Halogenated hydrocarbons, hydrocarbons with "activated" double bonds, and epoxides are conjugated with L-glutathione catalyzed by glutathione-*S*-transferase. The resulting conjugates can be further metabolized to mercapturic acids, polar excretion products (Fig. 2).

Most of the enzymes involved in drug metabolism occur as different isoenzyme forms with overlapping substrate specificities. For example, in the liver about 20 different cytochrome P-450 enzymes have been identified which have variable amino acid sequences. In some instances, only a few amino acids differ. Isoenzymes with different substrate specificities also exist for the other enzymes involved in drug metabolism. Not only are genetic differences in the enzyme proteins found in different species or in different organs of the same species, but also the different isoenzymes of one family can be induced (newly synthesized) by different exogenous or endogenous compounds. Enzyme induction is a very important process influencing drug metabolism. The different enzymes, the different substrate specificities, and the different inducibility of these enzymes complicate drug metabolism considerably.

C. Drug-Metabolizing Enzymes in Skin

I. General Remarks

Studies on drug metabolism in skin are extremely difficult because the organ is heterogeneous, the tissue is relatively resistant to homogenization, and the enzyme activities measurable are in general rather low. For example, in the studies on drug metabolism, dermis and epidermis are separated by simple scraping, by short-term heat treatment, by trypsinization, or by incubation with dithiothreitol.

Rough homogenization of skin leads to losses of some enzyme activities. Therefore, homogenization is often performed after freezing of the skin in liquid nitrogen. Nevertheless, the cytochrome P-450 spectra obtained with normal skin preparations often contain high absorptions at 420 nm, indicating enzyme alteration. Therefore, all the methods applied to determine drug-metabolizing enzyme activities in skin in vitro influence the results obtained more or less. In studies with monoclonal antibodies, specific enzymes can be identified in certain areas of the skin (BARON et al. 1983, 1986). However, enzyme activities are not measurable by this technique.

Although most studies carried out within the last 20 years conclusively demonstrate the presence of drug-metabolizing enzymes in skin, quantitative estimates are still difficult to obtain. This is mainly due to differences in species, ages, and skin samples used. Furthermore, drug-metabolizing enzymes have been studied in whole skin, epidermis, dermis, hair follicles, etc., the isolation procedure being most critical in obtaining reproducible results (see earlier in this section). Some authors have tried to overcome this problem by using cell culture systems. However, in cell culture various growth factors are added which may alter drug-metabolizing enzymes. On the other hand, enzyme activities may be based on wet weight, dry weight, or on protein content of whole homogenate, postmitochondrial supernatant, cytosol, or microsomes. In some studies, values are also based on DNA content or the numbers of cells. Depending on the substrate used, drug-metabolizing enzyme activities in skin are sometimes measurable only when induced by other xenobiotics. Therefore, exact numbers for constitutive drug-metabolizing enzyme activities in intact skin cannot be given yet. Furthermore, purification and characterization of cutaneous drug-metabolizing enzymes have not been carried out; this might be a very difficult task owing to the low activities present in skin.

II. Drug Oxidation by Cytochrome P-450

1. Aliphatic Oxidation

Only a few substrates have been tested to determine aliphatic oxidation reactions catalyzed by skin cytochrome P-450 (Table 1). Mainly, dealkylation reactions have been carried out, leading to products which can be determined by spectrofluorimetric methods. As shown in Table 1 dealkylation could be detected in various skin preparations, such as whole skin, epidermis, dermis, and isolated keratinocytes or the sebaceous cells of various species, the microsomal fractions being most active. Based on protein content, epidermis of hairless mice contains the highest activities, but based on wet weight, epidermal and dermal activities are equal (FINNEN et al. 1985). However, based on skin surface, epidermal activities are fivefold less than dermal activities (FINNEN et al. 1985). It has been calculated that whole skin contains about 1% of dealkylation activities of whole liver (FINNEN et al. 1985; GOERZ and MERK 1987).

Table 1. Aliphatic oxidation

Substrate	Product measured	Skin preparations used	Species examined[a]	Specific activities[b]	Studies
Ethoxy-coumarin	Hydroxy-coumarin	Whole skin strips	Mouse, rat	0.003–0.5 pmol/min per mg dry weight	MOLONEY et al. (1982b)
		Whole skin homogenate	Mouse, rat, cattle, rabbit	0.05–1 pmol/min per mg protein	FINNEN et al. (1985), DAMEN and MIER (1982)
		Whole skin postmitochondrial supernatant	Mouse, rat	0.3–3 pmol/min per mg protein	DAS et al. (1986c), FINNEN et al. (1984a), BICKERS et al. (1982b), MUKHTAR et al. (1982)
		Whole skin microsomes	Mouse, rat	0.3–80 pmol/min per mg protein	ASOKAN et al. (1986, 1985), RETTIE et al. (1986a), DAS et al. (1985), FINNEN et al. (1985, 1983), GOERZ et al. (1983, 1981), MUKHTAR and BICKERS (1983, 1982, 1981a), BICKERS et al. (1982b), MOLONEY et al. (1982c), POHL et al. (1976)
		Epidermal homogenate	Mouse	0.6 pmol/min per mg protein	FINNEN et al. (1985)
		Epidermal postmitochondrial supernatant	Mouse, rat	0.3–0.5 pmol/min per mg protein	FINNEN et al. (1984a), BICKERS et al. (1982b, c)
		Epidermal microsomes	Mouse, rat	0.3–1.6 pmol/min per mg protein	FINNEN et al. (1985), BICKERS et al. (1982a, b)
		Dermal homogenate	Mouse	0.3 pmol/min per mg protein	FINNEN et al. (1985)
		Dermal postmitochondrial supernatant	Mouse, rat	0.4 pmol/min per mg protein	FINNEN et al. (1984a), BICKERS et al. (1982b)
		Dermal microsomes	Mouse	0.8–1 pmol/min per mg protein	FINNEN et al. (1985), BICKERS et al. (1982b)
		Freshly isolated epidermal cells (keratinocytes)	Mouse	0.3–3 pmol/min per 10^6 cells	COOMES et al. (1984), POHL et al. (1984), FINNEN and SHUSTER (1985)

[a] Most studies with mice and rats were carried out either with neonatal or with adult hairless animals.
[b] Only activities of untreated skin are shown. The ranges given are either based on different values as reported by different authors or on conversion of original data published.

Table 1 (continued)

Substrate	Product measured	Skin preparations used	Species examined[a]	Specific activities[b]	Studies
		Freshly isolated skin cells (keratinocytes + sebaceous cells)	Mouse	5.7 pmol/min per 10^6 cells (228 pmol/min per mg DNA)	COOMES et al. (1983)
		Freshly isolated epidermal cells (basal keratinocytes)	Mouse	175 pmol/min per mg DNA	COOMES et al. (1983)
		Freshly isolated sebaceous cells	Mouse	1051 pmol/min per mg DNA	COOMES et al. (1983)
		Cultured epidermal cells (keratinocytes)	Mouse	0.1–0.5 pmol/min per mg protein	MUKHTAR et al. (1984a), BICKERS et al. (1982c)
		Whole hair follicles	Human	180 pmol/min per mg DNA	FINNEN and SHUSTER (1985)
		Cultured squamous carcinoma cells (postmitochondrial supernatant)	Human	0.1–0.8 pmol/min per mg protein	HUDSON et al. (1983)
Coumarin derivatives:					
Methoxy-	Hydroxy-coumarin	Whole skin microsomes	Mouse	3.9 pmol/min per mg protein	RETTIE et al. (1986a)
Ethoxy- (see above)	Hydroxy-coumarin	Whole skin microsomes	Mouse	7.2 pmol/min per mg protein	RETTIE et al. (1986a)
Propoxy-	Hydroxy-coumarin	Whole skin microsomes	Mouse	3.4 pmol/min per mg protein	RETTIE et al. (1986a)
Butoxy-	Hydroxy-coumarin	Whole skin microsomes	Mouse	2.5 pmol/min per mg protein	RETTIE et al. (1986a)

Ethoxyphenoxazone ("ethoxyresorufin")	Resorufin	Whole skin postmitochondrial supernatant	Mouse	0.2–0.5 pmol/min per mg protein	Das et al. (1986c), Finnen et al. (1984a)
		Whole skin microsomes	Mouse, rat	0.6–40 pmol/min per mg protein	Asokan et al. (1986, 1985), Rettie et al. (1986a, b), Finnen et al. (1985, 1983), Moloney et al. (1982c)
Phenoxazone derivates:					
Methoxy-	Resorufin	Whole skin microsomes	Mouse	8 pmol/min per mg protein	Rettie et al. (1986b)
Ethoxy- (see above)	Resorufin	Whole skin microsomes	Mouse	40 pmol/min per mg protein	Rettie et al. (1986b)
Propoxy-	Resorufin	Whole skin microsomes	Mouse	35 pmol/min per mg protein	Rettie et al. (1986b)
Butoxy-	Resorufin	Whole skin microsomes	Mouse	15 pmol/min per mg protein	Rettie et al. (1986b)
Benzyloxy-	Resorufin	Whole skin microsomes	Mouse	5 pmol/min per mg protein	Rettie et al. (1986b)
p-Nitroanisole	*p*-Nitrophenol	Whole skin postmitochondrial supernatant	Mouse	80 pmol/min per mg protein	Pannatier et al. (1981a)
p-Nitrophenetole	*p*-Nitrophenol	Whole skin postmitochondrial supernatant	Mouse	240 pmol/min per mg protein	Pannatier et al. (1981a, 1982)
Aminopyrine	Formaldehyde	Whole skin microsomes	Mouse, rat	2–6 pmol/min per mg protein	Mukhtar and Bickers (1982, 1981b), Goerz et al. (1981)
Aldrin	Aldrin epoxide	Whole skin microsomes	Mouse	4–5 pmol/min per mg protein	Rettie et al. (1986a)

Table 2. Aromatic oxidation

Substrate	Product measured	Skin preparations used	Species examined[a]	Specific activities[b]	Studies[c]
Benzo(a)pyrene	Hydroxybenzo(a)-pyrene[d]	Whole skin homogenate	Mouse, rat, human	0.01–3 pmol/min per mg protein	Lilienblum et al. (1986), Finnen et al. (1984b), Lawrence et al. (1984), Bickers et al. (1982d), Manil et al. (1981), Mukhtar and Bickers (1981a), Bickers and Kappas (1978), Akin and Norred (1976), Thompson and Slaga (1976a, b), Wiebel et al. (1975, 1971), Bickers et al. (1974), Poland et al. (1974), Benedict et al. (1973), Nebert et al. (1972, 1970), Schlede and Conney (1970)
		Whole skin post-mitochondrial supernatant	Mouse, rat	0.2–2 pmol/min per mg protein	Das et al. (1986c), Bickers et al. (1982b), Mukhtar et al. (1982), Chuang et al. (1978), Bürki et al. (1973)
		Whole skin microsomes	Mouse, rat	0.02–21 pmol/min per mg protein	Asokan et al. (1986, 1985), Rettie et al. (1986a), Das et al. (1985), Finnen et al. (1985, 1983), Goerz et al. (1983), Lesca (1983), Mukhtar and Bickers (1983), Bickers et al. (1982b), Kumar et al. (1982), Mukhtar and Bickers (1982), Rahimtula et al. (1982), Goerz et al. (1981), Manil et al. (1981), Mukhtar and Bickers (1981a), Vizethum et al. (1980), Pohl et al. (1976), Chan et al. (1972)
		Epidermal homogenate	Mouse, rat, human	0.02–4 pmol/min per mg protein	Finnen et al. (1984b), Puhvel and Ertl (1984), Mukhtar and Bickers (1981a), Pyerin and Hecker (1977), Slaga et al. (1977c), Akin and Norred (1976), Norred and Akin (1976), Thompson and Slaga

			(1976a), WIEBEL et al. (1975, 1973), BOWDEN et al. (1974), BRIGGS and BRIGGS (1973), GELBOIN et al. (1972, 1970), KINOSHITA and GELBOIN (1972)
Epidermal post-mitochondrial supernatant	Mouse, rat, human	0.02–9 pmol/min per mg protein	BICKERS et al. (1984, 1982b, c), AKIN and NORRED (1976), NORRED and AKIN (1976), BÜRKI et al. (1974, 1973)
Epidermal microsomes	Mouse, rat	0.5–42 pmol/min per mg protein	FINNEN et al. (1985), MUKHTAR et al. (1984b, c), BICKERS et al. (1982a, b), CAMUS et al. (1980), PYERIN and HECKER (1979, 1977), AKIN and NORRED (1976), NORRED and AKIN (1976)
Dermal homogenate	Mouse, rat, human	0.01–0.3 pmol/min per mg protein	FINNEN et al. (1984b), MUKHTAR and BICKERS (1981a), AKIN and NORRED (1976), THOMPSON and SLAGA (1976a), WIEBEL et al. (1975)
Dermal postmitochondrial supernatant	Mouse, rat	0.4–1.3 pmol/min per mg protein	BICKERS et al. (1982b), BÜRKI et al. (1974, 1973)
Dermal microsomes	Mouse, rat	1–2.8 pmol/min per mg protein	FINNEN et al. (1985), BICKERS et al. (1982b)
Freshly isolated skin cells (keratinocytes + sebaceous cells)	Mouse	118 pmol/min per mg DNA	COOMES et al. (1983)
Freshly isolated epidermal cells (basal keratinocytes)	Mouse	735 pmol/min per mg DNA	COOMES et al. (1983)
Freshly isolated sebaceous cells	Mouse	2377 pmol/min per mg DNA	COOMES et al. (1983)
Whole hair follicles	Human	1.7–5 pmol/min per mg DNA	HUKKELHOVEN et al. (1984)

[a] Most studies with mice and rats were carried out either with neonatal or with adult hairless animals.
[b] Only activities of untreated skin are shown. The ranges given are either based on different values as reported by different authors or on conversion of original data published. Lowest activities in human skin preparations.
[c] A number of reports on human AHH are not mentioned here, because the published data have been refuted by the authors (for details see FINNEN et al. (1983)).
[d] In most studies the fluorimetric test has been performed which was standardized with 3-hydroxybenzo[*a*]pyrene.

Table 2 (continued)

Substrate	Product measured	Skin preparations used	Species examined[a]	Specific activities[b]	Studies[c]
		Whole skin short-term culture	Mouse, rat, human	0.03–0.7 pmol/min per mg protein	Bickers and Kappas (1978), Alvares et al. (1973)
		Cultured epidermal cells (keratinocytes)	Mouse, human	0.08–3 pmol/min per mg protein	Das et al. (1986b), Lilienblum et al. (1986), Mukhtar et al. (1984a), Hosomi et al. (1983, 1982), Bickers et al. (1982c), Kuroki et al. (1981), Yuspa et al. (1976)
		Cultured skin fibroblasts	Mouse, human	0.2–9 pmol/min per mg protein	Lilienblum et al. (1986), Hosomi et al. (1983), Laaksonen et al. (1983), Kuroki et al. (1981), Gelboin and Wiebel (1971)
Other polycyclic aromatic hydrocarbons (PAH)	Hydroxy-PAH		see Table 8		
Diphenyloxazole	Hydroxydiphenyloxazole	Whole skin microsomes	Mouse, rat	6 pmol/min per mg protein	Rettie et al. (1986a), Rahimtula et al. (1982)
Coumarin	Hydroxycoumarin (umbelliferone)	Whole skin microsomes	Mouse	0.13 pmol/min per mg protein	Rettie et al. (1986a)
Aniline	Hydroxyaniline	Whole skin microsomes	Mouse	250 pmol/min per mg protein	Pohl et al. (1976)
Acetylaminofluorene (AAF)	1-, 3-, 5-, or 7-hydroxy-AAF (*N-OH*-AAF)	Whole fetal skin homogenate	Human	1.5 pmol/min per mg protein (0.2 pmol/min per mg protein)	Juchau et al. (1975)

2. Aromatic Oxidation

Even fewer aromatic compounds have been used as substrates for cytochrome P-450-dependent oxidation reactions in skin, although numerous studies determining hydroxylation of benzo[*a*]pyrene have been carried out. A sensitive test is based on the spectrofluorimetric measurement of the hydroxylated phenolic products formed. The test is normally standardized by 3-hydroxybenzo[*a*]pyrene. Benzo[*a*]pyrene hydroxylase oxidizes a great variety of polycyclic aromatic (aryl) hydrocarbons (PAH) and is therefore named "aryl hydrocarbon hydroxylase (AHH)." The activity measured in skin is due to isoenzymes of the cytochrome P-450 family which absorb near 448 nm (CO complex). Thus, it is also named "cytochrome P-448." It catalyzes not only aromatic, but also aliphatic oxidation. Most of the compounds used to measure dealkylation in skin (see Table 1) are actually substrates of AHH. This becomes clearly evident when AHH is induced (see Sect. C. VII).

AHH activity has been detected in almost all skin preparations of most species examined (Table 2) and is involved in skin carcinogenesis induced by PAH. The distribution of constitutive AHH activity in skin (Table 2) is almost the same as shown for enzymatic dealkylation (Table 1). Owing to the high sensitivity of the AHH test, activities can also be demonstrated in hair follicles and in skin fibroblasts (Table 2).

Experiments measuring the highest AHH activity in isolated sebaceous cells (COOMES et al. 1983) support earlier work using a histochemical fluorimetric technique to measure benzo[*a*]pyrene hydroxylation in intact skin (WATTENBERG and LEONG 1971). This is also consistent with the relatively high AHH activity found in vitro in the superficial layer of isolated dermis (WIEBEL et al. 1975). Nevertheless, based on protein content, whole dermis has lower AHH activities than whole epidermis, probably owing to "dilution" of cells of high activities with those of lower activities. But in some respects, e.g., in carcinogenesis, AHH activity in a certain cell type and not in the whole organ is crucial (CONNEY 1982).

III. Hydrolysis of Epoxides by Epoxide Hydrolase

Compared with AHH, epoxide hydrolase is present in skin in relatively high activities in all species examined (Table 3). Depending on the epoxide substrate used, different activities are found, indicating the existence of isoenzymes with different substrate specificities. As shown in Table 3 epoxide hydrolase activity is detectable in almost all parts of the skin as well as in cultured skin cells. The major activities of epoxide hydrolase are found in the microsomal fractions of skin preparations (Table 3). In studies with monoclonal antibodies, including human, the enzyme protein was found to be highest in the epidermis and in sebaceous glands (BARON et al. 1983, 1986).

Epoxide hydrolase activity in skin is also shown indirectly when examining the metabolism of PAH. A great number of diols of PAH have been identified in skin in vivo and in vitro, formed by hydrolysis of the respective epoxides (see Sect. D). Diols formed after hydrolysis of epoxides can form conjugates with glucuronic acid or with sulfate.

Table 3. Epoxide hydrolysis

Substrate	Product measured	Skin preparations used	Species examined[a]	Specific activities[b]	Studies
Benzo[*a*]pyrene 4,5-oxide	Benzo[*a*]pyrene 4,5-diol	Whole skin homogenate	Rat	20 pmol/min per mg protein	Manil et al. (1981)
		Whole skin microsomes	Mouse, rat, human	30–450 pmol/min per mg protein	Asokan et al. (1985), Das et al. (1985), Mukhtar and Bickers (1983, 1982, 1981a), Bickers et al. (1982b), Manil et al. (1981), Oesch and Schmassmann (1979), Oesch et al. (1978, 1977), Schassmann et al. (1976)
		Epidermal homogenate	Mouse	380 pmol/min per mg protein	Berry et al. (1977)
		Epidermal post-mitochondrial supernatant	Mouse, human	25–300 pmol/min per mg protein	Bickers et al. (1984, 1982c)
		Epidermal microsomes	Rat, human	50–350 pmol/min per mg protein	Mukhtar et al. (1984b), Bickers et al. (1982b), Oesch et al. (1978, 1977)
		Dermal microsomes	Rat, human	130–300 pmol/min per mg protein	Bickers et al. (1982b), Oesch et al. (1978, 1977)
		Whole hair follicles	Human	7 nmol/min per mg DNA	Hukkelhoven et al. (1982b)
		Cultured epidermal cells (keratinocytes)	Mouse	130 pmol/min per mg protein	Bickers et al. (1982c)
		Cultured skin fibrolasts	Human	80–130 pmol/min per mg protein	Oesch et al. (1980)

Phenanthrene 9,10-oxide	Phenanthrene 9,10-diol	Whole skin microsomes	Mouse, rat, human	1580–2530 pmol/min per mg protein	OESCH et al. (1978)
Benz[*a*]anthracene 5,6-oxide	Benz[*a*]anthracene 5,6-diol	Whole skin microsomes	Mouse, Rat, human	110–530 pmol/min per mg protein	OESCH et al. (1978)
7-Methylbenz[*a*]-anthracene 5,6-oxide	7-Methylbenz[*a*]-anthracene 5,6-diol	Whole skin microsomes	Mouse, Rat, human	120–380 pmol/min per mg protein	OESCH et al. (1978)
3-Methylcholanthrene 11,12-oxide	3-Methylcholanthrene 11,12-diol	Whole skin microsomes	Mouse, Rat, human	4–60 pmol/min per mg protein	OESCH et al. (1978)
Dibenz[a,h]anthracene 5,6-oxide	Dibenz[a,h]anthracene 5,6-diol	Whole skin microsomes	Mouse, Rat, human	2–21 pmol/min per mg protein	OESCH et al. (1978)
Stilbene oxide	Stilbene diol	Whole skin homogenate	Human	50 pmol/min per mg protein	O'NEILL et al. (1981)
		Whole skin microsomes	Rat, human	150–430 pmol/min per mg protein	MUKHTAR and BICKERS (1982, 1981a), O'NEILL et al. (1981)
		Epidermal microsomes	Human	140 pmol/min per mg protein	O'NEILL et al. (1981)
		Dermal microsomes	Human	170 pmol/min per mg protein	O'NEILL et al. (1981)
Styrene oxide	Styrene glycol	Epidermal homogenate	Mouse	50 pmol/min per mg protein	PYERIN and HECKER (1977, 1975)
		Epidermal microsomes	Mouse	300 pmol/min per mg protein	PYERIN and HECKER (1979, 1977, 1975)

[a] Most studies with mice and rats were carried out either with neonatal or with adult hairless animals.
[b] Only activities of untreated skin are shown. The ranges given are either based on different values as reported by different authors or on conversion of original data published. In general the activities decreased in the following order: human > mouse > rat.

Table 4. Examples of ester hydrolysis

Substrate	Product measured	Skin preparations used	Relative activities	Species	Studies
Betamethasone 17-valerate	Betamethasone	Whole skin homogenate	Low	Mouse	Cheung et al. (1985)
Betamethasone 21-valerate	Betamethasone	Whole skin homogenate	High	Mouse	Cheung et al. (1985)
N-Butyryloxymethyl-5-fluorouracil	5-fluorouracil	Whole skin strips	High	Human	Bundgaard et al. (1983), Møllgard et al. (1982)
Diflucortolone 21-valerate	Diflucortolone	Skin in vivo, whole skin strips	High	Rat, guinea pig, human	Täuber and Toda (1976)
Dithranol triacetate	Dithranol diacetate, -monoacetate, danthrone (via autoxidation)	Whole skin cytosol	High	Mouse, rat	Wiegrebe et al. (1984)
Fluocortin 21-butylester	Fluocortin	Skin in vivo, whole skin strips	Low	Guinea pig, human	Herz-Hübner and Täuber (1977)
Methylsulfinylmethyl 2-acetoxybenzoate	Salicylic acid	Whole skin strips	High	Mouse	Loftsson and Bodor (1981)
Metronidazole benzoate	Metronidazole	Whole skin strips	High	Human	Bundgaard et al. (1983)
p-Nitrobenzoate esters	*p*-Nitrobenzoic acid	Whole skin postmitochondrial supernatant and cytosol	High	Mouse	Pannatier et al. (1981b)
N-Pivaloyloxymethyl-5-fluorouracil	5-Fluorouracil	Whole skin strips	Low	Human	Møllgaard et al. (1982)
Vidarabine 5′-valerate	Vidarabine	Whole skin, epidermal and dermal strips, whole skin homogenate	High in epidermis, low in dermis (9/1)	Mouse	Yu et al. (1980a, b, 1979)

IV. Hydrolysis by Esterases

Esterases which split various esters, including esters of drugs, are measurable in all parts of the skin (Table 4). They are mainly present in the cytosolic fraction. Their substrate specificity is relatively low, but the reaction rates vary considerably. Based on the presence of esterases in skin, a number of ester pro-drugs have been developed which can be topically applied to the skin where they are hydrolyzed (Table 4). This pharmacologic concept is relatively new and can either be applied to treat skin diseases or for systemic therapy. However, the drugs released by the esterases of the skin are substrates for glucuronosyl transferase as well as for sulfotransferase of the skin.

V. Conjugation by Glucuronosyl Transferase and Sulfotransferase

Although the capacity of the skin to form glucuronic acid conjugates with xenobiotics or their derivatives has been known for a long time (HARPER and CALCUTT 1960), only a few reports deal with the measurement of drug-related glucuronosyl transferase activity in skin (Table 5). Compared with the oxidation reaction activities catalyzed by the cytochrome P-450 system, the activity of glucuronosyl transferase is relatively high. The tissue distribution of the enzyme in skin has not been tested, but the data available show that various skin cells contain glucuronosyl transferase activity (Table 5). Compared with liver, the activity of one isoenzyme seems to be lacking or extremely low in normal skin (Table 5). Furthermore, the activity usually measured after detergent activation (BOCK et al. 1983) does not necessarily mirror the activity in the skin in vivo.

That glucuronosyl transferase is active in the skin in vivo is indirectly demonstrated by the occurrence of glucuronidated drugs or derivatives in the skin in vivo (Table 5). On the other hand, β-glucuronidase is present in the skin in high activities (OHKAWARA et al. 1972, GOMEZ et al. 1977; MIER and VAN DEN HURK 1975). Therefore, following glucuronidation of drugs in the skin, the conjugates formed could be split in the same organ.

Sulfotransferase in the skin has not been systematically studied. However, analysis of drug metabolites in skin indicates that conjugation of drugs with sulfate is a rather minor metabolic pathway in the skin (Table 5), although it cannot be excluded that sulfate conjugates formed are immediately split by skin sulfatases.

VI. Conjugation by Glutathione-*S*-transferase

Glutathione-*S*-transferase is present in the skin of different species (Table 6). The activities measured either by chlorinated compounds or by epoxides as substrates are relatively high, but vary depending on the substrates used. Although tissue distribution has not been extensively examined, studies with monoclonal antibodies demonstrate that the enzyme is located in epidermis, hair follicles, and sebaceous glands (BARON et al. 1983, 1986). The results obtained with different substrates and antibodies against different isoenzymes suggest that skin, like liver,

Table 5. Glucuronidation and sulfation

Substrate	Product measured	Skin preparation used	Species examined	Specific activities[a]	Studies
1-Naphthol	1-Naphthol glucuronide	Whole skin homogenate	Mouse	9–30 nmol/min per g tissue wet weight	Lilienblum et al. (1986), Bock et al. (1982)
		Whole skin microsomes	Mouse, rat	0.8–13 nmol/min per mg protein	Moloney et al. (1982a), Bock et al. (1980)
		Cultured epidermal cells (keratinocytes)	Mouse	5 nmol/min per mg DNA	Lilienblum et al. (1986)
		Cultured skin fibroblasts	Mouse	8 nmol/min per mg DNA	Lilienblum et al. (1986)
3-Hydroxybenzo[*a*]-pyrene	3-Hydroxybenzo[*a*]-pyrene glucuronide	Whole skin homogenate	Mouse	0.4 nmol/min per g tissue wet weight	Lilienblum et al. (1986)
		Whole skin microsomes	Rat	0.08 nmol/min per mg protein	Bock et al. (1980)
Benzo[*a*]pyrene 7,8-diol	Benzo[*a*]pyrene 7,8-diol glucuronide	Whole skin homogenate	Mouse	0.1 nmol/min per g tissue wet weight	Lilienblum et al. (1986)
		Whole skin microsomes	Rat	0.003 nmol/min per mg protein	Bock et al. (1980)
o-Aminophenol	*o*-Aminophenol glucuronide	Whole skin strips	Mouse	6–9 nmol/2 h per 50 mg tissue dry weight	Dutton and Stevenson (1962)
		Whole skin homogenate	Mouse	1–4 nmol/30 min per 10 mg tissue dry weight	Dutton and Stevenson (1962)
o-Aminophenol (*p*-Nitrophenol)	*o*-Aminophenol glucuronide (*p*-nitrophenol glucuronide)	Cultured epidermal cells (keratinocytes)	Human	16 nmol/h per mg protein	Rugstad and Dybing (1975)
4-Methylumbelliferone	4-Methylumbelliferone glucuronide	Whole skin homogenate	Mouse	50 nmol/min per g tissue wet weight	Bock et al. (1982)
		Freshly isolated skin cells (keratinocytes + sebaceous cells)	Mouse	5 nmol/min per mg DNA	Coomes et al. (1983)
		Freshly isolated epidermal cells (keratinocytes)	Mouse	14 nmol/min per mg DNA	Coomes et al. (1983)
		Freshly isolated sebaceous cells	Mouse	39 nmol/min per mg DNA	Coomes et al. (1983)
7-Hydroxycoumarin	7-Hydroxycoumarin glucuronide	Whole skin strips	Mouse, rat	11–27 nmol/h per 100 mg tissue dry weight	Moloney et al. (1982b)
	7-Hydroxycoumarin	Whole skin strips	Mouse, rat	1–2 nmol/h per 100 mg tissue dry weight	Moloney et al. (1982b)

[a] Only activities of untreated skin are shown. The numbers given are as reported. The highest activities are always found in mice.

Table 6. Glutathione conjugation

Substrate	Skin preparations used	Species examined[a]	Specific activities[b]	Studies
1-Chloro-2,4-dinitrobenzene	Whole skin cytosol	Mouse	54 nmol/min per mg protein	DAS et al. (1985)
	Whole skin postmitochondrial supernatant	Mouse, rat, human	2–90 nmol/min per mg protein	BAARS et al. (1981), MUKHTAR et al. (1981)
	Epidermal cytosol	Rat	13 nmol/min per mg protein	MUKHTAR et al. (1984b)
	Epidermal postmitochondrial supernatant	Rat	54 nmol/min per mg protein	SUMMER and GÖGGELMANN (1980)
	Freshly isolated skin cells (keratinocytes + sebaceous cells)	Mouse	316 nmol/min per mg DNA	COOMES et al. (1983)
	Freshly isolated epidermal cells (basal keratinocytes)	Mouse	399 nmol/min per mg DNA	COOMES et al. (1983)
	Freshly isolated sebaceous cells	Mouse	1399 nmol/min per mg DNA	COOMES et al. (1983)
	Cultured skin fibroblasts	Human	30–90 nmol/min per mg protein	OESCH et al. (1980)
Benzo[*a*]pyrene 4,5-oxide	Whole skin cytosol	Rat	0.6–0.7 nmol/min per mg protein	MUKHTAR and BICKERS (1983, 1982, 1981a)
	Epidermal cytosol	Rat	1.2 nmol/min per mg protein	MUKHTAR et al. (1984b)
3-Methylcholanthrene 11,12-oxide	Whole skin cytosol	Mouse, human	0.03–0.2 nmol/min per mg protein	MUKHTAR and BRESNICK (1976)
Styrene oxide	Whole skin cytosol	Mouse, rat, human	0.4–3.2 nmol/min per mg protein	MUKHTAR and BICKERS (1982, 1981a), MUKHTAR and BRESNICK (1976)

[a] Most studies with mice and rats were carried out either with neonatal or with adult hairless animals.
[b] Only activities of untreated skin are shown. The ranges given are either based on different values as reported by different authors or on conversion of original data published. Lowest activities in human skin preparations.

contains a number of isoenzymes of glutathione-*S*-transferase. Glutathione conjugates have also been detected during the metabolism of PAH (Sect. D).

VII. Inducibility of Drug-Metabolizing Enzymes

Enzyme induction in skin has been extensively studied. Table 7 shows that pretreatment of animals with many xenobiotics applied either topically or systemically results in increased enzyme activities of drug metabolism. This could also be observed in humans after treatment of the skin with a number of xenobiotics.

The highest increase is observed when AHH is measured, indicating that certain isoenzymes are highly inducible (new protein synthesis) in skin. This is favored by the fact that after exposure of the skin to PAH, tetrachlorodibenzodioxin (TCDD), or Aroclor 1254, inducers of cytochrome P-448 in liver and other organs, the cytochrome P-450-CO complex determined in skin microsomes is highly increased and absorbs in the range 447–448 nm (BICKERS et al. 1982b, c; POHL et al. 1976). AHH, as measured by aliphatic dealkylation as well as by aromatic hydroxylation, increases severalfold, the enhancement being higher after topical application (Table 7). Drugs such as phenobarbital which induce isoenzymes absorbing near 450 nm are almost ineffective in inducing monooxygenase activity in the skin. The high inducibility of AHH in skin is also demonstrated by the large increases in PAH metabolite formation (see Sect. D).

In some areas of the skin, monooxygenase activity which was hardly detectable in controls becomes easily measurable after treatment with certain drugs (Table 7). This could also be demonstrated in intact skin. After treatment of animals with 3-methylcholanthrene, a dramatic increase in one of the cytochrome P-450 isoenzymes was observed in sebaceous glands and in hair follicles. In the latter it was not detectable without pretreatment (BARON et al. 1986). Similar results were obtained when bezo[*a*]pyrene metabolism was measured histochemically (BARON et al. 1986; WATTENBERG and LEONG 1971).

Epoxide hydrolase, in contrast to the monooxygenase system, is noninducible or only slightly inducible in the skin (Table 7). PAH and drugs with similar activity are ineffective, as are a number of epoxides which all significantly increase epoxide hydrolase activity in the liver. On the other hand, glucuronosyl transferase is inducible in skin by a number of drugs. Different enzyme activities are enhanced by different drugs used for pretreatment. This again suggests that different isoenzymes of glucuronosyl transferase are present in skin which only respond to defined inducing agents. Only limited information is available on induction of glutathione-*S*-transferase by drugs in the skin. The results obtained indicate that glutathione-*S*-transferase in the skin is not highly inducible.

D. Metabolism of Polycyclic Aromatic Hydrocarbons in the Skin as Related to Carcinogenicity

A number of PAH which are skin carcinogens are actively metabolized by drug-metabolizing enzymes of the skin (for reviews see GELBOIN 1980; CONNEY 1982;

Table 7 In vivo induction of drug-metabolizing enzymes

Compound applied[a]	Enzyme (activity) measured	Species examined	Inducing factor[b]	Studies
Polycyclic aromatic hydrocarbons (PAH): Pyrene, benzo[*a*]pyrene, benzanthracene, 3-methylcholanthrene, anthracene, dibenzanthracene, dimethylbenzanthracene, benzofluorene, phenanthrene, nitro-PAH	Aryl hydrocarbon hydroxylase, ethoxycoumarin deethylase, ethoxyresorufin deethylase	Mouse, rat	2–33 ×	ASOKAN et al. (1986, 1985), FINNEN et al. (1985, 1984a), MUKHTAR et al. (1984b, c), MUKHTAR and BICKERS (1983), BICKERS et al. (1982c, d, e), DAMEN and MIER (1982), MUKHTAR et al. (1982), RAUNIO and PELKONEN (1982), MANIL et al. (1981), MUKHTAR and BICKERS (1981a), CAMUS et al. (1980), PYERIN et al. (1980), VIZETHUM et al. (1980), PYERIN and HECKER (1979), BICKERS and KAPPAS (1978), CHUANG et al. (1978), SLAGA et al. (1977c), AKIN and NORRED (1976), POHL et al. (1976), Thompson and SLAGA (1976a, b), YUSPA et al. (1976), WIEBEL et al. (1975, 1973, 1971), BOWDEN et al. (1974), BÜRKI et al. (1974, 1973), BENEDICT et al. (1973), BRIGGS and BRIGGS (1973), CHAN et al. (1972), GELBOIN et al. (1972, 1970), GIELEN et al. (1972), KINOSHITA and GELBOIN (1972), NEBERT et al. (1972, 1970), SCHLEDE and CONNEY (1970)
	Cytochrome P-450 (control: 24 pmol per mg microsomal protein)	Mouse	2 ×	BICKERS et al. (1982c)
	NADPH-cytochrome P-450 reductase (control: 3–5 nmol/min per mg microsomal protein)	Rat	1 ×	VIZETHUM et al. (1980)
	Benzo[*a*]pyrene metabolism	Mouse, rat	1.5–19 ×	ASOKAN et al. (1986); DAS et al. (1986a), BICKERS et al. (1983), MUKHTAR and BICKERS (1983), BERRY et al. (1977), SLAGA et al. (1977a), WANG et al. (1974)

[a] In general single compounds or mixtures (as indicated) were topically applied (with some exceptions).
[b] The inducing factor (1 × = no induction) depends on the different compounds used, the different species examined, and the different enzymes measured in different skin preparations.

Table 7 (continued)

Compound applied[a]	Enzyme (activity) measured	Species examined	Inducing factor[b]	Studies
	Epoxide hydrolase	Mouse, rat	1–1.5 ×	Asokan et al. (1985), Mukhtar and Bickers (1983), Bickers et al. (1982c), Pyerin and Hecker (1979), Berry et al. (1977)
	Glucuronosyl transferase	Mouse, rat	1.2–3 ×	Moloney et al. (1982a), Dutton and Stevenson (1962)
	Glutathione-*S*-transferase	Mouse, rat	1 ×	Mukhtar and Bickers (1983), Mukhtar and Bresnick (1976)
	DT-diaphorase	Mouse	6–8 ×	Iversen and Digernes (1981)
Crude coal tar	Aryl hydrocarbon hydroxylase, ethoxycoumarin deethylase, ethoxyresorufin deethylase	Mouse, rat, human	1.2–11 ×	Das et al. (1986c, 1985), Finnen et al. (1984b), Hukkelhoven et al. (1984), Lawrence et al. (1984), Merk et al. (1984), Mukhtar et al. (1984b, 1982), Bickers et al. (1982e), Mukhtar and Bickers (1982), Bickers and Kappas (1978)
	Benzo[*a*]pyrene metabolism	Rat, human	1.5–9 ×	Van Cantfort et al. (1986), Mukhtar et al. (1984b)
	Epoxide hydrolase	Rat	1 ×	Mukhtar and Bickers (1982)
	Glutathione-*S*-transferase	Rat	1 ×	Mukhtar and Bickers (1982)
Cigarette smoke condensate	Aryl hydrocarbon hydroxylase	Mouse	10 ×	Norred and Akin (1976), Akin et al. (1975)
Naphthoflavone, benzoflavone	Aryl hydrocarbon hydroxylase, ethoxycoumarin deethylase	Mouse	2–23 ×	Coomes et al. (1984, 1983), Moloney et al. (1982c), Pyerin and Hecker (1979), Wiebel et al. (1975, 1973), Bowden et al. (1974), Gelboin et al. (1972, 1970), Wattenberg and Leong (1970)
	Benzo[*a*]pyrene metabolism	Mouse, rat	10–30 ×	Eling et al. (1986), Moloney et al. (1982b)
	Epoxide hydrolase	Mouse	1 ×	Berry et al. (1977)
	Glucuronosyl transferase	Mouse	1.5–3 ×	Coomes et al. (1983)
	Glutathione-*S*-transferase	Mouse	1.5–2 ×	Coomes et al. (1983)
Tetrachlorodibenzodioxin (TCDD)	Aryl hydrocarbon hydroxylase, ethoxycoumarin deethylase	Mouse, rat	6–61 ×	Puhvel and Ertl (1984), Lesca (1983), Raunio and Pelkonen (1982), Pohl et al. (1976), Poland et al. (1974)
	Cytochrome P-450	Mouse	2–3 ×	Pohl et al. (1976), Poland et al. (1974)
	Benzo[*a*]pyrene metabolism	Mouse	1.5–19 ×	Kao et al. (1984), Legraverend et al. (1980)

Polychlorinated biphenyls (Aroclor 1254)	Aryl hydrocarbon hydroxylase, ethoxycoumarin deethylase	Mouse, rat	2–24 ×	LILIENBLUM et al. (1986), MUKHTAR et al. (1984b), BICKERS et al. (1983, 1982b), RAHIMTULA et al. (1982), MUKHTAR and BICKERS (1981a), BICKERS et al. (1974)
	Diphenyloxazole hydroxylase	Rat	6 ×	RAHIMTULA et al. (1982)
	Cytochrome P-450	Rat	1.2–2 ×	BICKERS et al. (1982b)
	NADPH–cytochrome P-450 reductase	Rat	1 ×	MUKHTAR and BICKERS (1981a)
	Epoxide hydrolase	Rat	1 ×	BICKERS ET AL. (1982b), MUKHTAR and BICKERS (1981a), OESCH et al. (1978)
	Glucuronosyl transferase	Mouse	3 ×	LILIENBLUM et al. (1986)
	Glutathione-*S*-transferase	Rat	1 ×	MUKHTAR and BICKERS (1981a)
Hexachlorobenzene	Ethoxycoumarin deethylase	Mouse	2 ×	DAMEN and MIER (1982)
Dichlorodiphenyltrichloroethane (DDT)	Aryl hydrocarbon hydroxylase	Rat	1.5–2 ×	BICKERS et al. (1974)
Microscope immersion oil, jute batching oil, Kuwait crude oil	Aryl hydrocarbon hydroxylase	Mouse, rat	9–40 ×	KUMAR et al. (1982), RAHIMTULA et al. (1982), BICKERS et al. (1975), ALVARES et al. (1974)
Corticosteroids: Dexamethasone, clobetasol propionate, betamethasone valerate, fluocinonide, fluocinolone acetonide, fluocortolone, beclomethasone dipropionate	Aryl hydrocarbon hydroxylase, ethoxycoumarin deethylase, ethoxyresorufin deethylase	Mouse	2–5 ×	FINNEN et al. (1984a), Thompson and SLAGA (1976b), BRIGGS and BRIGGS (1973)
8-Methoxypsoralen	Aryl hydrocarbon hydroxylase	Rat	1 ×	BICKERS et al. (1982d)
Phenobarbitone	Aryl hydrocarbon hydroxylase, ethoxycoumarin deethylase	Mouse, rat	2–3 ×	DAMEN and MIER (1982), VIZETHUM et al. (1980)
Pregnenolone 16α-carbonitrile	Aryl hydrocarbon hydroxylase	Rat	2 ×	VIZETHUM et al. (1980)

THAKKER et al. 1985). Furthermore, most are potent inducers of AHH in the skin, thereby increasing their own metabolism as well as the metabolism of other skin carcinogens.

Benzo[*a*]pyrene which has been studied as a model compound is converted in skin in vivo and in vitro to a number of metabolites, e.g., phenols, quinones, diols, tetrols (Table 8). The metabolic pathways of benzo[*a*]pyrene which result in mutagenic and carcinogenic metabolites are shown in Fig. 3. The first step is the epoxidation of the 7,8 double bond which is stereospecific and yields two isomeric 7,8-epoxides (Fig. 3). Rearrangement to the phenols (7-hydroxybenzo[*a*]pyrene or 8-hydroxybenzo[*a*]pyrene) does not occur. In skin, the upper pathway of Fig. 3 is predominant, owing to the stereospecific attack of cytochrome P-450 enzymes involved. In the next step, epoxide hydrolase is involved which yields stable *trans*-7,8-diols (Fig. 3). These can be conjugated and excreted or reduced to catechols. However, a significant part of the 7,8-diols formed in skin is additionally epoxidized at the 9,10 double bond by the cytochrome P-450 enzyme system. Again this step is stereospecific and in the skin the main product is the *anti*-diol epoxide in which the 7-hydroxy group is *trans* to the 9,10-epoxide. (+)-BPDE-2 (Fig. 3) is highly reactive and is the major benzo[*a*]pyrene metabolite covalently bound to the amino group (N^2) of desoxyguanosine of cutaneous DNA in vivo (Table 8). It also forms some adducts with desoxyadenosine.

In skin carcinogenicity tests, however, (+)-BPDE-2, although more active than the other diastereoisomers, was less tumorigenic than the parent compound benzo[*a*]pyrene. This has been explained by the reactivity of this diol epoxide which can react with other tissue components before it reaches cellular DNA. But the precursor metabolite (−)-7β,8α-diol (Fig. 3) has higher or at least equal tumorigenic potency in skin, indicating that this is the proximate skin carcinogen. Based on specific DNA binding which correlates well with skin tumorigenicity, (+)-BPDE-2 (Fig. 3) is regarded as the ultimate carcinogenic metabolite of benzo[*a*]pyrene. These findings are consistent with the so-called bay region theory which has been well established in skin for many PAH (BRESNICK et al. 1977; BUENING et al. 1980; CHANG et al. 1982, 1983; CHOUROULINKOV et al. 1976, 1977, 1979a, b; HECHT et al. 1980; LEVIN et al. 1976a, b, 1977a, b, 1978a, b, 1979, 1980a, b; SLAGA et al. 1976, 1977b, d, 1978, 1979a, b, c; WISLOCKI et al. 1977; WOOD et al. 1977; for review see THAKKER et al. 1985).

The highly reactive diastereomeric bay region diol epoxides of PAH can be formed by monooxygenase, and also by co-oxidation during prostaglandin synthesis (for review see KRAUSS and ELING 1984). Although the enzymes involved are very active in skin (for review see RUZICKA and PRINTZ 1984), only a few reports dealing with this subject exist (ROGAN et al. 1978, 1983; SIVARAJAH et al. 1981; CAVALIERI and ROGAN 1984).

E. Conclusions

Numerous publications clearly demonstrate that the skin has a great number of drug-metabolizing enzymes present in many skin cells in variable activities. Qualitatively they are similar to those in other organs, but the quantitative turnover

Table 8. Formation of PAH metabolites

PAH	Metabolites and reaction products measured[a]	Skin preparations used	Species examined[b]	Studies
Benzo[*a*]pyrene	Polar metabolites, e.g., phenols (3-OH, 9-OH), dihydrodiols (4,5-diol, 7,8-diol, 9,10-diol), tetrols, quinones, conjugates (glucuronides, sulfates, glutathione conjugates)	Whole skin in vivo	Mouse, rat	MELIKIAN et al. (1986), WESTON et al. (1982a, b), NORDEN (1953), WEIGERT et al. (1948)
		Whole skin short-term culture	Mouse, rat, rabbit, marmoset, human	KAO et al. (1984, 1985), SHUGART and KAO (1984), SELKIRK et al. (1983), WESTON et al. (1983a), SMITH and HOLLAND (1981), MACNICOLL et al. (1980)
		Whole skin microsomes	Mouse, rat, hamster	ASOKAN et al. (1986), DAS et al. (1986a), BICKERS et al. (1983), MUKHTAR and BICKERS (1983), LEGRAVEREND et al. (1980), WANG et al. (1974)
		Epidermal homogenate	Mouse, human	VAN CANTFORT et al. (1986), COHEN et al. (1979), BERRY et al. (1977)
		Epidermal microsomes	Mouse, rat, human	DAS et al. (1986a), BICKERS et al. (1984), MUKHTAR et al. (1984b), BICKERS et al. (1983)
		Cultured epidermal cells (keratinocytes)	Mouse, rat, hamster, human	DAS et al. (1986b), MUKHTAR et al. (1984a), DIGIOVANNI et al. (1983b, 1982), HEIMANN and RICE (1983), HOSOMI et al. (1983, 1982), HUKKELHOVEN et al. (1982a, 1981), KUROKI et al. (1982, 1981, 1980), THEALL et al. (1981), PARKINSON and NEWBOLD (1980), FOX et al. (1975)
		Cultured skin fibroblasts	Mouse, human	HOSOMI et al. (1983), HUKKELHOVEN et al. (1982a), KUROKI et al. (1982, 1981), PARKINSON and NEWBOLD (1980), RÜDIGER et al. (1979), FOX et al. (1975), HUBERMANN and SACHS (1973), BROOKES and DUNCAN (1971), SELKIRK et al. (1983)
		Hair follicles in vitro	Human	HUKKELHOVEN et al. (1983a, b), VERMORKEN et al. (1979)

[a] Not all reported metabolites shown; conjugates of the metabolites named, but also of other mostly unkown metabolites.
[b] High qualitative and especially quantitative variations, depending on the species and the strains used.

Table 8 (continued)

PAH	Metabolites and reaction products measured[a]	Skin preparations used	Species examined[b]	Studies
	Specific adducts of benzo[*a*]pyrene 7,8-diol-9,10-oxide with DNA bases (major: desoxyguanosine; minor: desoxyadenosine)	Whole skin in vivo	Mouse, rat	Rojas et al. (1986), Das et al. (1985), Ashurst and Cohen (1981b), Vigny et al. (1980), Grover et al. (1976)
		Epidermis in vivo	Mouse, rat	Nakayama et al. (1984), Ashurst et al. (1983), Shugart et al. (1983), Alexandrov et al. (1982), Ashurst and Cohen (1982, 1981a), Baer-Dubowska and Alexandrov (1981), Cohen et al. (1979), Pereira et al. (1979), Koreeda et al. (1978), Daudel et al. (1975)
		Whole skin short-term culture	Human	Weston et al. (1983a)
		Epidermis short-term culture	Mouse	Shugart and Kao (1984)
		Cultured epidermal cells (keratinocytes)	Mouse, hamster, human	Nakayama et al. (1984), DiGiovanni et al. (1983b), Theall et al. (1981), Parkinson and Newbold (1980)
		Cultured skin fibroblasts	Human	Tejwani et al. (1982)
		Hair follicles in vitro	Human	Hukkelhoven et al. (1985, 1981), Vermorken et al. (1979)
Benzo[*a*]pyrene 7,8-diol	7,8,9,10-Tetrols	Whole skin and epidermal microsomes	Mouse	Das et al. (1986a)
	Specific adducts of benzo[*a*]pyrene 7,8-diol-9,10-oxide with DNA bases	Whole skin in vivo	Mouse	Das et al. (1986a), Brookes (1979), Grover et al. (1976)
		Freshly isolated keratinocytes	Mouse	Eling et al. (1986)
Benzo[*a*]pyrene 7,8-diol-9,10-oxide (BPDE)	Specific adducts of benzo[*a*]pyrene 7,8-diol-9,10-oxide with DNA bases	Whole skin in vivo	Mouse	Brookes (1979)
		Epidermis in vivo	Mouse	Pelling and Slaga (1982)
		Cultured skin fibroblasts	Human	Tejwani et al. (1982), Kaneko and Cerutti (1982)
7,12-Dimethylbenz[*a*]anthracene (DMBA)	Polar metabolites, e.g., phenols, dihydrodiols (1,2-diol; 3,4-diol; 5,6-diol;	Whole skin short-term culture	Mouse, rat, human	Weston et al. (1983b), MacNicoll et al. (1980)
		Epidermal homogenate	Mouse	DiGiovanni et al. (1980, 1977)

	8,9-diol; 10,11-diol), 7-hydroxymethyl-MBA, 12-hydroxymethyl-MBA, 7,12-dihydroxymethyl-BA	Cultured epidermal cells (keratinocytes)	Mouse, hamster	DiGiovanni et al. (1983b, 1980), Irmscher and Fusenig (1980), Hukkelhoven et al. (1980)
		Cultured skin fibroblasts	Mouse	Irmscher and Fusenig (1980), Hukkelhoven et al. (1980)
	Specific adducts of DMBA-3,4-diol-1,2-oxide with DNA bases	Whole skin in vivo	Mouse	Bigger et al. (1983, 1978), Vigny et al. (1981)
		Epidermis in vivo	Mouse	DiGiovanni et al. (1983a)
		Cultured epidermal cells (keratinocytes)	Mouse	DiGiovanni et al. (1985)
7-Methylbenz[*a*]-anthracene (MBA)	Polar metabolites, also diols	Whole skin short-term culture	Mouse	MacNicoll et al. (1980), Tierney et al. (1977)
	Specific adducts of MBA-3,4-diol-1,2-oxide with DNA bases	Whole skin in vivo	Mouse	Tierney et al. (1977)
Benz[*a*]anthracene (BA)	Polar metabolites, also diols	Whole skin short-term culture	Mouse	MacNicoll et al. (1980)
	Specific adducts of BA-3,4-diol-1,2-oxide with DNA bases	Whole skin in vivo	Mouse	Cooper et al. (1980)
Chrysene	Polar metabolites, also diols	Whole skin short-term culture	Mouse, rat, human	Weston et al. (1985), MacNicoll et al. (1980)
	Specific adducts of chrysene-1,2-diol-3,4-oxide with DNA bases	Whole skin short-term culture	Mouse, rat, human	Weston et al. (1985)
5-Methylchrysene	Polar metabolites, also diols, glucuronides and sulfates	Epidermis in vivo	Mouse	Melikian et al. (1983)
	Specific adducts of 5-methylchrysene-1,2-diol-3,4-oxide and 5-methylchrysene-7,8-diol-9,10-oxide with DNA bases	Whole skin in vivo, epidermis in vivo	Mouse	Melikian et al. (1983, 1982)
3-Methylcholanthrene	Polar metabolites, also diols	Whole skin short-term culture	Mouse	MacNicoll et al. (1980), Tierney et al. (1978)
Dibenzanthracene	Polar metabolites, also diols	Whole skin short-term culture	Mouse	MacNicoll et al. (1980)

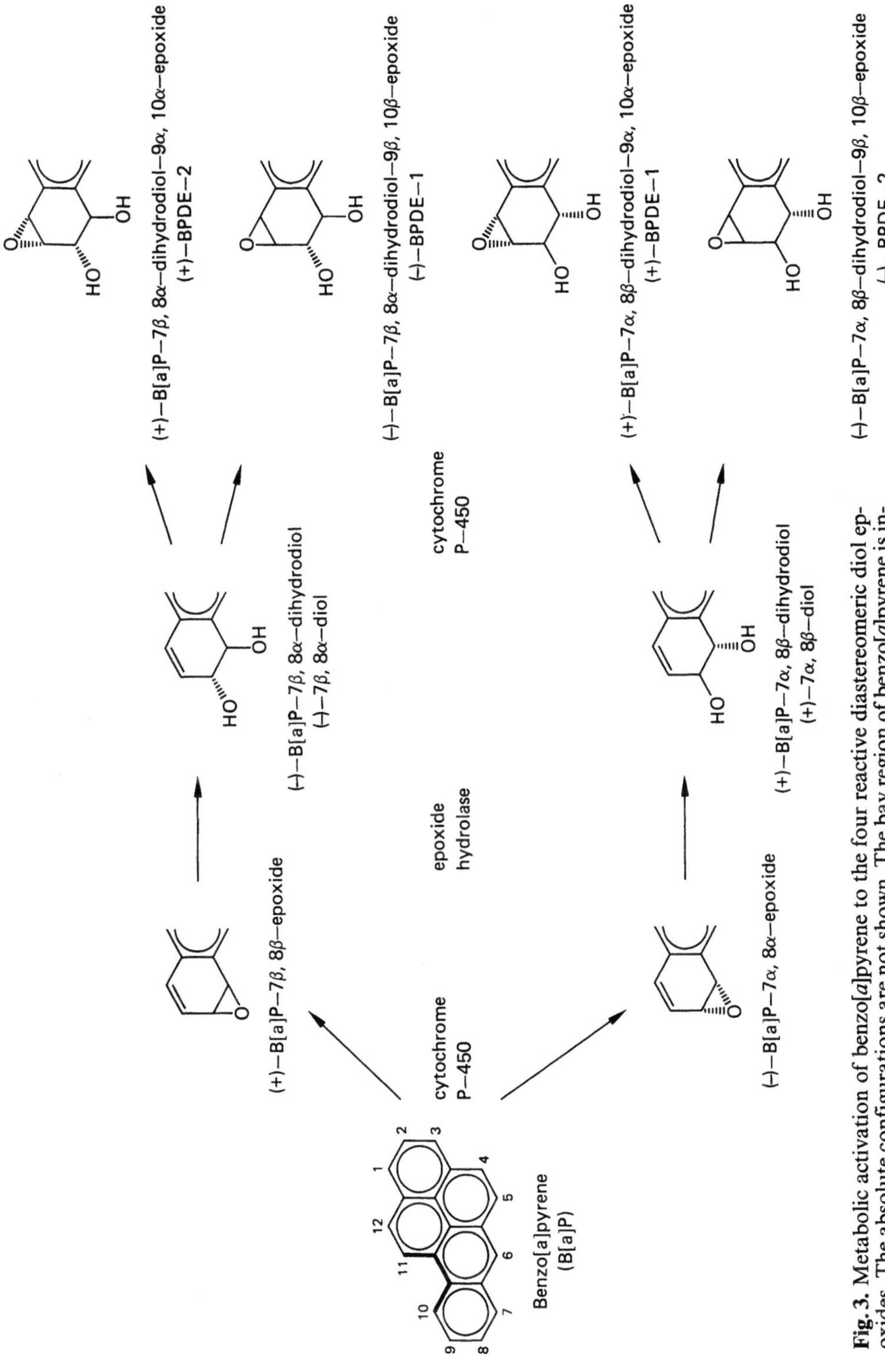

Fig. 3. Metabolic activation of benzo[*a*]pyrene to the four reactive diastereomeric diol epoxides. The absolute configurations are not shown. The bay region of benzo[*a*]pyrene is indicated by *bold lines*

of drugs in skin is extremely low. Therefore, skin does not substantially contribute to the metabolism of drugs systemically applied. But it cannot be excluded that after induction of these enzymes skin becomes involved in systemic drug metabolism.

On the other hand, after topical application drug metabolism is highly relevant. For example, it has been shown that after topical application of nitroglycerin to the skin the percutaneous first-pass effect is about 20% (WESTER et al. 1983). Depending on the drug used this effect can be higher or lower, decreasing or increasing the bioavailability of the drug. When a pro-drug is applied the effective metabolite may be released in the skin, exerting pharmacologic effects within the skin or elsewhere in the body.

On the other hand, it has been conclusively demonstrated that drug metabolism in skin is responsible for tumorigenicity of a number of chemical carcinogens. This is most important, because in the human population drug-metabolizing capacities in skin show great genetic variations, including inducibility (NEBERT 1979).

Acknowledgement. The excellent secretarial help of Mrs. Ch. Eckelmann during the preparation of this manuscript is gratefully acknowledged.

References

Akin FJ, Norred WP (1976) Factors affecting measurement of aryl hydrocarbon hydroxylase activity in mouse skin. J Invest Dermatol 67:709–712

Akin FJ, Chamberlain WJ, Chortyk OT (1975) Mouse skin tumorigenesis and induction of aryl hydrocarbon hydroxylase by tobacco smoke fraction. J Natl Cancer Inst 54:907–912

Alexandrov K, Becker M, Frayssinet C, Dubowska W, Guerry R (1982) Persistence of benzo[*a*]pyrenediolepoxide DNA adduct in mouse skin. Cancer Lett 16:247–251

Alvares AP, Kappas A, Levin W, Conney AH (1973) Inducibility of benzo[*a*]pyrene hydroxylase in human skin by polycyclic hydrocarbons. Clin Pharmacol Ther 14:30–40

Alvares AP, Bickers DR, Kappas A (1974) Induction of drug-metabolizing enzymes and aryl hydrocarbon hydroxylase by microscope immersion oil. Life Sci 14:853–860

Anders MW (1985) Bioactivation of foreign compounds. Academic, Orlando

Ashurst SW, Cohen GM (1981 a) In vivo formation of benzo[*a*]pyrenediolepoxide-deoxyadenosine adducts in the skin of mice susceptible to benzo[*a*]pyrene-induced carcinogenesis. Int J Cancer 27:357–364

Ashurst SW, Cohen GM (1981 b) The formation and persistence of benzo[*a*]pyrene metabolite-deoxyribonucleoside adducts in rat skin in vivo. Int J Cancer 28:387–392

Ashurst SW, Cohen GM (1982) The formation of benzo[*a*]pyrene-deoxyribonucleoside adducts in vivo and in vitro. Carcinogenesis 3:267–273

Ashurst SW, Cohen GM, Nesnow S, DiGiovanni J, Slaga TJ (1983) Formation of benzo[*a*]pyrene/DNA adducts and their relationship to tumor initiation in mouse epidermis. Cancer Res 43:1024–1029

Asokan P, Das M, Rosenkranz HS, Bickers DR, Mukhtar H (1985) Topically applied nitropyrenes are potent inducers of cutaneous and hepatic monooxygenases. Biochem Biophys Res Commun 129:134–140

Asokan P, Das M, Bik DP, Howard PC, McCoy GD, Rosenkranz HS, Bickers DR, Mukhtar H (1986) Comparative effects of topically applied nitrated arenes and their nonnitrated parent arenes on cutaneous and hepatic drug and carcinogen metabolism in neonatal rats. Toxicol Appl Pharmacol 86:33–43

Baars AJ, Mukhtar H, Zoetemelk CEM, Jansen M, Breimer DD (1981) Glutathione-*S*-transferase activity in rat and human tissues and organs. Comp Biochem Physiol [C]70:285–288

Baer-Dubowska W, Alexandrov K (1981) The binding of benzo[*a*]pyrene to mouse and rat skin DNA. Cancer Lett 13:47–52
Baron J, Kawabata TT, Redick JA, Knapp SA, Wick DG, Wallace RB, Jakoby WB, Guengerich FP (1983) Localization of carcinogen-metabolizing enzymes in human and animal tissues. In: Rydström J, Montelius J, Bengtsson M (eds) Extrahepatic drug metabolism and chemical carcinogenesis. Elsevier, Amsterdam, pp 73–88
Baron J, Voigt JM, Whitter TB, Kawabata TT, Knapp SA, Guengerich FP, Jakoby WB (1986) Identification of intratissue sites for xenobiotic activation and detoxication. In: Kocsis JJ, Jollow DJ, Witmer CM, Nelson JO, Snyder R (eds) Biological reactive intermediates. III. Mechanisms of action in animal models and human disease. Plenum, New York, pp 119–144
Benedict WF, Considine N, Nebert DW (1973) Genetic differences in aryl hydrocarbon hydroxylase induction and benzo[*a*]pyrene-produced tumorigenesis in the mouse. Mol Pharmacol 9:266–277
Berry DL, Bracken WR, Slaga TJ (1977) Benzo[*a*]pyrene metabolism in mouse epidermis: analysis by high pressure liquid chromatography and DNA binding. Chem Biol Interact 18:129–142
Bickers DR (1983) Drug, carcinogen and steroid hormone metabolism in the skin. In: Goldsmith LA (ed) Biochemistry and physiology of the skin. Oxford University Press, New York, pp 1169–1186
Bickers DR, Kappas A (1978) Human skin aryl hydrocarbon hydroxylase – induction by coal tar. J Clin Invest 62:1061–1068
Bickers DR, Kappas A (1980) The skin as a site of chemical metabolism. In: Gram TE, Vessel E, Garattini S (eds) Extrahepatic metabolism of drugs and foreign compounds. Spectrum, Jamaica, pp 295–318 (Monographs in pharmacology and physiology, vol V)
Bickers DR, Kappas A, Alvares AP (1974) Differences in inducibility of cutaneous and hepatic drug metabolizing enzymes and cytochrome P-450 by polychlorinated biphenyls and 1,1,1-trichloro-2,2-bis(*p*-chlorophenyl)ethane (DDT). J Pharmacol Exp Ther 188:300–309
Bickers DR, Eiseman J, Kappas A, Alvares AP (1975) Microscope immersion oils: effects of skin application on cutaneous and hepatic drug metabolizing enzymes. Biochem Pharmacol 24:779–783
Bickers DR, Dixit R, Mukhtar H (1982a) Hematoporphyrin photosensitization of epidermal microsomes results in destruction of cytochrome P-450 and in decreased monoxygenase activities and heme content. Biochem Biophys Res Commun 108:1032–1039
Bickers DR, Dutta-Choudhury T, Mukhtar H (1982b) Epidermis: a site of drug metabolism in neonatal rat skin. Studies on cytochrome P-450 content and mixed-function oxidase and epoxide hydrolase activity. Mol Pharmacol 21:239–247
Bickers DR, Marcelo L, Dutta-Choudhury T, Mukhtar H (1982c) Studies on microsomal cytochrome P-450 monooxygenases and epoxide hydrolase in cultured keratinocytes and intact epidermis from BALB/c mice. J Pharmacol Exp Ther 223:163–168
Bickers DR, Mukhtar H, Molica SJ, Pathak MA (1982d) The effect of psoralens on hepatic and cutaneous drug metabolizing enzymes and cytochrome P-450. J Invest Dermatol 79:201–205
Bickers DR, Wroblewski D, Dutta-Choudhury T, Mukhtar H (1982e) Induction of neonatal rat skin and liver aryl hydrocarbon hydroxylase by coal tar and its constituents. J Invest Dermatol 78:227–229
Bickers DR, Mukhtar H, Yang SK (1983) Cutaneous metabolism of benzo[*a*]pyrene: comparative studies in C57BL/6N and DBA/2N mice and neonatal Sprague-Dawley rats. Chem Biol Interact 43:263–270
Bickers DR, Mukhtar H, Dutta-Choudhury T, Marcello CL, Voorhees JJ (1984) Aryl hydrocarbon hydroxylase, epoxide hydrolase, and benzo[*a*]pyrene metabolism in human epidermis: comparative studies in normal subjects and patients with psoriasis. J Invest Dermatol 83:51–56
Bickers DR, Das M, Mukhtar H (1986) Pharmacological modification of epidermal detoxification systems. Br J Dermatol 115(Suppl 31):9–16

Bigger CAH, Tomaszewski JE, Dipple A (1978) Differences between products of binding of 7,12-dimethylbenz[*a*]anthracene to DNA in mouse skin and in a rat liver microsomal system. Biochem Biophys Res Commun 80:229–235

Bigger CAH, Sawicki JT, Blake DM, Raymond LG, Dipple A (1983) Products of binding of 7,12-dimethylbenz[*a*]anthracene to DNA in mouse skin. Cancer Res 43:5647–5651

Bock KW, von Clausbruch UC, Kaufmann R, Lilienblum W, Oesch F, Pfeil H, Platt KL (1980) Functional heterogeneity of UDP-glucuronyltransferase in rat tissues. Biochem Pharmacol 29:495–500

Bock KW, Lilienblum W, Pfeil H (1982) Functional heterogeneity of UDP-glucuronosyltransferase activities in C57BL/6 and DBA/2 mice. Biochem Pharmacol 31:1273–1277

Bock KW, Burchell B, Dutton GJ, Hänninen O, Mulder GJ, Owens IS, Siest G, Tephly TR (1983) UDP-glucuronosyltransferase activities. Biochem Pharmacol 32:953–955

Bowden GT, Slaga TJ, Shapas BG, Boutwell RK (1974) The role of aryl hydrocarbon hydroxylase in skin tumor initiation by 7,12-dimethylbenz[*a*]anthracene and 1,2,5,6-dibenzanthracene using DNA binding and thymidine-^{3}H incorporation into DNA as criteria. Cancer Res 34:2634–2642

Bresnick E, McDonald TF, Yagi H, Jerina DM, Levin W, Wood AW, Conney AH (1977) Epidermal hyperplasia after topical application of benzo[*a*]pyrene, benzo[*a*]pyrene diol epoxides and other metabolites. Cancer Res 37:984–990

Briggs MM; Briggs M (1973) Induction by topical corticosteroids of skin enzymes metabolizing carcinogenic hydrocarbons. Br J Dermatol 88:75–81

Brookes P (1979) The binding to mouse skin DNA of benzo[*a*]pyrene, its 7,8-diol and 7,8-diol-9,10-epoxides in relation to the tumorigenicity of these compounds. Cancer Lett 6:285–289

Brookes P, Duncan ME (1971) Carcinogenic hydrocarbons and human cells in culture. Nature 234:40–43

Buening MK, Levin W, Wood AW, Chang RL, Lehr RE, Taylor CW, Yagi H, Jerina DM, Conney AH (1980) Tumorigenic activity of benzo[*e*]pyrene derivatives on mouse skin and newborn mice. Cancer Res 40:203–206

Bürki K, Liebelt AG, Bresnick E (1973) Induction of aryl hydrocarbon hydroxylase in mouse tissues from a high and low cancer strain and their F_1 hybrids. J Natl Cancer Inst 50:369–380

Bürki K, Stoming TA, Bresnick E (1974) Effects of an epoxide hydrase inhibitor on in vitro binding of polycyclic hydrocarbons to DNA and on skin carcinogenesis. J Natl Cancer Inst 52:785–788

Bundgaard H, Hoelgaard A, Møllgaard B (1983) Leaching of hydrolytic enzymes from human skin in cutaneous permeation studies as determined with metronidazole and 5-fluorouracil pro-dugs. Int J Pharm (Amst) 15:285–292

Caldwell J, Jakoby WB (1983) Biological basis of detoxication. Academic, New York

Camus AM, Pyerin WG, Grover PL, Sims P, Malaveille C, Bartsch H (1980) Mutagenicity of benzo[*a*]pyrene 7,8-dihydrodiol and 7,12-dimethylbenz[*a*]anthracene 3,4-dihydrodiol in *S. typhimurium* mediated by microsomes from rat liver and mouse skin. Chem Biol Interact 32:257–265

Cavalieri EL, Rogan EG (1984) One-electron and two-electron oxidation in aromatic hydrocarbon carcinogenesis. In: Pryor WA (ed) Free radicals in biology, vol VI. Academic, Orlando, pp 323–369

Chan PC, Okamoto T, Wynder EL (1972) The possible role of riboflavin deficiency in epithelial neoplasia. III. Induction of microsomal aryl hydrocarbon hydroxylase. J Natl Cancer Inst 48:1341–1345

Chang RL, Levin W, Wood AW, Lehr RE, Kumar S, Yagi H, Jerina DM, Conney AH (1982) Tumorigenicity of bay-region diol-epoxides and other benzo-ring derivatives of dibenzo[*a,h*]pyrene and dibenzo[*a,i*]pyrene in mouse skin and in newborn mice. Cancer Res 42:25–29

Chang RL, Levin W, Wood AW, Yagi H, Tada M, Vyas KP, Jerina DM, Conney AH (1983) Tumorigenicity of enantiomers of chrysene 1,2-dihydrodiol and of the diastereomeric bay-region chrysene 1,2-diol-3,4-epoxides on mouse skin and in newborn mice. Cancer Res 43:192–196

Cheung YW, Po ALW, Irvin WJ (1985) Cutaneous biotransformation as a parameter in the modulation of the activity of topical corticosteroids. Int J Pharm (Amst) 26:175–189

Chouroulinkov I, Gentil A, Grover PL, Sims P (1976) Tumour-initiating activities on mouse skin of dihydrodiols derived from benzo[*a*]pyrene. Br J Cancer 34:523–532

Chouroulinkov I, Gentil A, Tierney B, Grover PL, Sims P (1977) The metabolic activation of 7-methylbenz[*a*]anthracene in mouse skin: high tumour-initiating activity of the 3,4-dihydrodiol. Cancer Lett 3:247–253

Chouroulinkov I, Gentil A, Tierney B, Grover PL, Sims P (1979a) Biological activities of dihydrodiols derived from two polycyclic hydrocarbons in rodent test systems. Br J Cancer 39:376–382

Chouroulinkov I, Gentil A, Tierney B, Grover PL, Sims P (1979b) The initiation of tumors on mouse skin by dihydrodiols derived from 7,12-dimethylbenz[*a*]anthracene and 3-methylcholanthrene. Int J Cancer 24:455–460

Chuang AHL, Mukhtar H, Bresnick E (1978) Effects of diethyl maleate on aryl hydrocarbon hydroxylase and on 3-methylcholanthrene-induced skin tumorigenesis in rats and mice. J Natl Cancer Inst 60:321–325

Cohen GM, Bracken WM, Iyer RP, Berry DL, Selkirk JK, Slaga TJ (1979) Anticarcinogenic effects of 2,3,7,8-tetrachlorodibenzo-*p*-dioxin on benzo[*a*]pyrene and 7,12-dimethylbenz[*a*]anthracene tumor initiation and its relationship to DNA binding. Cancer Res 39:4027–4033

Conney AH (1982) Induction of microsomal enzymes by foreign chemicals and carcinogenesis by polycyclic aromatic hydrocarbons. Cancer Res 42:4875–4917

Coomes MW, Norling AH, Pohl RJ, Müller D, Fouts JR (1983) Foreign compound metabolism by isolated skin cells from the hairless mouse. J Pharmacol Exp Ther 225:770–777

Coomes MW, Sparks RW, Fouts JR (1984) Oxidation of 7-ethoxycoumarin and conjugation of umbelliferone by intact, viable epidermal cells from the hairless mouse. J Invest Dermatol 82:598–601

Cooper CS, Ribeiro O, Hewer A, Walsh C, Pal K, Grover PL, Sims P (1980) The involvement of a "bay-region" and a non-"bay-region" diol-epoxide in the metabolic activation of benz[*a*]anthracene in mouse skin and in hamster embryo cells. Carcinogenesis 1:233–243

Damen FJM, Mier PD (1982) Cytochrome P-450-dependent-*O*-dealkylase activity in mammalian skin. Br J Pharmacol 75:123–127

Das M, Bickers DR, Santella RM, Mukhtar H (1985) Altered patterns of cutaneous xenobiotic metabolism in UVB-induced squamous cell carcinoma in SKH-1 hairless mice. J Invest Dermatol 84:532–536

Das M, Bickers DR, Mukhtar H (1986a) Epidermis: the major site of cutaneous benzo[*a*]-pyrene and benzo[*a*]pyrene 7,8-diol metabolism in neonatal BALB/c mice. Drug Metab Dispos 14:637–642

Das M, Mukhtar H, Del Tito BJ, Marcelo CL, Bickers DR (1986b) Clotrimazole, an inhibitor of benzo[*a*]pyrene metabolism and its subsequent glucuronidation, sulfation, and macromolecular binding in BALB/c mouse cultured keratinocytes. J Invest Dermatol 87:4–10

Das M, Asokan P, Don PSC; Krueger GG, Bickers DR, Mukhtar H (1986c) Carcinogen metabolism in human skin grafted onto athymic nude mice: a model system for the study of human skin carcinogenesis. Biochem Biophys Res Commun 138:33–39

Daudel P, Duquese M, Vigny P, Grover PL, Sims P (1975) Fluorescene spectral evidence that benzo[*a*]pyrene-DNA products in mouse skin arise from diol-epoxides. FEBS Lett 57:250–253

Del Tito BJ, Mukhtar H, Bickers DR (1984) In vivo metabolism of topically applied benzo[*a*]pyrene-4,5-oxide in neonatal rat skin. J Invest Dermatol 82:378–380

DiGiovanni J, Slaga TJ, Berry DL, Juchau MR (1977) Metabolism of 7,12-dimethylbenz-[*a*]anthracene in mouse skin homogenates analyzed with high-pressure liquid chromatography. Drug Metab Dispos 5:295–301

DiGiovanni J, Viaje A, Fischer S, Slaga TJ, Boutwell RK (1980) Biotransformation of 7,12-dimethylbenz[*a*]anthracene by mouse epidermal cells in culture. Carcinogenesis 1:41–49
DiGiovanni J, Miller DR, Singer JM; Viaje A, Slaga TJ (1982) Benzo[*a*]pyrene metabolism in primary cultures of mouse epidermal cells and untransformed epidermal cell lines. Cancer Res 42:2579–2586
DiGiovanni J, Nebzydoski AP, Decina PC (1983a) Formation of 7-hydroxymethyl-12-methylbenz[*a*]anthracene-DNA adducts from 7,12-dimethylbenz[*a*]anthracene in mouse epidermis. Cancer Res 43:4221–4226
DiGiovanni J, Sina JF, Ashurst SW, Singer JM, Diamond L (1983b) Benzo[*a*]pyrene and 7,12-dimethylbenz[*a*]anthracene metabolism and DNA adduct formation in primary cultures of hamster epidermal cells. Cancer Res 43:163–170
DiGiovanni J, Fisher EP, Aalfs KK, Prichett WP (1985) Covalent binding of 7,12-dimethylbenz[*a*]anthracene and 10-fluoro-7,12-dimethylbenz[*a*]anthracene to mouse epidermal DNA and its relationship to tumor-initiating activity. Cancer Res 45:591–597
Dutton GJ, Stevenson IH (1962) The stimulation by 3,4-benzpyrene of glucuronide synthesis in the skin. Biochim Biophys Acta 58:633–634
Eling T, Curtis J, Battista J, Marnett LJ (1986) Oxidation of (+)-7,8-dihydroxy-7,8-dihydrobenzo[*a*]pyrene by mouse keratinocytes: evidence for peroxyl radical- and monoxygenase-dependent metabolism. Carcinogenesis 7:1957–1963
Finnen MJ, Shuster S (1985) Phase 1 and phase 2 drug metabolism in isolated epidermal cells from adult hairless mice and in whole human hair follicles. Biochem Pharmacol 34:3571–3575
Finnen MJ, Shuster S, Lawrence CM, Rawlins MD (1983) Aryl hydrocarbon hydroxylase activity and psoriasis. Biochem Pharmacol 32:1707–1711
Finnen MJ, Herdman ML, Shuster S (1984a) Induction of drug metabolizing enzymes in the skin by topical steroids. J Steroid Biochem 20:1169–1173
Finnen MJ, Lawrence CM, Shuster S (1984b) Human skin aryl hydrocarbon hydroxylase. Br J Dermatol 110:339–342
Finnen MJ, Herdman ML, Shuster S (1985) Distribution and sub-cellular localization of drug metabolizing enzymes in the skin. Br J Dermatol 113:713–721
Fox HC, Selkirk JK, Price FM, Croy RG, Sanford KK, Cottler-Fox M (1975) Metabolism of benzo[*a*]pyrene by human epithelial cells in vitro. Cancer Res 35:3551–3557
Gelboin HV (1980) Benzo[*a*]pyrene metabolism, activation, and carcinogenesis: role and regulation of mixed-function oxidases and related enzymes. Physiol Rev 60:1107–1166
Gelboin HV, Wiebel FJ (1971) Studies on the mechanism of aryl hydrocarbon hydroxylase induction and its role in cytotoxicity and tumorigenicity. Ann NY Acad Sci 179:529–547
Gelboin HV, Wiebel F, Diamond L (1970) Dimethylbenzanthracene tumorigenesis and aryl hydrocarbon hydroxylase in mouse skin: inhibition by 7,8-benzoflavone. Science 170:169–171
Gelboin HV, Kinoshita N, Wiebel FJ (1972) Microsomal hydroxylases: induction and role in polycyclic hydrocarbon carcinogenesis and toxicity. Fed Proc 31:1298–1309
Gielen JE, Goujon FM, Nebert DW (1972) Genetic regulation of aryl hydrocarbon hydroxylase induction. II. Simple Mendelian expression in mouse tissue in vivo. J Biol Chem 247:1125–1137
Goerz G, Merk H (1987) Animal models for cutaneous drug-metabolizing enzymes. In: Maibach HI, Lowe NJ (eds) Models in dermatology, vol III. Karger, Basel, pp 93–105
Goerz G, Berger H, Tsambaos D (1981) A new animal model for the study of cutaneous drug-metabolizing enzymes. Arch Dermatol Res 270:367–369
Goerz G, Merk H, Bolsen K, Tsambaos D, Berger H (1983) Influence of chronic UV-light exposure on hepatic and cutaneous monoxygenases. Experientia 39:385–386
Gomez EC, Michaelover J, Frost P (1977) Cutaneous β-glucuronidase: cleavage of mycophenolic acid by preparations of mouse skin. Br J Dermatol 97:303–306

Gram TE, Vessel E, Garattini S (1980) Extrahepatic metabolism of drugs and foreign compounds. Spectrum, Jamaica New York (Monographs in pharmacology and physiology, vol V)
Grover PL, Hewer A, Pal K, Sims P (1976) The involvement of a diol-epoxide in the metabolic activation of benzo[*a*]pyrene in human bronchial mucosa and in mouse skin. Int J Cancer 18:1–6
Harper KH, Calcutt G (1960) Conjugation of 3,4-benzpyrenols in mouse skin. Nature 186:80–81
Hecht SS, Rivenson A, Hoffmann D (1980) Tumor-initiating activity of dihydrodiols formed metabolically from 5-methylchrysene. Cancer Res 40:1396–1399
Heimann R, Rice RH (1983) Polycyclic aromatic hydrocarbon toxicity and induction of metabolism in cultivated esophageal and epidermal keratinocytes. Cancer Res 43:4856–4862
Herz-Hübner U, Täuber U (1977) Metabolisierung von Fluocortin-butylester in der Haut von Meerschweinchen und Mensch. Arzneim Forsch 27:2226–2229
Hosomi J, Nemeto N, Kuroki T (1982) Metabolism of benzo[*a*]pyrene and other carcinogens in mouse epidermal keratinocytes in culture. Gann 73:879–886
Hosomi J, Nemeto N, Kuroki T (1983) Genetic differences in metabolism of benzo[*a*]pyrene in cultured epidermal and dermal cells of responsive and nonresponsive mice. Cancer Res 43:3643–3648
Huberman E, Sachs L (1973) Metabolism of the carcinogenic hydrocarbon benzo[*a*]pyrene in human fibroblast and epithelial cells. Int J Cancer 11:412–418
Hudson LG, Shaikh R, Toscano WA, Greenlee WF (1983) Induction of 7-ethoxycoumarin *O*-deethylase activity in cultured human epithelial cells by 2,3,7,8-tetrachlorodibenzo-*p*-dioxin (TCDD): evidence for TCDD receptor. Biochem Biophys Res Commun 115:611–617
Hukkelhoven MWAC, Irmscher G, Fusenig NE (1980) Metabolism of 7,12-dimethylbenzanthracene (DMBA) by mouse skin keratinocytes, fibroblasts, and carcinoma cells in culture. Arch Toxicol 44:181–195
Hukkelhoven MWAC, Vromans E, Vermorken AJM, Bleomendal H (1981) Enzyme induction in cultured human hair follicle cells. Mol Biol Rep 8:25–28
Hukkelhoven MWAC, Vromans E, Vermorken AJM, Bloemendal H (1982a) Formation of dihydrodiol metabolites of benzo[*a*]pyrene in cultured human and murine skin cells. Anticancer Res 2:89–94
Hukkelhoven MWAC, Vromans EWM, Vermorken AJM, Bloemendal H (1982b) A sensitive fluorometric assay for epoxide hydratase. FEBS Lett 144:104–108
Hukkelhoven MWAC, Dijkstra AC, Vermorken AJM (1983a) Rapid high-performance liquid chromatographic method for detection of interindividual differences in carcinogen metabolism. J Chromatogr 276:189–196
Hukkelhoven MWAC; Dijkstra AC, Vermorken AJM (1983b) Human hair follicles and cultured hair follicle keratinocytes as indicators for individual differences in carcinogen metabolism. Arch Toxicol 53:265–274
Hukkelhoven MWAC; Vromans LWM, van Pelt FNAM, Keulers RAC, Vermorken AJM (1984) In vivo induction of aryl hydrocarbon hydroxylase in human scalp hair follicles by topical application of a commercial coal tar preparation. Cancer Lett 23:135–143
Hukkelhoven MWAC, Bronkhorst AM, Vermorken AJM (1985) Covalent binding of BP-metabolites to DNA of cultured human hair follicle keratinocytes. Arch Toxicol 57:6–12
Irmscher G, Fusenig NE (1980) Metabolism of 7,12-dimethylbenzanthracene (DMBA) by mouse skin keratinocytes, fibroblasts, and carcinoma cells in culture. Arch Toxicol 44:181–195
Iversen OH, Digernes V (1981) Increased epidermal DT-diaphorase activity induced by some carcinogens. Virchows Arch [B] 36:133–138
Jakoby WB (1980) Enzymatic basis of detoxication, vol I, II. Academic, New York
Jakoby WB, Bend JR, Caldwell J (1982) Metabolic basis of detoxication – metabolism of functional groups. Academic, New York

Juchau MR, Namkung MJ, Berry DL, Zachariah PK (1975) Oxidative biotransformation of 2-acetylaminofluorene in fetal and placental tissues of humans and monkeys – correlations with aryl hydrocarbon hydroxylase activities. Drug Metab Dispos 3:494–501

Kaneko M, Cerutti PA (1982) Excision of benzo[*a*]pyrene diol epoxide I adducts from nucleosomal DNA of confluent normal human fibroblasts. Chem Biol Interact 38:261–274

Kao J, Hall J, Shugart LR, Holland JM (1984) An in vitro approach to studying cutaneous metabolism and disposition of topically applied xenobiotics. Toxicol Appl Pharmacol 75:289–298

Kao J, Patterson FJ, Hall J (1985) Skin penetration and metabolism of topically applied chemicals in six mammalian species including man: an in vitro study with benzo[*a*]-pyrene and testosterone. Toxicol Appl Pharmacol 81:502–516

Kinoshita N, Gelboin HV (1972) The role of aryl hydrocarbon hydroxylase in 7,12-dimethylbenz[*a*]anthracene skin tumorigenesis: on the mechanism of 7,8-benzoflavone inhibition of tumorigenesis. Cancer Res 32:1329–1339

Koreeda M, Moore PD, Wislocki PG, Levin W, Conney AH, Yagi H, Jerina DM (1978) Binding of benzo[*a*]pyrene 7,8-diol-9,10-epoxides to DNA, RNA, and protein of mouse skin occurs with high stereoselectivity. Science 199:778–781

Krauss RS, Eling TE (1984) Arachidonic acid-dependent cooxidation. Biochem Pharmacol 33:3319–3324

Kumar S, Antony M, Mehrotra NK (1982) Induction of benzo[*a*]pyrene hydroxylase in skin and liver by cutaneous application of jute batching oil. Toxicology 23:347–352

Kuroki T, Nemoto N, Kitano Y (1980) Metabolism of benzo[*a*]pyrene in human epidermal keratinocytes in culture. Carcinogenesis 1:559–565

Kuroki T, Hosomi J, Nemoto N (1981) Variation in metabolism of benzo[*a*]pyrene in epidermal keratinocytes and dermal fibroblasts of humans and mice. Gann Monogr 27:59–71

Kuroki T, Hosomi J, Munakata K, Onizuka T, Terauchi M, Nemoto N (1982) Metabolism of benzo[*a*]pyrene in epidermal keratinocytes and dermal fibroblasts of human and mice with reference to variation among species, individuals, and cell types. Cancer Res 42:1859–1865

Laaksonen AM, Mäntyjärvi RA, Hänninen OOP (1983) Fibroblast cultures of nude mouse skin as targets for transformation by chemical carcinogens. Med Biol 61:59–64

Lawrence CM, Finnen MJ, Shuster S (1984) Effect of coal tar on cutaneous aryl hydrocarbon hydroxylase induction and anthralin irritancy. Br J Dermatol 110:671–675

Legraverend C, Mansour B, Nebert DW, Holland JM (1980) Genetic differences in benzo[*a*]pyrene-initiated tumorigenesis in mouse skin. Pharmacology 20:242–255

Lesca P (1983) Modulating effects of 2,3,7,8-tetrachlorodibenzo-*p*-dioxin on skin carcinogenesis initiated by 7,12-dimethylbenz[*a*]anthracene in CF-1 Swiss mice. In: Rydström J, Montelius J, Bengtsson M (eds) Extrahepatic drug metabolism and chemical carcinogenesis. Elsevier, Amsterdam, pp 589–590

Levin W, Wood AW, Yagi H, Dansette PM, Jerina DM, Conney AH (1976a) Carcinogenicity of benzo[*a*]pyrene 4,5-, 7,8-, and 9,10-oxides on mouse skin. Proc Natl Acad Sci USA 73:243–247

Levin W, Wood AW, Yagi H, Jerina DM, Conney AH (1976b) (+)-*trans*-7,8-dihydroxy-7,8-dihydrobenzo[*a*]pyrene: a potent skin carcinogen when applied topically to mice. Proc Natl Acad Sci USA 73:3867–3871

Levin W, Wood AW, Chang RL, Slaga TJ, Yagi H, Jerina DM, Conney AH (1977a) Marked differnces in the tumor-initiating activity of optically pure (+)- and (−)-*trans*-7,8-dihydroxy-7,8-dihydrobenzo[*a*]pyrene on mouse skin. Cancer Res 37:2721–2725

Levin W, Wood AW, Wislocki PG, Kapitulnik J, Yagi H, Jerina DM, Conney AH (1977b) Carcinogenicity of benzo-ring-derivatives of benzo[*a*]pyrene on mouse skin. Cancer Res 37:3356–3361

Levin W, Thakker DR, Wood AW, Chang RL, Lehr RE,Jerina DM, Conney AH (1978a) Evidence that benzo[*a*]anthracene 3,4-diol-1,2-epoxide is an ultimate carcinogen on mouse skin. Cancer Res 38:1705–1710

Levin W, Wood AW, Chang RL, Yagi H, Mah HD, Jerina DM, Conney AH (1978b) Evidence for bay region activation of chrysene 1,2-dihydrodiol to an ultimate carcinogen. Cancer Res 38:1831–1834
Levin W, Buening MK, Wood AW, Chang RL, Thakker DR, Jerina DM, Conney AH (1979) Tumorigenic activity of 3-methylcholanthrene metabolites on mouse skin and in newborn mice. Cancer Res 39:3549–3553
Levin W, Buening MK, Wood AW, Chang RL, Kedzierski B, Thakker DR, Boyd DR, Gadaginamath GS, Armstrong RN, Yagi H, Karle JM, Slaga TJ, Jerina DM, Conney AH (1980a) An enantiomeric interaction in the metabolism and tumorigenicity of (+)- and (−)-benzo[*a*]pyrene 7,8-oxide. J Biol Chem 255:9067–9074
Levin W, Wood AW, Chang RL, Ittah Y, Croisy-Delcey M, Yagi H, Jerina DM, Conney AH (1980b) Exceptionally high tumor-initiating activity of benzo[*c*]phenanthrene bay-region diol-epoxides in mouse skin. Cancer Res 40:3910–3914
Lilienblum W, Irmscher G, Fusenig NE, Bock KW (1986) Induction of UDP-glucuronyltransferase and arylhydrocarbon hydroxylase activity on mouse skin and in normal and transformed skin cells in culture. Biochem Pharmacol 35:1517–1520
Loftsson T, Bodor N (1981) Improved delivery through biological membranes X: percutaneous absorption and metabolism of methylsulfinylmethyl 2-acetoxybenzoate and related aspirin prodrugs. J Pharm Sci 70:756–758
MacNicoll AD, Grover PL, Sims P (1980) The metabolism of a series of polycyclic hydrocarbons by mouse skin maintained in short-term organ culture. Chem Biol Interact 29:169–188
Manil L, van Cantfort J, Lapiere CM, Gielen JE (1981) Significant variation in mouse-skin aryl hydrocarbon hydroxylase inducibility as a function of the hair growth cycle. Br J Cancer 43:210–221
Melikian AA, La Voie EJ, Hecht SS, Hoffmann D (1982) Influence of a bay-region methyl group on formation of 5-methylchrysene dihydrodiol epoxide: DNA adducts in mouse skin. Cancer Res 42:1239–1242
Melikian AA, La Voie EJ, Hecht SS, Hoffmann D (1983) 5-Methylchrysene metabolism in mouse epidermis in vivo, diol epoxide-DNA adduct persistence, and diol epoxide reactivity with DNA as potential factors influencing the predominance of 5-methylchrysene-1,2-diol-3,4-epoxide-DNA adducts in mouse epidermis. Carcinogenesis 4:843–849
Melikian AA, Leszczynska JM, Hecht SS, Hoffmann D (1986) Effects of the co-carcinogen catechol on benzo[*a*]pyrene metabolism and DNA adduct formation in mouse skin. Carcinogenesis 7:9–15
Merk H, Rumpf M, Bolsen K, Wirth G, Goerz G (1984) Inducibility of arylhydrocarbon-hydroxylase activity in human hair follicles by topical application of liquor carbonis detergens (coal tar). Br J Dermatol 111:279–284
Mier PD, van den Hurk JJMA (1975) Lysosomal hydrolases of the epidermis. I. Glycosidases. Br J Dermatol 93:1–10
Møllgaard B, Hoelgaard A, Bundgaard H (1982) Pro-drugs as drug delivery systems XXIII. Improved dermal delivery of 5-fluorouracil through human skin via *N*-acyloxymethyl pro-drug derivatives. Int J Pharm (Amst) 12:153–162
Moloney SJ, Bridges JW, Fromson JM (1982a) UDP-glucuronosyltransferase activity in rat- and hairless mouse skin-microsomes. Xenobiotica 12:481–487
Moloney SJ, Fromson JM, Bridges JW (1982b) The metabolism of 7-ethoxycoumarin and 7-hydroxycoumarin by rat and hairless mouse skin strips. Biochem Pharmacol 31:4005–4009
Moloney SJ, Fromson JM, Bridges JW (1982c) Cytochrome P-450 dependent deethylase activity in rat and hairless mouse skin microsomes. Biochem Pharmacol 31:4011–4018
Mukhtar H, Bickers DR (1981a) Comparative activity of the mixed-function oxidases, epoxide hydratase, and glutathione-*S*-transferase in liver and skin of the neonatal rat. Drug Metab Dispos 9:311–314
Mukhtar H, Bickers DR (1981b) Aminopyrine-*N*-demethylase activity in neonatal rat skin. Biochem Pharmacol 30:3257–3260

Mukhtar H, Bickers DR (1982) Evidence that coal tar is a mixed inducer of microsomal drug-metabolizing enzymes. Toxicol Lett 11:221–227
Mukhtar H, Bickers DR (1983) Age-related changes in benzo[*a*]pyrene metabolism and epoxide-metabolizing enzyme activities in rat skin. Drug Metab Dispos 11:562–567
Mukhtar H, Bresnick E (1976) Glutathione-*S*-epoxidetransferase in mouse skin and human foreskin. J Invest Dermatol 66:161–164
Mukhtar H, Dixit R, Seth PK (1981) Reduction in cutaneous and hepatic glutathione contents, glutathione-*S*-transferase and aryl hydrocarbon hydroxylase activities following topical application of acrylamide to mouse. Toxicol Lett 9:153–156
Mukhtar H, Link CM, Cherniack E, Kushiner DM, Bickers DR (1982) Effect of topical application of defined hydroxylase and 7-ethoxycoumarin deethylase activities. Toxicol Appl Pharmacol 64:541–549
Mukhtar H, Del Tito BJ, Marcelo CL, Das M, Bickers DR (1984a) Ellagic acid: a potent naturally occurring inhibitor of benzo[*a*]pyrene metabolism and its subsequent glucuronidation, sulfation and covalent binding to DNA in cultured BALB/c mouse keratinocytes. Carcinogenesis 5:1565–1571
Mukhtar H, Del Tito BJ, Das M, Cherniack EP, Cherniack AD, Bickers DR (1984b) Clotrimazole, an inhibitor of epidermal benzo[*a*]pyrene metabolism and DNA binding and carcinogenicity of the hydrocarbon. Cancer Res 44:4233–4240
Mukhtar H, Das M, Del Tito BJ, Bickers DR (1984c) Epidermal benzo[*a*]pyrene metabolism and DNA-binding in BALB/c mice: inhibition by ellagic acid. Xenobiotica 14:527–531
Nakayama J, Yuspa SH, Poirier MC (1984) Benzo[*a*]pyrene-DNA adduct formation and removal in mouse epidermis in vivo and in vitro: relationship of DNA binding to initiation of skin carcinogenesis. Cancer Res 44:4087–4095
Nebert DW (1979) Genetic differences in the induction of monooxygenase activities by polycyclic aromatic compounds. Pharmacol Ther 6:395–417
Nebert DW, Bausserman LL, Bates RR (1970) Effect of 17β-estradiol and testosterone on aryl hydrocarbon hydroxylase activity in mouse tissues in vivo and in cell culture. Int J Cancer 6:470–480
Nebert DW, Benedict WF, Gielen JE, Oesch F, Daly JW (1972) Aryl hydrocarbon hydroxylase, epoxide hydrase, and 7,12-dimethylbenz[*a*]anthracene-produced skin tumorigenesis in the mouse. Mol Pharmacol 8:374–379
Noonan PK, Wester RC (1983) Cutaneous biotransformations and some pharmacological and toxicological implications. In: Marzulli FN, Maibach HI (eds) Dermatotoxicology, 2nd edn. Hemisphere, Washington, pp 71–90
Norden G (1953) The role of appearance, metabolism and disappearance of 3,4-benzpyrene in the epithelium of mouse skin. Acta Pathol Microbiol Scand [Suppl]96:1–87
Norred WP, Akin FJ (1976) Induction of aryl hydrocarbon hydroxylase and carbon monoxide binding hemoproteins in mouse epidermis by tobacco carcinogens. Biochem Pharacol 25:732–734
Oesch F, Schmassmann H (1979) Species and organ specificity of the *trans*-stilbene oxide induced effects on epoxide hydratase and benzo[*a*]pyrene monoxygenase activity in rodents. Biochem Pharmacol 28:171–176
Oesch F, Glatt H, Schmassmann H (1977) The apparent ubiquity of epoxide hydratase in rat organs. Biochem Pharmacol 26:603–607
Oesch F, Schmassmann H, Bentley P (1978) Specificity of human rat and mouse skin epoxide hydratase towards K-region epoxides of polycyclic hydrocarbons. Biochem Pharmacol 27:17–20
Oesch F, Tegtmeyer F, Kohl F-V, Rüdiger H, Glatt HR (1980) Interindividual comparison of epoxide hydratase and glutathione-*S*-transferase activities in cultured human fibroblasts. Carcinogenesis 1:305–309
Ohkawara A, Halprin KM, Taylor JR, Levine V (1972) Acid hydrolases in human epidermis. Br J Dermatol 83:450–459
O'Neill VA, Rawlins MD; Chapman PH (1981) Epoxide hydrolase activity in human skin. Br J Clin Pharmacol 12:517–521
Pannatier A, Jenner P, Testa B, Etter JC (1978) The skin as a drug-metabolizing organ Drug Metab Rev 8:319–343

Pannatier A, Testa B, Etter J-C (1981 a) Aryl ether *O*-dealkylase activity in the skin of untreated mice in vitro. Xenobiotica 11:345–350

Pannatier A, Testa B, Etter J-C (1981 b) Enzymatic hydrolysis by mouse skin homogenates: structure-metabolism relationships of *para*-nitrobenzoate esters. Int J Pharm 8:167–174

Pannatier A, Testa B, Etter J-C (1982) Effect of topically applied phenobarbital on *O*-dealkylase activity in mouse skin. Experientia 38:604–605

Parkinson EK, Newbold RF (1980) Benzo[*a*]pyrene metabolism and DNA adduct formation in serially cultured strains of human epidermal keratinocytes. Int J Cancer 26:289–299

Pelling JC, Slaga TJ (1982) Comparison of levels of benzo[*a*]pyrene diol epoxide diastereomers covalently bound in vivo to macromolecular components of the whole epidermis versus the basal cell layer. Carcinogenesis 3:1135–1141

Pereira A, Burns FJ, Albert RE (1979) Dose response for benzo[*a*]pyrene adducts in mouse epidermal DNA. Cancer Res 39:2556–2559

Pohl RJ, Philpot RM, Fouts JR (1976) Cytochrome P-450 content and mixed-function oxidase activity in microsomes isolated from mouse skin. Drug Metab Dispos 4:442–450

Pohl RJ, Coomes MW, Sparks RW, Fouts JR (1984) 7-Ethoxycoumarin *O*-deethylation activity in viable basal and differentiated keratinocytes isolated from the skin of the hairless mouse. Drug Metab Dispos 12:25–34

Poland AP, Glover E, Robinson JR, Nebert DW (1974) Genetic expression of aryl hydrocarbon hydroxylase activity. J Biol Chem 249:5599–5606

Puhvel SM, Ertl DC (1984) Decreased induction of aryl hydrocarbon hydroxylase activity in hyperproliferative hairless mouse epidermis. Br J Dermatol 110:29–35

Pyerin WG, Hecker E (1975) Epoxide hydrase activity in mouse skin epidermis. Z Krebsforsch 83:81–83

Pyerin WG, Hecker E (1977) On the biochemical mechanism of tumorigenesis in mouse skin. VIII. Isolation and characterization of epidermal microsomes and properties of their arylhydrocarbon monooxygenase and epoxide hydr(at)ase. Z Krebsforsch 90:259–279

Pyerin WG, Hecker E (1979) On the biochemical mechanism of tumorigenesis in mouse skin. IX. Interrelation between tumor initiation by 7,12-dimethylbenz[*a*]anthracene and the activities of epidermal arylhydrocarbon monooxygenase and epoxide hydratase. J Cancer Res Clin Oncol 93:7–30

Pyerin WG, Oberender HA, Hecker E (1980) Extent of skin tumour initiation in mice by 7,12-dimethylbenz[*a*]anthracene and induction of arylhydrocarbon monooxygenase are not causally related. Cancer Lett 10:155–162

Rahimtula AD, Payne JF, Martins I (1982) Hydrocarbon-based oils as inducers of cutaneous aryl hydrocarbon hydroxylase. Toxicol Lett 10:213–217

Raunio H, Pelkonen O (1982) Independent induction and inhibition of ornithine decarboxylase and aryl hydrocarbon hydroxylase activities in rat epidermis. J Invest Dermatol 79:246–249

Rettie AE, Willliams FM, Rawlins MD (1986 a) Substrate specificity of the mouse skin mixed-function oxidase system. Xenobiotica 16:205–211

Rettie AE, Williams FM, Rawlins MD, Mayers RT, Burke MD (1986 b) Major differences between lung, skin and liver in the microsomal metabolism of homologous series of resorufin and coumarin ethers. Biochem Pharmacol 35:3495–3500

Rogan E, Roth P, Katomski P, Benderson J, Cavalieri E (1978) Binding of benzo[*a*]pyrene at the 1,3,6 positions to nucleic acids in vivo on mouse skin and in vitro with rat liver microsomes and nuclei. Chem Biol Interact 22:35–51

Rogan EG, Hakam A, Cavalieri EL (1983) Structure elucidation of a 6-methylbenzo[*a*]pyrene-DNA adduct formed by horseradish peroxidase in vitro and mouse skin in vivo. Chem Biol Interact 47:111–122

Rojas M, Baer-Dubowska W, Alexandrov K (1986) Comparison of benzo[*a*]pyrene-DNA adduct levels in mouse and rat epidermis and dermis. Cancer Lett 30:35–39

Rüdiger HW, Marxen J, Kohl FV, Melderis H, von Wichert P (1979) Metabolism and formation of DNA adducts of benzo[*a*]pyrene in human diploid fibroblasts. Cancer Res 39:1083–1088

Rugstad HE, Dybing E (1975) Glucuronidation in cultures of human skin epithelial cells. Eur J Clin Invest 5:133–137
Ruzicka T, Printz M (1984) Arachidonic acid metabolism in skin: a review. Rev Physiol Biochem Pharmacol 100:121–160
Schlede E, Conney AH (1970) Induction of benzo[*a*]pyrene hydroxylase activity in rat skin. Life Sci 9:1295–1303
Schmassmann HU, Glatt HR, Oesch F (1976) A rapid assay for epoxide hydratase activity with benzo[*a*]pyrene 4,5-(K-region-)oxide as substrate. Anal Biochem 74:94–104
Selkirk JK, Nikbakht A, Stoner GD (1983) Comparative metabolism and macromolecular binding of benzo[*a*]pyrene in explant cultures of human bladder, skin, bronchus and esophagus from eight individuals. Cancer Lett 18:11–20
Shugart L, Kao J (1984) Effect of ellagic and caffeic acids on covalent binding of benzo[*a*]pyrene to epidermal DNA of mouse skin in organ culture. Int J Biochem 15:571–573
Shugart L, Holland JM, Rahn RO (1983) Dosimetry of PAH skin carcinogenesis: covalent binding of benzo[*a*]pyrene to mouse epidermal DNA. Carcinogenesis 4:195–198
Sivarajah K, Lasker JM, Elin TE (1981) Prostaglandin synthetase-dependent cooxidation of (±)-benzo[*a*]pyrene-7,8-dihydrodiol by human lung and other mammalian tissues. Cancer Res 41:1834–1839
Slaga TJ, Viaje A, Berry DL, Bracken W, Buty SG, Scribner JD (1976) Skin tumor initiating ability of benzo[*a*]pyrene 4,5-7,8- and 7,8-diol-9,10-epoxides and 7,8-diol. Cancer Lett 2:115–122
Slaga TJ, Buty SG, Thompson S, Bracken WM, Viaje A (1977a) A kinetic study on the in vitro covalent binding of polycyclic hydrocarbons to nucleic acids using epidermal homogenates as the activating system. Cancer Res 37:3126–3131
Slaga TJ, Bracken WM, Viaje A, Levin W, Yagi H, Jerina DM, Conney AH (1977b) Comparison of the tumor-initiating activities of benzo[*a*]pyrene arene oxides and diol-epoxides. Cancer Res 37:4130–4133
Slaga TJ, Thompson S, Berry DL, DiGiovanni J, Juchau MR, Viaje A (1977c) The effects of benzoflavones on polycyclic hydrocarbon metabolism and skin tumor initiation. Chem Biol Interact 17:297–312
Slaga TJ, Viaje A, Bracken WM, Berry DL, Fischer SM, Miller DR, Leclerc SM (1977d) Skin-tumor-initiating ability of benzo[*a*]pyrene-7,8-diol-9,10-epoxide (*anti*) when applied topically in tetrahydrofuran. Cancer Lett 3:23–30
Slaga TJ, Huberman E, Selkirk JK, Harvey RG, Bracken WM (1978) Carcinogenicity and mutagenicity of benz[*a*]anthracene diols and diol-epoxides. Cancer Res 38:1699–1704
Slaga TJ, Bracken WJ, Gleason G, Levin W, Yagi H, Jerina DM, Conney AH (1979a) Marked differences in the skin tumor-initiating activities of the optical enantiomers of the diastereomeric benzo[*a*]pyrene 7,8-diol-9,10-epoxides. Cancer Res 39:67–71
Slaga TJ, Gleason GL, DiGiovanni J, Sukumaran KB, Harvey RG (1979b) Potent tumor-initiating activity of the 3,4-dihydrodiol of 7,12-dimethylbenz[*a*]anthracene in mouse skin. Cancer Res 39:1934–1936
Slaga TJ, Huberman E, DiGiovanni J, Gleason G (1979c) The importace of the "bay region" diol-epoxide in 7,12-dimethylbenz[*a*]anthracene skin tumor initiation and mutagenesis. Cancer Lett 6:213–220
Smith LH, Holland JM (1981) Interaction between benzo[*a*]pyrene and mouse skin in organ culture. Toxicology 21:47–57
Summer K-H, Göggelmann W (1980) 1-Chloro-2,4-dinitrobenzene depletes glutathione in rat skin and is mutagenic in *Salmonella typhimurium*. Mutat Res 77:91–93
Täuber U, Toda T (1976) Biotransformation von Diflucortolonvalerianat in der Haut von Ratte, Meerschweinchen und Mensch. Arzneim Forsch 26:1484–1487
Tejwani R, Jeffrey AM, Milo GE (1982) Benzo[*a*]pyrene diol epoxide DNA adduct formation in transformable and non-transformable human foreskin fibroblast cells in vitro. Carcinogenesis 3:727–732
Thakker DR, Yagi H, Levin W, Wood AW, Conney AH, Jerina DM (1985) Polycyclic aromatic hydrocarbons: metabolic activation to ultimate carcinogens. In: Anders MW (ed) Bioactivation of foreign compounds. Academic, Orlando, pp 177–242
Theall G, Eisinger M, Grunberger D (1981) Metabolism of benzo[*a*]pyrene and DNA adduct formation in cultured human epidermal keratinocytes. Carcinogenesis 2:581–587

Thompson S, Slaga TJ (1976a) Mouse epidermal aryl hydrocarbon hydroxylase. J Invest Dermatol 66:108–111
Thompson S, Slaga TJ (1976b) The effects of dexamethasone on mouse skin initiation and aryl hydrocarbon hydroxylase. Eur J Cancer 12:363–370
Tierney B, Hewer A, Walsh C, Grover PL, Sims P (1977) The metabolic activation of 7-methylbenz[*a*]anthracene in mouse skin. Chem Biol Interact 18:179–193
Tierney B, Hewer A, Rattle H, Grover PL, Sims P (1978) The formation of dihydrodiols by chemical or enzymic oxidation of 3-methylcholanthrene. Chem Biol Interact 23:121–135
Van Cantfort J, Lorand T, Gielen JE, Lapiere CM (1986) Human epidermal blister: a convenient tissue for toxicological and genetic studies of benzo[*a*]pyrene metabolism. Arch Dermatol Res 278:324–328
Vermorken AJM, Goos CMAA, Roelofs HMJ, Henderson PTH, Bloemendal H (1979) Metabolism of benzo[*a*]pyrene in isolated human scalp hair follicles. Toxicology 14:109–116
Vermorken AJM, Goos CMAA, Hukkelhoven MWAC, de Bruyn CHMM (1985) Human hair follicles in metabolic studies. In: Maibach HI, Lowe NJ (eds) Models in dermatology, vol II. Karger, Basel, pp 189–208
Vigny P, Ginot YM, Kindts M, Cooper CS, Grover PL, Sims P (1980) Fluorescence spectral evidence that benzo[*a*]pyrene is activated by metabolism in mouse skin to a diol-epoxide and a phenol-epoxide. Carcinogenesis 1:945–950
Vigny P, Kindts M, Cooper CS, Grover PL, Sims P (1981) Fluorescence spectra of nucleoside-hydrocarbon adducts formed in mouse skin treated with 7,12-dimethylbenz[*a*]anthracene. Carcinogenesis 2:115–119
Vizethum W, Ruzicka T, Goerz G (1980) Inducibility of drug-metabolizing enzymes in the rat skin. Chem Biol Interact 31:215–219
Wang IY, Rasmussen RE, Crocker TT (1974) Metabolism of benzo[*a*]pyrene by microsomes from tissues of pregnant and fetal hamsters. Life Sci 15:1291–1300
Wattenberg LW, Leong JL (1970) Inhibition of the carcinogenic action of benzo[*a*]pyrene by flavones. Cancer Res 30:1922–1925
Wattenberg LW, Leong JL (1971) Tissue distribution studies of polycyclic hydrocarbon hydroxylase activity. In: Brodie BB, Gillette JR (eds) Concepts in biochemical pharmacology. Springer, Berlin Heidelberg New York, pp 422–430 (Handbook of pharmacology, vol 28/2)
Weigert F, Calcutt G, Powell AV (1948) The course of the metabolism of benzpyrene in the skin of the mouse. Br J Cancer 2:405–410
Wester RC, Noonan PK, Smeach S, Kosobud L (1983) Pharmacokinetics and bioavailability of intravenous and topical nitroglycerin in rhesus monkey: estimate of percutaneous first-pass metabolism. J Pharm Sci 72:745–748
Weston A, Grover PL, Sims P (1982a) Formation of the 11,12-diol as a metabolite of benzo[*a*]pyrene by rat skin in vivo. Biochem Biophys Res Commun 105:935–941
Weston A, Grover PL, Sims P (1982b) Metabolism and activation of benzo[*a*]pyrene by mouse and rat skin in short-term organ culture and in vivo. Chem Biol Interact 42:233–250
Weston A, Grover PL, Sims P (1983a) Metabolic activation of benzo[*a*]pyrene in human skin maintained in short-term organ culture. Chem Biol Interact 45:359–371
Weston A, Grover PL, Sims P (1983b) Formation of the 1,2-diol as a metabolite of 7,12-dimethylbenz[*a*]anthracene by rodent and human skin. Carcinogenesis 4:1307–1311
Weston A, Hodgson RM, Hewer AJ, Kuroda R, Grover PL (1985) Comparative studies of the metabolic activation of chrysene in rodent and human skin. Chem Biol Interact 54:233–242
Wiebel FJ, Gelboin HV (1977) Cutaneous carcinogenesis: metabolic interaction of chemical carcinogens with skin. In: Lee DHK (ed) Environmental physiology: reactions to environmental agents. Williams and Wilkins, Baltimore, pp 337–348 (Handbook of physiology, sect 9)
Wiebel FJ, Lentz JC, Diamond L, Gelboin HV (1971) Aryl hydrocarbon (benz[*a*]pyrene) hydroxylase in microsomes from rat tissue: differential inhibition and stimulation by benzoflavones and organic solvents. Arch Biochem Biophys 144:78–86

Wiebel FJ, Leutz JC, Gelboin HV (1973) Aryl hydrocarbon (benzo[*a*]pyrene) hydroxylase: inducible in extrahepatic tissues of mouse strains not inducible in liver. Arch Biochem Biophys 154:292–294
Wiebel FJ, Leutz JC, Gelboin HV (1975) Aryl hydrocarbon (benzo[*a*]pyrene) hydroxylase: a mixed-function oxygenase in mouse skin. J Invest Dermatol 64:184–189
Wiegrebe W, Retzow A, Plumier E, Ersoy N, Garbe A, Faro HP, Kunert R (1984) Dermal absorption and metabolism of the antipsoriatic drug dithranol triacetate. Arzneim Forsch 34:48–51
Wislocki PG, Chang RL, Wood AW, Levin W, Yagi H, Hernandez O, Mah HD, Dansette MP, Jerina DM, Conney AH (1977) High carcinogenicity of 2-hydroxybenzo[*a*]pyrene on mouse skin. Cancer Res 37:2608–2611
Wood AW, Levin W, Chang RL, Lehr RE, Schaefer-Ridder M, Karle JM, Jerina DM, Conney AH (1977) Tumorigenicity of five dihydrodiols of benz[*a*]anthracene on mouse skin: exceptional activity of benz[*a*]anthracene 3,4-dihydrodiol. Proc Natl Acad Sci USA 74:3176–3179
Wood AW, Chang RL, Levin W, Ryan DE, Thomas PE, Mah HD, Karle JM, Yagi H, Jerina DM, Conney AH (1979) Mutagenicity and tumorigenicity of phenanthrene and chrysene epoxides and diol epoxides. Cancer Res 39:4069–4077
Yu CD, Fox JL, Ho NFH, Higuchi WI (1979) Physical model evaluation of topical prodrug delivery – simultaneous transport and bioconversion of vidarabine-5′-valerate II: parameter determinations. J Pharm Sci 68:1347–1357
Yu CD, Fox JL, Higuchi WI, Ho NFH (1980a) Physical model evaluation of topical prodrug delivery – simultaneous transport and bioconversion of vidarabine-5′-valerate IV: distribution of esterase and deaminase enzymes in hairless mouse skin. J Pharm Sci 69:772–775
Yu CD, Gordon NA, Fox JL, Higuchi WI, Ho NFH (1980b) Physical model evaluation of topical prodrug delivery – simultaneous transport and bioconversion of vidarabine-5′-valerate V: mechanistic analysis of influence of nonhomogeneous enzyme distributions in hairless mouse skin. J Pharm Sci 69:775–780
Yuspa SH, Hennings H, Dermer P, Michael D (1976) Dimethyl sulfoxide-induced enhancement of 7,12-dimethylbenz[*a*]anthracene metabolism and DNA binding in differentiating mouse epidermal cell cultures. Cancer Res 36:947–951

CHAPTER 15

Skin Cancer (Excluding Melanomas)

F. MARKS

A. Human Skin Cancer[1]

I. History and Causative Factors

In human medicine, cancer of the skin is of outstanding importance for two reasons:

1. Among whites, epidermal cell carcinoma is by far the commenest form of malignant neoplasia, equaling in number of cases the incidence of all other cancer types combined; fortunately, skin carcinomas are also the easiest to prevent and to treat.

2. Human skin cancer was the first neoplastic disease shown to be caused by environmental influences.

As early as 1761, the British physician John HILL published a booklet in which he incriminated the "immoderate use of snuff" as being a cause of nasal mucosa carcinomas (HILL 1761). Sir Percivall Pott's famous work, which appeared 14 years later (POTT 1775), on the relationship between the high incidence of scrotal skin cancer in chimney sweeps and exposure to chimney soot, is considered as a landmark of medical science. This was followed by the first experimental induction of cancer by a chemical agent (coal tar) by YAMAGIWA and ICHIKAWA (1914), and the identification of aryl hydrocarbons as the carcinogenic "principles" of tar by KENNAWAY (1925). Although all skin cell types can become subject to malignant transformation, most human skin cancers arise from epidermis. Depending on morphological characteristics, one distinguishes between basal cell and squamous cell carcinomas. The exact orgin of these tumors (either from interfollicular epithelium or from hair follicles) was not entirely clear until recently. Now an exact histotypical characterization of skin tumors has become possible by determining the pattern of tissue-specific proteins such as cytokeratins and others (MOLL et al. 1982).

The great majority of human skin carcinomas can be attributed to environmental influences, i.e., the carcinogenic effect of ultraviolet irradiation, sunlight being one of the main causes (SCOTTO et al. 1983). The susceptibility to UV carcinogenesis is genetically determined and predominantly associated with the availability and effectiveness of light-protecting mechanisms of the tissue. Whites,

[1] For detailed reviews of this subject see GREITHER and TRITSCH (1957), MELCZER (1961), URBACH (1963, 1983), FOULDS (1975), ANDRADE et al. (1976), PRUNIERAS (1978), HELM (1979), LAERUM and IVERSEN (1981), LUGER and GESCHNAIDT (1983) and RIPOLL and WEBBER (1986).

especially of the Celtic type, are much more endangered than those with darker skin (URBACH et al. 1981). Tumors arise primarily on body sites which are consistently exposed to sunlight (head, neck, hands, forearms). Almost all squamous cell carcinomas and approximately two-thirds of all basal cell carcinomas occur at such sites. In addition to geographic and racial variations, three genetic disorders[2] are known which are connected with increased light sensitivity and an extraordinarily high incidence of skin cancer. These are xeroderma pigmentosum (XP), oculocutaneous albinism, and the nevoid basal cell carcinoma syndrome. In xeroderma patients and people with albinism, the development of cancer is directly related to disorders in light protection, showing defects either in DNA repair mechanisms or lack of protecting melanin (see Sect. A. II and URBACH et al. 1981). Human light-induced skin cancer can arise from several well-defined precancerous lesions of epidermis, with actinic keratosis being the commonest type (reviewed by PETERS and MÜLLER 1981).

There is no question that the fashionable trend to identify healthiness and sportsman-like flair with a suntanned skin has resulted in a dramatic increase of skin cancer over the past decades, at least as far as the white population is concerned. While sunlight "represents the most clear-cut etiological factor in human malignancy" (YUSPA 1983), the steadily increasing use of artificial suntanning devices, generally believed to work with a nondangerous type of long-wavelength UV light, has also led to concern among dermatologists, since the long-term effects of such irradiation cannot yet be predicted with certainty.

Besides UV light, ionizing radiation, burns, and several chemical agents have been shown to be carcinogenic for human skin, whereas the involvement of viruses in certain forms of the disease is highly probable, but not yet established beyond doubt. Combined or synergistic effects of these agents with UV light in skin carcinogenesis are frequently discussed (see RIPOLL and WEBBER 1986).

In general, a chronic (mainly occupational) exposure to chemical carcinogens is necessary to evoke skin cancer. Two groups of environmental chemical carcinogens especially involved in skin cancer can be distinguished, i.e., polycyclic aromatic hydrocarbons and inorganic arsenic. Skin cancer induction by other chemical agents has been shown in animal experiments, but is certainly of minor relevance to the human situation. Classical examples for occupational skin carcinogenesis by chemicals are the high incidence of skin cancer in workers chronically in contact with coal tar, soot, pitch, mineral oil products, or arsenic-containing substances such as certain pesticides or ores. The very long latency period (20 years and more) which generally precedes cancer formation makes it rather difficult to attribute a malignant skin disease to a distinct history of exposure, so that clear-cut statistical evaluations of the risk and the etiology of chemical skin carcinogenesis in humans are rare.

II. Ultraviolet Radiation Carcinogenesis

A correlation between benign and malignant skin disease and exposure to sunlight was suggested more than a century ago (UNNA 1884). The sun's ultraviolet

[2] For additional genetic disorders connected with an increased susceptibility to skin cancer see MCEVOY (1979).

radiation has indeed turned out to be the most important inducer of skin cancer in humans and the most abundant carcinogenic "agent" of all (reviewed in detail by KRIPKE et al. 1981; SCOTTO and FRAUMENI 1982; URBACH 1983; FRY and LEY 1984). For experimental reasons, the UV spectrum (200–400 nm) has been divided into three regions: the range 320–400 nm (UV-A), the range 290–320 nm (UV-B), and the range 200–290 nm (UV-C). The strongest biologic effects are observed with UV-B/UV-C, especially with light of wavelength 240–320 nm, which also has the highest carcinogenic potential. Fortunately, the most dangerous portion of sunlight, i.e., wavelength 240–290 nm, is absorbed by the ozone layer of the earth's atmosphere. A possible destruction of the ozone layer by environmental pollution may result in a considerable increase of skin cancer. It has been estimated that a reduction of the atmospheric ozone by only 15% may result in a 60% increase in the incidence of skin cancer (URBACH 1980). Although only weakly penetrating, UV radiation heavily damages living matter. In skin, the most exposed tissue of the body, cell death and inflammatory reactions (sunburn) occur after acute exposure. After chronic exposure, premature aging and disturbances of tissue differentiation (elastosis and keratosis) occur, as well as tumor formation (PARRISH 1983). Although UV light interacts with a wide variety of cellular constituents, the target most studied is cellular DNA which absorbs UV light between 240 and 300 nm with an absorption maximum at 258 nm. The most prominent and best-studied photochemical effect at the molecular level is the cyclobutane dimerization of adjacent pyrimidine residues in one DNA strand (Fig. 1) (CAMERMAN and CAMERMAN 1968). This reaction produces distortions of the DNA double helix which impair both transcription and replication (for review see JOHNSON 1978). It is generally assumed that biologic effects of UV light such as inhibition of DNA synthesis, mutations, neoplastic transformation, and cell death are mainly the result of this reaction.

A direct causal relationship between carcinogenesis and photodimerization of pyrimidine residues in DNA is indicated by several lines of evidence. For instance:

1. In carefully controlled experiments with the tropical fish *Poecilia formosa* (Amazon molly), the yield of tumors was found to be proportional to the number of pyrimidine dimers (HART et al. 1977);
2. In UV-irradiated human fibroblasts, symptoms of transformation were reversed by photoreactivation of a special enzyme which catalyzes the depolymerization of pyrimidine dimers (SUTHERLAND et al. 1980);

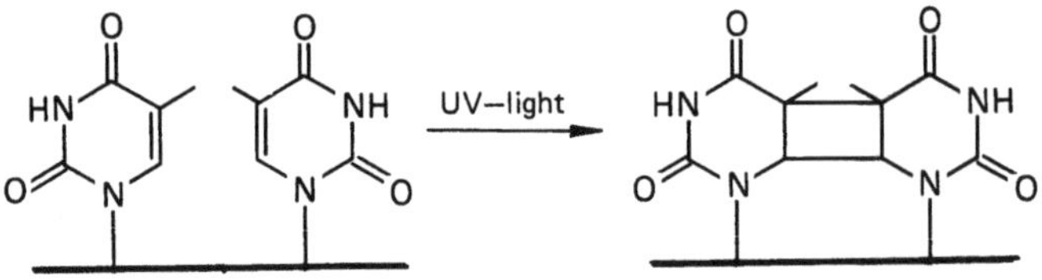

Fig. 1. Photodimerization of adjacent thymine residues in DNA

3. The action spectra for UV-induced lethality, transformation, and pyrimidine dimerization in hamster embryo fibroblasts were all identical within the narrow wavelength region 265–270 nm (DONIGER et al. 1981);
4. In xeroderma pigmentosum patients, a high skin cancer incidence correlates with an impaired ability of cells to remove pyrimidine dimers from their DNA (reviewed by JUNG 1978; GIANELLI 1978; URBACH et al. 1981).

The latter studies have turned out to be of special importance for a deeper understanding of the mechanisms involved in light-induced carcinogenesis of the skin.

Since life on earth depends on solar energy and most organisms are therefore constantly exposed to sunlight with all its hazardous effects, the development of protective mechanisms was one of the fundamental conditions for the evolution of terrestrial organisms. This is drastically shown by the high cancer incidence which occurs when such mechanisms fail, as in people with albinism or xeroderma pigmentosum patients. The most prominent symptom of xeroderma pigmentosum (which is a genetic disease affecting all cells of the body) is an extraordinarily high susceptibility of skin to sunlight carcinogenesis, resulting in the development of multiple tumors in light-exposed areas. This inherited genetic disorder occurs through a disturbed ability to repair UV-induced DNA lesions.

At present we know of four types of DNA repair mechanisms (for reviews see CARTER and ROSS 1983; THIELMANN 1984; FRIEDBERG 1985). The simplest is the so-called photoreactivation or light repair, an enzyme-catalyzed reaction by which the pyrimidine dimers are depolymerized in the presence of light. The light-dependent recovery of cells from UV-induced damage has been taken as a direct measure of the involvement of pyrimidine dimerization in a given biologic effect (SUTHERLAND et al. 1980; HART et al. 1977). The commonest light-independent repair process is the excision repair reaction, by which the damaged piece of DNA, together with approximately 100 adjacent nucleotides, is removed and replaced by an intact complementary strand. This reaction is catalyzed by several enzyme activities: a UV endonuclease activity, which recognizes the lesion and introduces nicks in the damaged strand (DNA incision); an exonuclease activity, which excises the DNA piece; a DNA polymerase, which forms the new strand, and a ligase, which closes the nick.

The so-called post-replication repair is another light-independent process which occurs during the S phase and fills the gaps which are left in the newly synthesized DNA strand when, in the course of DNA replication, the pyrimidine dimers of the template strand are "bypassed." Finally, single-strand breaks are sealed and gaps are filled by a mechanism called gap-filling repair. Single-strand breaks are characteristic of DNA damage induced by ionizing radiation rather than by UV.

The great majority (85%) of xeroderma pigmentosum patients are defective in DNA incision (CLEAVER 1968; SETLOW et al. 1969; CLEAVER and TROSKO 1970; THIELMANN et al. 1985), although in some cases post-replication repair (LEHMANN et al. 1975, 1977) and photoreactivation repair (SUTHERLAND et al. 1975) may also be impaired. A disturbance in post-replication repair is manifested as a decrease in the rate of replication of UV-damaged DNA; in such patients (XP variants), excision repair can be normal (LEHMANN et al. 1975; CLEAVER 1972). The defect of excision repair itself shows a considerable degree of genetic variability in that it can be due to at least nine different genetic alterations, as has been shown by fusion experiments with cells from different XP-patients (so-called complementation groups, see BOOTSMA et al. 1979; TAKEBE 1979; FISCHER et al. 1985). It is mainly the initial steps of the repair process (recognition, nicking, and excision) which seem to be impaired.

The situation becomes even more complicated by the finding that cells from xeroderma patients exhibit an extraordinarily high susceptibility not only to UV irradiation, but also to certain chemical carcinogens such as *N*-acetylaminofluorene or 4-nitroquinoline-*N*-oxide (so-called UV-like carcinogens), which do not give rise to pyrimidine dimerization, whereas the susceptibility to alkylating carcinogens as well as to X-rays does not differ from that of normal cells (for review see KRIPKE et al. 1981; THIELMANN et al. 1985).

In summary, this means that the clinical manifestations of one and the same neoplastic disease can be due to a great variety of genetic and metabolic alterations which impair the normal protective mechanisms of light- or chemical-induced DNA damage. In most, but certainly not all cases, such damage results in pyrimidine dimerization. How these alterations of DNA structure finally lead to the neoplastic phenotype is still unknown.

III. Ionizing Radiation Carcinogenesis

Actinic skin cancer is caused not only by UV light, but also by ionizing radiation such as X- and gamma-rays. Although quite rare, this type of skin carcinogenesis is historically important, having been known since 1902, only 7 years after the discovery of X-rays (FRIEBEN 1902). Similar to UV-induced cancer, X-ray cancer is preceded by precancerous lesions of skin (radiodermatitis) and manifested as squamous and basal cell carcinomas, but sometimes also as dermal fibrosarcomas, melanomas, or sweat gland tumors.

In almost all cases, the disease is associated with occupational or frequently repeated therapeutic exposure to ionizing radiation. A classical example is the higher incidence of skin cancer among radiologists in the early days of radiotherapy, i.e., before protective measures were taken. Chronic irradiation over a long period of time (10–30 years) seems to be obligatory since no increased skin cancer incidence has been found among Japanese who survived the atomic bomb explosions in Hiroshima and Nagasaki (Committee on the Biological Effects of Ionizing Radiation 1980). The biologic effects of (low level) ionizing radiation are different from that of UV light in that they include mainly DNA single-strand breaks instead of pyrimidine dimerization. Such damage is thought to lead to dominant or recessive mutations, neoplastic transformation, chromosome abnormalities, inhibition of cellular proliferation, and cell death, although the exact relationships are only poorly understood.

IV. Viruses as Causative Agents[3]

As far as viral tumorigenesis is concerned, the skin is again associated with a landmark in research, in that the infective nature of a tumorigenic agent was first demonstrated in 1907 by transferring human skin warts to another person by means of cell-free wart extracts (CUIFFO 1907). Nevertheless, a causative role of viruses in the etiology of *human malignant* skin tumors has not yet been unequivocally demonstrated. Skin warts (papillomas) are induced by papilloma viruses, a heterogeneous group of DNA viruses belonging to the family of Papovaviridae (for reviews see ORTH et al. 1977; ZUR HAUSEN 1981). Papilloma viruses are widely distributed in nature, and a rather large group is specialized on human hosts (human papilloma viruses, HPV). Among the human papilloma viruses, species with different carcinogenic potential seem to exist. Common (and plantar) warts of skin are induced by HPV-1 and HPV-2. In humans, a progression of such warts to carcinomas has never been observed. There are, however, three types of human papillomatosis which are sometimes associated with malignancy. The highest frequency of malignant development (20%–30%) is seen in epidermodysplasia verruciformis, a disseminated papillomatous skin disease (LUTZNER 1978). This progression of tumors from the benign to the malignant state occurs only in light-exposed papillomas and may be correlated with the appearance of specific human papilloma virus types (HPV-3 and HPV-5) which are different from HPV of common warts (ORTH et al. 1980). In addition, a correlation between HPV-5 infection and malignant progression of the disease was observed. Another example of a papillomatosis which sometimes gives rise to squamous cell carcinomas is condylomata acuminata (human anogenital warts) which is associated with HPV-6 and HPV-11 infections (reviewed by ZUR HAUSEN 1977, 1985; ZUR HAUSEN et al. 1984). By means of nucleic acid hybridization techniques, HPV-11 and distantly related virus types (HPV-16 and -18) can be identified in a number of human cervical, penile, and vulval carcinomas. According to ZUR HAUSEN et al. (1984), such observations "incriminate these agents as prime candidates for an etiological role in human genital cancer."

In addition to epidermodysplasia verruciformis and genital tumors, human laryngeal papillomas are consistently associated with HPV infectins, especially of the HPV-11 type (GISSMANN et al. 1983). Like genital warts, these tumors show a distinct tendency for malignant progression. In most of these papillomatoses, the progression to the cancerous state seems to require some "activating" mechanism such as UV, X-irradiation, or tissue injury, indicating a multistage mechanism of carcinogenesis.

In animals, the carcinogenic potential of papilloma viruses in skin and related epithelia is well established (see GROSS 1983). For example, the Shope papilloma virus system of the rabbit (ROUS and KIDD 1938; SYVERTON 1952) and the bovine papilloma viruses (JARRETT 1978; LANCASTER and OLSON 1978; JARRETT et al. 1980) are well-documented cases of carcinogenic potential. Again, the progression to malignancy seems to depend on additional "activators" such as chemical carcinogens, tissue injury, and alimentary factors.

---

[3] For reviews see ZUR HAUSEN (1981), GISSMANN (1984), and KIRCHNER (1986).

The possible role of other virus types in skin tumorigenesis is, although repeatedly discussed, still not very firmly established. Prime candidates are found among the herpes simplex viruses, which have been incriminated as being involved in the development of human genital cancer (penis, vulva, cervix) and of carcinomas in the lip region. The evidence for the role of these viruses is based on epidemiologic studies rather than on biochemical investigations (ZUR HAUSEN 1975) and is rather indirect and circumstantial.

B. Experimental Skin Carcinogenesis by Chemical Agents[4]

I. Historical Remarks

The skin of rodents (especially of mice) is the oldest and most favored model for studies on chemical carcinogenesis and for testing carcinogenicity based on a high sensitivity to all kinds of carcinogenic agents, the ease of manipulation, and the availability of different inbred strains with graded susceptibility. Only recently has there been competition from other models such as the liver, the bladder, and a variety of cell and tissue culture systems.

Studies on skin carcinogenesis have provided several landmarks in cancer research, beginning with the first experimental demonstration of chemical carcinogenesis by YAMAGIWA and ICHIKAWA in 1914. Another outstanding event was the introduction of the two-stage approach to carcinogenesis. The now generally accepted terms "initiation" and "promotion" were proposed in the mid-1940s by ROUS and co-workers (ROUS and KIDD 1941; FRIEDWALD and ROUS 1944) when they studied the augmentation of coal tar- and hydrocarbon-induced skin carcinogenesis by noncarcinogenic hyperplasiogenic agents such as chloroform and turpentine or by wounding. The "promoting" effect of skin wounds or irritants on mouse skin pretreated with coal tar or 3,4-benz[*a*]pyrene had, however, already been described almost 20 years earlier (DEELMAN 1924; TWORT and ING 1928).

Further important steps in the experimental investigation of skin carcinogenesis were the discovery of the strong tumor-promoting potency of croton resin (BERENBLUM 1941), the standardization of the two-stage approach (MOTTRAM 1944; BERENBLUM and SHUBIK 1947), and the isolation and identification of phorbol ester tumor promoters from croton oil by HECKER (for review see HECKER and SCHMIDT 1974) and VAN DUUREN and ORRIS (1965). Another step toward a clear-cut experimental approach was made by attempting to subdivide the stage of promotion into two different steps (BOUTWELL 1964; SLAGA et al. 1980; FÜRSTENBERGER et al. 1981), by introducing a new stage called "conversion" (FÜRSTENBERGER et al. 1985), and by manipulating the progression from the benign to the malignant state (HENNINGS et al. 1983). The whole process of carcinoma formation in skin can now be investigated by studying at least four separate and subsequent stages, again giving the skin model a clear advantage over other systems, with perhaps the exception of the rat liver model. This, however, does not question the steadily increasing practicability of other models, especially of in vitro

[4] For a comprehensive review of this subject see SLAGA (1984a).

systems suitable for the investigation of detailed problems at the cellular and molecular level.

Although skin cancer can, of course, be elicited by sufficient treatment with a single carcinogenic agent (solitary carcinogenesis), we shall discuss experimental skin carcinogenesis here according to the multistage approach in vivo. It must not be overlooked, however, that in their final result, one-stage protocols of skin carcinogenesis differ from multistage protocols in that the former yield mainly carcinomas and autonomously growing papillomas, whereas the latter lead first of all to reversibly growing papillomas with a progression to the malignant state being a late, low yield event (see Sect. B. III). These two types of experimental protocol also differ in strain susceptibility, as reviewed by YUSPA (1983). Regarding results obtained with skin cell cultures, the reader is referred to several comprehensive review articles and monographs (SLAGA et al. 1978; COLBURN 1979; YUSPA et al. 1980; FUSENIG 1981; HECKER et al. 1982; YUSPA 1983; SLAGA 1984b).

II. Initiation

According to the terminology of ROUS, initiation, the first step of multistage carcinogenesis, may be defined as the formation of "latent" tumor cells which do not give rise to a visible tumor, unless they are subsequently exposed to an agent or manipulation which induces tumor growth. Initiation of skin cells is generally brought about by a single, local, intragastric (BOUTWELL 1964; RITCHIE and SAFFIOTTI 1955; BERENBLUM and HARAN-GHERA 1957; GOERTTLER et al. 1980), or transplacental (GOERTTLER and LOEHRKE 1977) application of a chemical carcinogen in a very low "subthreshold" dose or by limited UV irradiation. Since successful induction of tumor growth can be accomplished at almost any time after initiation, the alterations caused by the initiating event in the tissue are regarded as virtually irreversible.

It is generally assumed, although not finally proven, that the key event of chemical carcinogenesis is the covalent binding of the carcinogen to certain target macromolecules in the cell (MILLER and MILLER 1971). Most carcinogens, especially those of the aromatic hydrocarbon type – these are mostly used in experimental skin carcinogenesis, although initiation can also be accomplished by many other chemicals, as listed by YUSPA (1983) and SLAGA (1984c) – are, however, chemically rather inert. To become reactive, they have to be metabolically activated, primarily by oxidative reactions (reviewed by LEVIN et al. 1982). The reactive metabolites have been called "ultimate" carcinogens, their immediate metabolic precursors "proximate" carcinogens (MILLER and MILLER 1976). The oxidative activation of aryl hydrocarbons requires NADPH and molecular oxygen and is catalyzed by the monooxygenase system. This enzyme complex is localized in the endoplasmic reticulum of most animal tissue cells, including skin cells (BICKERS 1983). The monooxygenases' physiologic role is the metabolism of endogenous hydrophobic compounds (steroid hormones, etc.) and xenobiotics (drug-metabolizing enzyme system). Generally, the enzyme-catalyzed reactions result in "detoxification" and formation of more hydrophilic metabolites which can be excreted. Depending on the substrate and the body site, however, metabolic intermediates may be more toxic than the parent compound. This is the case for chemical carcinogens.

Another enzyme which has recently been shown to bring about oxidative activation of carcinogens is prostaglandin endoperoxide synthetase. It is assumed that the carcinogens are co-oxidized during oxygenation of arachidonic acid to prostaglandin hydroperoxy endoperoxide and its reduction to the corresponding hydroxy endoperoxide (HONN et al. 1981). Although not yet proven, this reaction may also play a role in skin carcinogenesis, since epidermis contains a prostaglandin synthetase complex which might well be activated by certain carcinogenic compounds (see Sect. B. IV).

The formation of putative "proximate" and "ultimate" carcinogens from benzo[*a*]pyrene via the monooxygenase pathway is illustrated in Fig. 2. All carcinogenic aryl hydrocarbons contain a phenanthrene structure which forms a so-called "bay region" (see JERINA et al. 1980 and Figs. 2 and 3). It is predicted that dihydrodiol epoxides with the epoxy group in this region will exhibit the highest biologic activity of all isomeric compounds, owing to steric hindrance and increased chemical reactivity (JERINA et al. 1980). According to this concept, a 7,8-dihydrodiol-9,10-epoxide of benzo[*a*]pyrene (probably the (+)-enantiomer of the *anti*-isomer) would be the most important "ultimate" carcinogen derived from benzo[*a*]pyrene (for a more detailed discussion of this subject see SIMS 1980). The most potent skin carcinogen and most frequently used initiator in animal experiments is 7,12-dimethylbenz[*a*]anthracene (DMBA). Depending on the animal strain, only 5–100 nmol of this compound need be applied to skin for successful initiation.

The aryl hydrocarbon monooxygenase complex of epidermis catalyzes the metabolism and activation of DMBA in a manner quite similar to that of benzo[*a*]pyrene. The methyl groups of DMBA are, however, not only additional targets for hydroxylation, but introduce some steric strain on the molecule which decreases the stability of the aromatic system, and thus makes the hydrocarbon more reactive. This results in a rather bewildering number of metabolites and makes the investigation of DMBA biotransformation a difficult task (for review see DI GIOVANNI and JUCHAU 1980).

This hydrocarbon, and its hydroxymethyl derivatives, are the precursors of epoxides, dihydrodiols, and dihydrodiol epoxides. Again, the bay region dihydrodiol epoxides, i.e., the 3,4-dihydrodiol-1,2-epoxide (Fig. 3), are thought to be the most reactive species and, as such, probably the "ultimate" carcinogens.

Owing to their extreme instability, the bay region dihydrodiol epoxides of DMBA have not as yet been identified as metabolites in DMBA-treated animal skin. The observation, however, that in the usual initiation-promotion regimen of mouse skin carcinogenesis in the Sencar strain, the initiating potency of the 3,4-dihydrodiol surpasses that of DMBA and of all other DMBA derivatives tested (including non-bay region dihydrodiol epoxides), provides strong evidence for the validity of the bay region theory (Table 1). The 3,4-dihydrodiol may thus be regarded as a "proximate" carcinogen in mouse skin. Other data supporting this concept were provided by spectroscopic (VIGNY et al. 1981) and chromatographic (COOPER et al. 1980; BIGGER et al. 1983) analysis of DMBA-DNA adducts.

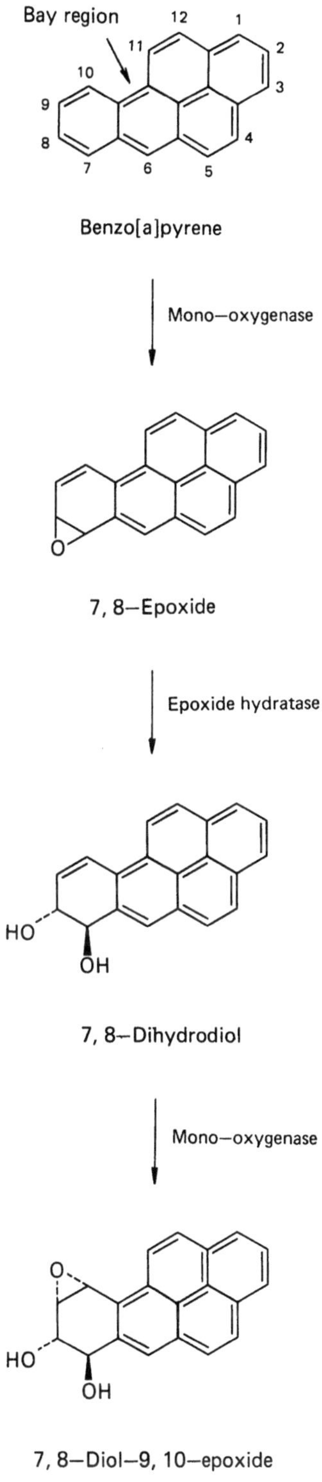

Fig. 2. Metabolic oxidation of 3,4-benzo[*a*]pyrene to *anti*-7,8-dihydrodiol-9,10-(ep)oxide, its putative "ultimate" carcinogen. The corresponding 4,5- and 9,10-dihydrodiols as well as the 9,10-dihydrodiol-7,8-(ep)-oxide are also formed in the course of the enzymatic reaction (not shown)

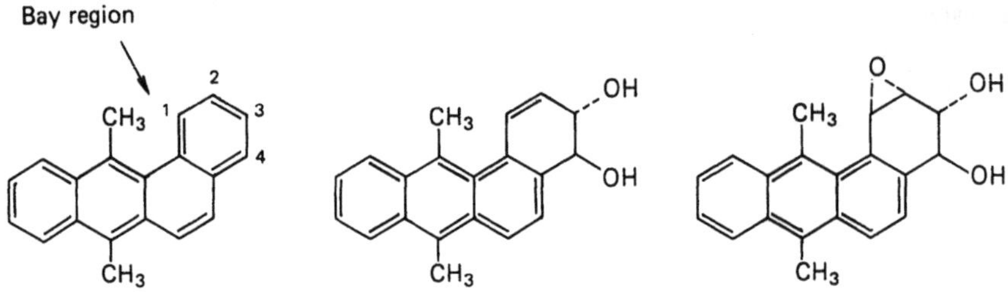

1 2 3

Fig. 3. 7,12-Dimethylbenz[*a*]anthracene (1) and two of its metabolically activated derivatives, the 3,4-dihydrodiol ("proximate" carcinogen, 2) and the 3,4-diol-1,2-(ep)oxide ("ultimate" carcinogen, 3). Corresponding products are thought to be derived from the different hydroxymethyl-dimethylbenz[*a*]anthracenes (not shown)

Table 1. Tumor-initiating activities of DMBA and various DMBA derivatives and metabolites in mouse skin[a] (DI GIOVANNU and JUCHAU, 1980)

Treatment	No. of mice[b]	Papillomas per mouse[c]	Mice with tumors[d] (%)
Control (only DMBA initiation)	30	0	0
Control (only TPA promotion)	29	0.10	6
DMBA	28	9.10	100
1-*OH*-DMBA	29	0.10	10
2-*OH*-DMBA	30	0.07	7
3-*OH*-DMBA	30	0.14	14
4-*OH*-DMBA	28	0.17	14
5-*OH*-DMBA	29	0.10	10
9-*OH*-DMBA	30	0.40	24
10-*OH*-DMBA	30	0.47	30
DMBA 3,4-dihydrodiol	29	22.84	100
DMBA 5,6-dihydrodiol	29	0.10	10
DMBA 8,9-dihydrodiol	29	0.21	18
DMBA 8,9-dihydrodiol-10,11-(ep)oxide	28	0.30	20
1-CH_3-DMBA	30	0.03	3
2-CH_3-DMBA	30	0.10	7
5-CH_3-DMBA	29	0.40	34
1-Fluoro-DMBA	30	0.03	3
2-Fluoro-DMBA	29	0.10	10
5-Fluoro-DMBA	28	0.20	15
11-Fluoro-DMBA	29	8.20	100

[a] A total of 30 mice (strain CD-1) were locally treated with 200 nmol of each compound (100 nmol 3,4-diol). 1 week later initiation was followed by twice-weekly applications of 17 nmol TPA (3.4 nmol for the 3,4-diol-treated animals) over a period of 30 weeks.
[b] Surviving at the 30th week after promotion.
[c] Total number of papillomas divided by total number of surviving mice.
[d] Percentage of surviving mice with tumors.

Employing the ^{32}P-post-labeling technique, SCHOEPE et al. (1986) have recently investigated the DNA binding of DMBA and DMBA metabolites applied to mouse skin in initiating doses. With DMBA, 12 adducts per 10^7 nuleotides were found. The binding of DMBA-3,4-dihydrodiol was 7–8 times higher, but showed practically the same pattern as that of DMBA.

The concept that oxidative metabolism of DMBA is necessary in order to express its initiating activity in skin is additionally supported by the finding that skin tumor initiation can be modified by a variety of compounds which interfere with aryl hydrocarbon metabolism (for reviews see WIEBEL 1980; DI GIOVANNI et al. 1980; SLAGA 1984c). Among these, certain benzoflavones and quercetin also decrease the tumor yield in the two-stage experiment, whereas myricetin, naringenin, and 17β-estradiol, another potent inhibitor of DMBA metabolism, do not show such an effect. On the other hand, 2,3,7,8-tetrachlorodibenzo-*p*-dioxin (TCDD), which is a strong inducer of aryl hydrocarbon monooxygenase activity in skin, has been found to be one of the most potent inhibitors of initiation by DMBA.

Such observations indicate that the relationship between oxidative metabolism and carcinogenic potential must be rather complex and not predictable a priori. This is not unexpected since aryl hydrocarbon metabolism involves both "activating" and "inactivating" pathways. It certainly depends on a distinct and delicate balance between both pathways, whether or not the carcinogenic potential of the hydrocarbon is expressed within the treated tissue, or whether it is abolished by "detoxification."

Ultimate carcinogens such as the bay region dihydrodiol epoxides, are expected to react spontaneously with all kinds of nucleophilic groups of cellular macromolecules, forming covalent bounds. Such alkylation (arylation) reactions are thought to be essential for the neoplastic transformation of the target cell. Considering the irreversibility of tumor formation (at least in the initiation stage) the close correlation between mutagenic and carcinogenic efficacy, and the high incidence of cancers in patients with defective DNA repair, DNA (i.e., mainly the hydroxyl and amino groups of the nucleobases) is generally assumed to be the critical target molecule for ultimate carcinogens. The DNA binding of carcinogens shows a high degree of specificity. For example, benzo[*a*]pyrene has been found to be almost exclusively bound to the 2-amino groups of guanine residues in epidermal DNA, whereas *N*-acetoxy-2-acetylaminofluorene prefers the C-8 position of guanine (see YUSPA 1983).

The concept of DNA binding as the essential step of chemical carcinogenesis has recently gained considerable support from studies on oncogenes (for review see MARKS 1987). Initially, oncogenes were identified as the transforming genetic principles of certain RNA tumor viruses. These viral oncogenes are derived from closely related genes of eukaryotic cells, the so-called proto-oncogenes. Today, approximately 50 different proto-oncogenes are known. Most, if not all of them, code for proteins required for the reception, transduction, and interpretation of endogenous mitogenic signals such as growth factors. The biochemical pathways of the "mitotic cascade" controlled by these proto-oncogene products are thought to play a key role in embryonic development, tissue regeneration, and wound repair (GOUSTIN et al. 1986; MARKS 1987). In tumor cells, either experimentally produced or of human origin, proto-oncogenes have been found to be mutated and/or overexpressed, resulting in a dysregulation of the mitotic cascade.

The transfection of oncogenes from tumor cells into proper recipient cells leads to neoplastic transformation of the latter, indicating the important role of oncogene mutation and expression in tumorigenesis. In human tumors, proto-oncogenes of the *ras* family have been repeatedly found to be activated, frequently by point mutation. In animal experiments, such point mutations can be induced by chemical carcinogens, whereby a direct relationship between the chemical type of the mutation (point of attack on DNA) and the reactivity of the carcinogen used was observed. The demonstration of a point-mutated and overexpressed Harvey *ras* proto-oncogene with transforming potential in papillomas and carcinomas in mouse skin initiated with DMBA (or other carcinogens) and promoted with the phorbol ester TPA indicates that, at least in animal experiments, skin carcinogenesis is also accompanied by an activation of a proto-oncogene (QUINTANILLA et al. 1986; BALMAIN and BROWN 1988). This oncogene activation is probably due to a direct attack of the carcinogenic agent on the genome, and results at least in papilloma formation.

Studies with keratinocyte cell cultures indicate that *ras*-containing Harvey murine sarcoma viruses can block epidermal cell differentiation (YUSPA et al. 1985), an effect which has been thought to provide a condition for the selection and clonal expansion of tumor cells occurring in the course of tumor promotion (YUSPA 1984). Moreover, epidermal cells carrying the viral H-*ras* oncogene develop into papillomas when grafted on athymic nude mouse recipients (ROOP et al. 1986). Although there is, thus, a good correlation between H-*ras* activation and papilloma formation, the precise role of H-*ras* in skin carcinogenesis is not known, since the function of the protein coded for by this oncogene is not fully clear. As yet, this protein has been only identified as a GTP-binding factor similar to, but not identical with, transducin or the regulatory subunit of adenylate cyclase. Neither in skin nor in other cells or tissues does activation of *ras* oncogenes appear to be sufficient for the development of malignant tumors (reviewed by MARKS 1987). Instead, the activation of at least one other type of proto-oncogene, as for example c-*myc* in addition to c-*ras*, has been found to be required for malignant progression in several cases. Whether the enhancement of malignant progression by treating mouse skin papillomas with carcinogenic agents (HENNINGS et al. 1983) is due to such an additional gene activation is still unknown.

III. The Induction of Tumor Development in Initiated Skin[5]

So long as the molecular events involved in initiation were not well enough understood and their analysis difficult, successful initiation could only be verified by inducing the initiated cells to multiply and to form visible tumors. It is one of the

[5] For detailed reviews of this subject see DIAMOND et al. (1980), BLUMBERG (1980), BOUTWELL (1964, 1974), VAN DUUREN (1978), HECKER (1978), SLAGA et al. (1978), SLAGA (1984a), HECKER et al. (1982), and MARKS (1981).

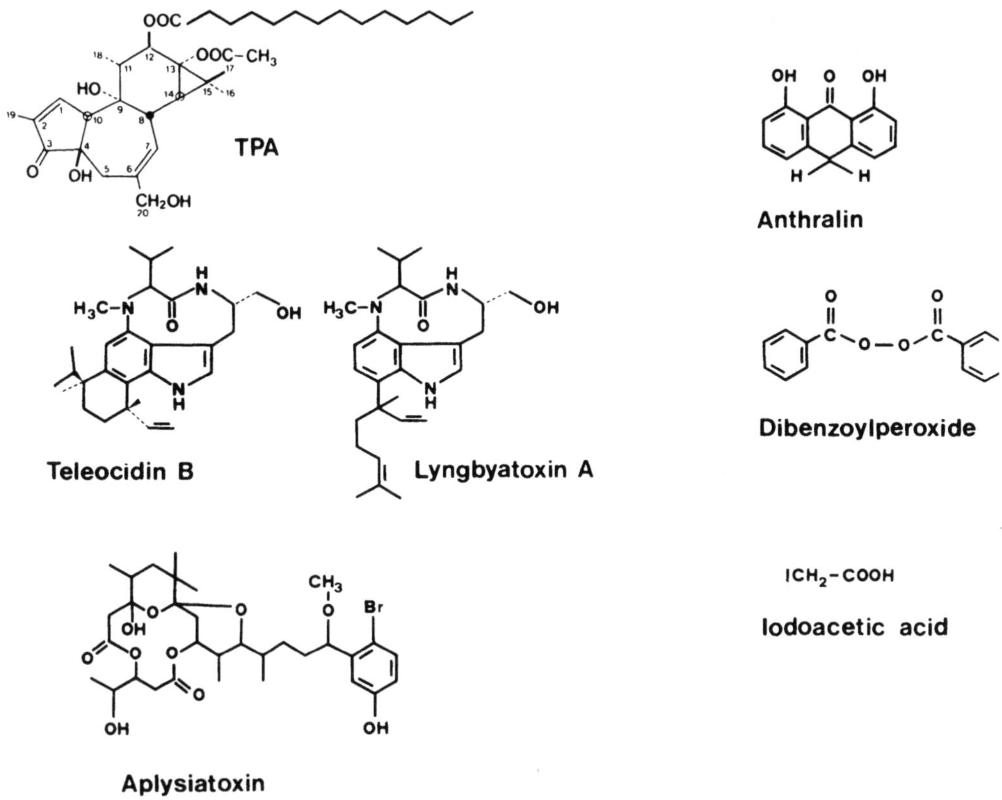

Fig. 4. Representative skin tumor promoters

great advantages of the multistage approach that this can be brought about by agents or manipulations which are by themselves practically noncarcinogenic, as long as the experimental conditions are carefully controlled.

Tumor development in initiated skin can be accomplished by wounding as well as by local application of a rather restricted number of chemical agents (tumor promoters) which all induce skin inflammation and epidermal hyperplasia (Fig. 4). The promoting agent must be repetitively applied *after* initiation over a period of several weeks. Most active in this respect are certain diterpene esters such as the phorbol ester 12-*O*-tetradecanoylphorbol-13-acetate (TPA) of plant origin (HECKER and SCHMIDT 1974) and alkaloids and polyacetate-type compounds such as teleocidin, lyngbyatoxin, and aplysiatoxin found in marine organisms (FUJIKI et al. 1984). Anthralin (dithranol) and some other phenolic compounds, detergents such as Tween 60, as well as iodoacetic acid, D-limonene, several fatty acid esters (BOUTWELL 1974), and organic peroxy compounds (SLAGA et al. 1981), belong to the group of "weak skin tumor promoters."

The most frequently used skin tumor promoter is the phorbol ester TPA (identical with phorbol-12-myristate-13-acetate, PMA), the active compound of croton oil, the seed oil of the tropical Euphorbiacea CROTON TIGLIUM (Fig. 4). Depending on the mouse strain, TPA induces tumor growth in initiated skin when given in a dose of 0.5–20 nmol per local application.

When TPA is given twice a week, the first tumors are seen after about 7 weeks and the maximal response is reached after 12–18 weeks. An extension of the time intervals between the applications leads to a drastic reduction of the tumor yield, showing that promotion is quickly reversible (BOUTWELL 1964).

Other phorbol esters and related compounds of the diterpene ester type have been isolated from different plant species (many of them belonging to the family Euphorbiaceae) or obtained by chemical modification of naturally occurring compounds, so that a whole series of pure agents with graded promoting potencies is currently available (HECKER 1978).

Studies on structure-activity relationships within the phorbol ester series indicate a distinct linkage between promoting potency and chemical structure which reflects the interaction of phorbol esters with a specific cellular binding site (for details see Sect. B. IV and HECKER 1978). There is no evidence for a metabolic activation of a phorbol ester tumor promoter.

Under the experimental conditions usually applied, the initiation-promotion approach in mouse skin yields mainly benign, reversibly growing tumors (papillomas, see BURNS et al. 1978, 1984), which are each of monoclonal origin (REDDY and FIALKOW 1983; TAGUCHI et al. 1984). It is generally assumed, although not finally proven, that in the course of the initiation-promotion regimen carcinomas arise directly from papillomas (malignant progression).

The rate of malignant progression is rather low (less than 10%), but can be increased by a second treatment with initiating agents following the promotion regime (HENNINGS et al. 1983). This indicates that an additional attack on the cellular genome might be required to achieve malignancy (see Sect. B. IV). The dose of the promoter and the duration of promoter treatment has no effect on the rate of malignant progression. On the other hand, carcinoma development practically does not occur in initiated skin without promotion. This indicates that promotion has to be regarded as a prerequisite for malignant progression, probably because carcinomas seem to develop from papillomas, the development of which, in turn, depends on promotion. It is easily conceivable that, if malignant progression is the result of a second genotoxic effect, the probability of this effect occurring will be much higher in papillomas (being the result of a clonal expansion of tumor cells) than in single initiated cells. If promoter treatment is stopped before malignant carcinomas appear, the whole process of skin tumorigenesis exhibits a high degree of remission, i.e., most of the papillomas disappear. A prevention or inhibition of promotion thus brings cancer development to a halt. The investigation of the mechanism of tumor promotion and the identification of tumor promoters in the environment may, therefore, provide a possibility of cancer prevention.

As mentioned already, the great majority of papillomas produced in mouse skin under the standard conditions of an initiation-promotion experiment do not show growth autonomy, but disappear after promoter treatment has been stopped, especially in the early phase of the experiment (BURNS et al. 1978, 1984). If the dose of the initiating carcinogen is increased by giving it repetitively above the "subthreshold level" of the two-stage experiment, more autonomously growing papillomas and carcinomas are generated, even in the absence of promotion (solitary carcinogenesis). This not only means that tumors induced by the two-stage approach differ from those induced by solitary carcinogenesis, but whether or not promoter treatment is required also depends on how the initial carcinogen treatment is carried out. A key element of promotion could therefore be a permanent proliferative stimulus in epidermis necessary to make the non-autonomous papillomas initiated by a single "subthreshold" dose of a carcinogen visible. In fact, all skin tumor promotoers are strong irritant mitogens, and the generation of sustained epidermal hyperplasia has been proposed as a prerequisite for promotion (Sisskin et al. 1982). It is still a matter of dispute whether a "promoting component" is also included in solitary carcinogenesis.

The fact that not every agent which induces epidermal hyperplasia is a good promoter has raised serious doubts about the concept that tumor promotion is the result of clonal expansion of initiated cells, owing to permanent stimulation of cellular proliferation. Moreover, it was found that certain skin mitogens, such as turpentine (Boutwell 1964), mezerein (Slaga et al. 1980), or the TPA derivatives 12-*O*-retinoylphorbol-13-acetate, RPA (Fürstenberger et al. 1981), and 12-*O*-tetradecatetra-2,4,6,8-enoylphorbol-13-acetate, Ti8 (Fürstenberger and Hecker 1986; Marks et al. 1979) which exhibit little or no tumor-promoting efficacy by themselves could, nevertheless, "complete" the tumorigenic effect of a limited treatment with a promoter such as croton oil or TPA. This means that, under certain experimental conditions, the long-term process of tumor induction in initiated skin can be subdivided into at least two stages, i.e., carried out by two successive treatments, each of which is insufficient to induce tumor development. Unfortunately, this approach has created some confusion in the nomenclature of multistage carcinogenesis.

Boutwell (1964) as well as Slaga et al. (1980) and Fürstenberger et al. (1981) have proposed the concept of "two-stage skin tumor promotion." The hyperplasiogenic compounds (such as mezerein or RPA, see Fig. 5) applied in stage II were called "incomplete" tumor promoters in order to distinguish them from "complete" promoters such as TPA. The two stages of promotion were also termed "conversion" and "propagation" (Boutwell 1964), implying that in stage I the initiated cell is converted into a tumor cell, whereas in stage II the proliferative propagation of such tumor cells is induced. Such a concept is based on the assumption that an initiated cell resembles a latent mutant, the neoplastic phenotype of which has to be expressed before clonal expansion can occur. Unfortunately, the term "conversion" has been used for the process leading from a benign to a malignant tumor (Hennings et al. 1983) although this has traditionally been called (malignant) progression (Foulds 1954).

Recently, Fürstenberger et al. (1985) suggested abandoning the concept of two-stage tumor promotion since stage I can also occur prior to initiation, whereas promotion is defined as completing tumorigenic processes started with initiation. These authors consider conversion to be a discrete stage of skin carcinogenesis and restrict the term promotion to what hitherto has been called stage II of promotion. According to this concept, one must not speak of "complete" and "incomplete" skin tumor promoters, but of promoters with (TPA) or without (Ti8, RPA, mezerein) an additional converting potency. For unknown reasons the converting potency of the phorbol ester tumor promoter TPA can be de-

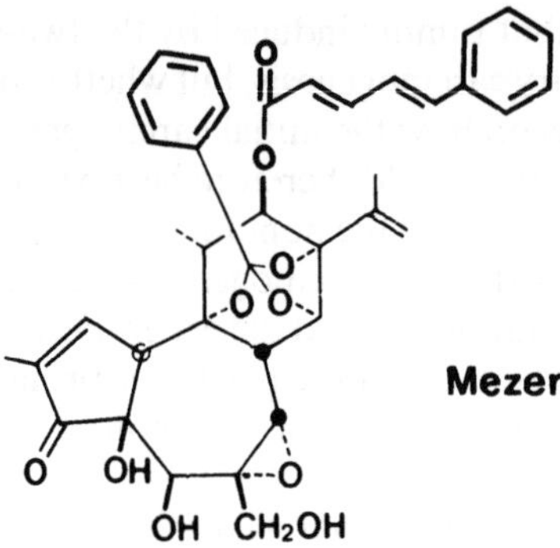

Mezerein

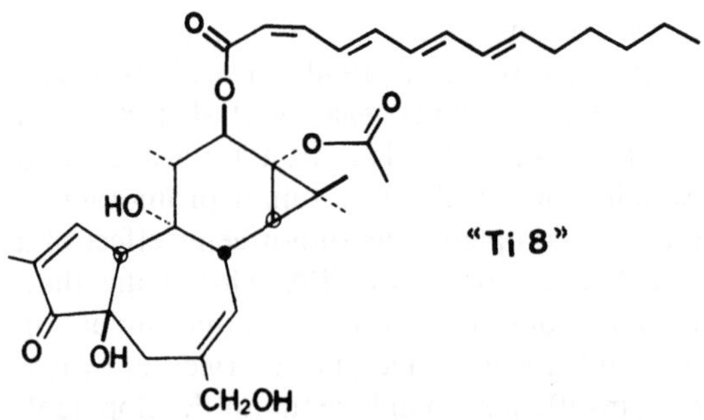

"Ti 8"

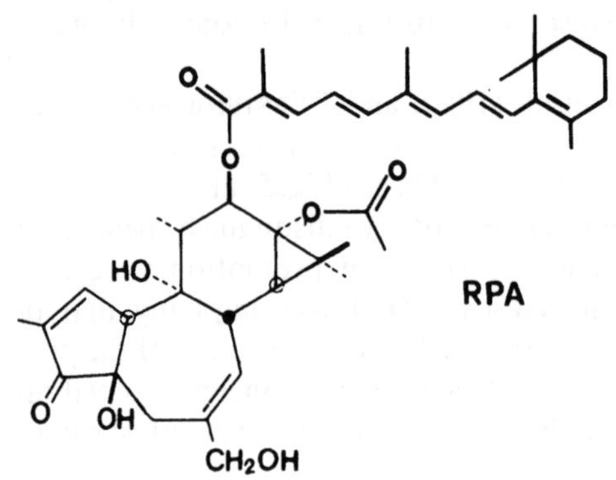

RPA

Fig. 5. "Incomplete" skin tumor promoters: mezerein, 12-*O*-tetradecatetra-2,4,6,8-enoylphorbol-13-acetate ($C_{14:4}$PA, Ti8), and 12-*O*-retinoylphorbol-13-acetate (RPA, PRA). The *trans* configuration of double bonds 6,7 and 8,9 in Ti8 is not definitely proven (Fürstenberger and Hecker 1986)

creased ("disarmed") to more than 90% either by inducing several conjugated double bonds into the C_{14} fatty acid side chain of the molecule or by replacing this side chain by a retinoyl residue (Fürstenberger et al. 1981). For the Sencar mouse strain, the 4-*O*-methyl ether of TPA has been found to provide a converting agent without promoting potency (Slaga 1984d), whereas, for other mouse strains, "pure" converting agents have not yet been found.

According to Fürstenberger et al. (1985) the terms conversion and promotion are first of all operationally defined as discrete stages of carcinogenesis observed under special experimental conditions. In the conversion stage the tissue is obviously sensitized for promotion, i.e., promotability is induced, if promotion is understood as a manipulation necessary to elicit papilloma development in initiated and converted skin. Considerable efforts have been made to replace these operational definitions by mechanistic ones. The results of this task are still fragmentary.

IV. The Mechanism of Skin Tumor Promotion

Since promotion is due to the development of chronic epidermal hyperplasia, in the course of which the clonal expansion of tumor cells occurs, the mechanisms involved in hyperplastic development of epidermis have to be regarded as important or even essential parameters of the promoting process. The development of a hyperplastic condition is the general response of epidermis to external injury. It is a rather dramatic event proceeding within 1–2 days and being accompanied by strong symptoms of inflammation and a sequence of characteristic biochemical events. This response has been termed "hyperplastic transformation" (MARKS et al. 1983a, b), implying that it is not simply the result of overshooting cellular proliferation, but is due to distinct cellular and molecular events which result in a temporary disturbance of tissue homeostasis, i.e., in a disquilibrium between the rates of cell gain and cell loss. Hyperplasia is indeed not an automatic consequence of epidermal hyperproliferation since under certain conditions, a strong hyperproliferative response can be evoked in mouse epidermis which does not result in hyperplasia and is not accompanied by irritation and those biochemical reactions observed in the course of hyperplastic transformation (MARKS et al. 1983a, b).

Two prominent events occur in mouse skin in the very early phase of hyperplastic transformation: an induction of the arachidonic acid cascade, resulting in the production of prostaglandins and metabolites produced along the lipoxygenase pathways (FÜRSTENBERGER and MARKS 1985), and an activation of the enzyme protein kinase C (PKC). The latter response is especially seen after application of phorbol esters, which have been found to be specific stimulators of PKC (CASTAGNA et al. 1982; ASHENDEL 1985) that mimic the effect of the second messenger diacylglycerol, the physiologic activator of the enzyme. Within the phorbol ester series, a good correlation exists between the degree of PKC stimulation and the induction of epidermal hyperplasia. While the function of PKC activation in the hyperproliferative process is still not clear, the role of arachidonic acid metabolites is better understood. In NMRI mouse skin the initial hyperplastic response depends critically on the accumulation of prostaglandin E_2 in epidermis, whereas, after repeated stimulation, this role seems to be taken over by prostaglandin $F_{2\alpha}$ (FÜRSTENBERGER and MARKS 1985). By themselves, prostaglandins are unable to evoke epidermal hyperproliferation; they act synergistically with the exogenous stimulus. In NMRI mouse skin, prostaglandins are exclusively involved in the induction of epidermal hyperproliferation, whereas the inflammatory response is probably mediated by arachidonic acid metabolites of the lipoxygenase pathway.

Other events in the course of hyperplastic transformation include an expression of the proto-oncogenes c-fos and c-myc (ROSE-JOHN et al. 1988), an increase of polyamine biosynthesis caused by an induction of the enzyme ornithine decarboxylase, desensitization of epidermal cells for antiproliferative endogenous signals such as catecholamine and chalone, and an impairment of intercellular communication (reviewed by MARKS et al. 1983 a, b). Several of the biochemical events observed in the course of hyperplastic transformation have been claimed as being of critical importance for skin tumor promotion (see, for example, WEEKS et al. 1984; TROSKO and CHANG 1984; FISCHER 1984; FISCHER and SLAGA 1985).

Such conclusions are mainly based on a correlation observed between the tumor-promoting potency of a given compound and its ability to induce a particular response in epidermis or in other cell types as well as on experiments with specific metabolic inhibitors. In the course of such studies certain strain differences were observed which probably reflect subtle differences in the biochemistry of the hyperplastic response. For example, prostaglandin synthase inhibitors were found to inhibit tumor promotion in NMRI mouse (FÜRSTENBERGER and MARKS 1985), but to stimulate promotion in Sencar mouse skin (FISCHER 1984), indicating a strain-specific difference in the balance between the cyclooxygenase and the lipoxygenase pathway and in the physiologic role of the arachidonic acid metabolites generated along these pathways (FISCHER et al. 1987).

The fact that phorbol ester tumor promoters mimic the effect of an intracellular second messenger, i.e., diacylglycerol, indicates that hyperplastic transformation of epidermis is normally controlled by receptor-mediated stimulation of the inositol trisphosphate/diacylglycerol cascade. The endogenous factors activating this signal-transducing pathway in skin may include transforming growth factor α (TGFα) and bradykinin.

The stimulation of diacylglycerol effects may also explain the wide variety of phorbol ester-induced effects of different tissues and cell types. Despite the fact that the diacylglycerol-like activity on phorbol esters correlates with efficacy to induce hyperplastic transformation in skin, the precise role of diaclyglycerol in epidermal physiology is not known, especially because the function of diacylglycerol-controlled protein phosphorylation catalyzed by PKC is still not clear. PKC activation may primarily be involved in epidermal cell maturation rather than in the induction of epidermal hyperproliferation. Hyperplastic transformation is, indeed, characterized not only by tissue growth, but, in addition, by profound changes in tissue differentiation. It has been repeatedly proposed that the initial event in the hyperplastic response is an acceleration of terminal differentiation of epidermal cells with hyperproliferation as a subsequent event (see, for example, POTTEN and ALLEN 1975). Moreover, the stimulation of terminal differentiation by phorbol ester tumor promoters has been proposed to provide a powerful selective pressure on initiated cells, which seem to be resistant to the stimulating effect of phorbol esters on maturation (YUSPA 1984; PARKINSON 1985).

Considering the tumor-promoting efficacy of skin wounding (see Sect. B. V) and in order to explain certain metaplastic processes (i.e., the de novo formation of hair follicles induced by tumor promoters in mouse tail skin, see SCHWEIZER and MARKS 1977; SCHWEIZER 1979) even retrodifferentiation of epidermal cells has been suggested as playing a critical role in tumor promotion (MARKS et al. 1978). In addition, the appearance of so-called dark cells, assumed to be primitive epidermal stem cells, in promoter-treated Sencar mouse epidermis, has been proposed as indicating a reversion of skin to a fetal phenotype, and therefore as a crucial step in multistage carcinogenesis (KLEIN-SZANTO 1984).

An important role in epidermal hyperproliferation, terminal differentiation, and tumor promotion has been attributed to active oxygen species and free radicals (CERUTTI 1985). Such concepts were put forward mainly to explain the tumor-promoting efficacy of peroxy compounds, the correlation between irritant and promoting activities (especially within the phorobol ester series), and the anti-promoting effect of superoxide dismutase, other oxygen scavengers, and anti-oxidants, and of anti-inflammatory drugs. This rapidly expanding field has been extensively reviewed (MARKS and FÜRSTENBERGER 1985; TROLL and WIESNER 1985). Retinoids (VERMA et al. 1979), protease inhibitors (TROLL and WIESNER 1983) and cyclosporin A, an immunosuppressive and antipsoriatic drug (GSCHWENDT et al. 1987) should be mentioned as potent inhibitors of skin tumor promotion. Other anti-promoting agents have also been listed (SLAGA et al. 1982; SLAGA 1980, 1984c, d).

Most symptoms of hyperplastic transformation induced by a single application of a skin irritant cease within 1–2 weeks. It appears as if the reversibility of promotion, as demonstrated by the decrease of tumor incidence on prolongation of the time interval between subsequent promoter treatments, is closely related to the remission rate of epidermal hyperplasia. Being a general response of skin to all kinds of chemical irritation and mechanical injury, hyperplastic transformation provides an important physiologic defense mechanism. Bearing this in mind, one hesitates to describe tumor promotion as the result of a special cytotoxic event, for instance comparable to the "genotoxic" effect of an initiator. Instead, one could understand promotion as an over-activation of a normal physiologic process.

V. Mechanistic Aspects of the Conversion Stage of Experimental Carcinogenesis

While skin tumor promotion can thus be explained as being due to a chronic hyperplasiogenic process occurring in epidermis, much less is known about the molecular events involved in the conversion stage of skin tumorigenesis. Conversion has been shown to depend on an induction of epidermal DNA synthesis by converting agents such as TPA (KINZEL et al. 1984, 1986) and to be subject to inhibition by retinoic acid and inhibitors of the arachidonic acid cascade (MARKS and FÜRSTENBERGER 1984). Thus, conversion seems to be closely related to promotion and hyperplastic transformation. On the other hand, the alteration induced by the converting agent (i.e., promotability of skin) exhibits a much longer half-life (10–12 weeks in NMRI mouse skin) than the events occurring in the course of hyperplastic transformation (2 weeks, FÜRSTENBERGER et al. 1985). Furthermore, reactions probably involved in hyperplasia and promotion, such as activation of PCK are apparently not critical for conversion (MARKS and FÜRSTENBERGER 1984).

The three-stage approach to skin carcinogenesis (Fig. 6) demonstrates that conversion, either before or after initiation (FÜRSTENBERGER et al. 1985), is required for initiated skin to respond to promoter application by tumor development. If promotion is explainable by clonal expansion of tumor cells in the course

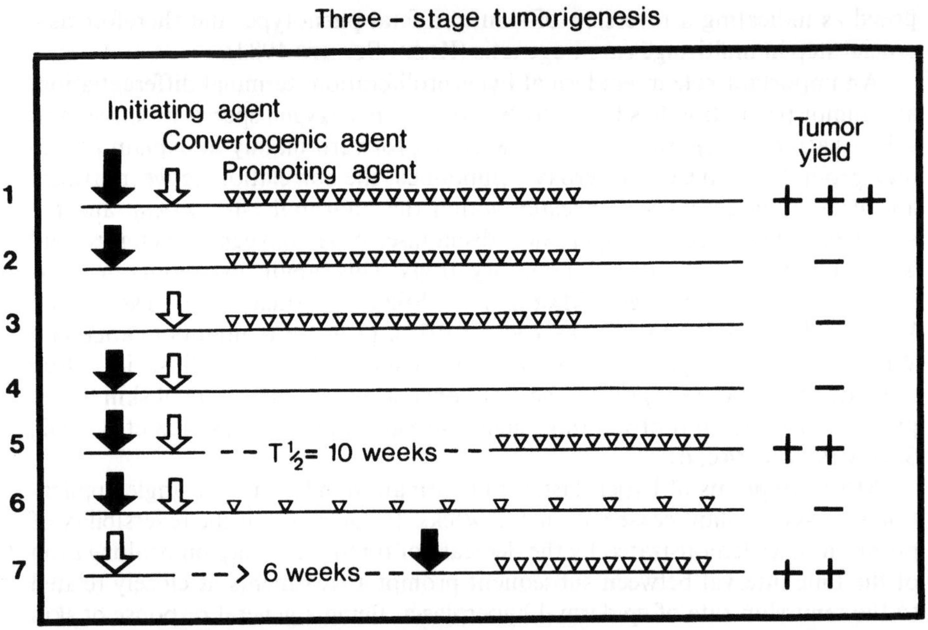

Fig. 6. Three-stage tumorigenesis in NMRI mouse skin. The scheme illustrates several features of the experimental approach. Papilloma formation depends on combined treatment with an initiating, a convertogenic (converting), and a promoting agent (*1*); a single application is sufficient for initiation and conversion whereas promotion requires repeated treatments (*1*); if one of the treatments is omitted, no papilloma development is observed (*2, 3, 4*); between conversion and promotion the experiment can be interrupted for several weeks (*5*); the tumor yield drops drastically when the time interval between subsequent promoter treatments is increased (*6*); the initiation-conversion sequence can be inverted (*7*)

of a chronic hyperproliferative reaction, this would mean that, in the absence of conversion, an initiated cell is unable to respond to the mitogenic effect of the promoter. It has been proposed (MARKS and FÜRSTENBERGER 1987) that initiation occurs in an immature stem cell population of the epidermis (as indicated, for instance, by the irreversible character of initiation) and that such stem cells may lack expression of one or another constituent (possibly an enzyme) required for the hyperplastic response. The role of the converting agent would then be to induce the formation of this constituent. On the other hand, an initiated cell may well respond to the mitogenic effect of a promoter, but, as already proposed by BOUTWELL (1964), may be unable to express its neoplastic phenotype in the absence of conversion. Such a hypothesis seems to be in conflict with the observation that conversion can be carried out prior to initiation (FÜRSTENBERGER et al. 1985). It could well be, however, that the converting agent creates a rather persistent situation in the tissue which allows spontaneous phenotypic expression of initiated cells, provided that initiation occurs within the time interval when the tissue is in the converted state.

At present, our knowledge is not sufficient for a final decision concerning these two hypotheses, which do not necessarily exclude each other. Moreover, we do not know whether conversion is a general phenomenon of neoplastic development or whether it is restricted to the mouse skin model in vivo. It must be emphasized, however, that converting potency is not a speciality of exotic chemical compounds such as the phorbol esters, but that skin wounding exhibits a strong convertogenic effect (FÜRSTENBERGER and MARKS 1983). This indicates that a converting agent induces pathways or mimics factors which are normally involved in the wound response. It appears as if a converting agent such as TPA makes the tissue "believe" that it is wounded. For conversion, the wound must be deep, i.e., include dermis and blood vessels. Superficial wounding such as removal of the horny layer of epidermis induces hyperplastic transformation, but does not exhibit convertogenic efficacy. This indicates a difference in the response of skin to superficial versus more extended wounding which may reflect the difference between promotion and conversion. Recently an intradermal injection of the putative wound hormones TGFα and TGFβ (transforming growth factors) has been found to provide a converting stimulus in initiated mouse skin (MARKS et al. 1988).

VI. Co-carcinogenesis

The term promotion is defined on an operational basis, i.e., the multistage protocol of carcinogenesis is characterized by the sequence initiation/conversion-promotion. Promotion has been regarded as a special case of co-carcinogenesis (HECKER 1985), although most authors prefer to restrict the latter term to experiments in which the effect of a carcinogen is augmented by *simultaneous* (instead of subsequent) administration of a non-carcinogenic agent (VAN DUUREN 1978; BOUTWELL 1964, 1974).

In addition to its precise operational definition, the term promotion implies a distinct conception of the underlying mechanisms, i.e., clonal expansion of tumor cells. Compared with promotion, the term co-carcinogenesis is poorly defined. It comprises effects which are, for instance, due to a facilitation of penetration and metabolic activation of the carcinogen or to a deterioration of protecting mechanisms such as DNA repair. Whereas co-carcinogenesis may, thus, be understood as an augmentation of initiation, promotion is defined as completing the carcinogenic process started with initiation/conversion. Several compounds such as catechol, pyrogallol, certain alkanes, and pyrene have been shown to be co-carcinogenic in mouse skin, rather than exhibiting promoting activity under the two-stage regimen.

VII. The Potential Importance of the Multistage Model for Prevention of Human Cancer

The multistage model of skin carcinogenesis is a highly artificial approach which allows the investigator to dissect the complex pattern of tumor development so that the process can be analyzed stage by stage. The argument that such a situation never occurs in real life and the model, therefore, is irrelevant for the human situation is frequently raised, but nevertheless trivial, since one should not demand more from an animal model than it can provide, i.e., an insight into biologic pathways along which tumors may develop (for a more extensive discussion of this subject see MARKS and FÜRSTENBERGER 1987). The most important results which have emerged from the skin model are:

1. Tumor development can be induced by a combination of treatments, each of which appears to be harmless when carried out alone;
2. The initiating effect of a carcinogenic agent on the tissue is irreversible, but remains silent provided the dose of the agent is low; it is most probably due to an alteration at the level of the cellular genome, perhaps involving proto-oncogene activation which occurs with a low degree of probability;
3. The developmental stages of skin carcinogenesis, i.e., conversion and promotion, proceed along physiologic pathways normally involved in defense against harmful external influences, wound response, and tissue regeneration; they facilitate the formation of papillomas from initiated cells;
4. The progression from the benign to the malignant state probably requires and additional effect on the cellular genome of a papilloma cell; the preceding clonal expansion of papilloma cells owing to promotion greatly increases the probability of such a genetic alteration.

The close relationship of conversion and promotion to physiologic processes raises the possibility that the development of tumors from initiated cells could be stimulated not only by exogenous (environmental) influences, but also by endogenous factors. This may help to explain the well-known effect of certain hormones on human cancer development.

Epidemiologic studies indicate that promoting influences may be important, or even critical, in the development of certain human cancers. The most striking evidence for this is the rather long latency period of most malignant diseases. In addition, risk reversibility and the synergistic effect of many carcinogenic agents show a striking parallelism to the situation in the skin model. For example, the carcinogenic effects of tobacco smoke and asbestos or ionizing radiation potentiate each other, as do the carcinogenic effects of initiators and promoters in the animal model. The risk of getting lung cancer decreases once smoking is stopped, showing that the process has a reversible component. In fact, this is again exactly what one would predict from the skin model, where the risk of getting skin carcinomas is drastically reduced when promotoer treatment is stopped before the end point of papilloma development is reached. Another practical impact of the multistage model is that it may help to develop measures for cancer chemoprevention, i.e., it may provide the opportunity for a patient to live with initiated or even fully transformed cells, provided one is able to inhibit the developmental stages of carcinogenesis. As mentioned already, conversion and/or promotion in mouse skin can be effectively suppressed by local treatment with vitamin A acid and its derivatives, by glucocorticosteroids, non-steroidal anti-phlogistic agents, antioxidants, and by the immunosuppressive drug cyclosporin A. It must be emphasized, however, that in other organs multistage carcinogenesis may proceed along other cellular pathways and may, therefore, be sensitive to other types of inhibitors. It may thus be impossible to design a generalized scheme for the chemoprevention of the post-initiation stages of tumor development by extrapolating the results obtained by studying experimental skin carcinogenesis. Rather, preventive measures have to be determined specifically for every tissue and tumor type.

References

Andrade R, Gumpert SL, Popkin GL, Rees TD (eds) (1976) Cancer of the skin. Saunders, Philadelphia

Ashendel CL (1985) The phorbol ester receptor: a phospholipid-regulated protein kinase. Biochim Biophys Acta 822:219–242

Balmain A, Brown K (1988) Oncogene activation in chemical carcinogenesis. Adv Cancer Res 51:147–183

Berenblum I (1941) The cocarcinogenic action of croton resin. Cancer Res 1:44–48

Berenblum I, Haran-Ghera N (1957) A quantitative study of the systemic initiating action of urethan (ethyl carbamate) in mouse skin carcinogenesis. Br J Cancer 11:77–84

Berenblum I, Shubik P (1947) A new, quantitative approach to the study of stages of chemical carcinogenesis in mouse skin. Br J Cancer 1:383–391

Bickers DR (1983) Drug, carcinogen, and steroid hormone metabolism in the skin. In: Goldsmith LA (ed) Biochemistry and physiology of the skin. Oxford University Press, New York, pp 1169–1186

Bigger CAH, Sawicki JT, Blake DM, Raymond LG, Dipple A (1983) Products of binding of 7,12-dimethylbenz[a]anthracene to DNA in mouse skin. Cancer Res 43:5647–5651

Blumberg P (1980) In vitro studies on the mode of action of the phorbol esters, potent tumor promoters. CRC Crit Rev Toxicol 8:153–234

Bootsma D, Zelle B, Keijzer W (1979) Deficient DNA repair in xeroderma pigmentosum. Stud Biophys 76:177–184

Boutwell RK (1964) Some biological aspects of skin carcinogenesis. Progr Exp Tumor Res 4:207–250

Boutwell RK (1974) The function and mechanism of promoters of carcinogenesis. CRC Crit Rev Toxicol 2:419–443

Burns FJ, Vanderlaan M, Snyder E, Albert RE (1978) Induction and progression kinetics of mouse skin papillomas. In: Slaga TJ, Sivak A, Boutwell RK (eds) Mechanisms of tumor promotion and cocarcinogenesis. Raven, New York, pp 91–96

Burns FJ, Albert RE, Altshuler B (1984) Cancer progression in mouse skin. In: Slaga TJ (ed) Mechanism of tumor promotion, vol II. CRC, Boca Raton Fl, pp 18–39

Camerman N, Camerman A (1968) Photodimer of thymine in ultraviolet-irradiated DNA: proof of structure by X-ray diffraction. Science 160:1451

Carter DM, Ross PM (1983) Cutaneous DNA damage and repair. In: Goldsmith LA (ed) Biochemistry and physiology of skin. Oxford University Press, New York, pp 292–317

Castagna M, Takai Y, Kaibuchi K, Sano K, Kikkawa V, Nishizuka Y (1982) Direct activation of calcium-activated, phospholipid-dependent proteinkinase by tumor-promoting phorbol esters. J Biol Chem 257:7847–7851

Cerutti PA (1985) Prooxidant states and tumor promotion. Science 227:375–381

Cleaver JE (1968) Defective repair of DNA in xeroderma pigmentosum. Nature 218:652–656

Cleaver JE (1972) Xeroderma pigmentosum: variants with normal DNA repair and normal sensitivity to UV light. J Invest Dermatol 58:124–128

Cleaver JE, Trosko JE (1970) Absence of excision of UV-induced cyclobutane dimers in xeroderma pigmentosum. Photochem Photobiol 11:547–550

Colburn NH (1979) The use of tumor promoter responsive experimental cell lines to study preneoplastic progression. In: Franks LM, Wigley CB (eds) Neoplastic transformation in differentiated epithelial cell systems in vitro. Pergamon, London, pp 112–134

Committee on the Biological Effects of Ionizing Radiation, Division of Medical Sciences, Assembly of Life Sciences and National Research Council (1980) The effects on populations of exposure to low levels of ionizing radiation. National Academy of Sciences of the United States of America, Washington DC

Cooper CS, Ribeiro O, Hewer A, Walsh C, Grover PL, Sims P (1980) Additional evidence for the involvement of the 3,4-diol-1,2-oxides in the metabolic activation of 7,12-DMBA in mouse skin. Chem Biol Interact 29:357–367

Cuiffo G (1907) Innesto positivo con filtrate di verruca volgare. Ital Mal Vener 48:12–17

Deelman HT (1924) Die Entstehung des experimentellen Teerkrebses und die Bedeutung der Zellregeneration. Z Krebsforsch 21:220–226
Diamond L, O'Brien TG, Baird WM (1980) Tumor promoters and the mechanism of tumor promotion. Adv Cancer Res 32:1–74
DiGiovanni J, Juchau MR (1980) Biotransformation and bioactivation of 7,12-dimethylbenz[a]anthracene (7,12-DMBA). Drug Metab Rev 11:61–101
DiGiovanni J, Slaga TJ, Berry DL, Juchau MR (1980) Inhibitory effects of environmental chemicals on polycyclic aromatic hydrocarbon carcinogenesis. In: Slaga TJ (ed) Modifiers of chemical carcinogenesis. Raven, New York, pp 145–168
Doniger J, Jacobson ED; Krell K, DiPaolo JA (1981) UV-light action spectra for neoplastic transformation, and lethality of Syrian hamster embryo cells correlate with that for pyrimidine dimer formation in cellular DNA. Proc Natl Acad Sci USA 78:2378–2382
Fischer E, Keijzer W, Thielmann HW, Popanda O, Bohnert A, Edler L, Jung EG, Bootsma D (1985) A ninth complementation group in xeroderma pigmentosum. Mutat Res 145:217–225
Fischer SM (1984) The role of prostaglandins in tumor promotion. In: Slaga TJ (ed) mechanisms of tumor promotion, vol II. CRC, Boca Raton, FL, pp 113–126
Fischer SM, Slaga TJ (eds) (1985) Arachidonic acid metabolism and tumor promotion. Nijhoff, Boston
Fischer SM, Fürstenberger G, Marks F, Slaga TJ (1987) Events associated with mouse skin tumor promotion with respect to arachidonic acid metabolism: A comparison between Sencar and NMRI mice, Cancer Res 87:3174–3179
Foulds L (1954) The experimental study of tumor progression: a review. Cancer Res 14:327–339
Foulds L (1975) Neoplasia of the skin. In: Foulds L, Neoplastic development, vol III. Academic Press, London, pp 17–108
Frieben A (1902) Demonstration eines Cancroids des rechten Handrückens, das sich nach langdauernder Einwirkung von Röntgenstrahlen entwickelt hat. Fortschr Röntgenstr 6:106–111
Friedberg EC (1985) DNA repair. Freeman, New York
Friedwald WF, Rous P (1944) The initiating and promoting elements in tumor production. J Exp Med 80:102–125
Fry RJM, Ley RD (1984) Ultraviolet radiation carcinogenesis. In: Slaga TJ (ed) Mechanisms of tumor promotion, vol II. Tumor promotion and skin carcinogenesis, pp 74–96. CRC, Boca Raton
Fürstenberger G, Hecker E (1986) Highly unsaturated irritant diterpene esters from *Euphorbia tirucalli* L. originating from Madagascar. J Nat Prod 49:386–398
Fürstenberger G, Marks F (1983) Growth stimulation and tumor promotion in skin. J Invest Dermatol 81:157s–161s
Fürstenberger G, Marks F (1985) Prostaglandins, epidermal hyperplasia and skin tumor promotion. In: Fischer SM, Slaga TJ (eds) Arachidonic acid metabolism and tumor promotion. Nijhoff, Boston, pp 50–72
Fürstenberger G, Berry DL, Sorg B, Marks F (1981) Skin tumor promotion by phorbol esters is a two-stage process. Proc Natl Acad Sci USA 78:7722–7726
Fürstenberger G, Kinzel V, Schwarz M, Marks F (1985) Partial inversion of the initiation-promotion sequence of multistage tumorigenesis in the skin of NMRI mice. Science 230:76–78
Fujiki H, Suganuma M, Tahira T, Yoshioka A, Nahayasu M, Endo Y, Shudo K, Takayama S, Moore RE, Sugimura T (1984) New classes of tumor promoters. In: Fujiki H, Hecker E, Moore RE, Sugimura T, Weinstein IB (eds) Cellular interactions by environmental tumor promoters. Japan Scientific Societies Press, Tokyo/VNU Science, Utrecht, pp 37–48
Fusenig NE (1981) In vitro systems for the study of skin carcinogenesis. In: Laerum OD, Iversen OH (eds) Biology of skin cancer, excluding melanomas. UICC Technical report no 13. UICC, Geneva, pp 120–123
Gianelli F (1978) Xeroderma pigmentosum and the role of DNA repair in oncogenesis. Bull Cancer (Paris) 65:323–334

Gissmann L (1984) Papillomaviruses and their association with cancer in animals and man. Cancer Surv 3:161–181
Gissmann L, Wolnik L, Ikenberg H, Koldovsky U, Schnürch HG, Zur Hausen H (1983) Human papilloma virus types 6 and 11 DNA sequences in genital and laryngeal papillomas and in some cervical cancers. Proc Natl Acad Sci USA 80:560–563
Goerttler K, Loehrke H (1977) Diaplacental carcinogenesis: tumor localization and tumor incidence in NMRI mice after diaplacental initiation with DMBA and urethane and postnatal promotion with the phorbol ester TPA in a modified 2-stage Berenblum Mottram experiment. Virchows Arch [A] 376:117–122
Goerttler K, Loehrke H, Schweizer J, Hesse B (1980) Positive two-stage carcinogenesis in female Sprague-Dawley rats using DMBA as initiator and TPA as promoter. Virchows Arch [A] 385:181–186
Goustin AS, Leof EB, Shipley GD, Moses HL (1986) Growth factors and cancer. Cancer Res 46:1015–1029
Greither A, Tritsch H (1957) Die Geschwülste der Haut. Thieme, Stuttgart
Gross L (1983) Carcinogenic viruses, vol I. Pergamon, Oxford, pp 48–83
Gschwendt M, Kittstein W, Marks F (1987) Cyclosporin A inhibits phorbol ester-induced cellular proliferation and tumor promotion as well as phosphorylation of a 100 kD protein in mouse epidermis. Carcinogenesis 8:203–207
Hart RW, Setlow RB, Woodhead AD (1977) Evidence that pyrimidine dimers in DNA can give rise to tumors. Proc Natl Acad Sci USA 74:5574–5578
Hecker E (1978) Structure-activity relationships in diterpene esters irritant and cocarcinogenic to mouse skin. In: Slaga TJ, Sivak A, Boutwell RK (eds) Mechanisms of tumor promotion and cocarcinogenesis. Raven, New York, pp 11–48
Hecker E (1985) Chemische Karzinogenese. In: Gross R, Schmidt CG (eds) Klinische Onkologie. Thieme, Stuttgart, pp 5.1–5.17
Hecker E, Schmidt R (1974) Phorbolesters – the irritants and cocarcinogens of *Croton tiglium* L. Fortschr Chem Org Naturst 31:377–467
Hecker E, Fusenig NE, Kunz W, Marks F, Thielmann HW (eds) (1982) Cocarcinogenesis and biological effects of tumor promoters. Raven, New York
Helm F (ed) (1979) Cancer dermatology. Lea and Febiger, Philadelphia
Hennings H, Shores R, Wenk ML, Spangler EF, Tarone R, Yuspa SH (1983) Malignant conversion of mouse skin tumours is increased by tumour initiators and unaffected by tumour promoters. Nature 304:67–69
Hill J (1761) Caution against the immoderate use of snuff. Founded on the human qualities of the tobacco plant. Baldwin and Jackson, London
Honn KV, Bockman RS, Marnett LJ (1981) Prostaglandins and cancer: a review of tumor initiation through tumor metastasis. Prostaglandins 21:833–864
Jarrett WFH (1978) Transformation of warts to malignancy in alimentary carcinoma of cattle. Bull Cancer (Paris) 65:191–194
Jarrett WFH, McNeil PE, Laird HM, O'Neil BW, Murphy J, Campo MS, Moar MH (1980) Papilloma viruses in benign and malignant tumors of cattle. In: Essex M, Todaro G, Zur Hausen H (eds) Viruses in naturally occurring cancer. Cold Spring Harbor Laboratory Press New York, pp 215–222
Jerina DM, Sayer JM, Thakker DR, Yagi H (1980) Carcinogenicity of polycyclic aromatic hydrocarbons: the bay-region theory. In: Pullman B, Ts'o POP, Gelboin H (eds) Carcinogenesis: fundamental mechanisms and environmental effects. Reidel, Dordrecht, pp 1–12
Johnson BE (1978) Formation of thymine containing dimers in skin exposed to UV light. Bull Cancer (Paris) 65:283–298
Jung EG (1978) Xeroderma pigmentosum: heterogenous syndrome and model for UV carcinogenesis. Bull Cancer (Paris) 65:315–322
Kennaway EL (1925) The identification of a carcinogenic compound in coal-tar. Br Med J 2:749–752
Kinzel V, Loehrke H, Goerttler K, Fürstenberger G, Marks F (1984) Suppression of the first stage of TPA-effected tumor promotion in mouse skin by nontoxic inhibition of DNA synthesis. Proc Natl Acad Sci USA 81:5858–5862

Kinzel V, Fürstenberger G, Loehrke H, Marks F (1986) Three-stage tumorigenesis in mouse skin: DNA synthesis as a prerequisite for the conversion stage induced by TPA prior to initiation. Carcinogenesis 7:779–782

Kirchner H (1986) Immunobiology of human papillomavirus infection. Progr Med Virol 33:1–41

Klein-Szanto AJP (1984) Morphological evaluation of tumor promoter effects on mammalian skin. In: Slaga TJ (ed) Mechanisms of tumor promotion, vol II. CRC, Boca Raton, pp 41–72

Kripke M, Urbach F, Witkop C (1981) Ultraviolet radiation carcinogenesis. In: Laerum OD, Iversen OH (eds) Biology of skin cancer, excluding melanomas. UICC technical report no 13. UICC, Geneva, pp 195–222

Laerum OD, Iversen OH (eds) (1981) Biology of skin cancer, excluding melanomas. UICC technical report no 13. UICC, Geneva

Lancaster WD, Olson C (1978) Demonstration of two distinct classes of bovine papilloma viruses. Virology 89:372–379

Lehmann AR, Kirk-Bell S, Arlett CF, Paterson MC, Lohmann PHM, deWeerd-Kastelein EA, Bootsma D (1975) Xeroderma pigmentosum cells with normal levels of excision repair have a defect in DNA synthesis after UV-irradiation. Proc Natl Acad Sci USA 72:219–223

Lehmann AR, Kirk-Bell S, Arlett CF, Harcourt SA, deWeerd-Kastelein EA, Keijzer W, Hall-Smith P (1977) Repair of UV light damage in a variety of human fibroblast cell strains. Cancer Res 37:904–910

Levin W, Wood A, Chang R, Ryan D, Thomas P, Yagi H, Thakker D, Yvas K, Boyd C, Chu SY, Conney AH, Jerina DM (1982) Oxidative metabolism of polycyclic aromatic hydrocarbons to ultimate carcinogens. Drug Metab Rev 13:555–580

Luger A, Gschnaidt G (eds) (1983) Dermatologische Onkologie. Urban and Schwarzenberg, Vienna

Lutzner MA (1978) Epidermodysplasia verruciformis. Bull Cancer (Paris) 65:169–182

Marks F (1981) Tumor-promoting phorbol esters. In: Laerum OD, Iversen OH (eds) Biology of skin cancer, excluding melanomas. UICC technical report no. 13. UICC, Geneva, pp 137–147

Marks F (1987) What's new in oncogenes and growth factors? Path Res Pract 182:831–848

Marks F, Fürstenberger G (1984) Multistage tumor promotion in skin. In: Fujiki H, Hecker E, Moore RE, Sugimura T, Weinstein IB (eds) Cellular interactions by environmental tumor promoters. Japan Scientific Societies Press, Tokyo/VNU Science, Utrecht, pp 273–287

Marks F, Fürstenberger G (1985) Tumor promotion in skin: are active oxygen species involved? In: Sies H (ed) Oxidative stress. Academic Press, London, pp 437–475

Marks F, Fürstenberger G (1987) Multistage carcinogenesis in animal skin: the reductionist's approach in cancer research. In: Iversen OH (ed) Theories of carcinogenesis. Hemisphere, Washington DC, pp 179–190

Marks F, Bertsch S, Grimm W, Schweizer J (1978) Hyperplastic transformation and tumor promotion in mouse epidermis: possible consequences of disturbances of endogenous mechanisms controlling proliferation and differentiation. In: Slaga TJ, Sivak A, Boutwell RK (eds) Mechanisms of tumor promotion and cocarcinogenesis. Raven, New York, pp 97–116

Marks F, Bertsch S, Fürstenberger G (1979) Ornithine decarboxylase activity, cell proliferation and tumor promotion in mouse epidermis in vivo. Cancer Res 39:41–4188

Marks F, Bertsch S, Fürstenberger G, Richter H (1983 a) Growth control in mouse epidermis – facts and speculations. In: Wright NA, Camplejohn RS (eds) Psoriasis: cell proliferation. Churchill Livingstone, Edinburgh, pp 173–188

Marks F, Fürstenberger G, Ganss M, Richter H, Seemann D (1983 b) Hyperplastic transformation, the response of skin to irritation. Br J Dermatol 109 (Suppl 25):18–21

Marks F, Fürstenberger G, Gschwendt M, Rogers M, Schurich B, Kaina B, Bauer G (1988) The wound response as a key element for an understanding of multistage carcinogenesis in skin. In: Feo F, Pani P, Columbano A, Garcea R (eds) Chemical carcinogenesis. Plenum, New York, pp 217–234

McEvoy BF (1979) Genodermatoses associated with malignancies. In: Helm F (ed) Cancer dermatology. Lea and Febiger, Philadelphia, pp 39–55

Melczer N (1961) Präcancerosen und primäre Krebse der Haut. Akadémiai Kiado, Budapest

Miller EC, Miller JA (1971) Chemical carcinogenesis: mechanisms and approaches to its control. J Natl Cancer Inst 47(3):V

Miller EC, Miller JA (1976) The metabolism of chemical carcinogens to reactive electrophiles and their possible mechanisms of actions in carcinogenesis. In: Searle CE (ed) Chemical carcinogens. American Chemical Society, Washington DC, pp 737–762 (ACS Monogr 173).

Moll R, Franke WW, Schiller DL, Geiger B, Krepler R (1982) The catalogue of human cytokeratins: patterns of expression in normal epithelia, tumors and cultured cells. Cell 31:11–24

Mottram JC (1944) A developing factor in experimental blastogenesis. J Pathol Bacteriol 56:181–187

Orth G, Breitburd F, Faure M, Croissant O (1977) Papilloma viruses: possible role in human cancer. In: Hiatt HH, Watson JD, Wisten JA (eds) Origin of human cancer. Cold Spring Harbor, New York, pp 1043–1068

Orth G, Faure M, Breiburd F, Croissant O, Jablonska S, Obalek S, Jarzabek-Chorzelska M, Rzesa G (1980) Epidermodysplasia verruciformis: a model for the role of papilloma viruses in human cancer. In: Essex M, Todaro G, Zur Hausen H (eds) Viruses in naturally occurring cancer, vol A. Cold Spring Harbor, New York, pp 259–282

Parkinson EK (1985) Defective responses of transformed keratinocytes to terminal differentiation stimuli. Their role in epidermal tumour promotion by phorbol esters and by deep skin wounding. Br J Cancer 52:479–493

Parrish JA (1983) Responses of skin to visible and ultraviolet radiation. In: Goldsmith LA (ed) Biochemistry and physiology of the skin. Oxford University Press, New York, pp 713–733

Peters J, Müller R (eds) (1981) Präkanzerosen und Papillomatosen der Haut. Springer, Berlin Heidelberg New York

Pott P (1775) The surgical works of Percivall Pott, vol 5. Haws, Clarke and Collins, London, pp 60–68

Potten CS, Allen TD (1975) The fine structure and cell kinetics of mouse epidermis after wounding. J Cell Sci 17:413–447

Prunieras M (ed) (1978) On skin carcinogenesis. Bull Cancer (Paris) 65:141–362

Quintanilla M, Brown K, Ramsden M, Balmain A (1986) Carcinogen-specific mutation and amplification of Ha-*ras* during mouse skin carcinogenesis. Nature 322:78–80

Reddy AL, Fialkow PJ (1983) Papillomas induced by initiation-promotion differ from those induced by carcinogen alone. Nature 304:69–71

Ripoll EA, Webber MM (1986) Skin cancer: Geographic distribution and risk factors. In: Webber MM, Sekely LI (eds) In vitro models for cancer research, vol III. Carcinomas of the mammary gland, uterus and skin. CRC, Boca Raton, pp 232–243

Ritchie AC, Saffiotti U (1955) Orally administered 2-acetyl-aminofluorene as an initiator and as a promoter in epidermal carcinogenesis in the mouse. Cancer Res 15:84–88

Roop DR, Lowy DR, Tambourin PE, Strickland J, Harper JR, Balaschak M, Spangler EF, Yuspa SH (1986) An activated Harvey *ras* oncogene produces beningn tumours on mouse epidermal tissue. Nature 323:822–824

Rose-John S, Fürstenberger G, Krieg P, Besemfelder E, Rincke G, Marks F (1988) Differential effects of phorbol esters on c-fos and c-myc and ornithine decarboxylase gene expression in mouse skin in vivo. Carcinogenesis 9:831–835

Rous P, Kidd JG (1938) The carcinogenic effects of papilloma viruses on the tarred skin of rabbits. I. Description of the phenomenon. J Exp Med 67:399–427

Rous P, Kidd JG (1941) Conditional neoplasms and subthreshold neoplastic states. A study of tumors in rabbits. J Exp Med 73:365–389

Schoepe KB, Friesel H, Schurdak ME, Randerath K, Hecker E (1986) Comparative DNA binding of 7,12-DMBA and some of its metabolites in mouse epidermis in vivo as revealed by the ^{32}P-postlabeling technique. Carcinogenesis 7:535–540

Schweizer J (1979) Neogenesis of functional hair follicles in adult mouse skin selectively induced by the tumor-promoting phorbol esters. Experientia 35:1651–1653
Schweizer J, Marks F (1977) The tumor promoter TPA induces the formation of new hair follicles in the epidermis of the mouse tail. Cancer Res 37:4195–4201
Scotto J, Fraumeni JF (1982) Skin cancer (other than melanoma). In: Schottenfeld D, Fraumeni JF (eds) Cancer epidemiology and prevention. Saunders, Philadelphia, pp 996–1011
Scotto J, Fears TR, Fraumeni JF (1983) Incidence of nonmelanoma skin cancer in the United States. US Department of Health and Human Services, pp 8–22, NIH Publ No 83-2433
Setlow RP, Regan JD, German J, Carrier WL (1969) Evidence that xeroderma pigmentosum cells do not perform the first step in the repair of UV damage to their DNA. Proc Natl Acad Sci USA 64:1035–2041
Sims P (1980) The metabolic activation of chemical carcinogens. Br Med Bull 36:11–18
Sisskin EE, Gray T, Barrett JC (1982) Correlation between sensitivity to tumor promotion and sustained epidermal hyperplasia of mice and rats treated with TPA. Carcinogenesis 3:403–408
Slaga TJ (1980) Antiinflammatory steroids: potent inhibitors of tumor promotion. In: Slaga TJ (ed) Modifiers of chemical carcinogenesis. Raven, New York, pp 111–126
Slaga TJ (ed) (1984 a) Mechanisms of tumor promotion, vol I–IV. CRC, Boca Raton
Slaga TJ (ed) (1984 b) Mechanisms of tumor promotion, vol III. Tumor promotion and carcinogenesis in vitro. CRC, Boca Raton
Slaga TJ (1984 c) Mechanisms involved in two-stage carcinogenesis in mouse skin. In: Slaga TJ (ed) Mechanisms of tumor promotion, vol II. Tumor promotion and skin carcinogenesis. CRC, Boca Raton, pp 1–16
Slaga TJ (1984 d) Multistage skin tumor promotion and specificity of inhibition. In: Slaga TJ (ed) Mechanisms of tumor promotion, vol II. CRC, Boca Raton, pp 189–196
Slaga TJ, Sivak A, Boutwell RK (eds) (1978) Mechanisms of tumor promotion and cocarcinogenesis. Raven, New York
Slaga TJ, Fischer SM, Nelson K, Gleason GL (1980) Studies on the mechanism of skin tumor promotion: evidence for several stages of promotion. Proc Natl Acad Sci USA 77:3659–3663
Slaga TJ, Triplett LL, Yotti LP, Trosko JE (1981) Skin tumor promoting activity of benzoyl peroxide, a widely used free radical generating compound. Science 213:1023–1025
Slaga TJ, Fischer SM, Weeks CE, Nelson K, Mamrack M, Klein-Szanto AJP (1982) Specificity and mechanism(s) of promoter inhibitors in multistage promotion. In: Hecker E, Fusenig NE, Kunz W, Marks F, Thielman HW (eds) Cocarcinogenesis and biological effects of tumor promoters. Raven, New York, pp 19–34
Sutherland BM, Rice M, Wagner EK (1975) Xeroderma pigmentosum cells contain low levels of photoreactivating enzyme. Proc Natl Acad Sci USA 72:103–107
Sutherland BM, Cimino JS, Delihas N, Shih AG, Oliver RP (1980) UV-induced transformation of human cells to anchorage-independent growth. Cancer Res 40:1934–1939
Syverton JT (1952) The pathogenesis of rabbit papilloma-to-carcinoma sequence. Ann NY Acad Sci 54:1126–1140
Taguchi R, Yokoyama M, Kitamura Y (1984) Intraclonal conversion from papilloma to carcinoma in the skin of $PgK\text{-}1^a/PgK\text{-}1^b$ mice treated by a complete carcinogen protocol or by an initiation-promotion regimen. Cancer Res 44:3779–3782
Takebe H (1979) Xeroderma pigmentosum: DNA repair defects and skin cancer. Gann Monogr Cancer Res 24:103–117
Thielmann HW (1984) Enzymology of DNA repair: a survey. In: Greim H, Jung R, Kramer M, Marquardt H, Oesch F (eds) Biochemical basis of chemical carcinogenesis. Raven, New York, pp 233–256
Thielmann HW, Edler L, Popanda O, Friemel S (1985) Xeroderma pigmentosum patients from the Federal Republic of Germany: decrease in post-UV colony-forming ability in 30 xeroderma pigmentosum fibroblast strains is quantitatively correlated with a decrease in DNA-incising capacity. J Cancer Res Clin Oncol 109:227–240

Troll W, Wiesner R (1983) Protease inhibitors as anticarcinogens and radioprotectors. In: Nygaard OF, Simic MG (eds) Radioprotectors and anticarcinogens. Academic Press, New York, pp 567–574
Troll W, Wiesner R (1985) The role of oxygen radicals as a possible mechanism of tumor promotion. Annu Rev Pharmacol Toxicol 25:509–528
Trosko JE, Chang CC (1984) Role of intercellular communication in tumor promotion. In: Slaga TJ (ed) Mechanisms of tumor promotion, vol IV. CRC, Boca Raton, pp 119–146
Twort CC, Ing HR (1928) Untersuchungen über krebserzeugende Agenzien. Z Krebsforsch 27:309–351
Unna P (1884) Histopathologie der Hautkrankheiten. Hirschwald, Berlin
Urbach F (ed) (1963) Conference on biology of cutaneous cancer. US Dept of Health, Education and Welfare, Bethesda (NCI Monogr 10)
Urbach F (1980) Ultraviolet radiation and skin cancer in man. Prev Med 9:227-
Urbach F (1983) Genese der Hautkarzinome. In: Luger A, Gschnaidt G (eds) Dermatologische Onkologie. Urban and Schwarzenberg, Vienna, pp 4–16
Urbach F, Witkop CJ, Laerum OD (1981) Skin cancer in man. In: Laerum OD, Iversen OH (eds) Biology of skin cancer excluding melanomas. UICC technical report no 13. UICC, Geneva, pp 56–58
Van Duuren BL (1978) Tumor-promoting and cocarcinogenic agents in chemical carcinogenesis. In: Searle CE (ed) Chemical carcinogens. American Chemical Society, Washington DC, pp 24–51 (ACS Monogr 173)
Van Duuren BL, Orris L (1965) The tumor-enhancing principles from *Croton tiglium* L. Cancer Res 25:1871–1875
Verma AK, Shapas BG, Rice HM, Boutwell RK (1979) Correlation of the inhibition by retinoids of tumor-promoter-induced mouse epidermal ornithine decarboxylase activity and of skin tumor promotion. Cancer Res 39:419–425
Vigny P, Kindts M, Cooper CS, Grover PL, Sims P (1981) Fluorescence spectra of nucleoside-hydrocarbon adducts formed in mouse skin treated with 7,12-DMBA. Carcinogenesis 2:115–119
Weeks CE, Slaga TJ, Boutwell RK (1984) The role of polyamines in tumor promotion. In: Slaga TJ (ed) Mechanisms of tumor promotion, vol. II. CRC, Boca Raton, pp 127–142
Wiebel FJ (1980) Activation and inactivation of carcinogens by microsomal monooxygenase: modification by benzoflavones and polycyclic aromatic hydrocarbons. In: Slaga TJ (ed) Modifiers of chemical carcinogenesis. Raven, New York, pp 57–84
Yamagiwa K, Ichikawa K (1914) Über die künstliche Epithelwucherung. Gann 8:11–15
Yuspa SH (1983) Cutaneous carcinogenesis: natural and experimental. In: Goldsmith LA (ed) Biochemistry and physiology of the skin. Oxford University Press, New York, pp 1115–1138
Yuspa SH (1984) Molecular and cellular basis for tumor promotion in mouse skin. In: Fujiki H, Hecker E, Moore RE, Sugimura T, Weinstein IB (eds) Cellular interactions by environmental tumor promoters. Japan Scientific Societies Press, Tokyo VNU Science, Utrecht, pp 315–326
Yuspa SH, Lichti U, Morgan D, Hennings H (1980) Chemical carcinogenesis studies in mouse epidermal cell cultures. In: Bernstein IA, Seiji M (eds) Biochemistry of normal and abnormal epidermal differentiation. University of Tokyo Press, Tokyo
Yuspa SH, Kilkenny AE, Stanley J, Lichti U (1985) Keratinocytes blocked in phorbol ester-responsive early stage of terminal differentiation by sarcoma viruses. Nature 314:459–462
Zur Hausen H (1975) Oncogenic herpes viruses. Biochim Biophys Acta 417:25–53
Zur Hausen (1977) Human papilloma viruses and their possible role in squamous cell carcinoma. Curr Top Microbiol Immunol 78:1–30
Zur Hausen H (1981) Viral carcinogenesis. In: Laerum OD, Iversen OH (eds) Biology of skin cancer, excluding melanomas. UICC technical report no 13. UICC Geneva, pp 228–237
Zur Hausen H (1985) Genital papillomavirus infections. Prog Med Virol 32:15–21
Zur Hausen H, Gissmann L, Schlehofer JR (1984) Viruses in the etiology of human genital cancer. Prog Med Virol 30:170–186

CHAPTER 16

Toxicology of Cosmetics

J. D. Middleton

A. Introduction

In order to establish the safety of a new cosmetic or toiletry product it is necessary to gather a large quantity of information from a wide variety of sources. This chapter gives an account of the type of information that is required and the sources from which much of the information can be obtained. The use of the information in establishing the safety of a product is discussed.

When any chemical comes into contact with the human body, through whatever route, the possibility exists of chemical or physicochemical interactions with the body resulting in a toxic effect. The science of toxicology is concerned with studying these interactions and predicting the extent to which any toxic effects manifested by a chemical or mixture of chemicals are likely to be a problem in humans exposed to anticipated levels.

A cosmetic or toiletry product can of course be regarded as a mixture of chemicals and the principles of toxicology to be applied are the same as those required in other industries such as pharmaceuticals, foods, agrochemicals or general chemicals. The problems of cosmetics and toiletries differ in emphasis from those in other industries in that in no other industry, apart from foods, is the product likely to be deliberately applied to the body by millions of people every day over periods of many years.

The food industry is closely controlled by governments with regard to chemicals which may or may not be used. The cosmetic industry is much less closely controlled. This situation has obvious advantages for the industry, but does bring with it a responsibility for the industry to ensure that its products are as safe as possible within the limits of current toxicological knowledge.

When one considers the degree of exposure of the population to cosmetics and toiletries it is perhaps remarkable that there have not been more problems of toxicity in the past. The small number of known instances of toxic effects attributable to cosmetics and toiletries testifies to the generally safe nature of the ingredients used. However, this is no cause to be complacent and the continual introduction of new ingredients, particularly those designed to have biological effects, means that toxicologists must be increasingly vigilant.

The primary objective of toxicology within the industry must therefore be to prevent the introduction of new ingredients under conditions where they may be harmful or to prevent the combination of otherwise harmless ingredients to produce products which may cause harm to the consumer. To achieve these objectives the toxicologist or the person responsible for product safety requires detailed information on the toxicity of a new product and its ingredients.

B. Safety Data Required

I. Scientific Requirements

There are a large number of possible toxic effects which could occur with cosmetics and toiletries if adequate safety standards were not met. Among the possible toxic effects which must be considered for products and ingredients are the following:

1. Skin irritation
2. Eye irritation
3. Skin sensitisation
4. Photoirritation
5. Photoallergy
6. Sensory effects
7. Systemic toxicity – single exposure effects
8. Systemic toxicity – repeated exposure effects
9. Teratology
10. Carcinogenicity
11. Mutagenicity

In order to arrive at a reasonable conclusion on the potential toxicity of a new product the toxicologist should have information on each of these effects for each ingredient in the product in addition to some information on the effects of the formulated product. For a product with more then 20 ingredients (excluding perfume ingredients) this may seem a daunting task. However, the task is not as impossible as it may seem at first sight. Many ingredients are in wide use and have a history of safe use. Once the safety data on an ingredient have been obtained, they can be used for many products provided the concentration in the product and the degree of human exposure is not greatly increased.

II. Legal Requirements

In most countries there is, at the time of writing, no list of specific data that are required before a cosmetic or toiletry product can be marketed. The usual situation is that there is a law stating that cosmetics must not cause harm and there may be, as in the European Economic Community, lists of potentially harmful ingredients which must not be used. "Positive" lists, that is lists of the only ingredients which are allowed for specific purposes such as preservatives or sunscreens, may become an increasing feature in future legislation. In some countries such as the United States and France there is an almost compulsory requirement to test products, but the tests to be carried out are not specified.

In the United States there is a requirement to label products as not tested for safety if no tests have been carried out and in France there is a requirement to maintain a dossier containing safety data on file for each product. A more detailed consideration of legislative requirements is beyond the scope of this chapter.

C. Sources of Safety Data

I. Scientific Literature

Before embarking on expensive toxicity testing the toxicologist should always make every effort to ascertain that there is no existing knowledge on the toxicology of the proposed ingredient. The scientific literature is one place that must be searched. The enormous task of searching the scientific literature has been made much easier in recent years with the introduction of computerised information retrieval systems. The correct use of these systems enables the toxicologist to obtain lists of references relating to the toxicology of the chemical under investigation.

The problem for the toxicologist in the cosmetics industry is that many ingredients are not pure chemicals or are not well defined chemically. This makes the use of the computer systems difficult and any information obtained may not be easily related to the possible effects of the chemical mixtures that the formulator is wishing to use.

II. Safety Data from Suppliers

If the scientific literature does not contain sufficient information on the toxicology of a product's ingredients, it is to be hoped that the manufacturers and suppliers of the ingredient will have further data. Chemical companies have an increasing legal requirement to obtain sufficient data to ensure the safety of workers handling a chemical. In addition manufacturing comsetic and toiletry companies are increasingly demanding safety data on a new chemical or product ingredient before considering purchase. In recent years it has therefore become increasingly likely that an ingredient supplier will have some safety data on the ingredient and these are usually made available to the toxicologist in the cosmetic company.

III. History of Safe Use

One of the most valuable sources of safety data of a cosmetic ingredient can be a history of safe use. However, it is important to consider carefully the conditions under which the ingredient has been used, the extent to which it has been used and the confidence one can have in the absence of reported adverse effects.

The prior use must relate to the same or greater degree of consumer exposure as is envisaged for the new product. The factors to be considered are the concentration of ingredient in the product, the method of use (e.g. whether the product is left on the skin or rinsed off) and the numbers and nature of the consumers who have used the product.

A history of prior use of products with similar formulations and methods of use to the new product can also be of value. In many cases such data may enable the toxicologist to decide that no further information on the new product is required. The degree to which a new product is allowed to differ from an existing known safe product before further investigation is required is a matter of judgment in each case. For example, if the only change were to lower the surfactant level in a shampoo then it might be considered reasonable to require no further data. However, if the surfactant were replaced by a new one then further investigations might be required.

In using information derived from a history of safe use great care must always be taken to ensure that there really is such a history and that any adverse effects are coming to the attention of the toxicologist. If a product is marketed in a country where consumers do not complain of adverse effects and where the medical profession has no contact with industry then the toxicologist may be unaware of any problems. Equally there must be a good system of recording complaints and ensuring that the toxicologist is aware of any adverse comments from consumers. Ideally the numbers of complaints of a biological nature should be related to the numbers of packs of a product sold. In this way it is possible to establish a normal and acceptble complaint rate for each product type in each country. Provided this rate is not exceeded then it is reasonable to assume that there is in fact a history of safe use.

IV. Safety Testing

If the sources of information considered in Sects. C. I–III have not provided sufficient information for a decision on safety to be made then some testing may be required. The extent of testing undertaken must depend upon the information already available and the method and extent of use of the final product.

All the possible toxic effects presented in Sect. B. I must be considered and a decision made whether or not testing is required for each of these. Examples of the tests that might be carried out for each effect are given in the following sections.

D. Toxic Effects

I. Skin Irritation

Skin irritation can be defined as a direct chemical attack on the skin or its biochemical and physiological processes resulting in a perceptible skin reaction. The degree of irritation is relative with all chemicals, including water, producing some irritation under the right exposure conditions. Everyone will react to an irritant if the exposure is sufficient.

Fully authenticated examples of skin irritation reactions to cosmetics are difficult to find in the scientific and dermatological literature. However, irritant reactions undoubtedly do occur in susceptible individuals. Ingredients such as aluminium salts in anti-perspirants and cationic surfactants are claimed to be the cause of occasional irritation problems (FISHER 1973).

The objective of skin irritation testing is to ensure that an insignificantly small number of people will react under the conditions of use of the product. The test methods available are not absolute. They give comparative results only and it is important to include a control product. The control should be a product with the same method of use and with a history of safe use. The history of safe use should be established as discussed in Sect. C. III. If no such control is available a standard irritant may have to be used.

Skin irritation studies are conventionally carried out in rabbits although other species including humans can be used. In recent years intensive research has been undertaken in an attempt to replace the animal models by in vitro systems. Frequently the methodology follows that published by the United States Government under the Federal Hazardous Substances Labelling Act (U.S FEDERAL REGISTER 1973) or by the French government (JOURNAL OFFICIEL DE LA REPUBLIQUE FRANCAISE 1971). In the rabbit the test materials are applied to intact and abraded skin sites under occlusion for 24 h and reactions of test and control products are compared after removal of the occlusive dressing. In Europe it is becoming a frequent practice to apply the test substance for 4 h. It is usual to use six rabbits for each test material. Occlusion is used as this hydrates the skin thereby increasing penetration and producing a visible reaction with many materials which might otherwise produce no reaction under the test conditions. Some reaction to either test or control product is required in order to make a valid comparison. Similar studies can be carried out in human volunteers with occlusive patch test techniques although the abraded skin sites are omitted. There are many methods for assessing skin irritation potential in human patch tests, including those published by FINKELSTEIN et al. (1963), KLIGMAN and WOODING (1967), PHILLIPS et al. (1972), UTTLEY and VAN ABBE (1973) and SMILES and POLLACK (1977). Before embarking on human studies care must be taken to ensure that there is a low degree of risk of other toxic side effects.

II. Eye Irritation

Some cosmetic products, for example eye make-up, face creams, toilet soaps or shampoos may accidentally enter the eye during use. In addition other products such as hand creams may be transferred to the eye on the fingers while rubbing the eye. In view of the potential damage to vision care must be taken to ensure that cosmetic products do not damage the ocular tissues.

Eye irritation studies are normally carried out in rabbits. The methodology involved is frequently based on modification of the method originally published by DRAIZE et al. (1944). Modifications in common use are those published by the United States Government under the Federal Hazardous Substances Act (U.S FEDERAL REGISTER 1973) and that published by the French Government (JOURNAL OFFICIEL DE LA REPUBLIQUE FRANCAISE 1971).

The test material is placed in one eye, normally in six rabbits, and reactions observed over a period the length of which may depend upon the degree of persistence of any reactions. As with skin irritation testing it is advisable to include control products to assist in the interpretation of results. The criterion of acceptability can then be that the test product is no worse than a control which is known to have been used without problems.

In order not to produce unnecessarily severe reactions in the animals it is normal practice to dilute products, particularly those containing surfactants, to a level where only slight and transient reactions are produced. The reactions to test and control products can then be compared. For example, adult shampoos are normally tested at a concentration of about 10% in water. Experience has shown that at this concentration most shampoos produce little irritation. As with skin irritation testing there have in recent years been strenuous efforts to find in vitro systems to predict eye irritation potential in humans in order to avoid the use of experimental animals altogether.

III. Skin Sensitisation

Skin sensitisation is probably the adverse effect most frequently attributed to cosmetics. The precise incidence of allergic reactions to cosmetics is difficult to determine although SUSKIND (1979) reported that from over 2000 cases of contact dermatitis 4% could be attributed to cosmetics. Similar figures have been quoted in a review by SKOG (1980).

Skin sensitisation is the result of an allergic process where contact with the causative chemical (the allergen) may induce a sensitised state. This process of induction is not accompanied by a visible skin reaction. A subsequent contact of the allergen by a sensitised person may then elicit a skin reaction. Not everybody will become sensitised by an allergen and there are many weak allergens to which only a very small number of people become allergic. An individual may contact a potential allergen regularly for many years before becoming sensitised or may become sensitised on the first contact. One characteristic of sensitisation is that once a person is sensitised only a very small concentration of the allergen may be required to elicit a reaction.

Skin sensitisation studies are normally carried out in guinea pigs although studies in humans are also performed, particularly in the United States. The ethics of such studies in humans are generally considered to be dubious in Europe. The guinea pig is the experimental animal of choice because its allergic responses to topically applied chemicals have been shown to be very similar to those of humans.

There are a number of published methods for determining the sensitisation potential of chemicals in the guinea pig. Perhaps the most widely used in Europe and the most sensitive is the test originally developed by MAGNUSSON and KLIGMAN (1969) and known as the maximisation test. A method developed by BUEHLER (1965) is used more frequently in the United States. A more recent method, specifically developed for cosmetics and toiletries has been published by KLECAK (1977) and is known as the open epicutaneous test.

All the methods have the same principle in that the test animals are treated with the experimental material during an induction phase. This may be by injection or by topical application and the objective is to sensitise the animals to the test material. After a rest period the animals are challenged with the test material under conditions which do not produce irritation. If the animals have become sensitised during the induction phase the challenge application should result in a skin reaction.

It is important that any reactions produced during the challenge application are due to sensitisation and not to irritation. In general irritant reactions occur at concentrations higher than those required to elicit responses in sensitised animals.

Preliminary experiments are therefore required in order to determine the threshold concentration for irritation so that the challenge can be carried out below this concentration. It is also essential to include in the experiment control animals which are treated identically to the test animals during induction except for the omission of test material, and challenged with the test material at the challenge phase. Only if there are reactions in the test group and not in the controls under these circumstances can it be concluded that the test material is a potential sensitiser.

There are numerous examples in the literature of sensitisaton reactions caused by cosmetic ingredients. However, the vast majority of these reports refer to single or very small numbers of cases and reports of significant numbers of people becoming allergic to a single cosmetic ingredient are few.

Two of the comsetic ingredients most frequently cited as being the cause of sensitisation are lanolin, and also the parahydroxybenzoate esters (parabens) used as preservatives (FISHER 1978; FREGERT 1981). Cases of allergy to these ingredients are undoubtedly seen relatively frequently in dermatological clinics. However, it appears that these cases are to a very large extent restricted to patients with dermatitis and cases of allergy to these ingredients in normal people are extremely rare. The implication is that sensitisation occurs more readily in damaged skin than in normal skin.

IV. Photoirritation

The phenomenon of photoirritation is analagous to that of skin irritation except that the presence of sunlight (frequently UV-A irradiation) is required to elicit the reaction. If the skin is exposed to sunlight after treatment with a photoirritant then a reaction may result. An example of photoirritation which has been encountered in the cosmetic industry is that produced by the methoxypsoralens found in undistilled oil of bergamot (MARZULLI and MAIBACH 1970; ZAYNOUN et al. 1977).

Tests of photoirritation of new chemicals or products are normally carried out on the clipped skin of rats or guinea pigs (VINSON and BORSELLI 1966) or on hairless mice (GLOXHUBER 1970). The animals are treated with test material and irradiated with ultraviolet lamps. Control animals are treated but not irradiated, or irradiated but not treated, in order to ensure that any reactions are not due to normal irritation or are not the result of the irradiation alone. Tests for photoirritation in humans have also been developed (KAIDBEY and KLIGMAN 1978).

V. Photoallergy

As with photoirritation the phenomenon of photoallergy is analogous to its non-light-mediated counterpart. Photoallergens are relatively uncommon, but two of the better known examples were discovered through their use in the cosmetic and toiletry industry. These were the germicide tetrachlorosalicylanilide (WILKINSON

1968) and the perfume ingredient 6-methylcoumarin (KAIDBEY and KLIGMAN 1980) neither of which is now used.

Tests for photoallergy are similar to those for conventional sensitisation potential except that the animals are irradiated after treatment during both the induction and challenge phases (VINSON and BORSELLI 1966; ICHIKAWA et al. 1981). Control animals which are not irradiated or not treated are necessary. Techniques for predicitve photoallergy testing in humans have also been published (KAIDBEY and KLIGMAN 1980).

VI. Sensory Effects

Until recently sensory effects of cosmetics and toiletries have not received attention from toxicologists and dermatologists. However, the phenomena of itching, stinging or skin tautness are not uncommonly the subject of consumer complaints. Little is known about these sensory phenomena and their causes although dermatologists have recently recognised that a proportion (given as approximately 10%) of the population may be more susceptible to facial stinging than the majority (FROSCH and KLIGMAN 1977). Stinging appears to occur more readily when the skin is wet and predictive tests have been developed using facial saunas with human volunteers.

This is a new area of product evaluation. The frequency and importance of sensory effects have not yet been propertly evaluated. These effects may have little toxicological significance and may be more important to marketing than to safety evaluation. However, it is possible that a consideration of sensory effects may become an important part of pre-market evaluation in the future.

VII. Systemic Toxicity: Single Exposure Effects

Systemic toxicity, i.e. toxicity within the body, may occur if a product or ingredient enters the body by penetration through the skin, by inhalation or by ingestion. A knowledge of the systemic toxicity following a single exposure, or acute toxicity, is of value if the toxicity of an ingredient is completely unknown and to take account of accidents such as might occur in a factory or if a child drinks a bottle of product.

Acute toxicity is of relatively little value in predicting the effects of repeated daily exposure over long periods as might be encountered in normal use of cosmetics. Acute toxicity is normally measured in terms of the LD_{50}, i.e. the dose required to kill 50% of experimental animals, normally rats or mice. The animals may be exposed to the test material by oral dosing, by inhalation or by application to skin.

Determinations or estimates of the LD_{50} values of manufactured cosmetic products are now rarely performed. Conventional products differ little in their LD_{50} values and determinations in individual products involves the unnecessary use of experimental animals.

Estimates of LD_{50} values are, however, still made when new ingredients are introduced, particularly when there is biological activity. A knowledge of the acute toxicity is necessary to know the handling precautions required in the factory and to enable the toxicologist to predict what is likely to happen should the product be misused by the consumer, for example, if a child is allowed to play with products or if aerosols are sprayed in confined spaces.

VIII. Systemic Toxicity: Repeated Exposure Effects

Cosmetics are used every day for many years by millions of people. It is therefore important to establish that small quantities of products or ingredients entering the body by any route will not cause toxic effects. Repeated exposure to a chemical may cause toxic effects at dose levels which are several orders of magnitude lower than the LD_{50}. It is therefore not possible to predict toxicity following repeated exposure (chronic toxicity or subacute toxicity) from acute toxicity data. An estimate of chronic toxicity is therefore required.

Before embarking on chronic toxicity studies some consideration should be given to the necessity of obtaining such information. Most cosmetic ingredients are used under conditions where the dose which consumers receive is several orders of magnitude lower than the dose required to produce toxic effects. It can therefore be considered unnecessary to obtain detailed knowledge of the systemic toxicology of a chemical at dose levels which will never be approached.

It is of course important to establish that the dose level encountered in normal use is in fact well below the level required to produce toxic effects. Simple range-finding experiments over reasonable periods of time such as 28 or 90 days should be adequate to obtain an estimate of the minimum toxic dose in experimental animals. This dose can then be related to consumer exposure and a decision made on whether or not a more detailed toxicological study is required.

Should a more detailed examination of chronic toxicity prove necessary the normal procedure would be to test new chemicals, possibly in a vehicle considered to be typical of the products in which it is likely to be used. These studies normally use rats as the experimental animal. The route of administration would usually simulate normal use and for many cosmetic ingredients would be by the skin. However, oral administration might be considered for toothpaste ingredients and inhalaton for aerosol ingredients. The experiments might last from as little as 28 days to a life span. In some studies the animals may breed during the experiment to assess any effects on fertility and reproduction.

The studies will normally be carried out at several dose levels, the objective being to determine the maximum dose at which not toxic effect is observed. Toxic effects will be sought from behavioural observations, blood and urine biochemistry during the study, post-mortem examination and microscopic examination of sections of all the major organs.

IX. Teratology

The birth of malformed offspring owing to exposure to toxic chemicals during pregnancy is obviously an effect to be avoided. As with chronic toxicity the time and costs involved preclude the testing of every product, but the toxicologist must make certain that any new ingredient is unlikely to be teratogenic.

In a teratology study the test compound is given to female animals during the early stages of pregnancy. Rabbits or rats are the normal experimental animals. The numbers of live and dead or resorbed foetuses and the type and frequency of any abnormalities are recorded and the data compared with that from control animals. Maternal toxicity can also be recorded.

X. Carcinogenicity

The possibility that cosmetic ingredients could cause skin cancer or other types of cancer following absorption into the body cannot be ignored. There is no evidence that any existing cosmetic ingredients cause cancer in humans, but epidemiological studies to prove that they do not are extremely difficult to perform. It is however possible to carry out studies for carcinogenicity in experimental animals and such studies have been carried out on cosmetic ingredients.

In a carcinogenicity study the test material must be dosed to the experimental animal (normally rats or mice) over a very long period which may be 2 years or more or for the life span of the animal. The test product will normally be added to the diet, but in some studies application to the skin may be carried out. The rate of cancer formation in treated animals is then compared with that of controls.

The great cost and length of time required for a study means that carcinogenicity studies cannot be carried out on new products and it is normally not possible for such studies to be carried out on new ingredients. Fortunately, in vitro studies which have some value in predicting carcinogenicity are now available. These studies depend upon investigating the effect of chemicals on DNA with resulting mutations in the cell population. These effects are thought to be involved in chemical carcinogenesis.

XI. Mutagenicity

Mutagenicity studies are primarily designed to predict the ability of a chemical to cause mutations in a cell population, and as such these studies might be used for predicting effects on mutation frequencies in mammalian species. However, the usual reason for carrying out mutagenicity studies is to predict the carcinogenic potential of the test material. Mutagenicity tests indicate a potential for carcinogenic activity, although not all mutagens are carcinogens.

There are many methods used in carrying out mutagenicity studies. The best known is the Ames test (Ames et al. 1975). In the Ames test for mutagenic potential in bacteria, strains of *Salmonella typhimurium* which require histidine for growth are exposed to the test chemical (with and without metabolic activation) and grown on a medium containing no histidine. If the chemical has caused a reverse mutation to the natural histidine-independent state the organism will grow. The number of colonies is therefore an indication of the mutagenicity of the test chemical.

The Ames test is only the first stage of an examination of potential for mutagenic activity. In order to assess the validity of positive results further tests may be carried out on mammalian cells in culture or by using in vivo tests to detect chromsome aberrations. Not all substances positive in the Ames test in bacteria induce mutagenic change in mammalian cells. Mutagenicity tests are reckoned to have a 90% reliability in detecting chemicals whose carcinogenic potential in experimental animals is known, particularly if two tests, one with bacterial and one with mammalian cells, are carried out.

E. Interpretation of Safety Data

Once the available safety data have been obtained and, if possible, any gaps filled by experimental studies, the toxicologist must then interpret the data to arrive at a recommendation on the likely safety of the final product or ingredient. This process of interpretation is often termed "risk assessment."

In making a risk assessment the toxicologist must consider the degree of exposure of the consumer to the material under investigation. To arrive at a conclusion on the degree of exposure the toxicologist must consider at least the following factors:

1. Concentration of ingredient in product and concentration of product contacting the consumer.
2. Quantity of product used at each application.
3. Frequency of application.
4. Duration of contact (is the product rinsed off?)
5. Nature of consumers (children, people with sensitive skin, etc.).
6. Projected number of consumers.
7. Quantity likely to enter the body.
8. Application on skin areas exposed to sunlight.
9. Foreseeable misuse which may increase exposure.

By considering all of these factors it should be possible for the toxicologist to arrive at a likely daily dose of the material under consideration. This daily dose can then be compared with the LD_{50} values or with the maximum no-effect levels in the chronic toxicity studies in order to arrive at a safety factor for the possibility of systemic toxicity. For example, it may be possible to show that the maximum no-effect level is 100 or more times the likely daily dose in which case it may be resonable to conclude that systemic toxic effects are unlikely.

For skin sensitisation or photoallergy the total dose may be less important than the concentration of a potential allergen. It may be possible to compare the concentration to which the consumer is exposed with the concentration required to induce sensitisation in the animal experiments and arrive at a safety factor in this way. Ideally of course known allergens should not be used at all, but if their use is desirable for technical or other reasons a consideration of safety factors may allow the toxicologist to conclude that there is unlikely to be a problem in normal use. The strength of the sensitisation (i.e. the proportion of animals or people sensitised) must also be considered when making these judgments.

The problem of interpreting skin and eye irritation results is different. The dose level of individual ingredients may be considered, but for final products which are applied undiluted to the skin and left there it is not possible to exaggerate exposure significantly in order to determine safety factors. The best method of interpreting both skin and eye irritation studies is to compare the level of reaction produced with that of products with the same degree of consumer exposure and which are known to be safe. Ideally these control products should be tested at the same time as the test products. The criterion of acceptability can then be that the test products do not produce reactions which are more severe than those of the controls.

The risk assessment based on available toxicity data must be combined with information obtained from any history of safe use of similar products or of products containing the ingredients in question before making a recommendation on the safety of a new product.

References

Ames BN, McCann J, Yamasaki E (1975) Methods for detecting carcinogens and mutagens with the *Salmonella*/mammalian microsome mutagenicity test. Mutat Res 31:347–364

Buehler EV (1965) Delayed contact hypersensitivity in the guinea pig. Arch Dermatol 91:171–177

Draize JH, Woodward G, Calvery HO (1944) Methods for the study of irritation and toxicity of substances applied topically to the skin and mucous membranes. J Pharmacol Exp Ther 82:377–390

Finkelstein P, Laden K, Miechowski W (1963) New methods for evaluating cosmetic irritancy. J Invest Dermatol 40:11–14

Fisher AA (1978) Contact dermatitis, 2nd edn. Lea and Febiger, Philadelphia

Fregert S (1981) Manual of contact dermatitis, 2nd edn. Munksgaard, Copenhagen

Frosch PJ, Kligman AM (1977) A method for appraising the stinging capacity of topically applied substances. J Soc Cosmet Chem 28:197–209

Gloxhuber C (1970) Prüfung von Kosmetik-Grundstoffen auf fototoxische Wirkung. J Soc Cosmet Chem 21:825–833

Ichikawa H, Armstrong RB, Harber LC (1981) Photoallergic contact dermatitis in guinea pigs: improved induction technique using Freund's complete adjuvant. J Invest Dermatol 76:498–501

Journal Officiel de la Republique Francaise (1971) 21 April, p 3862

Kaidbey KH, Kligman AM (1978) Identification of topical photosensitising agents in humans. J Invest Dermatol 70:149–151

Kaidbey KH, Kligman AM (1980) Photomaximisation test for identifying photoallergic contact sensitisers. Contact Dermatitis 6:161–169

Klecak G (1977) Screening of fragrance materials for allergenicity in the guinea pig. I. Comparison of four testing methods. J Soc Cosmet Chem 28:53–64

Kligman AM, Wooding WM (1967) A method for the measurement and evaluation of irritants on human skin. J Invest Dermatol 49:78–94

Magnusson B, Kligman AM (1969) The identification of contact allergens by animal assay. The guinea pig maximisation test. J Invest Dermatol 52:268–276

Marzulli FN, Maibach HI (1970) Perfume phototoxicity. J Soc Cosmet Chem 21:695–715

Phillips L, Steinberg M, Maibach HI, Akers WA (1972) A comparison of rabbit and human skin response to certain irritants. Toxicol Appl Pharmacol 21:369–382

Skog E (1980) Incidence of cosmetic dermatitis. Contact Dermatitis 6:449–451

Smiles KA, Pollack ME (1977) A quantitative human patch testing procedure for low level skin irritants. J Soc Cosmet Chem 28:755–764

Suskind RR (1979) Cutaneous reactions to cosmetics. J Dermatol (Tokyo) 6:203–209

U.S. Federal Register (1973) Section 1500:41, vol 38 (187)

Uttley M, van Abbe NJ (1973) Primary irritation of the skin, mouse ear test and human patch test procedures. J Soc Cosmet Chem 24:217–227
Vinson LJ, Borselli VF (1966) A guinea pig assay of the photosensitising potential of topical germicides. J Soc Cosmet Chem 17:123–130
Wilkinson DS (1968) Patch test reactions to certain halogenated salicylanilides and related compounds. Arch Dermatol 97:238–244
Zaynoun ST, Johnson BE, Frain-Bell W (1977) A study of oil of bergamot and its importance as a phototoxic agent. Contact Dermatitis 3:225–239

CHAPTER 17

Drug Sensitisation

J. R. GIBSON

A. Introduction

The study of drug sensitisation embraces a number of scientific disciplines and is relevant to virtually every branch of medical practice. It has been estimated that 5% of medical hospital admissions are due to adverse drug reactions and that among hospital in-patients 15% suffer from an ill effect due to drugs (PARKER 1975). It should be noted that many such unwanted occurrences are due to relative or absolute overdosing, side effects, secondary effects, drug interactions, intolerance and idiosyncrasy (DESWARTE 1980), but nonetheless drug allergy occurs in a significant number of patients and represents a considerable problem.

Drug allergy may be defined as an unpredictable and harmful reaction to a drug based on an immunological mechanism. However, in many instances presumptive allergy to a drug is difficult to prove as the availability of reliable test systems is somewhat limited. The problem is further compounded by the fact that patients may develop an immunological response following exposure without manifesting allergic symptoms on continued use or re-administration of the drug. In addition, even in patients who have recovered from a truly allergic reaction to a drug, it is often not known how long their allergic state will last or what percentage of these patients will react again to the same drug. This is due to incomplete knowledge of variables relating to patient and drug. It must also be remembered that biochemical and pharmacological phenomena may mimic immunological reactions, thus leading to further confusion (DESWARTE 1980) and that the formation of antibodies and antigen-antibody complexes may be secondary and not primary events.

For the purpose of this chapter it is necessary to concentrate on mechanisms of drug allergy which are relevant to the skin either because it is the portal of entry or because a significant part of the reaction occurs there. The analogy of a loaded gun aimed at a target is a useful one in this discussion, with the trigger likened to the antigenic determinants on the hapten-macromolecule complex; the gun and bullet representing the induction and elicitation of the reaction; and the target the site of damage in the skin.

The author considers it appropriate to blend some speculative ideas with well-established facts in order to create a more meaningful and stimulating review of an area in which large gaps exist in our knowledge. Throughout the text, an attempt will be made to discuss first subject matter which is firmly supported by scientific evidence and subsequently to introduce, where appropriate, topics which are far from well established conceptually. It must be recognised that even in the case of some of the commonest cutaneous drug reactions, e.g. the generalised exanthematous pattern, the pathogenesis is very poorly understood and open to much controversy, although it is true that other areas are relatively well delineated, e.g. allergic contact dermatitis and urticaria. An approach based on the four classical immunological reaction types (Coombs and Gell 1975) has been adopted in order to ensure general understanding and to provide a framework for discussion. It is appreciated that the high degree of integration between the various components of the immune system defies the rigidity of such classification and it has previously been proposed that type II and type III reactions should be combined as they share similar effector mechanisms (Henson 1982).

In this chapter, for the sake of simplicity, the concept of autoallergy will be discussed in the sections dealing with type II immune responses and all aspects of T-cell-mediated cytotoxicity will be included in type IV reaction patterns (Parish 1986). It should be noted that some common clinical entities may be caused by more than one immune mechanism, e.g. urticaria can be triggered via a type I or a type III reaction or a combination of both. Owing to limitations of space, many of the references chosen represent overviews of given conceptual areas and as such provide extensive citations of the relevant literature.

B. The Trigger

Substances with a molecular weight greater than 5000 have the potential to act as complete antigens whilst those of molecular weight less than 5000 are weakly immunogenic (Van Arsdel 1978; DeSwarte 1980). Virtually all drugs have a molecular weight of less than 1000 and in order to stimulate the immune system they must act as haptens and form complexes with macromolecules which are usually proteins, but may sometimes be carbohydrates (Levine 1966; DeSwarte 1980). This need to act as a hapten occurs both in cases of topical and systemic sensitisation to drugs. The bond between hapten and macromolecule must possess adequate stability and this is usually provided by covalent bonding although electrostatic bonding may occasionally suffice (Levine 1966).

Drugs which commonly or rarely cause allergic skin reactions have been reviewed (Arndt and Jick 1976; Bigby et al. 1986). The ability to act as a hapten requires a high degree of reactivity which is lacking in many drugs suspected of causing allergic reactions. In such cases drug metabolites or impurities are the likely culprits (DeSwarte 1980; de Weck 1971; Eisner and Shahidi 1972). Factors influencing a substance's ability to act as a successful hapten include water solubility and life span in the tissue system (Amos 1976). A number of metabolic reactions including hydroxylation may be needed to convert a relatively unreactive drug into a successful hapten (Amos 1976). However, knowledge in this field of research is far from complete. In cases where metabolic transformation of a drug is necessary to produce a haptenic molecule, allergy to that drug will depend on the individual's ability to effect the reaction needed (Amos 1976).

The antigenic determinant may be related both to the hapten itself and to the macromolecular structure (DeSwarte 1980). In addition to size, specific structural features such as rigidity appear to be required for antigenicity (Amos 1976).

Many types of macromolecules have been implicated as carriers for haptens. These include proteins synthesised locally within the liver and substances normally present in the skin. In some cases, the carrier may be a microorganism or a cell membrane derived from host tissues (DESWARTE 1980). Another possible mechanism by which a low molecular weight drug may become antigenic is by forming polymeric products which may be capable of inducing and eliciting hypersensitivity states (AMOS 1976).

In the case of photoallergic reactions it is likely that the role of light is simply that of photoactivation of the hapten, making it more avidly combine with the macromolecular carrier to create an antigenic substance (HARBER and BAER 1972; WILLIS and KLIGMAN 1968). Thereafter such reactions are similar conceptually to non-photoallergic reactions.

C. The Gun

I. Introduction

Allergy to drugs may occur owing to the classical reactions or combinations of reactions defined by COOMBS and GELL (1975) and detailed in Table 1. This classification represents an oversimplification of an integrated, complex and dynamic process, but it will be used in the absence of a widely accepted superior system. A general scheme of the evolution of allergic drug reactions is given in Fig. 1. In the induction of an immune response it is probable that the principles of the first steps in antigen handling are similar regardless of the portal of entry of the drug. Thus, the hapten-macromolecule complex is formed and subsequently it is processed and presented by antigen-presenting cells of the monocyte virgule macrophage series to T- or B-lymphocytes possessing the correct genetically predetermined antigenic receptors and often under the influence of helper and suppressor T-lymphocytes (PARISH 1986). It is not entirely clear where each stage of this chain reaction occurs although this will vary depending on the portal of entry. Thus, when appropriate, the skin, respiratory tract, gastrointestinal tract, blood, lymphatics and liver may all play a role with the final activity occurring in the lymph nodes or spleen. Thereafter the immunological process will vary in a qualitative manner, depending on whether antibody-mediated or cell-mediated immunity is the predominant response. Once more, factors governing this are far from clearly understood, but it is possible that the portal of entry may influence the type of response produced by a given drug: (a) by determining the type of macromolecular carrier linking with the drug which in turn may influence the affinity of the complex for either T- or B-lymphocytes; and (b) by influencing the subsequent route taken by the complex to the spleen or lymph nodes.

Phenomena which may at times apply to drug reactions are those of cross-reactivity between antigens with sufficient numbers of common antigenic determinants; tolerance induction (CHASE 1946; HJORTH and FREGERT 1979; MACHER and CHASE 1969; PARKER et al. 1983; POLAK 1977); interference in the immune response to one antigen by the presence of another antigen (KLIGMAN and EPSTEIN 1959) and the blocking of elicitation reactions by the presence of excessive amounts of monovalent haptens (AMOS 1976).

Table 1. Allergic reaction types, their association with specific skin manifestations and tests used in the detection of drugs responsible for sensitisation reactions (for reviews see ASSEM 1977; COOMBS and GELL 1975; DESWARTE 1980; GIRARD et al. 1976)

Reaction type (predominant effector mechanism)	Examples of drug involved	Skin manifestations	In vivo	References	In vitro	References
Type I anaphylactic (antibody-mediated)	Penicillin	Urticaria	Scratch, prick and intradermal tests	ADKINSON et al. (1971), PARKER (1972), LEVINE and ZOLOV (1969)	Histamine release from human leucocytes, human lung and monkey lung	ASSEM (1972a)
					Rat mast cell degranulation	PENCE and EVANS (1972)
					Human and rabbit basophil degranulation	MONERET-VAUTRIN et al. (1972)
					Radio-allergosorbent test (RAST)	BARNA et al. (1980), JUHLIN and WIDE (1972), SPATH et al. (1979)
Type II cytotoxic (antibody-mediated)	Penicillin	Purpura				
Type III immune complex (anti-mediated)	Penicillin; penicillamine	Necrotising vasculitis				
Type IV delayed hypersensitivity (T-cell-mediated)	Penicillin; neomycin	Allergic contact dermatitis	Patch test	AGRUP (1972), FISHER (1973), HJORTH (1977), HJORTH and FREGERT (1979)	Lymphocyte transformation	AMOS (1976), ASSEM (1972b), BARNA et al. (1980), DOBOZY et al. (1981), LING (1972), SUZUKI et al. (1978)
			Photopatch test	EPSTEIN (1977)	Macrophage/leucocyte migration inhibition	AMOS (1976), BENDIXEN and SOBORG (1969), BROSTOFF (1972), POLAK (1977)

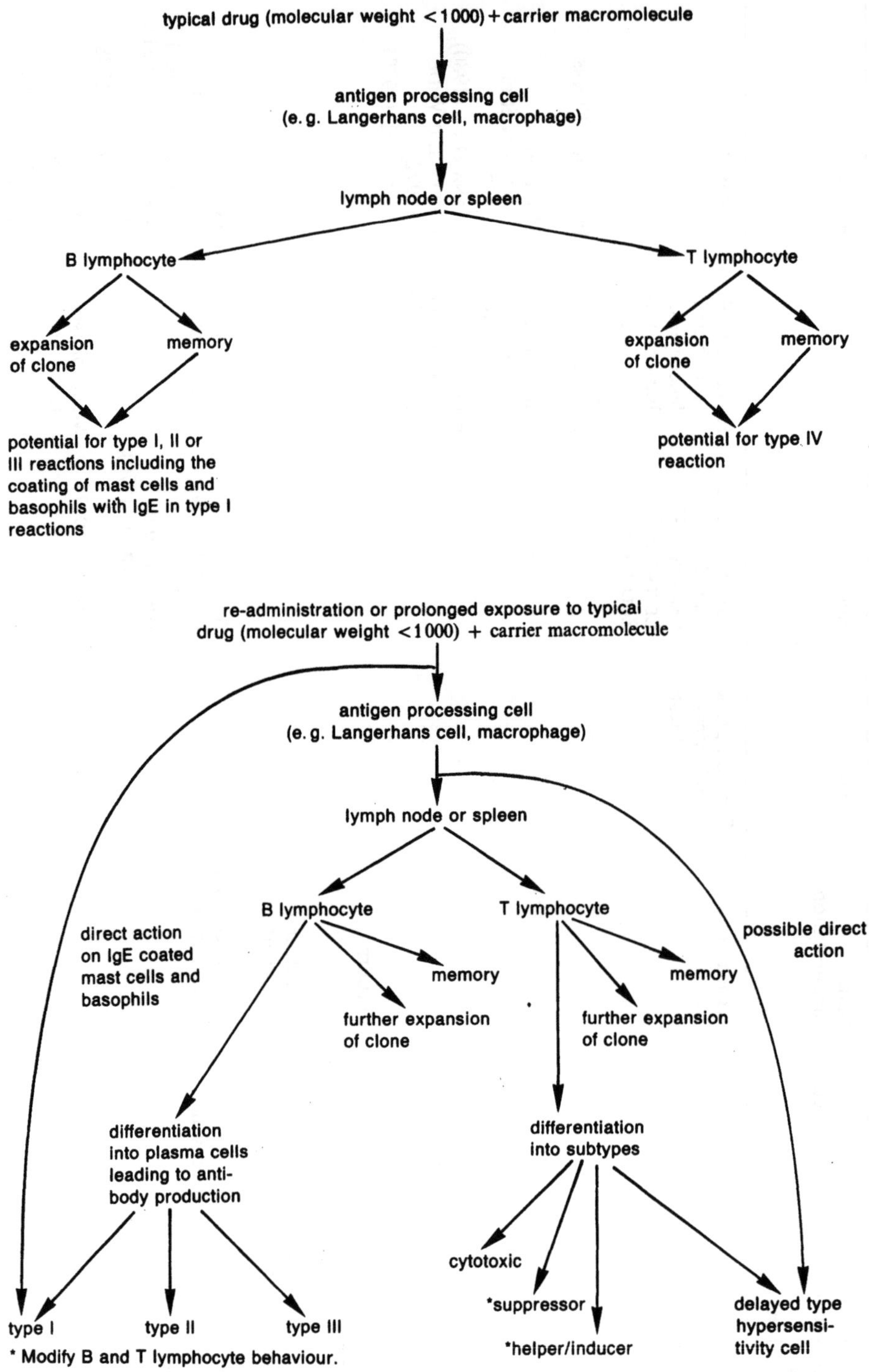

Fig. 1. General scheme of induction and elicitation of immune response in drug allergy

The portal of entry is an important aspect when considering the likelihood of inducing an immunological response to a drug. Thus, topical application to the skin constitutes the greatest risk, with parenteral administration occupying an intermediate position and the oral route being the safest (DESWARTE 1980). Once sensitisation has taken place, the intravenous route probably represents the most dangerous form of administration.

Apart from specific genetic factors which may influence the reaction of the host to a drug (DESWARTE 1980; WOOSLEY et al. 1978) certain disease states and their therapies may alter host susceptibility to drug allergy. Examples of such alterations are those associated with hypogammaglobulinaemia, impaired cellular hypersensitivity as in sarcoidosis, and impaired T-lymphocyte suppressor function (DESWARTE 1980; LAKIN et al. 1975). Much has been written about the immunological mechanisms involved in adverse reactions to drugs (ACKROYD 1975; WITTE and WEST 1982) and the following sections contain a synopsis of current opinions.

II. Type I

Anaphylaxis, urticaria and angio-oedema are the responses to type I reactions. Following the production of specific anti-drug antibodies of the IgE and occasionally IgG_4 class by plasma cells, the antibodies circulate and attach themselves to mast cells and basophils (PARISH 1986; WITTE and WEST 1982). On subsequent exposure to the drug-macromolecule complex which must be polyvalent and therefore capable of cross-linking two IgE antibody molecules on the cell membrane surface (WINTROUB et al. 1987), there follows mast cell activation and degranulation with the following consequences (SOTER and AUSTEN 1980; SOTER 1983):

1. Histamine release leading to increased venular permeability and enhanced chemotaxis of neutrophils and eosinophils.
2. The release of kallikrein which subsequently leads to the formation of bradykinin.
3. The release of eosinophil chemotactic factor of anaphylaxis (ECF-A) leads to accumulation of eosinophils.
4. Production and release of prostaglandin D_2 leading to vasodilatation and chemokinesis of neutrophils.
5. Leukotriene production and release. Slow reacting substance of anaphylaxis (SRS-A), which is now known to be a product of lipoxygenation of arachidonic acid, is produced. The recently defined components of SRS-A are the leukotrienes LTC_4, LTD_4, and LTE_4.
6. The generation and release of platelet-activating factors (PAFs).

The immediate effects of mast cell degranulation are those of erythema, weal and flare reactions, but IgE-mediated mast cell degranulation can also lead to late phase reactions (LEMANSKE and KALINER 1983) which peak at 6–8 h and are characterised by erythema and induration. A useful review of biochemical events following mast cell/basophil activation has been prepared by SIRAGANIAN (1983).

III. Type II

Type II immunological reactions which are relevant to the skin include: (a) drug-induced thrombocytopenia caused by the attachment of specific anti-drug antibody to drug antigen present on the surface of the thrombocyte; and (b) specific antibody deposition on drug antigens laid down on endothelial cells. In both instances, complement is activated, leading to damage either of the thrombocyte or of the endothelial cell. This gives rise to haemorrhages, some of which are manifested on the skin as purpura (WITTE and WEST 1982).

The likelihood that pemphigus and bullous pemphigoid are immunologically based diseases (GIGLI 1982; PARISH 1986) and the similarity between idiopathic and drug-associated types (LEVER and SCHAUMBURG-LEVER 1983; VAN JOOST et al. 1987) raises the possibility of antibody-mediated autoallergy as a factor in drug-associated pemphigus and pemphigoid. A theoretical model for such a mechanism has recently been proposed (VAN JOOST et al. 1987).

IV. Type III

Urticaria and allergic (leukocytoclastic) vasculitis are associated with type III reactions. Drugs may lead to type III reactions when specific anti-drug antibody forms complexes with the drug antigen or a cross-reacting antigen which are subsequently not removed from the circulation in an appropriate manner. The rate of removal of immune complexes depends on the following factors (JEGASOTHY 1983): (a) size, with larger complexes being cleared more quickly than smaller ones; (b) adequate functioning of the cells of the mononuclear phagocyte system (MPS) which at least in part appear to be dependent on receptors for the Fc portion of the IgG molecule and the third component of complement; and (c) the presence of massive amounts of complexes which may saturate the MPS, leading to impaired functioning.

The pathogenic potential of circulating immune complexes depends on their ability to activate complement and this in turn relates to the following factors (JEGASOTHY 1983; PARISH 1986): (a) the ratio of antigen to antibody, with moderate antigen excess providing greatest pathogenicity; (b) antigen valence, with multivalent antigens being most effective; and (c) antibody class – IgG and IgM activate complement by the classic pathway and IgA by the alternate pathway. IgE and IgD do not activate complement.

Deposition of immune complexes sometimes occurs at sites of previous tissue damage and it is quite possible that in some circumstances histamine release may facilitate the laying down of immune complexes by causing separation of the endothelial cell lining (GREAVES 1980). Following deposition and once the process of complement activation has commenced, the anaphylatoxins (C3a and C5a) can directly cause more histamine release (SOTER 1980) and thus increase immune complex deposition. Neutrophils are attracted into areas of damage owing to chemotactic substances which include the complement product C5a (JEGASOTHY 1983). Once present, neutrophils lead to a destructive process by releasing a variety of lysosomal enzymes (JEGASOTHY 1983).

V. Type IV

In many instances, the induction of a type IV T-cell-mediated immunological reaction leading to allergic contact dermatitis will be due to contact with the skin of the appropriate antigen. However, this may not always be the case and in some patients, delayed-type hypersensitivity reactions may occur in the skin when the initial site of induction of the immunological process may have been distant from the skin (EISEN 1959). In the usual case, the antigen which may be aided in its entry into the skin by mild, superficial trauma or inherent lipophilicity of the antigen itself (KRISTOFFERSON et al. 1982), will complex with a variety of macromolecules in the skin which are usually proteins (POLAK 1977). It is most likely that this process leads to a variety of antigenic determinants in many cases of allergic induction to drugs as each antigen may combine with several different carrier macromolecules (POLAK 1977). It is possible that a mixture of complexes is necessary for optimum induction/elicitation of allergic responses. These complexes are then processed by the antigen-presenting cells which in this case are the Langerhans' cells, and subsequently presented to T-lymphocytes (WOLFF and STINGL 1983). For appropriate presentation to T-lymphocytes to take place, it is necessary for the antigen-presenting cells to have on their surface type II antigens of the major histocompatibility complex – namely DR antigens (Ia-like) – and the antigen is presented to the T-lymphocyte in association with the DR antigen (WOLFF and STINGL 1983). The release of interleukin 1 from keratinocytes/macrophages and of interleukin 2 from T-lymphocytes appears to be an essential requirement for successful antigen presentation function. It is not entirely clear whether this presentation of antigenic material by the Langerhans' cell to the T-lymphocyte takes place in the epidermis or whether the Langerhans' cell transports the antigen to the regional lymph node before performing this transfer, but the latter possibility has more support (PARISH 1986; WOLFF and STINGL 1983).

Elicitation of a delayed-type hypersensitivity reaction in the skin following either subsequent skin contact with the drug antigen or systemic intake of antigen is due to infiltration of the affected skin area with specifically sensitised T-lymphocytes. The T-lymphocytes release lymphokines which attract macrophages into the area and also inhibit the macrophage from leaving the site of inflammation (PARISH 1986). Other lymphokines released by activated T-lymphocytes confer anti-antigen activity on previously non-immunocompetent lymphocytes and influence the behaviour of neutrophils, eosinophils and basophils (PARISH 1986). It would appear that, in the elicitation phase of the delayed-type hypersensitivity reaction taking place in allergic contact dermatitis, the Langerhans' cell may itself be a target for destruction (WOLFF and STINGL 1983). Specifically sensitised T-lymphocytes form only a very small percentage of the cellular infiltrate present in the delayed-type hypersensitivity reaction of contact dermatitis (PARISH 1986; POLAK 1977). It is clear that this small number of cells use their lymphokines to create the microenvironment described, in which antigen elimination can take place. Unfortunately, during this process, local tissue damage inevitably occurs. Further information on factors influencing the induction and elicitation of the contact dermatitis reaction can be found in the work of KLIGMAN (1966), ASHERSON et al. (1979), and THOMAS et al. (1980).

Although adequate proof does not exist, it may be considered reasonable to speculate that the commonly seen generalised exanthematous drug reaction patterns are at least partially due to cell-mediated immunity and there is some evidence to suggest that this mechanism may be relevant to the development of fixed drug eruptions (HINDSEN et al. 1987; MURPHY et al. 1985; SEHGAL and GANGWANI 1987). It is likely, but by no means certain, that both the induction and elicitation phases of such reactions follow systemic administration in virtually all cases. It may also be worth noting that drug-induced vasculitis is frequently of the lymphocytic pattern (LEVER and SCHAUMBURG-LEVER 1983), raising the possibility of a type IV reaction.

D. The Target

I. Introduction

One of the commonest sites for the manifestation of allergic drug reactions is the skin. The most frequent patterns of proven and presumed allergic drug reactions are generalised exanthematous rashes, fixed drug eruptions and urticaria following systemic administration (KAUPPINEN and STUBB 1984) and allergic contact dermatitis to topically applied agents. However, a wide range of less common reaction patterns occurs (KAUPPINEN and STUBB 1984). It is unlikely that the ease of clinical detection of reactions in the skin gives us a falsely heightened impression of the relative frequency of dermatological involvement in allergic reactions and it is most probable that the skin is genuinely a common target. Intuitively, it makes sense from an immunological point of view that this should be so because the skin experiences frequent contact with foreign elements, including drugs, and represents the first defensive barrier against them. Thus, it should be well equipped to handle this situation by being an immunologically highly reactive organ. There is in fact good evidence to support this view and it is probable that a large part of the skin's immunological responsiveness is due to the interrelationship between Langerhans' cells, T-lymphocytes and peripheral lymph nodes – the so-called skin-associated lymphoid tissue (SALT) system (STREILEIN 1983). In addition, co-factors, for example ultraviolet light, which enhance the opportunity for some types of allergic reactions, have ready access to the skin. In circumstances where the skin may be the site of metabolic transformation of the drug the opportunity for an allergic reaction occurring would be enhanced. Specific factors relating to type I–type IV drug reactions are discussed in the following sections.

II. Type I

The skin, along with the gastrointestinal tract and respiratory tract, has a high concentration of mast cells which act as one of the sentinel groups of the human defensive system (SOTER 1983). The presence of great numbers of mast cells with inflammatory potential must at least in part explain the relatively large number of type I reaction manifestations occurring in the skin. Additionally, apart from the blood-borne dissemination of inhaled or ingested specific allergens as a route to the mast cell, direct access by penetration of the skin may lead to local reactions as in contact urticaria (ODOM and MAIBACH 1977). It is well known, of course, that generalised type I reactions may also occur when the skin is the portal of entry (ROUPE and STRANNEGARD 1969).

III. Type II

The occurrence of haemorrhages as a result of thrombocytopenia due to type II reactions is probably no commoner in the skin than in other organs. However, purpura induced by type II reactions leading to endothelial damage may be commoner in the skin of the lower limbs owing to the effects of gravity compounding the local problems of microcirculation well known to occur in the legs.

In cases of drug-associated pemphigus and bullous pemphigoid, it could be speculated that the presence in the epithelium of molecules able to be involved in cross-reactions with antibodies directed against antigenic complexes or to act as macromolecular partners for drug haptens could be of critical importance.

IV. Type III

When the MPS fails to achieve adequate removal of circulating immune complexes for any reason (JEGASOTHY 1983) then a possible consequence is deposition of these products in the vasculature of a variety of organs ultimately leading to local damage and necrosis. There are a number of possible reasons why the skin should be a common site for such insults, but little hard evidence to confirm them. Possibilities include the following speculative thoughts:

1. The blood vessel types, diameters and anatomical position in the skin may in some way suit the size and other physical properties of some ciculating complexes. In particular there are a large number of relatively permeable post-capillary venules in the papillary dermis (GOWER et al. 1977).
2. The presence of mast cells/basophils which may release histamine and other vasoactive mediators, thus facilitating the deposition of immune complexes (GREAVES 1980).
3. Locally exposed and damaged tissues may act as cross-reacting antigens with antibody directed against a drug or may encourage the laying down of specific antigens at such sites. Both situations could lead to local immune complex formation. Such damaged tissues may occur more often in the skin owing to external trauma being commoner in such a superficial tissue. Additionally the skin of the lower leg may be especially susceptible to such damage owing to gravitational effects on the microvasculature (GREAVES 1980). These circumstances could also encourage the deposition of circulating immune complexes.
4. Cooling of the blood which leads to increased viscosity and reduction in the rate of blood flow readily occurs in the skin (GREAVES 1980). This could encourage the deposition of immune complexes.

V. Type IV

The skin is a common site for proven and presumed type IV drug reactions. The following factors are worth considering:

1. Owing to its function as a first defensive barrier, contact with antigen both in the induction and elicitation phases of the immunological response are commonplace and a contact dermatitis reaction is the usual clinical manifestation.
2. Whether or not the induction of immune reactivity to a specific antigen took place via the skin, the systemic administration of the specific antigen may lead to a flare of a contact dermatitis-like reaction or other manifestations of a presumed delayed-type hypersensitivity reaction (EKELUND and MOLLER 1969; HJORTH and FREGERT 1979; TAGAMI et al. 1983). The presence of suitable carrier macromolecules in the skin may contribute to the frequency of reactions occurring there (SUZUKI 1978).

E. The Detection of Drugs Responsible for Allergic Reactions

There is no doubt that a thorough clinical history and physical examination coupled with a comprehensive knowledge of the types of drugs likely to be associated with particular allergic reactions (BRUINSMA 1977; KUOKKANEN 1972; BIGBY et al. 1986) are the most important steps in reaching a correct diagnosis. All available diagnostic tests have limitations, but in some cases they may be of value. The most useful conclusions are likely to arise from a thoughtful series of investigations linked with the clinical picture. One of the most pressing problems is that the precise antigen derived from the offending drug is usually not known. A striking exception to this rule is that of penicillin where both the major determinant, penicilloyl, and several minor determinants have been identified (SCHNEIDER 1970) and thus much of the pioneering work in drug allergy has been done with the penicillins.

Many of the investigations which can be performed are more likely to allow one to deduce the allergic mechanisms involved rather than the offending drug and even the evaluation of mechanism may be difficult as some findings may be secondary and not primary events. Blood eosinophilia may be associated with type I or type IV drug reactions, but little is known of the precise role of the eosinophils in these reactions (SPRY 1980). The detection of elevated IgE levels raises suspicions of a type I reaction whereas evidence of immune complexes detected in the circulation by a variety of techniques (JEGASOTHY 1983) or found as deposits in tissues by immunofluorescence or immunoperoxidase techniques raises the possibility of a type III reaction although none of these methods will reveal the antigen involved. Similarly, tests of complement function such as Clq binding and CH50 may give further evidence of a type III reaction.

Several specific tests have been designed to detect the offending drug in allergic reactions and Table 1 summarises some of the methods used and also provides a bibliography for further reading. The re-administration of the suspected drug followed by the development of the previous reaction pattern is clearly the most reliable method, but is often inappropriate for ethical reasons.

Of the tests mentioned in Table 1, the ones involving human subjects – scratch, prick and intradermal tests for detecting penicillin allergy and patch and photopatch tests for detecting drug-induced contact and photoallegic contact dermatitis – are probably the most reliable and clinically relevant. False-positive and false-negative results can occur, but the use of adequate controls and thoughtul evaluation of data can yield valuable results. The radioallergosorbent test (RAST) is only of value in penicillin allergy and the other laboratory tests relating to type I reactions are of limited value.

The lymphocyte transformation test makes the assumption that the drug being tested bears the correct antigenic determinants and this may not be so as many drugs are metabolised to produce the reactive hapten. In addition, a positive result does not categorically incriminate a drug and a negative result does not exclude it as a cause of the reaction. Thus, this test has to be interpreted with caution. Similar comments apply to macrophage/leucocyte migration inhibition tests.

F. Management

An allergic reaction to a drug will usually necessitate the withdrawal of the offending agent although in certain circumstances where the drug is vital and the risk : benefit ratio favourable, cautious continuation of the medication may be appropriate. The precise therapy indicated will depend on the type and severity of the reaction and may include topical or systemic corticosteroids, β-adrenergic agents, anti-histamines and emollients. A detailed discussion of specific drug therapy of allergic drug reactions is not possible in this chapter, but the following two relevant techniques will be briefly mentioned. They would be most appropriately utilised only when there is no suitable alternative drug available:

1. Desensitisation to penicillin is performed by administering progressively increasing doses of the drug over short time intervals usually by the subcutaneous route. Once therapeutic doses have been reached, treatment with the drug is commenced. If therapy is discontinued for more than 48 h the patient should no longer be considered to be desensitised (DESWARTE 1980). Possible explanations for successful desensitisation in patients with proven penicillin allergy are the production of benzylpenicilloyl-specific "blocking" antibodies of the IgG class which prevent the drug from reacting with IgE, the depletion of mediators, penicillin degradation products acting as univalent haptens and the stimulation of suppressor T-lymphocyte function (DESWARTE 1980; PARISH 1986).
2. The use of a monovalent hapten to inhibit allergic reactions to penicillin produced in humans under experimental conditions was successfully employed by DE WECK and SCHNEIDER (1972) and subsequently used clinically (DE WECK 1979), but the value of this technique has not been fully established and it is unlikely to be of value where the minor antigenic determinants of penicillin are involved.

References

Ackroyd JF (1975) Immunological mechanisms in drug hypersensitivity. In: Gell PHG, Coombs RRA, Lachmann PJ (eds) Clinical aspects of immunology, 3rd edn. Blackwell, Oxford, pp 913–961
Adkinson NF, Thompson WL, Maddrey WC, Lichtenstein LM (1971) Routine use of penicillin skin testing on an inpatient service. N Engl J Med 285:22–24
Agrup G (1972) Patch testing in drug allergy. In: Dash CH, Jones HEH (eds) Mechanisms in drug allergy. Churchill Livingstone, Edinburgh, pp 135–138
Amos HE (1976) Allergic drug reactions. Arnold, London
Arndt KA, Jick H (1976) Rates of cutaneous reactions to drugs. JAMA 235:918–923
Asherson GL, Perera MACC, Thomas WR (1979) Contact sensitivity and the DNA response in mice to high and low doses of oxazolone: low dose unresponsiveness following painting and feeding and its prevention by pretreatment with cyclophosphamide. Immunology 36:449–459
Assem ESK (1972a) The passive sensitization of human lung as a test for drug allergy. In: Dash CH, Jones HEH (eds) Mechanisms in drug allergy. Churchill Livingstone, Edinburgh, pp 112–120
Assem ESK (1972b) IgE and other in vitro tests in the diagnosis and follow-up of drug allergy. In: Dash CH, Jones HEH (eds) Mechanisms in drug allergy. Churchill Livingstone, Edinburgh, pp 179–188
Assem ESK (1977) Tests for detecting drug allergy. In: Davies DM (ed) Textbook of adverse drug reactions. Oxford University Press, New York, pp 397–410
Barna BP, Gogate P, Deodhar SD, Moeder M (1980) Lymphocyte transformation and radioallergosorbent tests in drug hypersensitivity. Am J Clin Pathol 73:172–176
Bendixen G, Soborg M (1969) A leucocyte migration technique for in vitro detection of cellular (delayed type) hypersensitivity in man. Dan Med Bull 16:1–6
Bigby M, Jick S, Jick H, Arndt K (1986) Drug-induced cutaneous reactions: a report from the Boston collaborative drug surveillance program on 15438 consecutive inpatients, 1975 to 1982. JAMA 256:3358–3363
Brostoff J (1972) The leucocyte migration tests in drug allergy. In: Dash CH, Jones HEH (eds) Mechanisms in drug allergy. Churchill Livingstone, Edinburgh, pp 177–178
Bruinsma W (1977) A guide to drug eruptions. Krips Repro, Meppel, The Netherlands
Chase MW (1946) Inhibition of experimental drug allergy by prior feeding of the sensitizing agent. Proc Soc Exp Biol Med 61:257–259
Coombs RRA, Gell PGH (1975) Classification of allergic reactions responsible for clinical hypersensitivity and disease. In: Gell PGH, Coombs RRA, Lachmann PJ (eds) Clinical aspects of immunology, 3rd edn. Blackwell, Oxford, pp 761–781
DeSwarte RD (1980) Drug allergy. In: Patterson R (ed) Allergic diseases – diagnosis and management, 2nd edn. Lippincott, Philadelphia, pp 452–583
de Weck AL (1971) Immunological effects of aspirin anhydride, a contaminant of commercial acetylsalicylic acid preparations. Int Arch Allergy 41:393–418
de Weck AL (1979) Approaches to prevention and treatment of drug allergy. In: Turk JL, Parker D (eds) Drugs and immune responsiveness. Macmillan, London, pp 211–221
de Weck AL, Schneider CH (1972) Specific inhibition of allergic reactions to penicillin in man by a monovalent hapten. Int Arch Allergy 42:782–797
Dobozy A, Hunyadi J, Kenderessy ASZ, Simon N (1981) Lymphocyte transformation test in detection of drug hypersensitivity. Clin Exp Dermatol 6:367–372
Eisen HN (1959) Hypersensitivity to simple chemicals. In: Lawrence HS (ed) Cellular and humoral aspects of the hypersensitive states. Hoeber-Harper, New York, pp 89–122
Eisner EV, Shahidi NT (1972) Immune thrombocytopenia due to a drug metabolite. N Engl J Med 287:376–381
Ekelund AG, Moller H (1969) Oral provocation in eczematous contact allergy to neomycin and hydroxy-quinolines. Acta Derm Venereol (Stockh) 49:422–426
Epstein JH (1977) Photocontact allergy in humans. In: Marzulli FN, Maibach HI (eds) Dermatotoxicology and pharmacology. Hemisphere, Washington, pp 413–426
Fisher AA (1973) Contact dermatitis, 2nd edn. Lea and Febiger, Philadelphia

Gigli I (1982) Immunological aspects of skin diseases. In: Lachmann PJ, Peters DK (eds) Clinical aspects of immunology, 4th edn. Blackwell, Oxford, pp 790–821
Girard JP, Cattin S, Cuevas M (1976) Immunological mechanisms and diagnostic tests in allergic drug reactions. Ann Clin Res 8:74–84
Gower RG, Sams WM, Thorne EG, Kohler PF, Claman HN (1977) Leukocytoclastic vasculitis: sequential appearance of immunoreactants and cellular changes in serial biopsies. J Invest Dermatol 69:477–484
Greaves MW (1980) Pharmacological factors in initiation of cutaneous vasculitis. In: Wolff K, Winkelmann RK (eds) Vasculitis. Lloyd-Luke (Medical Books), London, pp 49–53
Harber LC, Baer RL (1972) Pathogenic mechanisms of drug-induced photosensitivity. J Invest Dermatol 58:327–342
Henson PM (1982) Antibody and immune-complex-mediated allergic and inflammatory reactions. In: Lachmann PJ, Peters DK (eds) Clinical aspects of immunology, 4th edn. Blackwell, Oxford, pp 687–709
Hindsen M, Christensen OB, Gruic V, Lofberg H (1987) Fixed drug eruption: an immunohistochemical investigation of the acute and healing phase. Br J Dermatol 116:351–360
Hjorth N (1977) Diagnostic patch testing. In: Marzulli FN, Maibach HI (eds) Dermatotoxicology and pharmacology. Hemisphere, Washington, pp 341–351
Hjorth N, Fregert S (1979) Contact dermatitis. In: Rook A, Wilkinson DS, Ebling FJG (eds) Textbook of dermatology, vol 1, 3rd edn. Blackwell, Oxford, pp 363–441
Jegasothy BV (1983) Immune complexes in cutaneous disease. Arch Dermatol 119:795–798
Juhlin L, Wide L (1972) IgE antibodies and penicillin allergy. In: Dash CH, Jones HEH (eds) Mechanisms in drug allergy. Churchill Livingstone, Edinburgh, pp 139–147
Kauppinen K, Stubb S (1984) Drug eruptions: causative agents and clinical types. Acta Derm Venereol (Stockh) 64:320–324
Kligman AM (1966) The identification of contact allergens by human assay. J Invest Dermatol 47:375–392
Kligman AM, Epstein WL (1959) Some factors affecting contact sensitization in man. In: Shaffer JH, Logrippo GA, Chase MW (eds) Mechanisms of hypersensitivity. Churchill Livingstone, London, pp 713–722
Kristofferson A, Ahlstedt S, Enander I (1982) Contact sensitivity in guinea pigs to different penicillins. Int Arch Allergy Appl Immunol 69:316–321
Kuokkanen K (1972) Drug eruptions. Acta Allergol 27:407–438
Lakin JD, Grace WR, Sell KW (1975) IgE antipolymyxin B antibody formation in a T-cell-depleted bone marrow transplant patient. J Allergy Clin Immunol 56:94–103
Lemanske RF, Kaliner MA (1983) Late phase allergic reactions. Int J Dermatol 22:401–409
Lever WF, Schaumburg-Lever G (1983) Eruptions due to drugs. In: Histopathology of the skin, 6th edn. Lippincott, Philadelphia, pp 259–270
Levine BB (1966) Immunochemical mechanisms of drug allergy. Annu Rev Med 17:23–38
Levine BB, Zolov DM (1969) Prediction of penicillin allergy by immunological tests. J Allergy 43:231–244
Ling NR (1972) Lymphocyte transformation in drug allergy. In: Dash CH, Jones HEH (eds) Mechanisms in drug allergy. Churchill Livingstone, Edingburgh, pp 171–175
Macher E, Chase MW (1969) Studies on the sensitization of animals with simple chemical compounds. J Exp Med 129:103–121
Moneret-Vautrin DA, Grilliat JP, Pupil P (1972) Basophil degranulation in drug allergy. In: Dash CH, Jones HEH (eds) Mechanisms in drug allergy. Churchill Livingstone, Edinburgh, pp 159–170
Murphy GF, Guillen FJ, Flynn TC (1985) Cytotoxic T lymphocytes and phenotypically abnormal epidermal dendritic cells in fixed cutaneous eruptions. Hum Pathol 16(12):1264–1271

Odom RB, Maibach HI (1977) Contact urticaria: a different contact dermatitis. In: Marzulli FN, Maibach HI (eds) Dermatotoxicology and pharmacology. Hemisphere, Washington, pp 441–453
Parish WE (1986) Clinical immunology and allergy. In: Rook A, Wilkinson DS, Ebling FJG, Champion RH, Burton JL (eds) Textbook of dermatology, vol 1, 4th edn. Blackwell, Oxford, pp 309–353
Parker CW (1972) Practical aspects of diagnosis and treatment of patients who are hypersensitive to drugs. In: Samter M, Parker CW (eds) International encyclopedia of pharmacology and therapeutics, section 75, vol I, Hypersensitivity to drugs, 1st edn. Pergamon Oxford, pp 367–394
Parker CW (1975) Drug therapy, drug allergy, (first of three parts). N Engl J Med 292:511–514
Parker D, Long PV, Turk JL (1983) A comparison of the conjugation of DNTB and other dinitrobenzenes with free protein radicals and their ability to sensitize or tolerize. J Invest Dermatol 81:198–201
Pence HL, Evans R (1972) In vitro methods of detecting drug hypersensitivity. Med Ann D C 41:431–436
Polak L (1977) Immunological aspects of contact sensitivity. In: Marzulli FN, Maibach HI (eds) Dermatotoxicology and pharmacology. Hemisphere, Washington, pp 225–288
Roupe G, Strannegard O (1969) Anaphylactic shock elicited by topical administration of bacitracin. Arch Dermatol 100:450–452
Schneider CH (1970) Immunochemistry of penicillin. In: Stewart GT, McGovern JP (eds) Penicillin allergy. Thomas, Springfield IL, pp 23–58
Sehgal VN, Gangwani OP (1987) Fixed drug eruption: current concepts. Int J Dermatol 26:67–74
Siraganian RP (1983) Histamine secretion from mast cells and basophils. Trends Pharmacol Sci 4:432–437
Soter NA (1980) Cutaneous necrotizing venulitis. In: Wolff K, Winkelmann RK (eds) Vasculitis. Lloyd-Luke (Medical Books), London, pp 173–182
Soter NA (1983) Mast cells in cutaneous inflammatory disorders. J Invest Dermatol 80(6 Suppl):22s–25s
Soter NA, Austen KF (1980) Biology of the mast cell and its products and of the complement system. In: Wolff K, Winkelmann RK (eds) Vasculitis. Lloyd-Luke (Medical Books), London, pp 54–67
Spath P, Huber H, Ludvan M, Roth A, Schwarz S, Zelger J (1979) Determinations of penicilloyl specific IgE antibodies for the evaluation of hypersensitivity against penicillin. Allergy 34:405–411
Spry CJF (1980) Eosinophilia and allergic reactions to drugs. Clin Haematol 9:521–534
Streilein JW (1983) Skin-associated lymphoid tissues (SALT): origins and functions. J Invest Dermatol 80(6 Suppl):12s–16s
Suzuki S, Asai Y, Hamada T, Mizoguchi Y, Yamada T, Morisawa S (1978) Drug-induced lymphocyte transformation in peripheral lymphocytes from patients with drug eruptions. Dermatologica 157:146–153
Tagami H, Tatsuta K, Iwatski K, Yamada M (1983) Delayed hypersensitivity in ampicillin-induced toxic epidermal necrolysis. Arch Dermatol 119:910–913
Thomas R, Edwards AJ, Watkins MC, Asherson GL (1980) Distribution of immunogenic cells after painting with the contact sensitizers fluorescein isothiocyanate and oxazolone. Different sensitizers form immunogenic complexes with different cell populations. Immunology 39:21–27
Van Arsdel PP (1978) Adverse drug reactions. In: Middleton E (ed) Allergy: principles and practice, vol 2. Mosby Year Book, London, pp 1133–1158
Van Joost T, Tank B, Stolz E, Habets JMW (1987) Causative agents and pathogenicity in iatrogenic cutaneous autoimmunity. Int J Dermatol 26:75–79
Willis I, Kligman AM (1968) The mechanism of photoallergic contact dermatitis. J Invest Dermatol 51:378–384

Wintroub BU, Stern RS, Arndt KA (1987) Cutaneous reactions to drugs. In: Fitzpatrick TB, Eisen AZ, Wolff K, Freedberg IM, Austen KF (eds) Dermatology in general medicine, 3rd edn. McGraw-Hill, New York, pp 1353–1366
Witte KW, West DP (1982) Immunology of adverse drug reactions. Pharmacotherapy 2:54–65
Wolff K, Stingl G (1983) The Langerhans cell. J Invest Dermatol 80(6 Suppl):17s–21s
Woosley RL, Drayer DE, Reidenberg MM, Nies AS, Carr K, Oates JA (1978) Effect of acetylator phenotype on the rate at which procainamide induces antinuclear antibodies and the lupus syndrome. N Engl J Med 298:1157–1159

Section C: Drugs and Diseases

CHAPTER 18

H_1- and H_2-Receptor Antagonists

D. A. A. OWEN

A. Introduction

Histamine is a naturally occurring substance found in most, if not all, tissues of the body. Histamine in tissues is stored predominantly in mast cells which are present in particularly large numbers in skin (KALINER 1979), where they are most evident around small blood vessels and nerves (COWEN et al. 1979; EADY et al. 1979).

B. Biological Actions of Histamine

Histamine was synthesised in 1907 by WINDAUS and VOGT and consequently has been available for study from that time. The first published biological studies were undertaken by DALE and LAIDLAW (1910, 1911) who identified the wide spectrum of action of histamine on both vascular and non-vascular smooth muscle and on the heart. The important capacity of histamine to cause secretion of acid in the stomach was discovered by POPIELSKI in 1920. Histamine also possesses other potentially very important properties, including a major influence on lymphocyte function (BUNCE et al. 1979) and as a neurotransmitter in the central nervous system (BUNCE et al. 1979).

I. Actions of Histamine on Skin

The powerful actions of histamine on vascular smooth muscle in general may be demonstrated in skin. The first known studies with histamine in humans were made by EPPINGER (1913) who described flushing of the skin after systemic administration as well as vasodilatation, wealing and itch when histamine was applied to a scratched skin surface. This complex response to local administration of histamine into human skin was confirmed by CARRIER (1922) and by HARMER and HARRIS (1926). The significance of the local cutaneous response to histamine became more widely appreciated after the extensive studies of LEWIS (1927) and LEWIS and GRANT (1924), who established a link between the cutaneous inflammatory response to mild or moderate injury and the reaction caused by local administration of histamine. These studies prompted LEWIS to attribute the cutaneous response to injury to the actions of an "H-like substance." LEWIS described the actions of histamine on human skin as the "triple response", comprising local oedema and redness at the site of injection surrounded by an area of flare. The local oedema and redness are due to histamine acting directly on blood vessels to increase microvascular permeability and to cause vasodilatation. The surrounding flare is ascribed to vasodilatation produced indirectly by stimulation of sensory nerve endings to cause a local axon reflex response. The term "triple response" ignores the sensory components of the cutaneous response to histamine which comprise itch and hyperalgesia (LYNN 1977). Thus, it has been appreciated for a long time that histamine is capable of eliciting vascular and sensory changes in skin. More recently, there has been additional interest in the possible role of histamine in epidermal cell function (AOYAGI et al. 1981) and in sebum production (LYONS and SHUSTER 1980).

The primary purpose of this chapter is to consider the pharmacology of the response to histamine in human skin. However, animal studies usually allow more comprehensive investigation, and consideration of animal studies will be made as background to the data derived from humans.

II. Action of Histamine on Blood Vessels

The changes in blood vessel function during exposure to histamine have been studied in many different tissues, organs and species. The direct actions of histamine on series-coupled blood vessels (MELLANDER and JOHANSSON 1968) are dilatation of resistance blood vessels to increase local blood flow, dilatation of precapillary sphincter vessels to increase functional capillary surface area, and to increase microvascular permeability and allow extravasation of macromolecules and water. These direct actions lead to increased capillary hydrostatic pressure and to distension of capacitance blood vessels so that local blood content increases (HADDY 1960; FLYNN and OWEN 1977). The important action of histamine to cause local oedema (wealing) occurs largely because of the increase in microvascular permeability although the increases in functional capillary surface area and capillary hydrostatic pressure also contribute to the oedema (WILLIAMS and MORLEY 1973; WILLIAMS and PECK 1977). Consideration of responses to histamine, particularly the oedema or wealing response, requires accurate appreciation of each of these contributory factors.

III. Histamine Receptors on Skin Blood Vessels

Numerous studies have established the presence of H_1-receptors in skin blood vessels in many laboratory species and in humans. Characteristic of these studies was the observation that part of the vascular response to histamine was refractory to doses of H_1-receptor antagonists which could be tolerated by subjects (BAIN et al. 1984). The use of larger doses of these antagonists in laboratory animals was questionable with respect to both the specificity of their action and the probability that higher doses might influence blood vessel function independently of H_1-receptor antagonism.

C. Animal Studies

Studies with histamine H_1- and H_2-receptor antagonists have been made in guinea pigs and monkeys. OWEN et al. (1980) developed techniques to distinguish between the direct actions of histamine on resistance blood vessels and on microvascular permeability and the consequent oedema formation and increased local skin blood content. Histamine caused dose-dependent and simultaneous dilatation of blood vessels; it also increased microvascular permeability and caused oedema. Studies with both histamine H_1- and H_2-antagonists (mepyramine and cimetidine), and with selective histamine receptor agonists (2,2-aminoethylpyridine and dimaprit) established the presence of both histamine H_1- and H_2-receptors on resistance vessels, each associated with dilatation, whereas increased microvascular permeability involved H_1-receptors only. Thus, oedema formation is due almost totally to the H_1-receptor-mediated increase in microvascular permeability although a very small part of the oedema could be attributed to the increase in capillary hydrostatic pressure caused by H_1- and H_2-receptor-mediated dilatation of resistance vessels. WOODWARD and OWEN (1982) confirmed the specificity of mepyramine and cimetidine as used in these experiments. Thus, the H_1-receptor antagonist alone produced a major reduction in the cutaneous blood vessel response to histamine. That part of the histamine response refractory to H_1-receptor antagonism reflects the presence of H_2-receptors on resistance vessels.

Other experiments in guinea pigs (CHENG et al. 1979) and monkeys (HUTCHCROFT et al. 1979) have also confirmed the predominant role of H_1-receptors in histamine-induced increases in vascular permeability in skin. In each of these studies, administration of an H_1-receptor antagonist (mepyramine or chlorpheniramine) caused substantial displacement to the right of the histamine doese-response curve. Addition of an H_2-receptor antagonist (metiamide or cimetidine) to the H_1-receptor antagonist led to significantly greater antagonism of histamine responses than the same dose of the H_1-receptor antagonists alone.

D. Human Studies

I. Local Administration of Histamine

Studies in humans must be made with less invasive techniques than are possible in laboratory animals. Responses to histamine have usually been expressed in terms of weal area (as an index of increased microvascular permeability) and erythema area (as an index of vasodilatation). The use of erythema area to measure vasodilatation does not adequately incorporate a component of the extent of dilatation, e.g. a small area of intense vasodilatation might involve a greater local increase in blood flow and content than a larger area of less intense dilatation; measurement of the response in terms of area alone would, in these circumstances, not be a true measure of the dilator response to histamine. Precise, objective and rapid measurement of skin blood flow in small areas will greatly facilitate studies on cutaneous blood vessels. The rate constants of histamine weal formation and disappearance were first measured in humans by COOK and SHUSTER (1980) giving $t_{1/2}$ = 5.4 and 87 min, respectively. This slow disappearance suggests a prolonged vasoactive effect.

The first comprehensive study of the effect of H_1-receptor antagonists on the vascular response to intradermal histamine in humans was made by BAIN et al.

(1948). They showed that whereas each of three H_1-receptor antagonists reduced the area of erythema and weal caused by intradermal histamine, a substantial part of these responses to histamine was refractory to the highest doses of the antagonists that their subjects could tolerate. It was not clear, at this time, whether the refractory responses were nevertheless H_1-receptor responses and that further, albeit unacceptable, doses of H_1-receptor antagonists were required or whether the refractory responses were independent of H_1-receptors. Recent studies with the newer non-sedative H_1-receptor antagonists terfenadine and astemizole show that histamine wealing can be almost totally inhibited, the response to compound 48/80 inhibited by 70%–80% and antigen-induced wealing, e.g. in response to intradermal injection of house dust mite, only by 50%–60% (Humphreys et al. 1987).

The first clear evidence of H_2-receptors on human skin blood vessels was reported by Marks and Greaves (1977), using chlorpheniramine as the H_1-receptor antagonist, and cimetidine as the H_2-receptor antagonist. Cimetidine, 200 mg alone or chlorpheniramine, 8 mg alone reduced the area of erythema and wealing caused by intradermal histamine or the mast cell degranulating agent compound 48/80. Combinations of chlorpheniramine and cimetidine caused a further, statistically significant displacement of the histamine and compound 48/80 dose-response curve than that caused by either antagonist alone. Although these studies showed clear evidence of both H_1- and H_2-receptors on the blood vessels, a substantial part of the histamine responses persisted in the presence of both types of antagonist. It would have been most valuable to determine whether increasing the dose of either or both antagonists – providing this could be tolerated by the subjects – would have caused further antagonism of histamine responses.

Subsequent studies using both H_1- and H_2-receptor antagonists have used different experimental designs, choices of antagonists, duration of antagonist treatment and route of antagonist administration, making comparison of findings difficult. Harvey and Schocket (1980) using hydroxyzine, 25 mg orally every 6 h or cimetidine, 300 mg orally every 6 h, or combinations of both antagonists, showed that hydroxyzine alone caused a very large reduction of the histamine-induced weal whereas cimetidine alone did not. The combination of hydroxyzine plus cimetidine almost abolished the histamine weal.

Hagermark et al. (1979) provided an interesting approach to the study of histamine receptors on skin vessels in humans by using intradermal administration of antagonists as well as intradermal administration of histamine. Although clearly ensuring the immediate presence of the antagonists in the vicinity of the histamine, this approach is vulnerable when compared with systemic administration of antagonists; it may lead to non-specific effects of high transient antagonist concentrations compounded by the possibility that the antagonists may not persist in skin. Despite the need to consider how either of these factors may influence the results, the study by Hagermark et al. (1979) provides evidence of both H_1- and H_2-receptors on skin vessels. Mepyramine reduced the weal and vasodilator responses to histamine in normal and lignocaine-treated skin, each in a dose-dependent manner. Either metiamide or cimetidine reduced the axon reflex response, which the authors attribute to H_1-receptor blockade although this interpretation does not adequately consider the potential local anaesthetic action of mepyramine.

SMITH et al. (1980) used chlorpheniramine, 8 mg (five doses at various intervals) and cimetidine, 300 mg (five doses at similar intervals) alone and in combination and measured the effect of these antagonists on the weal area following intradermal injection of histamine, 2 µg. Chlorpheniramine alone significantly reduced the area of the weal whereas cimetidine did not. The combination of chlorpheniramine plus cimetidine was more effective than chlorpheniramine alone. As in other studies, no evidence is provided about the maximum inhibition which can be achieved with higher doses of either antagonist alone or in combination.

NATHAN et al. (1981) considered the pharmacology of histamine-induced bronchoconstriction and dermal wealing in mild asthmatics. The antagonists used were chlorpheniramine, either 8 mg as a single dose or seven separate doses given every 6 h, and cimetidine, either 300 mg as a single dose or as seven separate doses every 6 h. In this study, chlorpheniramine as a single dose caused a very modest inhibition of the weal area induced by intradermal histamine, and a slightly greater inhibition after the multiple dosing. By contrast, cimetidine did not reduce responses to histamine whether given as a single dose or in multiple dosing and, unlike the findings in other studies, did not have an additive inhibitory effect when added to chlorpheniramine treatment. In this study, a substantial part of the histamine response persisted in subjects receiving both H_1- and H_2-receptor antagonists. The effect of H_1- and H_2-receptor antagonists in dermographic wealing has been studied by COOK and SHUSTER (1983) and KRAUSE and SHUSTER (1984a) by measuring the dermographic force-weal diameter response.

Studies with histamine receptor agonists have also supported the concept that human blood vessels possess both H_1- and H_2-receptors (ROBERTSON and GREAVES 1978). Thus, intradermal injection of 2-methylhistamine (a relatively selective H_1-receptor agonist) or of 4-methylhistamine (a relatively selective H_2-receptor agonist) each led to dose-dependent erythema and wealing although the extent of wealing was relatively modest and did not show dose dependence. The erythema caused by 2-methylhistamine was reduced, but not abolished, by chlorpheniramine and was not reduced by cimetidine whereas the weal was not inhibited by either antagonist. The erythema caused by 4-methylhistamine was reduced by either cimetidine or chlorpheniramine although the effect of cimetidine appeared to exceed that of chlorpheniramine. Neither antagonist reduced 4-methylhistamine-induced weal. The inhibitory action of the antagonists in combination was not investigated against responses to either antagonist.

Unpublished studies by M. J. BOYCE (1981) have shown that the very potent and highly selective H_2-receptor agonist, impromidine, causes very intense vasodilatation when injected intradermally in humans. The vasodilatation, as assessed by the intensity of the redness, exceeded that caused by histamine. A very small, but clear, area of wealing could be observed immediately over the injection area, probably reflecting increased permeability due to trauma in an area of intense vasodilatation (WILLIAMS and MORLEY 1973; WILLIAMS and PECK 1977) rather than evidence of an increase in microvascular permeability caused directly by the H_2-receptor agonist. L. J. COOK and S. SHUSTER (unpublished work) likewise showed a weal-and-flare response with intradermal cimetidine.

II. Evidence for Histamine Receptors on Skin Blood Vessels Following Systemic Administration of Histamine

Intravenous administration of histamine to humans has long been known to cause flushing due to dilatation of cutaneous blood vessels. The flushing, which occurs initially on the face and neck, shows clear dose dependence in terms of increasing intensity and also increasing distribution as it spreads from the neck to the trunk and eventually the legs as the dose of histamine increases. The initial clinical evaluation of burimamide, the first H_2-receptor antagonist, was designed to measure inhibition of histamine-induced acid secretion in subjects treated with mepyramine. Histamine caused profound flushing which was inhibited when subjects received burimamide (WYLLIE et al. 1972). These experiments provided clear evidence for the existence of H_2-receptors on human cutaneous blood vessels.

BOYCE and WAREHAM (1980) and BOYCE et al. (1980) have shown that flushing responses to histamine are readily reproducible. Pre-treatment with either chlorpheniramine, 10 mg intravenously, or cimetidine, 200 mg intravenously, reduced the flushing slightly although the two antagonists in combination were more effective than either alone. Increasing the dose of histamine, in subjects treated with both antagonists, re-established flushing. In a parallel study using hydroxyzine, 25 mg and cimetidine, 300 mg both four times a day, KALINER et al. (1981), showed that although neither antagonist treatment alone significantly changed the dose of histamine which caused flushing (or headache), after treatment with a combination of antagonists, 3–5 times more histamine was needed to elicit flushing. In these studies with systemic administration of histamine, there was no apparent difference in the response to histamine, or the pharmacology of the response, between asthmatic or normal subjects.

Intravenous infusions of impromidine also caused intense flushing in normal subjects. The intensity and distribution of the flushing is dose dependent and responses can be inhibited by treatment with cimetidine. Consistent with the competitive nature of H_2-receptor antagonism by cimetidine, impromidine flushing can be re-established after cimetidine treatment by increasing the dose of impromidine (BOYCE and WAREHAM 1980).

III. Histamine-Induced Pruritus

An important part of the response to local histamine is hyperalgesia or itching. Quantitative and objective measurement of itching is notoriously difficult, but merits substantial study as itching is a major distressing contribution to inflammatory skin disorders thought to involve histamine release. The available data for receptors involved in histamine-induce pruritus is more variable than the evidence for histamine receptors on blood vessels.

Evidence for H_1-receptors involved in pruritus has been reported by ARNOLD et al. (1979) using hydroxyzine, DAVIES and GREAVES (1980) using chlorpheniramine, and BOSS and BURTON (1981) using clemastine. In each of these studies, treatment with an H_1-receptor antagonist increased the threshold concentration of histamine to elicit itch. ARNOLD et al. (1979) considered that the anti-pruritic activity of hydroxyzine and other H_1-receptor antagonists might reflect their sedative action. In fact, their study demonstrated that hydroxyzine was far more effective as an anti-pruritic than a series of neuroleptics, consistent with activity of hydroxyzine as a consequence of H_1-receptor antagonism rather than sedation. HAGERMARK et al. (1979) found that intradermal mepyramine significantly reduced histamine-induced itch. It is impossible to distinguish between H_1-receptor antagonism and local anaesthesia as the mechanism of anti-pruritic activity of mepyramine under the conditions of their study. DAVIES et al. (1979) found that neither cimetidine, 400 mg nor chlorpheniramine, 4 mg reduced histamine induced itch although in combination the antagonists were a highly effective anti-pruritic treatment. Using nocturnal limb movement to measure itch as scratch MUSTON et al. (1979) showed that anti-pruritic drugs were mostly anxiolytic sedatives. Furthermore, it has recently been shown that terfenadine and astemizole which produce virtually complete H_1-receptor antagonism without sedation have no effect on pruritus (other than that due to urticaria where histamine is released in the skin), whereas the weak but sedative anti-histamines are anti-pruritic (KRAUSE and SHUSTER 1983). It can therefore be concluded that anti-histamines are not primarily anti-pruritic and act by an effect related to sedation.

DAVIES and GREAVES (1980) compared the pruritic activity of histamine receptor antagonists following intradermal injection. 2-Methylhistamine caused itch at a concentration approximately three times that needed for histamine. In subjects treated with chlorpheniramine, there was a small increase in the histamine concentration needed to cause itch whereas the concentration of 2-methylhistamine needed was substantially greater. The greater antagonism by chlorpheniramine of 2-methylhistamine than of histamine suggests that the histamine response may involve more than just H_1-receptors, presumably H_2-receptors. However, neither of the H_2-receptor agonists, 4-methylhistamine and dimaprit, caused itching at concentrations greatly in excess of the threshold concentration for histamine-induced itch and M. J. BOYCE (1981, unpublished work) has shown that impromidine does not cause itch. Neither low concentrations of 4-methylhistamine or dimaprit, nor a very high concentration of dimaprit decreased the threshold for histamine-induced itch. Low concentrations of the H_2-receptor agonist did not decrease the threshold concentration of 2-methylhistamine. Although this comprehensive study considered many combinations of agonist concentration, it did not adequately consider the relative potencies of the agonists at their receptor sites and the H_2-receptor agonists were used in inadequate concentrations when mixed with 2-methylhistamine.

Thus, although the pharmacology of histamine-induced pruritus remains unclear, very substantial evidence suggests a role for H_1-receptors, whereas the role of H_2-receptors in itch has been less well clarified and the agonist data particularly suggest that itch is independent of H_2-receptors. The itch of disease appears to be due to histamine only in the urticarial diseases (see Chaps. 31 and 32).

IV. Other Actions of Histamine in Skin

Histamines may increase epidermal cell production (e.g. HARPER et al. 1974) and cimetidine can reduce sebum production, but this may be an anti-androgenic effect.

V. Clinical Results with H_1- and H_2-Receptor Antagonists in Chronic Urticaria

The extensive experimental studies reviewed in this chapter provide overwhelming evidence for the involvement of H_1- and H_2-receptors in the acute inflammatory responses to exogenous histamine. The rational extrapolation of these findings to the use of combinations of antagonists in the treatment of urticaria has been demonstrated in studies based on rather small numbers of subjects and producing conflicting results, which were usually but not invariably positive (COOK and SHUSTER 1979, 1980). This prompted a larger multi-centre study with combined antagonists and confirms the benefit of adding an H_2-receptor antagonist (cimetidine) to the H_1-receptor antagonist (chlorpheniramine) in the treatment of urticaria (BLEEHEN et al. 1987). These findings, based on a substantial patient population would, if confirmed, offer the prospect of more effective treatment of urticaria than that achieved with H_1-receptor antagonists alone.

E. Newer Histamine Antagonists

The identification and successful development of some improved H_1-receptor antagonists is likely to exert an important influence on the treatment of histamine-induced inflammatory injuries in skin. In particular, these compounds have a greatly reduced central nervous system activity, usually because penetration into the central nervous system is substantially reduced relative to older H_1-receptor antagonists, and also a lesser anti-cholinergic activity.

Of these compounds, terfenadine (CHENG and WOODWARD 1982) astemizole (VAN WAUWE et al. 1981) and loratide (AHN and BARNETT 1986) have been introduced into clinical use and their efficacy demonstrated in urticarias (e.g. CERIO and LESSOF 1984; KRAUSE and SHUSTER 1984b, 1985b; FREDRIKSSON et al. 1986). However, it is already clear, that despite virtually complete H_1-receptor blockade, lesions of urticaria are only partially controlled, suggesting the role of mediators other than histamine (KRAUSE and SHUSTER 1984, 1985). It therefore remains to be seen whether additional effects are seen with the further H_1-receptor antagonists now under development.

F. Conclusions

The presence of both H_1- and H_2-receptors on skin blood vessels has been demonstrated. H_1-receptor antagonists have a long-established place in the therapy of urticarias although some patients respond poorly to these drugs. The recognition of H_2-receptors on cutaneous blood vessels provides a basis for the therapeutic use of H_2-receptor antagonists, although clinical trial results are still conflicting. The availability of the new non-sedative H_1-receptor antagonists has been a major advance.

References

Ahn HS, Barnett A (1986) Selective displacement of [^{3}H]-mepyramine from peripheral vs. central nervous system receptors by loratadine, a non-sedating antihistamine. Eur J Pharmacol 127:153–155

Aoyagi T, Adachi K, Halprin KM, Levine V, Woodyard CW (1981) The effect of histamine on epidermal outgrowth: its possible dual role as an inhibitor and stimulator. J Invest Dermatol 76:24–27

Arnold AJ, Simpson JG, Jones HE, Ahmed AR (1979) Suppression of histamine-induced pruritis by hydroxyzine and various neuroleptics. Am Acad Dermatol 1:509–512

Bain WA, Hellier FF, Warin RP (1948) Some aspects of the action of histamine antagonists. Lancet 2:964–969

Bleehen SS, Thomas SE, Greaves MW, Newton J, Kennedy CTC, Hindley F, Marks R et al. (1987) Cimetidine and chlorpheniramine in the treatment of chronic idiopathic urticaria: a multi-centre randomized double-blind study. Br J Dermatol 117:81–88

Boss M, Burton JL (1981) Lack of effect of the antihistamine drug clemastine on the potentiation of itch by prostaglandin E_1. Arch Dermatol 117:208–209

Boyce MJ, Wareham K (1980) Histamine H_1- and H_2-receptors in the cardiovascular system of man. In: Torsoli A, Lucchelli PE, Brimblecome RW (eds) H_2-antagonists. Excerpta Medica, Amsterdam

Boyce MJ, Balasubramanian V, Wareham K (1980) Cardiovascular effects in man of impromidine, a novel and specific histamine H_2-receptor agonist. Br J Pharmacol 70:157P–158P

Brittain RT, Daly MJ (1981) A review of the animal pharmacology of ranitidine – a new, selective histamine H_2-antagonist. Glaxo Symposium, London

Brown AE, Griffiths R, Harvey CA, Owen DAA (1986) Pharmacological studies with SK&F 93944 (temelastine), a novel histamine H_1-receptor antagonist with negligible ability to penetrate the central nervous system. Br J Pharmacol 87:569–578

Bunce KT, Owen DAA, Smith IR, Vickers MR (1979) Histamine. Int Rev Biochem 26:207–256

Carrier EB (1922) Studies on the physiology of capillaries. V. The reaction of the human capillaries to drugs and other stimuli. Am J Physiol 61:528–547

Cerio R, Lessof MH (1984) Treatment of chronic idiopathic urticaria with terfenadine. Clin Allergy 14:139–141

Cheng HC, Woodward JK (1982) Antihistaminic effect of terfenadine: a new piperidine-type antihistamine. Drug Dev Res 2:181–196

Cheng HC, Raevis OK, Munro NL, Woodward JK (1979) Cutaneous vascular histamine H_1-receptors in the guinea pig: the histamine skin wheal as a cutaneous vascular model. Arch Int Pharmacodyn Ther 249:241

Cook LJ, Shuster S (1979) Lack of effect of H_2 blockade in chronic urticaria. Br J Dermatol [Suppl 17]101:20–21

Cook LJ, Shuster S (1980) Histamine weal formation and absorption in man. Br J Pharmacol 69:579–585

Cook LJ, Shuster S (1983) The effect of H_1 and H_2 receptor antagonists on the dermographic response. Acta Derm Venereol (Stockh) 63:260–262

Cowen T, Trigg P, Eady RAJ (1979) Distribution of mast cells in human dermis-development of a mapping technique. Br J Dermatol 100:635–640

Dale HH, Laidlaw PP (1910) The physiological action of β-imidazolyethylamine. J Physiol (Lond) 41:318–344

Dale HH, Laidlaw PP (1911) Further observations on the action of β-imidazolylethylamine. J Physiol (Lond) 43:182–195

Davies MG, Greaves MW (1980) Sensory responses of human skin to synthetic histamine analogues and histamine. Br J Clin Pharmacol 9:461–465

Davies MG, Marks R, Horton RJ, Storari FE (1979) The efficacy of histamine antagonists as antipruritics in experimentally induced pruritis. Arch Dermatol Res 266:117–120

Durant GJ, Ganellin CR, Griffiths R, Harvey CA, Owen DAA, Sach GS (1984) Some newer H_1-receptor histamine antagonists. In: Ganellin CR, Schwartz JC (eds) Frontiers in histamine research. Pergamon, Oxford

Eady RAJ, Cowen T, Marshall TF, Plumer V, Greaves MW (1979) Mast cell population density, blood vessel density and histamine content in normal human skin. Br J Dermatol 100:623–633
Eppinger H (1913) Über eine eigentümliche Hautreaktion hervorgerufen durch Ergamin. Wien Klin Wochenschr 63:1414
Fredriksson T, Hersle K, Hjorth N, Mobacken H, Persson T, Saldo L, Salo O et al. (1986) Terfenadine in chronic urticaria: a comparison with clemastine and placebo. Cutis 38:128–130
Flynn SB, Owen DAA (1977) The effects of histamine on skeletal muscle vasculature in cats. J Physiol (Lond) 265:795–807
Haddy FJ (1960) Effect of histamine on small and large vessel pressures in dog forelimb. Am J Physiol 198:161–168
Hagermark O, Strandberg K, Bronneberg R (1979) Effects of histamine receptor antagonists on histamine-induced responses in human skin. Acta Derm Venereol (Stockh) 59:297–300
Harmer IM, Harris KE (1926) Observations on the vascular reaction in man in response to histamine. Heart 13:381–394
Harper RA, Flaskman BA, Chopra DP (1974) Mitotic response of normal and psoriatic keratinocytes in vitro to compounds known to effect intracellular cyclic AMP. J Invest Dermatol 62(4):384–387
Harvey RP, Schocket AL (1980) The effect of H_1- and H_2-blockade on cutaneous histamine response in man. J Allergy Clin Immunol 65:136–139
Humphreys F, Krause LB, Shuster S (1987) The effects of astemizole and indomethacin on weal and flare reactions to histamine, 48/80 and house dust mite antigen (Abstr). Br J Dermatol 116:435
Hutchcroft BJ, Moore EG, Orange RP (1979) The effects of H_1- and H_2-receptor antagonism on the response of monkey skin to intradermal histamine, reverse-type anaphylaxis, and passive cutaneous anaphylaxis. J Allergy Clin Immunol 63:376–382
Kaliner MA (1979) The mast cell – a fascinating riddle. N Engl J Med 301:498–499
Kaliner MJB, Sigler R, Summers R, Shelhamer JH (1981) Effects of infused histamine: analysis of the effects of H_1- and H_2-receptor antagonists on cardiovascular and pulmonary responses. J Allergy Clin Immunol 68:365–371
Krause LB, Shuster S (1983) Mode of action of H_1 antihistamines in itch. Br J Dermatol [Suppl 24]109:27
Krause LB, Shuster S (1984a) The effect of terfenadine on dermographic wealing. Br J Dermatol 110:73–79
Krause LB, Shuster S (1984b) H_1-receptor-active histamine not sole cause of chronic idiopathic urticaria. Lancet 2:929–930
Krause LB, Shuster S (1985a) A comparison of astemizole and chlorpheniramine in dermographic urticaria. Br J Dermatol 11:447–453
Krause LB, Shuster S (1985b) Minimal effect of complete H_1 receptor blockade on urticaria pigmentosa. Acta Derm Venereol (Stockh) 65:338–340
Lewis T (1927) The vessels of the human skin and their responses. Shaw, London
Lewis T, Grant RT (1924) Vascular reaction of the skin to injury. Heart 71:209–235
Lynn B (1977) Cutaneous hyperalgesia. Br Med Bull 33:103–108
Lyons F, Shuster S (1980) The suppression of sebaceous gland activity by H_2 receptor agonists. Br J Dermatol 102:730–731
Marks R, Greaves MW (1977) Vascular reactions to histamine and compound 48/80 in human skin: suppression by a histamine H_2-receptor blocking agent. Br J Clin Pharmacol 4:367–369
Mellander S, Johansson B (1968) Control of resistance, exchange, and capacitance functions in the peripheral circulation. Pharmacol Rev 20:117–196
Muston H, Felix R, Shuster S (1979) Differential effect of hypnotics and anxiolytics on itch and scratch. J Invest Dermatol 72:283
Nathan RA, Segall N, Schocket AL (1981) A comparison of the actions of H_1- and H_2-antihistamines on histamine-induced bronchoconstriction and cutaneous wheal response in asthmatic patients. J Allergy Clin Immunol 67:171–177

Ostoredo M, Arhand A, Vervloet D, Razzouk H, Charpin J (1980) Effets des antihistaminiques (H_1 et H_2) sur différents types de tests cutanés explorant l'allergie à médiation humorale ou cellulaire. Rev Fr Allergol 20:113–117

Owen DAA, Poy E, Woodward DF (1980) Evaluation of the role of histamine H_1- and H_2-receptors in cutaneous inflammation in the guinea-pig prodcued by histamine and mast cell degranulation. Br J Pharmacol 69:615–623

Popielski L (1920) βImidazolylaethylamine und die Organextrakte. I. β-Imidazolylaethylamine als mächtiger Erreger der Magendrüsen. Pflügers Arch Gesamte Physiol 178:214–236

Robertson I, Greaves MW (1978) Responses of human skin blood vessels to synthetic histamine analogues. Br J Clin Pharmacol 5:319–322

Smith JA, Mansfield LE, de Shazo RD, Nelson HS (1980) An evaluation of the pharmacologic inhibition of the immediate and late cutaneous reaction to allergen. J Allergy Clin Immunol 65:118–121

Thomas SE, Glenny H, Bleehen SS, Greaves MW, Pilgrim J, Rowell N, Fairiss G et al. (1985) Cimetidine and chlorpheniramine in the treatment of chronic urticaria – an interim analysis of the results from a multi-centre randomised double-blind study. In: Champion RH, Greaves MW, Kobza Black A, Pye RJ (eds) The urticarias. Churchill Livingstone, Edinburgh

Van Wauwe J, Avouters F, Niemegeers CJE, Janssens F, van Nueten JM, Janssens PAJ (1981) In vivo pharmacology of astemizole, a new type of H_1 anti-histaminic compound. Arch Int Pharmacodyn Ther 251(1):39–51

Williams TJ, Morley J (1973) Prostaglandins as potentiators of increased vascular permeability in inflammation. Nature 246:215–217

Williams TJ, Peck MJ (1977) Role of prostaglanin-mediated vasodilatation in inflammation. Nature 270:530–532

Woodward DF, Owen DAA (1982) Effect of H_1- and H_2-receptor antagonists on cutaneous inflammation evoked by histamine analogues and UV radiation. Eur J Pharmacol 77:103–112

Wyllie JH, Hesselbo T, Black JW (1972) Effects in man of histamine H_2-receptor blockade by burimamide. Lancet 2:1117–1120

CHAPTER 19

Clinical Pharmacology of Topical Steroids

D. N. BATEMAN

A. Introduction

The first report of the therapeutic use of topical hydrocortisone was published in 1952 by SULTZBERGER and WITTEN and many more potent synthetic steroids have been developed since. Ideally the pharmacological actions of topically applied steroids would be limited to the skin without systemic local adverse effects. Although this has not been achieved, there have been considerable advances both in potency of steroids available for clinical use and in the understanding of their pharmacological effects on the skin. Glucocorticoids are widely used in the treatment of skin disorders, but their specific indications are outside the scope of this chapter.

I. Structure-Activity Relationship

The basic steroid structure common to all the drugs under consideration is a 17-carbon atom rigid 4-ringed nucleus with three 6-membered and one 5-membered carbon rings (Fig. 1). Corticosteroids possess mineralocorticoid and glucocorticoid properties, and the relative potency of each depends on the ring structure and substitution. Thus, a double bond in the 1–2 position of hydrocortisone, to

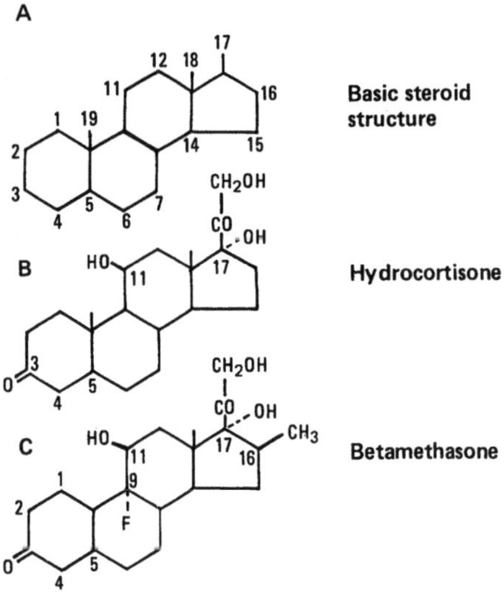

Fig. 1. The basic steroid structure (**A**) and the modifications to produce hydrocortisone (**B**) and a fluorinated corticosteroid, betamethasone (**C**)

give prednisolone, increases glucocorticoid activity about fourfold and reduces mineralocorticoid effects. Introduction of fluorine at the 9α position further increases glucocorticoid potency, but mineralocorticoid activity also increases. If a further group is substituted at the 16 position the mineralocorticoid activity is reduced again (e.g. triamcinolone 16α-hydroxy; dexamethasone 16α-methyl; betamethasone 16β-methyl).

In addition to synthesis of glucocorticoid with enhanced potency, efficacy will be increased by enhanced penetration into the skin, e.g. by improved lipid solubility. This is achieved by esterification of the steroid at the 17 or 21 position (e.g. betamethasone 17-valerate, hydrocortisone 21-acetate), and by masking the 16α-hydroxyl group; the acetonide group will simultaneously mask both 17 and 21 positions (e.g. fluocinolone acetonide). Increases in potency and absorption also increase the risk of adverse effects.

Glucocorticoids can be divided into four groups on the basis of clinical potency (Table 1) (MARTINDALE 1982). This classification takes into account both the inherent agonist activity of the drugs, and their ability to penetrate the skin. Though there is a differentiation between the very potent, group 1, and potent, group 2 agents, the difference between groups 3 and 4 is less well defined.

II. Mode of Application

The magnitude of effect of topically administered steroids will also depend on the vehicle in which the steroid is delivered (MILLER and MUNRO 1980). Vehicles may act as a carrier, increasing penetration into the stratum corneum, or by making the stratum corneum more permeable. Solubility of the glucocorticoid within the vehicle is enhanced by agents such as propylene glycol or ethanol. Very occlusive vehicles (e.g. ointments) increase stratum corneum hydration and therefore absorption and so may improve bioavailability (FRITZ and WESTON 1983). Inclusion of urea in the preparation also increases the skin penetration of the steroid; the effect of salicylic acid is less clear (MILLER and MUNRO 1980).

III. Skin Factors

The skin itself varies at different sites and the thickness of the stratum corneum appears to be particularly relevant, penetration being inversely proportional to its thickness (VICKERS 1963). Removal of the stratum corneum with adhesive tape was reported to increase the absorption of hydrocortisone from 1%–2% to over 50%, but MAIBACH (1976) was able to demonstrate only a 1%–3% increase in absorption in humans after tape stripping, suggesting that factors other than the stratum corneum are important for steroid penetration. Earlier evidence had also shown that fat solvents and keratolytics enhance steroid absorption by disrupting the stratum corneum (MALKINSON 1958).

The stratum corneum acts as a reservoir for the applied steroid, some evidence for which has been reported for up to 14 days after steroid administration under occlusion (VICKERS 1963). Absorption is also influenced by ambient temperature and humidity.

Table 1. Clinical potency of topical corticosteroids

Very potent: Group 1	
	Beclomethasone dipropionate 0.5%
	Clobetasol propionate 0.05%
	Diflucortolone valerate 0.3%
	Fluocinolone acetonide 0.2%
Potent: Group 2	
	Beclomethasone dipropionate 0.025%
	Betamethasone benzoate 0.025%
	Betamethasone dipropionate 0.05%
	Betamethasone valerate 0.1%
	Desonide 0.05%
	Desoxymethasone 0.25%
	Diflorasone diacetate 0.05%
	Diflucortolone valerate 0.1%
	Fluclorolone acetonide 0.025%
	Fluocinolone acetonide 0.025%
	Fluocinonide 0.05%
	Fluocortolone 0.5%
	Fluprednidene acetate 0.1%
	Flurandrenolone 0.05%
	Halcinonide 0.1%
	Hydrocortisone butyrate 0.1%
	Triamcinolone acetonide 0.1%
Moderately potent: Group 3	
	Clobetasone butyrate 0.05%
	Flumethasone pivalate 0.02%
	Fluocinolone acetonide 0.01%
	Fluocortin butylester 0.75%
	Fluocortolone 0.2%
	Flurandrenolone 0.0125 to 0.025%
	Hydrocortisone 1% with urea
Mild: Group 4	
	Dexamethasone 0.01%
	Hydrocortisone or hydrocortisone acetate 0.1%–1%
	Methylprednisolone 0.25%

B. Mode of Action

The precise mechanisms by which corticosteroids achieve their clinical effects in the wide variety of skin diseases in which they are used are unclear. A number of specific actions on the skin have been demonstrated and these are discussed in this chapter. The effects of corticosteroids on skin in vivo may result from a number of different actions on the different cell populations involved in the various diseases and no single unifying mechanism of action is apparent, although the action on phospholipase A_2 and lipocortin are obviously of importance (see Chap. 20).

I. Steroid Receptors

It is now clear that there are specific receptor proteins within the cells that mediate, at least in part, the actions of corticosteroids (JOHNSON et al. 1982). After binding to the receptor the steroid and receptor complex is translocated into the nucleus and changes in protein synthesis within the cell result following synthesis of specific mRNA. Such receptors have been demonstrated in the skin of mice (EPSTEIN and MUNDERLOH 1981), rats (SMITH and SHUSTER 1984a) and humans (EPSTEIN and BONIFAS 1982; LEIFERMAN et al. 1983). The typical features of the receptors are their high specificity and affinity, and low capacity for steroid binding.

The distribution of the receptor within the cells of the dermis and epidermis has also been studied. In rats 82% of the receptor sites are in the dermis and a similar situation appears to apply in humans (SMITH and SHUSTER 1984a; SMITH et al. 1983b). The receptors present in epidermis and dermis appear to have similar physicochemical properties, though differences in the association rate constant have been observed. However, these differences may reflect differences in protein content of the cytosol rather than a true difference in receptor affinity (SMITH and SHUSTER 1984a).

In rats dynamic changes in the numbers of skin cytosolic receptors have been shown. Adrenalectomy results in an increase in steroid receptors, with a maximum at 5 days after surgery and a return towards base-line by 15 days. Treatment of these animals with corticosterone resulted in a rapid dose-related decrease in the number of receptor sites (SMITH and SHUSTER 1984b), which was maximal within 30 min. These changes probably reflect both alterations in endogenous synthesis of the steroid receptor, and changes as a result of binding exogenous steroids administered acutely, the net effect being less available receptors for the ligand used in the receptor binding assay. There are no data on whether such changes occur in humans, but they would be compatible with the experimental evidence of tachyphylaxis to vasoconstriction (DU VIVIER and STOUGHTON 1975) and the clinical observation that patients become habituated to the effects of local steroid creams (ANONYMOUS 1977). The receptor concentration in different areas of human skin has also been shown to vary with the sites studied, the foreskin having the highest concentration. In addition an effect of age on the numbers of receptors on abdomen and breast skin was observed, the maximal number of receptors being found at the extremes of life (LEIFERMAN et al. 1983). The clinical implications of these observations are unclear.

It is imortant to stress that the binding between the receptor and the steroid is only one part of the sequence of events that occurs between drug administration and biological effect. Differences in bioavailability and local metabolism will also be important in determining the biological effect observed.

II. Inhibition of Prostaglandins

Corticosteroids act on prostaglandin production, and in this way exert an anti-inflammatory effect. Steroid inhibition of guinea pig peritoneal macrophage prostaglandin synthesis in vitro has been shown to parallel clinical efficacy (BRAY and GORDON 1978).

One mechanism by which corticosteroids achieve their effects on prostaglandin synthesis is by inducing synthesis of proteins that inhibit epidermal phospholipase A_2, which cleaves arachidonic acid from membrane phosphilipids (see Chap. 20). Arachidonic acid is the precursor for the synthesis of prostaglandins and leukotrienes and the degree of corticosteroid-induced inhibition of this enzyme in human skin is proportional to the enzyme's initial activity. Furthermore, the inhibitory effects of corticosteroids on phospholipase A_2 parallels clinical efficacy (NORRIS et al. 1984). In other sites the onset of steroid action in inhibiting prostaglandin synthesis is very rapid, occurring within 2 min (HORROBIN et al. 1976), suggesting that at least some of the glucocorticoid effect on prostaglandin synthesis is via a pathway that does not involve the action of steroids on the cell nucleus via the receptor system discussed in Sect. B. I.

III. Protein and Collagen Synthesis

Corticosteroids have been shown to exert a number of other actions including inhibition of skin fibroblast DNA synthesis, inhibition of protein synthesis and inhibition of collagen synthesis (JOHNSON et al. 1982; DURANT et al. 1986). SAARNI et al. (1980) demonstrated that in cultured human foetal skin fibroblasts there was inhibition of hyaluronic acid, sulphated aminoglycans and collagen synthesis. The steroid concentration required for inhibition of hyaluronic acid is much lower than that for the inhibtion of the other two synthetic pathways studied. PONEC et al. (1980) showed that inhibition of growth of human foreskin fibroblasts correlated with specific binding, probably to the glucocorticoid receptor, and with clinical efficacy. Corticosteroids interfere with wound healing, and it seems likely that the effects already mentioned will contribute to this. The major unwanted effect of topical corticosteroids is loss of skin collagen (SCARBOROUGH and SHUSTER 1960; SHUSTER et al. 1967; BLACK et al. 1973; DYKES and MARKS 1979).

IV. Immunosuppressant Effects

Corticosteroids are also used clinically as immunosuppressant agents and have been shown to affect both T-cell and monocyte functions. Topically applied corticosteroids also reduce the number of epidermal Langerhans' cells in mice. This effect was not observed after systemic administration, and may reflect the higher steroid concentrations obtained locally with topical therapy (HALLIDAY et al. 1986). Langerhans' cells which were exposed to short-term steroid did not alter their alloactivating capacity for allogeneic T-lymphocytes. If, in contrast, the Langerhans' cells were pulsed with purified protein derivative (PPD) as antigen there was a dose-dependent reduction of their stimulating capacity. The inhibition appeared, therefore, to be due to suppression of the presentation of the PPD in immunogenic form to sensitised T cells (BRAATHEN and HIRSCHBERG 1984). The significance of this effect is not clear.

V. Actions of Microsomal Oxidation

Corticosteroids induce the activity of certain drug-metabolising enzymes present in the skin. Studies in the hairless mouse have shown induction of aryl hydrocarbon hydroxylase (BRIGGS and BRIGGS 1973) and more recently FINNEN et al. (1984) using more specific substrates demonstrated a differential effect of fluorinated glucocorticoids on ethoxycoumarin *O*-dealkylase, which induced this enzyme, as compared with the non-fluorinated hydrocortisone and the androgen fluandrenolone, which did not. The effects of a range of steroids on this enzyme mirrored clinical activity, and the effect of clobetasol was maximum when applied at a concentration used clinically (0.05%). The inducing effect of clobetasol was also studied on ethoxyresorufin dealkylase activity, and the pattern of induction differed from that of the commonly studied inducing agent 3-methylcholanthrene. Corticosteroids may therefore have inducing properties which differ from those of polycyclic hydrocarbons. The clinical implications of these findings are still uncertain.

VI. Other Actions

Topical steroids inhibit histamine release from mast cells (STAHLE and HAGERMARK 1984). LAVKER et al. (1986) have found that mast cell numbers decline during therapy with potent steroids, and that after 6 weeks mast cells were not present in skin biopsies. These authors found that betamethasone was less potent than clobetasol-13-propionates and fluocinonide in reducing histamine levels in skin. They have therefore suggested a relationship between loss of mast cells and steroid structure, presumably reflecting differences in pharmacological activity on the mast cell. This pharmacological effect of steroids has been applied to the treatment of urticaria pigmentosa lesions (LAVKER et al. 1986).

Corticosteroids have a particular effect on the skin vasculature, producing vasoconstriction. The mechanism of this action is unclear, but in part probably results from inhibition of synthesis of vasodilator prostaglandins, together with enhancement of the vasoconstrictor response of vascular smooth muscle (JOHNSON et al. 1982). This vasoconstrictor effect of steroids is the basis for the in vivo vasoconstrictor test of glucocorticoid potency.

C. Assays of Glucocorticoid Activity

A number of assays have been developed in order to screen and assess corticosteroid action on the skin. These include in vitro methods based on steroid receptor assays or the effects of steroids on cultured fibroblasts mentioned already. In vivo methods have included inhibition of croton oil inflammation, ultraviolet-induced erythema (BURDICK et al. 1973), the response to histamine (REDDY and SINGH 1976) and irritant and contact dermatitis.

The most widely used in vivo assay of corticosteroid potency is the vasoconstrictor test, which has been refined by BARRY and WOODFORD (1977). The principle of this test is that glucocorticoids, when applied topically under occlusion, will produce blanching of the skin. The intensity, size and duration of this effect can be scored and relative potencies and duration of action thus determined. Despite modifications and refinements, even in experienced hands the ability of this assay to detect with precision likely clinical differences in potency remains in doubt. GIBSON et al. (1984) have studied the usefulness of the test, and related it to the ability of steroids to produce skin thinning (measured by ultrasound). They report that measurement of the area under the curve relating time and intensity of blanching will give the best prediction of clinical efficacy and skin-thinning potential. It does not seem possible from these studies to separate the unwanted skin thinning effect from the vasoconstrictor effect. The vasoconstrictor action of corticosteroids varies, depending on the site on the forearm to which the drugs are administered – higher scores being obtained nearer the elbow (KIRSCH et al. 1982). Thus, studies using this technique need to be carefully controlled.

D. Metabolism of Corticosteroids in Skin

Since corticosteroids are usually applied topically to diseased skin their local effect will be determined partly by their rate of removal from the site of action. Local metabolism within the skin is therefore of potential therapeutic importance. Conversion of cortisol to cortisone in the skin was demonstrated by MALKINSON et al. in 1959. Both dermis and epidermis are capable of metabolising cortisol to a number of metabolites and this is not a property just of the sebaceous glands, the principal site of androgen metabolism, since it occurs in skin from the sole of the foot (RONGONE 1977; PANNATIER et al. 1978). WHITEFIELD (1977) suggested that local toxic effects of some steroids on the skin were more likely because of its inability to metabolise 17-esters of synthetic steroids, and that these were thus particularly likely to produce the local toxic effects as a result of local accumulation. Studies on the metabolism of betamethasone-17-valerate on human skin in vitro have shown that metabolism of this ester does, however, occur with hydrolysis to betamethasone. the production of betamethasone in epidermis was approximately twice that in dermis per milligram tissue per hour, and was linear with time for up to 6 h (RAWLINS et al. 1979).

E. Toxic Effects of Glucocorticoids

Toxic effects of glucocorticoids can be local or sytemic from absorption of a topical preparation.

I. Local Toxicity

Used for short periods topical corticosteroids are remarkably free of toxicity. In a study of 2349 patients local problems such as irritation, bruising and dryness were rare and their incidence equal to that of the vehicle alone (AKERS 1980). The type of vehicle used may influence local irritation, ointments being least troublesome. If occlusive therapy is used the frequency of local complications increases and local bacterial or yeast colonisation may occur. Skin atrophy at the site of corticosteroid administration is also well recognised, particularly with occlusion. Epidermal thinning results over a short period of time and is predominantly due to reduction in cell size rather than cell number (DELFORNO et al. 1978). This effect is porportional to steroid dose, and is usually reversible.

Changes in the dermis result from effects of steroids on fibroblast function, as discussed elsewhere. It is generally thought that their mode of action is by impaired collagen synthesis and the possible effect on collagen breakdown is still unclear (PONEC et al. 1979; see Chap. 27). The clinical result of effects on dermis are a wrinkly, shiny thin skin, with telangectasia due to vessel dilatation and skin transparency. Striae and purpura are due to loss of collagen (SCARBOROUGH and SHUSTER 1960; SHUSTER 1979). The skin of the face and flexures appears to be particularly sensitive to the atrophic effects of glucocorticoids (FRITZ and WESTON 1983; ANONYMOUS 1977).

Steroids delay wound healing (EVANS et al. 1967) and penetration to local structures beneath the skin has been described with fat atrophy and even muscle wasting (JOHNS and BOWER 1970). Application of steroid creams to the eyelid may result in ocular adverse effects, including glaucoma and cataract (ANONYMOUS 1977). Inappropriate application of steroids at the site of infection may result in uncontrolled spread, as in impetigo, fungal infection and viral infections such as herpes and vaccinia. Scabies will also spread more rapidly if treated with steroids and fungal infections may worsen, and become more difficult to diagnose (IVE and MARKS 1968). It has been suggested that glucocorticoids can transform plaque to pustular psoriasis (BAKER 1976), but this may be a fortuitous association since most patients with severe psoriasis will have received corticosteroids. Habituation to the effects of topical steroids has been described (ANONYMOUS 1977), e.g. in perioral dermatitis (ANONYMOUS 1980) which reached a peak incidence in the early 1970s, but is less common now because of its recognition (WILKINSON et al. 1979). The papulo-pustular lesions of this condition usually respond to a slow reduction of steroid dose, and are exacerbated by rapid reduction (FRITZ and WESTON 1983). Hypertrichosis and depigmentation have been reported. Topical steroids may also interfere with patch testing (FISHER 1983).

II. Systemic Effects

Corticosteroids are absorbed from the skin, and if applied in sufficient amounts, particularly if given under occlusion, may result in a full-blown Cushing's syndrome (FRITZ and WESTON 1983). Adrenal suppression, as judged by a fall in plasma cortisol, was documented following 30 g daily application of betamethasone-17-valerate under occlusion (MUNRO 1976). This study showed that the effect was variable between individuals, both in respect of degree of adrenal suppression and rate of recovery. Patients with liver failure appear to be at increased risk (BURTON et al. 1974). Absorption of steroids is more of a problem in infants than in adults (WEST et al. 1981) in part because of their greater surface area : volume ratio. Potent fluorinated steroids produce far greater adrenal suppression than hydrocortisone. Thus, 75% of infants treated with 0.1% betamethasone valerate under occlusion had suppression of adrenocortical function on the second post-treatment day, but this was not observed in a group treated with 1% hydrocortisone (MUNRO 1976). However, the safety of topical hydrocortisone in this respect has been questioned (see SHUSTER 1985).

Growth retardation is a particular complication of intensive topical steroid therapy in children (FRITZ and WESTON 1983). In general, however, with the exceptions of children, patients with liver disease, highly potent topical corticosteroids, and excessive use, systemic effects are unlikely to be a major hazard (MILLER and MUNRO 1980).

F. Treatment Guidelines

Treatment guidelines for the use of topical corticosteroids include the use of glucocorticoid with a potency appropriate to the disorder being treated, and the site at which it occurs (MILLER and MUNRO 1980; FRITZ and WESTON 1983). Thus, higher potency is required for skin of the palms and soles and low potency for high risk patients such as children, and for the skin of the face, genital and perianal areas. A comparison of relative absorption, based on body area is: forearm 1; sole 0.14; palm 0.83; back 1.7; scalp 3.5; forehead 6; cheek 13; scrotum 42 (FRITZ and WESTON 1983).

Percutaneous absorption is increased by hydration; bathing prior to application or occlusive cover may be desirable. Application once daily is adequate for many patients. To avoid tachyphylaxis, which occurs with the vasoconstriction assay (DU VIVIER and STOUGHTON 1975) some recommend a rest period of 1–2 days every few days. To minimise the risk of atrophy after the first few weeks of treatment and once the disorder is controlled a weaker steroid should be used. The combination of steroids with other therapeutic substances such as antibiotics, anti-fungal agents and tars should be used only for clearly defined situations.

The vehicle in which the steroid is delivered is relevant to absorption, occlusive effect, irritation and patient acceptability; lotions and gels are useful for treating hairy skin, but may be irritant (FRITZ and WESTON 1983). In view of the wide range of steroid potencies and concentrations available (see Table 1) and the likelihood that all share a common mechanism of action, there is no indication for the dilution of potent steroids for clinical use. Furthermore, dilution may alter absorptive characteristics out of proportion to the diluent effect.

References

Akers WA (1980) Risks of unoccluded topical steroids in clinical trials. Arch Dermatol 116:786–788

Anonymous (1977) The hazardous jungle of topical steroids. Lancet 2:487–488

Anonymous (1980) Perioral dermatitis. Lancet 1:75–76

Baker H (1976) Corticosteroids and pustulae psoriasis. Br J Dermatol [Suppl 12]94:83–88

Barry BW, Woodford R (1977) Vasoconstrictor activities and bioavailabilities of seven proprietary corticosteroid creams assessed using a non-occluded multiple dosage regimen: clinical implications. Br J Dermatol 97:555–560

Black MM, Shuster S, Bottoms EJ (1973) Skin collagen and thickness in Cushing's syndrome. Arch Dermatol Forsch 246:365–368

Braathen LR, Hirschberg H (1984) The effect of short-term corticosteroids incubation on the alloactivating and antigen-presenting capacity of human epidermal Langerhans cells. Br J Dermatol 111:295–302

Bray MA, Gordon D (1978) Prostaglandin production by macrophages and the effect of anti-inflammatory drugs. Br J Clin Pharmacol 63:635–642
Briggs MM, Briggs M (1973) Induction by topical corticosteroids of skin enzymes metabolising carcinogenic hydrocarbons. Br J Dermatol 88:75–81
Bronstein SW, Bickers DR, Lamkin BC (1984) Bullous dermatosis caused by *Staphylococcus aureus* in locus minosis resistantiae. J Am Acad Dermatol 10:259–263
Burdick KH, Haleblian JK, Poulsen BJ, Cobner SE (1973) Corticosteroid ointments: comparison by two human bioassays. Curr Ther Res 15:233–242
Burton JL, Cunliffe WJ, Holti G, Wright V (1974) Complications to topical steroids in patients with liver disease. Br J Dermatol [Suppl 10]92:22–23
Delforno C, Holt PJA, Marks R (1978) Corticosteroid effect on epidermal cell size. Br J Dermatol 98:619–623
Durant S, Duval D, Homo-Delarche F (1986) Factors involved in the control of fibroblast proliferation by glucocorticoids: a review. Endocr Rev 7:254–269
Du Vivier A, Stoughton RB (1975) Tachyphylaxis to the action of topically applied steroids. Arch Dermatol 3:581–583
Dykes J, Marks R (1979) An appraisal of the methods used in the assessment of atrophy from topical corticosteroids. Br J Dermatol 101:599–609
Epstein EH, Bonifas JM (1982) Glucocorticoid receptors of normal human epidermis. J Invest Dermatol 78:144–146
Epstein EH, Munderloh NH (1981) Glucocorticoid receptors of mouse epidermis and dermis. Endocrinology 108:703–711
Evans CD, Harman RRM, Warin RP (1967) Varicose ulcers and use of topical corticosteroids. Br Med J 4:482
Finnen MJ, Herdman ML, Shuster S (1984) Induction of drug metabolising enzymes in the skin by topical steroids. J Steroid Biochem 20:1169–1173
Fisher AA (1983) Topical adrenal steroids and patch tests. Arch Dermatol 119:956
Fritz KA, Weston WL (1983) Topical glucocorticosteroids. Ann Allergy 50:68–76
Gibson JR, Kirsch JM, Darley CR, Harvey SG, Burke CA, Hanson ME (1984) An assessment of the relationship between vasoconstrictor assay findings, clinical efficacy and skin thinning effects of a variety of undiluted and diluted corticosteroiod preparations. Br J Dermatol [Suppl 27]3:204–212
Halliday GM, Knight BA, Muller HK (1986) Reduction of murine Langerhans cell ATPase staining following topical, but not systemic treatment with steroid and non-steroid immunosuppressants. Br J Dermatol 114:83–89
Horrobin DF, Mtabaji JP, Manku MS (1976) Physiological cortisol levels block the inhibition of vascular reactivity produced by prolcatin. Endocrinology 99:406–410
Ive FA, Marks R (1968) Tinea incognita. Br Med J 3:149–152
Johns AM, Bower BD (1970) Wasting of the napkin area after repeated use of fluorinated steroid ointments. Br Med J 1:347–348
Johnson LK, Longenecker JP, Baxter JD, Dallman MF, Widmaier EP, Eberhardt NL (1982) Glucocorticoid action: a mechanism involving nuclear and non-nuclear pathways. Br J Dermatol [Suppl 23]107:6–23
Kirsch J, Gibson JR, Darley CR, Barth J, Burke CA (1982) Forearm site variation with the corticosteroid vasoconstrictor assay. Br J Dermatol 106:495
Kligman AM, Willis I (1975) A new formula for depigmenting human skin. Arch Dermatol 111:40–48
Lavker RM, Schechter NM, Lazarus GS (1986) Effects of topical corticosteroids on human dermis. Br J Dermatol [Suppl 31]115:101–107
Leiferman KM, Schroeter A, Kirschner MK, Spelsberg TG (1983) Characterisation of the glucocorticoid receptor in human skin. J Invest Dermatol 81:355–360
Maibach HI (1976) In vivo percutaneous penetration of corticoids in man and unresolved problems in their efficacy. Dermatologica [Suppl 1]152:11–25
Malkinson FD (1958) Studies on the percutaneous absorption of C^{14} labelled steroids by use of the gas flow cell. J Invest Dermatol 31:19–26
Malkinson FD, Lee MW, Cutukovic I (1959) In vitro studies of adrenal steroid metabolism in the skin. J Invest Dermatol 32:101–108

Martindale (1982) The extra pharmacopoeia, 28th edn. Pharmaceutical Press, London
Miller JA, Munro DD (1980) Topical corticosteroids: clinical pharmacology and therapeutic use. Drugs 19:119–134
Munro DD (1976) The effect of percutaneously absorbed steroids in hypothalamic-pituitary-adrenal function after intensive use in inpatients. Br J Dermatol [Suppl 12]94:67–76
Munro DD, Clift DC (1973) Pituitary-adrenal function after prolonged use of topical corticosteroids. Br J Dermatol 88:381–385
Norris JFB, Iderton E, Yardley HJ, Summerly K, Forster S (1984) Utilisation of epidermal phospholipase A_2 inhibtion to monitor topical steroid action. Br J Dermatol [Suppl 27]3:195–203
Pannatier A, Jenner P, Testa B, Etter JC (1978) The skin as a drug-metabolising organ. Drug Metab Rev 8:319–343
Ponec M, Kempenaar SA, van der Meulen-van Harskamp GA, Bachra BN (1979) Effects of glucocorticoids on cultured human skin fibroblasts. IV. Specific decrease in synthesis of collagen but no effect on its hydroxylation. Biochem Pharmacol 28:2777–2783
Ponec M, Derkloet ER, Kempenaar JA (1980) Corticoids and human skin fibroblasts: intracellular specific binding in relation to growth inhibition. J Invest Dermatol 75:293–296
Rawlins MD, Shaw V, Shuster S (1979) The in vitro metabolism of betamethasone-17-valerate by human skin. Br J Pharmacol 66:441P
Reddy BSN, Singh G (1976) A new model for human bioassay of topical corticosteroids. Br J Dermatol 94:191–193
Rongone EL (1977) Cutaneous metabolism P 93–137. In: Marzulli FM, Maibach HI (eds) Dermatotoxicology and pharmacology. Wiley and Sons, New York (Advances in modern toxicology, vol 4)
Saarni H, Jalkanen M, Hopsu-Havu VK (1980) Effect of five anti-inflammatory steroids on collagen and glycoaminoglycan synthesis in vitro. Br J Dermatol 103:167–173
Scarborough H, Shuster S (1960) Corticosteroid purpura. Lancet 1:93–94
Shuster S (1979) The cause of striae distensae. Acta Derm Venereol (Stockh) [Suppl 89]59:161–169
Shuster S (1985) OTC sale of topical corticosteroids. Br Med J 2:967–968
Shuster S, Raffle EJ, Bottoms EJ (1967) Skin collagen in rheumatoid arthritis and the effects of corticosteroids. Lancet 1:525–527
Smith K, Shuster S (1984a) Characterisation and quantification of epidermal and dermal glucocorticoid receptors in the rat. J Invest Dermatol 82:44–48
Smith K, Shuster S (1984b) Effect of adrenalectomy and steroid treatment on rat skin cytosol glucocorticoid receptor. J Endocrinol 102:161–165
Smith K, Shuster S, Rawlins M (1983a) Characterisation of the glucocorticoid receptor in rat skin. J Endocrinol 96:229–239
Smith K, Shuster S, Walton S, Rawlins MD (1983b) Characterisation of the glucocorticoid receptor in human skin. Br J Dermatol 109:222–223
Stahle M, Hagermark D (1984) Effects of topically applied clobetasol-17-propionate on histamine release in human skin. Acta Derm Venereol (Stockh) 64:239–642
Sultzberger MB, Witten VH (1952) The effect of topically applied compound E in selected dermatoses. J Invest Dermatol 191:101–102
Vickers CFH (1963) Existence of reservoir in the stratum corneum. Arch Dermatol 88:20–23
West DP, Worobec S, Solomon LM (1981) Pharmacology and toxicology of infant skin. J Invest Dermatol 76:147–150
Whitefield M (1977) Topical steroids. Lancet 2:925
Wilkinson DS, Kirton V, Wilkinson JD (1979) Perioral dermatitis: a 12 year review. Br J Dermatol 101:245–257

CHAPTER 20

Glucocorticoids and Lipocortin

S. H. PEERS and R. J. FLOWER

A. Discovery of Lipocortin

I. Introduction

The steroids are the most useful anti-inflammatory drugs that we possess, effective against virtually any type of inflammation. The naturally occurring hormone hydrocortisone is still used in clinical medicine, but in many cases it has been superseded by synthetic analogues which have vastly increased duration of action and potency, making them particularly valuable to the dermatologist. As a group, the steroidal anti-inflammatories are amongst the most widely prescribed drugs in the world.

Our knowledge concerning their anti-inflammatory mechanism of action is still far from complete, but at least some of the effects of glucocorticoids appear to be the result of inhibition of phospholipase A_2 (PLA_2). This enzyme, which is abundant in most cells is important because it controls the liberation of arachidonic acid, the substrate of cyclo-oxygenase (which produces prostanoids) and lipoxygenase (which produces hydroxy fatty acids, the precursors of leukotrienes). Phospholipase A_2 also liberates lyso-PAF, the precursor of the potent lipid mediator platelet-activating factor (PAF; 1-*O*-acyl, 2-alkylglycero-3-phosphocholine). PLA_2 activation is therefore a key event in therelease from cells of a variety of mediators with potentially pro-inflammatory actions in the skin.

Glucocorticoids do not inhibit phospholipase directly, but act by the "classic pathway" of steroid action, i.e. occupation of a cytoplasmic steroid receptor, interaction of the receptor complex with the nucleus, formation of mRNA and subsequently, new protein synthesis. When exposed to glucocorticoids, many cells produce a protein with phosholipase inhibitory activity. Such protein (or proteins) were discovered independently in neutrophils (lipomodulin, HIRATA et al. 1980; HIRATA 1981), from macrophages and perfused lungs (macrocortin, BLACKWELL et al. 1980; CARNUCCIO et al. 1980) and in renal cells (renocortin, CLOIX et al. 1983; RUSSO-MARIE and DUVAL 1982). These proteins had molecular weights of 15000 (macrocortin, renocortin), 30000 (renocortin) and 40000 (lipomodulin, macrocortin) and similar physical properties. Other groups have since also isolated steroid-inducible proteins which appear to be related (GUPTA et al. 1984; VAN DE VELDE et al. 1985). Radioimmunoassay using a monoclonal antibody raised against lipomodulin also detected proteins of molecular weight 30 and 15000 (HIRATA et al. 1982), and the differing molecular weights of the proteins may be due to proteolytic breakdown. This family of proteins has been collectively renamed the *lipocortins* (DI ROSA et al. 1984). For reviews dealing with the

Table 1. Some research landmarks in the discovery of lipocortin

Year	Event	References
1974	Glucocorticoids (GCs) prevent prostaglandin (PG) generation in vivo: mechanism unknown	Herbaczynska-Cedro and Staszewska-Barczak (1974)
1975	GCs prevent release of PGs in organised tissues, but do not inhibit cyclo-oxygenase	Gryglewski et al. (1975)
1976	GCs prevent phospholipid deacylation depriving cyclo-oxygenase of substrate Inhibition of PLA_2 demonstrated	Hong and Levine (1976) Nijkamp et al. (1976)
1978	Inhibition of PG synthesis by GCs requires *de novo* RNA and protein synthesis	Danon and Assouline (1978)
1979	Inhibition of PG synthesis by steroids depends on synthesis of second messenger protein	Flower and Blackwell (1979), Russo-Marie et al. (1979)
	GC-induced inhibition of inflammation depends on "classic pathway"	Tsurufuji et al. (1979)
1980	Lipomodulin discovered in neutrophils	Hirata et al. (1980)
	Macrocortin detected in macrophages and lung	Blackwell et al. (1980), Carnuccio et al. (1980)
1982	GC action on renal cells depends on "second messenger protein" renocortin	Russo-Marie and Duval (1982)
	Anti-phospholipase proteins are anti-inflammatory	Blackwell et al. (1982)
1984	Anti-phospholipase proteins re-named lipocortin	DiRosa et al. (1984)
1986	Lipocortin sequenced and cloned	Wallner et al. (1986)

discovery of lipocortins, see Flower (1984, 1985), Flower et al. (1984) and Hirata et al. (1985). Table 1 is a summary of important research landmarks in this field.

Induction of lipocortin synthesis by glucocorticoids is dependent upon cytosolic glucocorticoid receptors. The order of potency of glucocorticoids in binding to the receptor parallels that of induction of lipocortin synthesis (Blackwell 1983), and lymphocytes from patients deficient in glucocorticoid receptors cannot synthesise lipocortin (Hirata et al. 1985). Lipocortin synthesis may be regulated by major histocompatibility complex (MHC) gene products (Hirata and Iwata 1983). Gupta and Goldman (1982) have also reported that sensitivity to glucocorticoids and lipocortin synthesis are related to the MHC genes in mice.

Lipocortin release from cells following exposure to hydrocortisone occurs within 1 h in vivo, but requires longer in vitro. In macrophages the release is followed by resynthesis of intracellular lipocortin within 2 h in vivo, but again the effect requires longer in vitro (4.5 h, Carnuccio et al. 1981). The sensitivity of rat peritoneal leucocytes to hydrocortisone depends upon their intracellular lipocortin content, such that when intracellular stores are depleted, the cells are refractory to the steroid (Carnuccio et al. 1981). It appears that steroids induce lipocortin release from cells (macrophages) as well as its synthesis. Activators of macrophages such as lipopolysaccharide, Ca^{2+} ionophore or the chemotactic peptide f-Met-Leu-Phe do not release lipocortin from cells in vitro, although agents which inhibit cell function by increasing cAMP such as PGI_2 or the stable cAMP analogue dibutyryl-cAMP inhibit steroid-induced release (BLACKWELL 1983). Inhibitors of intermediate metabolism such as 2,4-dinitrophenol and 2-deoxyglucose, and agents which inhibit protein or RNA synthesis or microtubule assembly also inhibit steroid-induced lipocortin release (BLACKWELL 1983).

Lipocortin may be phosphorylated on a tyrosine residue in vivo, possibly by protein kinase C (HIRATA 1981; HIRATA et al. 1984) and this is associated with increased arachidonate release from rabbit neutrophils, and also a decrease in its phospholipase inhibitory activity (HIRATA 1981). Alkaline phosphatase treatment of purified lipocortins tends to increase their anti-phospholipase activity, suggesting that the naturally occurring protein found in tissues is phosphorylated. This is particularly true of the protein of molecular weight 15000 (HIRATA et al. 1982). Phosphorylation-dephosphorylation of lipocortin in vivo therefore could provide fine control over phospholipase activity, and rather than being only an "anti-inflammatory protein", as first thought,lipocortin could be a key regulator of phospholipase in the "resting" cell as well.

II. Cloning and Expression of Lipocortin

Human lipocortin has now been cloned, expressed and sequenced, allowing studies to be made with recombinant material (hereafter referred to as rh-LC I). The gene was expressed in *Escherichia coli* (WALLNER et al. 1986) which produced a protein of molecular weight 37000 with phospholipase inhibitory activity. The protein is strongly polar, with 30% of its residues charged, these being interspersed with hydrophobic regions. The protein contains a single potential glycosylation site, and two potential phosphorylation sites. The rat and human proteins are very similar, with approximately 90% homology.

Two distinct lipocortin-like proteins have been purified from human placenta: lipocortin I, identical to the human recombinant lipocortin already described, and lipocortin II, also of molecular weight 37000 (HUANG et al. 1986). Lipocortin II has been partially sequenced, and shares about 50% homology with lipocortin I, the greatest homology being in the central region which is important for PLA_2 inhibitory activity. Lipocortin I contains repeats of a "consensus sequence" thought to be important in membrane binding (KRETSINGER and CREUTZ 1986); indeed the primary structure of lipocortin I, excluding its first 43 residues, may be considered four repeats of a single unit (SARIS et al. 1986).

Lipocortin I can be phosphorylated in vitro by protein kinase A, or at a site near its amino terminus by the protein kinase activity of the epidermal growth factor (EGF) receptor, and it seems that lipocortin I may be identical to the endogenous substrate for the EGF receptor/kinase, also known as p35 or calpactin II (PEPINSKY and SINCLAIR 1986; PEPINSKY et al. 1986).

III. Distribution of Lipocortins

Antibodies to lipocortins have been used to detect the presence of lipocortin in various tissues. Large amounts have been detected in lung, thymus, brain and spleen of both control and steroid-treated rats (FLOWER 1984), but the macrophage remains one of the richest sources known. Skin has not yet been examined.

Lipocortin or lipocortin-like proteins have now been purified from various tissues and cells, including of course neutrophils, macrophages, lung and renomedullary cells, where lipocortin was first described. Similar proteins have been isolated from or detected in other tissues including placenta (HUANG et al. 1986), mononuclear cells (ROTHHUT et al. 1987), T-lymphocytes (HIRATA and IWATA 1983), platelets (TOUQUI et al. 1986) and skin (FORSTER et al. 1984; NORRIS et al. 1984).

B. Properties of Lipocortin

I. Inhibition of Phospholipase A_2

By definition, the basic property of lipocortins is their ability to inhibit PLA_2. The most likely target in intact cells is a calcium-dependent membrane-bound enzyme, present in most cells in the plasma membrane and at least partially accessible to the cell exterior (YEDGAR et al. 1986). This enzyme is difficult to assay, but is similar both immunologically (OKAMOTO et al. 1985a) and in other aspects (such as pH optimum and sensitivity to inhibitors) to soluble phospholipases, in particular pig pancreatic PLA_2 (NIXON et al. 1985; OKAMOTO et al. 1985b; VOLWERK and DE HAAS 1982).

This enzyme is inhibited by lipocortins with [^{3}H]oleate-labelled *E. coli* or pure phospholipid vesicles as substrates (e.g. HIRATA 1981; BLACKWELL et al. 1982; ROTHHUT et al. 1983). The inhibition is dose dependent and requires a stoichiometric amount of lipocortin for maximal inhibition (HIRATA et al. 1982). The recombinant material (rh-LC I) inhibits binding of [^{125}I]PLA_2 to *E. coli* (FLOWER et al. 1986; PEERS et al. 1987) and also the release of [^{3}H]oleic acid from labelled *E. coli* (WALLNER et al. 1986), and it appears that a region contained within residues 83–212 is important for its activity (HUANG et al. 1987). It has been suggested that this inhibition could be due to lipocortin binding to substrate, and thereby blocking access to PLA_2, since the inhibition seen in some assays can be overcome by increasing the substrate concentration (AARSMAN et al. 1987; DAVIDSON et al. 1987; SCHLAEPFER and HAIGLER 1987). It has been reported that lipocortin does not interact directly with PLA_2 (SCHLAEPFER and HAIGLER 1987) and this situation awaits resolution.

Naturally occurring lipocortin inhibits PLA_2 present in cell membranes, such as that present in microsomal fractions of macrophages from rat (BLACKWELL 1983) and mouse (GHIARA et al. 1984). The release of [^{14}C]arachidonic acid from the membranes of pre-labelled rabbit neutrophils or cultured rat renomedullary cells is also inhibited by lipocortin (HIRATA et al. 1980; RUSSO-MARIE and DUVAL 1982), strongly suggesting inhibition of PLA_2. In addition, bradykinin-stimulated [^{3}H]arachidonic acid release from pre-labelled human fibroblasts is enhanced by antibodies to lipocortin, and inhibited by lipocortin itself (HIRATA et al. 1981).

II. Inhibition of Cellular Eicosanoid Synthesis

The anti-phospholipase proteins mimic closely the action of glucocorticoids in inhibition of eicosanoid synthesis, although without, of course, a lag phase. Inhibition of eicosanoid release by natural lipocortin has been noted in a variety of cells, including inhibition of thromboxane A_2 (TxA_2) from perfused guinea pig lung in response to histamine and leukotrienes, but not bradykinin (BLACKWELL et al. 1980), inhibtion of PGE_2 from phagocytosing leukocytes, and renomedullary cells (CARNUCCIO et al. 1980, 1981; DI ROSA and PERSICO 1979; PARENTE et al. 1984; RUSSO-MARIE and DUVAL 1982), and inhibition of leukotriene B_4 (LTB_4) release from leucocytes in response to heat-killed *Bordetella pertussis* (PARENTE et al. 1984). Lyso-PAF release also occurs owing to PLA_2 activity, and lipocortin has been shown to inhibit lyso-PAF release from resident leucoxytes in response to opsonised zymosan (PARENTE and FLOWER 1985).

rh-LC I has been shown to inhibit TxA_2 release from perfused guinea pig lung in response to leukotrienes, but not bradykinin (CIRINO et al. 1987), prostaglandin I_2 (PGI_2) release from human umbilical vein fragments (CIRINO and FLOWER 1987a) and rat peritoneal macrophages (CIRINO and FLOWER 1987b). Further studies are under way with this material.

Lipocortin appears to inhibit PLA_2 by interacting with the enzyme from the cell exterior, as might be expected with a protein of this size. The onset of action of lipocortin is rapid, but can be removed by washing (BLACKWELL et al. 1980). The action of lipocortin can also be terminated by exposure to proteolytic enzymes (HIRATA et al. 1980) or by neutralising antibodies (FLOWER 1984). Moreover, the neutralising antibody RM23 inhibits hydrocortisone-induced inhibition of prostaglandin synthesis (FLOWER 1984), suggesting not only that lipocortin can act from the cell exterior, but that it does so physiologically.

III. Anti-inflammatory and Other Effects of Lipocortin

Glucocorticoids are of course extremely effective anti-inflammatory agents. Lipocortins have also been shown to mimic glucocorticoid action in various models. Carrageenan-induced pleurisy or paw oedema is inhibited by lipocortin from macrophages (macrocortin) injected either locally or systemically (BLACKWELL et al. 1982; CALIAGNANO et al. 1985) and rh-LC I also inhibits paw oedema in response to carrageenan or PLA_2, but not to dextran or PAF (CIRINO et al. 1989). The effect of steroids against dextran-induced oedema may be mediated by a second steroid-induced protein, vasocortin (CALIAGNANO et al. 1985; CARNUCCIO et al. 1987; KOLTAI et al. 1987).

Potential anti-inflammatory actions of steroids which are mimicked by lipocortin include actions upon neutrophil chemotaxis (HIRATA et al. 1980) which may explain the reduced cell infiltration seen in various models. Lipocortin may be released from T cells, and be involved in the induction of suppressor cell formation (HIRATA and IWATA 1983), implicating its involvement in immunosuppressive actions of steroids. Lipocortin has also been shown to protect against sudden death in rats after coronary ligation (KOLTAI et al. 1983) and to prevent cerebral ischaemia in the same species (KOLTAI et al. 1984). Lipocortin or a closely related protein has been shown to cause differentiation of a histiocytic lymphoma cell line (HATTORI et al. 1983), and also to mediate glucocorticoid teratogenicity in an in vitro model (GUPTA et al. 1984). Obviously a great deal of work remains to discover which actions of steroids are mediated by lipocortin.

IV. Anti-lipocortin Antibodies and Disease

The case for an anti-inflammatory function for lipocortin is supported indirectly by findings that patients with the inflammatory diseases rheumatoid arthritis (RA) and systemic lupus erythematosus (SLE) have autoantibodies which cross-react with lipocortin (HIRATA et al. 1981; PODGORSKI et al. 1987).

In RA, PODGORSKI et al. (1987) reported that IgM anti-lipocortin antibodies were absent in non-steroid-treated subjects, but were markedly increased above control levels in patients requiring high maintenance steroids, and were also somewhat elevated in patients receiving low dose steroids. IgG antibodies to lipocortin showed a similar although less marked profile. In SLE, both IgM and IgG anti-lipocortin levels were increased in non-steroid-treated and in high maintenance steroid-treated patients, but were normal in those patients controlled with low dose steroids. It remains to be seen whether the antibodies are a feature of the disease or its treatment, but they may prove important in suggesting therapy. It is apparent however that such autoantibodies are not involved in all inflammatory diseases or a non-specific result of steroid treatment, since patients with polymyalgia rheumatica, and most disease-free controls do not have autoantibodies.

Circulating autoantibodies to lipocortin do not seem to explain the elevated epidermal PLA_2 activitiy seen in RA, since natural lipocortin or rh-LC I does not inhibit PLA_2 activity in skin from RA patients (DAWES et al. 1987), although it restores elevated PLA_2 activity in epidermis of psoriatic patients to control levels (FORSTER et al. 1983; ILCHYSHYN et al. 1984–1986).

It is not clear at the time of writing what the ultimate relevance of lipocortin to dermatology will be or what novel drugs this research will give rise to. However, the dermatologist relies heavily upon the glucocorticoids and any research work which sheds light upon their mode of action can be expected to have a long-term beneficial effect upon our understanding of the clinical effects of these powerful and useful drugs.

References

Aarsman AJ, Mynbeek G, Van Den Bosch H, Rothhut B, Prieur B, Comera C, Jordon L, Russo-Marie F (1987) Lipocortin inhibtion of extracellular and intracellular phospholipase A_2 is substrate concentration-dependent. FEBS Lett 219:76–180

Blackwell GJ (1983) Specificity and inhibition of glucocorticoid-induced macrocortin secretion from rat peritoneal macrophages. Br J Pharmacol 79:587–594

Blackwell GJ, Carnuccio R, di Rosa M, Flower RJ, Parente L, Persico P (1980) Macrocortin: a polypeptide causing the antiphospholipase effect of glucocorticoids. Nature 287:147–149

Blackwell GJ, Carnuccio R, di Rosa M, Flower RJ, Langham CSJ, Parente L, Persico P et al. (1982) Glucocorticoids induce the formation and release of anti-inflammatory and anti-phospholipase proteins into the peritoneal cavity of the rat. Br J Pharmacol 76:185–194

Caliagnano A, Carnuccio R, di Rosa M, Ialenti A, Moncada S (1985) The anti-inflammatory effect of glucocorticoid-induced phospholipase inhibitory proteins. Agents Actions 16:60–62
Carnuccio R, di Rosa M, Flower RJ, Pinto A (1980) Hydrocortisone-induced inhibitor of prostaglandin biosynthesis in rat leukocytes. Br J Pharmacol 68:14–16
Carnuccio R, di Rosa M, Flower RJ, Pinto A (1981) The inhibition by hydrocortisone of prostaglandin biosynthesis in rat peritoneal leukocytes is correlated with intracellular macrocortin levels. Br J Pharmacol 74:322–324
Carnuccio R, di Rosa M, Guerrasio B, Iuvone T, Sautebin L (1987) Vasocortin, a novel glucocorticoid-induced anti-inflammatory protein. Br J Pharmacol 90:443–445
Cirino G, Flower RJ (1987a) Human recombinant lipocortin I inhibits prostacyclin production by human umbilical artery. Prostaglandins 34:59–62
Cirino G, Flower RJ (1987b) The inhibitory effect of lipocortin on eicosanoid synthesis is dependent on divalent cation. Br J Pharmacol 92:521P
Cirino G, Flower RJ, Browning J, Sinclair LK, Pepinsky RB (1987) Recombinant human lipocortin I inhibits thromboxane release from guinea-pig isolated perfused lung. Nature 328:270–272
Cirino G, Peers SH, Flower RJ, Browning J, Pepinsky RB (1989) Human recombinant lipocortin I has acute local anti-inflammatory properties in the rat paw oedema test. Proc Natl Acad Sci USA (in press)
Cloix JF, Colard O, Rothhut B, Russo-Marie F (1983) Characterisation and partial purification of "renocortins", two polypeptides formed in renal cells causing the anti-phospholipase-like action of glucocorticoids. Br J Pharmacol 79:313–321
Danon A, Assouline G (1978) Inhibition of prostaglandin biosynthesis by corticosteroids requires RNA and protein synthesis. Nature 273:552–554
Davidson FF, Dennis EA, Powell M, Glenney JR (1987) Inhibition of phospholipase A_2 by "lipocortins" and calpactins. An effect of binding to substrate phospholipids. J Biol Chem 262:1698–1705
Dawes PT, Ilchyshyn A, Goulding NJ, Ilderton E, Shadforth MF, Yardley H, Hall ND (1987) Elevated epidermal PLA_2 activity in RA: evidence against a circulating lipocortin antibody. Br J Rheumatol [Suppl 1]26:62
Di Rosa M, Persico P (1979) Mechanism of inhibtion of prostaglandin biosynthesis by hydrocortisone in rat leukocytes. Br J Pharmacol 66:161–163
Di Rosa M, Flower RJ, Hirata F, Parente L, Russo-Marie F (1984) Nomenclature announcement. Anti-phospholipase proteins. Prostaglandins 28:441–442
Flower RJ (1984) Macrocortin and the antiphospholipase proteins. In: Weissman G (ed) Advances in inflammation research, vol 8. Raven, New York, p 1
Flower RJ (1985) Background and discovery of lipocortins. Agents Actions 17:255–262
Flower RJ, Blackwell GJ (1979) Anti-inflammatory steroids induce biosynthesis of a phospholipase A_2 inhibitor which prevents prostaglandin generation. Nature 278:456–459
Flower RJ, Wood JN, Parente L (1984) Macrocortin and the mechanism of action of the glucocorticoids. In: Otterness I, Capetola R, Wong S (eds) Advances inflammation research, vol 7. Raven, New York, p 61
Flower RJ, Moon-Parvin D, Peers SH, Pepinsky RB, Taylor RD (1986) A novel assay for assessing the binding of phospholipase A_2 to substrate. Br J Pharmacol 89:743P
Forster S, Ilderton E, Summerley R, Yardley HJ (1983) The level of phospholipase A_2 activity is raised in the univolved epidermis of psoriasis. Br J Dermatol 108:103–105
Forster S, Ilderton E, Norris JFB, Summerly R, Yardley HJ (1984) Characterization and activity of phospholipase A_2 in normal human epidermis and lesion-free epidermis of patients with psoriasis or eczema. Br J Dermatol 112:135–147
Ghiara P, Meli R, Parente L, Perscio P (1984) Distinct inhibition of membrane-bound and lysosomal phospholipase A_2 by glucocorticoid-induced proteins. Biochem Pharmacol 33:1445–1450
Gryglewski RJ, Panczenko B, Korbut R, Grodzinska L, Ocetkiewicz A (1975) Corticosteroids inhibit prostaglandin release from perfused mesenteric blood vessels of rabbit and from perfused lungs of sensitized guinea-pigs. Prostaglandins 10:343–355
Gupta C, Goldman A (1982) H-2 histocompatibility region: influence on the murine glucocorticoid receptor and its response. Science 216:994–996

Gupta C, Katsumo M, Goldman AS, Piddington R, Herold R (1984) Glucocorticoid-induced phospholipase A_2 inhibitory proteins mediate glucocorticoid teratogenicity in vitro. Proc Natl Acad Sci USA 81:1140–1143

Hattori T, Hoffman T, Hirata F (1983) Differentitation of a histiocytic lymphoma cell line by lipomodulin, a phospholipase inhibitory protein. Biochem Biophys Res Commun 111:551–559

Herbaczynska-Cedro K, Staszewska-Barczak J (1974) Adrenocortical hormones and the release of prostaglandin-like substances (PLS) (Abstr). 2nd Congress of the Hungarian Pharmacological Society, Budapest

Hirata F (1981) The regulation of lipomodulin, a phospholipase inhibitory protein, in rabbit neutrophils by phosphorylation. J Biol Chem 256:7730–7733

Hirata F, Iwata M (1983) Role of lipomodulin, a phospholipase inhibitory protein, in immunoregulation by thymocytes. J Immunol 130:1930–1936

Hirata F, Schiffmann E, Venkatasubramanian K, Salomon D, Axelrod J (1980) A phospholipase A_2 inhibitory protein in rabbit neutrophils induced by glucocorticoids. Proc Natl Acad Sci USA 77:2533–2536

Hirata F, Delcarmine R, Nelsom CA, Axelrod J, Schiffmann E, Warabi A, Deblas AL et al. (1981) Presence of autoantibody for phospholipase inhibitory protein(s) by radioimmuno-assay for lipomodulin. Biochem Biophys Res Commun 78:3190–3194

Hirata F, Notsu Y, Iwata M, Parente L, di Rosa M, Flower RJ (1982) Identification of several species of phospholipase inhibitory protein(s) by radioimmuno-assay for lipomodulin. Biochem Biophys Res Commun 109:223–230

Hirata F, Matsuda K, Notsu Y, Hattori T, Delcarmine R (1984) Phosphorylation at a tyrosine residue of lipomodulin in mitogen-stimulated murine thymocytes. Proc Natl Acad Sci USA 81:4717–4721

Hirata F, Notsu Y, Yamada R, Ishihara Y, Wano Y, Kunos I, Kunos G (1985) Isolation and characterisation of lipocortin (lipomodulin) Agents Actions 17:263–266

Hong SC, Levine L (1976) Inhibition of arachidonic acid release from cells as the biochemical action of anti-inflammatory steroids. Proc Natl Acad Sci USA 73:1720–1734

Huang K-S, Wallner BP, Mattaliano RJ, Tizard R, Burne C, Frey A, Hession C et al. (1986) Two human 35 kd inhibitors of phospholipase A_2 are related to substrates to $pp60^{v\text{-}src}$ and of the epidermal growth factor receptor/kinase. Cell 46:191–199

Huang K-S, McGray P, Mattaliano RJ, Burne C, Chow EP, Sinclair LK, Pepinsky RB (1987) Purification and characterisation of proteolytic fragments of lipocortin I that inhibit phospholipase A_2. J Biol Chem 262:7639–7645

Ilchyshyn A, Ilderton E, Kingsbury JA, Yardley HJ (1984) Evidence that the raised level of phospholipase A_2 in the uninvolved epidermis of psoriasis is caused by hyperphosphorylation of an inhibitor. Br J Dermatol 111:721–722

Ilchyshyn A, Ilderton E, Yardley HJ (1985) Further evidence that the raised level of phospholipase A_2 in lesion-free epidermis of psoriasis is caused by hyperpolarisation of an inhibitor. Br J Dermatol 112:135–147

Ilchyshyn A, Ilderton E, Pepinsky B, Yardley HJ (1986) Inhibition of phospholipase A_2 activity in extracts of psoriatic epidermis by lipocortin. 4th International Symposium on Psoriasis, Stanford, USA

Koltai M, Lepran I, Nemecz GY, Szekeres L (1983) The possible mechanism of protection induced by dexamethasone against sudden death due to coronary ligation in concious rats. Br J Pharmacol 79:327–329

Koltai M, Tosaki A, Adam G, Joo F, Nemecz G, Stezekeres L (1984) Prevention by macrocortin of global cerebral ischaemia in Sprague-Dawley rats. Eur J Pharmacol 105:347–350

Koltai M, Kovacs Z, Nemecz G, Mecs I, Szekeres L (1987) Glucocorticoid-induced low molecular mass anti-inflammatory factors which do not inhibit phospholipase A_2. Eur J Pharmacol 134:109–112

Kretsinger RH, Creutz CE (1986) Consensus in exocytosis. Nature 320:573

Nijkamp FP, Flower RJ, Moncada S, Vane JR (1976) Partial purification of rabbit aorta contracting substance-releasing factor and inhibition of its activity by anti-inflammatory steroids. Nature 263:479–482

Nixon JS, Wilkinson SE, Davis P, Bloxham DP (1985) The inhibitory profiles of hog pancreatic and human rheumatoid synovial cell phospholipase A_2. Agents Actions 17:299–301

Norris JFB, Ilderton E, Yardley HJ, Summerly R, Forster S (1984) Utilization of epidermal phospholipase A_2 inhibition to monitor topical steroid action. Br J Dermatol [Suppl 27]121:195–203

Okamoto M, Ono T, Tojo H, Yamano T (1985a) Immunochemical relatedness between secretory phospholipase A_2 and intracellular phospholipase A_2. Biochem Biophys Res Commun 128:788–794

Okamoto M, Tojo H, Ono T, Yamano T (1985b) Comparative study of intracellular and pancreatic phospholipase A_2. Adv Prostaglandin Thromboxane Leukotriene Res 15:135

Parente L, Flower RJ (1985) Hydrocortisone and "macrocortin" inhibit the zymosan-induced release of lyso-PAF from rat peritoneal leukocytes. Life Sci 36:1225–1231

Parente L, di Rosa M, Flower RJ, Ghiara P, Meli R, Persico P, Salmon JA, Wood JN (1984) Relationship between the anti-phospholipase and anti-inflammatory effects of glucocorticoid-induced proteins. Eur J Pharmacol 99:233–239

Peers SH, Taylor RD, Flower RJ (1987) A novel binding assay for phospholipase A_2. Biochem Pharmacol 36:4287–4291

Pepinsky RB, Sinclair LK (1986) Epidermal growth factor-dependent phosphorylation of lipocortin. Nature 321:81–84

Pepinsky RB, Sinclair LK, Browning JL, Mattaliano RJ, Tizard R, Burnes C, Frey A et al. (1986) Two human 35 kd inhibitors of phospholipase A_2 are related to substrates of $pp60^{v\text{-}src}$ and of the epidermal growth factor receptor kinase. Cell 46:191–199

Podgorski MR, Goulding NJ, Hall ND, Flower RJ, Maddison PJ, Pepinsky RB (1987) Autoantibodies to recombinant lipocortin in RA and SLE. Br J Rheumatol [Suppl 1]26:54–55

Rothhut B, Russo-Marie F, Wood J, di Rosa M, Flower RJ (1983) Further characterisation of the glucocorticoid-induced anti-phospholipase protein "renocortin". Biochem Biophys Res Commun 117:878–884

Rothhut B, Cumera C, Prieur B, Errasfa M, Minassion G, Russo-Marie F (1987) Purification and characterisation of a 32 kDa phospholipase A_2 inhibitory protein (lipocortin) from human peripheral blood mono-nuclear cells. FEBS Lett 219:169–175

Russo-Marie F, Duval D (1982) Dexamethasone-induced inhibition of prostaglandin production does not result from a direct action on phospholipase activities but is mediated through a steroid-inducible factor. Biochim Biophys Acta 712:177–185

Russo-Marie F, Paing M, Duval D (1979) Involvement of glucocorticoid receptors in steroid-induced inhibition of prostaglandin synthesis. J Biol Chem 254:8498–8504

Saris CJM, Tack BF, Kristensen T, Glenney JR, Hunter T (1986) The cDNA sequence for the protein-tyrosine kinase substrate p36 (calpactin 1 heavy chain) reveals a multi-domain protein with internal repeats. Cell 46:201–212

Schlaepfer DD, Haigler HT (1987) Characterisation of Ca^{2+}-dependent phospholipid binding and phosphorylation of lipocortin I. J Biol Chem 262:6931–6937

Touqui L, Rothhut B, Shaw AM, Fradin A, Vargaftig BB, Russo-Marie F (1986) Platelet activation – a role for a 40k anti-phospholipase A_2 protein indistinguishable from lipocortin. Nature 321:177–180

Tsurufuji S, Sugio K, Takemasa F (1979) The role of glucocorticoid receptor and gene expression in the anti-inflammatory action of dexamethasone. Nature 280:408–410

Van de Velde VJS, Bult H, Herman AG (1985) Dexamethasone and prostacyclin biosynthesis by serosal membranes of the rabbit peritoneal cavity. Agents Actions 17:308–309

Volwerk JJ, de Haas GH (1982) Pancreatic phospholipase A_2: a model for membrane-bound enzymes? In: Jost PC, Griffiths OH (eds) Lipid-protein interactions, vol 1. Wiley, New York, pp 69–149

Wallner BP, et al. (1986) Cloning and expression of human lipocortin, a phospholipase A_2 inhibitor with potential anti-inflammatory activity. Nature 320:77–81

Yedgar S, Reisfeld N, Dagan A (1986) Synthesis of cell impermeable inhibitor of phospholipase A_2. FEBS Lett 200:165–168

CHAPTER 21

Cutaneous Vasodilators

P. M. DOWD

A. Introduction

Vasodilator drugs are often employed in the management of symptomatic vasospastic disorders and diseases resulting in organic ischaemia of the skin. These drugs can be grouped into two large categories: (a) drugs acting on the sympathetic nervous system and (b) drugs with a direct vasodilator action. An understanding of the rationale for their use demands a knowledge of the innervation and pharmacology of the cutaneous vasculature.

B. Neurovascular Control in the Skin

Control of cutaneous blood flow was shown to be achieved by an α-adrenergic sympathetic vasoconstrictor mechanism (AHLQUIST 1948) with vasodilation occurring as a reflex action after cessation of α-adrenergic stimulation. Further understanding of the nature of adrenergic receptors (VON EULER 1972; ALTURA and HERSHEY 1967) led to the realisation that vasoconstriction in the skin was achieved through α_1-receptors, there being little if any evidence for the presence of α_2-adrenergic receptors in human skin. There is no direct evidence for a β-adrenergic vasodilator innervation of the cutaneous vasculature, but evidence for the presence of β-adrenergic receptors, at least in the digital cutaneous vasculature (COHEN and COFFMAN 1980), has led to the suggestion that there may be, albeit a weak, β-adrenergic innervation of the vasculature supplying the skin at this anatomical site.

C. Pharmacology of the Cutaneous Vasculature

It has long been recognised that the cutaneous vasculature possesses receptors for histamine, bradykin and serotonin. Stimulation of the first two results in vasodilation (LEWIS 1926; ERDÖS and WILDE 1970), and of the last in vasoconstriction. Hypoxia is a potent vasodilator stimulus, but it is difficult to differentiate between the direct effects of hypoxia itself and accumulation of vasoactive metabolites. However, perfusion studies (BUSSE and BASSENGEL 1984) showed that hypoxia per se was a potent vasodilator and that its effect was endothelium dependent and accompanied by increased 6 keto-PGF$_{1\alpha}$ production and inhibited by indomethacin, indicating that the vasodilatation was mediated by prostaglandin $I_2(PGI_2)$. Whether this mechanism operates in human skin has not yet been determined.

Acetylcholine has a vasodilator action mediated by the presence of muscarinic receptors which appear to be situated in the endothelial cells of the vasculature. It has recently been realised that when stimulated they effect release of endothelium-dependent relaxant factor (EDRF) which diffuses to the smooth muscle cells, causing it to relax (Furchgott and Zawadski 1980). The vasodilator activity of the prostaglandins E_1, E_2, I_2 and D_2, the lipoxygenase arachidonate derivatives 12- and 5-hydroxyeicosatetraenoic acids 5,12-dihydroxyeicosatetraenoic acid and the peptido leukotrienes has been demonstrated (Sondergaard and Greaves 1971; Higgs et al. 1979; Dowd et al. 1985; Cunningham et al. 1984; Camp et al. 1983). Platelet-activating factor (PAF) and platelet-derived growth factor (PDGF) are also potent vasodilators in human skin (Di Corletto et al. 1983). The neuropeptides substance P (Jorizzo et al. 1983), calcitonin gene-related peptide (CGRP) (Brain et al. 1986; Piotrowski and Foreman 1986) and the cytokine interleukin 1 (Dowd et al. 1988) have also been shown to produce vasodilatation in human skin.

Another class of agents, the adenine nucleotides, have also been shown to be vasodilators (Gordon 1982). ATP and to a lesser degree ADP produce an endothelium-dependent relaxation of smooth muscle. These nucleotides are in high concentration in the dense storage granules of platelets (Gordon 1987) and thus there is a potential for very high localised concentrations in the plasma following platelet degranulation.

The relative contributions of these varied pharmacological agents in the maintenance and control of cutaneous blood flow still awaits elucidation, but it is likely that, as in other vascular beds, substance P (Furchgott 1983; Moncada et al. 1986), CGRP (Brain et al. 1985), acetylcholine (Furchgott 1983) and PAF (D'Humieres et al. 1986) rely at least in part on the presence of an intact endothelium for their action and that their vasodilator activity may occur through secondary mediators such as endothelium-derived relaxant factor (EDRF) (Furchgott and Zwadski 1980) and PGI_2.

I. Drugs Acting on the Sympathetic Nervous System

Drugs acting as adrenergic neuronal blocking agents, drugs having both α_1- and α_2-receptor blocking activity, drugs having selective α_1-adrenergic receptor blocking activity and drugs acting centrally such as methyldopa have all been used to induce cutaneous vasodilatation with varying degrees of success.

1. Adrenergic Neurone-Blocking Agents

The ganglion blocking agents guanethidine, reserpine and α-methyldopa interfere with chemical mediation at post-ganglionic adrenergic nerve endings.

a) Guanethidine

The structural formula of guanethidine is

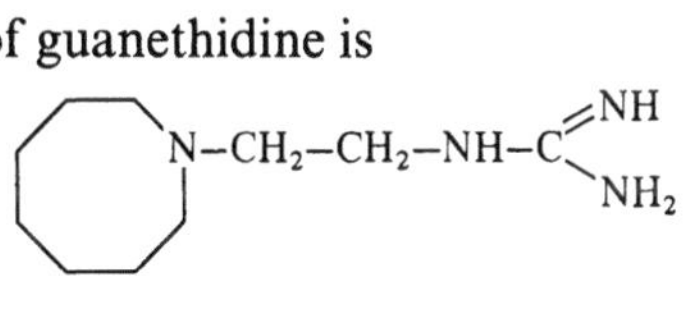

The guanidine grouping of this compound confers strong basic activity which is important in its ability to inhibit responses to stimulation of sympathetic nerves and indirectly acting sympathomimetic amines such as tyramine and amphetamine. The site of the inhibition is presynaptic and during chronic administration it is due to impaired release of neurotransmitter from peripheral adrenergic neurones (MITCHELL and OATES 1970).

Guanethidine is administered as the monosulphate. It is absorbed after both oral and parenteral routes of administration. Absorption after chronic oral administration varies from 3% to 30%, but appears to be relatively constant in a given patient. The percentage metabolised by hepatic microsomal enzymes to its two more polar and less active metabolites is higher after oral than after parenteral administration. Guanethidine is taken up and stored in adrenergic nerves irrespective of their anatomical site. The parent compound and its two polar metabolites are rapidly cleared from the circulation, but small amounts may remain in the body for as long as 14 days, probably the result of specific uptake in to adrenergic nerves and non-specific tissue uptake.

The accumulation of guanethidine in adrenergic nerves is essential for its action. Its effects are cumulative over extended periods, hence adverse effects may appear or progress for many weeks after a toxic dose and may not subside for several days after complete cessation of therapy. The most important toxic effect is hypotension, usually initially manifesting as postural hypotension which may be accentuated by hot weather, alcohol or exercise and accompanied by symptoms of cerebral and myocardial ischaemia. Fluid retention occurs and can lead to oedema. Guanethidine, by decreasing adrenergic myocardial effects, can decrease myocardial competence and together with fluid retention this may result in heart failure in patients with limited cardiac reserve. Increased gastrointestinal motility results in mild diarrhoea in a high percentage of patients. This is attributed to parasympathetic prominence after blockade of adrenergic fibres. This can often be controlled with anticholinergics or opioids. Guanethidine can cause severe hypotension in patients with phaeochromocytoma. The effect of guanethidine can be antagonised by tricyclic antidepressants (LE ROY et al. 1971) and chlorpromazine. Withdrawal of these agents from concurrent use with guanethidine can result in profound hypotension if the dose of guanethidine is not adjusted in advance. Guanethidine can sensitise to some directly acting sympathomimetics found in common cold remedies and can then result in hypertensive crises. It also augments responses to catecholamines in a large number of tissues and organs. Because it causes diminution in the release of transmitters, guanethidine reduces responses mediated by α and β-receptors about equally. Concentrations of catecholamines in the central nervous system are not altered by guanethidine as this polar drug does not penetrate the blood-brain barrier in sufficient concentration to exert significant effects.

α) Therapeutic Use. Guanethidine has been used in the management of Raynaud's phenomenon with some degree of success (PARKS et al. 1961), but postural hypotension has limited its usefulness.

b) Reserpine

Reserpine is a rauwolfia alkaloid whose structure is

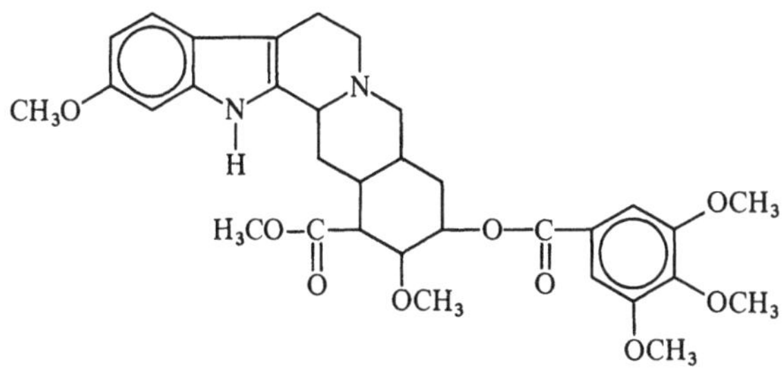

It is extracted from *Rauwolfia serpentina* (Benth), a climbing shrub of the Apocynaceae family indigenous to India. It is effective orally and by intra-arterial injection when the drug also has a transient sympathomimetic effect. Reserpine is an adrenergic neurone blocking agent with an inherent short vasodilator action unrelated to noradrenaline depletion.

Reserpine depletes stores of catecholamines and serotonin in many organs including the brain and adrenal medulla. Reduction of catecholamine stores occurs chiefly by intraneuronal deamination, is detectable after 1 h and is maximal at 24 h. Repeated doses even at intervals of about 1 week have a cumulative action as restoration of catecholamine concentration occurs only very slowly. Reserpine antagonises the uptake of noradrenaline by isolated chromaffin granules, apparently by inhibiting the ATP-Mg^{2+}-dependent uptake mechanism of the granule membrane, and is thought to act on noradrenaline synthesis in vivo at least in part by blocking dopamine uptake into storage granules containing the enzyme dopamine β-hydroxylase. Both inhibition and stimulation of tyrosine kinase are thought to occur and, with chronic administration, induction and increased levels of this enzyme and an increased turnover rate of noradrenaline occur. Chronic administration can result in supersensitivity to catecholamines.

α) Pharmacology. Reserpine causes a slowly developing fall in blood pressure, bradycardia and decreased peripheral resistance most marked in the skin and associated with increased cutaneous blood flow. Pressor responses are inhibited. Decreased cardiac output and postural hypotension occur and are produced in a manner analogous to that of guanethidine.

Reserpine acts centrally to produce sedation and indifference to environmental stimuli, and in this respect is similar to the phenothiazines. Presumably depletion of brain stores of catecholamines and serotonin occurs. Extrapyramidal effects can occur after prolonged administration of high doses. Gastrointestinal tone and motility are increased with abdominal cramps and diarrhoea and increased gastric acid secretion. Weight gain is common.

β) Therapeutic Use. Oral reserpine has been shown to increase capillary blood flow in the fingers in Raynaud's phenomenon (Coffman and Cohen 1971) and to reduce the vasoconstriction responses of the hand to the application of ice to the forehead (Kontos and Wasserman 1971). Intra-arterial injection also increases digital skin blood flow (Willerson et al. 1970). Oral administration produces amelioration of symptoms in up to 50% of patients with Raynaud's phenomenon and solitary intra-arterial injections have been reported to improve cutaneous blood flow and produce symptomatic relief for periods of days to weeks in Raynaud's phenomenon (NOBIN et al. 1978; NILSEN and JAYSON 1980). In view of its effects on the central nervous system, long-term therapy should be avoided in patients with depression.

γ) Adverse Effects. Vascular adverse effects include flushing and nasal congestion, the latter occasionally causing respiratory distress in neonates of mothers receiving reserpine. It is uncertain whether chronic administration of reserpine is associated with an increased incidence of carcinoma of the breast, as was reported in the Boston Collaborative Drug Surveillance Program, Editorial 1974.

2. α-Adrenergic Blocking Agents

These drugs bind selectively to the α-class of adrenergic receptors and thereby block sympathomimetic amine-initiated actions. Phenoxybenzamine, tolazoline, prazosin and indoramin are the drugs of this class which have been used in the management of cutaneous ischaemia.

a) Phenoxybenzamine

This haloalkylamine is closely related chemically to the nitrogen mustards. Its structural formula is

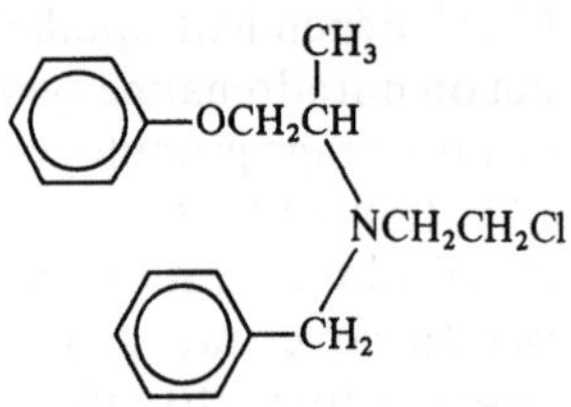

The tertiary amine cyclises to form a reactive ethylenimonium intermediate. The active metabolite is probably a highly reactive carbonium ion formed when the three-membered ring breaks. The slow onset of its action is probably related to the time required for the formation of these intermediates. It is considered that the arylalloylamine moiety is responsible for the specificity of action.

Phenoxybenzamine is effective on oral and parenteral administration, but is too irritant for other than intravenous injection. Only 20%–30% of orally administered phenoxybenzamine (phenoxybenzamine hydrochloride) appears to be absorbed in an active form. The drug is very lipid soluble at body pH and may accumulate in fat. Its metabolic fate is poorly understood. Radioisotope studies indicate that over 50% of intravenously administered phenoxybenzamine is excreted in 12 h and more than 80% in 24 h. Small amounts may remain in various tissues for several days. Many tissues contain α-receptors which are blocked by the direct action of this drug.

α) Pharmacology. Phenoxybenzamine increases the rate of turnover of noradrenaline in the periphery: this is associated with increased tyrosine hydroxylase activity probably owing to increased sympathetic activity occurring as a reflux response to α-adrenergic blockade. The amount of neurotransmitter released by each nerve impulse is increased owing to blockade of presynaptic α_2-receptors which mediate a negative feedback mechanism that inhibits the release of noradrenaline. In addition, the uptake of catecholamines into both adrenergic nerve terminals and extraneuronal tissues is inhibited. Phenoxybenzamine blocks the inhibitory action of adrenaline on insulin secretion and in the rat can block the effect of catecholamines on hepatic glycogenolysis. Central nervous system stimulation causing nausea, vomiting, hyperventilation, motor excitability and convulsions can occur. Mild to moderate sedation commonly results from slow intravenous infusion, and fatigue and lethargy may occur after oral treatment. The effects of endogenous and exogenous sympathomimetic amines mediated by α-adrenergic receptors are blocked, hence miosis is common in humans. Contractions of the nictitating membrane, arrector pili, retractor penis and the uterus of several species are inhibited. Motility of the non-pregnant human uterus is reduced, but transit time through the gastrointestinal tract is not appreciably altered.

Phenoxybenzamine effectively inhibits responses that are mediated by α-adrenergic agonists by binding covalently to α-receptors producing an irreversible blockade. It antagonises the α-receptor-mediated excitatory responses of smooth muscle and exocrine glands. Inhibition of the vasoconstrictor effect of noradrenaline occurs and inhibition of the vasoconstrictor effect of adrenaline may allow its vasodilator effects to predominate. However, even high doses of phenoxybenzamine may fail to block the pressor effect of noradrenaline because of its cardiac stimulant actions. The vasodilator effect of phenoxybenzamine is dependent on the degree of adrenergic vasomotor tone and the physiological state of different vascular beds.

Blockade develops slowly over the course of about 1 h; a single dose produces blockade with a half-life of 24 h in animals, and in humans a demonstrable effect persists for 3–4 days and effects of daily administration are cumulative for approximately 1 week. The effective blocking dose in normotensive individuals causes little change in systolic blood pressure although diastolic pressure may fall. As compensatory sympathetic vasoconstriction is blocked postural hypotension does, however, occur. Impairment of compensatory vasoconstriction also results in sensitisation to the hypotensive effects of agents producing vasodilation including hypercapnia, anaesthetic agents, morphine and meperamide. Rapid injection causes a precipitous fall in blood pressure probably involving factors other than α-adrenergic blockade.

A progressive increase in cardiac output and decrease in total peripheral resistance occurs in normal recumbent subjects, but changes in blood flow and resistance in different vascular beds vary widely with the conditions under which the drug is administered; the greater the degree of adrenergic vasomotor tone, the more pronounced the relevant effect. Cerebral flow is little affected and coronary flow increases with reflex cardiac stimulation. Resting blood flow is increased in muscle and cutaneous blood flow is enhanced in a cool environment, but little effect occurs in a warm environment, for, under these conditions α-receptor blockade produces little change. A shift of fluid from the interstitial to the vascular compartment occurs; this is the result of differential effects on precapillary and postcapillary resistance vessels. Pressor responses to sympathetic amines are blocked. Reflex tachycardia is often exaggerated because the pressor response is prevented and reflex vagal stimulation is minimised. Reflex tachycardia is also accentuated by enhanced release of noradrenaline and decreased inactivation of the amine owing to inhibition of neuronal and extraneuronal uptake mechanisms. Cardiac arrhythmias in the genesis of which catecholamines are implicated, are inhibited.

β) Adverse Effects. The toxic and adverse effects of phenoxybenzamine are largely those of α-adrenoceptor blockade including postural hypotension, reflex tachycardia and a sharp fall in blood pressure in the presence of hypovolaemia. With continued treatment, postural hypotension and palpitations may disappear, but reappear during exercise, after eating large meals or after consumption of alcohol, all of which promote vasodilatation.

γ) Therapeutic Use. Favourable clinical responses have been reported in Raynaud's phenomenon and acrocyanosis. Relief of vasospasm and reduction of sensitivity to cold have been reported in Raynaud's phenomenon with oral doses which produce incomplete adrenergic blockade (GIFFORD 1963). In cutaneous ischaemia occurring in association with limited flow in large vessels results of treatment with this agent have been disappointing and have not eliminated the need for surgery.

b) Tolazoline

This is a 2-substituted imidazoline, the structure of which is

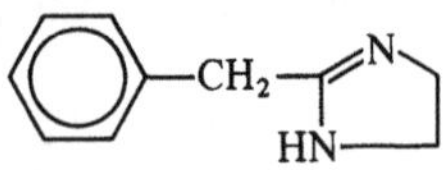

It produces α-adrenergic blockade by competitive inhibition: its effect is, therefore, unlike that of phenoxybenzamine, relatively transient. In addition it inhibits responses to serotonin.

Tolazoline (tolazoline hydrochloride) is well absorbed after both parenteral and oral administration. It is less effective when given orally because rapid renal excretion prevents accumulation during relatively slow absorption from the gut. It is excreted largely unchanged by an organic base transport in the renal tubule.

α) Pharmacology. Given intravenously, tolazoline produces vasodilatation and cardiac stimulation, its net effect is usually pressor. Vasodilatation leading to decreased peripheral resistance is predominantly due to a direct action on vascular smooth muscle in the dose now currently used. Pulmonary arterial pressure is usually reduced. As with phenoxybenzamine, cardiac stimulation can be associated with cardiac arrhythmias probably due to enhanced neural release of noradrenaline owing to presynaptic α_2-receptor blockade. The α-adrenergic receptor blocking action of tolazoline is similar in distribution to that of phenoxybenzamine, but the eye muscles are more resistant and mydriasis rather than miosis can result. Salivary, lacrimal, pancreatic and respiratory tract and eccrine secretion are stimulated possibly owing to a direct action on muscarinic cholinergic receptors. Gastric secretion of acid and pepsin are stimulated, hyperperistalsis and diarrhoea also occur: the latter effects are blocked by atropine.

β) Adverse Effects. The toxic and adverse effects are attributable to cardiac and gastrointestinal stimulation. Severe tachycardia, cardiac arrhythmias and anginal pain occur most frequently after parenteral administration and the drug has been implicated as precipitating myocardial infarction. Gastrointestinal stimulation may result in abdominal pain, nausea, vomiting, diarrhoea, exacerbation of peptic ulceration, piloerection, chilliness and apprehension.

γ) Therapeutic Use. Tolazoline has been used in situations similar to those in which phenoxybenzamine has been used with roughly similar effect. Favourable effects in Raynaud's phenomenon have been reported (Coffman and Cohen 1971).

c) Prazosin

Prazosin is a quinazoline derivative with the following structural formula

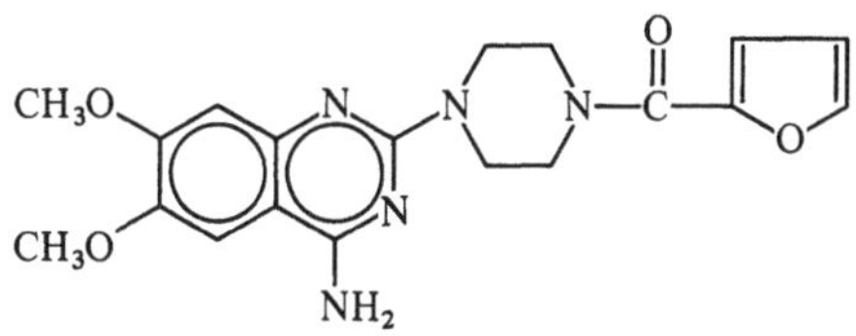

Prazosin (prazosin hydrochloride) is absorbed after oral administration. It undergoes first-pass metabolism by the liver. Peak plasma concentration occurs after 1–3 h and the plasma half-life is 2–3 h. About 90% of the drug is bound to α_1-acid glycoprotein and only a small percentage to albumin. Dealkylated metabolites are predominantly eliminated in the bile, a small amount being found unaltered in the urine.

α) Pharmacology. Prazosin has a very high affinity for α_1- and very low affinity for α_2-receptors. Its action is almost entirely due to vascular postsynaptic α_1-adrenergic receptor blockade. Its selectivity allows it to block noradrenaline-induced contraction of vascular smooth muscle without inactivation of its presynaptic α_2 activity, thus resulting in inhibition of further release of transmitter. Hence, it causes little or no tachycardia. Clinically effective doses reduce blood pressure and peripheral resistance whilst allowing normal cardiovascular responses to cold, exercise and carotid sinus pressure. Its hypotensive effect is greater when the patient is in an upright position and a mild reflex tachycardia can result. This effect often subsides with repeated dosage, suggesting the development of partial tolerance to the hypotensive effect. However, long-term administration does not result in marked tachyphylaxis.

β) Adverse Effects. Acute toxicity characterised by hypotension with sudden loss of consciousness can occur 30–90 min after an initial dose, especially if there is sodium depletion or concurrent administration of other anti-hypertensive medication. With limitation of the initial dose and administration just prior to bedtime, this effect may be avoided. Faintness and dizziness are frequently encountered. Less common adverse effects are headache, dry mouth, diarrhoea, nausea, constipation, weight gain, oedema, urinary urgency, nasal stuffiness, polyarthritis, eosinophilia, hypersensitivity, tachycardia and priapism. Occasionally plasma concentrations of glucose, free fatty acids and uric acid may also be elevated.

γ) Therapeutic Use. The results of its efficacy in Raynaud's phenomenon have been reported (HARPER and LE ROY 1982).

d) Indoramin

Like prazosin, indoramin is a selective α_2-adrenoceptor blocking agent. It is absorbed in the gut after oral administration. Its pharmacology and adverse effects are very similar to those of prazosin.

α) Therapeutic Use. Its efficacy in Raynaud's phenomenon has been reported (ROBSON et al. 1978).

II. Drugs with Direct Vasodilator Activity

Papaverine, isoxsuprine, nylidrin, cyclandelate and nicotinic acid (niacin) derivatives can all dilate vascular smooth muscle. Of these papaverine and nicotinic acid cause vasodilatation of the blush areas rather than of the extremities and the remainder do not significantly affect cutaneous blood flow. As studies have not shown a consistent increase in cutaneous blood flow with systemic administration of these drugs (OATTMAN and MANNICLE 1972; HANSTEEN and LORENTSEN 1974) there is no rational basis for their systemic use in vasospastic conditions and peripheral vascular insufficiency.

1. Nicotinic Acid Esters

Nicotinic acid esters (hexyl nicotinate, methyl nicotinate and tetrahydrofurfuryl nicotinate do, however, produce cutaneous vasodilatation on epicutaneous application (DOWD et al. 1987; GUY et al. 1985; BUNKER and DOWD 1987) and in appropriate concentration do cause increased blood flow of digital skin (BUNKER et al. 1988 a). In vivo the esters are believed to release nicotinic acid and this drug will therefore be briefly discussed. Nicotinic acid has the following chemical structure

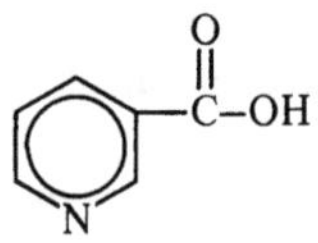

It is a vitamin extracted from the liver.

a) Pharmacology

Nicotinic acid functions as a vitamin after conversion to nicotinamide adenine dinucleotide (NAD) or nicotinamide adenine dinucleotide phosphate (NADP). NAD and NADP act as coenzymes for a wide variety of proteins that catalyse oxidation–reduction reactions. Nicotinic acid is metabolised by the formation of *N*-methyl nicotinamide and subsequently *N*-methyl-2-pyridone-5-carboxamide and *N*-methyl-4-pyridone-3-carboxamide. It can also be metabolised to its glycine peptide nicotinamic acid. The metabolic products are excreted in the urine.

b) Adverse Effects

No adverse local or systemic effects have so far been reported after repeated topical application of hexyl nicotinate for up to 6 weeks (BUNKER et al. 1988 b).

c) Therapeutic Use

After initially promising results in pilot studies epicutaneous application of a 2% lotion of hexyl nicotinate is currently under controlled investigation in the treatment of Raynaud's phenomenon and perniosis and appears to be an effective therapy in these disorders (BUNKER et al. 1988 b).

2. Glyceryl Trinitrate (Nitroglycerin)

Glyceryl trinitrate ointment is an effective epicutaneous vasodilator. Its structural formula is

$$\begin{array}{l} H_2C{-}O{-}NO_2 \\ \quad | \\ HC{-}O{-}NO_2 \\ \quad | \\ H_2C{-}O{-}NO_2 \end{array}$$

A small proportion of the glyceryl trinitrate is systemically absorbed. Other routes of administration do not result in a useful therapeutic effect on the peripheral vasculature. This drug will therefore be discussed with reference only to this method of administration.

Glyceryl trinitrate is a nitrate ester of nitric acid. It is a moderately volatile oily liquid. Glyceryl trinitrate is metabolised by reductive hydrolysis in the liver, a process catalysed by glutathione–organic nitrate reductase, with the formation of denitrated metabolites which are excreted.

a) Pharmacology and Mechanism of Action

Glyceryl trinitrate activates guanylate cyclase and increases cyclic GMP concentrations, it is believed by the formation of the reactive free radical nitric oxide. Cyclic GMP activates its dependent protein kinase with resultant phosphorylation of smooth muscle proteins and subsequent dephosphorylation of the myosin light chain and consequent relaxation of smooth muscle.

b) Adverse Effects

Acute toxicity which has not yet been reported, but which might occur after solely topical administration is characterised by decreased systemic and diastolic blood pressure and cardiac output with resulting pallor, weakness, dizziness and activation of compensatory sympathetic reflexes. Abrupt withdrawal from chronic exposure in industrial workers has been considered to cause ischaemic cardiac injury (LANGE et al. 1972).

c) Therapeutic Use

Glyceryl trinitrate (2%) ointment has been reported to improve digital temperature and blood flow and provide symptomatic improvement in patients with Raynaud's phenomenon (BEAUCHER 1979).

3. Calcium Channel Blocking Agents

Nifedipine, diltiazem and verapamil are the three calcium channel blocking agents currently in routine use. Nifedipine is a dihydropyridine, diltiazem a benzothiazepine and verapamil a benzene acetonitrile. Their structural formulae are illustrated here

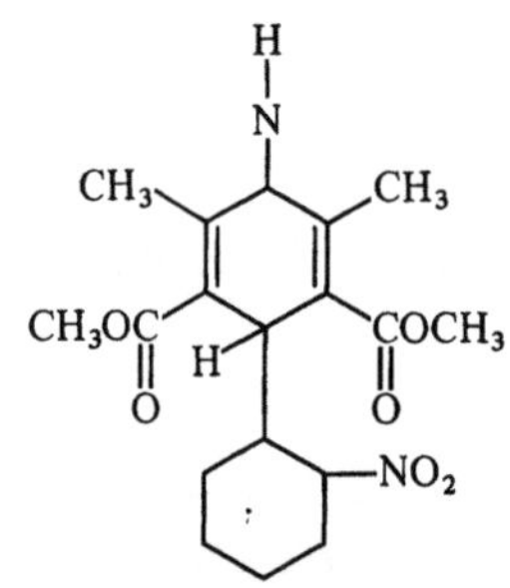

Nifedipine

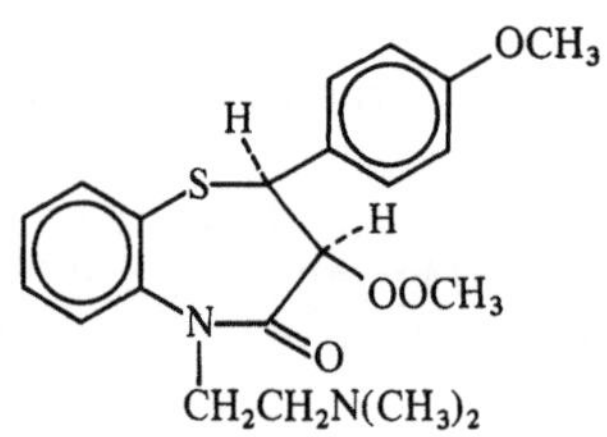

Diltiazem

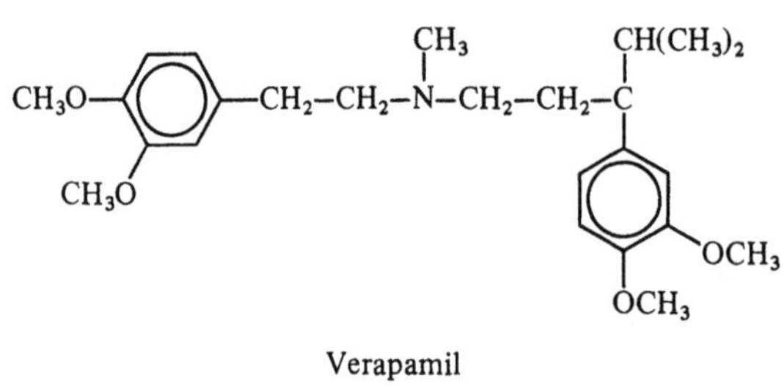

Verapamil

The L isomers of all these agents are 5–10 times more potent than the D isomers. Nifedipine is soluble only in organic solvents, such as propylene glycol and ethanol. Diltiazem and verapamil are water soluble. The calcium channel blocking agents are so called because they block the entrance of calcium into cells by inhibiting its movement through a membrane pore or "slow channel", the calcium flux through which contributes to the plateau phase of the cardiac action potential (KOHLHARDT et al. 1972).

Nifedipine is rapidly and completely absorbed after sublingual administration. Approximately 50% is absorbed after oral administration and peak concentrations are observed after 1–3 h. It is extensively (98%) plasma protein bound and is metabolised to inactive products with a half-life of 3–4 h. It is completely metabolised and excreted by the kidney.

Approximately 50% of diltiazem is absorbed after oral administration. Its noticeable effects reach a peak at 30 min and its half-life is 3–4 h. It is metabolised by deacetylation with subsequent O-a *N*-demethylation and is excreted by the kidney.

Verapamil is absorbed in high percentages after oral administration, but is extensively metabolised on first passage through the liver. Prolonged treatment appears to lead to decreased first-pass metabolism. Its effect becomes noticeable 1–2 h after administration and is maximal at 5 h. Its plasma half-life is 5 h, but this may be increased in children and the elderly on prolonged administration. This necessitates reduction of oral dosage by 80%. Verapamil can also be administered intravenously.

In the sinoatrial and ventricular nodes depolarisation is almost entirely dependent on the movement of calcium through the slow channel which is blocked by these drugs. Nifedipine does not affect the rate of recovery of the slow channel and does not affect conduction through the node. Verapamil blocks the rate of calcium flux through the slow channel, an effect which is enhanced as the frequency of the stimulus is linear, and also decreases the rate of recovery of the channel. Verapamil and diltiazem slow atrioventricular conduction and depress the rate of the sinus node pacemaker. All three drugs decrease coronary vascular resistance and increase coronary blood flow. At high doses, nifedipine increases red blood cell deformability.

a) Pharmacology

In vascular smooth muscle these agents block the voltage-sensitive contraction intensity channels which open in response to depolarisation of the membrane with efflux of sodium from the cell and influx of calcium into the cell down its electrochemical gradient (electromechanical coupling). They also block receptor-triggered phospholipase C-mediated inositol triphosphate stimulated release of intracellular calcium (pharmacomechanical coupling) from the sarcoplasmic reticulum (SOMLYO and SOMLYO 1968) which in turn results in an influx of extracellular calcium.

An increase in intracellular calcium results in a calcium-activated calmodulin-dependent activation of myosin light chain kinase with consequent myosin light chain phosphorylation and actin myosin interaction with the initiation of smooth muscle contraction. Mobilisation of calcium also results in activation of protein kinase C (PKC) and subsequent PKC-dependent protein phosphorylation, resulting in sustained smooth muscle contraction. The voltage-dependent calcium channels in vascular smooth muscle are blocked by lower concentrations of the drugs than are required to inhibit the intracellular mobilisation of calcium.

In vivo, nifedipine is the most and diltiazem the least potent vasodilator. Nifedipine given intravenously increases forearm blood flow with little effect on venous pooling, indicating a selective effect on arterial smooth muscle and little effect on the venous bed (ROBINSON et al. 1980). After oral administration, peripheral blood flow increases owing to arterial dilatation and does not affect venous flow. After both intravenous and oral administration there is increased cardiac output owing to decreased arteriolar resistance, with a sympathetic nervous system mediated increased heart rate and positive inotropic effect. In contrast to the effects in the peripheral vasculature, intravenous verapamil, by virtue of its direct negative chronotropic, dromotropic and inotropic actions, has more marked effect on the heart than does nifedipine and in patients with congestive cardiac failure can lead to a marked decrease in contractility and left ventricular function (CHEW et al. 1981). After oral administration of verapamil peripheral vascular resistance and blood pressure are reduced with no change in heart rate (THEROUX et al. 1980).

Intravenous diltiazem results in decreased peripheral vascular resistance and blood pressure with a reflex tachycardia and increased cardiac output. A direct negative inotropic effect subsequently reduces heart rate. After oral administration there is a sustained fall in both blood pressure and heart rate (KAHAN et al. 1982).

b) Adverse Effects

The major toxic effects of calcium channel blockers are excessive vasodilatation, negative inotropic effects with depression of the sinus node and atrioventricular node conduction disturbances. Nifedipine produces the most excessive vasodilatation, often with clinical flushing, peripheral oedema and dizziness or lightheadedness. Headaches, hypotension, nausea, vomiting and occasionally sedation occur in up to 20% of patients. They are often lessened with repeated administration and/or reduction of dosage. Aggravation of myocardial ischaemia owing to excessive hypotension and a coronary steal phenomenon may occur.

Verapamil is well tolerated orally – the adverse effects being similar to, but much less frequently encountered than those of nifedipine. After intravenous administration, bradycardia, transient asystole, hypotension and exacerbation of heart failure have been reported. Verapamil is contra-indicated in patients with ventricular, sinoatrial or atrioventricular disturbances and systolic blood pressure less than 90 mmHg. Verapamil can cause increased plasma digoxin concentration. Diltiazem appears to produce few adverse effects, but it does have effects similar to those of verapamil at high dose (above 360 mg daily).

c) Therapeutic Use

The more frequent peripheral vasodilatation effect of nifedipine makes it the calcium channel blocking agent with most significant effect on vasospastic conditions of the skin. In the author's opinion it is the first-line treatment for symptomatic Raynaud's phenomenon where it is effective orally (KAHAN et al. 1982; SMITH and McKendry 1982). It is also effective after sublingual administration. Oral nifedipine has also been demonstrated to be effective in the hitherto untreatable condition, severe idiopathic perniosis (Dowd et al. 1986). The modal effective dose is 60 mg in both these conditions. Occasionally and usually in the more mildly symptomatic patients adverse effects are so troublesome as to be intolerable and an effective dose cannot be achieved. In patients intolerant of the adverse effects of nifedipine, verapamil and diltiazem can be effective in the management of Raynaud's phenomenon and perniosis.

4. Prostaglandin I_2 (Prostacyclin)

Metabolism of the arachidonic acid derivative prostaglandin H_2 by prostacyclin synthetase results in the formation of the unstable compound prostaglandin I_2 (PGI_2) also known as prostacyclin. Its structure is

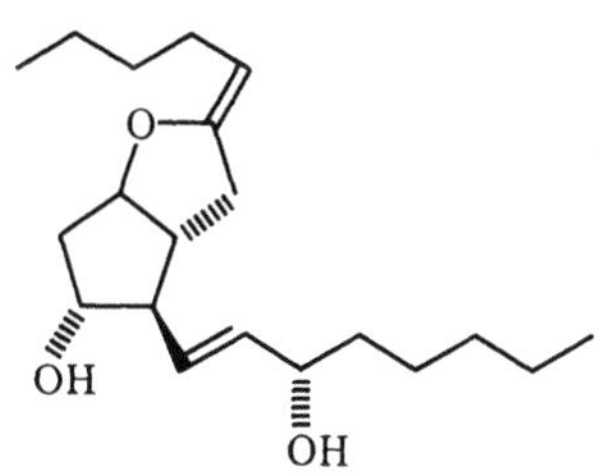

This compound was first discovered in vascular tissue (Moncada et al. 1976).

PGI_2 is hydrolysed nonenzymatically to a stable compound 6-keto-$PGF_{1\alpha}$. Prostacyclin is highly unstable and requires intravenous infusion after reconstitution in glycine buffer (pH 10.5), 10 ml buffer per 0.5 ml epoprostanol with further dilution in normal saline/glycine buffer in the proportion 5.25 : 1 by volume. Prostacyclin is spontaneously hydrolysed in blood, but not metabolised during first passage through the lungs. The first step appears to be the oxidation of the 15-OH group by 15-*OH*-dehydrogenase (PGDH) to a 15-keto compound which is subsequently reduced to the 13,14-dihydro derivative, a reaction catalysed by prostaglandin Δ13 reductase. Subsequently, the oxidation of the side chain results in the formation of the major metabolite 6-keto-$PGF_{1\alpha}$ which is excreted in the urine.

5. Synthetic Analogues

Prostacyclin has diverse effects on several organs and tissues. Vasodilatation occurs in most species and in most vascular beds, including the coronary, renal, mesenteric, skeletal muscle and pulmonary circulation. It relaxes all isolated vascular smooth muscle preparations and inhibits in vitro aggregation of human platelets at approximately 0.003–0.002 *M* concentrations and increases red blood cell deformability.

In the lungs PGI_2 is a mild bronchodilator and antagonises induced bronchoconstriction. PGI_2 relaxes uterine muscle from pregnant women and inhibits gastric secretion in response to food, histamine or gastrin. The acidity, pepsin content and volume of secretion are reduced and there is vasodilatation in the gastric mucosa. PGI_2 does not induce diarrhoea in animals or humans and inhibits the toxin-induced accumulation of intestinal fluid. In experimental models infusion of PGI_2 into the renal arteries of dogs increases renal blood flow with a resulting diuresis, natriuresis and kaliuresis with little effect on the glomerular filtration rate except in the presence of renal vasoconstriction. PGI_2 causes the release of renin and sensitises afferent nerve endings to the effects of chemical or mechanical stimuli.

a) Pharmacology

The mechanism of action of PGI_2 has been studied extensively in platelets where it has been shown that aggregation is inhibited by increasing the concentration of cyclic AMP with associated increase of intracellular calcium (Owen and Le Breton 1981). A similar effect has also been demonstrated in human leucocytes (Kirby et al. 1980).

Analogues of PGI_2 with greater stability and fewer side effects than the present compound have been developed. Of these iloprost has more potent effects on inhibition of platelet aggregation. Chronic toxic poisoning effects are not experienced as the drug is not administered long-term in ordinary clinical use.

b) Adverse Effects

Severe hypotension sometimes leading to loss of consciousness is the most serious manifestation of acute toxicity and responds to cessation of the infusion. Nausea, jaw pain, anorexia, facial flushing and fatigue are often also encountered. These are usually dose related and cease with termination of the infusion and there are no long-term adverse effects.

c) Therapeutic Use

Prostacyclin infusion at a dose of 4–12 ng kg^{-1} min^{-1} produces peripheral vasodilatation and increased hand and digital blood flow in humans (Dowd et al. 1982). For this reason, PGI_2 and its analogue iloprost, have been used in the treatment of Raynaud's phenomenon and cutaneous peripheral vascular insufficiency (Dowd et al. 1982; Belch et al. 1983; Yardoumian et al. 1988; Pardy et al. 1980). Increased peripheral cutaneous blood flow was associated with dramatic and often long-lasting relief of vascular insufficiency and decrease in the frequency, severity and duration of attacks of Raynaud's phenomenon.

References

Ahlquist RPM (1948) A study of the adrenotropic receptors. J Physiol 153:586–599
Altura BM, Hershey SG (1967) Pharmacology of neurohypophysical hormones and their synthetic analogues in the terminal vascular bed. Angiology 18:428–439
Beaucher WN (1979) Vasodilator drugs for peripheral vascular disease (letter). N Engl J Med 301:159
Belch JJF, Drury JK, Capell H, Forbes CD, Newman P, McKenzie F, Leiberman P, Prentice CRM (1983) Intermittent epoprostenol (prostacyclin) infusion in patients with Raynaud's syndrome: a double blind controlled trial. Lancet 1:313–315

Brain SD, Williams TJ, Tippins JR, Morris HR, MacIntyre I (1985) Calcitonin gene-related peptide is a potent vasodilator. Nature 313:54–56
Brain SD, Tippins JR, Morris HR, MacIntyre I, Williams TJ (1986) Potent vasodilator activity of calcitonin gene-related peptide in human skin. J Invest Dermatol 87:533–536
Bunker GB, Dowd PM (1987) Alterations in scalp blood flow after the epicutaneous application of 3% minoxidil and 0.1% hexyl nicotinate in androgenetic alopecia. Br J Dermatol 117:668–669
Bunker CB, Cooper F, Dowd PM (1983) A double blind placebo controlled trial of topical hexyl nicotinate on Reynaud's phenomenon. Brit J Dermatol (in press)
Bunker CB, Lanigan SW, Rustin MHA, Dowd PM (1988) The effects of topically applied hexyl nicotinate on the cutaneous blood in patients with Raynaud's phenomenon. Br J Dermatol 119:771–776
Busse R, Bassenge E (1984) Endothelium and hypoxic responses. Bibl Cardiol 38:21–34
Camp RD, Coutts AA, Greaves MW, Kay AB, Walport MJ (1983) Responses of human skin to intradermal injections of leukotrienes C_4, D_4 and B_4. Br J Pharmacol 80:497
Chew CVC, Hecht HS, Collett JT, McAllister RG, Singh BN (1981) Influence of severity of ventricular dysfunction on haemodynamic responses to intravenously administered verapamil in ischaemic heart disease. Am J Cardiol 47:917–922
Coffman JD, Cohen AS (1971) Total and capillary fingertip blood flow in Raynaud's phenomenon. N Engl J Med 285:259–263
Cohen BA, Coffman JD (1980) Beta adrenergic vasodilatation mechanism in the finger. Clin Res 28:161A
Cunningham FM, Woollard PM, Camp RD (1984) Skin monohydroxy fatty acids: correlation of biosynthesis with proinflammatory activity. Br J Dermatol 111:703–704
D'Humieres S, Russo-Marie F, Vargoftig BB (1986) PAF-acether-induced synthesis of prostacyclin by human endothelial cells. Eur J Pharmacol 131:13–19
Di Corletto PE, Dowen-Pope DF (1983) Cultured endothelial cells produce a platelet-derived growth factor-like protein. Proc Natl Acad Sci USA 80:1919–1923
Dowd PM, Martin MFR, Cooke ED, Bowcock SA, James R, Dieppe OA, Kirby JDT (1982) Treatment of Raynaud's phenomenon by intravenous infusion of prostacyclin (PGI_2). Br J Dermatol 106:81–89
Dowd PM, Kobza Black A, Woollard PM, Camp RDR, Greaves MW (1985) Cutaneousis responses to 12-hydroxy 5,8,10,14-eicosatetraenoic acid (12-HETE). J Invest Dermatol 84:537–544
Dowd PM, Rustin MHA, Lanigan S (1986) Nifedipine in the treatment of chilblains. Br Med J 293:923–924
Dowd PM, Whitfield M, Greaves MW (1987) Hexyl nicotinate-induced vasodilation in normal human skin. Dermatologica 174:239–243
Dowd PM, Camp RDR, Greaves MW (1988) Interleukin-1 is proinflammatory in human skin in vivo. Skin Pharmacol 1(1):30–37
Erdös EG, Wilde AF (eds) (1970) Bradykinin, kallidin and kallikrein. Springer, Berlin Heidelberg New York (Handbook of experimental pharmacology, vol 25)
Furchgott RF (1983) Role of endothelium in responses of vascular smooth muscle. Circ Res 53:557–573
Furchgott RF, Zawadski JV (1980) The obligatory role of endothelial cells in the relaxation of arterial smooth muscle by acetylcholine. Nature 288:373–376
Gifford RW (1963) Arteriospastic disorders of the extremities. Circulation 27:970–975
Gordon JL (1982) Vessel wall metabolism and vasoactive agents. In: Woolff N (ed) Biology and pathology of the vessel wall. Prager, Eastbourne, pp 41–56
Gordon JL (1987) Adenine nucleotides and the regulation of vascular tone. Thromb Haemost 58:251
Guy RH, Tur E, Bjerke S, Maibach HI (1985) Are there age and racial differences to methyl nicotinate-induced vasodilatation in human skin? J Am Acad Dermatol 12:1001
Hansteen V, Lorentsen (1974) Vasodilator drugs in the treatment of peripheral arterial insufficiency. Acta Med Scand [Suppl] 556:3–62
Harper FE, Le Roy EC (1982) Raynaud's phenomenon – an update on treatment. J Cardiovasc Med 7:282–290

Higgs GA, Cardinale PC, Moncada S, Vane JR (1979) Microcirculatory effects of prostacyclin (PGI_2 in the hamster cheek pouch). Microvasc Res 18:245–254
Jorizzo JL, Coutts AD, Eady RAJ, Greaves MW (1983) Vascular responses of human skin to injection of substance P and mechanism of action. Eur J Pharmacol 87:67–76
Kahan A, Weber S, Amor B, Saparta L, Hodana M, Degeorges M (1982) Etude controlée de la nifedipine dans le traitement du phénomène de Raynaud. Rev Rhum Mal Osteoartic 49:337–343
Kirby JDT, Lima DRA, Dowd PM, Kilfeather S, Turner P (1980) Prostacyclin increases cyclic nucleotide responsiveness of lymphocytes from patients with systemic sclerosis. Lancet 2:453
Kohlhardt M, Bauer B, Krause H, Fleckenstein A (1972) Differentiation of the transmembrane Na and Ca channels in mammalian cardiac fibres by the use of specific inhibitors. Pflügers Arch 335:309–322
Kontos HA, Wasserman AJ (1971) Effect of reserpine in Raynaud's phenomenon. N Engl J Med 285:259–263
Lange RL, Reid MS, Tresch DD; Kellan MH, Bernhard VM, Coolidge G (1972) Nonbatheromatous ischaemic heart disease following withdrawal from chronic industrial nitroglycerin exposure. Circulation 46:666–678
Le Roy EC, Downey JA, Cannon PJ (1971) Skin capillary blood flow in scleroderma. J Clin Invest 50:930–939
Lewis T (1926) The blood vessels of the human skin and their responses. Shaw, London
Mitchell JR, Oates JA (1970) Guanethidine and related agents in mechanism of the selective blockade of adrenergic neurons and its antagonism by drugs. J Pharmacol Exp Ther 172:100–107
Moncada S, Gryglewski R, Bunting S, Vane JR (1976) An enzyme isolated from arteries transforms prostaglandin endoperoxides to an unstable substance that inhibits platelet aggregation. Nature 263:663–665
Moncada S, Herman AS, Higgs AE (1986) Generation of prostacyclin an endothelium-derived relaxing factor from endothelial cells. In: Jolles J, Tegrand JY, Nurden A (eds) Biology and pathology of platelet-vessel wall interactions. Academic, London, pp 284–304
Nilsen KH, Jayson MIV (1980) Cutaneous microcirculation in systemic sclerosis and response to intra-arterial reserpine. Br Med J 280:1408–1411
Nobin BA, Nielsen SL, Eklov B, Lassen NA (1978) Reserpine treatment of Raynaud's disease. Ann Surg 187:12–16
Oattman JD, Mannicle JA (1972) Failure of vasodilator drugs in arteriosclerosis obliterans. Ann Intern Med 76:35–39
Owen NE, Le Breton GO (1981) Ca^{2+} mobilization in blood platelets as visualized by chlortetracycline fluorescence. Am J Physiol 241:613–619
Pardy BJ, Lewis DD, Eastcott HHG (1980) Preliminary experience with prostaglandins E_1 and I_2 in peripheral vascular disease. Surgery 88:826–832
Parks VJ, Sandison AG, Skinner SL, Whelan RF (1961) The mechanism of the vasodilator action of reserpine in man. Clin Sci 20:289–295
Piotrowski W, Foreman JC (1986) Some effects of calcitonin gene related peptide in human skin and on histamine release. Br J Dermatol 114:37–46
Robinson BF, Dobbs RJ, Kelsey CR (1980) Effects of nifedipine on resistance vessels, arteries and veins in man. Br J Clin Pharmacol 10:433–438
Robson P, Pearce V, Antcliffe AC, Hamilton M (1978) Double-blind trial of indoramin in digital artery disease. Br J Clin Pharmacol 6:88–91
Smith CD, McKendry RJR (1982) Controlled trial of nifedipine in the treatment of Raynaud's phenomenon. Lancet 2:1299–1301
Somlyo AV, Somlyo AP (1968) Electromechanical and pharmacomechanical coupling in vascular smooth muscle. J Pharmacol Exp Ther 159:129–145
Sondergaard J, Greaves MW (1971) Prostaglandin E effect on cutaneous vasculature and skin histamine. Br J Dermatol 84:424
Theroux P, Waters DD, DeBaisieux JC, Szlachcic J, Mizgala HF, Bourassa MG (1980) Haemodynamic effects of calcium ion antagonists after acute myocardial infraction. Clin Invest Med 3:81–85

von Euler US (1972) Synthesis, uptake and storage of catecholamines in adrenergic nervs. The effect of drugs. In: Blaschko H, Muscholl E (eds) Catecholamines. Springer, Berlin Heidelberg New York, pp 186–230 (Handbook of experimental pharmacology, vol 33)
Willerson JT, Thompson RH, Hookman P, Herdt J, Decker JL (1970) Reserpine in Ra ynand's disease and phenomenon short term response to intra-arterial injection. Ann Intern Med 72:17–27
Yardoumian DA, Isenberg DA, Rustin MHA, Dowd PM, Machin SJ (1988) Successful treatment of Raynaud's syndrome with Iloprost. Br J Rheumatol 27:220–226

CHAPTER 22

Fibrinolysis and Fibrinolytic Drugs

E. PANCONESI and T. LOTTI

A. Fibrinolysis

Fibrinolysis, the physiologic process which performs the task of demolishing fibrin, is dependent on various activation and inhibtion mechanisms, and is subject to functional fluctuations in biorhythm (FEARNLEY 1966). The fibrinolytic system can be simply defined as "the enzymatic breakdown of fibrin to fragments which are no longer able to form a coherent fibrin net" (VON KAULLA 1963). This system is part of a more substantial mechanism which creates in the cutaneous organ a dynamic balance between fibrinosynthesis and fibrinolysis, known as "the balance between fibrinolysis and fibrin deposition" (RYAN 1976). The equilibrium between fibrinosynthesis and fibrinolysis is a consequence of the activity of the inhibitor and activator systems of fibrinolysis. The lysis of fibrin takes place through a proteolytic reaction catalyzed by plasmin; this latter enzyme derives from the activation of plasminogen by agents called plasminogen activators. The principal components of the fibrinolytic system are outlined in the following sections.

I. Plasminogen

Plasminogen, the inactive precursor of plasmin, is a β-globulin present in plasma at a concentration of about 2 μM (ROBBINS and SUMMARIA 1970; LIJNEN and COLLEN 1982). It has not yet been ascertained where the synthesis of plasminogen takes place: in the liver (SAITO et al. 1980) and/or in the eosinophilic granulocytes (BARNHART and RIDDLE 1963). It has been reported in human and porcine uterine fluid (CASSLEN and OHLSSON 1981), in human saliva (MOUDY 1982), and in the basal layer of human epidermis (ISSEROFF and RIFKIN 1982, 1983).

Native plasminogen is comprised of a single polypeptide chain with molecular weight 92000 (KAPLAN et al. 1978; CASTELLINO and POWELL 1981).

II. Plasmin

Plasmin, the active enzyme of the fibrinolytic system, is an endopeptidase with powerful proteolytic activity; it has a molecular weight of about 81 000 (DANO et al. 1985). Plasmin acts by specific attack on the intra-peptide bonds in lysine, arginine, and many of their esters.

III. Plasminogen Activation

Plasminogen activation may follow two pathways, an intrinsic pathway in plasma (KAPLAN and AUSTEN 1972) or an extrinsic one in tissue (ASTRUP 1956), which are not alternatives, but coexist and generally occur simultaneously. Plasminogen activators have been divided into direct and indirect activators (MÜLLERTZ and LASSEN 1953). Fairly recently, there have been indications and then confirmation of the existence of two major types of plasminogen activators (distinguished by immunologic reactivity, functional differences, and differences in molecular weight): the *urokinase* plasminogen activator which was originally purified from urine, and the *tissue-type* plasminogen activator which was purified from tissue extracts.

1. Tissue-type Plasminogen Activator (t-PA)

t-PA is a serine proteinase $M_r = 74\,000$ which has been purified from different substances, including human plasma and vessel perfusates (COLE and BACHMANN 1977; BINDER et al. 1979; WALLEN 1977; WALLEN et al. 1982, 1983). PENNICA et al. (1983) suggested the molecular arrangement of t-PA and also established its exact amino acid sequence: t-PA is composed of a heavy A chain and light B chain (whereas in urokinase the A chain is the light one and B the heavy one); it is the B chain which has the active site.

Like all direct plasminogen activators, t-PA breaks the Arg – 560 to Val – 561 bond in the plasminogen molecule, thus generating the two structural chains of plasmin. t-PA is inhibited by dihisopropylfluorophosphate (DFP) (BINDER et al. 1979; RIJKEN and COLLEN 1981; ANDREASEN et al. 1984). *p*-Nitrophenyl-*p*-guanidinobenzoate and *p*-aminobenzamidine inhibit the incorporation of DFP into the active site (ANDREASEN et al. 1984). Macromolecular inhibitors of plasmin, such as aprotinin and soybean trypsin inhibitor, do not inhibit t-PA (ALLEN and PEPPER 1981).

2. Urokinase

A relatively high concentration of urokinase (u-PA) was discovered in human urine, allowing its purification (WHITE et al. 1966; OGAWA et al. 1975; VETTERLEIN and CALTON 1983). u-PA has a well-defined structure composed of two chains, a light A chain and heavy B chain, held together by a disulfide bond (GUNZLER et al. 1982; STRASSBURGER et al. 1983).

It has also been shown that there are three forms:

1. Pro-Urokinase (pro-UK), an active pro-enzyme characterized by the fact that it consists of a single amino acid chain with a high molecular weight (55 000);
2. Urokinase (HMW-UK) with approximately the same molecular weight (55 000) as pro-urokinase, but with two chains bound by a disulfide bond;
3. Urokinase (LMW-UK) with a low molecular weight (33 000) (BERNIK et al. 1974; NIELSEN et al. 1982; GUREVRICH et al. 1985).

It seems likely, therefore, that like the other serine proteinases, urokinase is released from cells as an inactive pro-enzyme (SKRIVER et al. 1984).

3. Endothelial Plasminogen Activator

The endothelial plasminogen activator is probably the same as the tissue-type activator (t-PA); it is easily soluble and is present in high quantities in the endotheliocytes of venules and veins and in minor quantities in those of the arteries (ASTRUP 1956; GUREVRICH et al. 1985).

4. Plasmatic Pro-activator

The plasmatic pro-activator, perhaps the same thing as pre-kallikrein or Fletcher factor, is the intermediate substance which, together with coagulation factor XII, activates plasminogen (GOLDSMITH et al. 1978; OGSTON et al. 1976).

5. Streptokinase – Activated Plasminogen Activator

Demonstration of a streptokinase – activated plasminogen activator dates from MÜLLERTZ and LASSEN'S (1953) experiments showing that streptokinase (a protein with molecular weight 47000, produced by group A β-hemolytic streptococci) was able to activate fibrinolysis.

6. "Activated" Macrophages Activator

"Activated" macrophages are also able to release a plasminogen activator (PLOW et al. 1971; PLOW and EDGINGTON 1973).

7. Plasminogen Activators in Neoplastic Cells

Numerous investigations have demonstrated the presence of plasminogen activators in many types of neoplastic cells (RIFKIN et al. 1974; DANØ and REICH 1978; LOTTI et al. 1985a; LOTTI and FABBRI 1985). These activators are predominantly urokinase and t-PA and are even considerd "tumor markers" (for review see DANØ et al. 1985); very few cultures of cells from neoplasms have been found to lack detectable activity of plasminogen activators (WILSON and DOWDLE 1978).

8. Erythrokinase

Erythrokinase is a direct plasminogen activator contained in the extra-stromal fraction of red blood cells (for review see NERI-SERNERI 1981).

9. Plasminogen Activators in Human Granulocytes

Substances with protease activity able to interfere with the fibrinolytic system have been identified in human granulocytes; they include a substance with direct fibrinogenolytic and fibrinolytic activity (HAWIGER 1969; SABA et al. 1967, 1968). Proteases are secreted actively by the leukocytes during phagocytosis, but not following contact with circulating immunocomplexes or endotoxins (GRANELLI-PIPERNO et al. 1977; REICH 1978).

10. Indirect Plasminogen Activators

Indirect plasminogen activators include: lysokinase, as yet imprecisely identified, present in tissue (ASTRUP 1956), streptokinase, which in low concentrations is not able to activate plasminogen into plasmin directly, and the so-called bacterial activators produced by *Escherichia coli, Pseudomonas aeruginosa* and *Staphylococcus* spp. (for review see NERI-SERNERI 1981).

IV. Inhibitors of Fibrinolysis

The fibrinolytic system is modulated by numerous factors with inhibiting activity which have been divided in antiactivators (of plasminogen) and antiplasmins. The antiactivators (plasmatic, placental, miometric, amniotic, endothelial, epidermal antiactivators, etc. LOSKUTOFF et al. 1983; TURNER et al. 1969 a, b) inhibit plasminogen activation, in contrast to the antiplasmins which specifically antagonize the proteolytic action of plasmin.

The antiplasmins are a group of substances able to inhibit the proteolytic effect of plasmin, and comprise the following:

1. Primary or α_2-antiplasmin, a protein with a molecular weight of about 70000 which reacts stoichiometrically with plasmin, thus inactivating it (COLLEN 1976);
2. α_2-Macroglobulin, a globulin with a molecular weight of about 80000 that inactivates plasmin through the formation of a plasmin-macroglobulin complex which mononuclear phagocytes quickly eliminate from the system (OGSTON et al. 1973);
3. α_1-Antitrypsin, a glycoprotein with a molecular weight of 56000 and less inhibiting activity on plasmin than that exercised by α_2-macroglobulin (CLEMMENSEN and CHRISTENSEN 1976).

The antiplasmin group also includes the so-called inhibitor of inter-α-trypsin, antithrombin III (minor inhibiting activity on plasmin) (CLEMMENSEN 1978), and C_1 inactivator (RANBY et al. 1982 a, b).

There is also a protease inhibitor, protease nexin, secreted by cultured human fibroblasts, which inhibits u-PA, plasmin, and thrombin (BAKER et al. 1980; SCOTT and BAKER 1983).

V. Plasminogen Activators in Physiologic Conditions

The presence of plasminogen activators in different types of cells indicates that these activators may play a role in numerous biologic processes: involution of the mammary gland, dehiscence of the ovarian follicle, thrombolysis (LARSSON et al. 1984; COLLEN 1980). The different distributions of u-PA and t-PA indicate the likelihood that the two types of activator have different functions.

Involution of the mammary gland after the period of milk secretion is an example of the role played by the plasminogen activators in physiolgic conditions (OSSOWSKI et al. 1979). The investigators have in fact observed a direct relationship between involution of the mammary gland in mice and u-PA activity in both glandular extracts and in short-term cultures of cells from the gland itself.

Another example of the function of plasminogen activators in the healthy organism is the destruction of the wall of the ovarian follicle during ovulation. It seems that t-PA, together with other factors, is mainly responsible for the dehiscence of the ovarian follicle (for review see DANØ et al. 1985), but it is probable that the low quantity of u-PA found in the follicle may be due to the presence of u-PA as a pro-enzyme at that site.

Thrombolysis is another biologic process in which the role of plasminogen activators has been studied, regarding both physiologic and pathologic conditions (PROWSE and CASH 1984). Both u-PA and t-PA can be demonstrated in the blood, but there are findings indicating that t-PA alone is the key enzyme in physiologic thrombolysis (RIJKEN et al. 1980; KRISTENSEN et al. 1984). Demonstration that cells in culture release t-PA in a partially or completely inactive single-chain form suggests that this may be the form best suited for therapy of thrombolysis (COLLEN et al. 1983, 1984a, b; GOLD et al. 1984).

VI. Plasminogen Activators in Inflammation

The possible role of proteolytic enzymes in inflammation has been under consideration for many years (for review see ASTRUP 1968). Renewed interest in this topic was aroused by the discovery that mice macrophages in culture stimulated with thioglycolate released plasminogen activators, while unstimulated macrophages did not. This release of plasminogen activators seems to be modulated by a vast range of agents, including asbestos, lymphokines (KLIMETZEK and SORG 1977; GORDON and COHN 1978), interferon, concanavalin A, and colony-stimulating factor (LIN and GORDON 1979). Production of u-PA by stimulated macrophages is inhibited by glycocorticoids, colchicine, and vincristine. Polymorphonuclear granulocytes also release u-PA; this production is induced by concanavalin A and inhibited by glycocorticoids (GRANELLI-PIPERNO et al. 1977).

The u-PA released by macrophages and granulocytes could contribute to macromolecular catabolism of tissue in inflammation through extracellular proteolysis, similar to what takes place in tumors.

VII. Plasminogen Activators in Neoplastic Conditions

For many years there has been the hypothesis of a relationship between coagulation and fibrinolytic activity in human tumor extracts (for review see ASTRUP 1956). However, only in the 1970s did research indicate that all the steps of the growth and spreading of tumor cells seemed to be influenced by components of the coagulation system and factors of the fibrinolytic system (SVANBERG et al. 1975; NEWSTEAD et al. 1976; YVEN and KWAAN 1983).

Fibrinolysis is strictly related to kininogenesis and the activation of the complement system which can be triggered by the enzymatic cascade of plasmatic coagulation factors as well as by leukocytes (PLOW and EDGINGTON 1975) and by endothelial cells able to synthesize vascular activators and inhibitors of plasminogen (LOSKUTOFF and EDGINGTON 1977).

There are fibrin deposits on the periphery of spontaneous and experimentally induced tumors (POGGI et al. 1977; HILGARD and HIEMEYER 1970) owing to reduced fibrinolytic activity and/or coagulation activation. Fibrin may be a support structure along which the tumor cells can grow and act (and migrate), promoting organization of granulation tissue, often found along the advancing edges of tumor tissue (AMBRUS et al. 1975).

The components of the cell surface responsible for interactions between neoplastic cells and fibrin are not well defined. However, fertile ground for investigation of the relationships between neoplastic transformation and fibrinolysis (VAHERI et al. 1978) has been opened up by the studies on protease degradation of fibronectin (a glycoprotein with a molecular weight of 440000 which is the principal surface-associated protein on the membrane of fibroblasts and many other cells) (NIEWIAROWSKI and GOLDSTEIN 1973).

Fibronectin is catabolized by plasmin, the active enzyme of the fibrinolytic system. Numerous findings indicate that plasminogen activation (resulting in formation of plasmin) is the commonest modulatory mechanism of extracellular proteolysis (BOGENMANN and JONES 1983).

Numerous oncological investigations have reported that cells undergoing neoplastic transformation produce plasminogen activators (for review see DANØ et al. 1985). Synthesis of these activators can be induced by physical carcinogens and co-carcinogenic substances (HAMILTON et al. 1976) and is enhanced in many spontaneous (human and other) and experimental tumor cells (CHRISTMAN and SILVERSTEIN 1977).

Plasmin is itself a powerful mitogen, even able to stimulate cell migration (POLLACK and RIFKIN 1975); furthermore, its proteolytic action on various substrates, such as fibrinogen, fibrin, and fibronectin, generates peptides which are active on vascular permeability, granulocytic chemotaxis, and cell growth (for review see DANØ et al. 1985).

The studies examined thus far demonstrate a correlation between increased synthesis of plasminogen activators, formation of plasmin, and malignant transformation (NAGY et al. 1977). Other studies demonstrate the role played by plasminogen activators independent of their effect on plasminogen and (in the absence of plasmin) on neoplastic growth (QUIGLEY 1979; WIGLER and WEINSTEIN 1976). The findings of these investigations were confirmed by DEL ROSSO et al. (1985) and DINI et al. (1985) when they demonstrated plasminogen-independent activity of urokinase in cultures of transformed human fibroblasts from Rous sarcoma virus.

VIII. Plasminogen Activation in the Skin

The skin is the site of continuous interaction between fibrinosynthesis and fibrinolysis. This equilibrium is largely controlled by the secretion of plasminogen activators by endothelial cells of the dermal venules, especially those of the middle and deep derma (RYAN 1973; LOTTI et al. 1979) and the presence in the epidermis of plasminogen activators (MORIOKA et al. 1984) and inhibitors of fibrinolysis (TURNER et al. 1969a, b). Although the role of *endothelial activators* seems well demonstrated in numerous studies dating as far back as the 1960s, studies to ascertain the presence of *epidermal activators* were done only a few years ago. The latter studies had a stimulating prologue in 1983 with ISSEROFF and RIFKIN's discovery of plasminogen in human basal keratinocytes.

Using anti-urokinase monoclonal antibodies, MORIOKA et al. (1984) demonstrated urokinase in human epidermal cells. These studies suggest that plasminogen activators play a physiologic role in the normal maturation of epidermal cells, proteolytically promoting detachment of the keratinocytes (GREEN 1977). It has also been shown that plasminogen activators are necessary for the growth of normal keratinocytes and that intracellular levels of plasminogen activators increase in direct proportion to the stage of differentiation of the keratinocytes (ISSEROFF et al. 1982). There have also been demonstrations (LOTTI et al. 1984b) of the existence of receptors for urokinase on the surface of human stratum corneum cells and on the surface of dermal fibroblasts.

IX. Physiopathology of the Cutaneous Fibrinolytic System

Disturbance of the equilibrium between fibrin deposition and its removal by the fibrinolytic system, for example because of an increase in inhibitors and/or activator deficit, can result in accumulation of fibrin in the derma, as seen in numerous dermatologic affections: bullous dermatoses of the pemphigoid group (RYAN 1976; PANCONESI et al. 1982), lichen planus (CRAWFORD and OGSTON 1975; LOTTI and FABBRI 1984), Schamberg's purpura (LOTTI et al. 1985b), and cutaneous vasculitis (CUNLIFFE 1968; FABBRI et al. 1981; TOKI et al. 1982; LOTTI et al. 1985c). There is instead an increase in fibrinolytic activity with excessive removal of fibrin from the microvessels which causes increased permeability of the dermal vessels in other diseases, such as urticaria (RYAN 1976; LOTTI et al. 1985c), Majocchi's purpura (LOTTI et al. 1985b), porphyria cutanea tarda (CHIMENTI et al. 1981), psoriasis (RYAN 1973, 1980; BRUNETTI and LOTTI 1983), pemphigus vulgaris (PANCONESI et al. 1982; FABBRI et al. 1985; HASHIMOTO et al. 1982, 1984), primary irritant dermatitis (LOTTI and FABBRI 1981), and pruritus aquagenicus (LOTTI et al. 1984a, 1986b). In particular, it has been suggested that plasminogen activators have a precise pathogenic role in pemphigus vulgaris where they are released by the keratinocytes following a specific interaction between antigen and anti-intercellular substance antibody, inducing acantholysis via plasmin (LOTTI et al. 1979; PANCONESI et al. 1982; HASHIMOTO et al. 1982; SINGER and LAZARUS 1983; FABBRI et al. 1985).

Investigations on psoriasis demonstrate that levels of plasminogen activators are higher in psoriatic epidermis than in the uninvolved epidermis of the same psoriatic subjects and the epidermis of healthy controls (FRÄKI et al. 1983; BRUNETTI and LOTTI 1983). Alterations in plasminogen activator activity seem to be correlated with disease activity (FRÄKI et al. 1983; LOTTI et al. 1986a). Only t-PA-dependent (and not u-PA-dependent) plasminogen activator activity has been demonstrated in psoriatic epidermis (LOTTI et al. 1988). Finally, the existence of u-PA receptors has been demonstrated by electron microscopic localization of ferritin–u-PA (LOTTI et al. 1984b) complexes on the cell surface of human keratinocytes.

X. Methods for Evaluation of Plasminogen Activators in the Skin

New methods for evaluating plasminogen activators in the skin are still being developed. In this section, we present and discuss the principal methods currently used in most laboratories, including ours.

1. Modified Autohistographic Fibrin Plate Assay with Monoclonal Antibodies Against t-PA and u-PA

The fibrin autohistograph was first described by TODD in 1959. This method has been modified so that 6-μm cryostated sections of the specimen are incubated under a uniformly thick layer of fibrin film composed of 50 mg plasminogen-rich fibrinogen mixed with 5 ml phosphate-buffered saline (PBS, pH 7.4) and 50 μl thrombin 25 U/ml mixed with 500 μl PBS. The slides are then placed in a humid chamber where they are incubated at 37 °C for 3 min and then at 4 °C for 16 h. The sections are stained with Harris' hematoxylin and examined by light microscopy (Fig. 1) (LOTTI et al. 1980).

Autohistographic control tests with plasminogen-free film are performed in all cases by heating the plasminogen-fibrin complex for 30 min at 80 °C to inactivate the plasminogen in the compound or by using plasminogen-free fibrinogen. This assay detects 10^{-16} mol u-PA, and is particularly sensitive to t-PA.

It has recently become possible to distinguish between u-PA and t-PA activity with this method further modified by specific monoclonal antibodies to u-PA and t-PA plasminogen activators (LOTTI et al. 1988c). A solution of 2 ml specific

Fig. 1. Fibrin plate assay; 6-μm section, Harris' hematoxylin, ×40

monoclonal antibodies (50 μg/ml) is put on the tissue section at room temperature; 5 min later the fibrin film is placed over the section, which is processed as specified.

The specific antibody reacts only with its specific substrate, u-PA or t-PA, inhibiting the specific substrate enzyme activity on plasminogen, thus allowing determination of the u-PA or t-PA part of overall cutaneous plasminogen activator activity. With this device, specific t-PA- and/or u-PA-dependent fibrinolytic activity can be selectively inhibited in the skin, at the anatomic sites where it is produced such that cutaneous t-PA and urokinase activity can be assessed individually. In perilesional skin of pemphigus subjects, epidermal fibrinolytic activity is u-PA dependent and perivascular fibrinolysis is t-PA dependent (LOTTI et al. 1987b, 1988).

2. Casein Plate Assay

Since fibrin is particularly sensitive to t-PA, casein may be used as substrate for plasmin. An assay similar to the fibrin plate method modified by using casein instead of fibrin was described by SAKSELA in 1981. In our experience with casein film (50 mg casein, plus 10 μg plasminogen, plus 500 μl PBS) on skin sections, we found that this assay showed a relatively very low sensitivity to plasminogen activator activity (LOTTI et al. 1987b).

3. Casein Substrate Assay

Casein substrate is useful for evaluating plasminogen activator activity with a spectrophotometric method (KLINE 1971; KLINE and REDDY 1977). With this method, the samples under examination (previously extracted with 2 *M* KSCN at room temperature after homogenization) are incubated with plasminogen in casein. The spectrophotometric absorbance (280 nm) is used as a measure of plasminogen activator activity. This assay has a low sensitivity for t-PA compared with the assays with fibrin (CAMIOLO et al. 1981). A selective inhibition of t-PA or u-PA activity in the extract sample under examination can be achieved with specific monoclonal antibodies; a concentration of 50 μg/ml inhibits 95% of the activity (LOTTI et al. 1987b).

4. Synthetic Substrate Assay

This assay allows evaluation of plasminogen activator activity in Triton X-100 skin extract samples by measuring the enzymatic activity of u-PA on the synthetic substrate pyro-Glu-Gly-Arg-*p*-nitroaniline, specific for u-PA (FRIBERGER 1982). The u-PA present in the specimen releases *p*-nitroaniline which is measured spectrophotometrically (405 nm). This assay allows detection of 2×10^{-7} mol u-PA (DANØ et al. 1985).

XI. Autohistographic Method for Evaluation of Inhibitors of Fibrinolysis in the Skin

TODD's fibrin slide autohistographic technique (1959) was modified by NOORDHOEK-HEGT and BRAKMAN (1974a, b) to determine the inhibition of fibrinolysis in tissue by incubating the section being investigated under a layer of fibrin over which another tissue section with fibrinolytic activity is placed.

LOTTI et al. (1983) proposed a method for detecting inhibition of fibrinolysis in tissue sections by a normal thread previously incubated in a fibrinolytically active solution (urokinase, streptokinase, t-PA, etc.) instead of the tissue section provided with fibrinolytic activity. Inhibitory capacity is evaluated with a low magnification microscope.

This technique may also be of value for critical evaluation of Todd's fibrin slide technique since it appears that substances which inhibit fibrinolysis greatly influence the size and shape of lysis zones produced by vascular plasminogen activators on plasminogen-rich fibrin film.

B. Drugs Affecting Plasminogen Activator Synthesis and Activity

I. Steroid Hormones

Steroid hormones, in particular glucocorticoids, have been reported to suppress urokinase-dependent activity in cultured cells and tissues from various mammalian species (VASSALLI et al. 1976, 1978; WERB 1978; NEUMAN and SORG 1983; RIFKIN 1978; JAKEN and BLACK 1981a, b; GRANELLI-PIPERNO and REICH 1983a, b; CAMIOLO et al. 1984). The response of t-PA activity to glucocorticoids appears to be more variable (CAMIOLO et al. 1984).

It has been suggested that glucocorticoids inhibit u-PA and t-PA synthesis by a cytoplasmatic molecule which in turn inhibits plasminogen activator gene transcription or plasminogen activator mRNA translation (RIFKIN 1978; RIFKIN and CROWE 1980). Estrogens increase u-PA activity in rat uterus (KATZ et al. 1976; PELTZ et al. 1983) and in human breast carcinoma (BUTLER et al. 1979; KATZENELLENBOGEN et al. 1984).

II. Polypeptide Hormones and cAMP

Calcitonin, prolactin, epinephrine, and luteinizing hormone have been found to regulate plasminogen activator activity in many cell types (DAYER et al. 1981; MIRA-Y-LOPEZ et al. 1983). cAMP and its derivatives have been reported to influence levels of intracellular and extracellular u-PA and t-PA (DAYER et al. 1981; LACROIX and FRITZ 1977).

III. Epidermal Growth Factor

Epidermal growth factor (EGF) was reported to increase plasminogen activator activity in an HeLa cell line and in cultured human foreskin fibroblasts (LEE and WEINSTEIN 1978; EATON and BAKER 1983).

IV. Retinoic Acid

Retinoic acid has been found to modulate plasminogen activator activity in a number of cultured cell types and in tissue explants (VASSALLI et al. 1976; MIRA-Y-LOPEZ et al. 1983). It seems that retinoic acid causes a parallel increase in intracellular and extracellular levels of plasminogen activators (WILSON and REICH 1978; SCHRODER et al. 1980), inducing de novo synthesis (WILSON and REICH 1978).

V. Phorbol Esters

Phorbol esters have been shown to induce plasminogen activator activity (WIGLER and WEINSTEIN 1976). In general, there is good correlation between the tumor-promoting activity and effect on plasminogen activator activity of the different phorbol esters (WIGLER and WEINSTEIN 1976; GROSS et al. 1982). Furthermore, it has been reported that the tumor promotion induced by phorbol esters is suppressed by various nonspecific protease inhibitors (TROLL et al. 1975). Phorbol esters seem to enhance de novo synthesis of intracellular u-PA and t-PA, causing parallel increases in intracellular and extracellular plasminogen activator levels (JAKEN and BLACK 1981 a, b).

C. Fibrinolytic Drugs Used in Thrombolytic Therapy

Urokinase (u-PA), streptokinase, and tissue-type plasminogen activator (t-PA) are efficacious in the treatment of thromboembolic diseases. Many other drugs have been shown to enhance plasmatic fibrinolysis, but the majority have no real clinical application. The anabolic steroid stanozolol, defibrotide (extract of bovine lungs or aorta), and glycosaminoglycans seem able to enhance plasmatic fibrinolysis (but to a lesser extent than u-PA, streptokinase, and t-PA); their use in dermatologic diseases has been proposed (DAVIDSON et al. 1972; OFOSU et al. 1984).

I. Urokinase

Urokinase (u-PA) is a two-chain plasminogen activator, deriving from an inactive single-chain pro-urokinase (WUN et al. 1982; NIELSEN et al. 1982); it activates transformation of the natural substrate plasminogen into the fibrinolytic enzyme plasmin. Urokinase is usually given intravenously with a loading dose of 4400 IU/kg (given over a period of 10 min, followed by a continuous infusion of 4400 IU kg/h (for 12 h or more) to accelerate resolution of pulmonary emboli or severe thrombophlebitis (SASAHARA et al. 1979). After this treatment, heparin or oral anticoagulants are usually administered. Febrile episodes and allergic reactions are rare.

II. Streptokinase

Streptokinase is a bacterial protein obtained from group C β-hemolytic streptococci capable for activating plasminogen. It has no enzymatic activity and acts by combining with plasminogen to form an equimolar complex with plasminogen activator activity (REDDY and MARKUS 1972; SUMMARIA et al. 1982). Streptokinase has been used successfully in the treatment of acute pulmonary embolism, deep vein thrombosis, and acute myocardial infarction (COLLEN 1980; ANDERSON et al. 1983).

The usual loading dose of streptokinase is 250000 IU, given intravenously over a 30 to 45-min period, followed by 100000 IU/h. During treatment thrombin time must be montored regularly. Heparin or oral anticoagulants should replace streptokinase after 2–5 days of treatment.

III. Tissue-type Plasminogen Activator (t-PA)

As stated in Sect. A. III. 1, t-PA is a M_r 74000 serine proteinase which consists of two disulfide-linked chains (WALLEN et al. 1978). t-PA preferentially activates plasminogen bound to fibrin, thus limiting fibrinolysis to the thrombus and avoiding systemic activation of plasminogen. It seems that, at least in vitro, t-PA is more efficient than urokinase or streptokinase in the dissolution of thrombi and that t-PA-induced fibrinolysis causes much less damage to other proteins (e.g., fibrinogen) than that induced by streptokinase and u-PA (MATSUO et al. 1981; MATTSSON et al. 1981).

IV. Stanozolol

Stanozolol (17-hydroxy-17-methylandrostanol[3,2*c*]pyrazole) is a synthetic anabolic steroid able to produce significant increases in plasmatic fibrinolysis (DAVIDSON et al. 1972; VERHEIJEN et al. 1985). Stanozolol is administered orally, 8–10 mg/day, for weeks or months.

V. Defibrotide

Defibrotide (NIADA et al. 1981; LOBEL and SCHROR 1985) is an extract of bovine lungs or aorta and is a polydesoxyribonucleotide. It causes release of plasminogen activators from blood vessels, reduces the hematic concentration of plasmin inhibitors, and causes an increase in prostacyclin levels (PGI_2) (LOBEL and SCHROR 1985). This drug does not alter the hemocoagulation parameters and has no anticoagulation activity. The drug is administered intravenously; the usual loading dose is 800 mg/day for 6–10 days (200 mg every 6 h), followed by intramuscular injections of 200 mg every 12 h for 15–50 days.

VI. Glycosaminoglycans

The glycosaminoglycans, or heparinoids, are a large class of polysaccharides which are able to stimulate fibrinolysis (HALSE 1962), probably releasing plasminogen activator from the endothelium (KLÖCKING and MARKWARDT 1984).

Pentosan polysulfate (BERGQUIST and LJUNGHER 1981), heparan sulfate, and dermatan sulfate (OFOSU et al. 1984), the most active substances in this group, can be administered (50–500 mg/day) either orally, subcutaneously, or intramuscularly (FISCHER et al. 1982).

D. Therapeutic Use of Fibrinolytic Drugs in Dermatologic Diseases

It has been suggested that fibrin deposits and loss of cutaneous fibrinolytic activity play a major role in the pathogenesis of many cutaneous diseases (CUNLIFFE 1968; RYAN 1976; HAUSTEIN 1969; LOTTI et al. 1979, 1987a; PANCONESI et al. 1982).

There is evidence that stimulation of plasmatic fibrinolytic activity improves the clinical condition of patients with cutaneous necrotizing vasculitis and Behçet's syndrome (CUNLIFFE 1968; CUNLIFFE and MENON 1969; MENON and CUNLIFFE 1969; DODMAN et al. 1973). Raynaud's phenomenon and systemic sclerosis have been successfully treated with stanozolol (JARRET et al. 1978). Stanozolol has also proved effective in diminishing cutaneous induration in patients with lipodermatosclerosis (LOTTI et al. 1987a). Ulcus cruris due to chronic venous hypertension following thrombophlebitis of the leg seems to respond well to treatments enhancing fibrinolysis (LOTTI et al. 1987a).

References

Allen RA, Pepper DS (1981) Isolation and properties of human vascular plasminogen activator. Thromb Haemost 45:43–50

Ambrus JL, Ambrus CM, Pikern L, Soldes S, Bros I (1975) Hematologic changes and thromboembolic complications in neoplastic disease and their relationship to metastasis. J Med 6:433–458

Anderson JL, Marshall HN, Bray BE, Lutz JR, Frederick PR, Vanowitz FG (1983) A randomized trial of intracoronary streptokinase in the treatment of acute myocardial infarction. N Engl J Med 308:1312–1318

Andreasen PA, Nielsen LS, Grondahl-Hansen J, Skriver L, Zeuthen J, Stephens R, Danø K (1984) Inactive proenzyme to tissue-type plasminogen activator from human melanoma cells identified after affinity purification with a monoclonal antibody. EMBO J 3:51–56

Astedt B (1979) No cross reaction between circulating plasminogen activator and urokinase. Thromb Res 14:535–539

Astedt B, Hamberg L, Wagner G, Richter P, Ploug J (1979) Purification of urokinase by a beta-naphthamidine affinity. Thromb Haemost 42:924–928

Astrup T (1956) Fibrinolysis in the organism. Blood 11:781–805

Astrup T (1968) Blood coagulation and fibrinolysis in tissue culture and tissue repair. Biochem Pharmacol [Suppl] 17:241–257

Baker JB, Lou DA, Simmer RL, Cunningham DD (1980) Protease nexin: a cellular component that links thrombin and plasminogen activator and mediates their binding to cells. Cell 21:37–45

Balian G, Click EM, Crouch E, Davidson JM, Bornstein P (1979) Isolation of a collagen-binding fragment from fibronectin and cold-insoluble globulin. J Biol Chem 254:1429–1432
Barnhart MI, Riddle JM (1963) Cellular localization of profibrinolysin (plasminogen). Blood 21:306–311
Bergquist D, Ljungher HA (1981) A comparative study of Dextran 70 and the sulphated polysaccharide in the prevention of post-operative thromboembolic complication. Br J Surg 68:449–451
Bernik MB, White WF, Oller EP, Kwaan HC (1974) Immunologic identity of plasminogen activator in human urine, heart, blood vessels and tissue culture. J Lab Clin Med 84:546–558
Binder BR, Spragg J, Austen KF (1979) Purification and characterization of human vascular plasminogen activator derived from blood vessel perfusates. J Biol Chem 254:1998–2003
Bogenmann E, Jones PA (1983) role of plasminogen in matrix breakdown by neoplastic cells. INGI 71:1177–1182
Brunetti L, Lotti T (1983) Study of cutaneous fibrinolytic activity after separation of epidermis from dermis. Ital Gen Rev Dermatol 22:111–116
Burnard KG, Clemenson C, Morland N (1980) Venous lipodermatosclerosis treatment by fibrinolytic enhancement and elastic compression. Br Med J 280:7–11
Butler WB, Kirkland WL, Jorgensen TL (1979) Induction of plasminogen activator by estrogen in a human breast cancer cell line (NCF-7). Biochem Biophys Res Commun 90:1328–1334
Camiolo SM, Thorsen S, Astrup T (1971) Fibrinogenolysis and fibrinolysis with tissue plasminogen activator, urokinase, streptokinase-activated human globulin and plasmin. Proc Soc Exp Biol Med 138:277–280
Camiolo SM, Markus G, Evers JL, Hobika GH, Depasquale JL, Beckley S, Grimaldi Jp (1981) Plasminogen activator content of neoplastic and benign human prostate tissue: fibrin augmentation of an activator activity. Int J Cancer 27:191–198
Camiolo SM, Markus G, Englander LS, Siuta MR, Hobika GH, Kohga S (1984) Plasminogen activator content and secretion in explants of neoplastic and benign human prostate tissue. Cancer Res 44:311–318
Casslen B, Ohlsson K (1981) Cyclic variation of plasminogen activator in human uterine fluid and the influence of an intrauterine device. Acta Obstet Gynecol Scand 60:97–101
Castellino FJ, Powell JR (1981) Studies on human proteases. Methods Enzymol 80:365–378
Chimenti M, Barontini A, Lotti T, Fabbri P (1981) Fibrinolytic activity and immunopathological findings in porphyria cutanea tarda. Ital Gen Rev Dermatol 18:163–170
Christman JK, Silverstein SC (1977) Plasminogen activators. In: Barrett AJ (ed) Proteinases in mammalian cells and tissues, vol 2. North-Holland, Amsterdam, pp 90–149
Clemmensen I (1978) Inhibition of urokinase by complex formation with human antithrombin III in absence and presence of heparin. Thromb Haemost 39:616–623
Clemmensen I, Christensen F 81976) Inhibition of urokinase by complex formation with human alpha$_1$-anti trypsin. Biochim Biophys Acta 429:591–599
Cliffton EE, Grossi CE (1956) Effect of human plasmin on toxic effects and growth of blood borne metastasis of Brown-Pearce carcinoma and V2 carcinoma of rabbit. Cancer 9:1147–1152
Cole ER, Bachmann FW (1977) Purification and properties of a plasminogen activator from pig heart. J Biol Chem 252:3729–3737
Collen D (1976) Identification and some properties of a new fast-reacting plasmin inhibitor in human plasma. Eur J Biochem 69:209
Collen D (1980) On the regulation and control of fibrinolysis. Eduard Kowalski Memorial Lecture. Thromb Haemost 43:77–89
Collen D (1983) Mechanism of inhibition of tissue-type plasminogen activator in blood. Thromb Haemost 50:678
Collen D, Stassen JM, Vestraete M (1983) Thrombolysis with human extrinsic (tissue-type) plasminogen activator in rabbits with experimental jugular vein thrombosis. Effect of

molecular form and dose of activator, age of thrombus and route of administration. J Clin Invest 71:368–376
Collen D, Stassen JM, Blaber M, Winkler M, Verstraete M (1984a) Biological and thrombolytic properties of proenzyme and active forms of human urokinase. III. Thrombolytic properties of natural and recombinant urokinase in rabbits with experimental jugular vein thrombosis. Thromb Haemost 52:27–30
Collen D, Stassen JM, Marafino BJ Jr, Builder S, de Cock F, Ogez J, Tajiri D et al. (1984b) Biological properties of human tissue-type plasminogen activator obtained by expression of recombinant DNA in mammalian cells. J Pharmacol Exp Ther 231:146–152
Crawford GMP, Ogston D (1975) The action of antithrombin III on plasmin and activators of plasminogen. Biochim Biophys Acta 391:189
Cunliffe WJ (1968) An association between cutaneous vasculitis and decreased blood fibrinolytic activity. Br J Dermatol 84:99–112
Cunliffe WJ (1978) An association between cutaneous vasculitis and decreased blood fibrinolytic activity. Lancet 1:226
Cunliffe WJ, Menon IS (1969) Blood fibrinolytic activity in diseases of the small blood vessels and the effect of low molecular weight dextran. Br J Dermatol 81:220
Danø K, Reich E (1978) Serine enzymes released by cultured neoplastic cells. J Exp Med 147:745–757
Danø K, Dabelsteen E, Nielsen LS, Kaltoft K, Wilson EL, Zeuthen J (1982) Plasminogen activating enzyme in cultured glioblastoma cells. An immunofluorescence study with monoclonal antibody. J Histochem Cytochem 39:1165–1170
Danø K, Andreasen PA, Grondahl-Hansen J, Kristensen P, Nielsen LS, Skriver L (1985) Plasminogen activators, tissue degradation and cancer. Adv Cancer Res 1(44):139–265
Davidson JF, Walker ID (1979) Synthetic fibrinolytic agents. Prog Cardiovasc Dis 21:375–396
Davidson JF, Lochead M, MacDonald GA, MacNicol GP (1972) Fibrinolytic enhancement by stanozolol: a double blind trial. Br J Haematol 22:543–559
Davies MC, Englert ME, de Renzo EC (1964) Interaction of streptokinase and human plasminogen. I. Combining of streptokinase and plasminogen observed in the ultracentrifuge under a variety of experimental conditions. J Biol Chem 239:2651–2656
Dayer JM, Vassalli JD, Bobbitt JL, Hull RN, Reich E, Krane SM (1981) Calcitonin stimulates plasminogen activator in renal tubular cells. J Cell Biol 91:195–200
Del Rosso M, Dini G, Fibbi G (1985) Receptors for plasminogen activator, urokinase, in normal and Rous sarcoma virus-transformed mouse fibroblasts. Cancer Res 45:630–636
Den Ottolander GJH, Leijnse B, Cremer-Elfrink HNJ (1967) Plasmatic and platelet antiplasmins and antiactivator. I. Thromb Diath Haemorrh 18:404
Dini G, Fibbi G, Pasquali F, del Rosso M (1985) Plasminogen activator: morphological evidence of binding internalization and delivery to lysosomes in 3T3 mouse fibroblast. Histochem J 17:333–341
Dodman D, Cunliffe WJ, Roberts BE, Sibbald R (1973) Clinical and laboratory double blind investigation on effects of fibrinolytic therapy in patients with cutaneous vasculitis. Br Med J 1:82–92
Eaton DL, Baker JB (1983) Phorbol ester and mitogens stimulate human fibroblast secretions of plasmin-activatable plasminogen activator and protease nexin, an antiactivator-antiplasmin. J Cell Biol 97:323–328
Fabbri P, Lotti T, Dindelli A, Petrini N, Cappugi P (1981) Studio della fibrinolisi tissutale e plasmatica nella vasculite cutanea necrotizzante. Ann Ital Dermatol Clin Sper 35:347–354
Fabbri P, Lotti T, Panconesi E (1985) Pathogenesis of pemphigus: role of the plasminogen activator in acantholysis. Int J Dermatol 24:422–426
Fearnley JD (1966) Fibrinolysis. Arnold, London, pp 16–31
Fischer AM, Merton RE, Marsh NA, Willans S, Gaffney PJ, Barrowcliffe TN, Thomas DP (1982) A comparison of pentosan polysulphate and heparin. II. Effects of subcutaneous injection. Thromb Hemost 47:109–113

Fräki JE, Djupsund BM, Hopsu-Havu VK (1978) Plasminogen activators of psoriatic scale extracts. Separation of two plasminogen activators by isoelectric focusing. Arch Dermatol Res 261:259–266
Fräki JE, Briggaman RA, Lazarus GS (1982) Uninvolved skin from psoriatic patients develops signs of involved psoriatic skin after being grafted onto nude mice. Science 215:685–687
Fräki JE, Lazarus GS, Gilgor RS (1983) Correlation of epidermal plasminogen activator activity with disease activity in psoriasis. Br J Dermatol 108:39–44
Friberger P (1982) Chromogenic peptide substrates. Their use for the assay of factors in the fibrinolytic and the plasma kallikrein-kinin systems. Scand J Clin Lab Invest [Suppl 162]42
Gelehrter TD, Barouski-Miller PA, Coleman PL, Cwikel BJ (1983) Hormonal regulation of plasminogen activator in rat hepatoma cells. Mol Cell Biochem 53–54:11–21
Gold HK, Fallon JT, Yasuda T, Leinbach RC, Kahe BA, Newall JB, Guerrero JL et al. (1984) Coronary thrombolysis with recombinant human tissue-type plasminogen activator. Circulation 70:700–707
Golds EE, Ciosek CP Jr, Hamilton JA (1983) Differential release of plasminogen activator and latent collagenase from mononuclear cell stimulated synovial cells. Arthritis Rheum 26:15–21
Goldsmith GH Jr, Saito H, Ratnoff OD (1978) The activation of plasminogen by Hageman factor (factor XII) and Hageman factor fragments. J Clin Invest 62:54–60
Gordon S, Cohn ZA (1978) Bacille Calmette-Guèrin infection in the mouse. Regulation of macrophage plasminogen activator by lymphocytes and specific antigen. J Exp Med 148:1175–1188
Granelli-Piperno A, Reich E (1983a) A study of proteases and protease-inhibitor complexes in biological fluids. J Exp Med 148:223–234
Granelli-Piperno A, Reich E (1983b) Plaminogen activators of the pituitary gland: enzyme characterization and hormonal modulation. J Cell Biol 97:1029–1037
Granelli-Piperno A, Vassalli JD, Reich E (1977) Secretion of plasminogen activator by human polymorphonuclear leukocytes. Modulation by glucocorticoids and other effectors. J Exp Med 146:1693–1706
Green H (1977) Terminal differentiation of cultured human epidermal cells. Cell 11:405–416
Gross JL, Moscatelli D, Jaffe EA, Rifkin DB (1982) Plasminogen activator and collagenase production by cultured capillary endothelial cells. J Cell Biol 95:974–981
Gunzler WA, Steffens GJ, Otting F, Kim SM, Frankus E, Flohe L (1982) The primary structure of high molecular mass urokinase from human urine. The complete amino-acid sequence of the A chain. Hoppe Seylers Z Physiol Chem 363:1155–1165
Gurevrich V, Pannel R, Kelley P (1985) Single-chain urokinase a 5500 M. W. zymogen precursor of urokinase with fibrin-dependent activation under physiological condition. In: Davidson JF, Cooccheri M, Donato MB (eds) progress in fibrinolysis, vol 3. Livingstone, Edinburgh, pp 352–355
Halse T (1962) Aktivierung der Fibrinolyse und Thrombolyse durch Polysaccharid. Schwefelsäureester (Heparin, Herparinoide). Arzneimittelforschung 12:574–451
Hamilton J, Vassalli JD, Reich E (1976) Macrophage plasminogen activator: induction by asbestos is blocked by anti-inflammatory steroids. J Exp Med 144:1689–1694
Hashimoto K, Shafran KM; Webber PS, Lazarus GS, Singer KH (1982) Anti-cell surface pemphigus autoantibody stimulates plasminogen activator activity of human epidermal cell A mechanism for the loss of epidermal cohesion and blister formation. J Exp Med 157:259–272
Hashimoto K, Singer KH, Shafran KM, Lazarus GS (1983) Incubation of human epidermal cell with pemphigus IgG results in an increase in plasminogen activator activity. J Invest Dermatol 81:424–429
Hashimoto K, Singer K, Lazarus GS (1984) Penicillamine induced pemphigus. Immunoglobulin from this patient induces plasminogen activator synthesis by human epidermal cells in culture: mechanism for acantholysis in pemphigus. Arch Dermatol 120:762–764

Haustein UF (1969) Die Lokalisation des Gewebsaktivators der Fibrinolyse bei Dermatosen. Arch Klin Exp Dermatol 234:182–194

Hawiger J (1969) Precipitation of soluble fibrin monomer complexes by lysosomal protein fraction of polymorphonuclear leucocytes. Proc Soc Exp Biol Med 131:349

Heisel M, Laug WE, Jones PA, Stowe SM (1984) Effect of X-irradiation on artificial blood vessel wall degradation by invasive tumor cells. Cancer Res 44:2441–2445

Hilgard P, Hiemeyer V (1970) Fibrinolysis, thromboplastin activity and localization of radioiodinated fibrinogen in experimental tumors. Eur J Cancer 6:157–158

Isseroff RR, Rifkin DB (1982) Plasminogen is present in the basal layer of epidermis. Clin Res 30:605A

Isseroff RR, Rifkin DB (1983) Plasminogen is present in the basal layer of epidermis. J Invest Dermatol 80:297–299

Isseroff RR, Fusening NE, Rifkin DB (1982) Plasminogen activator in differentiating mouse keratinocytes. J Invest Dermatol 80:217–222

Jaken S, Black P (1981 a) Regulation of plasminogen activator in 3T3 cells: effect of phorbol myristate acetate on subcellular distribution and molecular weight. J Cell Biol 90:727–731

Jaken S, Black P (1981 b) Correlation between a specific molecular weight form of plasminogen activator and metabolic activity of 3T3 cells. J Cell Biol 90:721–726

Jarrett PEM, Morland M, Browse NL (1978) Treatment of Raynaud's phenomenon by fibrinolytic enhancements. Br Med J 2:523–525

Jones PA, de Clerck A (1980) Destruction of extracellular matrices containing glycoproteins, elastin, and collagen by metastatic human tumor cells. Cancer Res 40:3222–3227

Kaltoft K, Nielsen LS, Zeuthen J, Danø K (1982) Monoclonal antibody that specifically inhibits a human M_r 52000 plasminogen activating enzyme. Proc Natl Acad Sci USA 79:3720–3723

Kaplan AP, Austen KF (1972) The fibrinolytic pathway of human plasma, isolation and characterization of plasminogen proactivator. J Exp Med 136:1378–1393

Kaplan AP, Castellino FJ, Collen D, Wiman B, Taylor FB Jr (1978) Molecular mechanisms of fibrinolysis in man. Thromb Haemost 39:263–283

Katz J, Troll W, Levy M, Filkins K, Russo J, Levitz M (1976) Estrogen dependent trypsin-like activity in the rat uterus. Localization of activity in the 12000 g pellet and nucleus. Arch Biochim Biophys 173:347–354

Katzenellenbogen BS, Norman MJ, Eckert RL, Peltz SW, Mangel WF (1984) Proactivites, estrogen receptor interactions, and plasminogen activator – inducing activities of tamoxifen and hydroxy tamoxifen isomers in MCF-7 human breast cancer cells. Cancer Res 44:112–119

Klimetzek V, Sorg C (1977) Lymphokine-induced secretion of plasminogen activator by murine macrophages. Eur J Immunol 7:185–187

Kline DL (1971) Significance of fibrinolysis occurring in patient with metastatic cancer. In: Bang NV, Beller FK, Deutsch E, Mammen EF (eds) Thrombosis bleeding disorders. Academic, New York, pp 358–360

Kline DL, Reddy KNN (1977) Proactivator and activator levels of plasminogen in plasma as measured by caseinolysis. J Lab Clin Med 89:1153–1158

Klöcking HP, Markwardt F (1984) Pentosan polysulphate-induced release of plasminogen activator. Haemostasis 14:121

Kristensen O, Larsson LI, Nielsen LS, Grøndahl-Hansen J, Andreasen PA, Danø K (1984) Human endothelial cells contain one type of plasminogen activator. FEBS Lett 168:33–37

Lacroix M, Fritz IB (1977) The control of the synthesis and secretion of plasminogen activator by rat Sertoli cells in culture. Mol Cell Endocrinol 9:227–236

Larsson LI, Skriver L, Nielsen LS, Grøndahl-Hansen J, Kristensen P, Danø K (1984) Distribution of urokinase-type plasminogen activator immunoreactivity in the mouse. J Cell Biol 98:894–903

Lee LS, Weinstein IB (1978) Epidermal growth factor, like phorbol esters, induced plasminogen activator in Hela cells. Nature 274:696–697

Lijnen HR, Collen D (1982) Interaction of plasminogen activators and inhibitors with plasminogen and fibrin. Semin Thromb Hemost 8:2–10
Lin HS, Gordon S (1979) Secretion of plasminogen activator by bone marrow-derived mononuclear phagocytes and its enhancement by colony-stimulating factor. J Exp Med 150:231–245
Ljungner H, Holmberg L, Kjeldgaard A, Nilsson IM, Astedt B (1983) Immunological characterization of plasminogen activator in the human vessel wall. J Clin Pathol 36:1046–1049
Lobel P, Schror K (1985) Selective stimulation of coronary vascular PG I_2 but not of platelet thromboxane formation by defibrotide in the platelet perfused heart. Naunyn Schmiedebergs Arch Pharmacol 331:125
Loskutoff DJ, Edgington TS (1977) Synthesis of a fibrinolytic activator and inhibitor by endothelial cells. Proc Natl Acad Sci USA 74:3903–3907
Loskutoff DJ, von Mourik JA, Erickson LA, Lawrence D (1983) Detection of an unusually stable fibrinolytic inhibitor produced by bovine endothelial cells. Proc Natl Acad Sci USA 80:2956–2960
Lotti T, Fabbri P (1981) Investigation on cutaneous fibrinolytic activity in cumulative irritant contact dermatitis. Ital Gen Rev Dermatol 18(2–3):195–203
Lotti T, Fabbri P (1984) Studio degli attivatori e degli inhibitori della fibrinolisi nel lichen planus. Ital Dermatol Venereol 19:97–103
Lotti T, Fabbri P (1985) Investigation on coagulation and fibrinolysis in Kaposi's sarcoma. In: Cerimele D (ed) Kaposi's sarcoma. Spectrum, New York, pp 117–130
Lotti T, Dindelli A, Barontini A, Fabbri P (1979) Fibrin deposits and fibrinolytic activity in certain dermatoses with immunological pathogenesis. III. Experimental results. Ital Gen Rev Dermatol 16(2–3):197–205
Lotti T, Dindelli A, Fabbri P (1980) Cutaneous fibrinolysis: a modified method of autohistographic identification of plasminogen activators in tissue. Technical note. Ital Gen Rev Dermatol 17:157–160
Lotti T, Brunetti L, Casigliani R, Romani A, Fabbri P (1983) A new technique for detecting inhibitors of fibrinolysis in skin section. Technical note. Ital Gen Rev Dermatol 20:105–109
Lotti T, Cappugi P, Lattari P, Panconesi E (1984a) Increased cutaneous fibrinolytic activity in a case of aquagenic pruritus. Int J Dermatol 23:61–62
Lotti T, Fabbri P, Panconesi E, Angiolini P, del Rosso M, Dini G, Fabbri G (1984b) Urokinase receptors on human keratinocytes. Ital Gen Rev Dermatol 21:101–104
Lotti T, Fabbri P, Brunetti L, Palombi C, Rossi P, Vallecchi C (1985a) Cutaneous plasminogen activator activity in melanoma basal cell and squamous cell carcinoma. Ital Gen Rev Dermatol 22:89–95
Lotti T, Battini ML, Zuccati G, Fabbri P (1985b) L'attività fibrinolitica cutanea nelle porpore infantili di Majocchi e Schamberg. G Ital Dermatol Venereol 2:39–45
Lotti T, Fabbri P, Panconesi E (1985c) Cutaneous fibrinolytic activity in urticaria and vasculitis. In: Champion RH et al. (eds) The urticarias. Livinstone, Edinburgh, pp 161–167
Lotti T, Brunetti L, Panconesi E (1986a) A novel action of anthralin (Letter). Arch Dermatol 122:748–749
Lotti T, Steinman HK, Grevaes MW, Fabbri P, Brunetti L, Panconesi E (1986b) Increased cutaneous fibrinolytic activity in aquagenic pruritus. Int J Dermatol 25:508–511
Lotti T, Fabbri P, Panconesi E (1987a) The pathogenesis of venous ulcers (Letter). J Am Acad Dermatol 16:877–879
Lotti T, Fabbri P, Bonan P, Panconesi E (1987b) Evaluation of the cutaneous plasminogen activators in pemphigus vulgaris. Technical note. Ital Gen Rev Dermatol 24:75–79
Lotti T, Bonan P, Panconesi E (1988) Epidermal plasminogen activator activity tPA dependent is a marker of disease activity in psoriasis (Letter). J Invest Dermatol 90:86–87
Low DA, Baker JB, Koonee WC, Cunningham DD (1981) Released protease-nexin regulates cellular binding internalization, and degradation of serine proteases. Proc Natl Acad Sci USA 78:2340–2344
Mackie M, Booth NA, Bennett B (1981) Comparative studies on human activators of plasminogen. Br J Haematol 47:77–90

Markus G, Werkheiser WC (1964) The interaction of streptokinase with plasminogen. I. Functional properties of the activated enzyme. J Biol Chem 239:2637–2643
Marrack D, Kubala M, Corry P, Leavens M, Howze J, Dewey W, Bale WF, Spar IL (1967) Localization of intracranial tumors. Comparative study with ^{131}-I-labeled antibody to human fibrinogen and neohydrin-^{203}Hg. Cancer 20:751–755
Matsuo O, Rijken DC, Collen D (1981) Thrombolysis by human tissue plasminogen activator and urokinase in rabbits with experimental pulmonary embolus. Nature 291:590–591
Mattsson C, Nyberg-Arrhenius V, Wallen P (1981) Dissolution of thrombi by tissue plasminogen activator urokinase and streptokinase in an artificial circulating system. Throm Res 21:535–545
Menon IS, Cunliffe WJ (1969) Phenformin and ethyloestrenol for Raynaud's disease. Lancet 2:1135–1149
Mira-Y-Lopez R, Reich E, Ossowski L (1983) Modulation of plasminogen activator in rodent mammary tumors by hormones and other effectors. Cancer Res 43:5467–5477
Miwa N, Takavanagi S, Suzuki A (1981) Comparative studies on two active enzyme forms of human urinary urokinase. I. Purification by serial column chromatography and homogeneity analyses of molecular weight and isoelectric point. Chem Pharm Bull (Tokyo) 29:463–671
Morioka S, Jensen PJ, Lazarus GS (1984) Localization of human keratinocyte plasminogen activator with anti-urokinase antibody (Abstr). J Invest Dermatol 82:393
Moudy GH (1982) The source of plasminogen activator in human saliva. Arch Oral Biol 27:33–37
Müllertz S, Lassen M (1953) An activator system in blood indispensable for formation of plasmin by streptokinase. Proc Soc Exp Biol Med 82:264–268
Nagy B, Ban J, Bardar B (1977) Fibrinolysis associated with human neoplasia. Production of plasminogen activator by human tumors. Int J Cancer 19:614–620
Neri-Serneri GG (1981) Malattie emorragiche e trombotiche. SEU Rome, pp 173–177
Neumann C, Sorg C (1983) Regulation of plasminogen activator secretion by interferon induction and proliferation in murine macrophages. Eur J Immunol 13:143–147
Newstead GL, Griffiths JD, Salsbury AJ (1976) Fibrinolytic activity of carcinoma of the colorectum. Surg Gynecol Obstet 143:61–64
Niada R, Mantovani M, Prino G, Pescador R, Berti F, Omini C, Foleo GC (1981) Antithrombotic activity of a polydesoxyribonucleotidic extracts from organs: a possible link with prostacyclin. Thromb Res 23:233
Nielsen LJ, Hansen JG, Skriver L, Wilson EL, Kalfoft K, Zeuthen J, Danø K (1982) Purification of zymogen to plasminogen activator from human glioblastoma cells by affinity chromatography with monoclonal antibody. Biochemistry 24:6410–6415
Niewiarowski S, Goldstein S (1973) Interaction of cultured human fibroblast with fibrin: modification by drugs and aging in vitro. J Lab Clin Med 82:605–610
Noordhoek-Hegt V, Brakmam P (1974a) Histochemical study of an inhibitor of fibrinolysis in the human arterial wall. Nature 248:75–76
Noordhoek-Hegt V, Brakman P (1974b) Inhibition of fibrinolysis by the human vascular wall related to the presence of smooth muscle cells. Haemostasis 3:118–128
Ofosu FA, Hodi GJ, Smith LM, Cerkus AL, Hirsh J, Blajchman MA (1984) Heparan sulfate and dermatan-sulfate inhibit the generation of thrombin activity in plasma by complementary pathways. Blood 54:727–747
Ogawa N, Yamamoto M, Katamine T, Tajima H (1975) Purification and some properties of urokinase. Thromb Diath Haemorrh 34:194–209
Ogston D, Bennet NB, Herbert RJ, Douglas AS (1973) The inhibition of urokinase by alfa-2-macroglobulin. Clin Sci 44:73–79
Ogston D, Bennet NB, Mackie M (1976) Properties of a partially purified preparation of a circulating plasminogen activator. Thromb Res 8:275
Ossowski L, Biegel D, Reich E (1979) Mammary plasminogen activator correlation with involution, hormonal modulation and comparison between normal and neoplastic tissue. Cell 16:929–940
Panconesi E, Fabbri P, Lotti T, Troiano G (1982) Cutaneous fibrinolytic activity in bullous diseases. Ital Gen Rev Dermatol 19:19–28

Peltz SW, Katzenellenbogen BS, Kneifel MA, Mangel WF (1983) Plasminogen activators in tissues of the immature and estrogen stimulated rat uetrus and in uterine luminal fluid: characterization and properties. Endocrinology 112:890–897
Pennica D, Collen D, Holmes WE, Kohr WJ, Harkin SRN, Yehar GA, Ward CA et al. (1983) Cloning and expression of human tissue type plasminogen activator c-DNA in *E. coli*. Nature 301:214–221
Plow EF, Edginton TS (1973) Discriminating neoantigenic differences between fibrinogen and fibrin derivatives. Proc Natl Acad Sci USA 70:1169–1175
Plow EF, Edginton TS (1975) An alternative pathway for fibrinolysis. I. The cleavage of fibrinogen by leukocyte proteases at physiologic pH. J Clin Invest 56:30–38
Plow EF, Hougie C, Edgington TS (1971) Neoantigenic expression engendered by plasmin cleavage of fibrinogen. J Immunol 197:1455–1463
Poggi A, Polentarutti N, Donati MB, de Gaetano G, Garattini S (1977) Blood coagulation changes in mice bearing Lewis lung carcinoma, a metastasizing tumor. Cancer Res 37:272–277
Pollack R, Rifkin D (1975) Actin-containing tables within anchorage-dependent rat embryo cells are dissociated by plasmin and trypsin. Cell 6:495–506
Prowse CV, Cash JD (1984) Physiologic and pharmacologic enhancements of fibrinolysis. Semin Thromb Hemost 10:51–60
Quigley JP (1979) Proteolytic enzymes of normal and malignant cells. In: Hynes RD (ed) Surfaces of normal and malignant cells. Wiley, New York, pp 247–285
Ranby M, Bergsdorf N, Pohl G, Wallen P (1982a) Isolation of two variants of native one-chain tissue plasminogen activator. FEBS Lett 146:289–292
Ranby M, Bergsdorf N, Nilsson T (1982b) Enzymatic properties of the one- and two-chain form of tissue plasminogen activator. Thromb Res 27:175–183
Reddy KNN, Markus G (1972) Mechanism of activation of human plasminogen by streptokinase. Presence of active center in streptokinase-plasminogen complex. J Biol Chem 247:1683–1691
Reich E (1978) Activation of plasminogen: a general mechanism for producing extracellular proteolysis. In: Berlin RD, Herrman H, Lepow IH, Tanzer JM (eds) Molecular basis of biological degradative processes. Academic, New York, pp 155–169
Rifkin DB (1978) Plasminogen activator synthesis by cultured human embryonic lung cells: characterization of the suppressive effect of corticosteroids. J Cell Physiol 97:421–428
Rifkin DB, Crowe RM (1980) Studies on the control of plasminogen activators production by cultured human embryonic lung cells: requirements for inhibition by corticosteroids. J Cell Physiol 105:417–422
Rifkin DB, Loeb JN, Moore G, Reich E (1974) Properties of plasminogen activators formed by neoplastic human cell in culture. J Exp Med 139:1317–1328
Rijken DC, Collen D (1981) Purification and characterization of the plasminogen activator secreted by human melanoma cells in culture. J Biol Chem 256:7035–7041
Rijken DC, Wijangaa RDS, Welbergen J (1980) Relationship between tissue plasminogen activator and the activators in blood and vascular wall. Thromb Res 18:815–830
Rijken DC, Wijangaa RDS, Welbergen J (1981) Immunological characterization of plasminogen activator activities in human tissues and body fluids. J Lab Clin Med 97:477–486
Robbins KC, Summaria L (1970) The structure of human plasminogen. Methods Enzymol 19:184–199
Robbins KC, Summaria L, Hsieh B, Shah RJ (1967) The peptide chains of human plasminogen to plasmin. J Biol Chem 242:2333–2342
Ryan TJ (1973) Inflammation, fibrin and fibrinolysis. In: Jarrett E (ed) The physiology and pathophysiology of the skin, vol 2. Academic, London, pp 445–477
Ryan TJ (1976) Microvascular injury. Saunders, Oxford, pp 317–319
Ryan TJ (1980) Microcirculation in psoriasis and tissue fluid. Pharmacol Ther 10:27
Saba HI, Roberts HR, Helion JC (1967) The anticoagulant activity of lysosomal cationic proteins from polymorphonuclear leukocytes. J Clin Invest 46:580
Saba HI, Roberts HR, Helion JC (1968) Antiheparin activity of lysosomal cationic proteins from polymorphonuclear leukocytes. Blood 31:369

Saito H, Hamilton SM, Tavill AS, Louis L, Ratnoff OD (1980) Production and release of plasminogen by isolated perfused rat liver. Proc Natl Acad Sci USA 77:6837–6840
Saksela O (1981) Radial caseinolysis in agarose: a simple method for detection of plasminogen activator in the presence of inhibitory substances and serum. Anal Biochem 111:276–282
Sasahara AA, Ho DD, Sharma GVRK (1979) When and how to use fibrinolytic agents. Drug Ther 9:111–128
Schroder EW, Chou IN, Black PH (1980) Effects of retinoic acid on plasminogen activator and mitogenic responses of cultured mouse cells. Cancer Res 40:3089–3094
Scott RW, Baker JB (1983) Purification of human protease nexin. J Biol Chem 258:10438–10444
Searls DB (1980) An improved colorimetric assay for plasminogen activator. Anal Biochem 107:64–70
Shepro D, Schleef R, Hechtman HB (1980) Plasminogen activator activity by cultured bovine aortic endothelial cells. Life Sci 26:415–422
Singer KH, Lazarus GS (1983) Molecular mechanism of pathogenesis. Clin Dermatol 1(2):196–221
Skriver L, Larsson LI, Kielberg U, Nielsen LS, Andreasen PB, Kristensen P, Danø K (1984) Immunocytochemical localization of urokinase type plasminogen activator in Lewis lung carcinoma. J Cell Biol 99:752–757
Soreq H, Miskin R (1983) Modulation in the levels and localization of plasminogen activator in differentiating neuroblastoma cells. Dev Brain Res 11:149–158
Strassburger W, Wollmer A, Pitts JE, Glover ID, Tickle IJ, Blundell TL, Steffens GJ et al. (1983) Adaptation of plasminogen activator sequences to known protease structures. FEBS Lett 157:219–223
Summaria L, Wohl RC, Boreisha IG, Robbins KC (1982) A virgin enzyme derived from human plasminogen. Specific cleavage of the arginyl-560-valy$_n$ peptide bond in the diisopropoxyphosphinyl virgin enzyme by plasminogen activators. Biochemistry 21:2056–2059
Svanberg L, Linell F, Pandolfi M, Asted B (1975) Plasminogen activators in ovarian tumours. Acta Pathol Microbiol Scand [A] 83:193–198
Thorsen S, Müllertz S (1974) Rate of activation and electrophoretic mobility of unmodified and partially degraded plasminogen. Effects of 6-aminohexanoic acid and related compounds. Scand J Clin Lab Invest 34:167–176
Todd AS (1959) The histological localization of fibrinolysis activator. J Pathol Bacteriol 78:281–283
Toki N, Tsushima H, Yamasaki M, Yamasaki R, Yamura T (1982) Isolation of tissue plasminogen activator from skin lesions with allergic vasculitis. J Invest Dermatol 78:18–23
Troll W, Rossman T, Katz J, Levitz M, Sugimura T (1975) Tumorigenesis in mouse skin: inhibition by synthetic inhibitors of proteases. In: Reich E, Rifkin DB, Show E (eds) Proteases and biological control. Cold Spring Harbor Laboratories, Cold Spring Harbor, pp 977–987
Turner RH, Kurban AK, Ryan TJ (1969 a) Inhibition of fibrinolysis by human epidermis. Nature 223:841–843
Turner RH, Kurban AK, Ryan TJ (1969 b) Fibrinolytic activity in human skin following epidermal injury. J Invest Dermatol 53:458–460
Vaheri A, Ruoslahti E, Mosher DF (1978) Fibroblast surface protein. Ann NY Acad Sci 312:111–116
Vassalli JD, Hamilton J, Reich E (1976) Macrophage plasminogen activator enzyme production by anti-inflammatory steroids, mitotic inhibitors, and cyclic nucleotides. Cell 8:271–281
Vassalli JD, Granelli-Piperno A, Reich E (1978) Macrophage plasminogen activator. In: Horton JE, Tarpley TM, Davies WF (eds) Mechanisms of localized bone loss. Information Retrieval, Washington, pp 201–212
Verheijen JH, Rijken DC, Changer GTG, Preston FE, Kluft C (1985) Decreased level of rapid tissue type plasminogen activator inhibition by oral administration of stanozolol.

In: Davidson JF, Donati MB, Coccheri S (eds) Progress in fibrinolysis. VII. Livingstone, Edinburgh, pp 102–104

Vetterlein D, Calton GJ (1983) Purification of urokinase from complex mixtures using immobilized monoclonal antibody against urokinase light chain. Thromb Haemost 49:24–27

Von Kaulla KN (1963) Chemistry of thrombolysis: fibrinolytic enzymes. Thomas, Springfield, pp 23–37

Wallen P (1977) Activation of plasminogen with urokinase and tissue activator. In: Paoletti R, Sherrys (eds) Thrombosis and urokinase. Academic, New York, pp 91–102

Wallen P, Kok P, Ranby M (1978) The tissue activator of plasminogen. In: Magnusson S, Ottesen B, Foltman B, Danø K, Neurath H (eds) Regulatory proteolytic enzymes and their inhibition. Pergamon, Oxford, pp 127–135

Wallen P, Bergsdorf N, Ranby M (1982) Purification and identification of two structural variants of bovine tissue plasminogen activator by affinity adsorption on fibrin. Biochim Biophys Acta 719:318–328

Wallen P, Pohl G, Bergsdorf M, Ranby M, Ny T, Jörnvall H (1983) Purification and characterization of a melanoma cell plasminogen activator. Eur J Biochem 132:681–686

Werb Z (1978) Biochemical actions of glucocorticoids on macrophages in culture specific inhibition of elastase, collagenase and plasminogen activator secretion and effects on other metabolic function. J Exp Med 147:1695–1712

White WF, Barlow GH, Hozen MN (1966) The isolation and characterization of plasminogen activators (urokinase) from human urine. Biochemistry 5:2160–2169

Wigler M, Weinstein TB (1976) Tumor promotor induces plasminogen activator. Nature 259:232–233

Wilson EL, Dowdle E (1978) Secretion of plasminogen activator by normal, reactive and neoplastic human tissue cultured in vitro. Int J Cancer 22:390–399

Wilson EL, Reich E (1978) Plasminogen activator in chick fibroblasts: induction of synthesis by retinoic acid synergism with viral transformation and phorbol ester. Cell 15:385–399

Wilson EL, Jacobs P, Dowdle EB (1983) The secretion of plasminogen activator by human myeloid leukemic cells in vitro. Blood 61:568–574

Wun TC, Ossowski L, Reich E (1982) A proenzyme form of human urokinase. J Biol Chem 257:7262–7268

Yuen P, Kwaan HC (1983) Fibrinolytic activity in human tumor tissue. Cancer Invest 1:369–378

CHAPTER 23

Non-steroidal Anti-inflammatory Agents and the Skin

M. W. GREAVES and S. SHUSTER

A. Introduction

The development of topical non-steroidal anti-inflammatory drugs (NSAIDs), devoid of the unwanted effects of corticosteroids but with equal or even greater potency, has been an unattained ideal for many years. Considerable impetus was given to the search by the discovery by VANE (1971) that aspirin and other NSAIDs owed their effectiveness, at least in part, to their in vitro ability to suppress enzymatic transformation of arachidonic acid to its prostaglandin metabolites PGD_2, PGI_2 and PGE_2 in a cell-free system, in concentrations which were of the same order of magnitude as those achieved therapeutically. This unique and specific inhibition occurs in almost all species and tissues examined. One useful by-product of Vane's findings has been that NSAIDs can be used to explore the role of prostaglandins in a wide range of pathophysiological processes.

Since the 1970s a number of inhibitors of rate-controlling enzymes involved in eicosanoid formation have emerged, including benoxaprofen (WALKER and DAWSON 1979; HARVEY et al. 1983) and, more recently, a chloronaphthalene derivative, lonapalene (LASSUS and FORSTROM 1985). Both these drugs are lipoxygenase inhibitors and both are reported to have effects on the lesions of psoriasis (KRAGBALLE and HERLIN 1983; LASSUS and FORSTROM 1985).

However, despite the increasing numbers of new drugs with selective inhibitory actions, the pharmacological complexity of inflammation, with its multiple interacting molecular pathways, and the evidence for nonenzymic formation of vasoactive agents, for example by free radical activity generated in anthralin inflammation (LAWRENCE and SHUSTER 1987), argue in favour of development of drugs which inhibit end-organ responses (GREAVES 1988). Since many of these topics are discussed in other sections, our aim in this chapter is briefly to draw together the more therapeutically practical aspects.

B. Aspirin, Indomethacin and Related Cyclo-Oxygenase Inhibitors

NSAIDs inhibit biosynthesis of prostaglandins by skin in vitro, but skin cyclo-oxygenase generally shows a low sensitivity to these drugs. *Aspirin*, for example, has an IC_{50} of 2 µm for platelet cyclo-oxygenase, but the corresponding value for a skin microsomal cyclo-oxygenase preparation is 60 µm (ZIBOH 1973; SMITH and WILLIS 1971). However, prior administration of aspirin in man does inhibit ex-

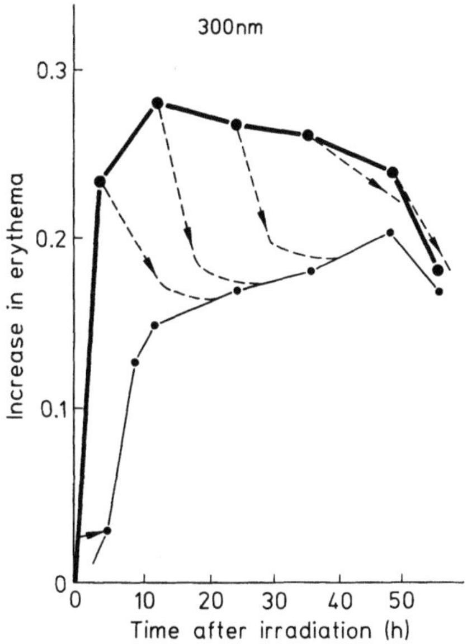

Fig. 1. The time course of UV-B erythema and the inhibitory effect of topical indomethacin on the early phase. The *upper curve* is the time course of UV-B erythema. The *lower curve* shows that when indomethacin is applied immediately after irradiation there is almost complete inhibition of the early phase, but little or no effect on the late phase. Consequently the inhibitory effect of indomethacin is dependent upon the time of application after UV-B exposure, as shown by the *arrow-marked dotted lines*, demonstrating that the second phase of the UV-B response is independent of the first. (Adapted from FARR and DIFFEY 1986)

perimental skin inflammation due to topical application of tetrahydrofurfuryl nicotinate (trafuril), and also reduces the concentrations of prostaglandins in the inflamed skin (PLUMMER et al. 1977). Oral aspirin, administered concurrently with H_1 and H_2 antihistamines, is reported to control the diarrhoea and systemic symptoms of mastocytosis (ROBERTS et al. 1980). These manifestations have been attributed to increased production of PGD_2 by the mast cell infiltrates, and aspirin reduces the urinary excretion of PGD_2 and its metabolites in parallel with clinical improvement. Apart from these special instances, aspirin is generally ineffective in inflammatory skin disorders, and indeed may be deleterious in some (see below).

Both topical and systemically administered *indomethacin* suppress the early phase of the erythemal reaction to ultraviolet B (UV-B) 290–320 nm radiation in human skin, a reaction known to be associated with increased concentrations of prostaglandins in the skin (KOBZA BLACK et al. 1978, 1980 a). Failure to suppress the later phase of UV-B erythema (Fig. 1) has been demonstrated by application of indomethacin at different times after UV-B irradiation (FARR and DIFFEY 1986) and may be explained by the appearance of proinflammatory lipoxygenase products at this stage (KOBZA BLACK et al. 1985). A similar inhibition of early UV-B erythema can been produced by topical application of other NSAIDs including flurbiprofen (KOBZA BLACK et al. 1980 b). Topical indomethacin has a prolonged inhibitory effect on cyclo-oxygenase activity in skin, and the early phase of UV-B erythema is inhibited significantly for up to 5 days after a single application (P. Farr, personal communication). There is no convincing evidence that indomethacin is capable of inhibiting the inflammation of inflammatory dermatoses, even though prostaglandins have been recovered in increased amounts from eczematous skin (GREAVES et al. 1970; GOLDYNE et al. 1973; BARR et al. 1984).

By contrast, indomethacin has a deleterious effect on a number of dermatoses. Idiopathic urticaria is frequently worsened by aspirin and other cyclo-oxygenase inhibitors. This is not due to increased release or augmentation of the effects of histamine (HUMPHREYS et al. 1987). Indomethacin applied topically *after* UV-B irradiation has been shown greatly to enhance the erythemal response in patients with certain photodermatoses and, in a clinically distinct subset of patients with polymorphic light eruption, will actually produce the urticated erythema of the disorder (FARR and DIFFEY 1988). Whether this is due to cyclo-oxygenase substrate diversion is not yet clear. There have been reports of worsening of some other dermatoses, for example dermatitis herpetiformis, but studies suggesting that psoriasis might be exacerbated by topical indomethacin (KATAYAMA and KAWADA 1981; ELLIS et al. 1983) have not been confirmed (GREEN and SHUSTER 1987).

C. Lipoxygenase Inhibitors

Increased concentrations of the 5-lipoxygenase product leukotriene B4 (LTB4) have been demonstrated in the skin lesions of psoriasis, allergic contact dermatitis and eosinophilic cellulitis, and the 12-lipoxygenase product 12-HETE has also been found in increased concentrations in psoriasis and UV-B erythema (KOBZA BLACK et al. 1985; HAMMARSTROM et al. 1975; BRAIN et al. 1984; WONG et al. 1984; BARR et al. 1984).

There have been strenuous efforts to develop drugs which inhibit lipoxygenase activity, mostly for use in asthma and arthritis, but most have failed because of low efficacy or toxicity.

I. Benoxaprofen

In 1979 benoxaprofen was reported to be an inhibitor of arachidonic lipoxygenase at concentrations of 10–15 µm/ml and several controlled clinical trials subsequently established its efficacy, given orally, in psoriasis, 75% of patients showing substantial improvement (ALLEN and LITTLEWOOD 1983; KRAGBALLE and HERLIN 1983). Unwanted effects included photosensitivity, onycholysis and hepatotoxicity, especially in elderly patients with impaired renal function, and the drug was withdrawn from the market. Since it has emerged that benoxaprofen is less active as a 5-lipoxygenase inhibitor than was originally believed (MASTERS and MCMILLAN 1984), its mode of action in psoriasis remains an intriguing enigma.

II. Lonapalene

The chloronaphthalene derivative lonapalene is a potent 5-lipoxygenase inhibitor with an IC_{50} ranging from 0.5 to 15 µm for the cyclo-oxygenase or 12-lipoxygenase enzymes. LASSUS and FORSTROM (1985) found 0.5% lonapalene ointment and fluocinolone acetonide 0.025% ointment to be equally effective in treatment of chronic plaque psoriasis. In a recent study, one of us (M.W.G.) found 2% lonapalene ointment to be clinically effective, and the improvement was associated with a selective decrease of LTB4 in the lesional skin (Camp et al. 1988). Although these findings should encourage evaluation of other 5-lipoxygenase inhibitors in psoriasis (lonapalene's proneness to cause local irritancy precludes its commercial development), double-blind studies of other topical lipoxygenase inhibitors have demonstrated little or no effect on psoriasis (G. Cauwenberg, personal communication).

D. Future Developments

Candidate mediators currently attracting attention in psoriasis and other inflammatory dermatoses include the leukotrienes, platelet-activating factor (PAF), the interleukins, free radical products of oxidation, proteases, kinins, neuropeptides, and amines. In the absence of conclusive evidence of the role of any of these, downgrading of end-organ response emerges as an increasingly attractive and pragmatic approach to drug development in this difficult area.

References

Allen BR, Littlewood SM (1983) Benoxaprofen: effect on cutaneous lesions of psoriasis. Br Med J 285:1241–1243

Barr RM, Brain S, Camp RDR et al. 1984) Human allergic and irritant contact dermatitis: levels of arachidonic acid and its metabolites in involved skin. Br J Dermatol 111:23–28

Brain S, Camp RDR, Dowd PM, Black AR, Greaves MW (1984) The release of leukotriene B4-like material in biologically active amounts from the lesional skin of patients with psoriasis. J Invest Dermatol 83:70–73

Camp R, Kobza Black A, Cunningham F, Mallet A, Greaves MW (1988) Pharmacological effects of topical lonapalene in psoriasis (abstract). J Invest Dermatol 90(4):550

Ellis CN, Fallon JD, Heezen JL, Voorhees JJ (1983) Topical indomethacin exacerbates the lesions of psoriasis. J Am Acad Derm 14:39–43

Farr PM, Diffey BL (1986) A quantitative study of the effect of topical indomethacin on cutaneous erythema induced by UVB and UVC erythema. Br J Dermatol 115:453–490

Farr PM, Diffey BL (1988) Augmentation of ultraviolet erythema by indomethacin in actinic prurigo: evidence of mechanism of photosensitivity. Photochem Photobiol 47:413–417

Goldyne ME, Winkelmann RK, Ryan RJ (1973) Prostaglandin activity in human cutaneous inflammation: detection by radioimmunoassay. Prostaglandins 4:737–749

Greaves MW (1988) Inflammation and mediators. Br J Dermatol 119:419–426

Greaves MW, Sondergaard JS, McDonald-Gibson W (1970) Recovery of prostaglandins in human cutaneous inflammation. Br Med J 2:258–260

Green CA, Shuster S (1987) Lack of effect of topical indomethacin on psoriasis. Br J Clin Pharmacol 24:381–384

Hammarstrom S, Hamberg M, Samuelsson B, Duell EA, Staviski M, Voorhees JJ (1975) Increased concentrations of non-esterified arachidonic acid, 12-L-hydroxy-5,8,10,14-eicosatetraenoic acid, prostaglandin E2, and prostaglandin F2 in epidermis of psoriasis. Proc Natl Acad Sci USA 72:5130–5135

Harvey J, Parish H, Hos PPK, Boot JR, Dawson W (1983) The preferential inhibition of 5-lipoxygenase product formation by benoxaprofen. J Pharm Pharmacol 35:44–45

Humphreys F, Krause LB, Shuster S (1987) The effects of astemizole and indomethacin on weal and flare reactions to histamine, 48/80 and house dust mite antigen. Br J Dermatol 116:435

Katayama H, Kawada A (1981) Exacerbation of psoriasis induced by indomethacin. J Dermatol (Tokyo) 8:323–327

Kobza Black A, Greaves MW, Hensby CN, Plummer NA, Warin AP (1978) Effects of indomethacin on prostaglandins E2, F2 alpha and arachidonic acid in human skin 24 h after UV-B and UV-C irradiation. Br J Clin Pharmacol 6:261–266

Kobza Black A, Fincham N, Greaves MW, Hensby CN (1980a) Time course changes in levels of arachidonic acid and prostaglandin D2, E2 and F2 in human skin following ultraviolet B irradiation. Br J Pharmacol 10:453–457

Kobza Black A, Greaves MW, Hensby CN (1980b) The anti-inflammatory and pharmacological effects of topically applied flurbiprofen on human skin 24 h after irradiation. Prostaglandins Med 5:405–413

Kobza Black A, Barr RM, Wong E et al. (1985) Lipoxygenase products of arachidonic acid in human inflamed skin. Br J Clin Pharmacol 20:185–190

Kragballe K, Herlin T (1983) Benoxaprofen improves psoriasis. A double-blind study. Arch Dermatol 119:548–552

Lassus A, Forstrom S (1985) A dimethoxynaphthalene derivative (RSO43179 gel) compared with 0.025% fluocinolone acetonide gel in the treatment of psoriasis. Br J Dermatol 113:103–106

Lawrence CM, Shuster S (1987) Effect of arachidonic acid on anthralin inflammation. Br J Clin Pharmacol 24:125–131

Masters DJ, McMillan RM (1984) 5-Lipoxygenase from human leucocytes (abstract). Br J Pharmacol 81:70

Plummer NA, Hensby CN, Kobza Black A, Greaves MW (1977) Prostaglandin activity in sustained inflammation of human skin before and after aspirin. Clin Sci Mol Med 52:615–620

Roberts LJ, Sweetman BJ, Lewis RA, Folarin VF, Austen KF, Oates JA (1980) Increased production of prostaglandin D2 in patients with systemic mastocytosis. N Engl J Med 303:1400–1404

Smith JB, Willis AL (1971) Aspirin selectivity inhibits prostaglandin production in human platelets. Nature (New Biol) 231:232–235

Vane JR (1971) Inhibition of prostaglandin synthesis as a mechanism of action for aspirin-like drugs. Nature (New Biol) 231, 232–235

Walker LR, Dawson W (1979) Inhibition of rabbit PMN lipoxygenase activity by benoxaprofen. J Pharm Pharmacol 31:278–279

Wong E, Greaves MW, O'Brien T (1984) Increased concentrations of immunoreactive leukotrienes in cutaneous lesions of eosinophilic cellulitis. Br J Dermatol 110:653–656

Ziboh VA (1973) Biosynthesis of prostaglandin E2 in human skin: subcellular localisation and inhbition by unsaturated fatty acids and anti-inflammatory drugs. J Lipid Res 14:377–384

CHAPTER 24

Immunosuppressive (Cytotoxic) and Immunostimulant Drugs

P. Goldsmith, J. L. Burton, and Richard C. D. Staughton

A. Introduction

Immunological mechanisms play a part in the pathogenesis of many dermatological diseases, but our knowledge of the production of immunological damage is limited, and until the recent introduction of cyclosporin most of the drugs used to modify these processes had an indirect effect on immune responses and inflammation. Thus many immunosuppressant drugs are cytostatic and some also have antiinflammatory effects. The beneficial effect of a drug on an autoimmune disease does not, therefore, necessarily imply that the drug is acting by a specific effect on the immune response. For instance, the cyclooxygenase inhibitor, indomethacin, will prolong survival of skin grafts in the guinea pig (Coppola et al. 1970).

Some diseases respond better to one immunosuppressant drug than another, but until there is better understanding of disease pathogenesis and the action of immunosuppressive drugs, it will not be possible to extrapolate from one disease to another. The use of these drugs in dermatology is still to some extent experimental, and more controlled studies are needed both to show which treatment is best for which disease and to provide clues to mode of action.

B. Corticosteroids

Synthetic glucocorticosteroids are the most widely used immunosuppressive agents in dermatology. They are potent and have the advantage of being rapid in onset of action.

I. Mode of Action

Corticosteroids have many effects (see Chaps. 19, 20 for discussion of steroid pharmacology and lipocortin and Christophers et al. 1988 for a review of their action on the skin) but their immunosuppressive action is not fully understood.

1. Leukocyte Distribution (Cooper and Penny 1985)

A single dose of cortisol will greatly reduce circulating lymphocytes, monocytes and eosinophils in 4–6 h, the effect lasting for about 24 h. This is due to an intravascular redistribution of cells. Corticosteroids also inhibit eosinophil chemotaxis and adherence, and thereby inhibit the effectiveness of eosinophils in hypersensitivity reactions. Following a single dose of cortisol the total granulocyte pool is increased by about 4000/mm^3 due to an increased bone marrow release and inhibition of white cell egress from the blood stream. Granulocytes are prevented from accumulating in the tissues and macrophage mobility is also inhibited, these effects combine to reduce the inflammatory response. Prevention of tissue granulocyte accumulation has an anti-inflammatory effect, as does the inhibition of macrophage mobility.

2. Humoral Effects

Short courses of moderate doses of steroids do not alter immunoglobulin levels, but high doses decrease the serum concentration of IgG, IgA, IgM and immune complexes, the lowest levels occurring after 2–3 weeks and persisting for at least 3 months. Turnover studies suggest that the effect is partly due to decreased synthesis.

3. Cell-mediated Immunity

Impaired lymphocyte transformation and release of lymphokines contribute to the impairment of immune activity.

II. Administration

Although the initial dose may need to be high, to balance the immunosuppressive effect aginst side effects a miserly dosage is used for maintenance. Once the condition has been controlled a single morning dose is used if possible on alternate days, or less frequently. Adrenal suppression can be reduced in this way, though the total dosage remains the same. Pulse therapy with large doses of steroid, e.g. 1000–2000 mg prednisolone or an equivalent intravenous dose, has been used for an additional immunosuppressive effect. The efficacy of this regimen has not been fully proven, although it may decrease serum immunoglobulins and cause disaggregation of immune complexes.

III. Adverse Effects

Adverse effects include cataracts and glaucoma, diabetes, suppression of the hypothalamic-pituitary-adrenal axis, growth retardation, osteoporosis, aseptic necrosis of bone, peptic ulcer, increased susceptibility to infection, weight gain, altered fat distribution and cushingoid appearance, collagen loss with striae and purpura, acne, steroid psychosis and hypercatabolism.

IV. Therapeutic Use

Systemic corticosteroids are used in many dermatological conditions, especially in certain autoimmune disorders.

1. Pemphigus and Pemphigoid

In limited disease 40–60 mg prednisolone per day is usually adequate, but in more severe disease 100–200 mg daily may be required. Other immunosuppressive drugs take several weeks to have their clinical effect (Pasricha 1975), and they are used to reduce the maintenance dose of steroids.

2. Dermatomyositis

Initially 60–120 mg of prednisolone is needed daily; thereafter the dose can be related to clinical response and muscle enzyme levels.

3. Systemic Lupus Erythematosus (Walport and Hughes 1981)

Severe systemic lupus erythematosus with renal involvement and failure to respond to antimalarials is the usual indication for corticosteroid therapy. Low doses of prednisolone may be adequate, but if renal damage progresses, 60 mg or more may be needed (Bresnahan 1979), together with other immunosuppressive drugs.

4. Pyoderma Gangrenosum

Moderate doses of prednisolone of 30–50 mg daily, with or without azathioprine, are used in patients with rapidly progressive pyoderma gangrenosum as well as for the treatment of associated diseases such as colitis.

C. Azathioprine

Azathioprine is the most commonly used immunosuppressive drug. It is metabolised to 6-mercaptopurine, which exerts strong cytotoxic and immunosuppressive effects. It is incorporated into nucleic acids, resulting in chromosome breaks and synthesis of fraudulent proteins with inhibition of RNA and DNA synthesis and cell division. Azathioprine suppresses both cell-mediated and antibody-mediated immune responses but has little effect on established transplant rejection or secondary responses (Perper et al. 1971), although it suppresses T cell effects if given early after transplantation (Currey 1971).

I. Pharmacokinetics

Azathioprine is well absorbed after oral administration, and peak plasma levels occur at about 2 h (Elion and Hitchings 1969). It is rapidly converted by erythrocytes to 6-mercaptopurine, which is further metabolised in the liver. The drug is rapidly cleared from the plasma by a combination of degradation, urinary excretion and tissue uptake, and the plasma half life of 6-mercaptopurine in adults is about 45 min (Leo et al. 1968). Azathioprine and 6-mercaptopurine are taken up by most tissues in the body (Elion et al. 1954).

II. Therapeutic Use

Azathioprine is the cytotoxic drug most commonly used as steroid-sparing agent in dermatological conditions.

1. Pemphigus and Pemphigoid

The combination of azathioprine and corticosteroids is thought to be safer than corticosteroids alone; it produces long-term remissions in most patients and possibly a cure in some.

2. Lupus Erythematosus

The place of azathioprine in the management of lupus erythematosus is controversial, and the improved survival of patients with severe lupus nephritis is probably due to its steroid-sparing effect (STEIJNBOCK et al. 1971; CADE et al. 1973a). With the exception of patients with marrow toxicity, azathioprine should be withdrawn slowly to prevent severe disease exacerbation (SHARON et al. 1973).

3. Dermatomyositis and Polymyositis

Although the early response of polymyositis to prednisolone alone and to prednisolone with azathioprine do not differ, after 3 years there is less muscular functional disability with combined therapy.

4. Other Dermatological Diseases

Psoriasis with and without arthropathy, pityriasis rubra pilaris, chronic dermatitis (atopic and other) polyarteritis nodosa, allergic vasculitis, erythema multiforme, pyoderma gangrenosum and Sézary syndrome have all been treated with azathioprine with varying degrees of success. There have been no large placebo-controlled trials (DANTZIG 1974; CORLEY et al. 1966).

III. Dosage

The recommended initial dosage is 2–3 mg/kg daily orally until remission occurs, after which the dose is matched against the clinical response. A smaller dose is used in patients with renal failure.

IV. Adverse Effects

The main problem is bone marrow depression with leukopenia, macrocytic anaemia, thrombocytopaenia, and increased susceptibility to infection. There may also be nausea, vomiting, anorexia, diarrhoea, stomal ulceration, jaundice secondary to bilary statis and/or hepatitis, retinopathy, arthralgia, pulmonary oedema, drug fever with rashes and serum sickness. There is a carcinogenic and teratogenic propensity.

V. Precautions

Patients need to be closely monitored for toxicitiy. Initially a full blood count is done weekly, then twice monthly for the 2nd and 3rd months of tratment, then monthly unless dosage alterations or other changes in therapy are necessary. Since the drug effect may continue for some days after the last dose, the drug should be stopped or the dosage reduced at the first sign of bone marrow depression until there is improvement. Abnormalities in liver function tests in the 1st week of treatment are usually transient, but if they progress, the drug should be stopped.

VI. Drug Interactions

The major catabolic pathway for azathioprine is via xanthine-oxidase. Allopurinol, which inhibits this enzyme, potentiates the effect of azathioprine by a factor of about three.

VII. Mutagenicity and Carcinogenicity

Azathioprine is mutagenic in animals and humans. Chromosomal abnormalities have been found in patients receiving azathioprine and disappear when the drug is stopped. Azathioprine is carcinogenic in animals, but the magnitude of the risk of neoplasia in humans is not known. The patients most at risk are those who have had transplants or previous treatment with alkylating agents (METCALF 1963; ARMSTRONG et al. 1970).

D. Methotrexate

Methotrexate is mainly used in dermatology for its cytotoxic effect on psoriasis (see Chap. 41).

I. Mode of Action

Methotrexate reversibly inhibits dihydrofolate reductase, leading to a reduction in the synthesis of purines, the conversion of deoxyuridylate to thymidylate, the synthesis of DNA and therefore inhibition of cell replication. Tissues with high rates of cellular proliferation such as neoplasms, psoriatic epidermis, bone marrow, intestinal mucosa, hair matrix and foetal cells are the most sensitive to the effects of methotrexate. The immunosuppressive and anti-inflammatory actions of the drug are due to inhibition of responses which depend on cell proliferation (SANTOS et al. 1974; BOREL 1968; GOIHMAN-YAHR et al. 1969).

II. Pharmacokinetics

Methotrexate can be given orally, intramuscularly or intravenously. It has a triphasic plasma disappearance. The initial phase lasts approximately 1 h and represents distribution and binding to tissue proteins, with highest concentrations in the kidneys, gall bladder, spleen, liver and skin. The second phase lasts about 3 h and is predominantly one of renal clearance. Of greater importance, especially with regard to toxicity, is the terminal half-life of 12 h, which is greatly extended should the renal clearance phase be slowed. Toxicity, especially to the gastrointestinal tract and bone marrow, will occur if the plasma concentration of the drug remains about 2×10 *M* for longer than 42 h and this is more important with regard to toxicity than the peak level of methotrexate (SPIRO and DEMIS 1969). Following systemic administration of a single dose of methotrexate, DNA synthesis in the psoriatic epidermis is diminished for 12–16 h.

III. Therapeutic Uses

Methotrexate produces beneficial effects in up to 75% of patients with severe recalcitrant psoriasis. It has also been used with effect in psoriatic arthropathy and dermatomyositis (ROENIGH 1972).

IV. Adverse Effects

There are two groups of adverse effects. (a) The expected of a cytostatic drug: bone marrow depression, leukopaenia, thrombocytopaenia and anaemia; gastrointestinal with stomatitis and ulceration of the entire gastrointestinal tract; immune depression with increased risk of infections, gonadal defects and faulty spermatogenesis, abortion and stillbirth. The erosions which occur in the lesions of psoriasis are usually a sign of overdosage. (b) Drug-specific changes are hepatic fibrosis and cirrhosis; renal failure, usually related to an overdose in patients with existing renal impairment; pulmonary infiltration occurring as a hypersensitivity reaction. Various other side effects are headaches, drowsiness, malaise and photosensitivity.

V. Hepatotoxicity

The use of methotrexate in psoriasis is limited chiefly by hepatotoxicity. Since toxicity is related to the time for which hepatocytes are exposed to methotrexate, rather than the drug concentration at each exposure (DAHL 1971; ROBINSON 1980), the drug is best given weekly in a single dose. It has been suggested that intramuscular is safer than oral administration, but there is no convincing evidence for this. Liver scans and function tests are of limited value in detecting methotrexate hepatotoxicity, and a liver biopsy is usually done before starting treatment and repeated after a cumulative dose of 1.5 g of methotrexate has been given, or earlier if there is evidence of liver abnormality. Methotrexate therapy should be stopped in patients with moderate fibrosis.

VI. Dosage

A test dose of 2.5 mg methotrexate is given orally, and, if no adverse effects develop, 5–25 mg can be given weekly or fortnightly for psoriasis, increments or reductions depending on response.

VII. Contraindications

Treatment is usually precluded by renal impairment, liver disease, alcoholism, peptic ulceration, mental instability or an inability to comply.

E. Cyclophosphamide

Cyclophosphamide is an alkylating agent derived from nitrogen mustard. It is converted by the active metabolites within 1 h of absorption (HILL et al. 1972).

I. Mechanism of Action

Cyclophosphamide is inert, but its metabolites inhibit cell division by cross-linking complementary DNA strands, interfering with the nucleic acid template for new DNA synthesis (ROBERTS 1971). Immune responses are suppressed mainly by depression of cell proliferation, though the capacity of the reticuloendothelial cells to retain antigen may also be impaired; cyclophosphamide can also kill small lymphocytes and suppress antibody production in relatively non-toxic doses by reducing the immunoglobulin synthesis (SANTOS 1974; LEVY 1973). A low dose (1.7 mg/kg daily) which has no effect on antibody production will suppress induction of delayed hypersensitivity. As with most immunosuppressives, reactions in previously sensitised subjects are usually unaffected, although unlike azathioprine, cyclophosphamide has some inhibiting effect on established immune responses (TOWNES et al. 1972). Several reports in animals and man suggest selective suppression of B lymphocytes, and prolongation of graft survival in man also suggests suppression of T cell activity (TURK and POWLTER 1972). In patients with rheumatoid arthritis, cyclophosphamide produces and equal decrease in the number of both B and T cells in the circulation, and in animals the suppression of delayed hypersensitivity is not dependent on its effect on B cells. Although cyclophosphamide has anti-inflammatory effects which are not entirely explained by its antimitotic activity, selective lymphopaenia does occur with doses too low to affect inflammatory reactions (DWYER et al. 1973; FAUCI et al. 1974; STEVENS and WILLOUGHBY 1969; PERPER et al. 1971; CURREY 1971).

II. Pharmacokinetics

Cyclophosphamide is well absorbed after oral administration. Hepatic metabolism is complex. Several active metabolites are formed, some of which are unstable with a half-life of 30 min, but they can be found in the serum for at least 24 h because of gradual release of drug from the tissues. The alkylating metabolites are bound to plasma proteins. About 15% of cyclophosphamide appears in the urine unchanged and a further 50%–60% is excreted by the kidney as metabolites (JAO et al. 1972; BAGLEY et al. 1973). Renal insufficiency therefore increases the effects of the drug.

III. Therapeutic Uses

Cyclophosphamide is used in organ transplantation as an alternative to azathioprine, particularly in patients with liver dysfunction (STARZL et al. 1971). It is also used as a steroid-sparing agent, e.g. in lupus erythematosus.

1. Pemphigus and Pemphigoid (GLICHMAN 1973)

Uncontrolled studies suggest that cyclophosphamide has a steroid-sparing effect. It has been claimed that as little as 50 mg daily can prevent relapse in patients in whom pemphigus and pemphigoid has been controlled with corticosteroids (PASRICHA 1975) and higher doses given every 10 days will suppress the disease with a lesser risk of side effects. The therapeutic effect takes several weeks to occur, but most regard azathioprine to be superior.

2. Lupus Erythematosus

Cyclophosphamide has been used for systemic lupus erythematosus and controlled trials have shown that given alone it produces no clinical improvement but given with corticosteroids it is better than steroids alone (STEINBERG et al. 1971, 1972; MEDVED and MAXWELL 1974; SCHULTZ and MENTER 1971).

3. Other Skin Diseases

There are case reports and uncontrolled studies of cyclophosphamide in many skin disorders, but in the absence of good comparative evidence, and because of the serious adverse effects, cyclosphosphamide is best reserved for patients refractory to corticosteroids and other less toxic agents.

IV. Dosage

Some 2–3 mg/kg daily is given orally until remission occurs; the dose should be less in patients with renal failure.

V. Adverse Effects

The adverse effects are much as would be expected of a cytotoxic drug, with leukopaenia occurring 8–15 days after a single dose, thrombocytopaenia, anaemia, increased risk of infection, alopecia, infertility, teratogenicity and carcinogenicity; there is also anorexia, nausea and vomiting, renal tubular necrosis, interstitial pulmonary fibrosis and inappropriate antidiuretic hormone syndrome. Miscellaneous effects are headache and dizziness, cardiotoxicity and hepatotoxicity (rarely). Sterile haemorrhagic cystitis due to chemical irritation by reactive metabolites occurs in 5%–10% of patients and may lead to bladder fibrosis. Its incidence can be reduced by administration of N-acetylcystine, high fluid intake and frequent voiding. The drug should be stopped with the first indication of dysuria or haematuria.

F. Gold

The two available preparations of gold are sodium aurothiomalate, which is given intramuscularly, and auranofin, the recently introduced oral preparation; in both preparations gold is attached by sulphur atoms to other groups.

I. Mechanism of Action

The anti-inflammatory, anti-arthritic and immunomodulating effects of gold compounds are not understood (GOTTLIEB 1976). Gold seems to concentrate in areas of inflammation, and therefore, as with chloroquine, in vitro assays may not reflect the in situ effects. Gold has long been thought to reduce inflammation by inhibiting lysozomal enzyme action, but there are other effects, including a decrease in cell-mediated immunity and histamine release after antigen challenge (BURGE et al. 1971; GOTTLIEB 1976).

II. Pharmacokinetics

The pharmacokinetics of gold are complicated. Following a single oral dose of auranofin approximately 25% is absorbed, with peak serum concentrations occurring within 2 h and with steady state concentrations after 9–12 weeks. After intramuscular injection, release is slow and irregular and peak serum concentrations occur in 3–6 h. With weekly injections steady-state serum concentrations are reached in 4–8 weeks. However, most studies have shown that whole blood, serum and total body gold concentrations attained with parenteral gold compounds do not correlate well with the therapeutic effects or toxicity. Gold is widely distributed through the body tissues. Elimination is slow in urine, and 30% is excreted in the faeces. Beneficial effects may occur after weeks or months and may persist long after discontinuation. Urinary excretion can be detected for as long as 12–15 months after a cumulative dose of 1 g parenteral gold compound; much smaller amounts of gold are retained during oral therapy with auranofin.

III. Therapeutic Uses

Gold is effective in patients with discoid lupus erythematosus resistant to other treatments and as a steroid-sparing agent in pemphigus (PENNEYS et al. 1976). Beneficial effects have been reported in some patients with psoriatic arthropathy, and gold is a second line agent in the treatment of rheumatoid arthritis.

IV. Adverse Effects

Oral auranofin seems to be less toxic and better tolerated than parenteral aurothiomalate, but gastrointestinal side effects are more common and it is less effective. The main adverse effects of gold therapy are thrombocytopaenia, agranulocytosis and anaemia; various rashes with pruritus, erythema, urticaria, lichenoid eruptions, exfoliative dermatitis and photosensitivity; stomatitis and vaginitis; diarrhoea, abdominal cramps, anorexia, dyspepsia, enterocolitis and toxic hepatitis; rarely, conjunctivitis, iritis and corneal ulceration; and uncommonly, headache, fever, alopecia, interstitial pneumonitis and fibrosis, and peripheral neuropathy. Gold should not be used in patients with renal impairment.

V. Dosage

For *auranofin* the usual initial dose is 6 mg daily, but 3 mg is better tolerated. Therapeutic effects generally appear after 3–4 months. If there has been no response, the dose may be increased to 9 mg daily after 6 months. For sodium *aurothiomalate* the initial dose is 10 mg i.m. followed by 25 mg a week later and 50 mg weekly subsequently until a total dose of 1 g has been given.

VI. Precautions

A full blood count and renal and liver function tests are done before therapy, and a blood count and urinalysis is done fortnightly during treatment.

G. Dapsone

I. Mechanism of Action

Dapsone (see Chap. 39) is a sulphonamide: it is bacteriostatic and is used in the treatment of leprosy and malaria. It probably works in a similar way to other sulphonamides by inhibiting folic acid synthesis in susceptible organisms. It also has an effect on the immune system (MILLIKAN and CONWAY 1974). Under certain experimental conditions dapsone can suppress the Arthus reaction, cell-mediated immunity, immune complex deposition, enzyme release from lysozymes, and complement activation by the alternative pathway, and, perhaps most importantly, it acts on the myeloperoxidase system of polymorphonuclear leukocytes (see Chap. 39).

II. Pharmacokinetics

Dapsone is rapidly absorbed after oral administration and peak serum concentrations occur in 2–8 h with a half-life of 20–30 h. Dapsone is 50%–90% bound to plasma proteins and is widely distributed in body tissues. It is metabolised in the liver and excreted in the urine, 20% of it unchanged.

III. Therapeutic Uses

1. Dermatitis Herpetiformis

Dapsone is the drug of choice for dermatitis herpetiformis. Its mode of action is not understood, but it may be relevant that a large proportion of inflammatory cells infiltrating the papillary dermis are polymorphonuclear leukocytes. Response is virtually invariable and rapid, improvement being obvious to the patient within 48 h, and can therefore be used as a therapeutic test. The suppressive dose varies greatly and is assessed by trial of reduction. Patients usually notice a relapse, with pruritus and sometimes blister formation, within 24–48 h. There is a poorly quantitated tendency to spontaneous regression, and most patients have to continue taking the drug. There is evidence that the dose required for control may be reduced following strict adherence to a gluten-free diet for a year or longer.

2. Other Uses

Dapsone is also used, often after a therapeutic trial, in patients suffering from a variety of conditions, some of which are too rare for formal trials to have been done, e.g. acute febrile neutrophilic dermatosis (Sweet's syndrome), subcorneal pustular dermatosis (Sneddon-Wilkinson), pyoderma gangrenosum, generalised pustular psoriasis especially in infancy, cutaneous vasculitis, polyarteritis nodosa, urticarial vasculitis, linear IgA disease, bullous disease of childhood and pemphigoid. It is rare for the response to dapsone to be as rapid or complete in these conditions as in dermatitis herpetiformis.

IV. Adverse Effects

The main adverse effect of dapsone is haemolysis and methaemoglobinaemia (usually dose-related and reversible), severe haemolysis occurring in patients with glucose-6-phosphate dehydrogenase deficiency; leukopaenia is rare. Peripheral motor neuropathy is rare and usually regresses with drug withdrawal. Dapsone can cause exfoliative dermatitis, toxic erythema and erythema multiforme. Patients may experience anorexia, abdominal pain, nausea and vomiting; cholestatic jaundice and toxic hepatitis occur rarely. Occasionally there may be blurred vision, tinnitus, fever, phototoxicity and drug-induced lupus.

V. Dosage

In the treatment of dermatitis herpetiformis the starting dose of dapsone is usually 50–100 mg daily; the dose is subsequently adjusted to the response and usually ranges from 25 to 400 mg/day.

H. Chloroquine

Chloroquine is a synthetic 4-amino-quinoline derivative. In addition to its antimalarial properties it has anti-inflammatory and immunosuppressive effects.

I. Mechanism of Action

The mechanism of action of chloroquine is not understood, and results in vitro conflict with in vivo findings. Chloroquine binds to the DNA template and stabilises it against heat denaturation (NEBLE et al. 1965). The drug has an inhibitory effect on protein synthesis (THOMPSON and BARTHOLOMEW 1964). In vitro it inhibits complement activation and destabilises lyzosomes. Its effect on delayed hypersensitivity reactions and immunoglobulin synthesis is controversial (SZILAGYNI et al. 1970). Chloroquine binds to melanin, but it is not known whether this explains its protective effect against ultraviolet radiation damage.

II. Pharmacokinetics

Chloroquine is rapidly absorbed from the bowel; peak plasma levels occur in 1–2 h and the drug is then widely distributed in tissues. Animal studies show concentrations of chloroquine in liver, spleen, kidney and lung to be at least 200–700 times higher than those in plasma. Chloroquine is also concentrated in the eye, particularly in the iris and choroid. The drug is partially metabolised, and it and its metabolites are slowly excreted by the kidneys; it has a half-life of 72–120 h.

III. Therapeutic Uses

Chloroquine is used as an adjunct to topical corticosteroids in the treatment of discoid lupus erythematosus (DUBOIS 1977) and as an adjunct to systemic corticosteroid and/or salicylate therapy in the treatment of systemic lupus erythematosus. It has also been used empirically in various photodermatoses.

IV. Dosage

One hundred and fifty milligrams of chloroquine is equivalent to 200 mg of chloroquine sulphate and to 250 mg of chloroquine phosphate. Chloroquine sulphate 200 mg b.i.d. is the usual starting dose, and the dose is reassessed after 6 weeks.

V. Adverse Effects

The main unwanted effect of chloroquine is retinopathy, which may not be reversible; blurred vision secondary to corneal oedema and deposits is reversible. There may be anorexia, nausea, vomiting, cramps and diarrhoea; pruritus, pigmentation of nails, skin and mucous membranes; bleaching of hair, exacerbation of existing psoriasis and, rarely, exfoliative dermatitis. Miscellaneous changes include headache, fatigue, irritability, hypotension and ECG changes. Nerve deafness is rare and occurs after prolonged high dosage; neutropenia and thrombocytopoenia are likewise rare.

VI. Precautions

Optic toxicity lilmits the use of chloroquine. Ophthalmological examinations, including slit lamp, fundoscopic and visual field tests, should be performed before starting the drug and periodically during treatment so that administration can be discontinued immediately if an abnormality is found.

J. Cyclosporin

Cyclosporin is a complex polypeptide isolated from the culture broth of the fungus *Tolypocladium inflatum* Gans, grown originally from specimens taken from a high treeless plateau in Norway. It is a powerful immunosuppressant.

I. Mechanism of Action

Cyclosporin is a novel immunosuppressive drug with a specific action on cytokine-driven T cell proliferation due in part to inhibition of synthesis of interleukin 2 (HESS et al. 1988). There is little effect of haemopoietic or phagocytic function; moreover, resting T cells are also unaffected, so that the increased susceptibility to infection usually seen with immunosuppressants is not observed. The drug is particularly active in suppressing T cell-dependent responses to new antigenic stimuli and has thus proved highly suitable for reducing the reaction to engrafted tissues (kidney, heart, liver and bone marrow). The therapeutic effects of cyclosporin in skin disease cannot yet be attributed to a single pharmacological action (SHUSTER 1988).

II. Pharmacokinetics

Cyclosporin has been employed orally and intravenously. Following oral administration it is slowly and incompletely (20%–30%) absorbed, reaching peak concentrations in the blood and plasma after 3 h with a half-life of about 19 h. Elimination is almost entirely biliary, with only 0.1% excreted in the urine.

III. Therapeutic Uses

1. Graft-Versus-Host Disease

Cyclosporin greatly decreases the severity of graft-versus-host disease (GVHD) after bone marrow transplantation. The incidence of chronic GVHD may be slightly increased, but this probably reflects the increased survival.

2. Psoriasis

The therapeutic effect of cyclosporin on psoriasis was found by chance in the treatment of patients with psoriatic arthropathy and was confirmed in both open and controlled studies (see Chap. 21). There is a report of response to intralesional injections, but a satisfactory preparation for topical usage has yet to be produced.

3. Dermatitis

Open studies using 2–5 mg/kg per day suggest a beneficial effect of cyclosporin in patients with dermatitis, particularly atopic, contact and photodermatitis (see Chap. 32), but as with psoriasis, nephrotoxicity remains a problem in more general use of the drug.

4. Other Skin Diseases

Studies in other skin diseases (see section of skin diseases in SECOND INTERNATIONAL CONGRESS ON CYCLOSPORINE 1988) are limited and as yet uncontrolled. There have been reports of efficacy of cyclosporin in lichen planus, pemphigus, pemphigoid, dermatomyositis, scleroderma, Behcet's syndrome, alopecia areata and alopecia totalis, all diseases with strong evidence of an immune component. The expected benefit in diseases where T helper lymphocytes are found has not been reported, and there are reports of serious exacerbations of Sézary syndrome; however, full studies of these diseases are awaited.

IV. Unwanted Effects

The main unwanted effect of cyclosporin is renal impairment, which is related to dose and duration. Patients with impaired renal function should not be treated, and regular studies of renal glomerular function are required during treatment. Hypertension also occurs and is presumed to be renal. The main unwanted effect in the skin is hirsutism.

K. Thalidomide

Thalidomide is a derivative of glutamic acid originally developed as a narcotic agent. Although there is some evidence that thalidomide acts as an immunosuppressive or lysozome-stabilising agent, the mechanism of action remains to be established. It has an antipruritic action (SHUSTER, personal communication) which is probably related to its sedative activity (see Chap. 7).

I. Therapeutic Uses

Thalidomide has been found to be of use in the treatment of discoid lupus erythematosus (HASPAR 1965; KNOP et al. 1983; NAAFS et al. 1985) which is resistant to chloroquine therapy, nodular prurigo (see Chap. 32), actinic prurigo and polymorphous light eruption. Its main use has been in the treatment of the lepra reaction.

II. Adverse Effects

The main adverse effect of thalidomide is foetal teratogenesis; it also produces dizziness and drowsiness, as well as a sensory peripheral neuropathy which may be irreversible.

III. Dosage

The starting dose is 100 mg in adults, given at night because of drowsiness, and if necessary increased cautiously. Dosages of 400 mg daily frequently produce the sensory abnormalities.

L. Clofazimine

Clofazimine is a famazine dye with antibacterial, antiinflammatory and immunosuppressant actions.

I. Mechanism of Action

Clofazimine binds preferentially to bacterial DNA, inhibiting replication and growth. Its anti-inflammatory action is attributed to inhibition of neutrophil mobility and lymphocyte transformation.

II. Pharmacokinetics

Clofazimine is poorly absorbed from the gut (unless administered as crystals in an oily base), and peak plasma concentrations occur after 6 h. The drug is widely distributed throughout the body and it has the very long tissue half-life of approximately 70 days. Clofazimine does not appear to be significantly metabolised, and most of the drug is excreted unchanged in the faeces.

III. Therapeutic Uses

Clofazimine is one of the agents recommended by the WHO in the treatment of multibacillary leprosy, and it is also used for dapsone-resistant leprosy. It is of particular importance in preventing erythema nodosum leprosum reactions which appear to be the result of damage by circulating immune complexes related to particularly sensitised tissues. Clofazimine has also been used in the treatment of discoid lupus erythematosus and pyoderma gangrenosum.

IV. Adverse Effects

Pink, brown and black discoloration of the skin is seen in 75%–100% of patients and resolves slowly after discontinuation of treatment. The red-brown discoloration of conjunctiva, corneal fluid and lacrimal fluid is also reversible. Abdominal pain, diarrhoea, nausea and vomiting occur in 60% of patients, occasionally with gastrointestinal bleeding and splenic infarction. There may be dizziness, drowsiness, fatigue and headache.

V. Dosage

The usual adult dosage is 50–100 mg daily; larger doses should be given only for short periods.

M. Colchicine

Colchicine is an alkaloid of the autumn crocus and has both anti-inflammatory and antimitotic effects. It appears to reduce the inflammatory response in gout by inhibiting polymorphonuclear leukocyte function, including metabolism, mobility and chemotaxis. Colchicine inhibits cell division in metaphase by interfering with the meiotic spindle. Dissolution of microtubule may be the common mechanism of the drug's antimitotic and anti-inflammatory effects. The antimitotic action is presumably responsible for toxic effects on proliferating tissue such as bone marrow, skin and hair.

I. Pharmacokinetics

Colchicine is rapidly absorbed after oral administration, peak plasma concentrations occurring in 30 min to 2 h. Colchicine is partially metabolised in the liver. The drug and metabolites re-enter the intestinal tract by biliary secretion, and unchanged drug may be reabsorbed from the intestine. This enteropathic recycling probably explains the intestinal manifestations of colchicine poisoning. Most of the drug is excreted in the faeces, with 10%–20% in the urine.

II. Therapeutic Uses

Colchicine was formerly the treatment of choice for gout and is now used in the prophylaxis and teratment of acute familial Mediterranean fever. In view of its action on polymorphonuclear leukocytes the drug has been tried in the neutrophilic dermatoses (SNEDDON-WILKINSON, subcorneal pustular dermatosis, Sweet's syndrome, pyoderma gangrenosum), necrotising vasculitis, Behcet's syndrome and pustular psoriasis; there are a number of favourable case reports.

III. Adverse Effects

Nausea, abdominal discomfort, vomiting and diarrhoea; agranulocytosis, thrombocytopaenia and aplastic anaemia with prolonged administration; loss of body and scalp hair and various dermatoses; peripheral neuropathy. It is not certain whether colchicine is teratogenic.

IV. Dosage

The dosages of colchicine used in dermatological disease have been on average 0.6 mg b.i.d. initially, usually increasing to 2 mg a day unless there are adverse effects.

V. Precautions

The most common side effects are gastrointestinal, and therapy should be stopped if abnormal symptoms occur and should not be instituted until these symptoms have disappeared (usually within 24–48 h). In long-term treatment, regular full blood counts should be done.

N. Levamisole

Levamisole is an antihelminthic with immunostimulant properties.

I. Mode of Action

Levamisole is toxic to a broad range of gastrointestinal and systemic nematodes, paralysing the worms through neuromuscular blockade and inhibition of the muscle enzyme fumerate reductase. Levamisole has been described as an immunostimulant drug, but may only restore a depressed immune response: it does not seem to alter the normal response. The drug stimulates phagocytosis and chemotaxis of polymorphs and macrophages; it also increases protein synthesis by hypofunctional T cells and improves lymphocyte cytotoxicity and E rosette formation. In vivo levamisole restores impaired delayed hypersensitivity (e.g. to tuberculin) in the elderly. The graft-versus-host reaction is accentuated in mice if the donor or recipient is pre-treated with levamisole. The drug has no effect on immunoglobulins and is not toxic to bacteria, viruses or fungi in therapeutic doses.

II. Pharmacokinetics

Levamisole is absorbed from the bowel, and peak blood levels are achieved within 2–4 h. Approximately 95% is metabolised in the liver, and any unchanged drug has an elimination half-life of 4–5 h in man.

III. Therapeutic Uses

Levamisole has no established place in dermatological therapy. However, it has been used experimentally with some reported success in the treatment of aphthous ulcers (Miller 1980), Behcet's disease, viral warts (Helin and Bergh 1975) and herpes simplex infections.

IV. Adverse Effects

Nausea, vomiting, abdominal pain, taste disturbances; euphoria; neutropaenia, thrombocytopaenia and an influenza-like reaction are reversible on drug withdrawal.

V. Dosage

The recommended dose of lavamisole is 150 mg per day, but side effects are fewer if one starts with a lower dose and increases it progressively. Regular blood counts should be done.

O. Other Cytostatic Drugs Used in Dermatology

Bleomycin has been used intralesionally for the treatment of warts, *5-fluorouracil* is used topically for the treatment of early stage epitheliomas, e.g. solar keratoses, and *hydroxyurea* has been employed for the treatment of psoriasis (see Chap. 33).

References

Aberer W, Wolffschreiner E, Stingl G, Wolff K (1987) Azathioprine in the treatment of pemphigus vulgaris. J Am Acad Dermatol 16(3):527–533

Armstrong MYK, Gleichmann E, Gleichmann H, Beldotti L, Andre-Schwarts J, Schwartz RS (1970) Chronic allogeneic diseases and the development of lymphoma. J Exp Med 132:417

Avinoviche R, Loewi G (1970) Comparison of the effects of two cytotoxic drugs and an anti-lymphocyte serum on immune and non-immune inflammation in experimental animals. Ann Rheum Dis 29:32

Axelrod L (1976) Glucocorticoid therapy. Medicine (Baltimore) 55:39–65

Bagley CM, Bostick FW, de Vita VT (1973) Clinical pharmacology of cyclophosphamide. Cancer Res 33:226

Barranco VP (1982) Dapsone: other indications. Int J Dermatol 21:513–520

Bernstein I, Lorinez A (1981) Sulphamides and sulphones in dermatology. Int J Dermatol 20:81–88

Biren CA, Barr RJ (1986) Immunological application of cyclosporin. Arch Dermatol 122:1028–1032

Black IL et al. (1964) Methotrexate therapy in psoriatic arthropathy. Double blind study on 21 patients. J Am Med Athol 189:743

Borel Y (1968) The effect of 6-mercaptopurine and methotrexate on passive delayed hypersensitivity reactions. Int Arch Allergy 33:583

Bresnahan B (1979) Systemic lupus erythematosus. Br J Hosp Med 2:16–25

Burge JJ et al. (1971) Inhibition of the alternative pathway of complement by gold in vitro. J Immunol 120:1625–1630

Cade R et al. (1973 a) Comparison of azathioprine, prednisolone and heparin alone or combined in the treatment of lupus nephritis. Nephron 10:37

Cade R et al. (1973 b) Tissue distribution of azathioprine. Gastroenterology 64:706

Champion RH (1981) Psoriasis and its treatment. Br Med J 282:343

Christophers E, Schopf E, Kligman AM, Stoughton RB (1988) Topical corticosteroid therapy. Raven, New York

Clemence PJ et al. (1974) Effects of cyclophosphamide on B and T lymphocytes in rheumatoid arthritis. Arthritis Rheum 17:347

Cooper DA, Penny R (1985) Glucocorticosteroid action on human immune and inflammatory responses. Clin Immunol Update Rev Physicians 1:1–31
Co-Operating Clinics of the American Rheumatological Association (1973) A controlled trial of gold salt therapy in rheumatoid arthritis. Arthritis Rheum 6:353–358
Coppola ED, Hernstreet GP, Villegas GR (1970) Pharmacological treatment. In: Bevrelli A, Moneco AP (eds) Organ and tissue transplantation. Excerpta Medica Foundation, Amsterdam, p 79
Corley CC et al. (1966) Azathioprine in the treatment of auto-immune disease. Am J Med 41:404
Currey HLF (1971) A comparison of immunosuppressive and anti-inflammatory agents in the rat. Clin Exp Immunol 9:879
Dahl B (1971) Liver damage due to methotrexate in patients with psoriasis. Br Med J 1:625–630
Dahl et al. (1972) Methotrexate toxicity in psoriasis: a comparison of different dose regimes. Br Med J 1:654–656
Dantzig PI (1974) Immunosuppressive and cytotoxic drugs in dermatology. Arch Dermatol 110:393
Dubois EL (1977) Anti-malarials in the management of discoid and systemic lupus erythematosus. Semin Arthritis Rheum 8:33–46
Dwyer JN et al. (1973) Selective suppression of T or B responses in man. Clin Res 21:576
Elion GB (1972) Significance of azathioprine metabolites. Proc R Soc Med 65:257
Elion GB, Hitchings GH (1959) The metabolism in vivo of anti-tumour derivatives of 6-mercaptopurine (Abstr.). Fed Proc 18:872
Elion GB et al. (1954) The fate of 6-mercaptopurine in mice. Ann NY Acad Sci 60:297
Fauci AS et al. (1974) Cyclophosphamide and lymphocyte subpopulations in Wegener's granulomatosis. Arthritis Rheum 17:335
Foley GE et al. (1961) Studies on the mechanism of action of cyclophosphamide. Evidence of activation in vitro and in vivo. Cancer Res 21:57
Glichman FS (1973) Pemphigus vulgaris treatment with cyclophosphamide. Arch Dermatol 107:467
Goihma-Yahr M et al. (1969) Effects of methotrexate upon experimental Misuda reaction. J Invest Derm 53:217
Goodman, Gilman. 7th edition. Chapter: Adrenocorticosteroids
Gottlieb NL et al. (1975) The influence of chrysotherapy on serum protein and immunoglobulin levels, rheumatoid factor and anti-epithelial antibody titres. J Lab Clin Med 86:962–972
Gottlieb NL (1976) Chrysotherapy. Bull Rheum Dis 27:912–917
Grant S, Forham P, Diramondo V (1965) Suppression of 17 hydroxycorticosteroids in plasma and urine by single and divided doses of triamcinome. N Engl J Med 273:115–118
Harper JI, Alan (1982) Use of colchicine in the treatment of Becet's disease. Int J Dermatol 21(9):551–554
Harper JI, Kendra JR, Desai S, Staughton RCD, Barrett AJ, Mobbs JR (1984) Dermatological aspects of the use of cyclosporine for prophylaxis of graft-versus-host disease. Br J Dermatol 110:469–474
Haspar MF (1965) Thalidomide in the treatment of lepra reaction. Clin Pharmacol Ther 6:303
Helin P, Bergh M (1975) Levamisole in the treatment of viral warts. N Engl J Med 291:1311
Hess AD, Esa AH, Colombani PM (1988) Mechanism of action of cyclosporine: effect on cells of the immune system and subsequent events in T cell activation. Transplant Proc [Suppl 2]20:29–40
Hill DL et al. (1972) Enzymatic metabolism of cyclophosphamide and nicotine production of toxic metabolites. Cancer Res 32:658
Horwitz DA (1973) Selected depletion of human circulating B lymphocytes by cyclophosphamide. Fed Proc 32:980
Hunsicker LE (1969) The pharmacology of anti malarials. Arch Intern Med 123:645–649

Hurd ER (1973) Immunosuppressive and anti-inflammatory properties of cyclophosphamide, azathioprine and methotrexate. Arthritis Rheum 16:84

Jao J et al. (1972) Phenobarbital effects on cyclophosphamide kinetics in man. Cancer Res 32:2761

Kersley GD (1968) Methotrexate in connective tissue disease, psoriasis and polyarteritis. Ann Rheum Dis 27:64

Knop J, Bonsmann E, Mapple R, Ludolph A, Matz DR, Mifsvol EJ, Macher E (1983) Thalidomide in the treatment of 60 cases of chronic discoid lupus erythematosus. Br J Dermatol 108:461–466

Krain LS et al. (1972) Cyclophosphamide in treatment of pemphigus vulgaris and bullous pemhigoid. Arch Dermatol 106:657

Lang PG (1979) Sulphones and sulphonimides in dermatology today. J Am Acad Dermatol 1:479–492

Leo et al. (1968) Clinical pharmacological observations on 6-mercaptopurine and 6-methylthiopurine ribonucleoside. Clin Pharmacol Ther 9:180

Levy J (1973) Comparative immunosuppressive effects of azathioprine and cyclophosphamide in the treatment of arthritis. Arthritis Rheum 15:444

Maldyk H, Chwalinski-Sadowski H (1968) Results obtained with cyclophosphamide treatment in psoriatic arthropathy and rheumatoid arthritis. Rheumatologica 6:III

Malkinson FD (1982) Colchine new uses of an old drug. Arch Dermatol 118:453–457

Medved A, Maxwell T (1974) Intermittent cyclosphosphamide in the treatment of pemphigus and pemphigoid. Can Med Assoc J 3:245

Melby JG (1974) Systemic corticosteroid therapy, pharmacology and endocrinological considerations. Ann Intern Med 81:505–512

Metcalf D (1963) The fate of parental preleukaemic cells in leukaemia-susceptible and leukaemia-resistant F_1 hybrid mice. Cancer Res 28:1774

Miller MF (1980) Levamisole in the treatment of Apthous ulcers. Drugs 20:131

Millikan LE, Conway FR (1974) Effect of drugs on the Pillener pathway – dapsone. J Invest Dermatol 64:541

Naafs B (1982) Thalidomide in the treatment of subacute cutaneous lupus erythematosus. Br J Dermatol 107:83–86

Naafs B et al. (1985) Thalidomide therapy: an open trial. Int J Dermatol 24:131–135

Neble HTR et al. (1965) Chloroquine: its mechanism of action upon immune phenomena. Arch Dermatol 92:720–725

Nettesheim P, Hamm (1970) Effect of immunosuppressive agents on retention of antigen in a mouse spleen. Proc Soc Exp Biol Med 133:696

Nicholls T et al. (1965) Diurnal variation in suppression of adrenal function by glucocorticoids. J Clin Endocrinol Metab 25:343–349

Pasricha JS (1975) Treatment of pemphigus with cyclophosphamide. Br J Dermatol 93:573

Penneys NS, Eaglestein WH, Frost P (1976) Management of pemphigus with gold compounds. Arch Dermatol 112:185–187

Perper RJ et al. (1971) The use of a standardised adjuvant arthritis assay to differentiate between anti-inflammatory and immunosuppressive drugs. Proc Soc Exp Biol Med 137:506

Philips FS et al. (1961) Cyclophosphamide and urinary bladder toxicity. Cancer Res 21:1577

Raynor A, Ashai AD (1980) Behcet's disease and treatment with colchicine. J Am Dermatol 2(5):396–400

Roberts JJ (1971) Evidence for the inactivation and repair of mammalian DNA template after alkylation by mustard gas and half mustard gas. Eur J Cancer 7:515

Robinson JK (1980) Methotrexate hepatotoxicity in psoriasis: consideration of liver biopsies at regular intervals. Arch Dermatol 116:413–415

Roenigh H (1972) Use of methotrexate in psoriasis. Arch Dermatol 105:363

Roenigh H, Mabic H (1983) Methotrexate. Semin Dermatol 2(4):231

Santos GW et al. (1974) Effects of selected cytotoxic agents on antibody production in man. Ann NY Acad Sci 114:404

Schultz EJ, Menter MA (1971) Treatment of discoid and subacute systemic lupus with cyclophosphamide. Br J Dermatol 85:60
Second International Congress on Cyclosporine (1988) Dermatology. Transplant Proc [Suppl 4]3:19–113
Sharon E et al. (1973) Exacerbation of systemic lupus erythematosus after withdrawal of azathioprine therapy. N Engl J Med 288:122
Shuster S (1988) Cyclosporine in dermatology. Transplant Proc [Suppl 4]3:19–22
Spiro JM, Demis DJ (1969) Treatment of psoriasis with minute divided oral doses of methotrexate. Arch Dermatol 99:459
Starzl TE et al. (1971) Cyclophosphamide and whole organ transplantation in human beings. Surg Gynaetol Obstet 133:981
Steinberg AD et al. (1971) Cyclophosphamide in lupus nephritis, a controlled trial. Ann Intern Med 75:165
Steinberg AD et al. (1972) Controlled trial of azathioprine, cyclophosphamide and placebo in lupus nephritis. Arthritis Rheum 15:456
Stejnbock M et al. (1971) Azathioprine in the treatment of systemic lupus erythematosus, a controlled trial. Arthritis Rheum 14:639
Stevens JE, Willoughby DA (1969) The anti-inflammatory effects of some immunosuppressive agents. J Pathol 2:367
Szilagyni T et al. (1970) The effect of chloroquine on the antigen antibody reaction. Acta Physiol Acad Sci Hung 38:411–417
Thompson GR, Batholomew L (1964) The effect of chloroquine on antibody production. Univ Mich Med Cent J 30:227–230
Townes AS et al. (1972) Controlled trial of cyclophosphamide in rheumatoid arthritis in 18 months double blind cross over study. Arthritis Rheum 15:129
Turk JL, Powlter LW (1972) Selective depletion of lymphoid tissue by cyclophosamide. Clin Exp Immunol 10:285
Walba A (1980) Therapeutic trial of oral colchicine in psoriasis. Acta Derm Venereol (Stockh) 60:515–520
Walport MJ, Hughes GRV (1981) Systemic lupus erythematosus. SLE Hosp Update 629–642
Weinstein GD (1984) Chemotherapy for psoriasis. Dermatol Clin 1(3):431
Weinstein GD et al. (1973) Archives Dermatol 108:36
Zvaifler NJ, Bernstein N (1962) Chloroquine deposition in ocular tissues. Arthritis Rheum 5:667

CHAPTER 25

Three Generations of Retinoids: Basic Pharmacologic Data, Mode of Action, and Effect on Keratinocyte Proliferation and Differentiation

R. Stadler

A. General Aspects

In the first half of this century, vitamin A and its natural derivatives: retinol, retinal (vitamin A aldehyde), and vitamin A acid, were recognized as playing an essential role in general growth, in the regulation of proliferation and differentiation of epithelial and non-epithelial tissues, in visual processes, and in the capacity of vertebrates to propagate (Wolbach and Howe 1925, 1933).

The hypervitaminosis A syndrome (Straumfjord 1942; Keddie 1948) and its resemblance to various dermatoses was the reason for the therapeutic application of vitamin A and later of all-*trans* retinoic acid (Fry and Schoch 1952; Beer 1962; Stüttgen 1962; Fulton et al. 1968; Frost and Weinstein 1969; Kligman et al. 1969; Fry et al. 1970; Bollag and Ott 1971; Orfanos et al. 1973). Compared with vitamin A, all-*trans*-retinoic acid proved to be highly efficient in patients with acne, psoriasis, genodermatoses, actinic keratoses, and basaliomas. Although the clinical results were encouraging, severe side-effects of the hypervitaminosis A syndrome restricted the further systemic administration of all-*trans*-retinoic acid. The fascinating prospect of treating dermatoses orally led to the development of various synthetic derivatives of vitamin A, the oral retinoids.

By modification of the all-*trans*-retinoic acid molecule which consists mainly of: (a) a cyclic end group; (b) a polyene side chain; and (c) a polar end group, more than 1500 synthetic retinoid derivatives have been developed in the last few years (Bollag 1981) with a partly improved therapeutic efficacy and fewer side-effects as compared with the parent substance (Orfanos 1980, 1982; Gollnick and Orfanos 1983). After 20 years of intensive laboratory and clinical retinoid research, there are now three generations of retinoids available for therapy and clinical research (Orfanos et al. 1985).

Four compounds from the first and second generations of retinoids are currently used clinically: tretinoin (all-*trans*-retinoic acid) and motretinide (Ro-11-1430), which are formulated for topical use; and isotretinoin (13-*cis*-retinoic acid, Ro 4-3780), and etretinate (Ro 10-9359), which are used systemically.

B. Synthesis of Retinoids

The synthesis of retinoids is performed according to well-established methods of modern polyene chemistry. This chemical procedure requires first a phosphonium

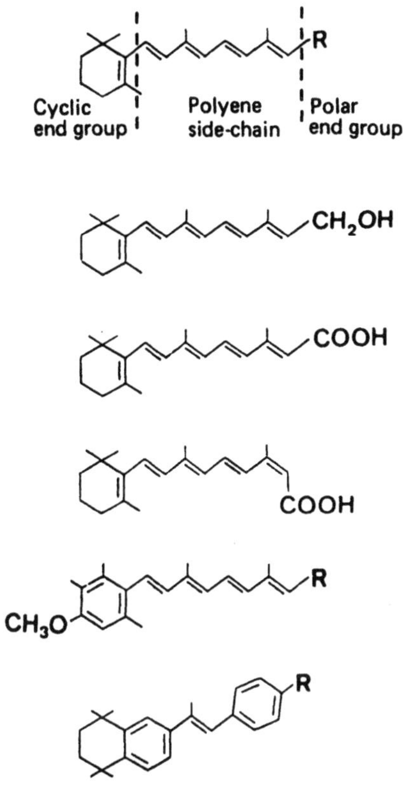

all-*trans*-Retinol

all-*trans*-Retinoic acid (tretinoin)

13-cis-Retinoic acid (isotretinoin)

Aromatic retinoids
R = COOH = free acid of etretinate
R = $COOC_2H_5$ = etretinate

Arotinoids
R = $COOC_2H_5$ = arotinoid ethyl ester
R = COOH = arotinoid acid
R = $SO_2C_2H_5$ = arotinoid ethylsulfone

Fig. 1. Retinoid derivatives

salt of an aldehyde linked to an end group R which can be acyclic, alicyclic, aromatic, or heterocyclic, and second a corresponding side chain component, e.g. the C_5 aldehyde ester or the C_{10} phosphonium salt. As coupling methods, the Wittig, Grignard, Reformatsky, or acetylide addition reactions can be employed.

Based on these chemical steps the synthetic retinoids are composed of three main building units: the cyclic end group, the polyene side chain, and the polar end group (Fig. 1). Modification of these units has led to an almost unlimited number of retinoid compounds (MAYER et al. 1978).

C. Therapeutic Index, Preclinical Evaluation

Retinoid analogs are selected for preclinical testing by several in vivo and in vitro assays. The most widely applied in vivo assay used the ability of various retinoids to cause regression of experimental skin papillomas in mice, as measured against the appearance of the hypervitaminosis A syndrome (BOLLAG 1974). Thus, a "therapeutic index" was defined as the ratio of the retinoid dose causing hypervitaminosis A to the dose producing a 50% papilloma regression in mice (Fig. 2).

	Chemical structure	Hypervitaminosis A (mg/kg)	Antipapilloma Effect (mg/kg)	Therapeutic Index
all - trans - Retinoic acid	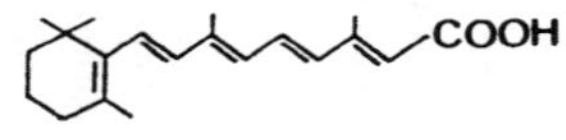	80	400	$\frac{80}{400} = 0.2$
13 - cis - Retinoic acid	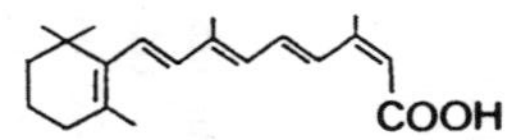	400	800	$\frac{400}{800} = 0.5$
Trimethylmethoxyphenyl (TMMP) analog of retinoic acid ethyl amide (Motretinide) Ro 11 - 1430	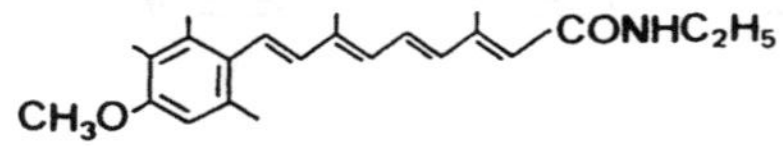	100	50	$\frac{100}{50} = 2.0$
Trimethylmethoxyphenyl (TMMP) analog of retinoic acid ethyl ester (Etretinate) Ro 10 - 9359	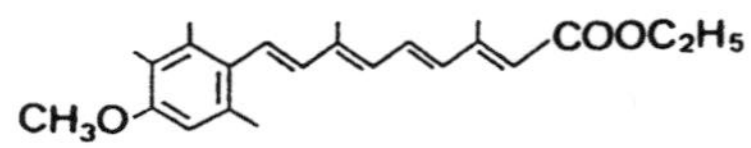	50	25	$\frac{50}{25} = 2.0$
Dichloromethylmethoxyphenyl (DCMMP) analog of retinoic acid ethyl ester Ro 12 - 7554	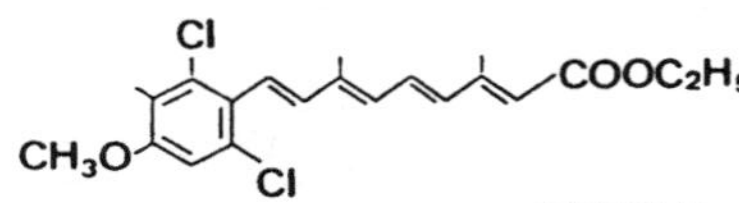	12	6	$\frac{12}{6} = 2.0$
Arotinoid ethyl ester Ro 13 - 6298	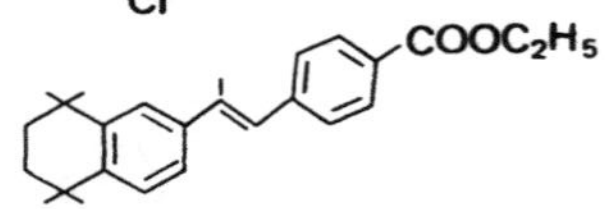	0.1	0.05	$\frac{0.1}{0.05} = 2.0$

Fig. 2. Therapeutic indices of some retinoids

However, this assay does not necessarily measure the retinoid's ability to influence cellular differentiation. Therefore, alternative in vivo and in vitro assay methods have been developed that measure the ability of retinoids to maintain cell differentiation in organ cultures of mouse epidermis or hamster tracheal epithelium (SPORN et al. 1975), hamster flank organ (TSAMBAOS et al. 1985), mouse prostate glands (LASNITZKI 1963), embryonic chicken epidermis (WILKOFF et al. 1976), cultured fibroblasts (DE LUCA et al. 1979), cultured neonatal mouse keratinocytes (STADLER et al. 1984), cultured chick foot skin (KISTLER 1985), and cultured promyelocytic leukemia cells HL-60 (BOLLAG 1985).

Based on these models, only some synthetic retinoid compounds of the first second, and third generations turned out to be of sufficient expected therapeutic value to justify screening in clinical pilot studies. Therefore, only selected retinoid analogs will be discussed further in this chapter.

D. First-Generation Retinoids

The first generation is composed of all-*trans*-retinoic acid, which proved efficient in the local therapy of acne, and its isomer 13-*cis*-retinoic acid (Fig. 1), which has by now become the drug of choice in the oral treatment of severe acne and may be beneficial in other acne-related diseases (PLEWIG et al. 1983, 1986).

I. Tretinoin

In 1962, the use of retinoic acid for topical treatment of keratinizing disorders was introduced (STÜTTGEN 1962; BEER 1962). Furthermore, topical tretinoin was found to be effective in the treatment of ichthyosis, psoriasis (FROST and WEINSTEIN 1969), actinic keratoses, and basal cell carcinoma (BOLLAG and OTT 1971). On the other hand, systemic administration of retinoic acid (SCHUMACHER and STÜTTGEN 1971; GÜNTHER 1973; RUNNE et al. 1973) did not always yield convincing therapeutic results. Increasing the orally applied dosage resulted in unbearable side effects, involving the liver, the bones, and the nervous system. Today, the clinical use of tretinoin is restricted to its local application. In this regard it has become the drug of choice in local therapy of acne, particularly of comedonic types.

1. Pharmacokinetics

All-*trans*-retinoic acid is not a normal die tary constituent. When given orally, this compound is absorbed directly into the portal blood (ZILE et al. 1982a) and circulates in the serum bound to albumin (ZILE et al. 1982b). Retinoic acid also requires a specific intracellular binding protein, cellular retinoic acid binding protein (CRABP). CRABP has been detected in various tissues of experimental animals and humans under normal and pathologic conditions (ONG and CHYTIL 1976a, b; SANI 1977, 1979). The first step in the metabolism of this drug is the isomerization of tretinoin to 13-*cis*-retinoic acid (isotretinoin), followed by their simultaneous glucuronidation at a constant rate (ZILE et al. 1982b). In various

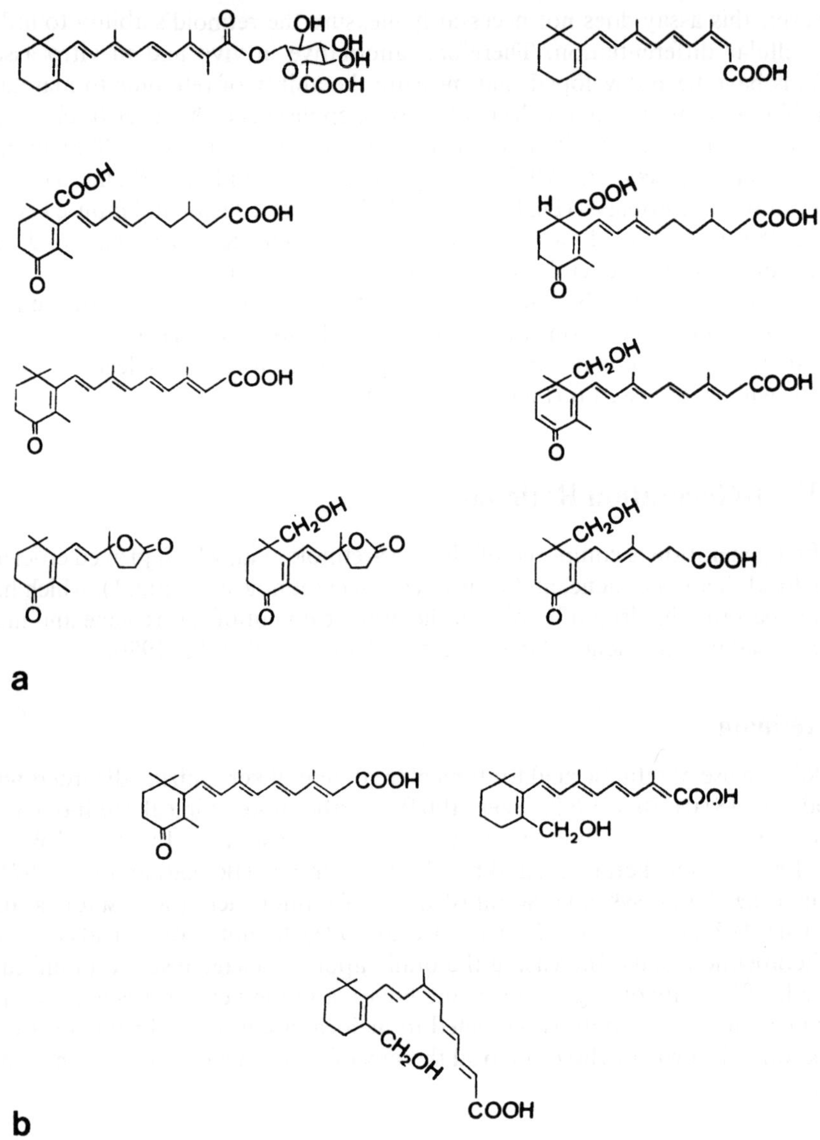

Fig. 3. a Urinary metabolites of retinoic acid. **b** Fecal metabolites of retinoic acid

rodents, between four and six major unknown retinoic acid metabolites have been identified, in addition to 5,6-epoxyretinoic acid (FROLIK 1981; DELUCA et al. 1981). Retinoic acid is not stored in organ depots, with a possible enteropathic recirculation, and is otherwise rapidly metabolized to polar substances, which are eliminated mainly in the bile and urine (KALIN et al. 1981). The major urinary and fecal metabolites of retinoic acid are shown in Fig. 3. When applied topically to humans, little, if any, tretinoin is absorbed systemically.

II. Isotretinoin

From the first generation of retinoids, isotretinoin, a geometric isomer of all-*trans*-retinoic acid, has attracted the greatest experimental and clinical attention. The drug was first used in psoriasis with some beneficial effect (ORFANOS et al. 1982; RUNNE et al. 1973). Specifically for isotretinoin, a dramatic inhibition of sebum production and reduction of the size of the sebaceous glands (GOMEZ 1981), and a reduced number of *Propionibacterium acnes* in the follicles (LEYDEN et al. 1982) have been shown. Isotretinoin is successfully used in nearly all types and grades of acne, including gram-negative folliculitis and rosacea (PECK et al. 1979, 1982; PLEWIG et al. 1982). As a result, after its release in the United States in 1982, some 500000 patients were treated in less than 20 months, including 120000 women of childbearing age. Several incidents of teratogenicity were reported among the large number of young females exposed to isotretinoin for various reasons (ORFANOS 1984). Obviously, the teratogenicity of isotretinoin has been underestimated and stricter instructions were issued by the Food and Drug Administration (MARWICK 1984); these are absolutely mandatory (ORFANOS 1984). The optimal dosage schedule recommended by the German multicenter group starts with an initial dose of 1.0 mg kg^{-1} day^{-1} during the first months of treatment, followed by dose reduction to 0.5–0.2 mg kg^{-1} day^{-1}, adapted to the individual response.

1. Pharmacokinetics

It is found that 99.9% of isotretinoin is bound in plasma, almost exclusively to albumin. According to the work of BRAZZELL and COLBURN (1982), displacement of isotretinoin from plasma protein binding sites by its metabolites does not appear to be significant. The bioavailability of isotretinoin has been shown to be influenced by food intake, the relative bioavailability of isotretinoin being 1.5–2 times greater when the dose is administered 1 h before, simultaneously with, or 1 h after a meal than when given during a complete fast (COLBURN et al. 1983a).

A great advantage of 13-*cis*-retinoic acid is its short metabolic elimination rate. The peak blood concentrations of isotretinoin, 400–950 µg/ml, were observed after a single oral dose of 3–5 µg/kg to normal subjects, and occurred 1–4 h after dosing. Isotretinoin exhibited an almost linear elimination rate with mean half-lives between 1.3 and 17.4 h. Interestingly enough, there is no significant change in this sequence of events after multiple doses. About 20–30% of isotretinoin is isomerized to tretinoin. The parent drug, its stereoisomer and oxidized metabolites undergo conjugation with glucuronic acid, biliary secretion, and enterohepatic recirculation. Oxidation occurs in the 4 position or the 5,6 position of the cyclohexenyl ring, forming the 4-*OH* or 5,6-epoxide intermediate. These units are converted to the major metabolite 4-oxoisotretinoin. The drug is then terminally eliminated in urine and feces after an interval of about 2–5 days (COLBURN et al. 1983b).

E. Second-Generation Retinoids

The second generation includes retinoids with an aromatic ring with methoxy, ethyl, and halogen groups at specific positions (Fig. 4). The aromatic retinoid etretinate (Tigason) with its broad application spectrum in dermatology is the standard agent (ORFANOS 1980). Among the more recent synthetic monoaromatics are the first metabolite of etretinate, the free aromatic acid etretin (Ro 10-1670), its isomer 13-*cis*-etretin (Ro-13-7652), the derivative demethyletretin (Ro-12-7310), and the chlorinated compound (Ro-12-7554). Today, three compounds are of practical clinical interest: (a) the parent aromatic compound etretinate, marketed as Tigason; (b) the main metabolite of etretinate, the free aromatic *p*-trimethylmethoxyphenylretinoic acid (TMMP-RA, Ro 10-1670) which is currently under clinical investigation; and (c) the aromatic analog of retinoic acid ethylamide (motretinide), marketed as Tasmaderm for local use.

I. Etretinate

This synthetic retinoid compound represents a breakthrough in modern oral dermatotherapy, like the steroids in the 1950s. It was found to be most effective and appropriate for broad-spectrum clinical use, especially for severe psoriasis conditions and keratinizing disorders. In addition, this retinoid analog turned out to be an agent with an extremely interesting preventive and therapeutic effect when administered sytemically in oncologic diseases, as well as in a series of dermatologic disorders (ORFANOS 1985).

1. Pharmacokinetics

In humans given a single oral dose of 100 mg etretinate, approximately 40% is absorbed after a lag time of 1.7–2.3 h. The absorption rate seems to be reduced in disease, e.g., psoriasis, and significantly increased by high fat food. Peak plasma concentrations of etretinate were 400–1600 ng/ml at 2–3 h, falling to 5–10

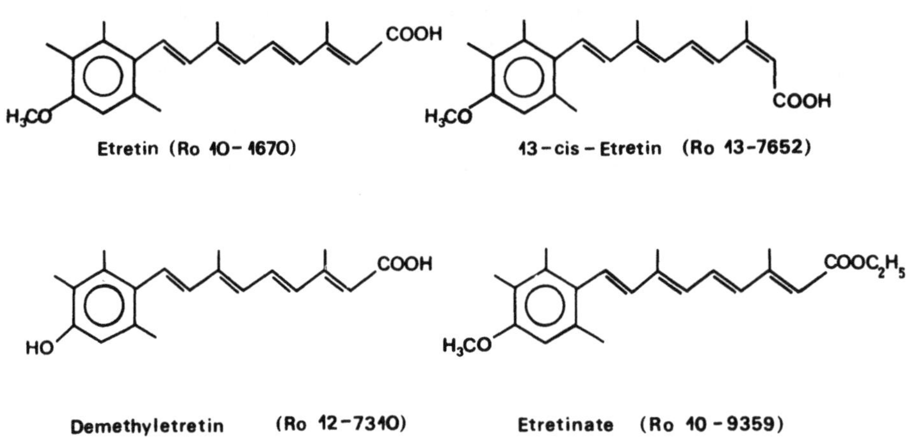

Fig. 4. Aromatic retinoids

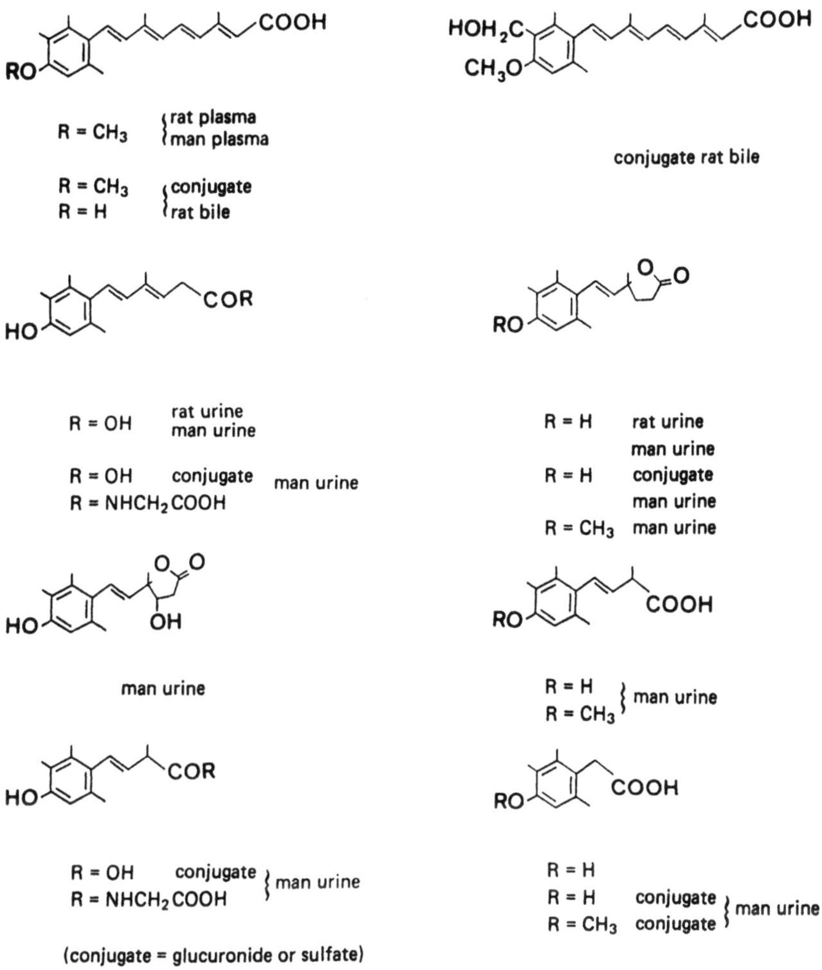

Fig. 5. Metabolites of the aromatic retinoid Ro 10-9359

ng/ml at 24–36 h, having a mean half-life of about 8 h (VAHLQUIST et al. 1981; PARAVICINI 1981). Etretinate and its major metabolite, Ro 10-1670, are bound almost exclusively to plasma proteins, etretinate being associated mainly with lipoproteins, while Ro 10-1670 is mainly bound to albumin (PARAVICINI 1981). Surprisingly, in multiple-dose studies, a prolonged terminal elimination phase was observed. Etretinate administered over a 2-week period was detectable in the plasma for more than 7 days after the last dose, and following long-term maintenance treatment for 6 months and 1 year, half-lives of 76 and 136 h, respectively, were observed. Mean plasma levels between 100 and 500 ng/ml were measured. After 140 days, etretinate plasma levels of 20–59 ng/ml were still detectable with a half-life of about 100 days.

This prolonged elimination rate is based on etretinate storage in adipocytes, primarily as the intact drug (ROLLMAN and VAHLQUIST 1983). Up to 1600 ng etretinate per gram subcutaneous tissue was detectable after long-term administration. Elimination from the fatty tissue is rather slow; even 18 months after cessation of therapy, significant plasma levels could be detected. However, the epidermis is not regarded as a storage pool for etretinate (PARAVICINI and BUSSLINGER 1984). Etretinate (Ro 10-9359) is immediately hydrolyzed in the gut, the gut wall, the blood, or the liver to the free aromatic acid (Ro 10-1670) and to its 13-*cis* isomer (Ro-13-7652) (MASSARELLA et al. 1983). These initial steps are followed by β-glucuronidation and β-oxidation of the side chain, forming at least 18 identified metabolites. They are all inactive and eliminated in the urine and feces (Fig. 5).

II. New Monoaromatics

As mentioned before, the aromatic retinoid etretin (Ro 10-1670) is the hydrolysis product of the ethyl ester, etretinate. The major pharmacokinetic advantage of etretin over the parent compound is the much faster elimination from the organism owing to the absence of storage in deep tissue compartments. The excretion rate is linear with an average elimination half-life of 50 h, which contrats with more than 80 days for etretinate (PARAVICINI et al. 1985). Based on these favorable pharmacokinetic data, etretin is expected to be introduced into therapy in the near future, simply because for women of childbearing age it allows the onset of pregnancy shortly after discontinuation of the drug without risk of fetal abnormalities. Preliminary in vitro and in vivo studies showed the therapeutic potency of Ro 10-1670 to be comparable to the parent compound. Its clinical efficacy is now under evaluation, the data available to us, however, suggest that a dose of 50 mg/day etretin appears somewhat less effective than 50 mg/day etretinate.

Another aromatic derivative of etretinate is the chlorinated compound Ro-12-7554, which was found to be very similar to the parent compound, but without any additional advantage (OTT and BOUNAMEAUX 1983).

F. Third-Generation Retinoids: Arotinoids (Polyaromatics)

The third-generation retinoids are synthesized by cyclization involving the polyene side chain, which leads to analogs with two and more aromatic rings in their molecules (Fig. 6) (LOELIGER et al. 1980; BOLLAG 1981). Various compounds of this group are most potent in minimal hormone-like doses.

I. Arotinoids with a Carbon-Containing Polar End Group

In recent years, the search for new retinoids, safer and more potent than the available compounds, has led to the development of arotinoids (LOELIGER et al. 1980). Two of the most active retinoids in this group are the arotinoid ethyl ester (Ro-13-6298) and its corresponding free acid (Ro-13-7410).

Experimentally, the arotinoid ethyl ester (Ro-13-6298) very marked anti-tumor, anti-metaplastic, differentiation-inducing, anti-keratinizing, anti-inflammatory, and immunomodulatory properties (BOLLAG 1981; HERCEND et al. 1981; ORFANOS and BAUER 1983; STADLER et al. 1984; TSAMBAOS et al. 1985). Prelimi-

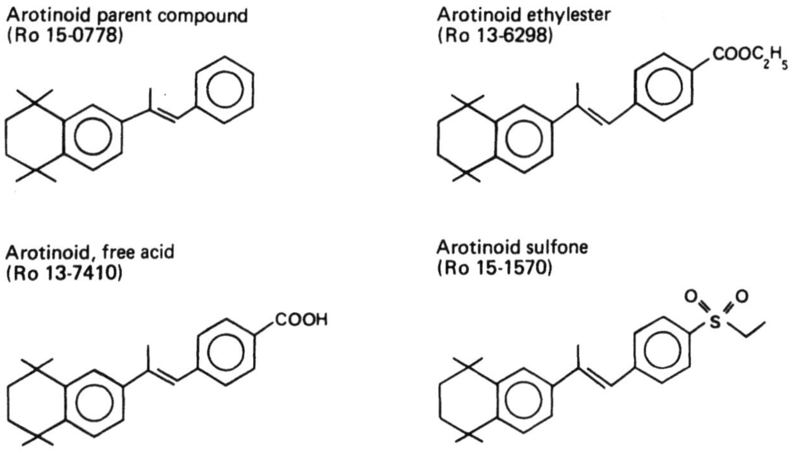

Fig. 6. New synthetic polyaromatics (arotinoids)

nary clinical experience with (Ro-13-6298) showed that this compound is highly effective in the treatment of severe cases of psoriasis and genodermatoses (TSAMBAOS and ORFANOS 1983; TSAMBAOS et al. 1983). Interestingly, psoriatic patients resistant to previous etretinate therapy responded to this analog; in addition, four patients with cutaneous T-cell lymphoma responded favorably. For therapeutic success, doses 1000 times lower than those of etretinate were administered.

Based on experimental data, arotinoid acid (Ro-13-7410) seems to be the active metabolite of the arotinoid ethyl ester. In these in vitro experiments concerning inhibition of keratinization, induction of cartilage degradation, and induction of differentiation, the free acid was more active than the ethyl ester. These experimental data are very promising for future oral dermatotherapy. Clinical pilot studies have been initiated.

1. Pharmacokinetics

The arotinoid (Ro-13-6298) as an ethyl ester seems to have pharmacokinetic properties similar to those of etretinate. It is rapidly metabolized to its corresponding free acid (Ro-13-7410), possibly its active metabolite, and is stored in a deep tissue compartment. Its elimination time is long.

II. Arotinoids with a Sulfur-Containing Polar End Group

Ro-15-1570 is the first derivative of sulfonated arotinoids (KLAUS et al. 1983). This compound is very active and inhibitory on clinically induced papillomas, carcinomas, and mammary carcinogenesis (BOLLAG 1985). Moreover, it is a potent anti-keratinizing compound and inhibits sebaceous gland differentiation (KISTLER 1985; STADLER 1987).

Compared with Ro-13-6298 and Ro-13-7410, its main advantage is that it lacks the bone toxicity in rodents, in spite of inducing the other signs of hypervitaminosis A on skin and mucous membranes (BOLLAG 1985). Therefore, this retinoid is the first analog with reduced sideeffects, at least in rodents, and would seem to be a candidate for clinical trial.

III. Arotinoid Ro-15-0778: The Parent Compound

Ro-15-0778 is an arotinoid lacking a polar end group. According to preliminary results, this compound has been reported to be a potent sebostatic drug, as seen in the hamster flank organ model. In contrast, preliminary experimental studies in vitro and clinical pilot studies in humans are rather disappointing and have shown Ro-15-0778 to be inactive in mouse keratinocyte cultures (STADLER 1987).

G. Retinoids, Intracellular Binding Proteins, and Mechanism of Action

The molecular mechanisms by which the retinoids may trigger their complex biologic effects are still largely unknown. According to our present knowledge, the biologic effect of the retinoids is based mainly on the influence exerted on a specific messenger RNA (FUCHS and GREEN 1981), gene expression/suppression (CHYTIL 1985), phosphorylation and dephosphorylation of proteins (COPE and BOUTWELL 1985), and on the glycolysis of glycoproteins (LOTAN et al. 1980), mediated mainly via two specific binding proteins. ONG and CHYTIL (1975a, b, 1976a, 1978) characterized them as cellular retinoic binding protein (CRBP) and cellular retinoic acid binding protein (CRABP), and they can be detected in various human and animal cell lines (ADACHI et al. 1981; ONG et al. 1982; DI GIOVANNA et al. 1985; GATES and KING 1983). They each have a molecular weight of 14600 and different immunologic properties (CHYTIL and ONG 1976). Available experimental data demonstrate that the retinol-CRBP-retinoid-CRABP complex is shifted from the cytosol to the nucleus and interacts with nuclear components or receptors (WIGGERT et al. 1977; TAKASE et al. 1979; LIAU et al. 1981; COPE et al. 1984).

For some cell lines, the effect of the retinoids via CRABP has been confirmed (LOTAN 1980; BICHLER and DAXENBICHLER 1982). We are probably dealing here with a specific biologic process, since only retinoids with a free carboxyl group on carbon atom 15 form a complex with CRABP, and thus have an antiproliferative effect. On the other hand, there are cell lines which show unchanged differentiation in the presence of potent retinoids, despite existing binding proteins in the cytosol (LOTAN 1980; SATO et al. 1980). This observation does not, in principle, cast doubt on the effect of retinoids via binding proteins, since, in an open biologic system, negative responses are possible as well. The exclusive significance of the binding proteins for retinoids is, however, questioned by findings demonstrating that the differentiation of cell lines may also be influenced by retinoids without detection of CRABP (LIBBY and BERTRAM 1982). In these cells, retinoids possibly exert their influence via membrane glycoproteins, phosphorylation of proteins, and induction of protein kinase (DE LUCA et al. 1979; LOTAN 1985; COPE

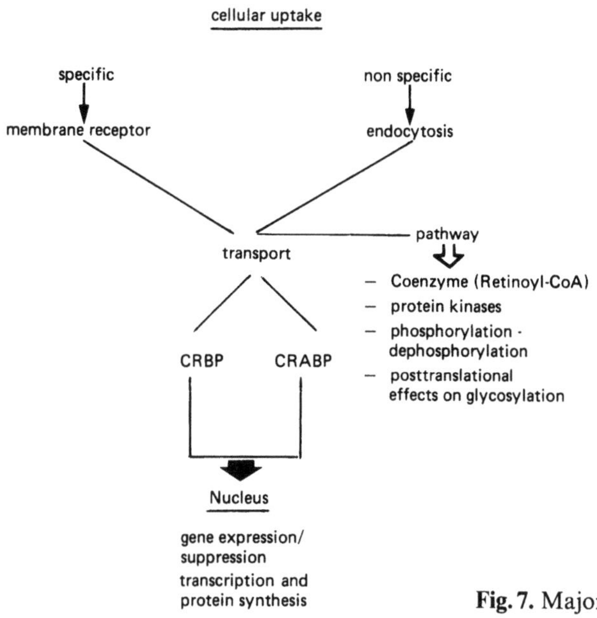

Fig. 7. Major hypothesis of retinoid action

and BOUTWELL 1985). The apparent differences between some in vivo and in vitro results probably stem from variations in technologies, cellular systems, and in the complex effects of the different retinoids themselves.

The cellular uptake of retinoids possibly takes place via a specific membrane receptor, as has already been described for vitamin A acid in murine brain (COPE and BOUTWELL 1983). However, a nonspecific uptake of retinoids via endocytosis with involvement of retinoid-inducible transglutaminase is feasible as well (LEVITSKI et al. 1980; YUSPA et al. 1982).

Two essential conclusions can be drawn from the findings available so far:

1. Retinoids belong to a group of substances whose effects at the cellular level cannot be completely explained by the hypothesis that they are similar to those of steroid hormones;
2. The effects of retinoids are extremely complex and cannot be explained by *one* molecular-biologic mechanism alone. Many years of intensive research will be necessary to make the scientific puzzle fit together to form a complete and plausible picture (Fig. 7).

H. Molecular-Biologic Effects of Synthetic Retinoids

In recent years, considerable efforts have been made to explain the extremely pleiotropic effect of synthetic retinoids in humans and animals. As a first target, retinoids influence epithelial cell tissue and, as a second target, mesenchymal dermal cells. These actions are comprised of: (a) regulation of epidermal proliferation and differentiation; (b) inhibition of experimental carcinogenesis; and (c) immunomodulating and anti-inflammatory effects, in particular, the response of immunocompetent cells such as peripheral blood lymphocytes is clearly modulated and the immigration of polymorphonuclear cells (PMNs) into the diseased epidermis is diminished. These effects, which are dose dependent and common for most retinoids, are summarized in Tables 1 and 2. Again, the apparent difference between some in vivo and in vitro results probably stems from variations in technologies, cellular systems, and in the complex effects of the different retinoids themselves.

Table 1. Epidermal effects of synthetic retinoids

Epidermal cell proliferation in vivo	Stimulation	GILLENBERG et al. (1980), ASHTON et al. (1984)
Epidermal cell proliferation in vitro	Stimulation	CHRISTOPHERS (1974), KLANN and MARCHOK (1982), HASHIMOTO et al. (1985), KUBILUS et al. (1981)
	Inhibition	MARCELO and TOMICH (1983), JETTEN (1984), STADLER et al. (1984), MADEN (1985)
Epidermal cell desquamation in vitro	Stimulation	MILSTONE et al. (1982), MILSTONE (1983)
Epidermal cell proliferation under pathologic conditions (e.g., psoriasis) in vivo	Inhibition	DIERLICH et al. (1979), KRUEGER et al. (1980)
Epidermal differentiation in vitro	Inhibition or modification	WILKOFF et al. (1976), YAAR et al. (1981), RICE et al. (1983), MARCELO and MADISON (1984), STADLER et al. (1984), GILFIX and GREEN (1984), JETTEN and SMITS (1985)
Epidermal differentiation in vivo	Modification	TSAMBAOS et al. (1980), ELIAS and WILLIAMS (1981, 1985)
Epidermal differentiation under pathologic conditions (e.g., psoriasis) in vivo	Normalization	STAQUET et al. (1983), KITAJIMA et al. (1983)
Epidermal cholesterol synthesis in vitro	Inhibition	PONEC (1984)
Epidermal cholesterol: phospholipid ratio	Change	PONEC (1985)
Epidermal glycolipid synthesis in vitro	Inhibition	OROZCIO-TOPETE et al. (1983)
Epidermal glycoprotein synthesis in human keratinocytes in vitro	Inhibition	ELIAS et al. (1983)
Epidermal glycoprotein synthesis in murine keratinocytes and fibroblasts in vitro	Stimulation	DE LUCA and YUSPA (1974), ADAMO et al. (1979), ELIAS and WILLIAMS (1985)
Epidermal glycoprotein synthesis in vivo	Inhibition	BIRKIN et al. (1985)
Epidermal DNA and RNA protein synthesis	Stimulation	FRITSCH et al. (1981), LOTAN et al. (1982)

Table 1 (continued)

Ornithine decarboxylase (ODC) activity in vivo	Inhibition	VERMA et al. (1979), HADDOCK-RUSSEL and HADDOX (1981), LICHTI and YUSPA (1985)
ODC activity and polyamine concentration under pathologic conditions (psoriasis) in vivo	Inhibition	KAPLAN et al. (1983), LOWE et al. (1982)
Epidermal cAMP content in vitro	Unchanged	STADLER et al. (1983)
Epidermal adenyl cyclase system in vitro	Stimulation	IIZUKA et al. (1985)
Epidermal surface antigens (pemphigus and bullous pemphigoid antigen)	Modulation	THIVOLET et al. (1984)
Epidermal growth factor receptors in vitro	Stimulation	JETTEN (1981)
Prostaglandin E_2 release in guinea pig skin	Reduction	ZIBOH et al. (1975)
Arachidonic acid metabolism in human epidermis	Uninfluenced	RUZIKA (1985)

Table 2. Dermal/mesenchymal effects of synthetic retinoids

Size of sebaceous glands	Reduction	PLEWIG (1980), LANDTHALER et al. (1981), TSAMBAOS (1984)
Sebaceous secretion and free fatty acids in sebum	Reduction	KING et al. (1982)
Collagenase and gelatinase activity in fibroblasts cultures in vitro	Reduction	BAUER et al. (1983)
Fibroblast proliferation in vitro	Inhibition	JETTEN et al. (1979)
cAMP-dependent protein kinase in fibroblasts	Stimulation	EVAIN BRION et al. (1985)
T-cell stimulation by mitogens	Inhibition	ORFANOS and BAUER (1983)
Cell-mediated cytotoxicity	Stimulation	LOTAN and DENNERT (1979)
Interleukin 2 production in T-helper cells	Stimulation	BLITSTEIN-WILLINGER (1983)
Langerhans' cells in the skin	Stimulation	TSAMBAOS and ORFANOS (1981)
Granulocytic motility in vitro	Inhibition	BAUER et al. (1982a, b), BAUER and ORFANOS (1984), CAMISA et al. (1982)
Lipoxygenase activity in leukotriene metabolism	Uninfluenced	KNIPPEL et al. (1984), RUZICKA (1985)
Release of leukotriene B_4 from rat granulocytes	Inhibition	BRAY (1984)
Arachidonic acid metabolites in the pathologically changed skin (psoriasis) in vivo	Reduction	VOORHEES (1983), WONG et al. (1983)
Chondrogenesis in vivo	Inhibition	TSAMBAOS et al. (1984)

J. Influence of Monoaromatic and Polyaromatic Retinoids on Neonatal Mouse Keratinocyte Cell Differentiation and Proliferation In Vitro

The successful clinical application of oral retinoids in keratinizing disorders (e.g., psoriasis, acne) is clearly based on their influence on cell differentiation of keratinizing epithelial tissues. They exert a keratolytic activity and reduce the thickness of the keratinized layers, leading to thinning of the skin.

In view of the lack of experimental test models to screen the effect of synthetic retinoids on keratinocyte differentiation, our group has in recent years used a well-defined neonatal mouse keratinocyte in vitro model system. Within 10–14 days, the mouse keratinocytes proliferate to a multilayer "epidermis" in vitro with ultrastructural and biochemical characteristics of terminal differentiation (MARCELO et al. 1978). In order to determine the influence of standard synthetic retinoids of the first and second generation and new synthetic monoaromatics and arotinoids (Figs. 6 and 8) on epidermal differentiation in vitro, keratin protein fractions corresponding to certain differentiation stages were extracted stepwise from the in vitro epidermis (STADLER et al. 1984).

These fractions are:

1. High salt soluble (S_2) proteins
2. Keratohyaline granule (R_2) macroaggregates
3. Nucleated viable cell (S_3) content
4. Non-cross-linked (S_4) keratins
5. Disulfide cross-linked (S_5) keratins
6. Cell envelope (R_5) proteins

a Etretin (free aromatic acid, Ro 10-1670)

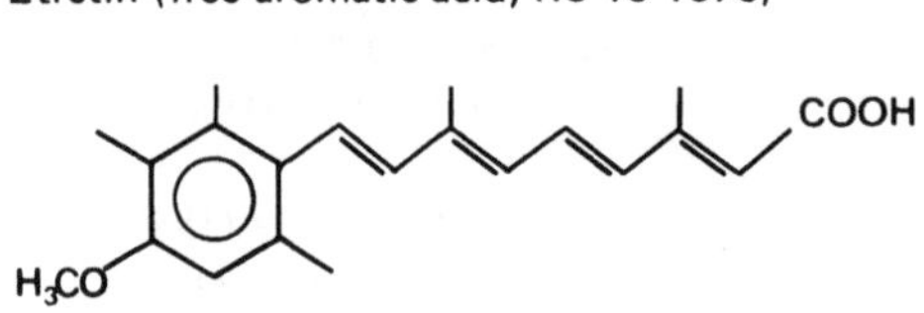

b 13-cis etretin (13-cis aromatic acid, Ro 13-7652)

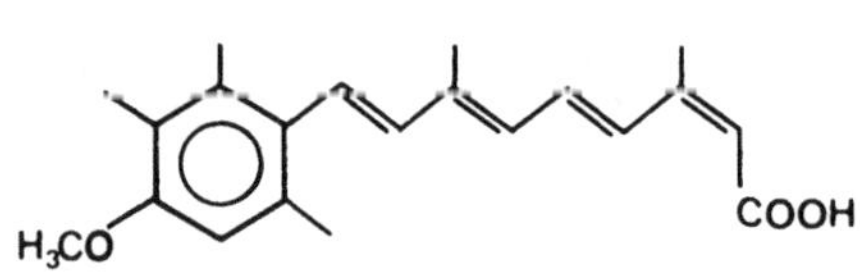

c Demethyl-etretin (demethyl aromatic acid (Ro 12-7310)

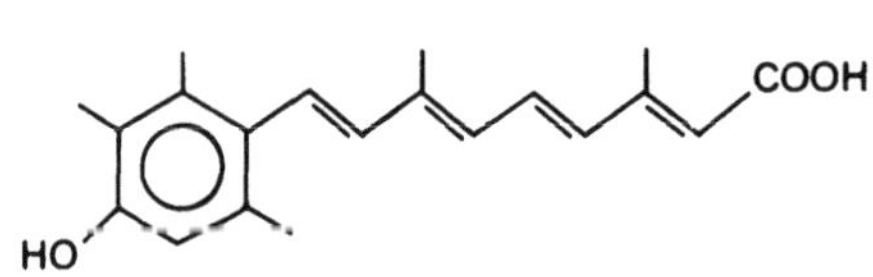

Fig. 8 a–c. In addition to the standard retinoids: all-*trans*-retinoic acid, 13-*cis*-retinoic acid, and etretinate (Tigason), a number of new monoaromatics have been synthesized, including (**a**) etretin (**b**) 13-*cis*-etretin, and (**c**) demethyl-etretin

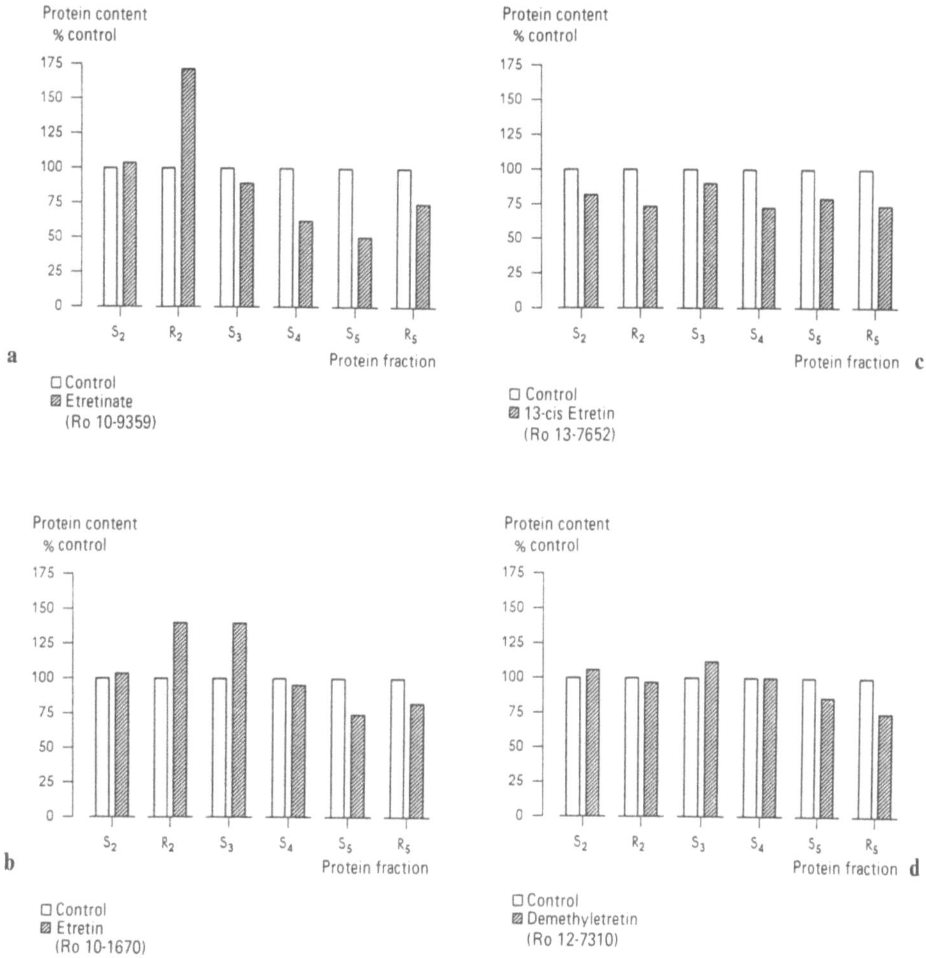

Fig. 9 a–d. The effect of (**a**) 14 μmol/ml etretinate (Ro 10-9359), (**b**) etretin (Ro-10-1670), (**c**) 13-*cis*-etretin (Ro-13-7652), and (**d**) demethyletretin (Ro-12-7310) on keratinocyte differentiation. All monoaromatics inhibited the non-disulfide (S_4) and disulfide cross-linked keratins (S_5) and envelope proteins. Etretinate and etretin are the most potent derivatives. The data are presented as percentages of the DMSO control

It was found that all retinoids, except the arotinoid parent compound Ro-15-0778, had anti-keratinizing potential.

In general, the second generation of retinoids exhibited a moderate anti-keratinizing effect, with etretin and etretinate being most potent (Fig. 9). When this response was analyzed at the level of keratin expression, a keratin of molecular weight 40000 was newly induced. Among the high salt soluble proteins (S_2) a protein of molecular weight 28000 is visualized (Fig. 10). This protein band is more strongly expressed under all retinoid derivatives other than arotinoid (Ro-15-0778). It probably reflects the expression of retinoid-treated neonatal mouse kera 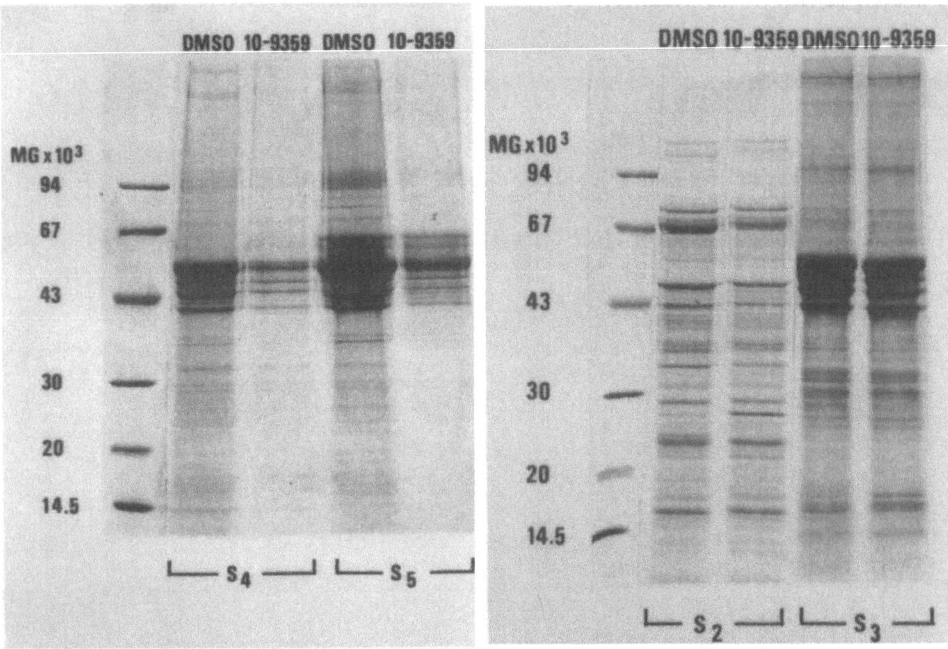 b

Fig. 10 a, b. SDS-PAGE of 12-day-old etretinate (Ro-10-9359) and etretin (Ro-10-1670) 14 mmol/ml-treated keratinocyte cultures. **a** non-disulfide (S_4) and disulfide cross-linked keratins (S_5) are inhibited by Ro-10-9359 and Ro-10-1670; **b** among the high salt soluble proteins (S_2), a protein of molecular weight 28 000 is strongly visualized and in the S_3 fraction, a protein of molecular weight 40 000 is newly expressed

atinocyte cultures in vitro, and could thus be of general importance for the retinoid effect in this culture system. It cannot be said with certainty whether this protein band represents the mouse CRABP protein. In any case, a CRABP complex with a molecular weight of 28 000 was recently isolated in the mouse brain (Cope and Boutwell 1983).

Among the third generation, all compounds except the parent arotinoid had anti-keratinizing activity, the arotinoid acid being the most potent analog, even in the picomolar range (Figs. 11 and 12). In this respect, the most significant inhibitory effects were observed at certain steps of the terminal differentiation process (disulfide cross-linked keratins and the cell envelope).

A striking example of the varying retinoid effects on epidermal differentiation was the influence on the keratohyaline granule fraction, corresponding to keratohyaline granular formation in the epidermis (filaggrin) (Dale 1985). Synthetic retinoids of the second generation (etretinate and etretin) and third generation (arotinoid ethyl ester, Ro-13-6298; arotinoid acid, Ro-13-7410; arotinoid sulfone, Ro-15-1570) stimulated these fractions, whereas other retinoids of the first generation and the arotinoid Ro-15-0778 did not (Fig. 13). On the basis of these observations and in correlation with clinical experience, particularly retinoids with known anti-psoriatic activity, such as etretinate (Ro 10-9359) and arotinoid ethyl

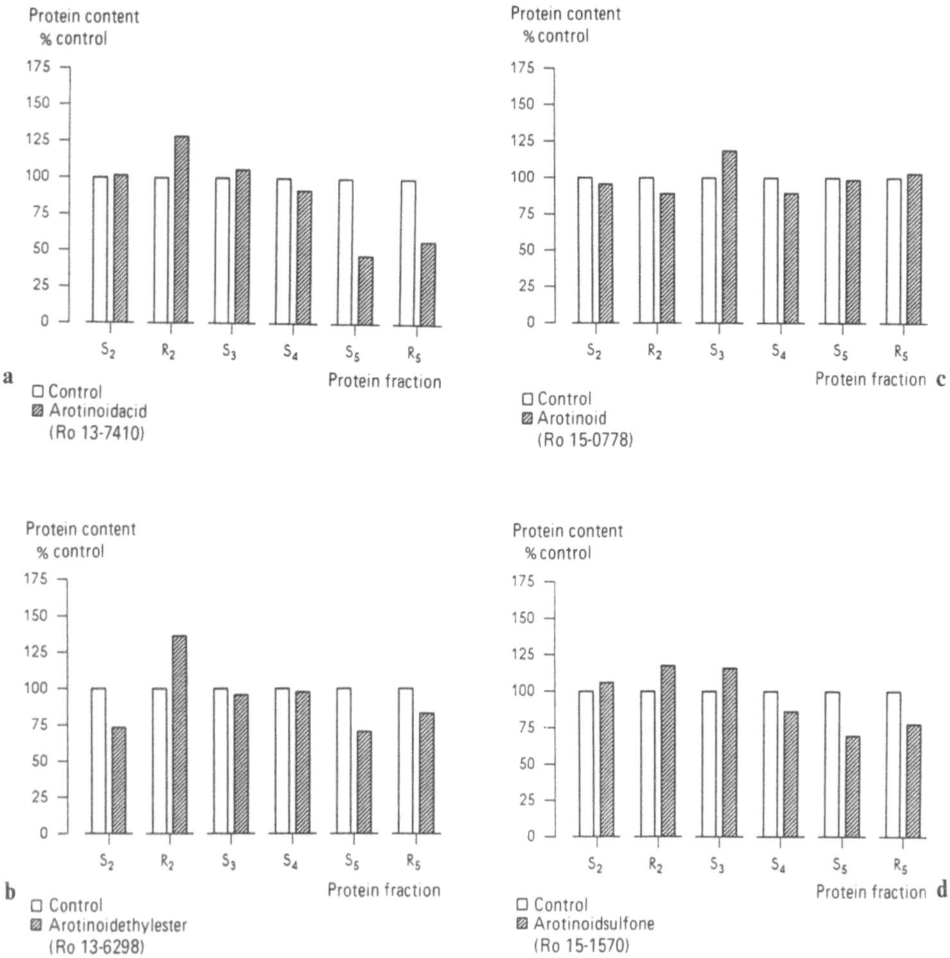

Fig. 11 a–d. Comparison of the effects of four arotinoids on keratinocyte differentiation; 12-day-old keratinocytes were serially extracted. All derivatives except arotinoid (Ro-15-0778) inhibited the disulfide cross-linked keratins (S_5) and envelope proteins (R_5)

ester (Ro-13-6298), biochemically increase the synthesis of keratohyaline granule-associated proteins (Fig. 14), whereas retinoids with low anti-psoriatic activity such as all-trans-retinoic acid and 13-*cis*-retinoic acid, inhibited this fraction (STADLER et al. 1985).

The last observation leads to the hypothesis that the stimulability of keratohyaline granule-associated proteins may represent a valuable marker for the anti-psoriatic effectiveness of synthetic retinoids. If this is true, it seems more useful to apply arotinoid acid or arotinoid sulfone instead of arotinoid Ro-15-0778 in future clinical studies on psoriasis.

Altogether, the findings obtained in a standardized neonatal mouse culture model indicate that synthetic retinoids have a modulating effect on the individual

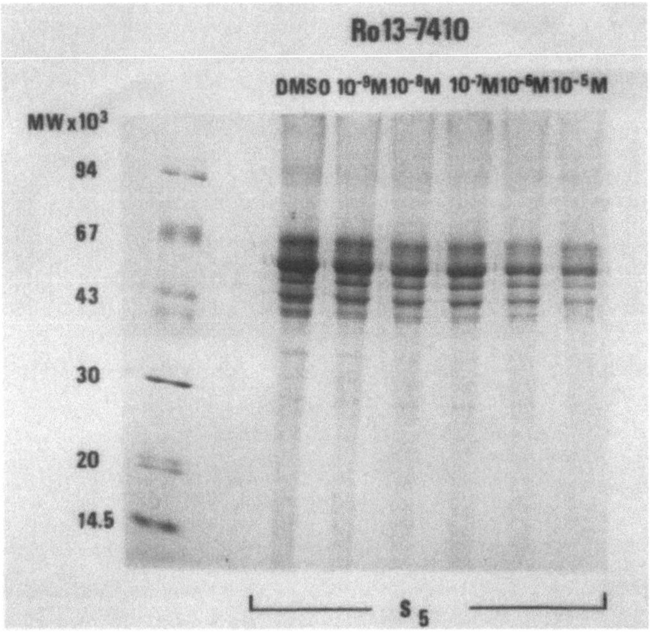

Fig. 12. SDS-PAGE of 12-day-old keratinocyte cultures treated with arotinoid acid (Ro-13-7410) 10^{-5}–10^{-9} mol/ml. Arotinoid acid inhibited disulfide cross-linked keratins (S_5) in nanomolar concentrations

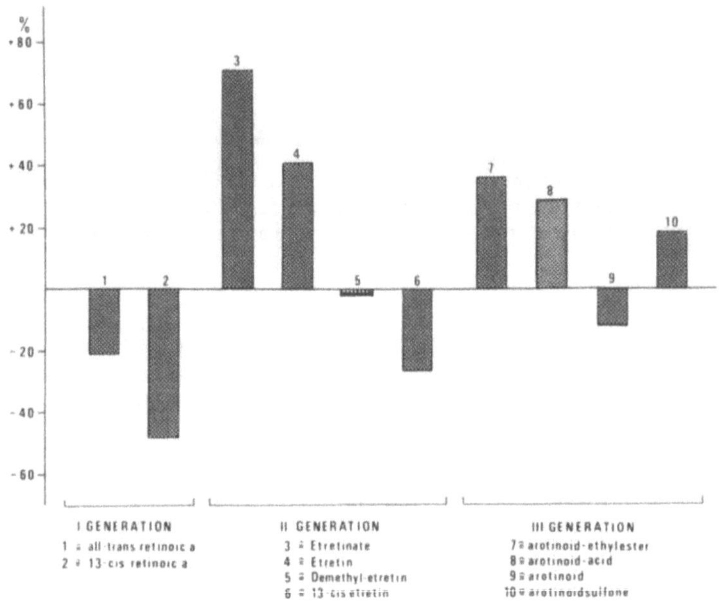

Fig. 13. Comparison of the effect of three generations of retinoids on keratohyaline granule-associated proteins (R_2). The data are presented as percentages of the DMSO control

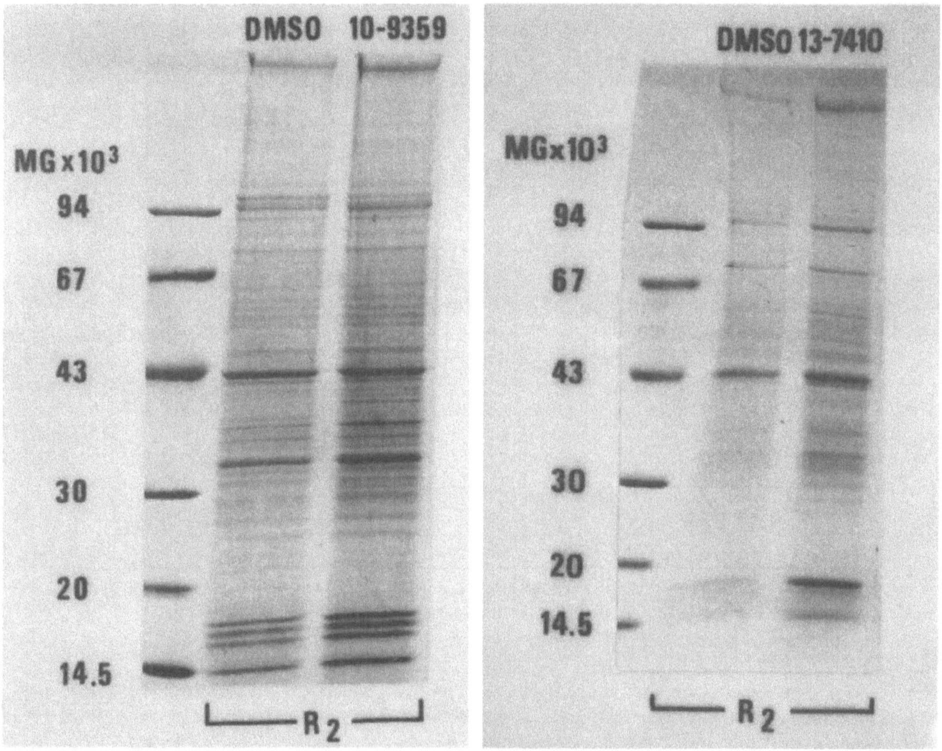

a b

Fig. 14a, b. SDS-PAGE of keratohyaline granule-associated proteins extracted from (**a**) etretinate (Ro 10-9359)- and (**b**) arotinoid acid (Ro-13-7410)-treated cultures with identical numbers of cells (3×10^6 cells per 60-mm Petri dish). 1 *M* KPO_4 buffer containing 0.22 *M* histidine preferentially extracts keratohyaline granule-associated proteins (R_2) from keratinocyte cultures. The R_2 fraction of etretinate and arotinoid acid is markedly stimulatd compared with the DMSO control lane

parameters of epidermal keratinization, and that the new arotinoids, except for arotinoid Ro-15-0778, are the hitherto most potent retinoid derivatives in this respect, the order of potency being: arotinoid acid > arotinoid sulfone > arotinoid ethyl ester > etretinate > etretin > 13-*cis*-etretin > demethyletretin.

In addition, compared with previous investigations with retinoids of the first and second generations, the arotinoids, with the exception of arotinoid Ro-15-0778, showed a 1000- to 2000-fold stronger antiproliferative effect in vitro (Figs. 15–18). In the model of the neonatal mouse keratinocyte culture, 10-day treatment with the lowest concentrations (picomolar or nanomolar) of arotinoid acid (Ro-13-7410, Fig. 15a, b), arotinoid sulfone (Ro-15-1570, Fig. 6a, b), or arotinoid ethyl ester (Ro-13-6298, Fig. 17a, b) led to a profound inhibition of proliferation, evident from the reduced incorporation of [^{3}H]thymidine into DNA or its labeling as measured by autoradiography. Of course, the in vitro and in vivo conditions cannot readily be compared; the present results nevertheless suggest that

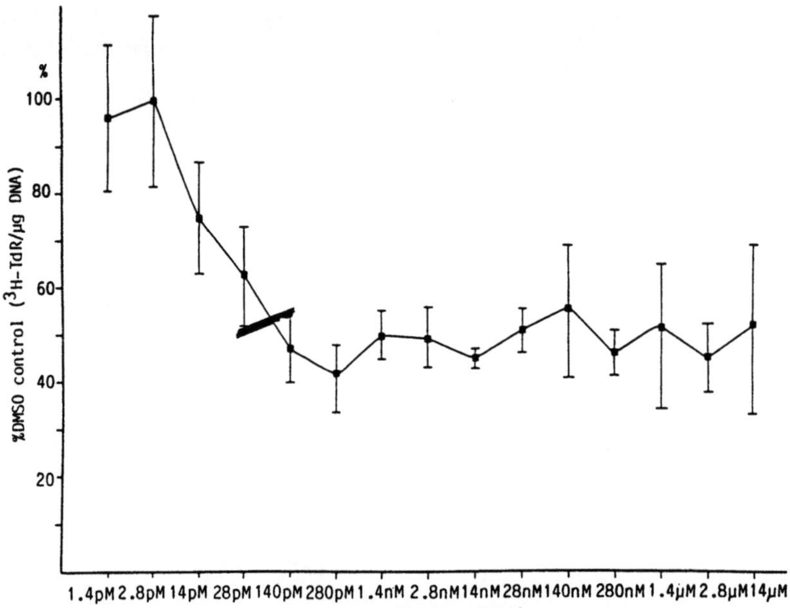

a

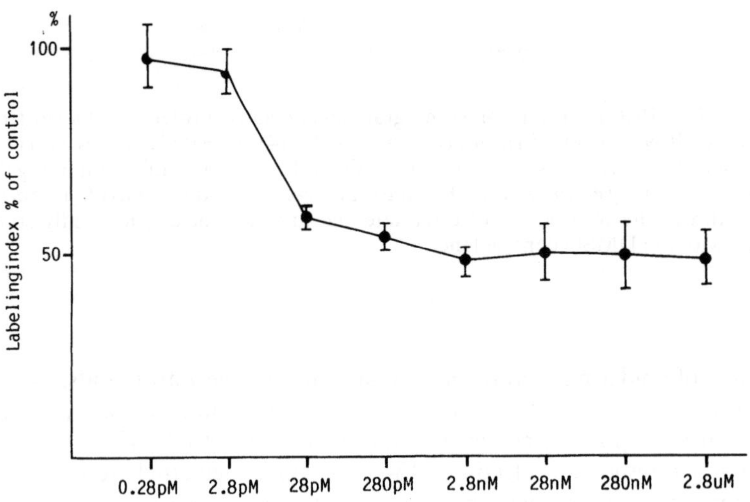

b

Fig. 15 a, b. Effect of Ro-13-7410 on epidermal keratinocyte proliferation. To quantitate keratinocyte proliferation [^{3}H]TdR incorporation into DNA and autoradiography was investigated. **a** The results were expressed as cpm [^{3}H]TdR label per µg DNA. The data are presented as percentages of the DMSO vehicle control, **b** for autoradiography, the data are expressed as percentage labeled cells. The labeled nuclei in a minimum of 60 fields were counted with a Zeiss phase microscope. *Bars* represent standard deviations

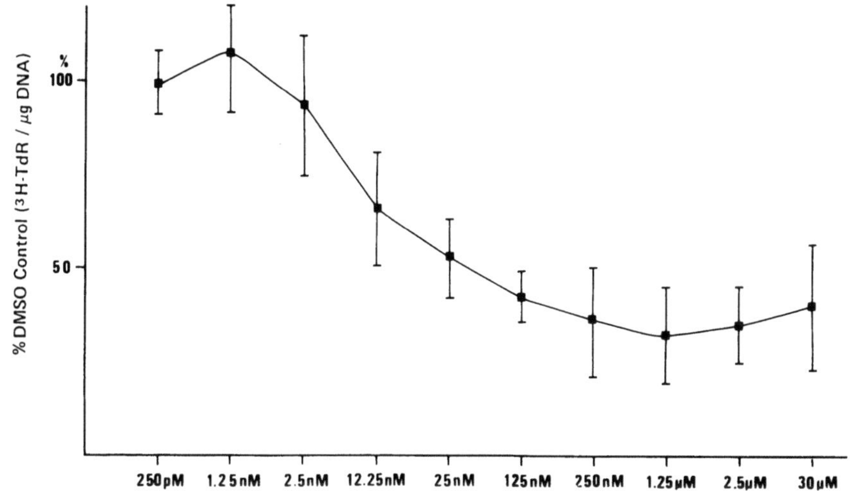

a

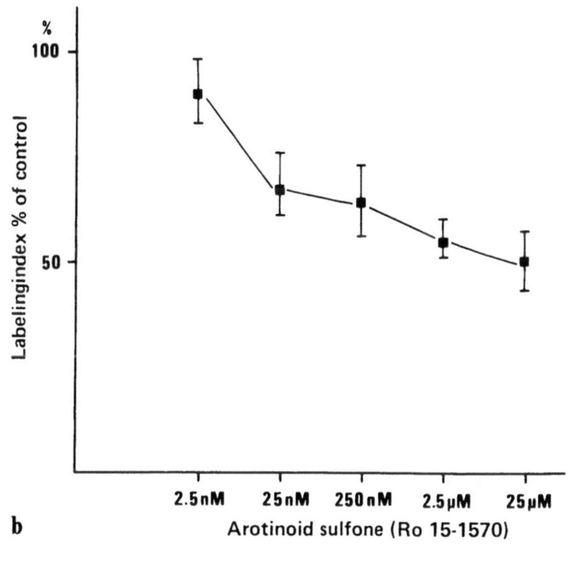

b

Fig. 16 a, b. Effect of Ro-15-1570 on epidermal keratinocyte proliferation

the new retinoids, especially the arotinoids, may exert potent effects on pathologically altered skin and could thus be of particular therapeutic value. Based on these data, the future of the retinoids appears most promising. The arotinoid analogs offer the exciting prospect of introducing synthetic retinoids with increased effectiveness and reduced side-effects.

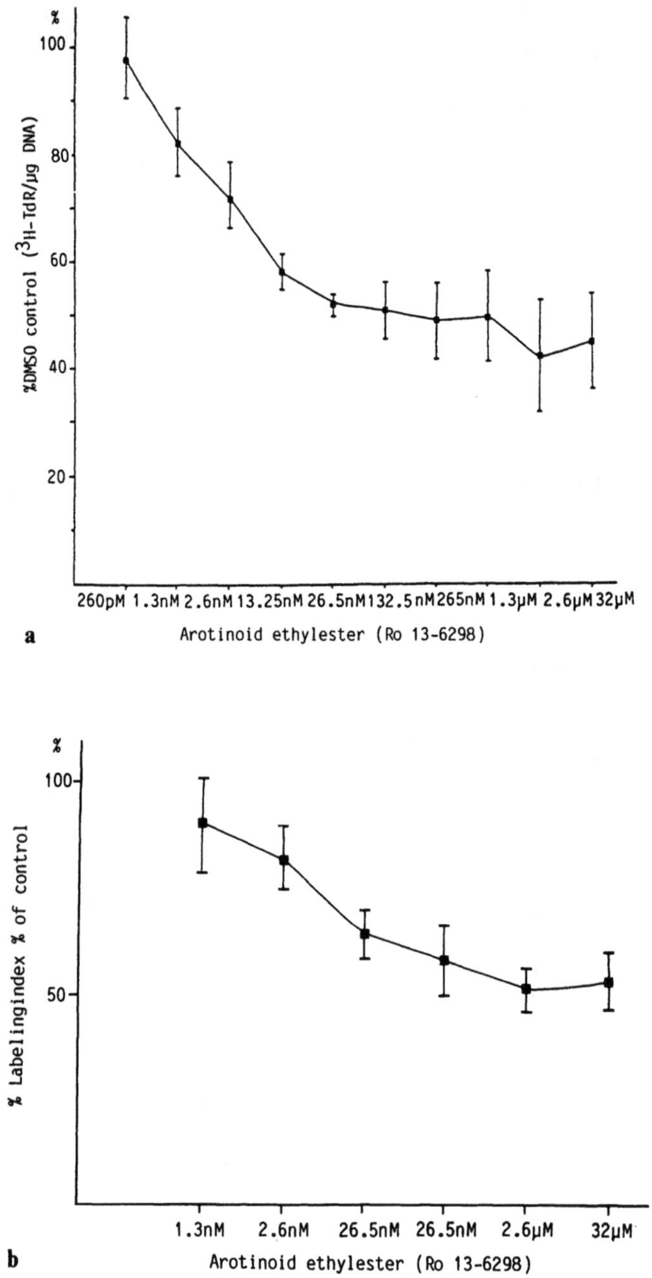

Fig. 17 a, b. Effect of Ro-13-6298 on epidermal keratinocyte proliferation

References

Adachi N, Smith JE, Sklan D, Goodman DS (1981) Radioimmunoassay studies of the tissue distribution and subcellar localization of cellular retinol-binding protein in rats. J Biol Chem 256:9471
Adamo S, de Luca LM, Akalovsky J, Bhat PV (1979) Retinoid-induced adhesion in cultured transformed mouse fibroblast. JNCI 62:1473
Ashton RE, Connor MJ, Lowe NJ (1984) Histological changes in the skin of the rhino mouse induced by retinoids. J Invest Dermatol 82:632
Bauer EA, Seltzer JL, Eisen AZ (1983) Retinoic acid inhibition of collagenase and gelatinase expression in human skin fibroblast cultures. Evidence for a dual mechanism. J Invest Dermatol 81:162
Bauer R, Orfanos CE (1984) Effects of synthetic retinoids on human peripheral blood lymphocytes and polymorphonuclears in vitro. In: Cunliffe WJ, Miller A (eds) Retinoid therapy. MTP Press, Lancaster, p 101
Bauer R, Schütz R, Orfanos CE (1982a) Granulozyten-Migration in vitro bei Psoriasis-Patienten unter aromatischem Retinoid. Z Hautkr 57:1247
Bauer R, Schütz R, Orfanos CE (1982b) Disordered motility of polymorphonuclear granulocytes under aromatic retinoid in psoriasis vulgaris. Inhibition of the cytoplasmic microtubular system? Arch Dermatol Res 273:166
Beer P (1962) Studies of the effects of vitamin A acid. Dermatologica 124:192
Bichler E, Daxenbichler G (1982) Retinoic acid-binding protein in human squamous cell carcinomas of the ORL region. Cancer 49:619
Birkin JA, Newson TP, Terras RS, Fry JR, Millard LG, Turner DR (1985) The effect of etretinate on glycoprotein biosynthesis and cell ultrastructure in the uninvolved epidermis of abnormal keratinizing skin during the initial 7–9 weeks of treatment. In: Saurat H (ed) Retinoids: new trends in therapy and research. Karger, Basel, p 183
Blitstein-Willinger E (1983) Effect of the aromatic retinoid etretinate (Tigason) on the human interleukin 2 production. Immunbiology 165:235
Bollag W (1974) Therapeutic effects of an aromatic retinoic acid analog on chemically induced skin papillomas and carcinomas in mice. Eur J Cancer 10:731
Bollag W (1981) Arotinoids. A new class of retinoids with activities in oncology and dermatology. Cancer Chemother Pharmacol 7:27
Bollag W (1985) New retinoids with potential use in humans. In: Saurat H (ed) Retinoids: new trends in therapy and research. Karger, Basel, p 274
Bollag W, Ott F (1971) Therapy of actinic keratoses and basal cell carcinomas with local application of vitamin A acid. Cancer Chemother Rep 55:59
Bray MA (1984) Retinoids are potent inhibitors of the generation of rat leukocyte leukotriene B_4-like activity in vitro. Eur J Pharmacol 98:61
Brazzell RK, Colburn WA (1982) Pharmacokinetics of the retinoids isotretinoin and etretinate – a comparative review. J Am Acad Dermatol 6:643
Camisa CH, Eisenstat B, Ragaz A, Weissmann G (1982) The effects of retinoids on neutrophil functions in vitro. J Am Acad Dermatol 6:620
Christophers E (1974) Growth stimulation of cultured post-embryonic, epidermal cells by vitamin A acid. J Invest Dermatol 63:450
Chytil F (1985) Retinoids, intracellular binding proteins and mechanism of action. In: Saurat H (ed) Retinoids: new trends in research and therapy. Karger, Basel, p 20
Chytil F, Ong DE (1979) Cellular retinol and retinoic acid-binding proteins in vitamin A action. Fed Proc 38:2510
Colburn WA, Gibson DM, Wiens RE, Hanigan JJ (1983a) Food increases the bioavailability of isotretinoin. J Clin Pharmacol 23:534
Colburn WA, Vane FM, Shorter HJ (1983b) Pharmacokinetics of isotretinoin and its major blood metabolite in man. Eur J Clin Pharmacol 24:689
Cope FO, Boutwell RK (1983) Retinoic acid binding to mouse brain plasma membrane. Characterization of two distinct receptors in adult and juvenile mice. Fed Proc 42:1317

Cope FO, Boutwell RK (1985) A role for retinoids and a tumor-promoting phorbol ester in the control of protein phosphorylation. In: Saurat H (ed) Retinoids: new trends in research and therapy. Karger, Basel, p 106
Cope FO, Knox K, Hall RJ (1984) Retinoid binding to nuclei and microsomes of rat interstitial cells. Nutr Res 4:289
Dale BA (1985) Filaggrin, the matrix protein of keratin. Am J Dermatopathol 7(1):65
DeLuca HF, Zile M, Sietsema WK (1981) The metabolism of retinoic acid to 5,6-epoxyretinoic acid, retinoyl-β-glucuronide and other polar metabolites. Ann NY Acad Sci 359:25
De Luca LM, Yuspa SH (1974) Altered glycoprotein synthesis in mouse epidermal cells treated with retinol acetate in vitro. Exp Cell Res 86:106
De Luca LM, Bhat P, Sasak W, Adamo S (1979) Biosynthesis of phosphoryl and glycosyl phosphoryl derivates in biological membranes. Fed Proc 38:2535
Dierlich E, Orfanos CE, Pullmann H, Steigleder GK (1979) Epidermale Zellproliferation unter oraler Retinoid-Therapie bei Psoriasis. Autoradiographische Befunde an befallener und nicht-befallener Haut. Arch Dermatol Res 264:169
Di Giovanna JJ, Fletcher RT, Chader GJ (1985) Qualitative and quantitative analysis of cytosol retinoid binding proteins in human skin. J Invest Dermatol 85:460
Elias PM, Williams ML (1981) Retinoids, cancer and the skin. Arch Dermatol 117:160
Elias PM, Williams ML (1985) Retinoid effects on epidermal differentiation. In: Saurat H (ed) Retinoids: new trends in research and therapy. Karger, Basel, p 138
Elias PM, Chung J-C, Orozcio-Topete R, Nemanic MK (1983) Membrane glycoconjugate visualization and biosynthesis in normal and retinoid-treated epidermis. J Invest Dermatol 81:815
Evain Brion D, Raynaud F, Plet A, Laurent P, Leduc B (1985) Effect of retinoids on cAMP-mediated events in human psoriatic fibroblasts. Growth and cAMP-dependent protein kinase. In: Saurat H (ed) Retinoids: new trends in therapy and research. Karger, Basel, p 189
Fritsch PO, Pohlin G, Längle U, Elias PM (1981) Response of epidermal cell proliferation to orally administered aromatic retinoid. J Invest Dermatol 77:287
Frolik CA (1981) In vitro and in vivo metabolism of all-*trans* and *cis*-retinoic acid in the hamster. Ann NY Acad Sci 359:37
Frost P, Weinstein GD (1969) Topical administrations of vitamin A acid for ichthyosiform dermatoses and psoriasis. JAMA 207:1863
Fry JR, Schoch MA (1952) Therapeutische Versuche bei Psoriasis mit Vitamin A, zugleich ein Beitrag zur A-Hypervitaminose. Dermatologica 104:80
Fry L, MacDonald A, McMinn RMH (1970) Effect of retinoic acid in psoriasis. Br J Dermatol 83:191
Fuchs E, Green H (1981) Regulation of terminal differentiation of cultured human keratinocytes by vitamin A. Cell 25:617
Fulton JE, Gross PR, Cornelius CE, Kligman AM (1968) Darier's disease. Treatment with topical vitamin A acid. Arch Dermatol 93:726
Gates RE, King LE (1983) Cytoplasmic vitamin A binding proteins in chick embryo dermis and epidermis. J Invest Dermatol 85:279
Gilfix BM, Green H (1984) Bioassay of retinoids using cultured human conjunctival keratinocytes. J Cell Physiol 119:172
Gillenberg H, Immel C, Orfanos CE (1980) Retinoid-Einfluß auf die Zellkinetik gesunder menschlicher Epidermis. Arch Dermatol Res 269:331
Gollnick H, Orfanos CE (1983) Klinisch-therapeutischer Index und Dosimetrie der oralen Behandlung mit aromatischem Retinoid. Hautarzt 34:605
Gomez EC (1981) Differential effect of 13-*cis*-retinoic acid and an aromatic retinoid (Ro 10-9359) on the sebaceous glands of the hamster flank organ. J Invest Dermatol 76:68
Günther S (1973) Vitamin A acid in treatment of oral lichen planus. Arch Dermatol 107:277
Haddock-Russel D, Haddox MK (1981) Antiproliferative effects of retinoids related to the cell cycle specific inhibition of ornithine decarboxylase. Ann NY Acad Sci 359:281

Hashimoto T, Dykes PJ, Marks R (1985) Retinoid-induced inhibition of growth and reduction of spreading of human epidermal cells in culture. Br J Dermatol 112:637
Hercend T, Bruley-Rosset M, Mathé G (1981) In vivo immunostimulating properties of two retinoids: Ro 10-9359 and Ro 13-6298. In: Orfanos CE et al. (eds) Retinoids. Advances in basic research and therapy. Springer, Berlin Heidelberg New York, p 21
Iizuka H, Okkuma N, Okkawara H (1985) Effects of retinoids on the cyclic AMP of pig skin epidermis. J Invest Dermatol 85:324
Jetten AM (1981) Action of retinoids and phorbol esters on cell growth and the binding of epidermal growth factor. Ann NY Acad Sci 359:200
Jetten AM (1984) Modulation of cell growth by retinoids and their possible mechanism of action. Fed Proc 43:134
Jetten AM, Smits H (1985) Regulation of differentiation of tracheal epithelial cells by retinoids. Ciba Found Symp 113:61
Jetten AM, Jetten MER, Shapiro SS, Poon JP (1979) Characterization of the action of retinoids on mouse fibroblast cell lines. Exp Cell Res 119:289
Kalin JR, Starling ME, Hill DL (1981) Disposition of all-*trans*-retinoic acid in mice following oral doses. Drug Metab Dispos 9:196
Kaplan RP, Russell DH, Lowe NJ (1983) Etretinate therapy for psoriasis: clinical responses, remission times, epidermal DNA and polyamine responses. J Am Acad Dermatol 8:95
Keddie F (1948) Use of vitamin A in the treatment of cutaneous diseases. Arch Dermatol Syph 58:64
King K, Jones DH, Daltrey DG, Cunliffe WF (1982) A double blind study of the effects of 13-*cis*-retinoic acid on acne, sebum excretion rate and microbial population. Br J Dermatol 107:583
Kistler A (1985) Structure-activity relationship of retinoids on the differentiation of cultured chick foot skin. In: Saurat H (ed) Retinoids: new trends in research and therapy. Karger, Basel, p 309
Kitajima Y, Tsuneda Y, Furuta H, Mori S (1983) Effects of oral retinoid on the SDS-PAGE profile of stratum corneum keratin in patients with psoriasis. J Dermatol 10:383
Klann RC, Marchok AC (1982) Effects of retinoic acid on cell proliferation and cell differentiation in a rat tracheal epithelial cell line. Cell Tissue Kinet 15:473
Klaus M, Bollag W, Huber P, Küng W (1983) Sulfur-containing arotinoids, a new class of retinoids. Eur J Med Chem Ther 18:425
Kligman AM, Fulton JE, Plewig G (1969) Topical vitamin A acid in acne vulgaris. Arch Dermatol 99:469
Knippel J, Bauer R, Orfanos CE (1984) No influence of three synthetic retinoids (Ro 10-1670, Ro 4-3780, Ro 13-6298) on lipoxygenase activity in two in vitro systems. In: Cunliffe WJ, Miller A (eds) Retinoid therapy. MTP Press, Lancaster, p 345
Krueger GG, Shelby NJ, Hansen CD, Taylor MD (1980) Comparison of labeling indices of skin involved and uninvolved with psoriasis; placebo and oral retinoid Ro 10-9359 vs. time. Clin Res 28:21
Kubilus J, Rand R, Baden H (1981) Effects of retinoic acid and other retinoids on the growth and differentiation of 3T3 supported human keratinocytes. In Vitro 17:786
Landthaler M, Kummermehr J, Wagner A, Nikolowski J, Plewig G (1981) Effects of 13-*cis*-retinoic acid on sebaceous glands in humans. In: Orfanos CE et al. (eds) Retinoids. Advances in basic research and therapy. Springer, Berlin Heidelberg New York, p 259
Lasnitzki I (1963) Growth pattern of the mouse prostate gland in organ culture and its response to sex hormones, vitamin A and 3-methyl-cholanthrene. NCI Monogr 12:381
Levitski A, Willingham M, Pastan J (1980) Evidence for participation of transglutaminase in receptor mediated endocytoses. Proc Natl Acad Sci USA 77:2706
Leyden JJ, McGinley KJ, Webster GF (1982) Prolonged reductions of *Propionibacterium acnes* in acne conglobata patients treated with 13-*cis*-retinoic acid. J Invest Dermatol 78:350
Liau G, Ong DE, Chytil F (1981) Interaction of the retinol cellular retinol-binding protein complex with isolated nuclei and nuclear compounds. J Cell Biol 91:63

Libby P, Bertram J (1982) Lack of intracellular retinoid-binding proteins in a retinol sensitive cell line. Carcinogenesis 3:481
Lichti U, Yuspa SH (1985) Inhibition of epidermal terminal differentiation and tumour promotion by retinoids. In: Nugent J, Clark S (eds) Retinoids, differentiation and disease. Pitman, London, p 77
Loeliger P, Bollag W, Mayer H (1980) Arotinoids, a new class of highly active retinoids. Eur J Med Chem 15(1):9
Lotan R 81980) Effects of vitamin A and its analogs (retinoids) on normal and neoplastic cells. Biochem Biophys Acta 605:33
Lotan R (1985) Mechanism of inhibition of tumor cell proliferation by retinoids. In: Saurat H (ed) Retinoids: new trends in research and therapy. Karger, Basel, p 97
Lotan R, Dennert G (1979) Stimulatory effects of vitamin A analogs on induction of cell-mediated cytotoxicity in vivo. Cancer Res 39:55
Lotan R, Neumann G, Lotan D (1980) Relationships among retinoid structure, inhibition of growth, and cellular retinoic acid-binding protein in cultured S91 melanoma cells. Cancer Res 40:1097
Lotan R, Fischer J, Meromsky L, Modlave K (1982) Effects of retinoic acid on protein synthesis in cultured melanoma cells. J Cell Physiol 113:47
Lowe NJ, Kaplan R, Breeding J (1982) Etretinate treatment for psoriasis inhibits epidermal ornithine decarboxylase. J Am Acad Dermatol 6:697
Maden M (1985) Retinoids and the control of pattern in regenerating limbs. Ciba Found Symp 113:123
Marcelo CL, Madison KC (1984) Regulation of the expression of epidermal keratinocyte proliferation and differentiation by vitamin A analogs. Arch Dermatol Res 276:381
Marcelo CL, Tomich J (1983) Cyclic AMP, glucocorticoid, and retinoid modulation of in vitro growth. J Invest Dermatol 81:64s
Marcelo CL, Kim YA, Jeffrey LK, Voorhess JJ (1978) Stratification, specialization, and proliferation of primary keratinocyte cultures. J Cell Biol 79:356
Marwick CH (1984) More cautionary labelling appears on isotretinoin. JAMA 252:3208
Massarella JW, Colburn WA, Vane FM, Bugge C, Rodriguez L, Cunningham W (1983) The pharmacokinetic profile of Ro 10-9359 during chronic dosing in patients with severe psoriasis. Data on file. Hoffman-LaRoche, Nutley
Mayer H, Bollag W, Hanni R, Ruegg R (1978) Retinoids, a new class of compounds with prophylactic and therapeutic activities in oncology and dermatology. Experientia 34:1105
Milstone LM (1983) Population dynamics in cultures of stratified squamous epithelia. J Invest Dermatol 81:69s
Milstone LM, McGuire J, La Vigne JF (1982) Retinoic acid causes premature desquamation of cells from confluent cultures of stratified squamous epithelia. J Invest Dermatol 79:253
Ong DE, Chytil F (1975 a) Specificity of cellular retinol-binding protein for compounds with vitamin A activity. Nature 255:74
Ong DE, Chytil F (1975 b) Retinoic acid binding protein in rat tissue. J Biol Chem 250:6113
Ong DE, Chytil F (1976 a) Changes in levels of cellular retinol- and retinoic acid-binding proteins of liver and lung during perinatal development. Proc Natl Acad Sci USA 73:3976
Ong DE, Chytil F (1976 b) Presence of retinol and retinoic acid binding proteins in experimental tumors. Cancer Lett 2:25
Ong DE, Chytil F (1978) Cellular retinol-binding protein from rat liver. Purification and characterization. J Biol Chem 253:4551
Ong DE, Crow JA, Chytil F (1982) Radioimmunochemical determination of cellular retinol and retinoic acid-binding proteins in cytosol of rat tissues. J Biol Chem 257:13385
Orfanos CE (1980) Oral retinoids – present status. Br J Dermatol 103:473
Orfanos CE (1982) Oral retinoid in psoriasis; current clinical experiences and possible mechanisms of action. In: Farber EM, Cox AJ, Nall L, Jacobs PH (eds) Psoriasis. Grune and Stratton, New York, p 197
Orfanos CE (1984) Teratogenität von Isotretinoin. Hautarzt 35:503

Orfanos CE (1985) Retinoids in clinical dermatology; an update. In: Saurat H (ed) Retinoids: new trends in research and therapy. Karger, Basel, p 314
Orfanos CE, Bauer R (1983) Evidence for anti-inflammatory activities of oral synthetic retinoids and experimental findings and clinical experience. Br J Dermatol [Suppl 25]109:55
Orfanos CE, Schmidt HW, Mahrle G, Gartmann H, Lever WF (1973) Retinoic acid in psoriasis: its value for topical therapy with and without corticosteroids. Br J Dermatol 88:167
Orfanos CE, Stadler R, Gollnick H, Tsambaos D (1985) Current developments of oral retinoid therapy with three generations of drugs. Curr Probl Dermatol 13:33
Orozcio-Topete R, Chung JC; Elias PM (1983) Epidermal glycoconjugate biosynthesis in organ and cell culture; effect of retinoids. J Invest Dermatol 80:316
Ott F, Bounameaux Y (1983) Pilot study of a new retinoid, Ro 12-7554, in psoriasis and in some congenital disorders of keratinization. Dermatologica 167:52
Paravicini U (1981) Pharmacokinetics and metabolism of oral aromatic retinoids. In: Orfanos CE et al. (eds) Retinoids. Advances in basic research and therapy. Springer, Berlin Heidelberg New York, p 13
Paravicini U, Busslinger A (1984) Etretinate and isotretinoin, two retinoids with different pharmacokinetic profiles. In: Miller A, Cunliffe WJ (eds) Retinoid therapy. MTP Press, Lancaster, p 11
Paravicini M, Camenzind M, Gower M, Geiger JM, Saurat JH (1985) Multiple dose phamarcokinetics of Ro 10-1670, the main metabolite of etretinate (Tigason). In: Saurat H (ed) Retinoids: new trends in research and therapy. Karger, Basel, p 289
Peck GL, Olsen TG, Yoder FW, Strauss JS, Downing DT, Pandya M, Butkus D, Arnaud-Battandier J (1979) Prolonged remissions of cystic and conglobate acne with 13-*cis*-retinoic acid. N Engl J Med 300:329
Peck GL, Olsen TG, Butkus D, Pandya M, Arnaud-Battandier J (1982) Isotretinoin versus placebo in the treatment of cystic acne. J Am Acad Dermatol 6:735
Plewig G (1980) Der Einfluß des aromatischen Retinoids Ro 10-9359 und der 13-*cis*-Retinsäure Ro 4-3780 auf die Talgdrüsen des Syrischen Hamsters. Arch Dermatol Res 268:329
Plewig G, Nikolowski J, Wolff HH (1982) Action of isotretinoin in acne rosacea and gram-negative folliculitis. J Am Acad Dermatol 6:766
Plewgi G, Gollnick H, Meigel W, Wokalek H (1983) 13-*cis*-Retinsäure zur oralen Behandlung der Acne conglobaton – Ergebnisse einer multizentrischen Studie. Hautarzt 32:634
Plewig G, Braun-Falco O, Klövekorn W, Luderschmidt C (1986) Isotretinoin zur örtlichen Behandlung von Akne und Rosazea sowie tierexperimentelle Untersuchungen mit Isotretinoin und Arotinoid. Hautarzt 37:138
Ponec M (1984) Retinoid-induced changes in lipid synthesis in cultured human epidermal keratinocytes. Br J Dermatol [Suppl 27]111:242
Ponec M (1985) Retinoid-induced effects on lipid synthesis in human epidermal keratinocytes cultured at a low and high calcium level. In: Saurat H (ed) Retinoids: new trends in research and therapy. Karger, Basel, p 175
Rice RH, Cline PR, Coe LE (1983) Mutually antagonistic effects of hydrocortisone and retinyl acetate on envelope competence in cultured malignant human keratinocytes. J Invest Dermatol 81:176s
Rollman O, Vahlquist A (1983) Retinoid concentrations in skin, serum and adipose tissue of patients treated with etretinate. Br J Dermatol 109:439
Runne U, Orfanos CE, Gartmann H (1973) Perorale Applikation zweier Derivate der Vitamin A-Säure zur internen Psoriasis-Therapie. 13-*cis*-beta Vitamin A-Säure und Vitamin A-Säureacethylamid. Arch Dermatol Forsch 247:171
Ruzicka T (1985) Effect of retinoids on arachidonic acid metabolism in human epidermis and polymorphonuclear leukocytes. In: Saurat H (ed) Retinoids: new trends in research and theapy. Karger, Basel, p 180
Sani BP (1977) Localization of retinoic acid-binding protein in muscle. Biochem Biophys Res Commun 75(1):7

Sani BP (1979) Retinoic acid-binding protein: a plasma membrane component. Biochem Biophys Res Commun 91:502
Sato M, Hiragun A, Mitsui H (1980) Preadipocytes possess cellular retinoid-binding proteins and their differentiation is inhibited by retinoids. Biochem Biophys Res Commun 95:1839
Schumacher A, Stüttgen G (1971) Vitamin A-Säure bei Hyperkeratosen, epithelialen Tumoren und Akne. Dtsch Med Wochenschr 96:1547
Sporn MB, Clamon GH, Dunlop NM, Newtow DL, Smith JM, Saffiotti V (1975) Activity of vitamin A analogues in cell cultures of mouse epidermis and organ cultures of hamster trachea. Nature 253:47
Stadler R (1987) Kultivierung humaner und muriner Keratinozyten als in vitro Epidermis-Modell. Einfluß synthetischer Retinoide und humaner Interferone auf Proliferation und Differenzierung. *Habilitationsschrift*, Freie Universität Berlin
Stadler R, Marcelo C, Voorhees JJ, Orfanos CE (1983) Effect of a new retinoid, arotinoid, Ro 13-6298 on in vitro keratinocyte proliferation, differentiation and cellular cAMP content. J Invest Dermatol 80:356A
Stadler R, Marcelo CL, Voorhees JJ, Orfanos CE (1984) Effect of a new retinoid, arotinoid (Ro 13-6298), on in vitro keratinocyte proliferation and differentiation. Acta Derm Venerol (Stockh) 64:405
Stadler R, Marcelo CL, Orfanos CE (1985) Einfluß synthetischer Retinoide auf die Synthese histidinreichen Proteins in vitro – Ein Parameter antipsoriatischer Wirksamkeit? In: Mahrle G, Ippen H (eds) Dermatologische Therapie. perimed, Erlangen, p 144
Staquet MJ, Faure MR, Reano A, Viac J, Thivolet J (1983) Keratin polypeptide profile in psoriatic epidermis normalized by treatment with etretinate (aromatic retinoid Ro 10-9359). Arch Dermatol Res 275:124
Straumfjord JV (1942) Lesions of vitamin A deficiency: their local character and chronicity. Northwest Med 41:229
Stüttgen G (1962) Zur Lokalbehandlung von Keratosen mit Vitamin A-Säure. Dermatologica 124:65
Takase S, Ong D, Chytil F (1979) Cellular retinol-binding protein allows specific interaction of retinol with the nucleus in vitro. Proc Natl Acad Sci USA 76:2204
Thivolet CH, Hintner HH, Stanley JR (1984) The effect of retinoic acid on the expression of pemphigus and pemphigoid antigens in cultured human keratinocytes. J Invest Dermatol 82:329
Tsambaos D (1984) Effekte von Arotinoidsäure auf die Embryogenese, Proliferation und Differenzierung der Epidermis und auf die Talgdrüsenaktivität. *Habilitationsschrift*, Freie Universität Berlin
Tsambaos D, Orfanos CE (1981) Ultrastructural evidence suggesting an immunomodulatory activity of oral retinoid. Its effect on dermal components in psoriasis. Br J Dermatol 104:37
Tsambaos D, Orfanos CE (1983) Antipsoriatic activity of a new synthetic retinoid. The arotinoid Ro 13-6298. Arch Dermatol 119:746
Tsambaos D, Mahrle G, Orfanos CE (1980) Epidermal changes by oral excess of aromatic retinoid in guinea pigs. Arch Dermatol Res 267:141
Tsambaos D, Gollnick H, Orfanos CE (1983) Orales Arotinoid bei Psoriasis, kongenialer Ichthyose, Lichen, Palmoplantarkeratosen, M Darier (15 Patienten). Hautarzt [Suppl 6]34:40
Tsambaos D, Zimmermann B, Orfanos CE (1984) Effects of retinoids on chondrogenesis and epidermogenesis in vitro. In: Cunliffe WJ, Miller A (eds) Retinoid therapy. MTP Press, Lancaster, p 119
Tsambaos D, Stadler R, Hilt K, Zimmermann B, Orfanos CE (1985) Effects of arotinoid ethyl ester on epithelial differentiation and proliferation. Ciba Found Symp 113:97
Vahlquist A, Michaelson G, Kober A, Sojholm J, Palmskog G, Petterson U (1981) Retinoid-binding proteins and the plasma transport of etretinate (Ro 10-9359) in man. In: Orfanos CE et al. (eds) Retinoids. Advances in basic research and therapy. Springer, Berlin Heidelberg New York, p 109

Verma A, Shapas B, Rice H, Boutwell R (1979) Correlation of the inhibition by retinoids of tumor promotor-induced ornithine decarboxylase activity and of skin tumor promotion. Cancer Res 39:419

Voorhess JJ (1983) Leukotrienes and other lipoxygenase products in the pathogenesis and therapy of psoriasis and other dermatoses. Arch Dermatol 119:541

Wiggert B, Russel P, Lewis M, Chader JG (1977) Differential binding to soluble nuclear receptors and effects on cell viability of retinol and retinoic acid in cultured retinoblastoma cells. Biochem Biophys Res Commun 79:218

Wilkoff LJ, Peckham JC, Dulmadge EA, Mowry RW, Chopra DP (1976) Evaluation of vitamin A analogs in modulating epithelial differentiation of 13-day chick embryo metatarsal skin explants. Cancer Res 36:964

Wolbach SB, Howe PR (1925) Tissue changes following deprivation of fat-soluble A vitamin. J Exp Med 42:753

Wolbach SB, Howe PR (1928) Vitamin A deficiency in the guinea pig. Arch Pathol [Lab Med]5:239

Wolbach SB, Howe PR (1933) Epithelial repair in recovery from vitamin A deficiency. J Exp Med 57:511

Wong E, Barr RM, Brain SD, Olins LA, Greaves MW (1983) The effect of etretinate on cyclo-oxygenase and lipoxygenase products of arcachidonic acid in psoriatic skin. Br J Dermatol 109:703

Yaar M, Stanley JR, Katz SI (1981) Retinoic acid delays the terminal differentiation of keratinocytes in suspension culture. J Invest Dermatol 76:363

Yuspa S, Ben T, Steinert P (1982) Retinoic acid induces transglutaminase activity but inhibits cornification in cultured epidermal cells. J Biol Chem 257:9906

Ziboh VA, Price B, Fulton J (1975) Effects of retinoic acid on prostaglandin biosynthesis in guinea pig skin. J Invest Dermatol 65:370

Zile MH, Inhorn RC, DeLuca HF (1982a) Metabolism in vivo of all-*trans*-retinoic acid. J Biol Chem 257:3544

Zile MH, Inhorn RC, DeLuca HF (1982b) Metabolites of all-*trans*-retinoic acid in bile: identification of all-*trans*- and 13-*cis*-retinoyl glucuronides. J Biol Chem 257:3537

CHAPTER 26

Hypolipidaemic Agents in the Treatment of Xanthomata

N. E. MILLER

A. Introduction

Animal studies (KODAMA et al. 1981) and the resolution of xanthomata in many hyperlipidaemic patients after plasma lipid reduction (HESSEL et al. 1976) have confirmed that the association between plasma lipids and xanthoma formation is causal. Accordingly, the treatment of hyperlipidaemia is an important component of the management of xanthomatous patients. Not all types of hyperlipidaemia produce xanthomata, and the nature of the lesions varies according to the lipoprotein class (HESSEL et al. 1976). The lipoprotein patterns found in association with different forms of xanthomata are presented in Table 1 and the causes in Table 2 (LEWIS 1976).

B. Treatment of Hyperlipidaemia

Secondary hyperlipidaemias are corrected by treatment of the primary condition. Lipoprotein lipase deficiency and apolipoprotein CII deficiency can be controlled by dietary fat restriction, supplemented if necessary by medium-chain triglycerides (absorbed without chylomicron formation). Other types of primary hyper-

Table 1. Lipoprotein patterns associated with different types of cutaneous xanthomata

Xanthomata	Lipoprotein phenotype[a]					
	I	IIa and IIb	III	IV	V	LpX[b]
Eruptive	1		2	2	1	3
Tuberous		2[c]	1			3
Planar		2[c]				
Palmar crease			1	2		3
Palpebral[d] (xanthelasma)		1	2	2		3

1 Usual; 2 occasional; 3 rare.

[a] WHO classification (BEAUMONT et al. 1970). Lipoprotein(s) elevated: I chylomicrons; IIa LDL; IIb VLDL and LDL; III IDL; IV VLDL; V chylomicrons and VLDL. Elevated HDL levels (hyperalphalipoproteinaemia) do not produce xanthomata.

[b] Lipoprotein X: an abnormal lipoprotein which accounts for the hypercholesterolaemia of chronic cholestasis.

[c] When due to severe familial hypercholesterolaemia only (see Table 2).

[d] Commonly associated with normal plasma total lipid levels.

Table 2. Causes of hyperlipoproteinaemias

Phenotype	Familial	Other
I	Lipoprotein lipase deficiency Apolipoprotein CII deficiency	Insulin-dependent diabetes mellitus Paraproteinaemia [a]
IIa, IIb	Familial hypercholesterolaemia [b] Familial combined hyperlipidaemia [c]	Polygenic (multifactorial) hypercholesterolaemia [c] Myxoedema Nephrotic syndrome Alcohol Paraproteinaemia [a]
III	Broad-beta disease	Myxoedema Paraproteinaemia [a]
IV	Familial hypertriglyceridaemia Familial combined hyperlipidaemia [c]	Insulin-resistant diabetes mellitus Nephrotic syndrome Alcohol Oestrogens Glycogen storage disease Paraproteinaemia [a]
V	As for type I Familial hypertriglyceridaemia	As for types I and IV

[a] May produce xanthomata without hyperlipidaemia.
[b] Produces tendon xanthomata in addition to cutaneous forms.
[c] The commonest forms of hypercholesterolaemia; produce xanthelasma only (never other cutaneous or tendon xanthomata).

Table 3. Drug selection for familial and polygenic hyperlipoproteinaemias

Disorder	Cholestyramine, colestipol	Nicotinic acid	Clofibrate	Probucol	Gemfibrozil	Mevinolin
Familial hypercholesterolaemia	1	2		2		2
Familial hypertriglyceridaemia		2	2		1	
Broad-beta disease		2	2		1	
Familial combined hyperlipidaemia	2	2			1	2
Polygenic hypercholesterolaemia	1	2		2	2	2

1 First choice; 2 second choice.

lipidaemia and polygenic hyperlipidaemia may be controlled by weight reduction and dietary fat modification (LEWIS 1976). Many patients, however, require lipid-lowering drugs (Table 3). Newer drugs include specific inhibitors of cholesterol synthesis (e.g. mevinolin) and drugs related to clofibrate (e.g. gemfibrozil, fenofibrate).

I. Cholestyramine

Cholestyramine has its major effect on the plasma low density lipoprotein (LDL) concentration, which it usually reduces by 20%–35% in heterozygous familial hypercholesterolaemia, familial combined hyperlipidaemia, polygenic hypercholesterolaemia and some cases of primary biliary cirrhosis. It is ineffective in homozygous familial hypercholesterolaemia. A nonabsorbable anion exchange resin, cholestyramine releases chloride in exchange for bile acids, thereby reducing bile acid reabsorption in the terminal ileum; reduced feedback suppression of cholesterol of 7α-hydroxylase in the liver increases the conversion of cholesterol to bile acids. Although cholesterol synthesis increases concomitantly, this is insufficient to prevent a decrease in hepatocyte cholesterol, which in turn increases the receptor-mediated uptake and catabolism of LDL (Shepherd et al. 1980). The plasma concentration of very low density lipoprotein (VLDL), and therefore also of triglyceride, may increase. The dosage is 4–12 g three times daily. Gastrointestinal side effects include flatus, nausea, constipation and, rarely, vomiting. Very large doses may reduce the absorption of fat-soluble vitamins. Absorption of cardiac glycosides, thyroxine, phenylbutazone, vancomycin and paracetamol is also reduced. Reduction of vitamin K absorption may potentiate oral anticoagulants.

II. Colestipol

This alternative bile acid-sequestering resin has the same mode of action and side effects as cholestyramine (Miller et al. 1973); it is given at the dose of 5–10 g three times daily.

III. Nicotinic Acid

Nicotinic acid, administered as the sodium salt at doses of 1–3 g three times daily, frequently reduces the plasma concentrations of both VLDL and LDL (raising that of high density lipoprotein, HDL) in all forms of primary endogenous hyperlipidaemia. It is absorbed rapidly after oral administration and is metabolised by *N*-methylation. The primary mode of action is to inhibit lipolysis in adipose tissue, which in turn diminishes hepatic VLDL synthesis by reducing delivery of free fatty acids to the liver (Grundy et al. 1981). Reduction of LDL concentration reflects reduced LDL production from VLDL. Effects have also been observed on cholesterol metabolism. The most frequent side effects are transient flushing, with pruritus or a burning sensation, owing to prostaglandin-mediated vasodilation during drug absorption. These effects can be reduced by commencing with a small dose (300 mg/day) and gradually increasing to the therapeutic dose over 2–6 weeks, by taking the drug during a meal, or by taking 300 mg aspirin 30 min before each dose (Kane et al. 1981). Other side effects include rashes, dyspepsia, hyperuricaemia, impaired glucose tolerance and reversible liver damage.

IV. Clofibrate

Clofibrate, ethyl-*para*-chlorophenoxyisobutyrate, is particularly effective in reducing the elevated levels of intermediate density lipoprotein in broad-beta disease (familial type III hyperlipoproteinaemia). VLDL levels are also consistently reduced in familial hypertriglyceridaemia. The effect on LDL concentration is less predictable (ROSE et al. 1976), but clofibrate can be of value in familial combined hyperlipidaemia or polygenic hypercholesterolaemia. Because in addition to a reduced incidence of non-fatal myocardial infarction a WHO trial (1980) showed an increase in non-cardiovascular mortality, clofibrate should be reserved for hyperlipidaemic patients unresponsive to other drugs. The dose is 0.5–1.0 g 2–3 times daily; much lower doses must be used in renal disease (see later in this section). After hydrolysis the free acid circulates almost entirely bound to plasma albumin. The major route of elimination is renal excretion of the glucuronide. Many effects on lipid metabolism have been documented (e.g. MILLER et al. 1976), but the primary mode of action remains uncertain. Side effects include impotence, slight transient elevation of serum transaminases, and myositis with increased serum creatine kinase. Bile cholesterol saturation is also increased, owing to increased biliary cholesterol secretion, raising the risk of cholelithiasis (WHO 1980). Clofibrate potentiates the coumarin anti-coagulants, phenindione and the sulphonylureas. It also competes with frusemide for albumin binding sites. When patients with nephrotic syndrome are treated with clofibrate and frusemide, the combination of hypoalbuminaemia, delayed elimination of clofibrate and displacement from albumin can produce very high plasma concentrations of free *p*-chlorophenoxyisobutyrate, producing severe myositis. Very small doses should be given initially to all patients with renal disease (e.g. 500 mg/week), gradually increasing the dose while monitoring serum creatinine kinase activity.

V. Probucol

This sulphur-containing biphenolic compound lowers the VLDL, LDL and HDL cholesterol concentrations (NOSEDA et al. 1980). It is not clear whether increased fractional catabolism or decreased synthesis of LDL is the mode of action (NESTEL and BILLINGTON 1981). Cholesterol synthesis is diminished. With 500 mg twice daily, probucol is very effective in some cases of familial hypercholesterolaemia (heterozygous and homozygous) and polygenic hypercholesterolaemia. Other disorders do not respond well; the reason for this is not known. Diarrhoea may occur. Cholesterol-fed monkeys have developed syncope and ventricular arrhythmias with probucol, but cardiac effects have not been established in humans (BARBER et al. 1981).

VI. Drug Combinations

In resistant cases of familial hypercholesterolaemia combinations of two drugs may be more effective than either drug alone; e.g. a bile acid binding resin plus nicotinic acid (KANE et al. 1981) or clofibrate (MILLER et al. 1973).

References

Barber MJ, Gilmour FR, Mueller TM, Zipes DP (1981) Electrophysiological effects of probucol on monkey ventricle. Circulation 64:IV–123

Beaumont JL, Carlson LA, Cooper GR, Fejfar Z, Fredrickson DS, Strasser T (1970) Classification of hyperlipidaemis and hyperlipoproteinaemias. Bull WHO 43:891–915

Grundy SM, Mok HYI, Zech L, Berman M (1981) Influence of nicotinic acid on the metabolism of cholesterol and triglycerides in man. J Lipid Res 22:24–36

Hessel LW, Vermeer BJ, Polano MK, de Jonge H, de Pagter HAT, van Get CM (1976) Primary hyperlipoproteinaemia in xanthomatosis. Clin Chim Acta 69:405–416

Kane JP, Malloy MJ, Tun P, Phillips NR, Freedman DD, Williams ML, Rowe JS, Havel RJ (1981) Normalization of low density lipoprotein levels in heterozygous familial hypercholesterolemia with a combined drug regimen. N Engl J Med 304:251–258

Kodama H, Nagao Y, Arakawa K, Tada J, Nohara N (1981) Cholesterol synthesis and esterification in experimental xanthoma tissues. J Lipid Res 22:1033–1041

Lewis B (1976) The hyperlipidaemias. Blackwell, Oxford

Miller NE, Clifton-Bligh P, Nestel PJ, Whyte HM (1973) Controlled clinical trial of a new bile acid-sequestering resin colestipol, in the treatment of hypercholesterolaemia. Med J Aust 1:1223–1227

Miller NE, Mjos OD, Oliver MF (1976) Effects of *p*-chlorophenoxyisobutyrate on free fatty acid mobilization from canine subcutaneous adipose tissue in situ. Clin Sci 51:107–110

Nestel PJ, Billington T (1981) Effects of probucol on low density lipoprotein removal and high density lipoprotein synthesis. Atherosclerosis 38:203–209

Noseda G, Lewis B, Paoletti R (1980) Diet and drugs in atherosclerosis. Raven, New York

Rose HG, Haft GK, Juliano J (1976) Clofibrate-induced low density lipoprotein elevation. Atherosclerosis 23:413–427

Shepherd J, Packard CJ, Bicker S, Lawrie TDV, Morgan HG (1980) Cholestyramine promotes receptor-mediated low density lipoprotein catabolism. N Engl J Med 302:1219–1222

WHO (1980) WHO cooperative trial on primary prevention of ischaemic heart disease using clofibrate to lower serum cholesterol: mortality follow-up. Lancet 2:379–385

CHAPTER 27

Drugs Acting on Dermal Connective Tissue

B. NUSGENS and C. M. LAPIÈRE

A. Introduction

The dermis is a complex tissue composed of differentiated fibroblasts embedded in an extracellular matrix containing various macromolecules: collagens, proteoglycans, elastin and glycoproteins. Interactions between these various constituents and with the cells determine the architectural organisation and the function of the supporting network. The different steps and their potential controls operating in the synthesis, degradation and organisation of these molecules are described in Chap. 9. Each of these steps can potentially be modified or modulated by exogenous agents. The discovery of pharmacological agents active in some of them and the knowledge of their mode of action has helped to elucidate physiological mechanisms. Much of the information about the pharmacology of the dermis is related to fibroblasts and mainly one of their biosynthetic products, collagen. This protein will be considered as a model compound to monitor the activity of drugs on the tissue. Some information will also be briefly presented about elastin, proteoglycans and glycoproteins in the dermis and in tissues other than skin.

We define as drugs a broad variety of chemical or physiological compounds capable of modifying any one of the components of the connective tissues in vitro and/or in vivo. These molecules will be considered in classes according to the metabolic step at which they act or are supposed to exert their activity.

B. Drugs Acting at the Level of Transcription and Translation

A large number of drugs probably act by regulating the transcription of the genome or at the translational level. The demonstration of promoter sequences in the collagen genes and of specific binding of repressors or activators provide additional opportunities for regulation of the expression of the gene (HATAMOCHI et al. 1986).

I. Glucocorticoids

The successful use of glucocorticoids as anti-inflammatory agents is tempered by numerous unwanted side effects among which are inhibition of growth, osteoporosis and skin atrophy due to loss of collagen (SHUSTER et al. 1967; BLACK et al.

1973). The reduction of collagen in connective tissues by glucocorticoids results from a selective depression of collagen synthesis (MCNELIS and CUTRONEO 1978; PONEC et al. 1979) related to a specific decrease in the cellular concentration of translatable procollagen mRNA as demonstrated in the dermis (ROKOWSKI et al. 1981), in calvaria (OIKARINEN and RYHANEN 1981) and in cultured human skin fibroblasts (OIKARINEN et al. 1983). The mechanisms mediating this decrease are unknown although recent data suggest that it could result from the activation of an intracellular regulation system rather than a direct effect of the steroid-receptor complex on the expression of the collagen genes (OIKARINEN et al. 1983). An increased degradation of the mRNAs in the presence of cortisol has also been reported in human skin (HAMALAINEN et al. 1985) although no such increased degradation was observed with dexamethasone on chick skin (COCKAYNE et al. 1986). VERBRUGGEN and ABE (1982) observed a lower ratio of type III to type I procollagen synthesised by fibroblasts in culture in the presence of corticosteroids. Rather than a net decrease in type III collagen synthesis, SHULL and CUTRONEO (1986) suggest a coordinated decrease in type I and type III collagen synthesis, with selective changes in the extracellular processing of procollagens and/or differential degradation of these two molecules. The simultaenous reduced activity of enzymes involved in collagen synthesis observed in vitro (OIKARINEN 1977) and in human skin (OIKARINEN and HANNUKSELA 1980) could represent an example of hormonal integrated regulation of protein synthesis.

Although growth and function of fibroblasts are generally depressed by anti-inflammatory corticosteroids these hormones can potentiate the mitogenic effect of polypeptide growth factors (GOSPODAROWICZ and MORAN 1974; IVANOVIC and WEINSTEIN 1981). At low concentration, they are required to sustain the growth of fibroblasts in chemically defined media (WALTHALL and HAM 1981). The large number of factors influencing the activity of glucocorticoids on fibroblasts in culture explains the variability of their effect in vitro (for review see PONEC 1984).

Hydrocortisone and dexamethasone inhibit the production of collagenase by human skin in organ culture (KOOB et al. 1974). A similar effect has also been demonstrated for collagenase and elastase secreted by macrophages in vitro (WERB 1978). Dexamethasone selectively decreases collagenase production by human skin fibroblasts at a pre-translational level (BAUER et al. 1985) and by synovial fibroblasts without affecting the half-life of mRNA (BRINCKERHOFF et al. 1986). The high collagenolytic activity induced by interactions between dermal fibroblasts and keratinocytes in reconstituted skin in vitro is also suppressed by dexamethasone and cortisol (YOSHIZATO et al. 1986).

Anti-inflammatory steroids decrease the amount and the synthesis of sulphated and non-sulphated glycosaminoglycans in vivo and in vitro (MANTHORPE et al. 1980). Hyaluronic acid proves to be the most affected in fibroblast culture (SAARNI 1978). Glucocorticoids enhance (10- to 30-fold) fibronectin mRNA level and synthesis by normal fibroblasts without altering the biosynthesis of the 140 kdalton fibronectin receptor or the splicing patterns of mRNA (OLIVER 1986). The retractile properties displayed by fibroblasts embedded in a three-dimensional collagen lattice are inhibited by corticosteroids (COULOMB et al. 1984; VAN STORY-LEWIS and TENENBAUM 1986) by a yet undefined mechanism.

II. Sex Hormones

At physiological concentration oestrogen specifically increases collagen synthesis by fibroblasts in culture without modifying cell multiplication (HOSOKAWA et al. 1981). Collagen degradation is decreased by oestrogen in bone (CRUESS and HONG 1975) and uterus (WOESSNER 1969). Induction of collagenase activity in cultured macrophages is largely inhibited by oestrogen and/or progesterone (WAHL 1977). Hyaluronic acid (not proteoglycans) is increased by topical application of oestradiol in vivo with a concomitant increase in hyaluronic synthetase activity (UZUKA et al. 1981). When studying sex hormone regulation of fibroblast activity, one has to keep in mind that receptors for these hormones are not expressed to the same extent in genital and non-genital skin (HERFERT et al. 1980).

III. Photochemotherapy

Photochemotherapy can induce dermal modifications since a part of the incident UV-A penetrates through the epidermis into the dermis. Histological (PIERARD and ACKERMAN 1979) and ultrastructural studies (KUMAKIRI et al. 1977) demonstrate modifications of the structure of the dermis and discrete remodelling of the elastin network (WOLFF et al. 1976). Mechanical properties of the superficial dermis are alterd (PIERARD et al. 1977). Fractionated treatment of skin fibroblasts in culture with 8-methoxypsoralen and UV-A is responsible for their photoinactivation (POHL and CHRISTOPHERS 1979). In the human no significant modification has been observed in urinary hydroxyproline excretion, hydroxyproline and hydroxylysine content in skin or enzymes involved in collagen synthesis (PIERARD et al. 1977; BLUMENKRANTZ et al. 1979; VAATAINEN et al. 1980). A small, but significant, increase in circulating immunoreactive amino terminal precursor sequences of procollagen type III in psoriatic patients after a few weeks of PUVA therapy (D. PIERARD, CH. M. LAPIERE and B. NUSGENS 1984, unpublished work) supports an increased synthesis of this collagen in irradiated skin.

IV. Antimitotics and Antibiotics

High doses of cyclophosphamide inhibit the synthesis of collagen in granulation tissue (WIE and BECK 1981). This effect is partly reversible (HANSEN et al. 1978). The reduced body weight of the treated animals and a reduced level of serum albumin suggest that the drug does not specifically affect collagen, but protein synthesis in general.

Bleomycin is an antibiotic agent producing a dose-dependent pulmonary fibrosis and skin lesions. Procollagen synthesis by skin fibroblasts is significantly increased when cultured in the presence of bleomycin and this depends predominantly on an enhanced synthesis of type I procollagen (CLARK et al. 1980). A selective increase in the rates of transcription of the genes coding for procollagen α_1 (I), fibronectin and elastin was demonstrated in interstitial pulmonary fibrosis induced by bleomycin (RAGHOW et al. 1985).

V. Growth Factors

Growth factors may be defined as polypeptides stimulating cell multiplication and metabolism through their binding to specific cell surface receptors. Much of the stimulus for the numerous studies of this expanding field arises from the homologies between growth factors, their receptors and oncogenes and proto-oncogenes and their presumed involvement in cancer (for review see GOUSTIN et al. 1986). Chapters 3 and 17, Vol. I, are devoted to growth factors, and the discussion here will be restricted to their activity on dermal cells.

1. Epidermal Growth Factor

EGF is a potent mitogenic peptide (6 Kdalton) isolated from mouse submaxillary gland and human urine (urogastrone) (COHEN and CARPENTER 1975) acting on various epithelial and mesenchymal cell cultures or organ cultures through interaction with a surface receptor exhibiting a tyrosine kinase function, internalisation of the EGF-receptor complexes, fusion with lysosomes and degradation (for review see COHEN 1983). Recent in situ hybridisation analyses indicate that RNA complementary to cloned EGF probes are present in a large variety of tissues (RALL et al. 1985). In fibroblast culture, EGF causes a dose-dependent stimulation of protein synthesis, but collagen synthesis remains unaffected (HUEY et al. 1980) or is reduced (STEINMANN et al. 1982).

Several recent reports have demonstrated that wound healing in vivo is accelerated by EGF (BUCKLEY et al. 1985; LAATO et al. 1986) or urogastrone (FRANKLIN et al. 1986) by an accumulation of mesenchymal cells in granulation tissue together with an increased collagen biosynthesis. The addition of EGF to skin fibroblasts embedded in a three-dimensional collagen lattice resulted in a stimulation of both non-collagen proteins and collagen synthesis as observed in vivo (COLIGE et al. 1987).

2. Fibroblast Growth Factors

FGFs, a family of growth factors (14–18 Kdalton), acidic and basic, isolated from brain (GOSPODAROWICZ et al. 1984) stimulate the proliferation of a variety of mesodermal cells in vitro. The FGF receptor has been identified and characterised (NEUFELD and GOSPODAROWICZ 1985). While fibroblasts are clearly stimulated by FGF in vitro, the effect of FGF in vivo on the proliferation of fibroblasts in the dermis is less clear (KATOH et al. 1985).

3. Platelet-Derived Growth Factor

PDGF, stored in the α-granules of platelets, selectively stimulates the growth of connective tissue cells and induces a chemotactic response in fibroblasts and smooth muscle cells. When platelets adhere to traumatised tissues and release PDGF, the connective tissue cells respond by replication. This effect needs the cooperation of other serum factors, lipoproteins, somatomedins, etc. (for review see ROSS 1985).

4. Insulin and Insulin-Like Growth Factors

Insulin (plus dexamethasone and EGF) is a required ingredient of a serum-free medium to sustain the growth of fibroblasts (WALTHALL and HAM 1981). When cultured in low concentrations of glucose, fibroblasts are responsive to insulin (SHAW and AMOS 1973). The concentration of glucose can influence the amount and the type of collagen secreted and some post-translational modification of this protein (VILLEE and POWERS 1977). The IGFs, a family ancestrally related to pro-insulin, comprise IGF-I (human somatomedin C) and IGF-II (human somatomedin A and rat multiplication-stimulating activity). IGF-I, one of the important growth factors found in serum and plasma, is produced in response to growth hormone and has been shown to stimulate the incorporation of sulphate into proteoglycans. The human genes for IGF-I and IGF-II have been cloned. The IGF-II gene is closely linked to the human insulin gene. Distinct receptor molecules for each IGF have been described. The IGF-I receptor shows homology with the insulin receptor (for review see GOUSTIN et al. 1986).

5. Cytokines

Mononuclear cells secrete a variety of factors capable of regulating several fibroblast functions such as collagen synthesis and collagenase production. These factors, among which interleukin 1 (IL-1), interleukin 2 (IL-2) and gamma interferon (IFN-γ) are important regulatory molecules involved in the inflammatory response in many diseases (for review see FREUNDLICH et al. 1986). The specific inhibitory activity of IFN-γ on fibroblast collagen production was used to reduce the excessive production of collagen by scleroderma fibroblasts in culture (ROSENBLOOM et al. 1986), opening a new therapeutic approach to this disease. Other mediators of inflammation are active on fibroblasts in vitro, e.g. histamine (HATAMOCHI et al. 1985), serotonin (BOUCEK et al. 1972) and prostaglandins (KRANE 1982).

6. Tumour Angiogenesis Factor and Angiogenesis Inhibitors

TAF, isolated from tumours, vitreous humour of foetal calf eyes or corpus luteum, induces the proliferation of blood vessels in vivo and endothelial cells in vitro (FOLKMAN and HAUDENSCHILD 1980).

7. Transforming Growth Factors

The TGFs may be defined as factors capable of stimulating the anchorage-independent growth in soft agar of non-transformed cells. Two peptides were isolated from conditioned medium of sarcoma cells. TGF-α shows a significant homology with EGF, binds to EGF receptor, elicits the same biological responses and, alone, stimulates soft agar colony formation weakly. It has been demonstrated in a variety of non-neoplastic tissues. TGF-β promotes the growth of normal cells in soft agar, alone or in the presence of EGF or TFG-α. It has been demonstrated both to stimulate and inhibit cell proliferation and differentiation, depending upon the type of cell. A major non-neoplastic storage site of TGF-β is found in human platelets, indicating that it may be an important mediator in wound healing (for review see ROBERTS and SPORN 1986).

VI. Oncogenes

Transfection of fibroblasts with oncogenes induces significant modifications in the expression of the genes of type I and type III collagen (SCHMIDT et al. 1985; SETOYAMA et al. 1986). The expanding study of oncogenes, proto-oncogenes, growth factors and their receptors has led to identification of specific genes important in growth regulation and differentiation in normal cells, and also the involvement of growth factors in the aetiology of cancer.

VII. Precursor Sequences of Procollagens

Amino terminal precursor sequences of bovine procollagen type I (p-N-I) and type III (p-N-III) display a specific inhibitory effect on collagen synthesis in calf (NUSGENS et al. 1978), normal human (WIESTNER et al. 1979) and scleroderma (KRIEG et al. 1978) skin fibroblasts in culture. Both p-N-I and p-N-III inhibit the translation of mRNA procollagen type I in a cell-free system (PAGLIA et al. 1979).

The carboxypropeptide also displays a specific inhibitory effect on collagen synthesis and induces a reduced amount of mRNA for type I collagen. Such a regulation acts most probably at a pre-translational level (WU et al. 1986). Using a synthetic peptide of 22 amino acids homologous with a highly conserved region of the carboxypropeptide of pro-α_2, AYCOCK et al. (1986) observed an inhibition of both type I procollagen and fibronectin synthesis without alteration in the steady state levels of their respective mRNA, suggesting a post-transcriptional regulation. Synthetic peptides containing the cleavage site for procollagen peptidase are potent inhibitors of this enzyme (MORIKAWA et al. 1980) and could be of significance in the regulation of procollagen processing and regulation of synthesis.

C. Drugs Acting on Hydroxylation

In the absence of adequate hydroxylation either by substrate substitution (amino acid analogues) or deficiency (lack of O_2), cofactor chelation (iron chelators) or deficiency (vitamin C), the collagen molecules are under-hydroxylated, do not form a stable triple helix, accumulate inside the cells and become susceptible to intracellular degradation (for review see PROCKOP et al. 1979).

I. Amino Acid Analogues

Proline and lysine analogues in vitro are incorporated into collagen polypeptides and depress the function of the hydroxylases. In vivo, these amino acid analogues are toxic and inefficient in modifying collagen synthesis, perhaps owing to the large size of the proline pool in the body and the lower affinity of the analogues for their tRNAs (MADDEN et al. 1973).

II. Iron Chelators

Chelation of iron by 1,10-phenanthroline, α,α'-dipyridyl or desferrioxamine inhibits the hydroxylases and decreases collagen secretion in vitro. Direct assay for prolyl hydroxylase activity of various tissues of 1,10-phenanthroline-treated animals does not show any inhibition of the enzyme (CHVAPIL et al. 1974).

III. Vitamin C

In vivo, ascorbate deficiency leads to scurvy in species dependent on exogenous sources of the vitamin, such as guinea pigs and humans. In vitro, contradictory reports have been published concerning the effect of ascorbate deficiency on fibroblasts in culture. Ascorbate has been found to have no effect in some cell lines while in human skin fibroblasts it increases collagen synthesis. In the absence of added ascorbate, the secreted collagen polypeptide has normal amounts of hydroxlysine and is mildly deficient in hydroxyproline (MURAD et al. 1981). The persistence of prolyl hydroxylation in the absence of ascorbate has suggested the existence in serum-enriched culture medium of a reducing factor which can partially substitute for ascorbate (MATA et al. 1981). Ascorbate stimulates the deposition of glycosaminoglycans into the insoluble matrix of normal skin fibroblasts (EDWARD 1986).

D. Drugs Acting on Secretion

I. Cytoskeleton-Disruptive Drugs

Drugs interfering with the function of microtubules and microfilaments (colchicine, vinblastine, cytochalasin B) produce in fibroblasts an intracellular accumulation of procollagen by inhibiting secretion. A low concentration of colchicine increases the synthesis and release of collagenase and prostaglandins in synovial cell culture (HARRIS and KRANE 1971; ROBINSON et al. 1973) in human skin fibroblasts (BAUER and VALLE 1982) and in granulation tissue (CHVAPIL 1975). The activity of lysyl oxidase is not altered by colchicine (KUIVANIEMI et al. 1986). This drug has been used as an anti-fibrotic agent in liver cirrhosis (ROJKIND et al. 1973) and in scleroderma (ALARCON-SEGOVIA et al. 1974) with some clinical improvement, but without significant variation of the urinary excretion of hydroxyproline, an index of collagen catabolism (HARRIS et al. 1975). Cytoskeleton-disruptive drugs also inhibit the contractile properties of fibroblasts towards a three-dimensional collagen lattice (BELL et al. 1979) or a fibrin clot (AZZARONE and MACIEIRA-COELHO 1984).

II. Canavanin

Canavanin is an analogue of arginine which inhibits the normal proteolytic processing of the signal and precursor sequences of the procollagen molecules and their secretion (SCHEIN et al. 1980).

III. Tunicamycin

This antibiotic inhibits the addition of a *N*-linked oligosaccharide by the dolichol pathway to procollagen and other glycoproteins in culture. It reduces the secretion of procollagen and fibronectin (HOUSLEY et al. 1980) and the processing of procollagen (DUKSIN and BORNSTEIN 1977).

IV. Ionophores

The intracellular mobilisation of Ca^{2+} by the ionophore A23187 specifically inhibits collagen production by human fibroblasts, suggesting a modulatory role of intracellular Ca^{2+} in collagen metabolism similar to the phorbol esters which directly activate protein kinase C (FLAHERTY and CHOJKIER 1986). The Na^{+} ionophore, monensin and nigericin causes an impairment of procollagen and fibronectin secretion in vitro (UCHIDA et al. 1979) and a reduction of lysyl oxidase (KUIVANIEMI et al. 1986). These drugs provided interesting information about the subcellular organisation of post-translational modifications of proteoglycans (HOPPE et al. 1985).

E. Drugs Acting on Cross-Linking

Components interfering with the activity of lysyl oxidase either as enzyme inhibitor, chelators of Cu^{2+} or by interaction with the aldehydes and/or the Schiff bases depress covalent cross-linking. Inversely, activators of the enzyme or exogenous cross-linkers can increase the tensile strength of the extracellular matrix.

I. Lathyrogens

β-Aminopropionitrile is the commonest lathyrogen. When administered to growing animals it induces an extreme fragility of various connective tissues and acts by irreversible inhibition of lysyl oxidase (PINNELL and MARTIN 1968). The collagen molecules synthesised in its presence are defective in aldehydes; they polymerise but do not form adequate cross-links. Elastin is also affected and becomes extractable (SYKES and PARTRIDGE 1974). Lathyrogenic nitriles have been used for preventing arterial lesions in the hypertensive rat (IWATSUKI et al. 1977) and retraction of the burned oesophagus during healing (DAVIS et al. 1972) and in scleroderma (KEISER and SJOERDSMA 1967). Side effects in the human, e.g. neurotoxicity, have prevented its extended use as a therapeutic agent (BARROW et al. 1974).

II. Penicillamine

Penicillamine is a chelating agent used in the treatment of Wilson's disease and heavy metal intoxication. It is also used in the treatment of cystinuria. Penicillamine binds specifically to free aldehydes on collagen and prevents their subsequent participation in the formation of intra- and intermolecular cross-links

(SIEGEL 1977). At high dosage ($>10^{-4}$ *M*), lysyl oxidase activity is also reduced in vitro, perhaps by chelation of Cu^{2+}. In addition to preventing the cross-linking of newly synthesised collagen, penicillamine seems to labilise some Schiff base cross-links (for review see NIMNI 1977). It has been demonstrated that in the presence of D-penicillamine at pharmacological doses in organ culture the synthesis of collagen is inhibited and prolyl hydroxylase activity is decreased by chelation of Fe^{2+} (UITTO 1969). In contrast, administration of D-penicillamine to rats does not reduce the enzyme activity (HALME et al. 1970). D-Penicillamine seems to stimulate [^{35}S]sulphate binding to the proteoglycans in skin, bone and aorta (JUNKER et al. 1981).

III. Copper

Cu^{2+} ions are required for lysyl oxidase activity. The effect of copper deficiency is largely documented in animals (for review see SIEGEL 1979). Dietary copper deficiency or zinc-induced reduction of tissue copper levels seem to reduce lysyl oxidase activity. Elastin can be extracted from tissue of copper-deficient pigs (SANDBERG et al. 1971). Patients with Menkes' kinky hair syndrome have decreased intestinal copper absorption and their skin fibroblasts in culture display a reduced lysyl oxidase activity (ROYCE et al. 1980) although their copper content is higher than control cells. Administration of copper to these patients in attempts to correct their deficiency leads to increased copper accumulation in various tissues and toxic effects (BERATIS et al. 1980).

IV. Sex Hormones

Extractability of collagen from the dermis is markedly reduced by oestrogen treatment, perhaps by a stimulation of lysyl oxidase activity (SANADA et al. 1978).

V. Flavonoids

It has recently been demonstrated that the protective activity of (+)-catechin on collagen degradation by collagenase in vitro is due to its spontaneous oxidation products (WAUTERS et al. 1986).

VI. Catechol Analogues

Analogues of catechol, dopamine, dopa or D,L-serine-(2,3,4-trihydroxybenzyl)hydrazide inhibit the formation of hydroxylysine-derived intermolecular collagen cross-links and the activity of lysyl hydroxylase in vitro (MURRAY et al. 1977).

VII. Radiotherapy

At low doses in vitro, X-rays increase intermolecular cross-links in collagen (VAN CANEGHEM and LAPIÈRE 1975). A low doses, in vivo, X-rays decrease the amount of collagen in skin (OHUCHI and TSURUFUJI 1970) while at higher doses β-rays produce fibrosis (OHUCHI et al. 1979).

VIII. Coagulation Factor XIIIa

This enzyme, also called fibrin-stabilising factor, is a transglutaminase that can cross-link fibrin, fibronectin (MOSHER 1975) and collagen (SORIA et al. 1975) and establish a complex meshwork of these proteins. Factor XIIIa has been shown to induce attachment of deficient type I collagen attachment cells in association with fibronectin (PAYE and LAPIÈRE 1986). Factor XIIIa decreases collagen synthesis and increases collagen degradation by normal and scleroderma skin fibroblasts in vitro (PAYE et al. 1986). This observation might account for the beneficial effect observed in some scleroderma patients (DELBARRE et al. 1981).

F. Drugs Acting on Degradation

A large number of agents are able to induce collagenase activity in cultured fibroblasts (cytochalasin B, colchicine, concanaval in A, IL-1, phorbol esters, uric acid, proteases and growth factors). This induction most probably involves various mechanisms.

I. Diphenylhydantoin

DPH at doses used in the treatment of epilepsy can induce hyperplasia of the gingiva. Studied in organ culture, DPH was shown to inhibit collagen degradation (BERGENHOLTZ and HANSTROM 1979) and to increase the proteoglycan content of granuloma (HIRAMATSU et al. 1979). Although DPH has no direct inhibitory activity on purified collagenase, the addition of DPH to skin fibroblast cultures results in a marked reduction of collagenase activity and immunoreactive protein concentration which suggests that it inhibits synthesis and/or secretion of collagenase. By this property DPH is effective as a therapeutic agent in some patients with recessive dystrophic epidermolysis bullosa (RDEB) (BAUER et al. 1980).

II. Prostaglandins

The addition of PGE_2 to human skin fibroblasts in culture elevates the concentration of intracellular cAMP and inhibits collagen production, perhaps by increasing intracellular degradation (BAUM et al. 1980). Similar results were observed with various agents such as beta-agonists, isoproterenol and epinephrine, or cholera toxin which raise intracellular cAMP by independent mechanisms. Although increased cyclic nucleotides appear to increase the level of collagenase in macrophages and in synovial tissue (WAHL et al. 1977) no such increased collagenase activity could be detected in skin fibroblasts. In isolated adherent rheumatoid synovial cells, indomethacin inhibits PGE_2 production and stimulates collagenase production (DAYER et al. 1976).

III. Vitamin A and Retinoids

Vitamin A plays an important role in growth and differentiation of epidermal and mesodermal structures. Retinol is known to cause degradation of proteoglycans in cartilage (POOLE 1975) and inhibit chondrogenic differentiation (ZIMMERMAN and TSAMBAOS 1985). Retinoids inhibit the production of collagenase and PGE_2 in cultured synovial cells (BRINCKERHOFF and HARRIS 1981) and in human skin fibroblasts. Treatment of stimulated fibroblasts with all-*trans*-retinoic acid results in decreased collagenase mRNA levels without modification of their half-life, suggesting regulation at the transcriptional level (BRINCKERHOFF et al. 1986). These compounds could therefore be therapeutically useful in RDEB (BAUER 1984). The inhibitory activity is not effective on type IV collagenolytic activity of melanoma cells, perhaps owing to the small concentration of the binding protein in these cells (OIKARINEN and SALO 1986). It is worth noting that retinoic acid has been shown to stimulate collagenase activity of mucosal keratinocytes at concentrations two or three orders of magnitude lower than those required to inhibit fibroblast collagenase secretion (WELLS and BIRKEDAL-HANSEN 1985). It is perhaps related to a process of dedifferentiation. Reduced levels of type I procollagen mRNA induced by retinoids in human skin fibroblasts have also been described (OIKARINEN et al. 1985). Studies in vivo are in conflict with the effects reported in vitro (KLIGMAN et al. 1984) for collagen synthesis. Increased synthesis of oversulphated heparan sulphate and hyaluronic acid have been observed in vitro (SHAPIRO and MOTT 1981).

IV. Enzyme Replacement Therapy

Most inherited lysosomal storage diseases are related to the lack of activity of a defined hydrolase. In fibroblast culture, deficient activity can be corrected by adding the missing enzyme. Enzyme replacement therapy for this group of disorders can be performed in vivo by transplantation of fibroblasts (for review see MCKUSICK et al. 1978).

V. Elastase Inhibitor

Elastase can induce emphysema by destruction of elastin. It is prevented by instillation of α_1-antitrypsin (for detailed discussion see SANDBERG et al. 1981). Chemical inhibitors have also been described and one of them displays its activity in vivo (KLEINERMAN et al. 1980). Danazol, a mild androgenic drug increasing the hepatic production of α_1-antitrypsin, might prove interesting in preventing elastin degradation (GADEK et al. 1980).

G. Conclusions

Except for some pharmacological agents whose activity has also been demonstrated in vivo, many compounds active on fibroblasts have mostly been studied in vitro and mainly in cell culture. The in vivo situation may be very different owing to the existence of numerous regulation mechanisms and the control exerted by interactions with the organised connective tissue matrix. In vitro skin-equivalent models, attempting to introduce such cell-matrix interactions, but lacking other environmental regulation mechanisms represent new tools in pharmacology.

Much work is needed to develop the pharmacology of the connective tissue. Diseases and trauma frequently lead to the formation of a scar that is often more harmful to the organism than the process that has led to its formation (LAPIÈRE 1977). Anti-fibrotic drugs more specific than corticosteroids at high doses are urgently needed to fill a gap in the therapeutic arsenal of modern medicine. A large amount of knowledge about the connective tissues is available to develop them on a rational basis.

References

Alarcon-Segovia D, Ibanez G, Kershenobich D, Rojkind M (1974) Treatment of scleroderma by modification of collagen metabolism. A double blind trial with colchicine-placebo. J Rheumatol 1:97–102

Aycock RS, Raghow R, Stricklin GP, Seyer JM, Kang AH (1986) Post-transcriptional inhibition of collagen and fibronectin synthesis by a synthetic homolog of a portion of the carboxyl terminal propeptide of human type I collagen. J Biol Chem 261:14355–14362

Azzarone B, Macieira-Coelho A (1984) Role of cytoskeletal elements in the retractile activity of human skin fibroblasts. Exp Cell Res 155:299–304

Barrow MV, Simpson CF, Miller EJ (1974) Lathyrism: a review. Q Rev Biol 49:101–128

Bauer EA (1984) Inhibition of collagenase expression in normal and recessive dystrophic epidermolysis bullosa fibroblast cultures by retinoic acid (Abstr 33). Dermatologica 169:234

Bauer EA, Valle KJ (1982) Colchicine induced modulation of collagenase in human skin fibroblast cultures. I. Stimulation of enzyme synthesis in normal cells. J Invest Dermatol 79:398–402

Bauer EA, Cooper TW, Tucker DR, Esterly NB (1980) Phenytoin therapy of recessive dystrophic epidermolysis bullosa: clinical trial and proposed mechanism of action on collagenase. N Engl J Med 303:776–781

Bauer EA, Kronberger A, Valle KJ, Jeffrey JJ, Eisen AZ (1985) Glucocorticoid modulation of collagenase expression in human skin fibroblast cultures. Evidence for pretranslational inhibition. Biochim Biophys Acta 825:227–235

Baum BJ, Moss J, Breul SD, Berg RA, Crystal RG (1980) Effect of cyclic AMP on the intracellular degradation of newly synthesized collagen. J Biol Chem 255:2843–2847

Bell E, Ivarsson B, Merrill C (1979) Production of a tissue like structure by contraction of collagen lattices by human fibroblasts of different proliferative potential in vitro. Proc Natl Acad Sci USA 76:1274–1278

Beratis NG, Yee M, Labadie GU, Hirschhorn K (1980) Effect of copper on Menkes' and normal cultured skin fibroblasts. Dev Pharmacol Ther 1:305–317

Bergenholtz A, Hanstrom L (1979) The effect of diphenylhydantoin upon the biosynthesis and degradation of collagen in cat palatal mucosa in organ culture. Biochem Pharmacol 28:2653–2660

Black MM, Shuster S, Bottoms E (1973) Skin collagen and thickness in Cushing's syndrome. Arch Dermatol Forsch 246:365–368

Blumenkrantz N, Weismann K, Asboe-Hansen G (1979) Lack of dermal changes in uninvolved skin after photochemotherapy of psoriasis: biochemical studies. Acta Derm Venereol (Stockh) 59:149–151

Bohlen P, Baird A, Esch F, Ling N, Gospodarowicz D (1984) Isolation and partial molecular characterization of pituitary fibroblast growth factor. Proc Natl Acad Sci USA 81:5364–5368

Boucek RJ, Speropoulos AJ, Noble NL (1972) Serotonin and ribonucleic acid and collagen metabolism of fibroblasts in vitro. Proc Soc Exp Biol Med 140:599–603
Brinckerhoff CE, Harris ED (1981) Modulation by retinoid acid and corticosteroids of collagenase production by rabbit synovial fibroblasts treated with phorbol myristate acetate or polyethylene glycol. Biochim Biophys Acta 677:424–431
Brinckerhoff CE, Plucinska IM, Sheldon LA, O'Connor GT (1986) Half life of synovial cell collagenase mRNA is modulated by phorbol myristate acetate but not by all-*trans*-retinoic acid or dexamethasone. Biochemistry 25:6378–6384
Buckley A, Davidson JM, Kamerath CD, Wolt TB, Woodward SC (1985) Sustained release of epidermal growth factor accelerates wound repair. Proc Natl Acad Sci USA 82:7340–7344
Chvapil M (1975) Pharmacology of fibrosis: definitions, limits and perspectives. Life Sci 16:1345–1362
Chvapil M, McCarthy D, Madden JW, Peacock EE (1974) Effect of 1,10-phenanthroline and desferrioxamine in vivo on prolyl hydroxylase and hydroxylation of collagen in various tissues of rats. Biochem Pharmacol 23:2165–2173
Clark JG, Starcher BC, Uitto J (1980) Bleomycin induced synthesis of type I procollagen by human lung and skin fibroblasts in culture. Biochim Biophys Acta 631:359–370
Cockayne D, Sterling KM, Shull S, Mintz KP, Illeyne S, Cutroneo KR (1986) Glucocorticoids decrease the synthesis of type I procollagen mRNAs. Biochemistry 25:3202–3209
Cohen S (1983) The epidermal growth factor (EGF) Cancer 51:1787–1791
Cohen S, Carpenter G (1975) Human epidermal growth factor: isolation and chemical and biological properties. Proc Natl Acad Sci USA 72:1317–1321
Colige A, Nusgens B, Lapière CM (1987) The effect of EGF on human skin fibroblasts is modulated by the extracellular matrix. Arch Dermatol Res 280:S42–S46
Coulomb B, Dubertret L, Bell E, Touraine R (1984) The contractility of fibroblasts in a collagen lattice is reduced by corticosteroids. J Invest Dermatol 82:341–344
Cruess RL, Hong KC (1975) Effect of cortisone on collagenolytic activity in the rat. Proc Soc Exp Biol Med 148:887–892
Davis WM, Madden JW, Peacock EE (1972) A new approach to the control of esophageal stenosis. Ann Surg 176:469–476
Dayer JM, Krane SM, Russell RGG, Robinson DR (1976) Production of collagenase and prostaglandins by isolated adherent rheumatoid synovial cells. Proc Natl Acad Sci USA 73:945–949
Delbarre F, Godeau P, Thivolet J (1981) Factor XIII treatment for scleroderma. Lancet 2:204
Duksin D, Bornstein P (1977) Impaired conversion of procollagen to collagen by fibroblasts and bone treated with tunicamycin, an inhibitor of protein glycosylation. J Biol Chem 252:955–962
Edward M (1986) Ascorbate induced changes in glycosaminoglycan synthesis and distribution of normal and SV40-transformed fibroblasts. J Cell Sci 85:217–229
Flaherty M, Chojkier M (1986) Selective inhibition of collagen synthesis by the Ca^{2+} ionophore A23187 in cultured human fibroblasts. J Biol Chem 261:12060–12065
Folkman J, Haudenschild C (1980) Angiogenesis in vitro. Nature 288:551–556
Franklin TJ, Gregory H, Morris WP (1986) Acceleration of wound healing by recombinant human urogastrone (epidermal growth factor). J Lab Clin Med 108:103–108
Freundlich B, Bomalaski JS, Neilson E, Jimenez SA (1986) Regulation of fibroblasts proliferation and collagen synthesis by cytokines. Immunol Today 7:303–307
Gadek JE, Fulmen JD, Gelfand JA, Franck MM, Petty TL, Crystal RG (1980) Danazol-induced augmentation of serum alpha 1 antitrypsin levels in individuals with marked deficiency of this antiprotease. J Clin Invest 66:82–87
Gospodarowicz D (1975) Purification of a fibroblast growth factor from bovine pituitary. J Biol Chem 250:2515–2520
Gospodarowicz D, Moran JS (1974) Stimulation of division of sparse and confluent 3T3 cell population by a fibroblast growth factor, dexamethasone and insulin. Proc Natl Acad Sci USA 71:4584–4588

Gospodarowicz D, Cheng J, Lui GM, Baird A, Bohlen P (1984) Isolation of brain fibroblast growth factor by heparin Sepharose affinity chromatography: identity with pituitary fibroblast growth factor. Proc Natl Acad Sci USA 81:6963–6967
Goustin AS, Leof EB, Shipley GD, Moses HL (1986) Growth factors and cancer. Cancer Res 46:1015–1029
Halme J, Uitto J, Kahanpaa K, Karhunen P, Lindy S (1970) Protocollagen proline hydroxylase activity in experimental pulmonary fibrosis of rats. J Lab Clin Med 75:535–541
Hamalainen L, Oikarinen J, Kivirikko KI (1985) Synthesis and degradation of type I procollagen mRNAs in cultured human skin fibroblasts and the effect of cortisol. J Biol Chem 260:720–725
Hansen TM, Andreassen TT, Lorenzen I (1978) Reversibility of the effects of cyclophosphamide on collagen: biochemical studies on skin and granulation tissue and determination of thermal stability of tail tendons of rats. Acta Pharmacol Toxicol (Copenh) 42:103–109
Harris ED, Krane SM (1971) Effects of colchicine on collagenase in cultures of rheumatoid synovium. Arthritis Rheum 14:669–684
Harris ED, Hoffmann GS, McGuire JL, Strosberg JM (1975) Colchicine: effects upon urinary hydroxyproline excretion in patients with scleroderma. Metabolism 24:529–535
Hatamochi A, Fugiwara K, Weki H (1985) Effects of histamine on collagen synthesis by cultured fibroblasts derived from guinea pig skin. Arch Dermatol Res 277:60–64
Hatamochi A, Paterson B, de Crombrugghe B (1986) Differential binding of a CCAAT DNA binding to the promoters of the mouse alpha 2 (I) and alpha 1 (III) collagen genes. J Biol Chem 261:11310–11314
Herfert J, Wienker TF, Ropers HH (1980) The presence of androgen binding receptors in genital and non genital skin fibroblasts. Hum Genet 53:271–273
Hiramatsu M, Hatakeyama K, Abe I, Minami N (1979) The effect of diphenylhydantoin sodium on acid mucopolysaccharide and collagen contents of cotton pellet granulomas in rats. J Periodontal Res 14:173–177
Hoppe W, Glossi J, Kresse H (1985) Influence of monensin on biosynthesis, processing and secretion of proteodermatan sulfate by skin fibroblasts. Eur J Biochem 152:91–97
Hosokawa M, Ishii M, Inoue K, Yao C, Takeda T (1981) Estrogen induces different responses in dermal and lung fibroblasts: special reference to collagen. Connect Tissue Res 9:115–120
Housley TJ, Rowland FN, Ledger PW, Kaplan J, Tanzer ML (1980) Effects of tunicamycin on the biosynthesis of procollagen by human fibroblasts. J Biol Chem 255:121–128
Huey J, Narayanan AS, Jones K, Page RC (1980) Effect of epidermal growth factor on the synthetic activity of human fibroblasts. Biochim Biophys Acta 632:227–233
Ivanovic V, Weinstein B (1981) Glucocorticoids and benzo[*a*]pyrene have opposing effects on EGF receptor binding. Nature 293:404–406
Iwatsuki K, Cardinale GJ, Spector S, Udenfriend S (1977) Reduction of blood pressure and vascular collagen in hypertensive rats by beta-aminopropionitrile. Proc Natl Acad Sci USA 74:360–362
Junker P, Helin G, Lorenzen IB (1981) Effect of D-penicillamine on collagen, glycosaminoglycans, DNA and RNA of granulation tissue and connective tissue of skin, bone and aorta in rats. Acta Pharmacol Toxicol (Copenh) 48:300–310
Katoh Y, Kodama K, Ishikawa T (1985) In vivo effects of epidermal and fibroblast growth factors on DNA replication in mouse skin. Exp Cell Res 161:111–116
Keiser HR, Sjoerdsma A (1967) Studies on beta-aminopropionitrile in patients with scleroderma. Clin Pharmacol Ther 8:593–602
Kleinerman J, Ranga V, Rynbrandt D, Sorensen J, Powers JC (1980) The effect of the specific elastase inhibitor, alanyl alanyl prolyl alanine chloromethylketone, on elastase-induced emphysema. Am Rev Respir Dis 121:381–387
Kligman LH, Duo CH, Kligman AM (1984) Topical retinoic acid enhances the repair of ultraviolet damaged dermal connective tissue. Connect Tissue Res 12:139–150
Koob TJ, Jeffrey JJ, Eisen AZ (1974) Regulation of human skin collagenase activity by hydrocortisone and dexamethasone in organ culture. Biochem Biophys Res Commun 61:1083–1088

Krane SM (1982) Collagenases and collagen degradation. J Invest Dermatol 79:83S–85S
Krieg T, Horlein D, Wiestner M, Muller PK (1978) Aminoterminal extension peptides from type I procollagen normalize excessive collagen synthesis of scleroderma fibroblasts. Arch Dermatol Res 263:171–180
Kuivaniemi H, Ala-Kokko L, Kivirikko KI (1986) Secretion of lysyl oxidase by cultured human skin fibroblasts and effects of monensin, nigericin, tunicamycin and colchicine. Biochim Biophys Acta 883:326–334
Kumakiri M, Hashimoto K, Willis I (1977) Biologic changes due to long wave ultraviolet irradiation on human skin: ultrastructural study. J Invest Dermatol 69:392–400
Laato M, Niinikoski J, Lebel L, Gerdin B (1986) Stimulation of wound healing by epidermal growth factor: a dose dependent effect. Ann Surg 203:379–381
Lapière CM (1977) Les médications antiscléreuses. Med Hyg 35:2677–2678
Madden JW, Chvapil M, Carlson EC, Ryan JN (1973) Toxicity and metabolic effects of 3,4-dehydroproline in mice. Toxicol Pharmacol 26:426–437
Manthorpe R, Garbarsch C, Lorenzen I (1980) Long term effect of glucocorticoid on connective tissue of aorta and skin. Morphological and biochemical studies of tissues from rabbits with intact and injured aortas. Acta Endocrinol (Copenh) 95:271–281
Mata JM, Assad R, Peterkosfsky B (1981) An intramembranous reductant which participates in the proline hydroxylation reaction with intracisternal prolylhydroxylase and unhydroxylated procollagen in isolated microsomes from L-929 cells. Arch Biochem Biophys 206:93–104
Maugh T (1981) Angiogenesis inhibitors link many diseases. Science 212:1374–1375
McKusick VA; Neufeld E, Kelley TE (1978) The mucopolysaccharide storage diseases. In: Stanbury IB, Wyngaarden JB, Frederickson DS (eds) The metabolic basis of inherited disease, 4th edn. McGraw-Hill, New York
McNelis B, Cutroneo KR (1978) A selective decrease of collagen peptide synthesis by dermal polysomes isolated from glucocorticoid treated newborn rats. Mol Pharmacol 14:1167–1175
Morikawa T, Tuderman L, Prockop DJ (1980) Inhibitors of procollagen *N*-protease. Synthetic peptides with sequences similar to the cleavage site in the pro-alpha I chain. Biochemistry 19:2646–2650
Mosher DF (1975) Cross-linking of cold insoluble globulin by fibrin stabilizing factor. J Biol Chem 250:6614–6621
Murad S, Grove D, Lindberg KA, Reynolds G, Sivarajah A, Pinnell SR (1981) Regulation of collagen synthesis by ascorbic acid. Proc Natl Acad Sci USA 78:2879–2882
Murray J, Lindberg K, Pinnell S (1977) In vitro inhibition of collagen cross-links by catechol analogs. J Invest Dermatol 68:146–150
Neufeld G, Gospodarowicz D (1985) The identification and partial characterization of the fibroblast growth factor receptor of baby hamster kidney cells. J Biol Chem 260:13860–13868
Nimni ME (1977) Mechanism of inhibition of collagen cross-linking by penicillamine. Proc R Soc Med 70:65–72
Nusgens B, Gielen J, Lapière CM (1978) Inhibition of collagen synthesis by the aminoterminal extension of procollagen. J Invest Dermatol 70:230
Ohuchi K, Tsurufuji S (1970) Degradation and turnover of collagen in the mouse skin and the effect of whole body X-irradiation. Biochim Biophys Acta 208:475–481
Ohuchi K, Chang LF, Tabachnick J (1979) Radiation fibrosis of guinea pig skin after beta irradiation and an attempt at its suppression with proline analogs. Radiat Res 79:273–288
Oikarinen A, Salo T (1986) Effects of retinoids on type IV collagenolytic activity in melanoma cells. Acta Derm Venereol (Stockh) 66:346–348
Oikarinen H, Oikarinen AI, Tan EML, Abergel RP, Meeker CA, Chu ML, Prockop DJ, Uitto J (1985) Modulation of procollagen gene expression by retinoids. Inhibition of collagen production by retinoic acid accompanied by reduced type I procollagen messenger ribonucleic acid levels in human skin fibroblast cultures. J Clin Invest 75:1545–1553

Oikarinen J (1977) Effect of cortisol acetate on collagen biosynthesis and on the activities of prolyl hydroxylase, lysyl hydroxylase, collagen galactosyltransferase and collagen glucosyltransferase in chick embryo tendon cells. Biochem J 164:533–539
Oikarinen J, Hannuksela M (1980) Effects of hydrocortisone-17-butyrate, hydrocortisone and clobetasol-17-propionate on prolyl hydroxylase activity in human skin. Arch Dermatol Res 267:79–82
Oikarinen J, Ryhanen L (1981) Cortisol decreases the concentration of translatable type I procollagen mRNA species in the developing chick embryo calvaria. Biochem J 198:519–524
Oikarinen J, Pihlajaniemi T, Hamalainen L, Kivirikko KI (1983) Cortisol decreases the cellular concentration of translatable procollagen mRNA species in cultured human skin fibroblasts. Biochim Biophys Acta 741:297–302
Oliver N (1986) Molecular mechanism of glucocorticoid enhanced fibronectin gene expression. Coll Relat Res 6:441
Paglia LM, Wilczek J, Diaz de Leon L, Martin GR, Horlein D, Muller P (1979) Inhibition of procollagen cell free synthesis by aminoterminal extension peptides. Biochemistry 18:5030–5034
Paglia LM, Wiestner M, Duchene M, Ouellette LA, Horlein D, Martin GR, Muller PK (1981) Effects of procollagen peptides on the translation of type II collagen messenger ribonucleic acid and on collagen biosynthesis in chondrocytes. Biochemistry 20:3523–3527
Paye M, Lapière CM (1986) The lack of attachment of transformed embryonic lung epithelial cells to collagen I is corrected by fibronectin and FXIII. J Cell Sci 86:95–107
Paye M, Etievant C, Michiels F, Pierard D, Nusgens B, Lapière CM (1986) Characterization of a spontaneously transformed pulmonary embryonic rat (PER) epithelial cell line. J Cell Sci 86:83–93
Pierard GE, Ackerman A (1979) Histopathology of remodelling induced by PUVA within superficial dermis. Br J Dermatol 100:251–256
Pierard GE, de la Brassinne M, Lapière CM (1977) Effects of long term photochemotherapy on the dermis. J Invest Dermatol 68:249–250
Pinnell SR, Martin GR (1968) The cross-linking of collagen and elastin: enzymatic conversion of lysine in peptide linkage to alpha-aminoadipic-8-semialdehyde (allysine) by an extract from bone. Proc Natl Acad Sci USA 61:708–714
Pohl J, Christophers E (1979) Photoinactivation of skin fibroblasts by fractionated treatment with 8-methoxypsoralen and UVA. J Invest Dermatol 73:176–179
Ponec M (1984) Effects of glucocorticoids on cultured skin fibroblasts and keratinocytes. Int J Dermatol 23:11–24
Ponec M, Kempenaar JA, van der Meulen-van Harskamp GA, Bachra BN (1979) Effects of glucocorticosteroids on cultured human skin fibroblasts. IV – Specific decrease in the synthesis of collagen but no effect on its hydroxylation. Biochim Pharmacol 28:2777–2783
Poole AR (1975) Immunocytochemical studies of the secretion of a proteolytic enzyme cathepsin D, in relation to cartilage breakdown. In: Burleigh PMC, Poole AR (eds) Dynamics of connective tissue macromolecules. North-Holland, Amsterdam, pp 357–406
Prockop DJ, Kivirikko KI, Tuderman L, Guzman NA (1979) The biosynthesis of collagen and its disorders. N Engl J Med 301:13–23
Raghow R, Lurie S, Seyer JM, Kang AH (1985) Profiles of steady state levels of messenger RNAs coding for type I procollagen, elastin and fibronectin in hamster lungs undergoing bleomycin induced interstitial pulmonary fibrosis. J Clin Invest 76:1733–1739
Rall LB, Scott J, Bell GI, Crawford RJ, Penschow JD, Niall HD, Coghlan J (1985) Mouse prepro-epidermal growth factor synthesis by the kidney and other tissues. Nature 313:228–231
Roberts AB, Sporn MB (1986) Transforming growth factors. Cancer Surv 4:683–694
Robinson DR, Smith H, Levine L (1973) Prostaglandin (PG) synthesis by human synovial cultures and its stimulation by colchicine. Arthritis Rheum 16:129–136
Rojkind M, Uribe M, Kersenobich D (1973) Colchicine and the treatment of liver cirrhosis. Lancet 7793:38–39

Rokowski RJ, Sheehy J, Cutroneo KR (1981) Glucocorticoid mediated selective reduction of functioning collagen messenger ribonucleic acid. Arch Biochem Biophys 210:74–81
Rosenbloom j, Feldman G, Freundlich B, Jimenez SA (1986) Inhibition of excessive scleroderma fibroblast collagen production by recombinant gamma interferon. Association with a coordinate decrease in type I and type III procollagen messenger RNA levels. Arthritis Rheum 29:851–856
Ross R (1985) Platelets, platelet-derived growth factor, growth control, and their interactions with the vascular wall. J Cardiovasc Pharmacol 7:S186–S190
Royce PM, Camakaris J, Danks DM (1980) Reduced lysyl oxidase activity in skin fibroblasts from patients with Menkes' syndrome. Connect Tissue Res 7:205–211
Saarni H (1978) Cortisol effects on the glycosaminoglycan synthesis and molecular weight distribution in vitro. Biochem Pharmacol 27:1029–1032
Sanada H, Shikata J, Hamamoto H, Ueba Y, Yamamuro T, Takeda T (1978) Changes in collagen cross-linking and lysyl oxidase by estrogen. Biochim Biophys Acta 541:408–414
Sandberg LB, Zeikus RD, Coltrain IM (1971) Tropoelastin purification from copper deficient swine: a simplified method. Biochim Biophys Acta 236:542
Sandberg LB; Soskel NT, Leslie JG (1981) Elastin structure, biosynthesis and relation to disease states. N Engl J Med 304:566–579
Schein J, Harsch M, Cywinski A, Rosenbloom J (1980) Effect of the arginine analog, canavanin, on the synthesis and secretion of procollagen. Arch Biochem Biophys 203:572–579
Schmidt A, Setoyama C, de Crombrugghe B (1985) Regulation of a collagen gene promoter by the product of viral *mos* oncogene. Nature 314:286–289
Setoyama C, Hatamochi A, Peterkofsky B, Prather W, de Crombrugghe B (1986) V-*fos* stimulates expression of the alpha 1 (III) collagen gene in NIH 3T3 cells. Biochem Biophys Res Commun 136:1042–1048
Shapiro SS, Mott DJ (1981) Modulation of glycosaminoglycan biosynthesis by retinoids. Modulation of cellular interactions by vitamin A and derivatives (retinoids). Ann NY Acad Sci 359:306–321
Shaw SN, Amos H (1973) Insulin stimulation of glucose entry in chick fibroblast and Hela cells. Biochem Biophys Res Commun 5:357–365
Shull S, Cutroneo KR (1986) Glucocorticoids change the ratio of type III to type I procollagen extracellularly. Coll Relat Res 6:295–300
Shuster S, Raffle EJ, Bottoms E (1967) Skin collagen in rheumatoid arthritis and the effects of corticosteroids. Lancet 1:525–527
Shuster S, Black MM, MacVitie E (1975) The influence of age and sex on skin thickness, skin collagen and density. Br J Dermatol 93:639–643
Siegel RC (1977) Collagen cross-linking. Effect of D-penicillamine on cross-linking in vitro. J Biol Chem 252:254–259
Siegel RC (1979) Lysyl oxidase. Int Rev Connect Tissue Res 8:73–118
Soria J, Soria C, Boulard C (1975) Fibrin stabilizing factor (FXIII) and collagen polymerization. Experientia 31:1355–1357
Steinmann BU, Abe S, Martin GR (1982) Modulation of type I and type III collagen production in normal and mutant human skin fibroblasts by cell density, prostaglandin E_2 and epidermal growth factor. Coll Relat Res 2:185–195
Sykes BC, Partridge SM (1974) Salt soluble elastin from lathyritic chicks. Biochem J 141:567–572
Uchida N, Smilowitz H, Tanzer ML (1979) Monovalent ionophores inhibit secretion of procollagen and fibronectin from cultured human fibroblasts. Proc Natl Acad Sci USA 76:1868–1872
Uitto J (1969) Effect of D-penicillamine on collagen biosynthesis in organ culture. Biochim Biophys Acta 194:498–503
Uzuka M, Nakajima K, Ohta S, Mori Y (1981) Induction of hyaluronic acid synthetase by estrogen in the mouse skin. Biochim Biophys Acta 673:387–393
Vaatainen N, Oikarinen A, Kuutti-Savolainen ER (1980) The effects of long-term local PUVA treatment on collagen metabolism in human skin. Arch Dermatol Res 269:99–104

Van Caneghem P, Lapière CM (1975) Influence des rayons X sur la vitesse de dépolymérisation des fibres de collagène lathyrique reconstituées in vitro. C R Soc Biol (Paris) 16:242–245
Van Story-Lewis PE, Tenenbaum HC (1986) Glucocorticoid inhibition of fibroblast contraction of collagen gels. Biochem Pharmacol 35:1283–1286
Verbruggen LA, Abe S (1982) Glucocorticoids alter the ratio of type III type I collagen synthesis by mouse dermal fibroblasts. Biochem Pharmacol 31:1711–1715
Villee DB, Powers ML (1977) Effect of glucose and insulin on collagen secretion by human skin fibroblasts in vitro. Nature 268:156–157
Wahl LM (1977) Hormonal regulation of macrophage collagenase activity. Biochem Biophys Res Commun 74:838–845
Wahl LM, Olsen CE, Sandberg AL, Mergenhagen SE (1977) Prostaglandin regulation of macrophage collagenase production. Proc Natl Acad Sci USA 74:4955–4958
Walthall BJ, Ham RG (1981) Multiplication of human diploid fibroblasts in a synthetic medium supplemented with EGF, insulin and dexamethasone. Exp Cell Res 134:303–311
Wauters P, Eeckhout Y, Vaes G (1986) Oxidation products are responsible for the resistance to the action of collagenase conferred on collagen by (+)-catechin. Biochem Pharmacol 35:2971–2973
Wells BR, Birkedal-Hansen H (1985) Retinoic acid stimulates degradation of interstitial collagen fibrils by rat mucosal keratinocytes in vitro. J Dent Res 64:1186–1190
Werb Z (1978) Biochemical actions of glucocorticoids on macrophages in cultures. Specific inhibition of elastase, collagenase and plasminogen activator secretion and effects on other metabolic functions. J Exp Med 147:1695–1712
Wie H, Beck EI (1981) Synthesis and solubility of collagen in rats during recovery after high dose cyclophosphamide administration. Acta Pharmacol Toxicol (Copenh) 48:294–299
Wiestner P, Krieg T, Horlein D, Glanville RW, Fietzek P, Muller PK (1979) Inhibiting effect of procollagen peptides on collagen biosynthesis in fibroblast cultures. J Biol Chem 254:7016–7023
Woessner JF (1969) Inhibition by oestrogen of collagen breakdown in the involuting rat uterus. Biochem J 112:637–645
Wolff K, Gschnait F, Honigsmann H, Konrad K, Stingl G, Wolff-Schreiner E, Fritsch P (1976) Oralphotochemotherapy: results, follow-up and pathology. In: Farber EM, Cox AJ (eds) Psoriasis. Yorke, New York, p 300
Wu CH, Donovan CB, Wu GY (1986) Evidence for pretranslational regulation of collagen synthesis by procollagen propeptides. J Biol Chem 261:10482–10484
Yoshizato K, Nishikawa A, Taira T, Koganei Y, Yamamoto N, Kishi J, Hayakawa T (1986) A reconstituted skin: dermal-epidermal interactions in collagenolysis and cell morphology. Biomed Res 7:219–231
Zimmermann B, Tsambaos D (1985) Retinoids inhibit the differentiation of embryonic mouse mesenchymal cells in vitro. Arch Dermatol Res 277:98–104

CHAPTER 28

Fungal Skin Infections

R. J. HAY

A. Introduction

The dermatophyte or ringworm infections are confined to stratum corneum or keratinised structures derived from epidermis, such as nail or hair, and these superficial infections are considered in this chapter. The natural course of infection is variable and in many instances spontaneous remission occurs. Infections with *Trichophyton rubrum*, the commonest cause of dermatophytosis, and *T. concentricum*, found in parts of the Far East, Pacific islands and South America, are often chronic. Chronic dermatophytosis is seldom highly inflammatory. The most severely inflamed lesions in humans are cattle ringworm infections (*T. verrucosum*) where granuloma and pustule formation (kerion) may occur; second infections are rare presumably because of an enduring immune response. Clearance of dermatophyte lesions has been correlated with delayed-type hypersensitivity and poor T-lymphocyte responses correlate with chronicity. Other factors include the thickness and rate of turn-over of keratinised tissue.

Superficial *Candida* infections mainly involve the mucous membranes or occluded skin surfaces. They are caused by species of the yeast genus *Candida*, primarily *C. albicans*, although other species such as *C. guilliermondii* may sometimes cause infection. *C. albicans* is a common commensal in the human mouth, gastrointestinal tract and vagina. Factors known to increase the rate of carriage include age, antibiotic therapy, immunosuppression and endocrine disease. Under certain circumstances both these and other factors are associated with the development of clinical lesions of candidosis. Immunosuppression, debility, iron deficiency and diabetes mellitus have all been associated with this change. Local conditions including occlusion of colonised sites by dressings or dentures may also promote invasion. Candidosis of the oral or vaginal mucosa is very common.

Infections around nail folds (paronychia) must be distinguished from and may co-exist with bacterial infections. Rarely nail plate infection may occur either in conjunction with paronychia or in isolation. Persistent superficial candidosis in the mouth, vagina, nails or multiple sites (chronic mucocutaneous candidosis) remains a major therapeutic problem.

Tinea versicolor is a mild infection, particularly common in the tropics, caused by the lipohilic yeast *Malassezia furfur*. Yeast forms of this organism are known to occur on the skin surface of normal individuals. Usually, but not invariably, the appearance of disease is associated with the development of pseudohyphae amongst clusters of *M. furfur* yeasts. The reasons for this transformation are complex and ill understood. Heat, sun exposure, lipid composition of sebum and immunosuppression have all been implicated. Tinea versicolor is usually symptomless. Discrete or confluent hypo- or hyper-pigmented scaling patches are found on the trunk, upper arms or face. Spontaneous improvement may occur, but generally untreated infections are persistent.

Other superficial infections include the infections caused by *Hendersonula toruloidea* and *Scytalidium hyalinum* which closely mimic dry type dermatophytosis caused by *T. rubrum*, onychomycosis caused by non-dermatophytes such as *Scopulariopsis brevicaulis* and the tropical infections, tinea nigra, black piedra and white piedra.

B. Management of Fungal Skin Infections

Management of fungal skin infections is concerned with specific drug therapy and alteration of predisposing factors. The latter is usually more important in superficial *Candida* infections than with the dermatophytes or tinea versicolor. A wide variety of compounds with varying degrees of anti-fungal activity are used in the treatment of superficial fungal infections, but in recent years more specific drugs have been developed (ROBERTS 1980).

I. Dyes and Keratolytic Agents

A number of dyes have weak anti-fungal activity. Best known of these are magenta (Castellani's paint) which also contains resorcinol, potassium permanganate (1 in 5000) and 20% aluminium chloride which are also bacteriostatic.

Keratolytics increase the shedding of fungal organisms. The most widely used preparation is benzoic (6%) and salicylic acid (3%) ointment (benzoic acid compound or Whitfield's ointment). Its main use is for common dermatophyte infections: it is only weakly active against *Candida*, but can be used for erythrasma.

II. Specific Anti-fungal Agents

There are now a large number of substances with specific anti-fungal activity. *Tolnaftate* is a topical anti-fungal drug which, in vitro, has fungicidal activity at lower concentrations than other non-imidazole compounds. It is less active against arthrospores and prolonged treatment is required in chronic dermatophytosis (PLEMPEL 1976). It is used against dermatophytes and tinea versicolor. *Undecylenic acid*, a short chain fatty acid, is an effective anti-dermatophyte compound in a concentration as low as 0.005%. The sodium, zinc or calcium salts retain this activity (LYDDON et al. 1980) and ointments are based on an undecylenate-undecyclenic acid formula, e.g. zinc undecylenate 20% and undecylenic acid 5%. Their clinical effect is similar to tolnaftate. *Haloprogin* (2,4,5-trichlorophenyl-γ-iodopropyl ether) is a broad-spectrum agent with comparable in vitro activity to tolnaftate against certain dermatophytes and activity against *C. albicans* (minimum inhibitory concentration, MIC 0.2 mg/l; HARRISON et al. 1970). Like tolnaftate the fungicidal and fungistatic concentrations of the drug are very close. Haloprogin is readily absorbed through human skin, approximately 11% of the topically applied dose entering the urine within 5 days of application at 4 mg/cm^2 (BARTEK et al. 1972). Given as a 1% cream or lotion, its effect is comparable to the other available topical anti-fungals. It is also active against gram-positive bacteria.

III. Polyene Anti-fungal Agents

These highly active compounds are derived from streptomycetes (MEDOFF and KOBAYASHI 1980). They include amphotericin B (*Streptomyces nodosus*), nystatin (*S. albidus* or *S. noursei*), natamycin (*S. natalensis*) and candicidin (*S. griseus*). All the polyenes have a macrolide ring of carbon atoms stabilised by a lactone containing a conjugated double bond system. Each molecule contains a large number of hydroxyl groups. Recent additions include partricin and mepartricin. All the polyenes are amphipathic and are poorly soluble in water or non-polar organic substances, tending to form a suspension in aqueous solvents. Their mechanism of action is by alterations in permeability of the fungal sterol membrane (*Kinsky* 1961) with potassium leakage, cell permeability and lysis. Interaction between the drug and sterols in the membrane leads to particle aggregation and crater formation (NOZAWA et al. 1974). The ratio of cholesterol to phospholipid in cell membranes is probably a major factor in determining the sensitivity of fungi to polyenes (*Brajtburg* et al. 1974).

The polyenes are particularly active against yeasts such as *C. albicans*. MIC's against *C. albicans* varies from 3.0 mg/l for nystatin to 0.5 mg/l for amphotericin B. Synergistic effects can be demonstrated between amphotericin B and flucytosine (*Medoff* et al. 1971).

The polyenes are poorly absorbed via the oral route and in superficial infections parenteral therapy is not given, apart from low dose intravenous amphotericin B in some cases of chronic mucocutaneous candidosis. Parenterally administered amphotericin B is excreted in bile, only a small proprtion of the does entering the urine. After topical therapy only small quantities of the drugs enter the bloodstream and toxicity is therefore minimal. Resistance to the polyenes is rare (WOODS et al. 1974). Both amphotericin B and nystatin are effective for *Candida* infections, but relatively inactive against the dermatophytes and *M. furfur*. They are given in ointment, cream, pessary, tablet or suspension form (1%–2%) as appropriate. Natamycin is active against dermatophytes as well as yeasts.

IV. Imidazole Anti-fungal Agents

The best known of the tritylimidazole and phenethylimidazole derivatives are clotrimazole, miconazole, econazole, ketoconazole, tioconazole and isoconazole. Like the polyenes these compounds are poorly soluble in water, but can be dissolved in organic solvents. They are active against a wide range of fungi, including dermatophytes, *Candida* spp. and *M. furfur*. MIC values for clotrimazole for the common fungi *C. albicans, Torulopsis glabrata, Aspergillus fumigatus* and *Trichophyton* spp. are all between 0.5 and 10 mg/l (HOLT 1974). Comparison of different imidazoles is difficult because the marked variation in MIC is obtained by using different inoculum sizes, media, solvents and assay systems. However, clinically significant resistance to these drugs appears to be rare (HOLT and AZMI 1978).

The sequence of changes which occur when a susceptible organism is exposed to an imidazole involve an increase of membrane permeability, damage to cytoplasmic membrane structures and lipid deposition in the cell wall (RAAB 1980). One site of specific activity is cell membrane phospholipid. There is also inhibition of protein and RNA synthesis (IWATA et al. 1973). Ergosterol synthesis is inhibited by interference with the conversion of lanosterol to ergosterol (VAN DEN BOSSCHE 1974). Some imidazole compounds such as levamisole have immunostimulant properties.

Miconazole, clotrimazole, econazole, bifonazole, isoconazole and sulconazole are given topically as creams, ointments or pessaries in 1%–2% concentration. Absorption from the skin is variable for dermatophyte or *Candida* infections and tinea versicolor (RAAB 1980). Less than 0.5% of the applied dose of clotrimazole is excreted in the urine and when 50 mg/cm^2 isoconazole is applied to skin in vitro 3 mg/ml active drug reaches the dermis (RAAB 1980). Early studies suggest that toical ketoconazole is effective in seborrhoeic eczema and dandruff (FARR and SHUSTER 1985). Contact sensitisation to the imidazoles is uncommon and has been described with miconazole (SAMSOEN and JELEN 1977). Clotrimazole is now unavailable in tablet form, but ketoconazole is the main imidazole drug which is active in superficial infections after oral administration.

V. Oral Anti-fungal Drugs Used in Superficial Infections

1. Ketoconazole

The newest addition to the orally absorbed anti-fungal agents active against cutaneous fungi is an imidazole compound, ketoconazole. Less is known about the clinical efficacy of this drug although it is clearly both active and useful.

Ketoconazole is a phenylimidazole with a broad spectrum of activity against pathogenic moulds and yeasts. MIC values range from 0.5 to 3.5 mg/l for *C. albicans* and from 0.5 to 25 mg/l for dermatophytes (R. J. HAY and Y. M. CLAYTON 1980, unpublished work) which is higher than that reported for other imidazoles. Its inhibitory activity against *Candida* may be enhanced by leucocytes (BORGERS and VAN DEN BOSSCHE 1981).

Its mode of action is partially understood. It causes accumulation of lanosterol and other C_{14} sterols, due to interference in their conversion to ergosterol in the cytoplasmic membrane (VAN DER BOSSCHE et al. 1980). There is early disorganisation of intracellular organelles and plasma membrane as well as release of oxidative enzymes (BORGERS and VAN DEN BOSSCHE 1981). There is a low incidence of toxicity in the therapeutic range in animals, but there is a low incidence of foetal bone abnormalities with 80 mg/kg in rats (HEEL 1981 b) and liver dysfunction in rats and dogs. In humans, side effects are seen in less than 2% of those given therapeutic doses. The commonest is nausea which often disappears with continued treatment. Headach, pruritus and faintness have also been described. Liver enzyme abnormalities may occur, but almost always return to normal without stopping therapy. Their significance is unknown, but heaptitis has been ascribed to ketoconazole (HEIBERG and SVEJGAARD 1981). The incidence appears to be approximately 1 in 10000 treated cases, although it has been reported most frequently in patients with onychomycosis. It is commoner in women and in middle-aged or elderly patients. The peak incidence occurs betwen 16 and 60 days of therapy (LEWIS et al. 1984). A number of deaths from fulminant hepatic necrosis have now been reported. The development of severe hepatotoxicity is idiosyncratic, but follows the occurrence of liver function abnormalities, particularly hepatocellular enzyme changes. Later the alkaline phosphatase levels rise, serum bilirubin becomes abnormal and frank jaundice develops. Frequent monitoring of liver function is appropriate in patients receiving ketoconazole, the drug should be stopped if significant liver function abnormalities or hepatic symptoms develop. At higher doses (in excess of 400–600 mg daily) ketoconazole has androgen blocking activity and may produce side effects such as gynaecomastia in some individuals (DEFELICE et al. 1981).

Most of orally administered ketoconazole in humans is excreted via the biliary system, while only about 13% of the dose enters urine; 4 days after an oral dose 70% has been excreted, 57% via the faeces. Between 20% and 65% of excreted drug is not metabolised (HEEL 1981 a). Peak serum concentrations of drug 2 h after a single 200-mg tablet range from 3 to 5 mg/l. In rats given ^{3}H-labelled ketoconazole significant concentrations of the drug have been found in sebaceous glands and hairs from treated animals show inhibitory activity against *Trichophyton mentagrophytes* (HEEL 1981 a). Poor absorption may prevent an adequate clinical response in some patients. In certain patients with chronic mucocutaneous candidosis drug resistance has developed (RYLEY et al. 1984).

Ketoconazole is active against dermatophytes, *Candida* and *M. furfur*. However, its clinical efficacy bears little relation to the MIC values obtained in the laboratory. In dermatophytosis eradication of infection has been noted in patients with infections of the feet, body, scalp as well as the nails. It may be more effective than griseofulvin in *T. rubrum* infections (LEGENDRE and STELTZ 1980). However, in early studies involving scalp ringworm, *Microsporum canis* infections showed a response rate of only 29% (HEEL 1981 c). Patients with infection of the palms, finger-nails or body who have already received griseofulvin and failed may respond to ketoconazole whereas sole and toe-nail infections often persist (HAY and CLAYTON 1982 b). Generally, comparative studies of ketoconazole versus griseofulvin in dermatophytosis have shown similar efficacy (WISHART 1983; HAY et al. 1988).

In superficial candidosis ketoconazole has been used to treat oral, vaginal, nail and chronic mucocutaneous infections. The drug has been found to induce clinical and mycological remission in up to 80% of patients with chronic mucocutaneous candidosis; early relapse within 6 months is usually confined to the mouth and responds well to treatment (HAY and CLAYTON 1982 a). In vaginal candidosis a 3-day course of 400 mg ketoconazole appears to be effective (FREGOSO-DUENAS 1980). Although it is not significantly more effective than comparable topical agents (e.g. clotrimazole) ketoconazole was preferred by patients in one study (BINGHAM 1984).

The initial studies suggested that the response in tinea versicolor was slow although up to 92% of patients attain clinical and mycological cure in a mean period of 4 weeks of treatment. However, such long periods of treatment may not be necessary and much shorter courses such as 5 days (HAY and MIDGELEY 1984), or even shorter, appear to be effective.

The normal dose in human fungal infections is 200 mg, increasing to 400–600 mg daily if necessary. In view of its side effects ketoconazole is normally reserved for chronic dermatophytosis which is unresponsive to griseofulvin, chronic persistent candidosis and resistant tinea versicolor. It is also useful for a number of systemic or deep infections ranging from candida oesophagitis to paracoccidioidomycosis. The role of lipophilic yeasts in the pathogenesis of seborrhoeic dermatitis and dandruff has once again been proposed (SHUSTER 1984) and following this oral ketoconazole was found to be effective in seborrhoeic dermatitis and dandruff (FORD et al. 1984). Although oral therapy is probably unwarranted, topical imidazole treatments (FARR and SHUSTER 1985) are being evaluated in this disease. Preliminary studies with ketoconazole also suggest that the common localisation of psoriasis in the scalp may be similarly related to lipophilic yeast infections (FARR et al. 1985).

2. Griseofulvin

Griseofulvin is a coumarin-containing compound produced by species of *Penicillium* such as *P. griseofulvum* and *P. nigricans*. Following the publication by GENTLES (1958) of the cure of experimental dermatophyte infections in the guinea pig, the development of griseofulvin as a treatment for human dermatophytosis proceeded rapidly. Griseofulvin combines with RNA as well as ribosomes (EL-NAKEEB and LAMPEN 1964) and disrupts mitosis after metaphase, probably by disorganisation of spindle microtubules (WILSON 1975). It impairs cell wall formation in young hyphae, possibly by interference with chitin biosynthesis. The drug is mainly active against dermatophytes, MIC values for *T. rubrum* ranging from 0.5 to 6.5 mg/l. Resistance in treated patients is uncommon (ARTIS et al. 1981) although dermatophytes such as *M canis* adapt to serially increased concentrations of griseofulvin in vitro, an effect reversed by passage of infection of the guinea pig (ROSENTHAL and WISE 1960).

Griseofulvin is soluble in water and poorly absorbed after oral administration. Absorption is enhanced by reducing the particle size (ATKINSON et al. 1962) and possibly the presence of lipid. Peak blood levels 4 h after equal doses of the original and the microcrystalline forms were 0.3–1.4 and 0.8–3.0 mg/l, respectively (CROUNSE 1963). Griseofulvin concentrations in stratum corneum are ten or more times those in serum (EPSTEIN et al. 1972). It is not clear how much enters this layer directly with the keratinocytes or indirectly in sweat or extracellular fluid. The drug is also present in nail and hair keratin in active amounts. Griseofulvin is eliminated in urine (50% of the oral dose) and to a lesser extent faeces. The major urinary metabolite is 6-desmethylgriseofulvin. Side effects are rare, mostly nausea and headache; abdominal pain and urticaria are less frequent and acute intermittent porphyria is rare in predisposed subjects. Absorption and metabolism are impaired by phenobarbitone (BUSFIELD et al. 1963) and there is an interaction with warfarin (UDALL 1969).

Griseofulvin is only effective in dermatophyte infections and is usually given for infections of hair or nail or when topical therapy has failed. Daily dosage in adults is 500–1000 mg given in one or two doses or 10 mg/kg in children. In treating large numbers of children with endemic scalp ringworm, modified dosage regimens of spaced large doses (50 mg/kg) are often effective. Treatment is usually continued for 4–6 weeks or longer, depending on response. Nail infections, particularly those involving toe-nails, may respond poorly even to 18 months of therapy.

3. Other Oral Anti-fungal Drugs

Other orally active anti-fungal agents have little value in the management of superficial mycoses although flucytosine is the treatment of choice in chromomycosis. Two new triazoles, itraconazole and fluconazole as well as an allylamine, terbinafine, are under evaluation in dermatomycosis.

However, these new compounds do have potential value and are likely to become available shortly. Itraconazole is a triazole which is active in vitro against the main pathogens involved in superficial fungal infections – the dermatophytes, *Candida* species, and *Malassezia furfur* (HAY et al. 1987). Clinical studies have shown that it will induce remissions in tinea corporis/cruris within 2 weeks and tinea pedis within 4 weeks with a dose of 100 mg daily. It is active in vaginal candidosis in total doses of 400–600 mg (SANZ SANZ and PALACIO HERNANZ 1987) given in a single or three divided doses. It is effective in pityriasis versicolor in a total dose of 1000 mg. Clinical trials have also shown that it is useful in sporotrichosis and chromomycosis, as well as certain systemic mycoses such as histoplasmosis and cryptococcosis (HAY et al. 1987). Itraconazole accumulates in the stratum corneum and has a prolonged skin surface retention time (HEYKANTS et al. 1987). At present it has not produced significant side effects. It does not inhibit human cytochrome P-450-dependent pathways such as adrenal androgen biosynthesis and has not been found to cause hepatitis. Its mode of action is thought to be similar to that of ketoconazole. Itraconazole is absorbed in low concentrations, producing serum levels of 300–400 ng/ml 2 h after a 100-mg dose, but is quickly bound to plasma proteins and in cellular compartments.

Fluconazole is another triazole anti-fungal which is highly active in vitro and in model systems against yeast infections such as those caused by *Candida* species and *Cryptococcus neoformans* (RICHARDSON et al. 1985). It is rapidly absorbed, penetrates body cavities such as peritoneum and cerebrospinal fluid, the skin and urine. Only a small fraction ($<1\%$) is bound to plasma proteins. It has been shown to be rapidly active in both vaginal and oral candidosis. In vaginal candidosis a single dose of 150 mg will induce remissions in over 85% of those treated. It has been found to be effective in chronic atrophic oral candidosis, chronic mucocutaneous candidosis and oral candidosis in immunocompromised patients (HAY 1988). Preliminary studies indicate that fluconazole is active in dermatophytosis as well (RUPING et al. 1987). In systemic infections the main indications at present are candidosis (DUPONT and DROUHET 1988) and cryptococcosis, particularly where the latter occurs in patients with the acquired immunodeficiency syndrome. The normal dose for superficial infections is 50 mg daily. Fluconazole appears to act by inhibition of cytochrome P-450-dependent synthesis of the fungal cell membrane. It has not been shown to produce serious adverse reactions in humans. Modification of the dose will be required for patients with renal impairment as fluconazole is largely excreted by the kidneys.

The other new antifungal agent is terbinafine, an allylamine compound (Ryder 1985). Terbinafine has the unusual feature of being fungicidal in vitro to many fungi at concentrtaions close to the minimum inhibitory concentration (MIC). For many dermatophytes the MIC is less than 0.05 mg/l. Terbinafine is less active against *Candida* species although it can produce clinical responses in vaginal candidosis. Most of the studies carried out at present with terbinafine have involved the treatment of dermatophytosis. The drug is active against most species and has been shown in one study of "dry-type" *T. rubrum* infections to produce a clinical and mycological remission rate of 75% after 30 days with doses of 125 mg twice daily compared with 40% for griseofulvin at 250 mg twice daily (Savin 1986). Terbinafine is also active in onychomycosis. The doses of terbinafine used are 125–250 mg twice daily. This drug inhibits squalene epoxidase which is active in the biosynthesis of the fungal cell membrane. It is well absorbed, but avidly bound to plasma proteins and is distributed particularly to skin and adipose tissue (Jensen 1988). At present no serious adverse effects have been recorded with the drug.

C. Conclusion

There have been few comparative studies between individual topical anti-fungal drugs. For instance Clayton and Knight (1976) showed that miconazole and clotrimazole were equally effective in superficial infections and the activity of Whitfield's ointment was very similar (Clayton and Connor 1973). Likewise there have also been few comparative studies of the new griseofulvin and azoles.

Newer anti-fungal drugs are still required, especially as many of the more specific topical agents are no more effective than Whitfield's ointment. Perhaps of greater importance would be an understanding of why chronic infections, particularly of toe-nails and webs, are only partially affected by current therapy with drugs to which they appear sensitive in vitro. The contribution of host factors such as site of infection, immune response and penetration of drug into thick keratin may be important. In this respect new topical methods of treating onychomycoses including a nail formulation of 28% tioconazole (Hay et al. 1985) and 1% bifonazole incorporated into urea paste (Nolting 1984) may be advantageous. Studies with the latter, for instance, indicate that it is possible to obtain clinical and mycological remission of nail disease and in some cases the patient is able to apply the treatment, using a simple instruction sheet. The new oral anti-fungals, have not been completely assessed, but show considerable promise provided that they can be shown to be non-toxic once larger numbers of patients have received treatment.

References

Artis WM, Odle MB, Jones HE (1981) Griseofulvin resistant dermatophytosis correlates with in vitro resistance. Arch Dermatol 117:16–19

Atkinson RM, Bedford C, Child KJ, Tomich EG (1962) Effect of particle size on blood griseofulvin levels in man. Nature 193:588–589

Bartek MJ, LaBudde JA, Maibach HJ (1972) Skin permeability in vivo: comparison in rat, rabbit, pig and man. J Invest Dermatol 58:114–123

Bingham JS (1984) Single blind comparison of ketoconazole 200 mg oral tablets and clotrimazole 100 mg vaginal tablets and 1% cream in treating acute vaginal candidosis. Br J Venereol Dis 60:175–177

Borgers M, van den Bossche H (1981) The mode of action of antifungal drugs. In: Levine HB (ed) Ketoconazole in the management of fungal disease. Addis, Balgoweah, Australia, pp 25–47

Brajtburg JK, Price HD, Medoff G, Schlessinger D, Kobayashi GS (1974) The molecular basis for the selective toxicity of amphotericin B for yeast and filipin for animal oils. Antimicrob Agents Chemother 5:377–382

Busfield D, Child KJ, Atkinson RM, Tomich EG (1963) An effect of phenobarbitone on blood levels of griseofulvin in man. Lancet 2:1042–1043

Clayton YM, Connor BL (1973) Comparison of clotrimazole cream, Whitifield's ointment and ystatin ointment for topical treatment of ringworm infections, pityriasis vesicolar, erythrasma and candidiasis. Br J Dermatol 89:297–303

Clayton YM, Knibht AG (1976) A clinical double blind trial of topical minconazole and clotrimazole against superficial fungal infection and erythrasma. Clin Exp Dermatol 1:225–229

Crounse RG (1963) Effective use of griseofulvin. Arch Dermatol 87:176–178

DeFelice R, Johnson DG, Galgiani IM (1981) Gynaecomastia with ketoconazole. Antimicrob Agents Chemother 19:10173–10174

Dupont B, Drohet E (1988) Fluconazole in the management of oropharyngeal candidosis in a predominantly HIV antibody positive group of patients. J Med Vet Mycol 26:67–71

El-Nakeeb MA, Lampen JO (1964) Formation of complexes of griseofulvin and nucleic acids of fungi and its relation to griseofulvin sensitivity. Biochem J 92:59–60

Epstein WL, Shah VP, Riegelman S (1972) Griseofulvin levels in stratum corneum. Study after oral administration in man. Arch Dermatol 106:344–348

Farr PM, Shuster S (1985) Treatment of seborrhoeic dermatitis with topical ketoconazole. Lancet 2:1271–1272

Farr PM, Krause LB, Marks JM, Shuster S (1985) The response of scalp psoriasis to ketoconazole. Lancet 2:921–922

Ford GP, Farr PM, Ive FA, Shuster S (1984) The response of seborrhoeic dermatitis to ketoconazole. Br J Dermatol 111:603–607

Fregoso-Duenas F (1980) Ketoconazole in vulvovaginal candidosis. Rev Infect Dis 2:620–626

Gentles JC (1958) Experimental ringworm in guinea pigs: oral treatment with griseofulvin. Nature 182:476–477

Harrison EF, Zwadyk P, Bequette RJ, Hamlow EE, Tavormina PA, Zygmunt WA (1970) Haloprogin: a topical antifungal agent. Appl Microbiol 19:746–750

Hay RJ (1988) Overview of studies of fluconazole in oropharyngeal candidosis. Rev Infect Dis (in press)

Hay RJ, Clayton YM (1982a) The treatment of patients with chronic mucocutaneous candidosis and *Candida* onychomycosis with ketoconazole. Clin Exp Dermatol 7:155–162

Hay RJ, Clayton YM (1982b) Treatment of chronic dermatophyte infections. The use of ketoconazole in griseofulvin treatment failures. Clin Exp Dermatol 7:611–617

Hay RJ, Midgeley G (1984) Short course ketoconazole in pityriasis versicolor. Clin Exp Dermatol 9:571–573

Hay RJ, Mackie RM, Clayton YM (1985) Tioconazole nail solution – an open study of its efficacy in onychomycosis. Clin Exp Dermatol 10:111–115

Hay RJ, Dupont B, Graybill JR (eds) (1987) Rev Infect Dis 9[Suppl 1]
Hay RJ, Clayton YM, Griffiths WAD, Dowd PM (1988) A comparative double-blind study of ketoconazole and griseofulvin in dermatophytosis. Br J Dermatol (in press)
Heel RC (1981 a) Pharmacokinetic properties. In: Levene HB (ed) Ketoconazole in the management of fungal disease. Addis, Balgowlah, Australia, pp 67–73
Heel RC (1981 b) Toxicology and safety studies. In: Levene HB (ed) Ketoconazole in the mangement of fungal disease. Addis, pp 74–76
Heel RC (1981 c) Dermatomycoses.In: Levene HB (ed) Ketoconazole in the management of fungal disease. Addis, Balgowlah, Australia, pp 79–88
Heiberg JK, Svejgaard E (1981) Toxic heptatitis during ketoconazole therapy. Br Med J 283:825–826
Heykants J, Michiels M, Meuldermans W, Monbaliu J, Lavrijsen K, Van Peer A, Levron JC, Woestenborghs R, Cauwenbergh G (1987) The pharmacokinetics of itraconazole in animals and man: an overview. In: Fromtling RA (ed) Recent advances in the discovery, development and evaluation of antifungal drugs. Telesymposium Proceedings Prous, Barcelona
Holt RJ (1974) Recent developments in antimycotic chemotherapy. Infection 2:95–107
Holt RJ, Azmi A (1978) Miconazole-resistant *Candida*. Lancet 1:50
Iwata K, Yamajuchi H, Hiratani T (1973) The mode of action of clotrimazole. Sabouraudia 11:158–168
Jensen JC (1988) Clinical pharmacokinetics of terbinafine (Lamisil). Clin Exp Dermatol (in press)
Kinsky SC (1961) the effect of polyene antibiotics on permeability in *Neurospora crassa*. Biochem Biophys Res Commun 4:353–357
Legendre R, Steltz M (1980) A multi-center double-blind comparison of ketoconazole and griseofulvin in the treatment of infections due to dermatophytes. Rev Infect Dis 2:578–581
Lewis JH, Zimmerman JH, Benson GD, Ishak KG (1984) Hepatic injury associated with ketoconazole therapy – an analysis of thirty three cases. Gastroenterology 86:503–513
Lyddon FE, Gundersen K, Maibach HI (1980) Short chain fatty acids in the treatment of dermatophytoses. Int J Dermatol 19:24–28
Medoff G, Kobayashi GA (1980) The polyenes. In: Speller DCE (ed) Antifungal chemotherapy. Wiley, Chichester, pp 3–33
Medoff G, Comfort M, Kobayashi GS (1971) Synergistic action of amphotericin B and 5-fluorocytosine against yeast-like organisms. Proc Soc Exp Biol Med 138:571–574
Nolting S (1984) Non traumatic removal of the nail and simultaneous treatment of onychomycosis. Dermatologica [Suppl 1]169:117–120
Nozawa Y, Kitajima Y, Sekiya T, Ito Y (1974) Ultrastructural alterations induced by amphotericin B in the plasma membrane of *Epidermophyton floccosum* as revealed by freeze-etch electron microscopy. Biochim Biophys Acta 367:32–38
Plempel M (1976) Probleme der Theapie mit modernen Antimykotika. MMW [Suppl 1]118:19–23
Raab WPE (1980) The treatment of mycoses with imidazole derivatives. Springer, Berlin Heidelberg New York
Richardson K, Brammer KW, Marriott MS, Troke PF (1985) Activity of UK 49858, a *bis*-triazole derivative against experimental infections with *Candida albicans* and *Trichophyton mentagrophytes*. Antimicrob Agents Chemother 27:832–835
Roberts SOB (1980) Treatment of the superficial and subcutaneous mycoses. In: Speller DCE (ed) Antifungal chemotherapy. Wiley, Chichester, pp 225–283
Rosenthal SA, Wise RS (1960) Studies concerning the development of resistance to griseofulvin by dermatophytes. Arch Dermatol 81:684–689
Ruping KU, Stary A, Tronnier H (1987) A comparative study of fluconazole and ketoconazole in the treatment of fungal infection by systemic agents. 17th World Congress of Dermatology, Berlin 1987. Abstracts, part I, p 452
Ryder NS (1985) Specific inhibition of fungal sterol biosynthesis by SF 86-327, a new allylamine antimycotic agent. Antimicrob Agents Chemother 27:252–256

Ryley JF, Wilson RG, Barrett-Bee KJ (1984) Azole resistance in *Candida albicans*. J med Vet Mycol 22:53–64
Samsoen M, Jelen G (1977) Allergy to Daktarin gel. Contact Dermatitis 3:351–352
Sanz Sanz F, Palacio Hernanz A (1987) Randomised comparative trial of three regimens of itraconazole for treatment of vaginal mycoses. Rev Infect Dis [Suppl 1]9:S139–142
Savin R (1988) Successful treatment of chronic tinea pedis (moccasin type) with terbinafine. Clin Exp dermatol (in press)
Shuster S (1984) The aetiology of dandruff and the mode of action of therapeutic agents. Br J Dermatol 111:235–242
Udall JA (1969) Drug interference with warfarin therapy. Am J Cardiol 23:143
Van den Bossche H (1974) Biochemical effects of miconazole on fungi. Biochem Pharmacol 23:887–889
Van den Bossche H, Willemsens G, Coole W, Cornelissen F, Lauwers F, van Cutsem JM (1980) In vitro and in vivo effects of the antimycotic drug, ketoconazole, on sterol synthesis. Antimicrob Agents Chemother 17:922–928
Wilson L 81975) Action of drugs on microtubules. Life Sci 17:303–310
Wishart JM (1980) A double blind trial of ketoconazole versus griseofulvin treatment for dermatophyte infections. Aust J Dermatol 24:40–44
Woods RA, Bard M, Jackson IE, Drutz DJ (1974) Resistance to polyene antibiotics and correlated sterol changes in two isolates of *Candida tropicalis* from a patient with an amphotericin B-resistant funguria. J Infect Dis 129:53–58

CHAPTER 29

Bacterial Infections

M. I. WHITE

A. Introduction

Bacterial infections of the skin are common and not usually serious. Complications do, however, arise if any delay occurs in the treatment of debilitated patients. Recently bacteria of low invasive potential have become recognised as causes of infection. The treatment of common conditions will be reviewed and rarer examples cited if they have been the subject of special study.

B. Normal Skin Flora

The positive and negative effects of the wide variety of microbes associated with the external surfaces of healthy people have been reviewed by MACKOWIAK (1982). The most comprehensive recent monograph on the microbiology of the human skin is that by NOBLE (1981). In this book he discusses the difficulties inherent in an apparently simple classification of bacteria on the skin into three classes, temporary residents, contaminants and residents. Additionally, he summarises recent work on the interactions between skin bacteria which include stimulation, inhibition and the important exchange of genetic material.

Infants' skin is colonised immediately after birth and within a short time the composition is similar to that of adults (SARKANY and GAYLARDE 1968). Both aerobic and anaerobic bacteria are found and the principal genera are *Staphylococcus, Propionibacterium* and *Cornynebacterium*. Additionally there are species of yeasts and fungi. The various anatomical sites have different proportions and total numbers of organisms.

C. Skin Surface Defences

From comparative studies (JENKINSON 1980) the complexity of the surface ecosystem is becoming appreciated. The bacteria derive essential nutrients on the skin surface, but lipids, for example, may be inhibitory or stimulatory. Physical factors such as temperature, humidity, pH and the partial pressure of gases also affect bacterial growth.

The intact skin is a formidable barrier to bacterial invasion. Host defences include the secretions of the sweat and sebaceous glands. Immunoglobulins and lysosyme have been demonstrated on the surface. In a healthy person definable factors such as poor hygiene or friction from clothing should be sought. Certain occupations such as meat handlers have a predilection to infection (BARNHAM and KERBY 1981), but although the precise organism causing infection may vary the key factor is frequently trauma.

D. Skin Flora in Disease States

Bacteria may cause disease both by direct tissue invasion and by toxins. For example, strains of *Staphylococcus aureus* are usually the predominant organisms in boils while other strains, by the production of a toxin, may cause the dramatic staphylococcal scalded skin syndrome even though no focus of infection is apparent. Increasingly it is being appreciated that the coagulase-negative staphylococci on the skin may be pathogenic – especially after the insertion of surgical prostheses such as ventriculo-atrial shunts (HOLT 1972).

When swabs are taken from an apparently infected skin lesion the result of culture is usually a mixed growth. The proportions of bacteria found on culture may not always reflect the proportions in vivo because of overgrowth of one species such as *Pseudomonas aeruginosa*. Many attempts have been made to quantitate skin bacteria (MARPLES 1981), but these methods are not yet realistic routine procedures. In most clinical situations the assessment must be made both from the Gram's stain report and from the result of culture.

The abstract concept of colonisation as distinct from infection has practical implications. Even when an organism generally considered pathogenic is recovered from an ulcer this is not necessarily an indication for specific treatment if the ulcer is healing well. The major exception to this generalisation is *Streptococcus pyogenes* Lancefield group A which so readily causes cellulitis.

E. General Principles of Treatment

As in every other clinical situation, safe and effective chemotherapy ought to be based on the isolation and identification of the pathogen and studies to determine its susceptibility to anti-microbial agents. In practice, however, there is usually a need or at least a demand for therapy before all these procedures can be completed – and therapy is usually commenced on an empirical basis. In hospital-acquired infections and in patients who are immune-compromised it is mandatory to send appropriate samples for bacteriological investigation before commencing treatment since this will ensure that continued therapy is rationally based.

I. Non-specific Measures

In certain circumstances debridement, removal of crusting or lancing are appropriate. Measures to combat nasal carriage of pathogenic staphylococci have been advocated in situations of recurring infection on the face. Any underlying and predisposing condition should be sought and treated.

II. Topical

In the treatment of skin infections topical antibiotics or antiseptics may be useful adjuncts. Their use, however, is limited by toxicity, sensitisation, staining and odour. As with any drug applied to the skin systemic absorption occurs and this may cause undesirable effects such as ototoxicity from neomycin products. The topical use of an antibiotic may lead to the development of a delayed hypersensitivity reaction; a generalised dermatitis may then develop if the drug is subsequently used systemically (CRONIN 1980).

Perhaps the main reason for limiting the topical use of important antibiotics is the ease with which bacteria develop drug resistance. Patients may then not only harbour, but also disperse multi-resistant strains (NAIDOO and NOBLE 1978).

Although antiseptics used topically may also cause allergy or systemic toxicity they are important everyday drugs in soaks or ulcer dressings. Aqueous solutions of the traditional antiseptics: potassium permanganate, silver nitrate, hydrogen peroxide and sodium hypochlorite, tulle dressings incorporating chlorhexidine and products containing iodine or polynoxylin are invaluable, but will not cure established infection.

III. Systemic

It is common practice to estimate the minimum inhibitory concentration of an antibiotic for the infecting organism. It is much more difficult however to predict the essential level or to measure the achieved level in infected tissues. Likewise, the knowledge of the free, as distinct from the protein-bound antibiotic level in serum, will not predict the antibiotic actually available at the site of infection. Additional factors influencing antibiotic tissue concentration are blood flow to the tissue site, transport systems controlling tissue penetration and the effects of disease on local binding sites. Thus, a measured serum "half-life" must be interpreted with great caution in clinical situations. Disease, such as walled-off abscesses, may result in decreased tissue penetration (RYAN 1978) and necrotic material may alter the effectiveness of the antibiotic (BRYANT and HAMMOND 1974).

In any individual patient the decision to use systemic antibiotic requires consideration of the patient, the disease and the drug. The risks of treatment must be assessed and balanced against the morbidity and possible mortality. The dosage, route of administration and duration of treatment must be determined and amended as necessary. In some circumstances, e.g. mixed bacterial infections, combinations of agents may be required. Hypersensitivity reactions limit the choice of antibiotic in an individual patient. Co-existing hepatic or renal disease may result in reduced excretion of a drug and enhanced toxicity. Interaction with other drugs may occur and result in diminished effect or increased toxicity.

Many antibiotics cause nausea and gastrointestinal symptoms. The development of diarrhoea must not be ignored, but must prompt investigation by sigmoidoscopy and stool culture. The list of antibiotics causing antibiotic-associated colitis is long (PITTMAN 1979). The recognition that toxigenic *Clostridium difficile* can cause pseudomembranous entero-colitis has led to improved management of this condition with vancomycin. Superinfection by other organisms may occur and troublesome oral infection or perianal inflammation due to *Candida albicans* is frequent. Other side effects such as marrow toxicity, skin eruptions of varying severity and the possibility of teratogenic effects should be borne in mind.

F. Antibiotics

I. Penicillins

Since the isolation of the 6-aminopenicillanic acid nucleus various semi-synthetic penicillins have been produced and the penicillins may now be considered in groups: natural penicillins; β-lactamase-resistant penicillins; ampicillin-like agents; and the broad-spectrum anti-pseudomonal penicillins. The β-lactam bond area of the nucleus is a structural analogue of a component of muramic acid, a major bacterial cell wall constituent. β-Lactams also affect target sites in cell division (TOMASZ 1979; TIPPER 1979).

Since the oral absorption of benzylpenicillin is poor and renal excretion rapid, the serum half-life is short (about 20 min). Adequate serum levels may be obtained by the frequent administration of parenteral doses and the use if necessary of probenecid. Formulations such as procaine penicillin have been developed for intramuscular use and are more reliable than oral phenoxymethylpenicillin (penicillin V).

The anti-bacterial activity of various penicillins has been summarised (WISE 1982). The principal use in dermatology for benzylpenicillin is in infections due to Lancefield group A β-haemolytic streptococci. The drug has low toxicity, but hypersensitivity reactions may be serious. Haemolytic anaemia has been reported and massive infusions of sodium benzylpenicillin may cause hypokalaemia and hypernatraemia. Hypersensitivity reactions range from urticaria to arthritis and anaphylaxis.

The isoxazolyl penicillins: oxacillin, cloxacillin, dicloxacillin and flucloxacillin resist hydrolysis by the exocellular β-lactamase produced by *Staphylococcus aureus* and *Staphylococcus epidermidis*. Methicillin is also in this group, but since it is not acid stable and not absorbed by the oral route it has been less important clinically. These drugs are all less effective against organisms such as β-haemolytic streptococci, but are the antibiotics of choice for *S. aureus* infections. The pharmacological properties of the drugs have been reviewed (NEU 1982), and the stability of these drugs compared with various cephalosporins has been studied (BASKER et al. 1980). Flucloxacillin has pharmacokinetic advantages (SUTHERLAND et al. 1970).

The adverse reactions are those of penicillins generally, but neutropenia and elevated hepatic enzymes have been described. Resistance to methicillin has been identified in staphylococci by in vitro testing on specific media at the low temperature of 30 °C. This finding however does not inevitably mean that the clinical results with other antibiotics in the group will be inadequate (LOWBURY et al. 1977) although such organisms are frequently resistant to the isoxazolyls. A further type of resistance of *S. aureus* to penicillins was described by SABATH et al. (1977). The dissociation of minimal inhibitory concentration and the minimal bactericidal concentration may be noted at 24 h. How often this phenomenon is responsible for antimicrobial therapeutic failure is unknown.

The bacteriological properties of ampicillin, the ampicillin esters, and analogues such as amoxycillin are similar. All are destroyed by staphylococcal β-lactamase. The oral absorption of amoxycillin is greater than that of ampicillin and it might be expected that diarrhoea would be less likely following treatment with the former. Ampicillin is active against gram-negative bacilli which do not produce hydrolysing enzymes. A recent innovation has been the combination of amoxycillin and potassium clavulanate. This latter is a β-lactamase inhibitor. The anti-bacterial activity of the combination is broad-spectrum and although clinical experience is limited it has proved effective in respiratory and urinary tract infections. In addition to the original anti-pseudomonal penicillin carbenicillin, acylureidopenicillins have been developed. Azlocillin has useful activity against *P. aeruginosa* and, like mezlocillin, has broad-spectrum activity.

II. Macrolides

Erythromycin inhibits protein synthesis at the level of the 50 S bacterial ribosome in a wide variety of bacteria (OLEINICK and CORCORAN 1969). It is a macrocyclic lactone ring (macrolide) inactivated by acid. Modifications of the base to produce salts or esters has yielded clinically active oral preparations as well as intravenous preparations. The use of the latter is limited because of the high incidence of associated phlebitis.

The anti-microbial activity of erythromycin is wide and it may be used as an alternative antibiotic in patients hypersensitive to penicillin. The bioavailability of erythromycin and erythromycin stearate is diminished by food. The absorption is maximum at 4 h, but the half-life is short, 2 h. The drug is concentrated in the liver and excreted in the bile, and a proportion may be *N*-demethylated in the liver. Erythromycin estolate is acid stable. It is absorbed as an ester and this drug particularly has been implicated in the major side effect of this group – cholestatic jaundice (LUNZER et al. 1975). Dose-related gastrointestinal symptoms are the most troublesome side effect of the group and hypersensitivity is rare.

Originally reserved as "second-line" anti-staphylococcal drugs their use in skin infections used to be frowned upon. However, with the development of other anti-staphylococcal drugs erythromycin may now be used in dermatology for pyodermas, erythrasma and acne as well as for superinfected skin lesions.

III. Polymyxins

Polymyxin B and polymyxin E (colistin) are branched cyclic decapeptides which destroy cytoplasmic membranes of bacteria (SUD and FEINGOLD 1972). They may be used topically in infected ulcers and have a synergistic effect with neomycin and some other antibiotics such as penicillin.

IV. Aminoglycosides

This group of antibiotics has a central hexose (or aminocyclitol) nucleus joined via a glycosidic linkage to amino sugars. Although active against staphylococci their greatest usefulness is in infections caused by gram-negative aerobic bacilli. In cases suspected of septicaemia a synergistic combination with an anti-staphylococcal penicillin may be used until blood culture results are available.

V. Cephalosporins

The basic 7-aminocephalosporonic acid nucleus has been modified with the generation of large numbers of antibiotics for oral and parenteral use. Later drugs are stable to β-lactamases and have a wider gram-negative spectrum. Several of these, cefuroxime, cefoxitin and cefotaxime, are active against penicillin-resistant gonococci. Patients who have had a hypersensitivity to penicillin may also be allergic to the cephalosporins.

VI. Tetracyclines

Tetracyclines have a broad spectrum of activity against bacteria, rickettsiae, mycoplasmas and chlamydiae. They are bacteriostatic and interfere with protein synthesis at the ribosomal level. The absorption of tetracycline is increased if the drug is taken half an hour before food and excretion is largely in the urine, but additionally in bile. The drug has been widely used with success in acne vulgaris and rosacea. The absorption of doxycycline and minocycline is much higher and the longer half-life permits twice daily dosage in most infections. These drugs are much more lipid soluble (Brogden 1975), but have not yet been fully evaluated in long-term use.

Tetracyclines should be avoided in patients with acute and chronic renal failure, pregnancy (both because of the risk of hepatotoxicity and because of deposition in foetal teeth and bones), patients with systemic lupus erythematosus and children under 8 years of age.

Gastrointestinal side effects are commonest and superinfection of the gastrointestinal tract by resistant gram-negative bacilli, or staphylococci as well as by *C. albicans* may occur. Acute benign intracranial hypertension has been reported rarely (Walters and Gubbay 1981) and photosensitivity is less common than with the earlier tetracyclines. The dose of concomitant anticoagulant treatment may have to be reduced and an interaction with oral contraceptives has been reported (Bacon and Shenfield 1980; Hudson and Callen 1982).

VII. Anti-tuberculous Drugs

Isoniazid, the hydrazide of isonicotinic acid, is bactericidal for replicating mycobacteria. Absorption may be diminished by antacids and the drug is well distributed in all body fluids. Isoniazid is metabolised in the liver by acetylation and hepatotoxicity may be a major adverse effect. Peripheral neuropathy is dose related and is preventable by pyridoxine.

Rifampicin is a complex molecule which inhibits bacterial DNA-dependent RNA polymerase. It is able to kill dormant organisms which have short periods of active metabolism or growth (MITCHISON 1979). It is well absorbed orally, but may be less well absorbed if *para*-aminosalicylic acid (PAS) is given simultaneously. The drug is metabolised in the liver and there is an enterohepatic circulation. Rifampicin induces liver enzymes so that the doses of other drugs may require alteration. Liver function tests should be monitored since hepatotoxicity may occur.

Ethambutol is active only against dividing organisms. Although metabolised by the liver the drug will accumulate in renal insufficiency. In higher doses the drug may cause optic neuritis. In general, the initial loss of colour discrimination is reversible.

Streptomycin must be administered parenterally; it is excreted predominantly by the kidney. It is an aminoglycoside and both vestibular and auditory ototoxicity occur, as can nephrotoxicity. Great caution must therefore be taken when prescribing for the elderly. PAS and cycloserine are second-line drugs, but may be required if, for example, in vitro studies show resistant organisms.

VIII. Metronidazole

Metronidazole is a nitroimidazole drug active against obligate anaerobic bacteria such as *Bacteroides fragilis* and *Bacteroides melaninogenicus*. These may be isolated from abscesses and leg ulcers. Whilst their exact role in pathogenicity in dermatological practice has not been established, specific treatment may at times be necessary (TALLY et al. 1978).

G. Clinical Situations

I. Infections

1. Impetigo

The use of the clinical term impetigo applied to crusted lesions has caused confusion between British and American writers. In the more humid environment of parts of the United States streptococcal pyodermas are more frequent, perhaps arising from insect bites. Since cultures from such lesions may produce a mixed growth of haemolytic streptococci and penicillin-resistant staphylococci, oral antibiotic treatment is appropriate with, for example, erythromycin.

In the United Kingdom, localised crusted lesions – typically on the face of children – are principally due to *S. aureus* (NOBLE et al. 1974). Although they are commonly successfully managed by the use of topical antibiotics or antiseptics, it is difficult to recommend a particular regimen since the trials of treatment have frequently been inconclusive. The untreated disease may persist for 6–60 days and treatment is rarely commenced in the first few days. It is thus difficult to separate the effects of treatment from spontaneous healing. Additionally, the base of the topical preparation may be as important as the antibiotic component. A systemic antibiotic, such as flucloxacillin, is required if the lesions are extensive or the patient debilitated. If a nephritogenic streptococcus is suspected the antibiotic of choice is penicillin or erythromycin.

Bullous impetigo is a localised form of the staphylococcal scalded skin syndrome and is mediated by a toxin. Flucloxacillin or an alternative anti-staphylococcal antibiotic for 10 days is recommended, the route depending on the urgency of the illness.

2. Furunculosis

Superficial pustules and folliculitis result from superficial damage to the skin and quickly resolve. A more painful inflamed nodule, the furuncle, results from deeper infection. If adjacent hair follicles are affected a much larger lesion, the carbuncle, may form with multiple draining centres. Penicillinase-resistant antibiotics are the agents of choice, but other precautions must be taken to prevent auto-inoculation. Collections of pus should be drained and all contaminated clothing should be adequately laundered – a precaution commonly neglected.

Topical neomycin or chlorhexidine prescribed to reduce the nasal carriage of bacteria may be of some benefit in patients with recurrent folliculitis in the beard area. The use of antiseptic hand washing does reduce the transmission of infection in hospital outbreaks.

Obesity, diabetes mellitus, blood dyscrasias and disorders of neutrophil function predispose patients to these infections as does the use of immunosuppressive drugs. It remains to be seen if detailed studies will detect more subtle defects in host defence mechanisms (CATES and QUIE 1979).

3. Ecthyma

This is an ulcerative form of pyoderma characterised by crusted ulcers surrounded by a halo of erythema and caused by group A β-haemolytic streptococci (KELLY et al. 1971). As with staphylococcal infections the identification and elimination of external factors is the most important aspect of treatment. A minimum of 10 days penicillin is required, if necessary parenterally.

4. Cellulitis

Infections of the skin by group A β-haemolytic streptococci have a characteristic demarcated edge and can extend rapidly with lymphangitis or produce bacteraemia. In one series of 293 patients with bacteraemia 53 had a preceding cellulitis. Another 39 patients developed this potentially lethal condition as a consequence of secondarily infected skin lesions such as leg ulcers, burns and skin diseases. Benzylpenicillin is the antibiotic of choice. If after 24–48 h there has been an adequate response oral antibiotics may be substituted, either penicillin V or amoxycillin for at least 10 days. Necrotising fasciitis is a rapidly developing complication and extensive surgical resection of affected areas is recommended (SEAL et al. 1980).

5. "Gram-negative" Infections

In certain situations such as chronic paronychia, stasis ulceration and chronic otitis externa there may be significant infections by *Pseudomonas* species. It would be rare for such patients to require intensive antibiotic therapy since the bacteria disappear as the original problem is controlled (NOBLE and SAVIN 1971).

The syndrome of "gram-negative" folliculitis (LEYDEN et al. 1973) has been attributed to the antibiotic therapy of acne vulgaris. Folliculitis due to *Escherichia coli* has also been attributed to the prolonged use of a hexachlorophene preparation (KENNEDY 1979). *Proteus mirabilis* has been implicated in non-staphylococcal axillary abscesses (SHEHADEH and KLIGMAN 1963).

6. Erythrasma

Erythrasma is a superficial bacterial infection due to *Corynebacterium minutissimum*. As a model of superficial skin infection it demonstrates several features: the responsible organism forms part of the normal flora of healthy people; it may be chronic; it responds to both topical and systemic treatment; there may be coexisting fungal infection (SOMERVILLE 1972; SCHLAPPNER et al. 1979).

The treatment of this has been the subject of several good clinical studies of which SOMERVILLE et al. (1971) and PITCHER et al. (1979) are examples. Nevertheless there are suggestions that relapse occurs more frequently after topical treatment than by systemic erythromycin.

7. Tuberculosis

Skin lesions caused by *Mycobacterium tuberculosis* and the other mycobacteria are comparatively rare. The classifications of cutaneous tuberculosis vary, but dividing the tuberculoses, in which bacilli can be demonstrated in the skin, from the tuberculides has practical advantages (WILKINSON 1979). Nevertheless, the distinction is not absolute as the cases of lichen scrofulosorum discussed by SMITH et al. (1976) illustrate. Here the skin lesions of a tuberculide responded to chemotherapy.

The regimens for respiratory tuberculosis have been extensively assessed and are continually revewied. Therapy for extra-pulmonary tuberculosis has not been so rigorously tested, but the principles of treatment apply to all disease patterns. It is essential to establish the total extent of the disease, to estimate the patient's general health, nutrition and immunity as well as to consider public health aspects. In the first 2 months of treatment rifampicin plus isoniazid plus either ethambutol or streptomycin are recommended. Subsequently rifampicin and isoniazid are continued for a further 7 months. Pyridoxine should be given concurrently with isoniazid to prevent peripheral neuropathy when the daily dose of isoniazid exceeds 5 mg/kg body weight (HORNE 1982).

8. Miscellaneous

The diversity of primary bacterial infections of the skin is enormous. Certain occupations increase the risk of acquiring infection and immune deficiency predisposes individual patients. A useful summary is the chapter by ROBERTS and ROOK (1979) or the monograph by NOBLE (1981).

II. Diseases Exacerbated by Bacteria

1. Atopic Dermatitis

Viruses are the principal cause of serious clinical infection in patients with atopic dermatitis. Nevertheless, since the discovery in the nineteenth century that bacteria are present in the exudates of eczema there has been discussion of their role in the pathogenesis of the disease. Two studies which include personal observations and extensive reviews of the relevant literature are those of STORCK (1948) and ANDERSON and HEILESEN (1951). A further major study was reported by RAJKA in 1963. One-third of his 70 adult patients with atopic dermatitis showed immediate and delayed hypersensitivity reactions to staphylococcal products. The effect of a staphylovaccine was assessed and although some patients showed a subsequent decrease in the immediate reaction, the experiments did not result in a clinically useful vaccine.

The nasal carriage rate of staphylococci in this condition is high and the lesional skin is almost invariably colonised by *S. aureus* (ALY et al. 1977; BIBEL et al. 1977). The difficulty of recognising secondary infection in atopic dermatitis has been highlighted by LEYDEN et al. (1974). After studying the clinical response to antibiotic therapy they concluded that *S. aureus* at a density of 10^6 organisms/cm^2 was likely to aggravate the primary lesion. The mechanisms of such aggravation are unknown. The human cutaneous reaction to teichoic acid, a bacterial cell wall component, has been studied by MARTIN et al. (1967) and the reaction to staphylococcal protein A by WHITE and NOBLE (1980). This latter protein has interesting properties including a non-specific reaction with IgG. The immediate weal and flare is succeeded by a later induration which persists, suggesting a lymphocyte-mediated delayed hypersensitivity reaction. Antibody to *S. aureus* of the IgE class has been detected in the serum of patients with chronic eczema and the hyper-immunoglobulinaemia E syndromes. Whether this indicates a hypersensitivity reacton to staphylococci in these patients awaits further investigation (SCHOPFER et al. 1979).

It has proved extremely difficult to evaluate the role of topical steroid-antibiotic combinations. Recently, in a study of the effect of topical corticosteroid alone on the microbial flora of normal human skin no significant quantitative change could be demonstrated, although a rise in microbial count could be detected after occlusion of the skin (CHAN et al. 1982). The effects of a broad-spectrum steroid-antibiotic combination in experimental *S. aureus* infection of healthy subjects suggested that the omission of any one component decreased the effectiveness of the mixture (MARPLES et al. 1973 b).

In studies of children with atopic dermatitis LEYDEN and KLIGMAN (1977) assessed the clinical response as well as the alterations in the total aerobic flora following topical treatment. They advocate the use of steroid-antibiotic combinations to produce a rapid reduction in the density of potentially pathogenic bacteria. The usefulness of systemic cloxacillin has also been documented (EAGLSTEIN et al. 1977), but again this study was very short term. It is to be hoped that this problem will soon be studied with long-term follow-up of patients to determine, for example, if recurrences are reduced.

The development of an allergic contact dermatitis to a topical antibiotic, such as neomycin, is less important than the development of a generalised hypersensitivity to penicillin. However, topical antibiotics have been implicated in other complications such as the development of pyoderma due to resistant *S. aureus* (MARPLES and KLIGMAN 1969) and the loss of the natural antibacterial action of the skin following the topical application of neomycin (LACEY 1969).

2. Psoriasis

The percentage of patients carrying *S. aureus* on psoriatic plaques was found to be 30% by ALY et al. (1976). This is an intermediate between previous estimates of 50% and 26%, but certainly higher than the carriage rate in their control population. The count on plaques, averaged from 40 subjects was 3×10^3 colony forming units/cm^2. Where the effect of occlusion has been studied the absolute numbers of *S. aureus* increased and the proportion of the total aerobic flora due to that bacterium also increased (MARPLES et al. 1973 a). It seemed that occluded skin was less responsive to subsequent treatment. Another factor which has been noted to increase the carriage rate of *S. aureus* was the use of emollients in the course of photochemotherapy treatment (WEISSMANN and NOBLE 1980). In this study one of the ten patients developed pustular lesions requiring antibiotic treatment.

3. Napkin Dermatitis

The treatment of all forms of napkin dermatitis involves more frequent changing of nappies and, if possible, in severe cases the avoidance of nappies. In some children the rash may be a manifestation of atopic dermatitis, psoriasis or seborrhoeic dermatitis. *C. albicans* infection may be the cause, but, in the remainder, friction and maceration are more important than microorganisms (LEYDEN and KLIGMAN 1978). At times, specific therapy may be required against staphylococci or *C. albicans*. Topical applications which combine nystatin, chlorhexidine and hydrocortisone are popular, but should not be used before the condition has been accurately diagnosed nor used as a substitute for the more tedious chores of parenthood.

4. Leg Ulcers

Venous stasis is the commonest cause of ulceration in the leg and attention to this is the most important therapeutic responsibility. Gentle cleansing and a simple sterile gauze dressing should be sufficient. However, frank infection does occur. The most serious infections are due to group A β-haemolytic streptococci, but *S. aureus* or *E. coli* infections also occur. There Is no role for prophylactic antibiotic treatment and in established infection systemic antibiotic treatment will be required. There is no agreement about the best topical application if infection is suspected or has occurred. New formulations and novel agents are continually being introduced. Allergic contact dermatitis to medicaments may develop insidiously to cause delayed healing and ultimately a generalised dermatitis.

5. Hidradenitis Suppurativa

This is an unusual but severe chronic disease in which repeated suppuration of the skin of the axillae and groins results in scarring and sinus formation. Both aerobic and anaerobic bacteria are important. Long-term tetracycline is of benefit and it has been suggested that infection with *Streptococcus milleri*, a frequent commensal in the gastrointestinal tract, can cause treatable exacerbations in this disease (Highet et al. 1980).

6. Miscellaneous

Crusting or ulceration is frequent in many chronic skin conditions. Any compress may soothe, but observers have recorded that improved healing occurs if antibacterial measures are taken. In particular, the bullous disorders and pyoderma gangrenosum heal better with meticulous dressings (Hickman 1977). In contrast, the use of topical medicaments in miliaria is theoretically indicated, but not practically helpful (Holzle and Kligman 1978).

III. Infection in Immune-Compromised Patients

Patients may have an increased susceptibility to infection because of disease or therapy, such as cytotoxic drugs. The routine management of such patients should include measures to minimise the risk of infection and any infective episode should be treated promptly and vigorously. The final choice of antibiotic will be determined by the sensitivities of isolated organisms and the resistance patterns of organisms prevalent in each hospital. The parenteral administration of a semi-synthetic penicillin active against *S. aureus*, an aminoglycoside agent against gram-negative bacteria and azlocillin to combat *P. aeruginosa* is one approach, although inactivation may occur if these drugs are admixed prior to infusion. As in every other clinical situation the prescription will be individually tailored for the patient by the experience of the clinician, having due regard to the results of the microbiological tests.

References

Aly R, Maibach HI, Mandel A (1976) Bacterial flora in psoriasis. Br J Dermatol 95:603–606
Aly R, Maibach HI, Shinefield HR (1977) Microbial flora of atopic dermatitis. Arch Dermatol 113:780–782
Anderson EK, Heilesen B (1951) On the occurrence and typing of *Staphylococcus pyogenes* in weeping eczema. Acta Derm Venereol (Stockh) 31:679–703
Bacon JF, Shenfield GM (1980) Pregnancy attributable to interaction between tetracycline and oral contraceptives. Br Med J [Clin Res] 280:293
Barnham M, Kerby J (1981) Skin sepsis in meat handlers. J Hyg (Lond) 87:465–476
Basker MJ, Edmondson RA, Sutherland R (1980) Comparative stabilities of penicillins and cephalosporins to staphylococcal β-lactamase and activities against *Staphylococcus aureus*. J Antimicrob Chemother 6:333–341
Bibel DJ, Greenberg JH, Cook JL (1977) *Staphylococcus aureus* and the microbial ecology of atopic dermatitis. Can J Microbiol 23:1062–1068
Brogden RN (1975) Minocycline: a review of its antibacterial and pharmacokinetic properties and therapeutic use. Drugs 9:251–291

Bryant RE, Hammond D (1974) Interaction of purulent material with antibiotics used to treat *Pseudomonas* infections. Antimicrob Agents Chemother 6:702–707
Cates KL, Quie PG (1979) Neutrophil chemotaxis in patients with *Staphylococcus aureus* furunculosis. Infect Immun 26:1004–1008
Chan H-L, Aly R, Maibach HI (1982) Effect of topical corticosteroid on the microbial flora of human skin. J Am Acad Dermatol 7:346–348
Cronin E (1980) Contact dermatitis. Livingstone, Edinburgh
Eaglstein WH, Feinstein RJ, Halprin KM, Bergstresser PR, Mertz PM (1977) Systemic antibiotic therapy of secondarily infected dermatitis. Arch Dermatol 113:1378–1379
Hickman JG (1977) New look at pyoderma gangrenosum. Cutis 20:209–219
Highet AS, Warren RE, Staughton RCD, Roberts SOB (1980) *Streptococcus milleri* causing treatable infection in perineal hidradenitis suppurativa. Br J Dermatol 103:375–382
Holt RJ (1972) The pathogenic role of coagulase negative staphylococci. Br J Dermatol [Suppl 8]86:42–47
Holzle E, Kligman AM (1978) The pathogenesis of miliaria rubra. Br J Dermatol 99:117–137
Horne N (1982) Tuberculosis. Med Int 1:978–983
Hudson CP, Callen JP (1982) The tetracycline – oral contraceptive controversy. J Am Acad Dermatol 7:269
Jenkinson DMcE (1980) The topography, climate and chemical nature of the mammalian skin surface. Proc R Soc Edinburgh [Biol Sci]79:3–22
Kelly C, Taplin D, Allen AM (1971) Streptococcal ecthyma. Arch Dermatol 103:306–310
Kennedy C (1979) Gram-negative folliculitis of the face. Clin Exp Dermatol 4:123–124
Lacey RW (1969) Loss of the antibacterial action of skin after toical neomycin. Br J Dermatol 81:435–439
Leyden JJ, Kligman AM (1977) The case for steroid-antibiotic combinations. Br J Dermatol 96:179–187
Leyden JJ, Kligman AM (1978) The role of micro-organisms in diaper dermatitis. Arch Dermatol 114:56–59
Leyden JJ, Marples RR, Mills OH, Kligman AM (1973) Gram-negative folliculitis – a complication of antibiotic therapy in acne vulgaris. Br J Dermatol 88:533–538
Leyden JJ, Marples RR, Kligman AM (1974) *Staphylococcus aureus* in the lesions of atopic dermatitis. Br J Dermatol 90:525–538
Lowbury EJL, Lilly HA, Kidson S (1977) "Methicillin-resistant" *Staphylococcus aureus*: reassessment by controlled trial in burns unit. Br Med J 1(6068):1054–1056
Lunzer MR, Huang SN, Ward KM, Sherlock S (1975) Jaundice due to erythromycin estolate. Gastroenterology 68:1284–1291
Mackowiack PA (1982) The normal microbial flora. N Engl J Med 307:83–93
Marples RR (1981) Newer methods of quantifying skin bacteria. In: Maibach H, Aly R (eds) Skin microbiology. Springer, Berlin Heidelberg New York, pp 45–49
Marples RR, Kligman AM (1969) Pyoderma due to resistant *Staphylococcus aureus* following topical application of neomycin. J Invest Dermatol 53:11–13
Marples RR, Heaton CL, Kligman AM (1973 a) *Staphylococcus aureus* in psoriasis. Arch Dermatol 107:568–570
Marples RR, RR, Rebora A, Kligman AM (1973 b) Topical steroid-antibiotic combinations. Arch Dermatol 108:237–239
Martin RR, Crowder JG, White A (1967) Human reactions to staphylococcal antigens. J Immunol 99:269–275
Mitchison DA (1979) Basic mechanisms of chemotherapy. Chest [Suppl] 76:771–781
Naidoo J, Noble WC (1978) Acquisition of antibiotic resistance by *Staphylococcus aureus* in skin patients. J Clin Pathol 31:1187–1192
Neu HC (1982) Antistaphylococcal penicillins. Med Clin North Am 66:51–60
Noble WC (1981) Microbiology of human skin, 2nd edn. Lloyd-Luke, London
Noble WC, Savin JA (1971) "Gram-negative" infections of the skin. Br J Dermatol 85:286–289
Noble WC, Presbury D, Connor BL, Savin JA, Taplin D, Lansdell L (1974) Prevalance of streptococci in lesions of impetigo. Br J Dermatol 91:115

Oleinick NL, Corcoran JW (1969) Two types of binding of erythromycin to ribosomes from antibiotic-sensitive and resistant *Bacillus subtilis* 168. J Biol Chem 244:727–735
Pitcher DG, Noble WC, Seville RH (1979) Treatment of erythrasma with miconazole. Clin Exp Dermatol 4:453–456
Pittman FE (1979) Antibiotic-associated colitis – an update. Adverse Drug React Bull 75:268–271
Rajka G (1963) Studies in hypersensitivity to molds and staphylococci in prurigo Besnier (atopic dermatitis). Acta Derm Venereol (Stockh) [Suppl 54]43
Roberts SOB, Rook A (1979) Bacterial infections. In: Rook A, Wilkinson DS, Ebling FJG (eds) Textbook of dermatology. Blackwell, Oxford, pp 541–605
Ryan DM (1978) Implanted tissue cages – a critical evaluation of their relevance in measuring tissue concentrations of antibiotics. Scand J Infect Dis [Suppl] 13:58–62
Sabath LD, Wheeler N, Laverdiere M, Blazevic D, Wilkinson BJ (1977) A new type of penicillin resistance of *Staphylococcus aureus*. Lancet 1(8009):443–447
Sarkany I, Gaylarde CC (1968) Bacterial colonisation of the skin of the newborn. J Pathol Bacteriol 95:115–122
Schlappner OLA, Rosenblum GA, Rowden G, Phillips TM (1979) Concomitant erythrasma and dermatophytosis of the groin. Br J Dermatol 100:147–151
Schopfer K, Baerlocher K, Price P, Krech U, Quie PG, Douglas SD (1979) Staphylococcal IgE antibodies, hyperimmunoglobulinemia E and *Staphylococcus aureus* infections. N Engl J Med 300:835–838
Seal D, Leppard B, Widdowson J, McGill J, Tormey P (1980) Necrotising fasciitis due to *Streptococcus pyogenes*. Br Med J 280(6229):1419–1420
Shehadeh NH, Kligman AM (1963) The effect of topical antibacterial agents on the bacterial flora of the axilla. J Invest Dermatol 40:61–71
Smith NP, Ryan TJ, Sanderson KV, Sarkany I (1976) Lichen scrofulosorum. Br J Dermatol 94:319–325
Somerville DA (1972) A quantitive study of erythrasma lesions. Br J Dermatol 87:130–137
Somerville DA, Noble WC, White PM, Seville RH, Savin JA (1971) Sodium fusidate in the treatment of erythrasma. Br J Dermatol 85:450–453
Storck H (1948) Experimentelle Untersuchungen zur Frage der Bedeutung von Mikroben in der Ekzemgenese. Dermatologica 96:177–262
Sud IJ, Feingold DS (1972) Effect of polymyxin B on antibiotic resistant *Proteus mirabilis*. Antimicrob Agents Chemother 1:417–421
Sutherland R, Croydon EAP, Rolinson GN (1970) Flucloxacillin, a new isoxazolyl penicillin, compared with oxacillin, cloxacillin and dicloxacillin. Br Med J 4(5733):455–460
Tally FP, Goldin BR Sullivan N, Johnston J, Gorbach SL (1978) Antimicrobial activity of metronidazole in anaerobic bacteria. Antimicrob Agents Chemother 13:460–465
Tipper DJ (1979) Mode of action of β-lactam antibiotics. Rev Infect Dis 1:39–54
Tomasz A (1979) The mechanism of the irreversible antimicrobial effects of penicillins: how the beta-lactam antibiotics kill and lyse bacteria. Annu Rev Microbiol 33:113–137
Walters BNJ, Gubbay SS (1981) Tetracycline and benign intracranial hypertension: report of five cases. Br Med J [Clin Res] 282:19–20
Weissman A, Noble WC (1980) Photochemotherapy of psoriasis: effects on bacteria and surface lipids in uninvolved skin. Br J Dermatol 102:185–193
White MI, Noble WC (1980) Skin response to protein A. Proc R Soc Edinburgh [Biol Sci]79:43–46
Wilkinson DS (1979) Tuberculosis of the skin. In: Rook A, Wilkinson DS, Ebling FJG (eds) Textbook of dermatology. Blackwell, Oxford, pp 677–699
Wise R (1982) Penicillin and cephalosporins: antimicrobial and pharmacological properties. Lancet 2(8290):140–143

CHAPTER 30

Herpes Virus Infections

D. Brigden

A. Introduction

Virology is a rapidly advancing science which is now beginning to pay clear dividends in terms of our understanding of the relationship betwen viruses and their host cells and in providing more rational ways of developing major therapeutic advances. Since the first viruses to be readily tackled by chemotherapy are members of the Herpesviridae this chapter will confine itself to a discussion of the biology of this fascinating group and our attempts at the therapeutic control of the human diseases these cause.

B. Classification

Humans are commonly infected with four members of the group: herpes simplex virus (HSV), varicella zoster virus (VZV), Epstein-Barr virus (EBV) and cytomegalovirus (CMV). All these four viruses have the ability to remain latent following the primary infection. HSV and VZV both migrate to the sensory nerve ganglia where they remain dormant until an as yet essentially unknown stimulus produces reactivation. EBV and CMV are able to remain latent in lymphocytes and probably other tissue cells.

C. Herpes Simplex

There are two main strains of this virus which behave in many ways in a similar manner. They share some homologous genetic information, but can readily be distinguished by restriction endonuclease analysis. HSV type I is usually transmitted during early childhood and the primary infection normally causes a stomatitis which frequently passes unnoticed. HSV type II is generally considered to be transmitted by venereal contact and thus only usually infects humans when sexual activity begins. The major exception to this is when a new-born infant becomes infected during passage down the birth canal, leading to a severe life-threatening infection carrying an approximate 50% mortality. This firm distinction between venereally acquired HSV type II and childhood acquired HSV type I is by no means absolute. The primary infection in both cases lasts for around 20 days following an incubation period of usually around 1 week. The characteristic sequence of lesions runs from a macular papular lesion through a vesicle, a pustule, ulceration or crusting and finally healing. The more severe primary infections can be associated with systemic illness. Virus can usually be cultured from lesions for around 2 weeks.

I. Latency and Recurrence

The observed clinical phenomenon of a frequently recurrent cold sore is a manifestation of the reactivation of HSV from its latent state in the sensory nerve cell nucleus. The most widely accepted hypothesis is that the primary infecting virus travels up the sensory nerve neurone and becomes latent in the cell nucleus at the nerve ganglion. The evidence for this has been assembled by COOK and STEVENS (1973). When reactivation occurs new virus particles travel down the neurone again and are discharged from the nerve endings where a clinical recurrence may take place if the local conditions are right. This appears to occur without host cell death as there is very little evidence of sensory loss despite numerous recurrences at the same site. This hypothesis would suggest that viral DNA remains in the cell nucleus and acts as a template for further viral replication at a later date. If this is the case then anti-viral drugs are unlikely to eliminate the latent viral DNA and thus a course of anti-viral therapy would not alter the likelihood of recurrence at a later date. An alternative hypothesis would suggest that reinfection of the cell nucleus has to take place, either of the same cell (a variant of the hypothesis) or of an adjacent cell. If either of these situations occurs then it could theoretically be possible to reduce or eliminate the latent virus in the ganglion and thus reduce or eliminate recurrences. As will be seen later the nature of latency in humans also has significance for the development of virus drug resistance.

What do the experimental facts tell us? FIELD et al. (1979) were able to protect mice from establishing latent infections by giving acyclovir (50 mg/kg by intraperitoneal injection) from the day before virus injection for 10 days. They were unable, however, to affect the virus once latency had been established. KLEIN et al. (1979) using topical acyclovir therapy were also able to protect mice completely when treatment was started 3 h after the infection. Protection was only partial if treatment was delayed until 24 h after infection. Systemic treatment was, however, ineffective if delayed at all. BLYTH et al. (1980) demonstrated that acyclovir given systemically would prevent recurrence induced by tape-stripping, but were unable to influence the established latent viral load. These results strongly suggest that once the virus has entered the nerve fibres it is not possible to prevent the virus integrating with the cellular DNA. Furthermore, FIELD et al. (1979) were unable to eradicate latency by relatively long-term treatment in animals. In a particularly elegant model BLYTH et al. (1980) induced several cycles of virus replication during treatment with acyclovir. No reduction in virus latency was demonstrable. It can be concluded from these experiments that whilst it is possible to prevent the establishment of latency, treatment has to be started before the virus enters the neurone and this happens within hours of virus inoculation. Furthermore, latency once established is not susceptible to elimination. Whether these findings are directly transferable to the human situation is debatable and will only be determined following the use of effective anti-herpetic drugs in humans.

D. Varicella Zoster

The primary infection with this virus causes chickenpox (varicella). This is usually a mild infection of young children although it can be severe and sometimes fatal in adults, neonates and individuals who are immune-compromised by disease or therapy. This virus also establishes latency in the sensory nerve ganglia and can reactivate later in life. The reactivated virus travels down the sensory nerve supplying one or more dermatomes and produces an attack of shingles (zoster). This usually only happens once though second attacks of zoster are well documented. The reasons why this herpes virus produces this rather different type of secondary infection are completely unknown. The factors that precipitate reactivation are also largely unknown although the chance of getting shingles is greatly increased in the immune-compromised subject.

E. Therapy

Progress in anti-herpes therapy has been bedevilled by wild claims of activity based on anecdotal reporting or on poorly designed clinical studies. Amongst the therapies that have been claimed to be effective, but for which there is no acceptable scientific evidence, are herpes virus vaccines, methisoporinol, diethyl ether, photodynamic inactivation, deoxyglucose and lysine. In two cases therapies have been shown to be actually harmful, namely idoxuridine in herpes encephalitis and cytosine arabinoside in herpes zoster when the drugs in question have been subjected to controlled trial.

I. Idoxuridine

Despite these disappointing early results, exciting progress has been made in recent years in the therapy of herpes virus infections. KAUFMAN et al. (1962) first showed that idoxuridine was useful in the treatment of herpetic keratitis. Since that time it has become the mainstay of therapy in this disease. JUEL-JENSEN and MACCULLUM (1972) were the first to demonstrate that idoxuridine might be useful as a topical preparation in herpes zoster and this has been confirmed in a well-conducted trial by WILDENHOFF et al. (1979). Unfortunately idoxuridine has been limited to topical use because of its systemic toxicity.

II. Adenine Arabinoside

PAVAN-LANGSTON and DOHLMAN (1972) were the first to demonstrate the efficacy of adenine arabinoside in human herpetic keratitis. Adenine arabinoside has also been demonstrated to be effective when used systemically in life-threatening herpes virus infections. In a multi-centre trial in herpes encephalitis, WHITLEY et al. (1978) showed that the mortality in this disease was reduced from about 70% to about 30%. Unfortunately, the analysis of this trial has come under criticism because of the small number of placebo-treated patients. Further follow-up in an open study has confirmed that the mortality of brain biopsy-positive herpes encephalitis is about 40% when treated with a full course of adenine arabinoside (WHITLEY et al. 1981). WHITLEY and ALFORD (1981) have shown that adenine arabinoside is effective in the treatment of neonatal herpes and WHITLEY et al. (1976) have demonstrated its effectiveness in herpes zoster in immune-compromised patients. WHITLEY et al. (1980) have reviewed the pharmacological and therapeutic properties of adenine arabinoside.

III. Trifluorothymidine

Trifluorothymidine was also demonstrated by WELLINGS et al. (1972) to be effective in the treatment of herpes keratitis; however, it has been considered too toxic for systemic use. A very full review of this drug has been published by HEIDELBERGER and KING (1979).

IV. Acyclovir

All the drugs mentioned so far essentially need to be phosphorylated and then inhibit DNA metabolism by varying mechanisms. However, a new anti-herpes drug was reported by ELION et al. (1977) which has far greater specificity for virus-infected cells. In radioactive tracer experiments these workers demonstrated that acyclovir was largely converted to the mono-, di- and triphosphates in Vero cells infected with HSV whereas uninfected cells or cells infected with vaccinia virus did not show significant phosphorylation. There was a 40-fold greater concentration of acyclovir triphosphate in the infected cells than in the uninfected cells. FURMAN et al. (1979) using isolated HSV and cellular DNA polymerases showed that the apparent k_i, of acyclovir triphosphate for the HSV DNA polymerase was 0.08 μM (approximately 1/30th of the concentration for the α-DNA polymerase of the host cell (2.1 μM). They were also able to show that acyclovir triphosphate not only inhibits the DNA polymerase activity by competition with deoxyguanosine triphosphate, but is also capable of preventing further DNA synthesis by being incorporated as a DNA chain terminator.

Extensive evaluation of acyclovir has now taken place in humans, although much still remains to be done. Human pharmacokinetic studies have been reported by DE MIRANDA et al. (1979) and BRIGDEN et al. (1980, 1981 a). Essentially the drug has been shown to have a half-life of about 3 h with the majority of the drug being excreted unchanged through the kidneys. Renal clearance is by both glomerular filtration and tubular secretion. Absorption from the gut following oral dosing is only partial although mean peak plasma levels of 4 μM are achieved which is at least ten-fold the in vitro ED_{50} level for HSV.

1. Herpes Simplex Infection

The first clinical results to be reported were from trials in the eye. JONES et al. (1979) showed that acyclovir eye ointment would prevent early recrudescence of a dendritic ulcer after minimal wipe debridement. Double-blind controlled trials in dendritic ulcers have since been reported and are reviewed by BRIGDEN et al.

(1981 b). Essentially they demonstrated that acyclovir is either as good as or superior to idoxuridine, adenine arabinoside and trifluorothymidine.

Several anecdotal reports have appeared claiming efficacy of acyclovir in severe herpes virus infections; however, data from controlled trials are now available. MITCHELL et al. (1981) reported the results of a small trial of intravenous acyclovir at a dose of 250 mg/m^2 8-hourly for 7 days in immune-compromised patients with mucocutaneous herpes simplex infections. This trial demonstrated the dramatic effect in reducing virus shedding a clear effects on the cessation of pain and other clinical parameters. A larger study, of which MITCHELL et al. (1981) was a part, was presented by MEYERS et al. (1982) and has demonstrated the effect on clinical parameters in a more statistically meaningful way.

A recent study conducted by SKOLDENBERG et al. (1984) has demonstrated that intravenous acyclovir at a dose of 10 mg/kg 8-hourly for 10 days reduces the mortality in herpes encephalitis from 50% in patients on vidarabine to 19% in patients on acyclovir.

The effect of intravenous acyclovir used as prophylaxis following bone marrow grafting to prevent herpes simplex recurrence during the period of severe immune suppression was clearly demonstrated by SARAL et al. (1981) and HANN et al. (1983). In the latter studies the acyclovir-treated arm recovered bone marrow function more rapidly than the placebo-treated arm, presumably because the bone marrow suppressive influence of the herpes infection was removed. GLUCKMAN et al. (1983), PRENTICE (1983), and WADE et al. (1984) have all demonstrated that oral acyclovir can also achieve protection in these patients provided they are able to take and absorb the adminstered drug.

MINDEL et al. (1982) and COREY et al. (1983) have shown that intravenous acyclovir in doses of either 250 mg/m^2 or 5 mg/kg given 8-hourly for 5 and 7 days respectively produced a highly significant reduction in the time of virus shedding and the rate of healing in severe initial genital herpes. NILSEN et al. (1982) have demonstrated a highly satisfactory response in primary genital herpes with orally administered acyclovir (200 mg at 4-hourly intervals for 7 days) and these findings have been confirmed by BRYSON et al. (1983).

Recurrent genital herpes and herpes labialis have presented a difficult challenge to the clinical trialists as in these diseases virus replication is over within a day or two, usually before patients present themselves at clinics. Consequently, the early trials were rather disappointing although COREY et al. (1982, 1983) were able to demonstrate some activity with topical ointment and NILSEN et al. (1982) clearly showed effects with oral dosing in recurrent genital herpes. Improved results have been seen when patients were instructed to initiate oral therapy as soon as lesions of a recurrent attack appeared (REICHMAN et al. 1984). Similar findings have been demonstrated in herpes labialis with acyclovir cream (5% in an aqueous cream base) which has been shown to abort some 40% of attacks whereas with the base alone only 10% fail to progress to disease (FIDDIAN et al. 1983).

Patients suffering frequent severe recurrences present a special problem in that the disease can be a major imposition on their lives, leading to severe emotional trauma. The results of trials of suppressive therapy, i.e. continuous therapy with the object of preventing recurrences, have produced dramatic results. KINGHORN et al. (1983), MINDEL et al. (1984), and STRAUS et al. (1984) have all reported a marked reduction in the incidence of attacks without evidence of adverse effects over periods of 3–6 months. If studies currently underway over long periods confirm the effect and lack of adverse reactions and if these patients do not run into problems of resistance development, then this seems to be the way this relatively small number of patients with severe disease are likely to be managed in the future.

2. Herpes Zoster Infection

Intravenous acyclovir has been shown by BALFOUR et al. (1983) to prevent the dissemination of varicella zoster in immune-compromised patients with shingles. Intravenous acyclovir has also been shown to be effective in the treatment of herpes zoster in patients with normal immune responses. PETERSLUND et al. (1981) showed that the drug significantly improved the rate of healing of skin lesions and shortened the duration of pain in the acute phase of zoster. They also demonstrated that the effect was greater in the patients suffering a more severe attack, namely the elderly, those with fever and those with a short duration of pain before the start of therapy. These results have been confirmed by BEAN et al. (1982) and McGILL et al. (1983). MCKENDRICK et al. (1984) have recently shown that oral acyclovir at a dose of 800 mg five times per day is effective in reducing the acute pain and time to healing of herpes zoster.

V. Bromovinyldeoxyuridine

Another nucleoside analogue, bromovinyldeoxyuridine, which probably works through the same mechanisms as acyclovir, has been reported by DE CLERQ et al. (1980) and they claim in a few anecdotal cases that the drug is effective when given by mouth in the treatment of varicella zoster infections in the immune-compromised. Unfortunately, problems with metabolism and potential toxicity appear to be going to limit the usefulness of this interesting development.

VI. Interferon

Interferon is also being evaluated in herpes infections and indeed MERIGAN et al. (1978) have shown that it is effective in varicella zoster infections. It would seem, however, that the biochemical approach to therapy of the herpes virus infections is likely to be both more effective and less expensive than the use of this biological mediator.

F. Drug Resistance

It is now well established that drug resistance can be easily demonstrated in tissue culture by passing herpes simplex through high concentrations of the various anti-viral drugs. FIELD et al. (1980) showed that resistant strains to acyclovir could be mutants at either the locus coding for the thymidine kinase or for the DNA polymerase. The addition to culture medium of drugs which have to be activated by a virally coded thymidine kinase (i.e. acyclovir or bromovinyldeoxyuridine) results in thymidine kinase-deficient or -defective mutants, whereas the use of less selective agents such as phosphonoacetic acid results in DNA polymerase mutants. There is a fair amount of cross-resistance although at present the extent of this is still somewhat unclear.

Despite the relative ease of resistance production in vitro we have to ask whether this is likely to be a problem in vivo and in humans. FIELD and DARBY (1980) showed that thymidine kinase-deficient mutants have reduced pathogenicity which might suggest that resistance development could be beneficial. This type of mutant is the usual one seen when mutants are selected with acyclovir; however, the thymidine kinase-defective mutants and the mutants at the DNA polymerase locus do not seen to have this property.

If we accept the most likely hypothesis for the latent state, namely that the template remains constant, then it can be assumed that even if a resistant strain grows up during therapy the next recurrence would originate from the constant template and would again be sensitive. Close monitoring by DEKKER et al. (1983) of some of the acyclovir clinical trials would suggest that at least in the treatment of recurrences in normal individuals resistance is unlikely to be a problem. It is interesting that the biological reasons which make the eradication of latency unlikely also mean that the development of resistance is unlikely to be a major problem.

In what situations could resistance be a problem? The possibilities warranting careful monitoring are:

1. The treatment of primary infection. If a resistant strain developed before the establishment of latency, then it is possible that latency might be established with a resistant mutant. This also of course carries the corollary that it might be possible to prevent latency establishment. Based on the animal experiments mentioned previously this seems unlikely.
2. The acquisition of a primary infection from an individual on treatment who has developed a resistant mutant. This seems a genuine likelihood although the risk might be reduced by the relatively dramatic reduction of virus shedding in patients treated with acyclovir (see Sect. E. IV). It is possible also that should this occur with a thymidine kinase-deficient mutant the subject may not develop recurrences. Information on this aspect, unfortunately, will take a long time to acquire.
3. The severely immune-suppressed patient where viral replication is continuing in non-neuronal sites. BURNS et al. (1982), CRUMPACKER et al. (1982) and others have reported the development of viruses resistant to acyclovir following a course of therapy. It seems likely that virus in these severely immune-deficient patients could continue to replicate in peripheral tissues and thus be more subject to the selection of drug-resistant mutants.

G. Conclusion

The age of specific anti-viral therapy has arrived. Whilst I have dealt with recent advances in herpes simplex and varicella zoster virus infections, other areas are beginning to look promising. There is new evidence that acyclovir may be of value in Epstein-Barr virus infection and new drugs are being developed with activity against cytomegalovirus. Amantadine is a start in the therapy of influenza. Interferon is certainly able to prevent experimentally induced infections with rhinovirus in human volunteers and is probably effective in injections with papilloma viruses and hepatitis B virus. These early beginnings should act as a spur to developing further knowledge and hopefully to a steady increase in our therapeutic options.

References

Balfour HH, Bean B, Laskin OL, Ambinder RF, Meyers JD, Wade JC, Zaia JA et al. (1983) Acyclovir halts progression of herpes zoster in immunocompromised patients. N Engl J Med 308:1448–1453

Bean B, Braun C, Balfour HH (1982) Acyclovir therapy for acute herpes zoster. Lancet 2:118–121

Blyth WA, Habour DA, Hill TJ (1980) Effect of acyclovir on recurrence of hepres simplex skin lesions sin mice. J Gen Virol 48:417–419

Brigden D, Fowle A, Rosling A (1980) Acyclovir, a new antiherpetic drug: early experience in man with systemically administered drug. In: Collier LH (ed) Developments in antiviral therapy. Academic, Oxford, pp 53–62

Brigden D, Bye A, Fowle ASE, Rogers H (1981 a) Human pharmacokinetics of acyclovir (an antiviral agent) following rapid intravenous injection. J Antimicrob Chemother 7:399–404

Brigden D, Fiddian P, Rosling AE, Ravenscroft T (1981 b) Acyclovir – a review of the preclinical and early clinical data of a new antiherpes drug. Antiviral Res 1:203–212

Bryson Y, Dillon M, Lovett M, Acuna G, Taylor S, Cherry J, Johnson L et al. (1983) Treatment of first episodes of genital herpes simplex virus infection with oral acyclovir. N Engl J Med 308:916–921

Burns WH, Saral R, Santos GW, Laskin OL, Leitman PS, McLaren C, Barry DW (1982) Isolation and characterisation of resistant herpes simplex virus after acyclovir therapy. Lancet 1:421–423

Cook ML, Stevens JG (1973) Pathogenesis of herpetic neuritis and ganglionitis in mice: evidence of intra-axonal transport of infection. Infect Immun 7:272–288

Corey L, Nahmias AJ, Guinan ME, Benedetti J, Critchlow C, Holmes KK (1982) A trial of topical acyclovir in genital herpes simplex virus infections. N Engl J Med 306:1313–1319

Corey L, Fife KH, Benedetti K, Winter CA, Fahnlander A, Connor JD, Hintz MA, Holmes KK (1983) Intravenous acyclovir for the treatment of primary genital herpes. Ann Intern Med 98:914–921

Crumpacker CS, Schnipper LE, Marlowe SI, Kowalsky PN, Hershey BJ, Levin MJ (1982) Resistance to antiviral drugs of herpes simplex virus isolated from a patient treated with acyclovir. N Engl J Med 306:343–346

De Clercq E, Degreef H, Wildiers J, de Jonge G, Drochmans A, Deschamps J, de Somer P (1980) Oral (E)-5-(2-bromovinyl)-2-deoxyuridine in severe herpes zoster. Br Med J 281:1178

Dekker C, Ellis MN, McLaren C, Hunter G, Rogers J, Barry DW (1983) Virus resistance in clinical practice. J Antimicrob Chemother 12B:137–152

De Miranda P, Whitley RJ, Blum MR, Keeney RE, Barton N, Cochetto DM, Good S et al. (1979) Acyclovir kinetics after intravenous infusion. Clin Pharmacol Ther 26:718–728

Elion GB, Furman PA, Fyfe JA, de Miranda P, Beauchamp L, Schaeffer HJ (1977) Selectivity of action of an antiherpetic agent 9-(2-hydroxyethoxymethyl) guanine. Proc Natl Acad Sci USA 74:5716–5720

Fiddian AP, Yeo JM, Stubbings R, Dean D (1983) Successful treatment of herpes labialis with topical acyclovir. Br Med J 286:1699–1701
Field HJ, Darby G (1980) Pathogenicity in mice of strains of herpes simplex virus which are resistant in vitro and in vivo. Antimicrob Agents Chemother 17:209–216
Field HJ, Bell SE, Elion GB, Nash AA, Wildy P (1979) Effects of acycloguanosine treatment on acute and latent herpes simplex infections in mice. Antimicrob Agents Chemother 15:554–561
Field HJ, Darby G, Wildy P (1980) Isolation and characterisation of acyclovir resistant mutants of herpes simplex virus. J Gen Virol 49:115–124
Furman PA, St Clair MH, Fyfe JA, Rideout JL, Keller PM, Elion GB (1979) Inhibition of herpes simplex virus-induced DNA polymerase activity and viral DNA replication by 9-(2-hydroxyethoxymethyl) guanine and its triphosphate. J Virol 32:72–77
Gluckman E, Lotsberg J, Devargie A, Zhao A, Zhao XM, Melo R, Gomez-Morales M, Nebout T et al. (1983) Prohylaxis of herpes infections after bone marrow transplantation by oral acyclovir. Lancet 2:706–708
Hann IM, Prentice HG, Blacklock HA, Ross MGR, Brigden D, Rosling AE, Burke C et al. (1983) Acyclovir prophylaxis against herpes virus infections in severely immunocompromised patients: randomised double blind trial. Br Med J 287:384–388
Heidelburger C, King DH (1979) Trifluorothymidine. Pharmacol Ther 6:427–442
Jones BR, Coster DJ, Fison PN, Thompson GM, Cobo LM, Falcon MG (1979) Efficacy of acycloguanosine against herpes simplex corneal ulcers. Lancet 1:243–244
Juel-Jensen BE, MacCullum FO (1972) Herpes simplex, varicella and zoster. Heinemann, London
Kaufman HE, Martola EL, Dohlamnm CH (1962) Use of 5-iodo-2′-deoxyuridine (IDU) in treatment of herpes simplex keratitis. Arch Ophthalmol 68:235–239
Kinghorn GR, Barton IG, Potter CW, Fiddian AP (1983) Oral acyclovir prophylaxis of recurrent genital herpes (Abstr 185). 5th International Meeting of the International Society for STD Research, Seattle
Klein RJ, Freidman-Kein AE, de Stefano E (1979) Latent herpes simplex virus infections in sensory ganglia of hairless mice prevented by acycloguanosine. Antimicrob Agents Chemother 15:723–729
McGill J, MacDonald DR, Fall C, McKendrick GDW, Copplestone A (1983) Intravenous acyclovir in acute herpes zoster infection. J Infect 6:157–161
McKendrick MW, McGill JI, Bell AM, Hickmott E, Burke C (1984) Oral acyclovir for herpes zoster. Lancet 2:925
Merigan TC, Rand KH, Pollard RB, Abdallah PS, Jordon GW, Fried RP (1978) Human leukocyte interferon for the treatment of herpes zoster patients with cancer. N Engl J Med 298:981–987
Meyers JD, Wade JC, Mitchell CD, Aral R, Leitman PS, Durak DT, Levin MJ et al. (1982) Multicentre collaborative trial of intravenous acyclovir for the treatment of mucocutaneous HSV infection in the compromised host. Am J Med 73(1A):229–235
Mindel A, Adler MW, Sutherland S, Fiddian AP (1982) Intravenous acyclovir treatment for primary genital herpes. Lancet 1:697–700
Mindel A, Weller IVD, Faherty A, Sutherland S, Hindlay D, Fiddian AP, Adler MW (1984) Prophylactic oral acyclovir in recurrent genital herpes. Lancet 2:57–59
Mitchell CD, Bean B, Gentry SR, Groth KE, Boen JR, Balfour HH (1981) Acyclovir therapy for mucocutaneous herpes simplex infections in immunocompromised patients. Lancet 1:1389–1392
Nilsen AE, Aasen T, Halsos AM, Kinge BR, Tjøtta EAL, Wikström K, Fiddian AP (1982) Efficacy of oral acyclovir in the treatment of initial and recurrent genital herpes. Lancet 2:571–573
Pavan-Langston D, Dohlman CH (1972) A double blind clinical study of adenine arabinoside therapy of viral keratoconjunctivitis. Am J Ophthalmol 74:81–88
Peterslund NA, Seyer-Hansen K, Ipsen J, Esmann V, Schonheyder H, Juhl H (1981) Acyclovir in herpes zoster. Lancet 2:827–830
Prentice HG (1983) Use of acyclovir for prophylaxis of herpes infections in severely immunocompromised patients. J Antimicrob Chemother 12B:153–159

Reichman RC, Badger GJ, Mertz GJ, Corey L, Richmann DD, Connor JD, Redfield D et al. (1984) Treatment of recurrent genital herpes simplex infection with oral acyclovir: a controlled trial. JAMA 251:2103–2107
Saral R, Burns WH, Laskin OL, Santos GW, Leitmann PS (1981) Acyclovir prophylaxis of herpes simplex virus infections, a randomised double-blind, controlled trial in bone marrow transplant recipients. N Engl J Med 305:63–67
Skoldenberg B, Forsgren M, Alestig K, Bergstrom T, Burman L, Dahlquist E, Forkman A et al. (1984) Acyclovir versus vidarabine in herpes simplex encephalitis. Lancet 2:707–711
Straus SE; Takiff HE, Seidlin M, Bachrach S, Lininger L, di Giovanna JJ, Western KA et al. (1984) Suppression of frequently recurring genital herpes: a placebo-controlled double-blind trial of oral acyclovir. J Med 310:1545–1550
Wade JC, Newton B, Flournoy N, Meyers JD (1984) Oral acyclovir for prevention of herpes simplex virus reactivation after marrow transplant. Ann Intern Med 100:823–828
Wellings PC, Audry PN, Borz FH, Jones BR, Brown DC, Kaufman HE (1972) Clinical evaluaton of trifluorothymidine in the treatment of herpes simplex corneal ulcers. Am J Ophthalmol 73:932–942
Whitley RJ, Alford CA (1981) Parenteral antiviral chemotherapy of human herpesviruses. In: Nahmias AJ, Dowdle WR, Shinazi RF (eds) The human herpesviruses. Elsevier, New York, pp 478–490
Whitley RJ, Chien LT, Dolin R, Galasso GJ, Alford C (1976) Adenine arabinoside therapy of herpes zoster in the immunosuppressed NIAID collaborative antiviral study. N Engl J Med 294:1193–1199
Whitley RJ, Soong S, Dolin R, Galasso GJ, Chien LT, Alford C (1978) Adenine arabinoside therapy of biopsy-proved herpes simplex encephalitis. N Engl J Med 297:289–294
Whitley R, Alford C, Hess F, Buchanan R (1980) Vidarabine: a preliminary review of its pharmacological properties and therapeutic use. Drugs 20:267–282
Whitley RJ, Soong S, Kirsch MS, Karchmer AW, Dolin R, Galasso G, Dunnick J, Alford CA (1981) Herpes simplex encephalitis. Vidarabine therapy and diagnostic problems. N Engl J Med 304:313–318
Wildenhoff KE, Ipsen J, Esmann V, Ingemann-Fensen J, Hjelm Poulsen J (1979) Treatment of herpes zoster with idoxuridine ointment, including a multivariate analysis of symptoms and signs. Scand J Infect Dis 11:1–9

CHAPTER 31

The Urticarias

A. Kobza Black

A. Introduction

Urticaria is characterised by wealing (localised oedema) with itching and erythema and results from localised vasodilatation and transudation of fluid from the dermal microvasculature. Deep spread of the fluid produces angio-oedema. Histologically there is dermal oedema, dilated small blood vessels and lymphatics and a sparse perivascular mixed cellular infiltrate (Ackerman 1968). The urticarial lesions usually subside within hours as the extravasated fluid is absorbed into the lymphatics. There are many precipitating causes of urticaria (Table 1) which induce a chain of events in the skin of susceptible subjects, with release of mediators which increase vascular permeability and thus cause wealing (Table 2).

B. Histamine

Histamine when injected intradermally into human skin reproduces the manifest signs of an urticarial response, inducing a short-lived erythema, weal and flare (Lewis 1927). There are at least two subclasses of histamine receptors, H_1 and H_2. Both classes appear to be present on human skin blood vessels (Marks and Greaves 1977) and are involved in the production of erythema and wealing

Table 1. Precipitating causes of urticaria

Drug reactions
Foods
Infections
Physical stimuli
Dermographism
Cold
Cholinergic (sweat gland activation)
Solar
Pressure
Localised heat
Aquagenic
Vibration
Complement activation
$C\bar{1}$ inhibitor deficiency
Contactants

Table 2. Potential mediators of urticaria

Cell-derived		Plasma-derived
Cell	Mediator	
Mast cell	Histamine Arachidonic acid metabolites Prostaglandins Leukotrienes Platelet-activating factor Eosinophil and neutrophil chemotactic factors	Kinins Fibrin degradation products Anaphylatoxins
Platelets	Platelet factor 4	
Eosinophils	Eosinophil granule proteins	
Neutrophils and mast cell	Proteases	
Neurones	Neuropeptides Acetylcholine	

Table 3. Increased histamine concentrations in urticarial lesions

Urticaria	Sample	References
Chronic idiopathic	Suction blister fluid	KAPLAN et al. (1978)
	Skin biopsy	PHANUPHAK et al. (1980)
Dermographism	Blood	ROSE (1941)
	Dermal perfusate	SONDERGAARD and GREAVES (1970)
Cholinergic	Blood	KAPLAN and BEAVEN (1976)
Cold contact	Blood	KAPLAN et al. (1975), BENTLEY-PHILLIPS et al. (1976)
	Suction blister fluid	MISCH et al. (1982)
Solar	Blood	SOTER et al. (1979), HAWK et al. (1980)
Pressure	Suction blister fluid	KAPLAN et al. (1978)
Aquagenic	Blood	SIBBALD et al. (1981), DAVIS et al. (1981)
Heat	Blood	GRANT et al. (1981), ATKINS and ZWEIMAN (1981)
Vibratory angio-oedema	Blood	KAPLAN and BEAVEN (1976), TING et al. (1983)

(ROBERTSON and GREAVES 1978). The production of itching by histamine is predominantly due to H_1-receptor activation (DAVIES and GREAVES 1980).

I. Mast Cell Release

Histamine release has been demonstrated in most forms of urticaria (Table 3). Histamine can be released from dermal mast cells and circulating basophils by a variety of immunological and non-immunological processes (MACGLASHAN et al. 1983). Though most of the attention has focused on IgE-mediated mast cell histamine release, there is little evidence of an immunological process in most cases of idiopathic urticaria (CHAMPION et al. 1969). However, IgE-mediated histamine release occurs more commonly in acute urticaria due to food and drugs. In some physical urticarias there is evidence of a passive transfer factor in the serum, which when characterised appears to be IgE-like: in dermographism (AOYOMA 1971; NEWCOMB and NELSON 1973); in cold urticaria (HOUSER et al. 1970; KAPLAN et al. 1975; AKIYAMA et al. 1981); in cholinergic urticaria (ILLIG and HEINICKE 1967). It is possible that the physical stimulus alters a normal protein constituent of skin in such a way that it becomes antigenic and reacts with an IgE autoantibody (KAPLAN et al. 1981).

Non-immunological histamine release from mast cells can be produce by a variety of exogenous agents including morphine, codeine, (+)-tubocurarine, compound 48/80 and polymyxin B. Endogenous agents with a similar ability include anaphylatoxins C3a, C4a, C5a, eosinophil granule proteins and substance P.

II. Mast Cell Numbers and Stability

Mast cell number was found to be normal in the skin of patients with chronic urticaria (JUHLIN 1979), though in some patients skin histamine content was raised (JUHLIN 1979; KAPLAN et al. 1978). Spontaneous histamine release from skin slices from these non-urticated areas has been demonstrated (PARISH 1985).

In physical urticarias normal mast cell numbers were found in a preliminary study in normal or wealed skin of cold, solar, pressure, aquagenic and cholinergic urticaria and dermographic patients (JAMES et al. 1980). In urticated skin there was a variable degree of mast cell degranulation, but there was no statistically significant increase compared with skin that was not wealed.

III. Basophil Responses

The peripheral basophils of patients with chronic idiopathic urticaria showed decreased release of histamine by heterologous anti-IgE antibody compared with normal controls (GREAVES et al. 1974). It is possible that an in vivo desensitisation of the basophils had occurred to IgE, as these cells responded normally to chemical releasers of histamine (KERN and LICHTENSTEIN 1976).

IV. Reactivity of Blood Vessels

Patients with idiopathic, chronic, heat and cold urticaria and dermographism had a similar reaction to intradermal injections of histamine to that of normal controls (JUHLIN and MICHAELSSON 1969a). However, in idiopathic chronic urticaria there was also a delayed reaction, which was thought to be due to a leakage of plasma with activation of kallikrein, to which these patients were very sensitive.

V. Involvement of Other Mediators

That other mediators, whether mast cell derived or not, are likely to be involved is suggested by a number of observations:

1. Suppression of histamine release in cold urticaria did not produce a concomitant improvement of erythema and oedema, though pruritus was improved (KEAHEY and GREAVES 1980).
2. Measurement of the rate of disappearance of histamine-induced oedema suggested that it was slower than would be accounted for by histamine alone (COOK and SHUSTER 1980).
3. Though treatment with a potent H_1 anti-histamine (astemizole) of patients with chronic idiopathic urticaria almost totally abolished weals following intradermal injections of histamine, spontaneously occurring urticarial lesions were only partially improved (KRAUSE and SHUSTER 1984).

C. Mediators Derived from Arachidonic Acid

Arachidonic acid is a 20-carbon polyunsaturated fatty acid constituent of cell membrane phospholipid. After its release, mainly by the enzyme phospholipase, it is metabolised via two main pathways. Products of the cyclo-oxygenase pathway include the primary prostaglandins (PGE_2, PGD_2, PGF_2, and PGI_2) and thromboxane A_2. The products of the lipoxygenase pathway include the monohydroxy fatty acids 12- and 15-hydroxyeicosatetraenoic acids (12- and 15-HETE), and leukotrienes B_4, C_4, D_4, and E_4 (LTB_4, LTC_4, LTD_4, and LTE_4). The leukotrienes C_4, D_4, and E_4 are components of the pharmacological slow reacting substance of anaphylaxis (SRS-A), released during immediate allergic reactions.

I. Arachidonate Cyclo-oxygenase Products

All nucleated cells are capable of generating prostaglandins, but there is preferential production of various prostaglandins by different types of cells. The predominant cyclo-oxygenase product of the human mast cell is PGD_2 (LEWIS et al. 1982) while leucocytes produce mainly PGE_2, and vascular endothelium PGI_2. When injected into human skin, they are all capable of inducing sustained erythema and oedema (lasting longer than 1 h). Prostaglandin E_2 is the most potent (FLOWER et al. 1976), followed by PGI_2 (HIGGS et al. 1979) and PGD_2 (FLOWER et al. 1976). In animal skin, histamine- and bradykinin-induced vascular permeability is potentiated by prostaglandin E_2 (JOHNSTON et al. 1976) and PGI_2 (FORD-HUTCHINSON et al. 1978). Prostaglandin D_2 potentiates the vasopermeability effects of histamine in rat skin (FLOWER et al. 1976), but this has not been confirmed in human skin. Prostaglandin D_2 injection in high concentration (1 μg) into human skin caused polymorphonuclear leucocyte infiltrate at 30 min and 6 h after injection (SOTER et al. 1983).

Increased amounts of PGD_2 have recently been recovered in venous blood draining cold urticarial reactions (KOBZA BLACK et al. 1985), in concentrations that might potentiate the effects of histamine. Non-mast-cell-derived PGE_2, perhaps derived from the inflammatory cells recruited into cold urticarial reactions, was found in increased concentrations in suction blister fluid obtained from the cold-challenged urticated skin in the minority (2/9) of cold urticaria patients (MISCH et al. 1982). This subgroup was no different clinically, therefore the significance of this finding is uncertain.

However, treatment of urticaria with non-steroidal anti-inflammatory drugs, which block cyclo-oxygenase activity and prostaglandin generation, is not effective. Indeed, in some patients with chronic idiopathic urticaria (MOORE ROBINSON and WARIN 1967) and physical urticaria (DOEGLAS 1975), aspirin exacerbates their urticarial reactions. The suggestion that this is due to redirection of arachidonate metabolism, via the lipoxygenase pathway to form vasoactive leukotrienes, remains to be proven.

II. Arachidonate Lipoxygenase Products

Leukotrienes C_4 and D_4 are potent vasoactive compounds and, when injected into human skin in nanomolar amounts, produce an immediate weal-and-flare reaction (CAMP et al. 1983; SOTER et al. 1983). At the 1–3 nmol dose of LTC_4, LTD_4 and LTE_4 injected into the forearm by SOTER et al. (1983), the erythema persisted up to 6 h and oedema up to 2 h. The histological appearance at 2 h showed capillary and venular dilations with little or no evidence of a cellular infiltrate. Though LTC_4 and LTD_4 have been identified in rat mast cells and from human lung (LEWIS et al. 1980), these compounds have not been recovered from human skin mast cells, or in increased concentrations in urticaria. Intradermal injections of LTC_4 produced the same degree of wealing in normal controls as in patients with chronic urticaria (JUHLIN and HAMMARSTROM 1982).

Leukotriene B_4 is one of the most powerful neutrophil chemoattractant substances tested in vitro (FORD-HUTCHINSON et al. 1980). When injected into human skin (1.6 nmol) an initial weal-and-flare reaction is followed by a papule, persisting for 6 h and characterised by a predominantly polymorphonuclear cell infiltrate (SOTER et al. 1983; CAMP et al. 1983). 12-Hydroxyeicosatetraenoic acid (12-HETE) is a less powerful leucocyte chemoattractant substance in vitro (PALMER et al. 1980). Topical application to human skin (2 ng–20 μg) after 6 h induces erythema accompanied by a dermal mixed neutrophil cell infiltrate (DOWD et al. 1985). Thus, these compounds could be involved in the cellular response in the delayed urticarial reactions. In extracts of pressure lesions from a small number of patients with delayed pressure urticaria, chemotactic activity was found in LTB_4 and 12- and 15-HETE, fractions though this was not statistically significant compared with controls (CZARNETZKI et al. 1984). The role of the products of arachidonate metabolism in the initiation and perpetuation of the urticarial reaction needs to be elucidated. This would be helped by the development of specific lipoxygenase enzyme inhibitors or antagonists.

D. Neutrophil and Eosinophil Chemotactic Factors

The mast cell releases these factors on stimulation. The presence of these factors has been studied in venous blood draining areas of physical urticaria.

In cold urticaria, following immersion of the forearm in ice water, these chemotactic factors appeared with a similar time course to that of histamine. The neutrophil chemotactic factor was a single high molecular weight factor (molecular weight greater than 750000) (WASSERMAN et al. 1977). The eosinophil chemotactic activity consisted of two intermediate molecular weight factors (1000–3000) and a low molecular weight factor (300–700). The intermediate molecular weight factors were also chemoattractant to mononuclear cells, and all factors deactivated eosinophils at subchemoattractant concentrations (WASSERMAN et al. 1982). There was however absence of infiltrating white cells at the urticated sites up to 24 h later (CENTER et al. 1979). An impaired chemotactic response of the neutrophils harvested from the challenged arm was found 5 min–4 h after challenge. This suggested that deactivation of the leucocytes to the neutrophil chemotactic factor had taken place (CENTER et al. 1979).

In cholinergic urticaria induced experimentally by exercise (SOTER et al. 1980) and in solar urticaria (SOTER et al. 1979), neutrophil and eosinophil chemotactic factors were also demonstrated. In chronic urticaria defective release of eosinophil chemotactic factor from peripheral blood was demonstrated (CZARNETZKI et al. 1976). The significance of this finding is uncertain.

E. Platelet-Activating Factor

Platelet-activating factor (PAF) is an ether-linked phospholipid, acetylglycerylether phosphorylcholine (AGEPC). It is released from a wide range of inflammatory cell types, including platelets, eosinophils, neutrophils, macrophages, basophils, mast cells and endothelial cells (HENSON and LYNCH 1983). It has a wide variety of biological functions including platelet aggregation, chemoattraction of eosinophils, polymorphonuclear leucocytes and mononuclear cells, and induction of vascular permeability (PAGE et al. 1983 a). In human skin intradermal injections of PAF induce an immediate weal-and-flare reaction lasting 1 h (ARCHER et al. 1984). In a proportion of subjects, a delayed response consisting of an indurated papule occurs 3–6 h later, persisting up to 48 h. Histological examination showed a perivascular predominantly neutrophil cellular infiltrate at 4 h. After 24 h, a predominantly lymphocytic and histocytic infiltrate was accompanied by marked dermal vessel damage (ARCHER et al. 1985). PAF thus reproduces most of the characteristics of an immediate and delayed allergic skin response and may be implicated in urticaria of the immediate and delayed types.

Recently, a platelet-activating factor-like lipid (PAF-IL) has been recovered from venous blood draining cold urticated skin tissue (GRANDEL et al. 1985). The exact pathophysiological interrelationship to the urticaria remains to be determined. The cold could have induced local release of PAF-IL, which itself causes urticaria. Alternatively, PAF-IL release may be a secondary phenomenon released by inflammatory cells attracted to the site of the cold urticarial reaction.

F. Platelet Factor 4

Since the discovery of PAF as a product of IgE-dependent basophil activation (BENVENISTE et al. 1972), more attention has been focused on the role of platelets in the inflammatory response. In addition to its vasodilating properties, intradermal injections of PAF induce a local accumulation of platelets (PAGE et al. 1983 b). Since platelets mechanically obstruct vessels and release a range of mediators, including prostaglandin E, the monohydroxyeiocosatetraenoic acids 5- and 12-HETE, thromboxane A_2 and PAF, this activation may contribute to tissue injury and further accumulation of inflammatory cells.

Platelet aggregation has been observed in one patient with cold urticaria after several applications of ice onto the skin (EADY et al. 1981). Platelet factor 4 is a platelet-specific polypeptide which is released on platelet stimulation, and has been used as a measurement of in vivo platelet activation (KAPLAN and OWEN 1981). Platelet factor 4 has been demonstrated in the blood draining a cold urticated area in five patients, reaching a peak 20 min after challenge, paralleling the clinical reaction (WASSERMAN and GINSBERG 1984). This factor may contribute directly to the urticarial reaction by its ability to induce basophil histamine release (BRINDLEY et al. 1983) or may merely reflect the release of other platelet products, including PAF.

G. Eosinophil Granule Proteins

Though eosinophils are present in small numbers in histological sections of chronic idiopathic urticaria, they may be present in transit and may not be recognised in histological sections, as was suggested in atopic eczema (LEIFERMAN et al. 1985). In delayed urticarial reactions such as urticarial vasculitis (RUSSELL-JONES et al. 1985) and in delayed pressure urticaria (WINKELMANN et al. 1986) larger numbers of eosinophils are present. It is postulated that they localise at the site of urticarial reactions, from release of mast cell-derived chemotactic factors for eosinophils such as eosinophil chemotactic factors and 12-HETE.

Two constituents of the eosinophil granules, eosinophil major basic protein (MBP) and eosinophil cationic protein (ECP) have been implicated in urticarial lesions. Purified MBP, which accounts for over 50% of the granule constituents, produces non-cytolytic histamine release from human basophils and rat mast cells (O'DONNELL et al. 1983). Immunoreactive MBP has been demonstrated in 12/18 biopsy lesions of chronic urticaria either in blood vessels, dermis, or connective tissue fibres (PETERS et al. 1983). Elevated concentrations of MBP have been demonstrated in 38% of patients with chronic urticaria (WASSOM et al. 1981), even though these patients have had normal eosinophil counts. This raises the possibility that extracellular MBP seen in blood vessels and connective tissue may be derived from degranulated circulating blood eosinophils, in addition to eosinophils infiltrating the urticarial lesions. MBP- and ECP-rich fractions injected intradermally into human skin produce a weal-and-flare reaction, which may be partially related to the ability of the compounds to release histamine (LEIFERMAN et al. 1984).

Immunocytochemical studies, using a monoclonal antibody EG_2 which recognises ECP only from activated and degranulating eosinophils, showed large numbers of activated and degranulating eosinophils and sometimes extracellular ECP in semi-thin sections of urticaria vasculitis and chronic idiopathic urticaria with a dense perivascular infiltrate (RUSSELL-JONES et al. 1985). From these findings, it is possible that eosinophils have an important role in the pathogenesis of the immediate and, later, the delayed phase of the urticarial reaction.

H. Proteases

The human skin contains a number of proteases that are probably involved in acute inflammation (FRAKI et al. 1983). The chymotrypsin-like and trypsin-like neutral proteases are derived from dermal mast cells and, when injected into skin, produce increased vascular permeability (SEPPA 1980; FRAKI 1977). The mechanism of oedema formation is not known, but may be related to proteolytic hydrolysis of fibronectin, an important component of endothelial basement membranes (VARTIO et al. 1981). Trypsin-like protease is released in cold urticarial reactions (FRAKI et al. 1983). Plasma protease inhibitors are reduced in various forms of urticaria. In cold urticaria, α_1-anti-trypsin (DOEGLAS and BLEUMINK 1975) and anti-chymotrypsin (EFTEKHARI et al. 1980) are reduced. In cholinergic urticaria, anti-chymotrypsin (EFTEKHARI et al. 1980) and in dermographism (BREATHNACH et al. 1983) and angio-oedema (DOEGLAS and BLEUMINK 1975), α_1-anti-trypsin levels have been found to be reduced. This deficiency may explain the successful treatment of angio-oedema with the synthetic protease inhibitor, tranexamic acid (THOMPSON and FELIX-DAVIES 1978). In longitudinal studies of patients with cold urticaria and angio-oedema up to 10 years, the deficiency of the protease inhibitor α_1-anti-trypsin persisted in a subpopulation of these patients and was unrelated to disease activity (DOEGLAS and BLEUMINK 1985). This suggested an underlying deficiency of this protease inhibitor and not a consumption deficiency during the urticarial process. Under normal conditions, there would be enough protease inhibitor to inactivate proteolytic enzymes derived from mast cells, or other sources, but the inhibitors might not be able to cope with increased proteolytic enzymes released from mast cells activated by the precipitating stimulus. In addition, lack of protease inhibitor may act through relative inability to inhibit kinin or kinin-like mediators formed in plasma (see Sect. L. I).

J. Neuropeptides

Neuropeptides consist of a group of oligopeptides, which are putative neurotransmitters in the central nervous system. It has been assumed that tissue damage releases chemical mediators that stimulate the sensations of itch and pain which are important components of the inflammatory response, including urticaria. Some of these chemical mediators may include neuropeptides, opioid peptides and substance P.

A wide range of opioid peptides, including Leu-enkephalin, Met-enkephalin and β-endorphin have an important role in the pharmacology of pain. They may however regulate the peripheral sensations of itch and pain. Enkephalin has been shown to enhance histamine-induced pruritus (HAGERMARK and FJELLNER 1980), and endorphin to inhibit substance P release from sensory nerve terminals (JESSEL and IVERSEN 1977).

The undecapeptide, substance P, is released from sensory nerve endings (GAMSE et al. 1980). Intradermal injections of substance P into human skin induce an immediate response of itching, erythema and wealing, the erythema being mainly due to an axon reflex (LEMBECK and HOLZER 1979). Some of these actions may have been partially due to histamine release from the cutaneous mast cell (JORIZZO et al. 1983). Capsaicin is a substance which after topical application can block the effector side of the axon reflex, possibly by depleting the nerve terminals of substance P (CARPENTER and LYNN 1981). The involvement of the sensory nerves in heat and cold urticaria is suggested by the fact that repeated topical applications of capsaicin prevented the urticarial response following appropriate challenge (TOTH-KASA et al. 1983).

The calcitonin gene-related peptide (CGRP) is a novel neuropeptide which is co-localised in sensory neurones with substance P. Human CGRP when injected intradermally into human skin causes intense microvascular dilation (BRAIN et al. 1985). It may thus be a potential mediator in urticarial reactions.

K. Acetylcholine

Acetylcholine is released on stimulation of post-ganglionic sympathetic nerves that innervate sweat glands. Cholinergic urticaria develops in association with stimulation of these nerves (HERXHEIMER 1956). The abnormality at the nerve ending receptor level, as well as its relationship to the urticarial reaction remains unclear. There may be impaired choline esterase activity, leading to an accumulation of acetylcholine (MAGNUS and THOMPSON 1954) or a lowered threshold to acetylcholine at the receptor level. There is, however, no evidence of generalised increase of sensitivity of cholinoceptors in the autonomic system of patients with cholinergic urticaria (MURPHY et al. 1984). An increase of specific muscarinic receptors in a lesion of cholinergic urticaria has been demonstrated in one patient (SHELLEY et al. 1983). It is well established that mast cell-derived mediators are released during cholinergic urticaria (SOTER et al. 1980). Cholinergic impulses can cause mast cell degranulation (CHO and OGLE 1979). This may occur in cholinergic urticaria, either by a direct effect or mediated by an immune mechanism, possibly IgE, as passive transfer has been demonstrated in cholinergic urticaria (ILLIG and HEINICKE 1967).

L. Mediators Derived from Plasma

I. Kinins

Kinins are pharmacologically active oligopeptides of which bradykinin, a nonapeptide, is the major effector agent. The kinins are widely distributed in tissues, and are generated in response to a wide range of stimuli, including complement activation and contact with negatively charged particles (see Chap. 19, Vol. I). Briefly, Hageman factor, a proenzyme, is activated by one of these stimuli to an active enzyme, the pre-kallikrein activator. This in turn activates pre-kallikrein to kallikrein, which then cleaves its substrate kininogen to form bradykinin. Kallikrein catalyses a positive feedback mechanism that converts more of the Hageman factor to pre-kallikrein activator. Kallikrein is inhibited by various substances including C_1 esterase inhibitor. Bradykinin is rapidly destroyed by kininases.

Bradykinin has potent pharmacological properties, which makes it a potential mediator of urticaria. Intradermal injection into human skin induces increased vascular permeability. The response does not show tachyphylaxis on repeated injections (Greaves and Shuster 1967). Bradykinin also interacts with prostaglandings in a complex manner. It can activate prostaglandin synthesis (Lembeck et al. 1976). Prostaglandin E_1 potentiates the pain caused by subdermal infusions of bradykinin (Ferreira 1972), or by application of bradykinin onto previously scarified human skin (Greaves and McDonald-Gibson 1973).

Kinin-like activity, demonstrated by bioassay, has been recovered from dermal perfusates in various forms of urticaria. Low levels were detected in perfusates obtained from dermographic skin (following application of tetrahydrofurfuryl nicotinate) (Winkelmann et al. 1965) and lesions of cold urticaria (DeLaus and Winkelmann 1968). However, the identity of this kinin-like activity was tentative. In contrast, no increased kinin-like activity was found in samples obtained from dermographic skin, either in dermal perfusates (Greaves and Sondergaard 1970) or in suction blister fluid raised on urticated skin (Plummer et al. 1977) compared with control uninvolved skin. There is no definite evidence that kinins are involved in physical urticaria, but this awaits confirmation when more sensitive methods are used to assay kinins.

Intradermal injections of a crude kallikrein preparation in normal controls and patients with chronic urticaria caused an immediate weal-and-flare response, followed by a palpable delayed response maximal at 24 h (Juhlin and Michaelsson 1969a). Patients with chronic, but not acute or physical urticaria, produced a significantly greater response than normal controls. The delayed response may have been mediated by the kallikrein-like system, with leakage of plasma caused by bradykinin activating kallikrein, to which the patients with chronic urticaria were particularly sensitive.

M. Hereditary Angio-oedema (HAE)

This condition is a rare genetic disease, inherited in an autosomal dominant manner. There is no HLA association (Robson et al. 1979). Clinically it is characterised by episodic non-prurititic angio-oedema of the skin and mucous membranes, lasting 2–3 days. The face, extremities, larynx, pharynx and gastrointestinal tract may be affected, sometimes following minor trauma. The condition may present as abdominal pain, and airway obstruction due to laryngeal oedema may be fatal (Ruddy et al. 1972).

HAE is due to deficiency of $C\bar{1}$ inhibitor activity ($C\bar{1}$ INH) (Donaldson and Evans 1963). The $C\bar{1}$ INH can be measured antigenically by radial immunodiffusion. Functional estimation can be made by techniques such as the ability of $C\bar{1}$ INH to split *N*-acetyl-L-tyrosine ethyl ester (Levy and Lepow 1959) or the ability of $C\bar{1}$ INH to block $C\bar{1}$ in immune haemolysis (Gigli et al. 1968). Two biochemical forms of HAE exist (Rosen et al. 1965). In 85% of the patients, both the immunochemical and functional concentrations are low, being approximately 5%–30% of normal. In the other 15% the antigenically determined $C\bar{1}$ INH levels are normal or high, but functionally determined concentrations are low. There are no clinical distinguishing features of these two biochemical forms of HAE.

C$\bar{1}$ INH, previously known as α_2-neuroaminoglycoprotein, has molecular weight of approximately 100000. It is a plasma protein which inhibits a number of plasma proteases, including the activated first component of complement (C$\bar{1}$r and C$\bar{1}$s) of the classical pathway, kallikrein, plasmin and factors XIa and XIIa of the intrinsic coagulation system (RUDDY et al. 1972).

It is thought that tissue injury causes activation of the Hageman factor, resulting in the generation of plasmin and kallikrein, which can activate C1 (DONALDSON 1968). This activation may lead to an exhaustion of the available reduced or inactive C$\bar{1}$ INH, with unopposed cleavage of C4 and C2, the natural substrates of C$\bar{1}$ INH. During an episode of angio-oedema the levels of C4 and C2 are decreased, with the low levels of C4 persisting between attacks. For a reason that remains unclear, the levels of C3 are not decreased during or between episodes. In view of the multiple effects of C$\bar{1}$ INH, it is likely that other systems, apart from the complement pathway, play a role in HAE.

The mediator that causes the oedema remains under debate. DONALDSON et al. (1969) have demonstrated a cleavage product of C2, a kinin-like fragment, that enhances vasopermeability. That this may be significant is suggested by the fact that intradermal injection of C$\bar{1}$ INH into patients with HAE produces a larger weal than in normal subjects, but no weal formation in C2-deficient patients (KLEMPERER et al. 1968). However, the identity of the peptide is an assumption. Alternatively, CURD et al. (1980) have demonstrated that in suction blisters raised on lesions of HAE, increased kallikrein activity was present compared with control subjects. Plasma kallikrein injected into subjects with HAE showed an increased permeability response compared with normal control subjects, suggesting increased bradykinin formation when C$\bar{1}$ INH was decreased (JUHLIN and MICHAELSSON 1969b). It is possible that both C2 fragment and bradykinin are involved.

Treatment of HAE is based on the biochemical abnormalities. Prophylactic treatment can be given with attenuated androgens such as danazol (GELFAND et al. 1976). Though this treatment can raise the levels of C$\bar{1}$ INH by stimulation of synthesis, lower doses than that required to restore the C$\bar{1}$ INH levels to normal are clinically effective, suggesting a different mechanism of action (WARIN et al. 1980). Though the prophylactic use of the androgenic steroids is effective, it can cause masculinisation of female foetuses, delay puberty in children and potentially cause liver damage in adults. The consumption of C$\bar{1}$ INH may be reduced by inhibiting the activation of other enzymes with which it reacts. ε-Aminocaproic acid, or its derivative tranexamic acid, inhibits the activation of plasmin from plasminogen and both compounds have been used successfully prophylactically to prevent attacks (FRANK et al. 1972). The side-effects of ε-aminocaproic acid are dose related and include myalgia and depp venous thrombosis. Tranexamic acid has fewer side effects. The recent availability of C$\bar{1}$ INH concentrate enables more effective therapy for acute attacks (ANGOSTONI 1980; LOGAN and GREAVES 1984).

I. Acquired C1 Inhibitor Deficiency

A small number of such patients have been described, with associated angio-oedema (GELFAND et al. 1979). The angio-oedema closely resembles that of HAE. Many of the patients, however, had a lymphoproliferative disorder, but occasionally lupus erythematosus, adenocarcinoma or cold urticaria. Most of these cases were associated with a paraprotein, cryoglobulin, an autoantibody which presumably initiates $C\bar{1}$ INH activation and $C\bar{1}$ INH consumption. Although $C\bar{1}$ INH, C2 and C4 levels are low as in HAE, a distinguishing feature is that C1 ($C\bar{1}q$) is very low in the acquired disease, but is normal or only minimally depressed in HAE.

In acquired cold urticaria in 20 patients, NILSSON and BACK (1984) found a functional deficiency of $C\bar{1}$ esterase activity, though the concentration measured by immunoassay was normal. Prekallikrein and kallikrein activity in the plasma was found to be normal. The significance of these findings remains to be evaluated.

N. Fibrin and Fibrinolysis

The relationship between urticaria and fibrinolysis has been reviewed by RYAN (1977). The release of histamine by the mast cell stimulates the endothelium to secrete plasminogen activators which act on plasminogen in the blood to form plasmin (RYAN et al. 1971). Plasmin lyses the fibrin film that lines blood vessels, with resultant increase in vascular permeability (RYAN et al. 1971). Plasmin diffuses into the surrounding tissue and fibrinolytic activity is increased in the acute urticarial lesions (BIANCHINI et al. 1983). Plasmin also contributes to the inflammatory response by cleaving the third component of complement to yield the vasoactive C3a, activating the kinin system to form bradykinin (RUDDY et al. 1972) and forming vasoactive fibrin degradation products (MALOFIEJEW 1972).

In the more persistent inflammatory response such as cutaneous vasculitis, fibrinolytic activity is inhibited or exhausted (CUNLIFFE et al. 1971; BIANCHINI et al. 1983). In the delayed persistent weals found in pressure urticaria, though there is no frank vasculitis, fibrinolytic activity in the blood is decreased (RYAN et al. 1971) and fibrin deposition is one of the features of the histology of pressure urticaria (JAMES et al. 1980). Fibrin may be increased in other forms of delayed persistent weals, and account partially for their persistence.

O. Complement

The complement system consists of a complex series of serum proteins which on activation results in the ordered and sequential formation of several highly specialised proteins (see Chap. 23, Vol. I). Activation of the classical and alternate pathways converge with activation of C3 and proceed through further steps to activation of C8 and C9. Biologically active fragments are formed, including the anaphylatoxins C3a, C4a, C5a. All these fragments when injected into human skin induce erythema and oedema, C5a being the most active (GORSKI et al. 1979).

This property may be mediated by histamine release from mast cells (JOHNSON et al. 1975). C5a is powerfully chemotactic to neutrophils in vitro and in vivo (WARD and HILL 1970) though neither C3a nor C4a are chemotactic (GORSKI et al. 1979). In the serum, C3a, C4a, and C5a are rapidly cleaved to their des-arginine forms by carboxypeptidase N, with only C5a retaining biological activity. The role of complement fragments in urticaria is largely inferred, except in HAE. Here a vasoactive fragment is formed from C2 (DONALDSON et al. 1969), which is not an anaphylatoxin.

Anaphylatoxins are not thought to be involved in chronic urticaria which is rarely accompanied by hypocomplementaemia (CHAMPION and HIGHET 1982). Urticarial vasculitis is a syndrome with persistent urticarial lesions showing histological features of leucocytoclastic vasculitis, angio-oedema, polyarthralgia and other systemic features (SANCHEZ et al. 1982). A proportion of these patients have hypocomplementaemia, C3 deposition in the blood vessels and circulating immune complexes. Until direct measurements of anaphylatoxins are made, the role of complement must be inferred.

P. Cellular Infiltration

The acute urticarial reaction is self-limiting, the clinical response resolving within 24 h. In theory, after all immediate reactions the release of mast cell mediators with chemotactic properties could cause an accumulation of leucocytes, leading to a persistent reaction. The mast cell mediators capable of attracting circulating polymorphonuclear leucocytes, including eosinophil chemotactic factors, neutrophil chemotactic factor, PGD_2, 12-HETE, LTB_4 and PAF, have been discussed under the relevant headings. Mononuclear cells are also attracted by 12-HETE (DOWD et al. 1985) and by a mast cell granule-derived inflammatory mediator of anaphylaxis (TANNENBAUM et al. 1980). The mediators released could be removed by the increased blood flow draining an urticarial site, degraded by their respective enzymes and further release from the mast cell, regulated by such factors as prostaglandin E (LICHTENSTEIN et al. 1972). Tachphylaxis of the blood vessel response also occurs (GREAVES and SHUSTER 1967). In chronic idiopathic urticaria, there is usually a sparse perivascular mononuclear leucocyte infiltrate (NATBONY et al. 1983), though dense perivascular mixed cellular infiltrates can occur (RUSSEL-JONES et al. 1983).

The less frequently observed delayed urticarial reactions of pressure urticaria (RYAN et al. 1968), delayed dermographism (KALZ et al. 1950) and delayed cold urticaria (SARKANY and TURK 1965) clinically resemble cutaneous late phase reactions. These cutaneous late phase reactions can follow immunological and non-immunological mast cell degranulation, but have mostly been studied after the IgE-mediated response (SOLLEY et al. 1976). Clinically the lesions follow an immediate urticarial response, are maximal at 6–8 h, are erythematous, burning and indurated and are beginning to resolve at 24 h. Though the delayed urticarial reactions are similar they do not always follow a clinically obvious immediate urticarial response, and the lesions persist for longer than 24 h.

Histologically in the late cutaneous reaction there is a mixed cellular infiltrate of neutrophils, eosinophils, lymphocytes and macrophages, the relative proportion of cells appearing to be influenced by the type of stimulus and the timing of the biopsy during the reaction (LEMANSKE and KALINER 1983). There is no evidence of immune complex deposition or complement activation (SOLLEY et al. 1976). The histology of pressure urticaria has been studied and is consistent with a delayed cutaneous reaction (RYAN et al. 1968).

References

Ackerman AB (1978) Superficial perivascular dermatitis. In: Histologic diagnosis of inflammatory diseases. Lea and Febiger, Philadelphia, pp 180–181

Akiyama T, Ushijima N, Anan S, Takahashi I, Yoshida H (1981) A case of cold urticaria belonging to the IgE class. J Dermatol 8:139–143

Angostoni A, Bergamaschini L, Martignoni GC, Cicardi M, Marasi B (1980) Treatment of acute attacks of hereditary angio-oedema with $C\bar{1}$ inhibitor concentrate. Ann Allergy 44:299–301

Aoyama H (1971) IgE as a dermographism-inducing principle of urticaria factitia. Nippon Hifuka Gakkai Zasshi 81:266–271

Archer CB, Page CP, Paul W, Morley J, MacDonald D (1984) Inflammatory characteristics of platelet activating factor (PAF-acether) in human skin. Br J Dermatol 110:45–50

Archer CB, Page CP, Paul W, Morley J, MacDonald DM (1985) Histamine independent role for PAF-acether as a potential mediator of urticarial vasculitis. In: Champion RH, Greaves MW, Kobza Black A, Pye RJ (eds) The urticarias. Livingstone, Edinburgh, pp 156–160

Atkins PC; Zweiman B (1981) Mediator release in local heat urticaria. J Allergy Clin Immunol 68:286–289

Bentley-Phillips CB, Kobza Black A, Greaves MW (1976) Induced tolerance in cold urticaria caused by cold-evoked histamine release. Lancet 2:63–66

Benveniste J, Henson PM, Cochrane CG (1972) Leucocyte dependent histamine release from rabbit platelets: the role of IgE, basophils and a platelet activating factor. J Exp Med 136:1356–1377

Bianchini G, Lotti T, Fabri P (1983) Fibrin deposits and fibrinolytic activity in Schonlein-Henoch syndrome. Int J Dermatol 22:103–106

Brain SD, Williams TJ, Tippins JR, Morris HR, MacIntyre I (1985) Calcitonin gene related peptide is a potential vasodilator. Nature 313:54–56

Breathnach SM, Allen R, Milford-Ward A, Greaves MW (1983) Symptomatic dermographism: natural history, clinical features, laboratory investigations and response to therapy. Clin Exp Dermatol 8:463–476

Brindley LL, Sweet JM, Goetzl EJ (1983) Stimulation of histamine release from human basophils by human platelet factor 4. J Clin Invest 72:1218–1223

Camp RDR, Coutts AA, Greaves MW, Kay AB, Walport MJ (1983) Responses of human skin to intradermal injections of leukotrienes C_4, D_4 and B_4. Br J Pharmacol 80:497–502

Carpenter SA, Lynn C (1981) Vascular and sensory responses of human skin to mild injury after topical treatment of capsaicin. Br J Pharmacol 73:755–758

Center DM, Soter NA, Wasserman SI, Austen KF (1979) Inhibition of neutrophil chemotaxis in association with experimental angio-oedema in patients with cold urticaria. Clin Exp Immunol 35:112–118

Champion RH, Highet AS (1982) Investigation and management of chronic urticaria and angio-oedema. Clin Exp Dermatol 7:291–300

Champion RH, Roberts SOB, Carpenter RG, Roger JH (1969) Urticaria and angio-oedema. A review of 544 patients. Br J Dermatol 81:588–597

Cho CH, Ogle CW (1979) Cholinergic-mediated gastric cell degranulation with subsequent histamine H_1 and H_2 receptor activation in stress ulceration in rats. Eur J Pharmacol 55:23–33

Cook J, Shuster S (1980) Histamine weal formation and absorption in man. Br J Pharmacol 69:579–585
Cunliffe WJ, Dodman B, Holmes RL, Foster RA (1971) Local fibrinolytic activity in patients with cutaneous vasculitis. Br J Dermatol 84:420–423
Curd JG, Prograis LJ, Cochrane CG (1980) Detection of active kallikrein in induced blister fluids of hereditary angio-oedema patients. J Exp Med 12:742–747
Czarnetzki BM, Kern F, Lichtenstein LM (1976) Defective release of eosinophil chemotactic factor from peripheral leucocytes in patients with chronic urticaria. J Invest Dermatol 67:276–278
Czarnetzki BM, Meentken J, Rosenbach T, Pokropp A (1984) Clinical, pharmacological and immunological aspects of delayed pressure urticaria. Br J Dermatol 111:315–323
Davies MG, Greaves MW (1980) Sensory response of human skin to synthetic histamine analogues and histamine. Br J Clin Pharmacol 9:461–465
Davis RS, Remigio LK, Schoket AL, Bock SA (1981) Evaluation of a patient with both aquagenic and cholinergic urticaria. J Allergy Clin Immunol 68:479–483
DeLaus FV, Winkelmann RK (1968) Kinins in cold urticaria. Arch Dermatol 98:67–74
Doeglas HMG (1975) Reactions to aspirin and food additives in patients with chronic urticaria, including the physical urticarias. Br J Dermatol 93:135–143
Doeglas HMG, Bleumink E (1975) Protease inhibitors in plasma of patients with chronic urticaria. Arch Dermatol 111:979–985
Doeglas HMG, Bleumink E (1985) Plasma protease inhibitors in chronic urticaria. In: Champion RH, Greaves MW, Kobza Black A, Pye RJ (eds) The urticarias. Livingstone, Edinburgh, pp 59–69
Donaldson VH (1968) Mechanism of activation of $C\bar{1}$ esterase inhibitor in hereditary angio-neurotic plasma in vitro. The role of Hageman factor, a clot-promoting agent. J Exp Med 127:411–429
Donaldson VH, Evans RR (1963) A biochemical abnormality in hereditary angio-neurotic oedema. Am J Med 35:37–44
Donaldson VH, Ratnoff OD, Dias Da Silva W, Rosen FS (1969) Permeability-increasing activity in hereditary angio-neurotic edema plasma: II. Mechanism of formation and partial characterization. J Clin Invest 48:642–653
Dowd PM, Kobza Black A, Woollard PM, Camp R, Greaves MW (1985) Cutaneous responses to 12-hydroxy-5,8,10,14-eicosatetraenoic acid (12-HETE). J Invest Dermatol 84:537–541
Eady RAJ, Keahey TM, Sibbald RG, Kobza Black A (1981) Cold urticaria with vasculitis: report of a case with light and electron microscopic, immunofluorescence and pharmacological studies. Clin Exp Dermatol 6:355–366
Eftekhari N, Milford Ward A, Allen R, Greaves MW (1980) Protease inhibitor profiles in urticaria and angio-oedema. Br J Dermatol 103:33–39
Ferreira SH (1972) Prostaglandins, aspirin-like drugs and analgesia. Nature [New Biol] 240:200–203
Flower RJ, Harvey EA, Kingston WP (1976) Inflammatory effects of prostaglandin D_2 in rat and human skin. Br J Pharmacol 56:229–233
Ford Hutchinson AW, Walker JR, Davidson EM, Smith MJH (1978) PGI_2: a potential mediator of inflammation. Prostaglandins 16:253–258
Ford-Hutchinson AW, Bray MA, Doig MV, Shipley ME, Smith MJH (1980) Leukotriene B, a potent chemokinetic and aggregating substance released from polymorphonuclear leukocytes. Nature 286:265–266
Fraki JE (1977) Human skin proteases. Effect of separated proteases on vascular permeability and leukocyte emigration in skin. Acta Derm Venereol (Stockh) 57:393–398
Fraki JE, Schecter NMS, Lazarus GS (1983) Human skin proteases as inflammatory mediators. Br J Dermatol [Suppl 25]109:72–76
Frank MM, Sergent JS, Kane MA, Alling D (1972) Epsilon amino caproic acid therapy of hereditary angio-neurotic oedema. A double-blind study. N Engl J Med 286:808–812
Gamse R, Holzer P, Lembeck F (1980) Increase of substance P in primary afferent neurones and impairment of neurogenic plasma extravasation by capsaicin. Br J Pharmacol 68:207–213

Gelfand JA, Sherins RJ, Alling DW, Frank MM (1976) Treatment of hereditary angio-oedema with danazol: reversal of clinical and biochemical abnormalities. N Engl J Med 295:1444–1448
Gelfand JA, Boss GR, Conley CL, Reinhart RMM (1979) Acquired $C\bar{1}$ esterase deficiency and angio-oedema: a review. Medicine (Baltimore) 58:321–328
Gigli I, Ruddy S, Austen KF (1968) The stoichiometric measurement of the serum inhibitor of the first component of complement by the inhibition of immune haemolysis. J Immunol 100:1154–1164
Gorski JP, Hughli TE, Muller-Eberhard HJ (1979) The third anaphylatoxin of the human complement system. Proc Natl Acad Sci USA 76:5299–5302
Grandel KE, Farr RS, Wanderer AA, Eisenstadt TC, Wasserman SI (1985) Association of platelet-activating factor with primary acquired cold urticaria. N Engl J Med 313:405–409
Grant JA, Findlay SR, Thueson DO, Fine DP, Krueger GG (1981) Local heat urticaria: angio-oedema: evidence for histamine release without complement activation. J Allergy Clin Immunol 67:75–77
Greaves MW, McDonald-Gibson W (1973) Itch: role of prostaglandins. Br Med J 111:608–609
Greaves MW, Shuster S (1967) Responses of skin blood vessels to bradykinin, histamine and 5-hydroxytryptamine. J Physiol (Lond) 193:255–267
Greaves MW, Sondergaard J (1970) Urticaria pigmentosa and factitious urticaria. Direct evidence for release of histamine and other smooth muscle-contracting agents in dermographic skin. Arch Dermatol 101:418–425
Greaves MW, Plummer VM, McLaughlan P, Stanworth DR (1974) Serum and cell bound IgE in chronic urticaria. Clin Allergy 4:265–271
Hagermark O, Fjellner B (1980) Enhancement of histamine-induced pruritus by enkephalin and morphine. J Invest Dermatol 74:459 a
Hawk JLM, Eady RAJ, Challoner AVJ, Kobza Black A, Keahey TM, Greaves MW (1980) Elevated blood histamine levels and mast cell degranulation in solar urticaria. Br J Clin Pharmacol 9:183–186
Henson PM, Lynch JM (1983) Cellular orgins of PAF. In: Beneviste J, Arnoux B (eds) Platelet-activating factor. Elsevier, Amsterdam, pp 75–82 (INSERM Symposium no 23)
Herxheimer A (1956) The nervous pathway mediating cholinergic urticaria. Clin Sci 15:195–205
Higgs EA, O'Grady J, Thrower PA, Moncada S (1979) Prostacyclin: inflammatory effects in human skin (Abstr). 4th International Prostaglandin Conference, May 27–31, Washington
Houser DD, Arbesman CE, Ito K, Wicher K (1970) Cold urticaria: immunologic studies. Am J Med 49:23–33
Illig L, Heinicke (1967) Zur Pathogenese der cholinergischen Urticaria IV. Arch Klin Exp Dermatol 229:360–371
James MP, Eady RAJ, Kobza Black A, Hawk JLM, Greaves MW (1980) Physical urticaria: a microscopical and pharmacological study of mast cell involvement. J Invest Dermatol 74:451
Jessel TM, Iversen LL (1977) Opiate analgesics inhibit substance P release from rat trigeminal nucleus. nature 268:549–551
Johnson AR, Hugli TE, Muller-Eberhard HJ (1975) Release of histamine from rat mast cells by the complement peptides C3a and C5a. Immunology 28:1067–1080
Johnston MG, Hay JB, Movat HZ (1976) The modulation of enhanced vascular permeability by prostaglandins through alterations in blood flow (hyperaemia). Agents Actions 6:705–711
Jorizzo JL, Coutts AA, Eady RAJ, Greaves MW (1983) Vascular responses of human skin to injections of substance P and mechanism of action. Eur J Pharmacol 87:67–76
Juhlin L (1979) Urticaria and mast cells in the skin. In: Pepys J, Edwards AM (eds) The mast cell: its role in health in disease. Pitman, London, pp 612–616

Juhlin L, Hammarstrom (1982) Effects of intradermally injected leukotriene C_4, and histamine in patients with urticaria, psoriasis and atopic dermatitis. Br J Dermatol [Suppl 23]107:106–110
Juhlin L, Michaelsson G (1969a) Cutaneous reactions to kallikrein, bradykinin and histamine in healthy subjects and in patients with urticaria. Acta Derm Venereol (Stockh) 49:26–36
Juhlin L, Michaelsson G (1969b) Vascular reactions in hereditary angio-oedema. Acta Derm Venereol (Stockh) 49:20–25
Kalz F, Bower CM, Prichard H (1950) Delayed and persistent dermographism. Arch Dermatol 61:772–779
Kaplan AP, Beaven MA (1976) In vivo studies of the pathogenesis of cold urticaria, cholinergic urticaria and vibration-induced swelling. J Invest Dermatol 67:327–332
Kaplan AP, Bray L, Shaff RE, Horakova Z, Beaven MA (1975) In vivo studies of mediator release in cold urticaria and cholinergic urticaria. J Allergy Clin Immunol 55:394–402
Kaplan AP, Horakova Z, Katz KI (1978) Assessment of tissue fluid histamine levels in patients with urticaria. J Allergy Clin Immunol 61:350–354
Kaplan AP, Garofalo J, Sigler R, Hauber T (1981) Idiopathic cold urticaria: in vitro demonstration of histamine release upon challenge of skin biopsies. N Engl J Med 18:1074–1077
Kaplan KL, Owen J (1981) Plasma levels of β-thromboglobulin and platelet factor 4 as indices of platelet activation in vivo. Blood 57:199–202
Keahey TM, Greaves MW (1980) Cold urticaria: dissociation of cold evoked histamine release and urticaria following cold challenge. Arch Dermatol 116:174–177
Kern F, Lichtenstein LM (1976) Detective histamine release in chronic urticaria. J Clin Invest 57:1369–1377
Klemperer MR; Donaldson VH, Rosen FS (1968) Effect of C1 esterase on vascular permeability in man: studies in normal and complement-deficient individuals and in patients with hereditary angio-neurotic oedema. J Clin Invest 47:604–611
Kobza Black A, Heavey DJ, Barr RM, Barrow SE, Chappell CG, Greaves MW, Dollery CT (1985) The release of histamine and eicosanoids in cold urticaria. J Invest Dermatol 84:433
Krause LB, Shuster S (1984) H_1-receptor-active histamine not the sole cause of chronic idiopathic urticaria. Lancet 2:929–930
Leiferman KM; Loegering DA, Gleich GJ (1984) Production of wheal-and-flare reactions by eosinophil granule protein. J Invest Dermatol 82:414
Leiferman KM; Ackerman SJ, Sampson HA, Haughen, HS, Venecie PY, Gleich GJ (1985) Dermal deposition of eosinophil-granule major basic protein in atopic dermatitis. N Engl J Med 313:282–285
Lemanske RF, Kaliner MA (1983) Late phase allergic reactions. Int J Dermatol 22:401–409
Lembeck F, Holzer P (1979) Substance P as a neurogenic mediator of antidromic vasocilation and neurogenic plasma extravasation. Nauyn Schmiedebergs Arch. Pharmacol 310:175–183
Lembeck F, Popper H, Juan H (1976) Release of prostaglandins by bradykinin as an intrinsic mechanism of its algesic effect. Nauyn Schmiedebergs Arch Pharmacol 294:69–73
Levy LR, Lepow IH (1959) Assay and properties of serum inhibitor of $C\bar{1}$ esterase. Proc Soc Exp Biol Med 101:608–611
Lewis RA, Austen KF, Drazen JM, Clark DA, Marfat A, Corey EJ (1980) Slow reacting substance of anaphylaxis: identification of leukotrienes C-1 and D from human and rat sources. Proc Nat Acad Sci USA 77:3710–3714
Lewis RA, Soter NA, Diamond PT, Austen KF, Oates JA, Roberts LJ (1982) Prostaglandin D_2 generation after activation of rat and human mast cells with anti-IgE. J Immunol 129:1627–1631
Lewis T (1927) The blood vessels of the human skin and their responses. Shaw, London, pp 80–90

Lichtenstein LM, Gillespie E, Bourne HR, Henney CS (1972) The effects of a series of prostaglandins on in vitro models of allergic response and cellular immunity. Prostaglandins 2:519–528
Logan RA, Greaves MW (1984) Hereditary angio-oedema: treatment with C1 esterase inhibitor. J R Soc Med 77:1046–1048
MacGlashan DW, Schleimer RP, Peters SP, Schulman ES, Adams GK, Sobotka AK, Newball HH, Lichtenstein LM (1983) Comparative studies of human basophils and mast cells. Fed Proc 42:2504–2509
Magnus IA, Thompson RHS (1954) Cholinesterase activity of human skin. Br J Dermatol 66:163–173
Malofiejew M (1972) The biological and pharmacological properties of some fibrinogen degradation products. Scand J Haematol [Suppl]13:303
Marks R, Greaves MW (1977) Vascular reactions to histamine and compound 48/80 in human skin: suppression by a histamine H_2 receptor blocking agent. Br J Clin Pharmacol 4:367–369
Misch KJ, Black AK, Barr R, Hensby CN, Mallet AI (1982) Histamine and non-histamine pharmacological activity in cold urticaria. J Invest Dermatol 18:329
Moore-Robinson M, Warin RP (1967) The effect of salicylates in urticaria. Br Med J 4:262–264
Murphy GM, Smith SE, Smith SA, Greaves MW (1984) Autonomic function in cholinergic urticaria and atopic eczema. Br J Dermatol 110:581–586
Natbony SF, Phillips ME, Elias JM, Godfrey HP, Kaplan AP (1983) Histologic studies of chronic idiopathic urticaria. J Allergy Clin Immunol 71:177–183
Newcomb RW, Nelson H (1973) Dermatographia mediated by immunoglobulin E. Am J Med 54:174–180
Nilsson T, Back O (1984) On the role of C1-esterase inhibitor in cold urticaria. Acta Derm Venerol (Stockh) 64:197–202
O'Donnell MC, Ackerman SJ, Gleich GJ, Thomas LL (1983) Activation of basophil and mast cell histamine release by eosinophil granule major basic protein. J Exp Med 157:1981–1991
Page CP, Paul W, Archer CB, MacDonald DM, Morley J (1983 a) PAF-acether, a mediator of acute and persisting inflammation to non-allergic and allergic stimuli. In: Beneviste J, Arnoux B (eds) Platelet-activating factor. INSERM symposium 23. Elsevier, Amsterdam
Page CP, Paul W, Morley J (1983 b) Continuous monitoring of plasma protein extravasation and cell accumulation in vivo. Br J Derm 109:103–105
Palmer RMJ, Stephney RG, Higgs GA, Eakins KE (1980) Chemokinetic activity of arachidonic acid lipoxygenase products on leucocytes of different species. Prostaglandins 20:411–418
Parish WE (1985) Possible relevance of changes in mast cells and neutrophils to perpetuation of chronic urticaria. In: Champion RH, Greaves MW, Kobza Black A, Pye RJ (eds) The urticarias. Churchill Livingstone, London, pp 70–85
Peters MS, Schroeker AL, Kephart GM, Gleich GJ (1983) Localization of eosinophil granule major basic protein in chronic urticaria. J Invest Dermatol 81:39–43
Phanuphak P, Schocket AL, Arroyave CM, Kohler PF (1980) Skin histamine in chronic urticaria. J Allergy Clin Immunol 65:371–375
Plummer NA, Hensby CN, Kobza Black A, Greaves MW (1977) Prostaglandin activity in sustained inflammation of human skin before and after aspirin. Clin Sci Mol Med 52:615–620
Robertson I, Greaves MW (1978) Responses of human skin blood vessels to synthetic histamine analogues. Br J Clin Pharmacol 5:319–322
Robson EB, Lachmann PJ, Hobart MJ, Johnston WA (1979) Linkage studies in hereditary angio-oedema. J Med Gen 16:347–350
Rose B (1941) Studies on blood histamine in cases of allergy. J Allergy 12:327–334
Rosen FS, Charache P, Pensky J, Donaldson V Hereditary angio-neurotic oedema: two genetic variants. Science 148:957–958

Ruddy S, Gigli I, Austen KF (1972) The complement system of man (third of four parts). New Engl J Med 287:592–596
Russell Jones R, Bhogal B, Dash A, Schifferli J (1983) Urticaria and vasculitis: a continuum of histological and immunopathological changes. Br J Dermatol 108:695–703
Russell Jones R, Tai PC, Spry CJF, Eady RAJ, Kobza Black A, Greaves MW (1985) Criteria for the diagnosis of vasculitis and identification of activated eosinophils. In: Champion RH, Greaves MW, Kobza Black A, Pye RJ (eds) The urticarias. Churchill Livingstone, London, pp 149–155
Ryan TJ (1977) Urticaria and fibrinolysis. Clin Exp Dermatol 2:177–182
Ryan TJ, Shim-Young N, Turk JL (1968) Delayed pressure urticaria. Br J Dermatol 80:485–490
Ryan TJ, Nishioka K, Dawber RPR (1971) Epithelial-endothelial interaction in the control of inflammation through fibrinolysis. Br J Dermatol 84:501–515
Sanchez NP, Winkelmann RK, Schroeter AL, Dicken CH (1982) The clinical and histopathologic spectrums of urticarial vasculitis: study of forty cases. J Am Acad Dermatol 7:599–605
Sarkany I, Turk JL (1965) Delayed hypersensitivity to cold. Proc R Soc Med 58:622–623
Seppa HEJ (1980) The role of chymotrypsin-like proteinase of rat mast cells in inflammatory vasopermeability and fibrinolysis. Inflammation 4:1
Shelley WB, Shelley ED, Ho AKS (1983) Cholinergic urticaria: acetyl choline receptor dependent immediate-type hypersensitivity reaction to copper. Lancet 1:843–846
Sibbald RG, Kobza Black A, Eady RAJ, James M, Greaves MW (1981) Aquagenic urticaria: evidence of cholinergic and histaminergic basis. Br J Dermatol 105:297–302
Solley GO, Gleich GJ, Jordon RE, Schroeter AL (1976) The late phase of the immediate weal and flare skin reactions. J Clin Invest 58:408–420
Sondergaard J, Greaves MW (1970) The responses of the skin to tetrahydrofurfuryl nicotinate (Trafuril) studied by continuous skin perfusion. Br J Dermatol 82:14–18
Soter NA, Wasserman SI, Pathak MA, Parrish JA, Austen KF (1979) Solar urticaria: release of mast cell mediators into the circulation after experimental challenge. J Invest Dermatol 72:282
Soter NA, Wasserman SI, Austen KF, McFadden ER (1980) Releaes of mast cell mediators and alterations in lung function in patients with cholinergic urticaria. N Engl J Med 302:604–608
Soter NA, Lewis RA; Corey EJ, Austen KF (1983) Local effects of synthetic leukotrienes (LTC_4, LTD_4 and LTB_4) in human skin. J Invest Dermatol 80:115–119
Tannenbaum S, Ortel H, Henderson W, Kaliner M (1980) The biological activity of mast cell granules.I. Elicitation of inflammatory responses in rat skin. J Immunol 125:325–335
Thompson RA, Felix-Davies DD (1978) Response of "idiopathic" recurrent angioneurotic oedema to tranexamic acid. Br Med J 2:608
Ting S, Reimann BEF, Rauls DO, Mansfield LE (1983) Nonfamilial, vibration-induced angio-oedema. J Allergy Clin Immunol 71:546–551
Toth-Kasa I, Jancso G, Obal F, Husz S, Simon N (1983) Involvement of sensory nerve endings in cold and heat urticaria. J Invest Dermatol 80:34–36
Vartio T, Seppa H, Vaheri A (1981) Selective susceptibility of soluble and matrix fibrinectin to degradation by the tissue proteinases, mast cell chymase and cathepsin G. J Biol Chem 256:471
Ward PA, Hill JH (1970) C5 chemotactic fragments produced by an enzyme in lysosomal granules. J Immunol 104:535–543
Warin AP, Greaves MW, Gatecliff M, Williamson DM, Warin RP (1980) Treatment of hereditary angio-oedema by low dose attenuated androgens: dissociation of clinical response from levels of C1 esterase inhibitor and C4. Br J Dermatol 103:405–409
Wasserman SI, Ginsberg MH (1984) Release of platelet factor 4 into the blood after cold challenge of patients with cold urticaria. J Allergy Clin Immunol 74:275–279
Wasserman SI, Soter NA, Center DM, Austen KF (1977) Cold urticaria: recognition and characterization of neutrophil chemotactic factor which appears in serum during experimental cold challenge. J Clin Invest 60:189–196

Wasserman SI, Austen KF, Soter NA (1982) The functional and physiochemical characterization of three eosinophilotactic activities released into the circulation by cold challenge of patients with cold urticaria. Clin Exp Immunol 47:570–578
Wassom DL, Loegering DA, Solley GO, Moore SB, Schooley RT, Fauci AS, Gleich GJ (1981) Elevated serum levels of the eosinophil granule major basic protein in patients with eosinophilia. J Clin Invest 67:651–661
Winkelmann RK, Kobza Black AK, Dover J, Greaves MW (1986) Pressure urticaria: a histopathological study. Clin Exp Dermatol 11:139–147
Winkelmann RK, Wilhelmj CM, Horner FA (1965) Experimental studies on dermographism. Arch Dermatol 92:436–442

CHAPTER 32

Eczema

S. SHUSTER

A. Introduction

Treatment of eczema can occasionally be directed at causative mechanisms; more often it is symptomatic as dictated by acuteness, extent, severity, site and clinical type.

B. Treatment Directed at Causative Mechanisms

I. Asteatotic Eczema

Despite its name, the mechanism of this type of eczema has not been established and its inclusion in this section owes more to convenience than correctitude. Nevertheless, the condition often occurs in the elderly, particularly after frequent bathing and use of detergents, and it responds to less frequent and milder bathing, together with the regular application of emollients, common constituents of which are glycerol lanolin, vegetable oil and waxes. There have, however, been few therapeutic comparisons of the various emollients and many different preparations continue to be used as they are in the ichthyoses (see Chap. 41).

II. Atopic Eczema

The view that atopic eczema is provoked by a variety of environmental and dietary antigens, most particularly the house-dust mite and cows milk, is widely held but has still to be established. Reports of response to reduction of putatively causal antigens have not been consistent (ATHERTON et al. 1978; CANT et al. 1986; MUNKVAD et al. 1984), and there have been too few controlled therapeutic studies; this important field remains interesting but confused (ATHERTON 1988). Therapy is therefore directed at the cause of exacerbation (infection, scratch, dehydration of the stratum corneum) or is symptomatic, as described below.

III. Contact Eczema

Both immune and irritant forms of contact eczema may be controlled by avoidance of the responsible chemical. Symptomatic therapy with corticosteroids and emollients is used initially and in relapses when complete avoidance cannot be achieved, and in atopic patients, in whom hand eczema is more easily provoked by irritants. There is no good evidence that barrier creams are effective.

IV. Infected Eczema

Many eczemas become infected, most often by pyogenic bacteria, occasionally by yeasts and viruses (see Chap. 29). Staphylococcal infection is a particular problem with atopic eczema and may precipitate a relapse; corticosteroid-antimicrobial combinations are believed to be more effective than either type of agent alone (LEYDEN and KLIGMAN 1977). Antibiotics, most often erythromycin, are usually given systemically until the organism is identified and its sensitivity established (DAVID and CAMBRIDGE 1986). Because of the risk of resistance, it is important to avoid the topical use of antibiotics which may be required for the systemic treatment of infections; likewise drugs such as the penicillins and sulphonamides which readily sensitise the skin. There are a number of antibacterial agents suitable for topical use, e.g. clioquinol, neomycin and more recently mupirocin, as well as the many antiseptics. Microbial growth on the skin is mainly related to water content of the stratum corneum (see NOBLE 1981) an presumably this explains the greater tendency of eczematous skin to infection. It may also explain the action of many topical antiseptics which desiccate the stratum corneum; this may partly explain the effect of potassium permanganate soaks in acute weeping eczema. Infection with the herpes simplex virus may occasionally provoke woresening of atopic eczema and responds to acyclovir (DAVID and LONGSON 1985). Eczema, particularly of the flexures, may become infected by yeasts and respond to the appropriate antifungal agent.

V. Seborrhoeic Eczema

It has now been shown that seborrhoeic eczema, and dandruff, its minor manifestation (SHUSTER 1984; FORD et al. 1984), is due to an infection by *Pityrosporum ovale* (see Chap. 28) and responds to appropriate antimycotic agents, of which ketoconazole (2% cream or shampoo) has been studied most extensively, although other antimycotic agents with an antipityrosporal activity are also effective (see SHUSTER and BLATCHFORD 1988). The condition usually recurs 4–8 weeks after treatment with ketoconazole has stopped, presumably because the organism has not been entirely eradicated, and recolonises the skin; therapy has therefore to be used intermittently as indicated by the relapse.

The seborrhoeic dermatitis of AIDS responds likewise to antipityrosporal therapy, although other infective dermatoses may be confused with pityrosporal dermatitis. There are early reports that *Pityrosporum* may be the causal organism in the flexural eczema of infancy (napkin psoriasis), but confirmation is required that this disorder responds to antipityrosporal therapy; likewise the possible ancillary role of *Pityrosporum* in atopic eczema (WAERSTED and HJORTH 1985) and scalp psoriasis (FARR et al. 1985).

VI. Stasis Eczema

In addition to the symptomatic therapy described below, relief of stasis surgically, or more often by postural drainage, compression stocking or well-applied bandage, will reduce or completely control the disorder. Contact dermatitis to topically applied medicaments is particularly common (PARAMSOTHY et al. 1988), and sensitisers must be detected and avoided.

C. Symptomatic Treatment

The various mediators of inflammation which have been found in the skin of eczematous lesions, notably the prostaglandins and histamine (see Chap. 29 in Part I of this volume), do not appear to be part of the pathological process, in that eczema does not respond to inhibitors of prostaglandin synthesis nor to the antihistamines. The main symptomatic treatment uses the anti-inflammatory effect of corticosteroids.

I. Corticosteroids

(see Chaps. 19, 20 and CHRISTOPHERS et al. 1988)

Corticosteroids are invariably effective, but the way they are used and the associated treatments depend on the stage and type of disease. As a rule systemic administration is reserved for initial treatment of extensive or severe disease or when a rapid effect is required in localised disease. There are many different topical corticosteroids but they differ only in potency; they can be selected for use on that basis alone, since, despite claims to the contrary, there is no evidence that loss of skin collagen, their main unwanted toxic effect in skin (SHUSTER et al. 1967; Chap. 27 herein) is related other than to their therapeutic potency. Although it is generally believed that the therapeutic response of dermatoses to topical corticosteroids is subject to tachyphylaxis, and that this may be overcome by changing to a different corticosteroid, there is no convincing evidence for either notion. Likewise there is no hard evidence to support the common view that stopping corticosteroids may induce a "rebound" relapse to a state worse than that before treatment. The use of corticosteroids is limited by toxicity, of which systemic absorption is particularly important in children in whom there is higher surface area volume ratio. With prolonged treatment there is loss of skin collagen, telangiectasia and facial rosacea and a predilection to infection. These unwanted effects occur with all corticosteroid preparations apparently in proportion to their biological potency. The debate about the toxicity of 1% hydrocortisone (SHUSTER 1985 a, b; GREAVES 1985) arises largely because the magnitude and frequency of toxicity has not been well studied.

II. Tar Applications

Tar applications are used mainly for chronic eczemas, particularly atopic eczema. The active principle is not known, nor is the mode of action. The source of the tars varies (e.g. coal, wood or fish), and the concentrations generally vary from 2% to 10%, used as a solution or paste. Some preparations are combined with a corticosteroid or incorporated in an occlusive bandage.

III. Emollients

It is believed that emollients produce their effects by their occlusive action, which leads to hydration of the stratum corneum (see Chap. 41). The hydrated stratum corneum is more flexible and cracks less, with consequent reduction of pruritus, excoriation, infection and the further development of eczema. Bathing hydrates the stratum corneum and some emollients are designed for use by dispersal in the bath. Patient preference is surprisingly variable and seems to depend on feel, particularly of greasiness, and disappearance of the material. Urea is included in some preparations to enhance hydration of the stratum corneum osmotically.

IV. Antihistamines

Although there have been relatively few controlled studies (e.g. Frosch et al. 1984) it is generally accepted that antihistamines (see Chaps. 7, 18 herein and Chap. 18 of Part I) have no direct effect on eczematous dermatitis. The antipruritic action of the older non-selective antihistamines in eczema has been shown not to be due to histamine receptor antagonism, but to a central property related to their sedative effect (see Chaps. 7, 18); thus the newer non-sedative antihistamines have no effect on the itch of the eczematous dermatoses. Of the sedative antihistamines, trimeprazine is often used in children and promethazine in adults. The main use of these drugs is for nocturnal pruritus, since their daytime use is limited by sedation. Alleviation of itch, especially in children, helps greatly to reduce scratch-evoked worsening of the rash. The benzodiazepines are also effective but should only be used for short periods, and thalidomide is effective but toxic. Not all sedatives are antipruritic, and barbiturates make itch worse (see Chap. 7; Shuster 1981; Krause and Shuster 1983).

V. Cyclosporin A

Although well-controlled studies of the effect of Cyclosporin A on eczematous dermatitis have not yet been published, early reports and our own studies show the treatment to be extremely effective. The mechanism of action of the drug has not yet been elucidated, though its effects on T lymphocyte function are thought likely to be involved through inhibition of interleukin 2 production (Hess et al. 1988). From our own studies, relapse, unlike in psoriasis, occurs rapidly, usually within a few days after Cyclosporin A is stopped. The therapeutic place of the drug has therefore still to be established; in particular, whether safe dosage regimens can be devised.

VI. Cytostatic Drugs

Cytostatic drugs are used for severe persistent erythroderma and exfoliative dermatitis (see Chap. 24), but there have been few critical studies of their use, and it is not clear whether they have any advantage over systemic corticosteroids.

VII. Photochemotherapy (see Chap. 41)

The use of methoxypsoralen and UVA (PUVA) for atopic eczema has not been as well studied as it has for psoriasis (see Chaps. 33, 38) and this treatment is employed mostly for severe or recalcitrant disease. The mode of action of PUVA is discussed in Chap. 38.

VIII. Grenz Ray Therapy

Grenz ray therapy is not often used but is said to have been effective in chronic eczemas, particularly of the hands and feet.

D. Specific Eczematous Dermatoses

I. Lichenified Eczema

The usual treatment for lichenified eczema is occlusion with tar or corticosteroid preparations and oral antipruritic drugs at night.

II. Nodular Prurigo

Occlusion therapy together with nocturnal administration of antipruritic drugs is used for nodular prurigo as for lichenified eczema. Thalidomide is also effective (VAN DEN BROEK 1980), probably because of the antipruritic activity associated with its central depressant action (DALY and SHUSTER, unpublished; Chaps. 1, 18 herein): if the scratching of nodular prurigo and lichenification can be reduced or eliminated, the rash will regress with no other therapy.

III. Erythroderma and Exfoliative Dermatitis

Systemic corticosteroid therapy is the mainstay for erythroderma and exfoliative dermatitis; cytostatic drugs are also used, and Cyclosporin A may well prove useful. Many of the patients require antibiotics. In addition to the primary disorder, there are many systemic complications of this extensive dermatosis (SHUSTER 1967; SHUSTER and MARKS 1970), some of which, e.g. increased capillary permeability, hypothermia and heart failure, require urgent treatment.

IV. Atopic Eczema

The treatment for atopic eczema is mainly symptomatic; the role of specific antigens and the therapeutic effect of their removal remains unproven. Drugs believed to act on the mast cell, such as the chromones, have not proved useful (ATHERTON et al. 1982), possibly because of poor absorption. However, iontophoretically delivered nedocromil was without effect on antigen-induced weal reactions (HUMPHREYS and SHUSTER 1987). There have been relatively few studies of topical chromones, and it remains difficult to assess reports of benefit (ARYANAYAGAM et al. 1985). The main treatment is corticosteroids, most often topically, occasionally systemically. The role of skin infection is particularly important (Leyden and Kligman 1977; David and Cambridge 1986), and exacerbations have often to be controlled by antibiotics, or acyclovir in the much less common herpes simplex infections (David and Longson 1985); the role of *Pityrosporum* (Waersted and Hjorth 1985) requires further study. Exacerbation of itch, initiating scratch, excoriation, infection and worsening of the eczema, is particularly important in children, and sedative antihistamines are used for this at night (see above). Coarse woollen clothing worsens the rash by friction. Because of the associated ichthyotic skin and a tendency of the rash to occur when the stratum corneum dries and becomes brittle – e.g. in a centrally, heated environment without appropriate humidification – emollients are used to maintain its flexibility (see Chap. 41). The efficacy of Cyclosporin A is clear, but because of toxicity its possible therapeutic use requires further study. A degree of improvement is reported after oral evening primrose oil (Wright and Burton 1982) and attributed to its content of gamma-linoleic acid, which, it is suggested, may be relatively deficient in the skin of patients with atopic eczema (Manku et al. 1984). However, both effect and mechanism require further substantiation.

References

Aryanayagam M, Barlow JJG, Graham P, Hall-Smith SP, Harris JM (1985) Topical sodium cromoglycate in the management of atopic eczema: a controlled trial. Br J Dermatol 112:343–348
Atherton DJ (1988) Diet and atopic eczema. Clin Allergy 18:215–228
Atherton DJ, Sewell M, Soothill JF, Wells RS, Chilvers CED (1978) A double blind cross over controlled trial of an antigen avoidance diet in atopic eczema. Lancet 1:401–403
Atherton DJ, Soothill JF, Elvidge J (1982) A controlled trial of oral sodium cromoglycate in atopic eczema. Br J Dermatol 106:681–685
Cant AJ, Bailes JA, Marsden RA, Hewitt D (1986) Effect of material dietary exclusion on breast-fed infants with eczema. Br Med J 293:231–233
Christophers E, Schöpf E, Kligman AM, Stoughton RB (1988) Topical corticosteroid therapy. Raven, New York
David TJ, Cambridge GC (1986) Bacterial infection and atopic eczema. Arch Dis Child 61:20–23
David TJ, Longson M (1985) Herpes simplex infection in atopic eczema. Arch Dis Child 60:338–343
Farr PM, Krause LB, Marks JM, Shuster S (1985) The response of scalp psoriasis to ketoconazole. Lancet 2:921–922
Ford GP, Farr PM, Ive FA, Shuster S (1984) The respnse of seborrhoeic dermatitis to ketoconazole. Br J Dermatol 111:603–607
Frosch JP, Schwanitz HJ, Macher E (1984) A double-blind trial of H_1 and H_2 receptor antagonists in the treatment of atopic dermatitis. Arch Dermatol Res 276:36–40
Greaves MW (1985) Over the counter sale of topical corticosteroids: evidence versus anecdote. Br Med J 291:276–277
Hess AD, Esa AH, Colombani PM (1988) Mechanisms of action of cyclosporine: effects on cells of the immune system and sub-cellular events in T cell activation. Transplant Proc [Suppl 2]20:29–40
Humphreys F, Shuster S (1987) The effect of nedocromil on weal reactions in human skin. Br J Clin Pharmacol 24:405–408
Krause L, Shuster S (1983) The mechanism of action of antipruritic drugs. Br Med J 287:1199–1200
Leyden JJ, Kligman AM (1977) The case for steroid antibiotic combinations. Br J Dermatol 96:179–187

Manku MS, Horrobin DF, Morse NL, Wright S, Burton JL (1984) Essential fatty acids in the plasma phospholipids of patients with atopic eczema. Br J Dermatol 110:643–648
Munkvad M, Davidsen L, Hoj L, Poulsen CO, Secher L, Svejgaard E, Bandgard A, Larsen PO (1984) Antigen free diet in adult patients with atopic dermatitis: a double blind study. Acta Derm Venerol (Stockh) 64:524–528
Noble WC (1981) Microbiology of human skin, 2nd edn. Lloyd-Luke, London
Paramsothy Y, Collins M, Smith AG (1988) Contact dermatitis in patients with leg ulcers. Contact Dermatitis 18:30–36
Shuster S (1967) Systemic effects of skin disease. Lancet 1:907–912
Shuster S (1981) Reason and the rash. Proc R Inst GB 53:136–163
Shuster S (1984) The aetiology of dandruff and mode of action of therapeutic agents. Br J Dermatol 111:235–242
Shuster S (1985a) Over the counter sale of topical corticosteroids: the need for debate. Br Med J 291:38–39
Shuster S (1985b) Over the counter sale of topical corticosteroids. Br Med J 291:967–968
Shuster S, Blatchford NR (1988) Seborrhoeic dermatitis and dandruff – a fungal disease. R Soc Med Int Congr Symp Ser 132:1–54
Shuster S, Marks JM (1970) Systemic effects of skin disease. Heinemann, London
Shuster S, Raffle EJ, Bottoms E (1967) Skin collagen in rheumatoid arthritis and the effects of corticosteroids. Lancet 1:525–527
Van den Broek H (1980) Treatment of prurigo nodularis with thalidomide. Arch Dermatol 116:571–572
Waersted A, Hjorth N (1985) *Pityrosporum orbiculare* – a pathogenic factor in atopic dermatitis of the face, scalp and neck? Arch Derm Venereol (Stockh) [Suppl] 114:146–148
Wright S, Burton JL (1982) Oral evening primrose oil improves atopic eczema. Lancet 2:1120–1122

CHAPTER 33

Treatment of Psoriasis

J. MARKS

A. Introduction and Aetiological Factors

The very fact that so many treatments are still in use for psoriasis indicates how far from ideal all of them are. There is also a lack of good trials upon which selection of treatment can be based for the different forms of the disease. Most remedies were introduced on an empirical basis, but, with increasing knowledge, their application has become more rational. The fact that the fundamental abnormality in psoriasis is still not known has hampered progress in this respect, but even with this knowledge, there are still a number of levels at which the therapeutic attack can logically be made.

I. Genetics

There is an inherited susceptibility in all who get the rash and when the underlying abnormality is eventually discovered (presumably an enzyme defect), it may turn out to follow an autosomal dominant pattern although monozygotic twins are not invariably concordant for the rash (FARBER and NALL 1971). There is an increased risk of psoriasis in people with certain HLA antigens (MARCUSSON et al. 1981) particularly Cw6 and DR7.

II. Precipitating Factors

What it is that precipitates an attack is completely unknown in most cases. The exceptions, though interesting, have few therapeutic implications.

1. Streptococcal Infection

It is worthwhile treating repeated *Streptococcus pyogenes* infections in children suffering from guttate psoriasis as this may prevent subsequent attacks.

2. Koebner Phenomenon

Although psoriasis often occurs at the site of skin injury, trauma is an unusual precipitating factor, and scope for prevention is negligible.

3. Stress

The view much held by patients and some doctors that psoriasis is due to stress has never been proved and in any case, scope for prevention is very limited.

4. Lithium

Administration of lithium has been reported to precipitate or exacerbate attacks of psoriasis (SKORAN and THORMANN 1981) though the relevance of this to psoriasis as a whole is not known.

III. Abnormalities in Skin and Other Organs

A specific arthropathy occurs in about 7% of psoriatics patients, but its relationship to the rash is not one of cause and effect. Many abnormalities of skin, blood and other organs have been found in psoriasis. Most are epiphenomena; some result from the disease and return to normal when the rash is cleared, and treatment directed towards their correction is irrelevant to the rash.

1. Increased Epidermal Proliferation

The discovery that psoriatic epidermis was proliferating at about ten times the normal rate (WEINSTEIN and FROST 1968) led to the use of anti-mitotic drugs, particularly methotrexate, for treating psoriasis. Moreover, treatments which had previously been used empirically have since been found to decrease epidermal proliferation, though whether this is the mechanism of their action in psoriasis is not known. The precise cytokinetic abnormality is a matter of debate, but it seems that there is an increase in the number of proliferating cells rather than a shortening of the cell cycle time (WRIGHT 1980). Further elucidation of this phenomenon may help with finding new drugs and drug combinations for psoriasis, as well as with working out the best regimens for using the ones in existence. Cyclic AMP decreases epidermal cell division in vitro and a decrease in cAMP was proposed as a mechanism of the kinetic abnormality. Neither this mechanism nor the alleged beneficial effect of therapeutic regimens with phosphodiesterase inhibitors to which it led have been confirmed. The role of vitamin D_3 in cell proliferation has led to the use of the vitamin in psoriasis and topical treatment with 1,24-dihydroxycholecalciferol is reported to have had some effect (KATO et al. 1986).

2. Leukotriene Production

It has been suggested that the polymorphonuclear cell infiltration and leukotriene production in lesions of psoriasis may be close to the primary defect (DOWD et al. 1983) and the improvement following benoxaprofen treatment (ALLEN and LITTLEWOOD 1983; KRAGBALLE and HERLIN 1983) would fit in with this.

3. Immunological Abnormalities

Discovery that cyclosporin A works in psoriasis has raised the possibility of a primary immunological defect (ELLIS et al. 1986). There is, however, no evidence that the drug's action in psoriasis is an immunological one.

B. Choice of Treatment

Choice of treatment has to take into account many factors of which the most important are: (a) clinical type of psoriasis; (b) extent of rash; (c) site of rash; (d) other medical conditions; (e) age and sex of patient; (f) desire for treatment and personal preference; (g) occupation and hobbies; (h) history of response to treatment in previous attacks; (j) treatment facilities in hospital and at home; (k) distance of home and work from treatment centre, and ability to travel; (l) local medical and nursing preference and expertise. The main problem is chronic plaque psoriasis and the methods used to clear the existing rash are discussed first: what little can be done to alter the subsequent natural history of exacerbation and remission is discussed later.

C. Clearance of Chronic Plaque Psoriasis

I. Topical Preparations

1. Corticosteroids

These are clean and much used, but not very effective. They have some place short-term if plaque psoriasis is sore or irritated and, in patients who are relatively intolerant of dithranol (anthralin), they can be used to reduce dithranol inflammation without otherwise reducing its effectiveness.

2. Tar

This has been much used, particularly since the time of GOECKERMAN (1925) who used coal tar paste combined with ultraviolet light (UV-B, 290–320 nm). It contains thousands of different chemicals (GRUPPER 1971) and tar from different sources varies considerably in composition. It is not known which of its components are effective and refined preparations are as a rule less potent therapeutically. Coal tar preparations are more messy and smell worse than dithranol, but they stain the skin and clothes less and are easier to apply. Tar was introduced as an empirical treatment, long before it was shown to suppress DNA synthesis (WALKER et al. 1978) though whether this is the mechanism of its action is not known.

3. Dithranol

Dithranol (anthralin 1,8-dihydroxyanthrone) is a simple substance relative to coal tar. It binds to DNA (see Chap. 34) and reduces epidermal proliferation (WRIGHT 1980), but it is not known whether this is its mode of action. Its stability is pH sensitive and most preparations use salicyclic acid as stabiliser. Its main inconveniences are that it stains and burns. Until recently most dermatologists who used dithranol followed the regimen of Ingram (1953). After a bath containing coal tar the patient is irradiated with ultraviolet light (UV-B) and the dithranol in a paste containing salicylic acid (Lassar's paste BNF) is then applied to the lesions and left on for 24 h when the whole process is repeated. By using this regimen it is possible to clear in-patients with chronic plaque psoriasis in a mean of 21 days. Although devised apparently for very good reasons which centred round the fact that dithranol irritates non-lesional skin more than psoriatic lesions and is therefore most efficiently used if the application stays confined to the psoriatic plaques, the Ingram regimen has many disadvantages. The paste is thick and therefore difficult to apply and to remove, and it takes a skilled nurse up to an hour a day to apply: the most effective concentration varies from patient to patient (usually 0.025%–1%) and for the most part it is by intelligent guessing that the starting concentration is chosen.

Attempts to find more acceptable and easily usable preparations of dithranol (Seville 1975; Whitefield 1981; Chalmers et al. 1982) have been more successful than attempts to find analogues with an improved therapeutic rate. It has been shown that dithranol (mostly 0.5–8%) left on for 0.25–4 h a day is as effective as that left on for 24 h (Schaefer et al. 1980; Marsden et al. 1983; Kunze and Runne 1984). Despite the shorter time of application, burning of non-lesional skin is still a problem, either because of the higher concentrations used or their less accurate application in cream and ointment bases which are used because they are easier to deal with. The optimal way of using this form of dithranol treatment has not been fully worked out or its effectiveness (including remission time) fully compared with the Ingram regimen. In some centres short-contact dithranol treatment is replacing the Ingram regimen and this is a great advance, because of the reduction in skilled nurse time and the fact that it is easily carried out by patients in their own homes. It has not proved possible to separate the irritant and staining properties of dithranol from its therapeutic effect though a number of free radical scavengers inhibit its inflammatory action, and are being investigated for possible clinical application (Finnen et al. 1984; Lawrence et al. 1987).

II. Photochemotherapy (PUVA)

This treatment (Parrish et al. 1974) depends upon the photoactivation by ultraviolet light (wavelength 320–440 nm, U-VA) of the photosensitising drug 8-methoxypsoralen which reacts with DNA in the skin (see Chap. 38). The drug is usually given orally because response is less predictable with topical applications. The drug itself is non-toxic and the effects of the treatment are, with minor exceptions, confined to the skin and eyes. UV-A is usually provided by an artificial source and treatment is given up to four times a week. The irradiation is given 2 h after the drug, to correspond to maximal blood and skin levels. It is much more acceptable than dithranol to most patients and more people can work while having it.

The dose of UV-A must be carefully adjusted to prevent burning the skin and goggles worn to protect the eyes. Absolute contra-indications are pregnancy and aphakia. There are theoretical reasons for believing susceptibility to skin cancer and changes of ageing to be increased. As the latent period may be long, it will be some years before it is known whether they it occur and whether the PUVA regimen used is relevant. Reprots that cancer is already occurring are suspect because the observations were uncontrolled. PUVA with whole-body radiation is usually given when more than 20% of the body surface is involved.

1. Psoralen Baths

Psoralen is given by mouth in most instances and psoralen ointments, paints, etc. defeat one of the main purposes of PUVA, namely to produce a clean treatment. Tripsoralen baths followed by UV-A are effective and acceptable, certainly in Finland (SALO et al. 1981) and their advantages and disadvantages are worth a formal comparison.

III. How to Choose Between Tar, Dithranol and PUVA

Few dermatologists use tar and dithranol with equal enthusiasm and direct comparisons have not been made. In one study with tar it was reported that the psoriasis was 80% clear after a mean of 24 days and 90% clear after a mean of 30 days (GRUPPER 1971). Although this is not very different from the results with dithranol by the Ingram regimen (e.g. ROGERS et al. 1979) studies of dithranol have in general been carried out more rigorously. The side effects and inconvenience of the two are similar and both are not very acceptable to most patients although short-term dithranol regimens are a great advance in patient convenience and acceptability. Both treatments are safe. Overall in the United Kingdom dermatologists use dithranol more often than tar. A few patients who cannot be cleared with tar can be cleared with dithranol and vice versa, but irritated psoriasis is often equally intolerant of both.

The obvious advantage of PUVA is that it is clean. A randomised controlled trial (ROGERS et al. 1979) of one PUVA regimen showed that it worked in a slightly greater proportion of patients than the Ingram regimen and although it took longer to clear the rash, the actual number of treatments required was fewer and the socio-economic cost was also less (MARKS et al. 1979). Thus, apart from geographical and domestic factors which prevent patients from attending hospital for PUVA, it would be the treatment of choice were it not for the uncertainty about long-term toxicity. For this reason it has been recommended that it should for the time being be used as first-line treatment only in the more elderly patients (BRIFFA et al. 1981). Nevertheless it is often in the younger patients that the cosmetic disability of the disease and its messy treatments cause greatest distress, and PUVA should not be withheld purely on the grounds of youth if the psoriasis is extensive, severe, or unresponsive to tar of dithranol.

D. Prevention of Recurrence of Chronic Plaque Psoriasis

Clearing of an attack with dithranol, tar of PUVA does not seem to alter the natural history of psoriasis and the chances of a recurrence in these patients are approximately 50% and 75% at 6 months and 1 year, respectively. It has been shown by a randomised controlled trial (BRIFFA et al. 1981) that if PUVA is continued after the rash is clear, the risk is decreased significantly to 20% at 6 months and 25% at 1 year. Findings of a non-controlled, non-randomised multi-centre trial (HENSELER et al. 1981) contradict these findings, but because of the nature of the trial are less likely to be correct. PUVA is also effective prophylactically after dithranol treatment (MARKS et al. 1984) and dithranol itself has some effect prophylactically applied regularly to the whole skin once it is clear (J. MARKS, unpublished work).

E. Psoriasis at Special Sites

I. Face

Fortunately this is unusual. Dithranol and tar are rarely acceptable though PUVA is satisfactory as long as the eyes are suitably shielded. Topical corticosteroids are effective, but must be used for short periods of time because the face is an area where the adverse effects of topical corticosteroids seem to be most severe and noticeable. Psoriatic patients with seborrhoeic dermatitis-like rash of the face respond to topical ketoconazole (FARR et al. 1985).

II. Scalp

Dithranol and tar are effective and are used as pomades which are easy to apply and to wash out. Topical corticosteroid applications for the scalp are cleaner and more soothing. For removal of ointment and debris a detergent shampoo such as Teepol is used.

III. Flexures

Sore, macerated psoriasis of the flexures can be treated with weak dithranol paste, but this may be uncomfortable, in which case topical corticosteroids are preferable. PUVA is unsatisfactory because of the inaccessibility of the flexures to the UV-A.

IV. Palms and Soles

If dithranol and tar fail, topical corticosteroids can be used under polythene occlusion. PUVA is more effective, but the response is slow (MURRAY et al. 1980) although this can be hastened by etretinate (LAWRENCE et al. 1984). It is possible to treat palms and soles with limited irradiation from small UV-A units.

V. Nails

Neither pitting nor onycholysis respond to topical treatment or to PUVA. Very painful or cosmetically disabling nail psoriasis can be treated with corticosteroid injections into the nail base, but the effect is short-lived.

F. Systemic Treatments (Excluding PUVA)

These are used only in severe forms of the disease such as erythrodermic, generalised pustular and arthropathic psoriasis. Exceptionally, extensive plaque psoriasis which has not responded to topical applications or to PUVA, as well as psoriasis which responds well, but recurs rapidly, may have to be treated in this way.

I. Corticosteroids

Response is related to potency and there is no proven advantage in any particular corticosteroid (or adrenocorticotropic hormone). Both refractoriness during treatment and worsening after treatment are believed to occur by many clinicians, but the evidence is neutral. Corticosteroids have no place in the treatment of chronic plaque psoriasis though if given for arthropathic psoriasis the skin lesions may benefit. In erythrodermic psoriasis corticosteroids (usually prednisolone) are the most effective agents for urgent reduction of the acute inflammation and for prevention or reversal of cardiovascular and thermoregulatory complications and, as in other forms of erythroderma, can sometimes save lives (Marks 1982). Treatment of this phase of psoriasis may well have to be followed by treatment more appropriate to severe chronic psoriasis, e.g. methotrexate.

II. Anti-mitotics

These have a place in the management of erythrodermic and generalised pustular psoriasis and in severe resistant and rapidly recurring chronic plaques. Methotrexate is very effective and is the drug in this group which is most often used, but no trials have been done to compare it with other cytostatic drugs or to assess the effectiveness of drug combinations. All have the disadvantages that they may depress the bone marrow and have adverse effects on the germ cells, so that pregnancy must be avoided.

1. Methotrexate

Although it is generally assumed that methotrexate acts locally to inhibit DNA synthesis in the epidermal cells, the fact that topical applications of methotrexate are ineffective does raise the question of a more central mechanism. The main long-term toxic effect is hepatic fibrosis and is dose related (Dahl et al. 1972; Zachariae et al. 1980). It should not normally be given to anyone with liver disease. To avoid the more immediately serious side effects of bone marrow depression and gastrointestinal bleeding which are occasionally fatal, it is necessary to start with a small dose, e.g. 5 mg weekly. A weekly oral dose of 5–15 mg is usually adequate for maintenance. More frequent doses should never be given because of increased toxicity and adequate supervision is essential (Roenigk et al. 1982). There appears to be no advantage in various split dose and parenteral regimens.

2. Hydroxyurea

Its main adverse effect is on the bone marrow where its effects include macrocytosis. Its disadvantage is that it works in a relatively small proportion of patients. It may be used in patients who are unable to take methotrexate.

III. Aromatic Retinoids

Drugs of this group are still being assessed in psoriasis. Etretinate seems to be lacking in serious side effects except for those on the foetus, so that pregnancy must not be allowed to occur in a woman on the drug, or for 2 years after stopping. The importance of changes that occur in serum lipids and liver enzymes is not known, but is unlikely to be great and bone changes are rare. However, a number of annoying complications such as dry mouth and hair loss occur. On its own etretinate has an effect in psoriasis though it is still not known how great this effect is or what type of psoriasis it is best used in, as relatively few cases of each type have been studied. The exception to this is pustular psoriasis of palms and soles where the rash clears quickly in a high proportion of patients who can tolerate the drug. Relapse occurred quickly after stopping the drug, though continuing on a lower dose at clearance was possible with few side effects and had some clinical effect (White et al. 1986).

There is some evidence that etretinate used in conjunction with PUVA reduces the dose of UV-A needed to clear the rash (Frisch et al. 1978). Our own studies throw doubt on its usefulness in practice except for pustular psoriasis of palms and soles: etretinate led to a slight and insignificant reduction in the dose of PUVA and the time to clear chronic plaque psoriasis (Parker et al. 1984) and there was a significant and larger reduction of dose of PUVA required to clear pustular psoriasis of the palms and soles (Lawrence et al. 1984). However, the unwanted effects, especially alopecia, made treatment unacceptable to many of the patients.

IV. Cyclosporin A

Cyclosporin A is very effective in low dosage for clearing psoriasis. Relapse occurs when the drug is withdrawn, particularly in severe cases, but we have shown that, in disease of only moderate severity, the relapse rate is not statistically different from that after clearing with dithranol or PUVA (Marks et al. 1988). Adverse effects in the short term appear to be negligible, but in the longer term most patients experience some rise in blood pressure and serum creatinine. It is too early to assess the possible role of cyclosporin A in psoriasis, but it may be useful in patients in whom short courses of treatment, e.g. once or twice a year, are followed by good periods of remission.

We do not know how cyclosporin A works in psoriasis and have no reason to suppose its well-known immunological action is particularly relevant here. Other possible mechanisms have been investigated systematically in our patients (Higgins et al. 1988): we have found no decrease in inflammatory response to such stimulators as UV-B and dithranol, no decrease in response to histamine or

48/80 and no decrease in immediate hypersensitivity reactions: some delayed hypersensitivity reactions are in fact enhanced. As would be expected, very few psoriatics and others on cyclosporin A can be sensitised to DNCB using doses that sensitise 95% of the population. Evidence on cell turnover is conflicting, but our own pilot study suggests that the abnormally high turnover rate of psoriasis is not corrected by cyclosporin A.

V. Other Treatments

The non-steroidal anti-inflammatory drug benoxaprofen, which inhibits leukotriene production, appeared to improve psoriasis in small but controlled studies (ALLEN and LITTLEWOOD 1983; KRAGBALLE and HERLIN 1983). In view of the findings of increased leukotriene production (DOWD et al. 1983) confirmation of this observation is keenly awaited. Other inhibitors of prostaglandin synthesis have not been shown to be effective. Likewise, claims for an effect of phosphodiesterase inhibitors have never been confirmed. A few studies of renal dialysis appear to show an effect which is of theoretical rather than practical importance, and the same applies to the use of vitamin D_3 analogues.

References

Allen B, Littlewood S (1983) The aetiology of psoriasis: clues provided by benoxaprofen. Br J Dermatol [Suppl 25]109:126–129

Briffa DV, Greaves MW, Warin AP, Rogers S, Marks J, Shuster S (1981) Relapse rate and long-term management of plaque psoriasis after photochemotherapy and dithranol. Br Med J 282:937–940

Chalmers RJG, Marks JM, Shuster S (1982) A novel wax stick preparation of anthralin. Acta Derm Venereol (Stockh) 62:181–182

Dahl MGC, Gregory MM, Scheuer PJ (1972) Methotrexate hepatotoxicity in psoriasis – comparison of different dose regimens. Br Med J 1:654–656

Dowd PM, Brain S, Kamp R, Black AK, Fincham NJ, Greaves MW (1983) The detection of leukotriene B_4-like material in psoriasis. Br J Dermatol 108:225–226

Ellis CN, Gorsulowsky DC; Hamilton TA, et al. (1986) Cyclosporin improves psoriasis in a double blind study. JAAMA 256:3110–3116

Farber E, Nall ML (1971) Genetics of psoriasis. twin study. In: Farber E, Cox A (eds) Psoriasis. Stanford University Press, Stanford, pp 7–13

Farr PM, Krause LB, Marks JM, Shuster S (1985) The response of scalp psoriasis to ketoconazole. Lancet 2:921–922

Finnen MJ, Lawrence CM, Shuster S (1984) Inhibition of dithranol inflammation by free-radical scavengers. Lancet 2:1129–1130

Fritsch P, Honigsmann H, Jaschke E, Wolff K (1978) Augmentation of oral methoxsalen photochemotherapy with an oral retinoic acid derivative. J Invest Dermatol 7090:178–182

Goeckerman WH (1925) The treatment of psoriasis. Northwest Med 24:229–231

Griffiths CEM, Powles AV, Leonard JM, et al. (1986) Clearance of psoriasis with low dose cyclosporin. Br Med J 293:731–732

Grupper C (1971) The chemistry, pharmacology and use of tar in psoriasis. In: Farber E, Cox A (eds) Psoriasis. Stanford University Press, Stanford, pp 347–356

Harper JI, Kendra JR, Desai S, Straughton RCD, Barrett AJ, Hobbs JR (1984) Dermatological aspects of the use of cyclosporin A for prophylaxis of graft-versus-host disease. Br J Dermatol 110:469–474

Henseler T, Wolff K, Honigsmann H, Christophers E (1981) Oral 8-methoxypsoralen photochemotherapy of psoriasis. Lancet 1:853–857
Higgins EM, Munro C, Rees J, Humphreys F, Farr PM, Ramsay B, Marks JM, Friedmann PS, Shuster S (1988) Effects of cyclosporin on physiological and pharmacological reactions in skin. Paper presented at meeting of European Society for Dermatological Research, Munich
Ingram JT (1953) The approach to psoriasis. Br Med J 2:591–594
Kato TE, Rokugo M, Terui T, Thagamighi H (1986) Successful treatment of psoriasis with topical application of active vitamin D_3 analogue 1,24-dihydroxycholecalciferol. Br J Dermatol 115:431–433
Kragballe K, Herlin T (1983) Benoxaprofen improves psoriasis. Arch Dermatol 119:548–552
Kunze J, Runne U (1984) Left-right comparison of psoriasis by short contact ("minutes") therapy with anthralin (dithranol). Acta Derm Venereol [Suppl] (Stockh) 113:161–165
Lawrence CM, Marks J, Parker S, Shuster S (1984) A comparison of PUVA-etretinate and PUVA-placebo for palmo-plantar pustular psoriasis. Br J Dermatol 110:221–226
Lawrence CM, Shuster S, Collins M, Bruce JM (1987) Reduction of anthralin inflammation by potassium hydroxide and Teepol. Br J Dermatol 116:171–177
Marcusson JA, Johannesson A, Moller E (1981) HLA-A,B,C and DR antigens in psoriasis. Tissue Antigens 17:525–529
Marks J (1982) Erythroderma and its management. Clin Exp Dermatol 7:415–422
Marks J (1986) Psoriasis. Br Med J 293:509
Marks J, Rogers S, Shuster S (1979) Socioeconomic benefits of psoralen/ultraviolet therapy. Lancet 2:464
Marks JM, Lawrence CM, Corbett M, Coburn PR, Parker S, Shuster S (1984) The influence of prophylactic photochemotherapy with PUVA on relapse rate in psoriasis cleared with anthralin. Br Med J 188:95–96
Marks JM, Friedmann PS, Shuster S, Munro C, Higgins EM (1988) Response and relapse of moderate psoriasis treated with cyclosporin A. Paper presented at meeting of European Society for Dermatological Research, Munich
Marsden JR, Coburn P, Marks J, Shuster S (1983) Response to short term application of dithranol in psoriasis. Br J Dermatol 108:243
Murray D, Corbett MF, Warin AP (1980) A controlled trial of photochemotherapy for persistent palmo plantar pustulosis. Br J Dermatol 102:659–663
Parker S, Coburn P, Lawrence CM, Marks J, Shuster S (1984) A randomised double blind comparison of PUVA-etretinate and PUVA-placebo in the treatment of chronic plaque psoriasis. Br J Dermatol 110:215–220
Parrish J, Fitzpatrick T, Tananbaum L, Pathak M (1974) Photochemotherapy of psoriasis with oral methoxsalen and longwave ultraviolet light. N Engl J Med 291:1207–1212
Roenigk H, Auerback R, Maibach H, Weinstein G (1982) Methotrexate guide lines revised. J Am Acad Dermatol 6:145–155
Rogers S, Marks J, Shuster S, Briffa DV, Warin A, Greaves MW (1979) Comparison of photochemotherapy and dithranol in the treatment of chronic plaque psoriasis. Lancet 1:455–458
Salo OP, Lassus A, Taskinen J (1981) Trioxsalen bath plus UVA treatment of psoriasis. Acta Derm Venereol (Stockh) 61:551–554
Schaefer H, Farber E, Goldberg L, Schalla W (1980) Limited application period for dithranol in psoriasis. Preliminary report on penetration and clinical efficacy. Br J Dermatol 102:571–573
Seville RH (1975) Simplified dithranol treatment for psoriasis. Br J Dermatol 93:205–208
Skoran I, Thormann J (1979) Lithium compound treatment and psoriasis. Arch Dermatol 115:1185–1187
Walker JF, Stoughton RD, de Quay PR (1978) Suppression of epidermal proliferation by UVL, coal tar and anthralin. Br J Dermatol 99:89–98
Weinstein GD, Frost P (1968) Abnormal cell proliferation in psoriasis. J Invest Dermatol 50:254–259

White SI, Puttick L, Marks JM (1986) Low dose etretinate in the maintenance of remission of palmar plantar pustular psoriasis. Br J Dermatol 115:577–582
Whitefield M (1981) Pharmaceutical formulations of anthralin. Br J Dermatol [Suppl]105:28–32
Wright NA (1980) The kinetics of human epidermal cell populations in health and disease. In: Rook A, Savin J (eds) Recent advances in dermatology, vol 5. Livingstone, Edinburgh, pp 317–343
Zachariae H, Kragballe K, Sagaard H (1980) Methotrexate-induced cirrhosis. Studies including serial liver biopsies during continued treatment. Br J Dermatol 102:407–412

CHAPTER 34

Anthralin

B. SHROOT and H. SCHAEFER

A. Introduction

Since the pioneering work of UNNA (1916) anthralin (dithranol, 1,8-dihydroxy-anthr-9-one) has occupied an important place in the clinical management of psoriasis. A synthetic compound consisting of a simple tricyclic skeleton, anthralin rapidly superseded chrysarobin, a natural product, introduced by SQUIRE (1877). Side effects of chrysarobin, which is mainly 3-methlyanthralin, were soon noted by CROCKER (1888) who warned physicians of the severe irritation which occurs in the non-involved skin which comes into contact with the drug. Mahogany-coloured staining of skin and linen also accompanied treatment with these drugs and it is probably as a direct consequence of these two major drawbacks, irritation and staining, that anthralin did not enjoy great popularity in the first 30 years of its use. European interest was rekindled when INGRAM (1953, 1954) introduced a stiff anthralin paste which enabled a skilled nurse to limit the application of the drug to the lesion. Since that time anthralin studies have been focused on a search for effective treatment schedules, either by changing the dose of the drug or combining it with light and/or other anti-psoriatic therapies (SEVILLE 1981; SHROOT et al. 1985). Within this framework fits all the painstaking research which has been directed towards the discovery of stable readily manageable derivatives and formulations of anthralin which ideally should be usable by the patient at home.

B. Chemistry

I. Anthrone-Anthranol Tautomerism

The key to the anthralin problem lies in having a better understanding of the chemistry of the molecule. It is of basic importance to realize that although anthralin can theoretically exist in two forms (Fig. 1), under normal conditions it is the keto form which predominates and this anthrone is a molecule which is highly susceptible to oxidation.

A key intermediate in the oxidation process is thought to be the radical formed by hydrogen radical abstraction at C-10 (WIEGREBE et al. 1981). As shown in Fig. 2 this radical may react with itself to form a dimeric substance or with molecular oxygen to form a hydroperoxide which can further decompose to furnish a 9,10-anthraquinone.

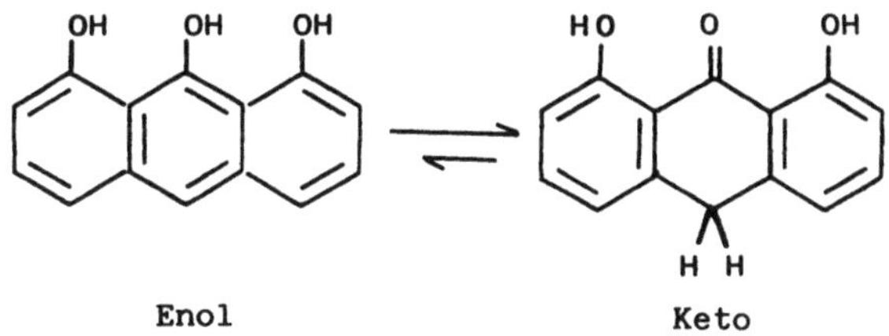

Fig. 1. Keto-enol tautomerism of anthralin (SHROOT 1987)

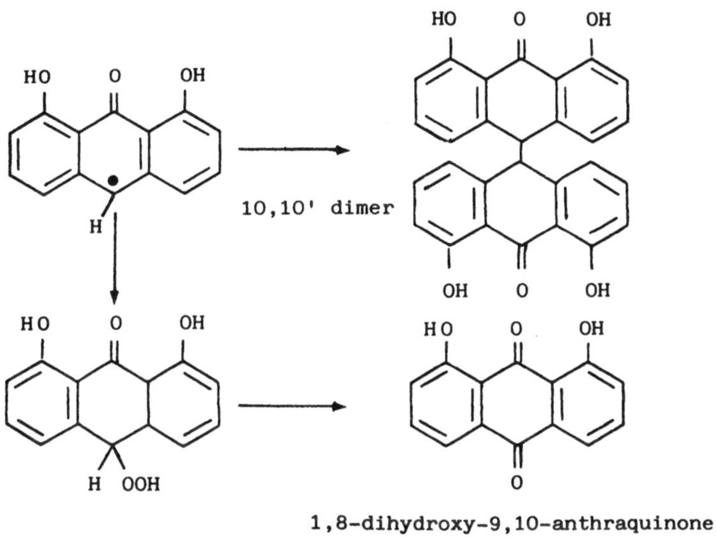

Fig. 2. Possible scheme for the free radical-mediated decomposition of anthralin (SHROOT 1987)

II. Chemical Assay

The UV-visible spectra of anthralin, the dimer and the quinone are shown in Fig. 3 and it is to be noted that they all absorb in the range 300–420 nm in chloroform solution. This is of relevance for two reasons: first, none of these three substances are highly coloured and second, the similarity in spectral characteristics severely restricts the use of UV spectroscopy for specific assays of anthralin in formulations or in metabolic studies.

Thus the violet-brown products which are responsible for the staining of skin and linen are possibly formed through further oxidation of the dimer. CARON and SHROOT (1981) used the United States Pharmacopeia UV assay for several commercial and standard preparations of anthralin and compared this with a high pressure liquid chromatographic method. Only the latter technique permits simultaneous detection and quantification of anthralin, the dimer and the quinone (Table 1).

Formulations which were apparently stable by simple UV absorbance considerations contained as little as 50% free anthralin. This technique also permits

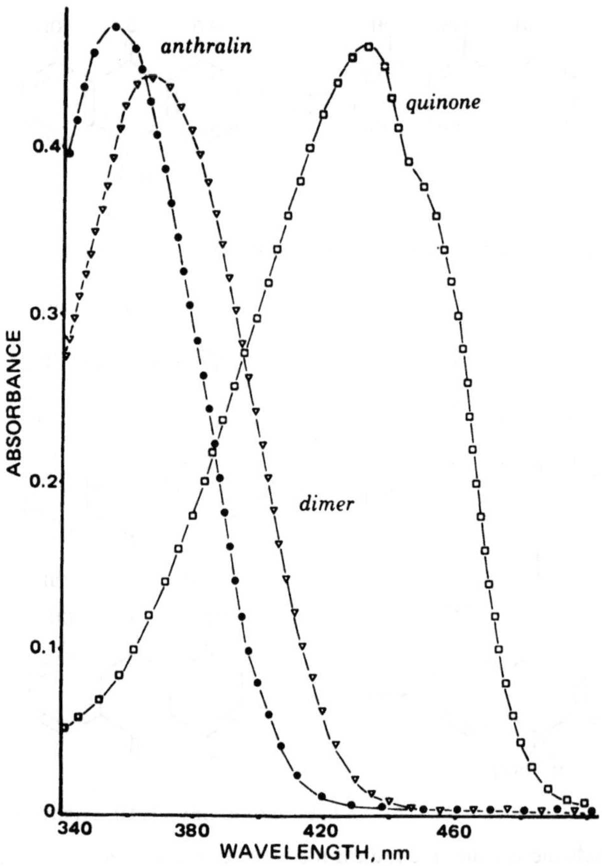

Fig. 3. UV-visible spectra at 10 μg/ml in chloroform of anthralin (*circles*), the dimer (*triangles*) and the quinone (*squares*). Molar absorption coefficients ε at 345 and 432 nm are, respectively, 10650 and 100 $l\,mol^{-1}\,cm^{-1}$ for anthralin, 16970 and 220 $l\,mol^{-1}\,cm^{-1}$ for the dimer, and 1680 and 10890 $l\,mol^{-1}\,cm^{-1}$ for the quinone (CARON and SHROOT 1981; SHROOT 1987)

Table 1. High pressure liquid chromatographic assay procedure for anthralin, dimer and quinone

Extraction	Analytical conditions (normal phase)	Compound	Detection wavelength (nm)	Retention time (min)
Cream				
1. Extract with chloroform	SiO_2 (5 μm), 250 × 4 mm	Anthralin	365	3.70
	Mobile phase (vol.-%):	Quinone	436	3.35
2. Evaporate to dryness	Isooctane (44.5)	Dimer	365	4.52
	Diisipropyl ether (49.9)			
3. Redissolve in chromatography elution solvent	Methanol (5.4)			
	Acetic acid (0.2)			
	Flow rate = 1 ml/min			
4. Inject	Temperature 30 °C			

quality control of unformulated anthralin in its crystalline state. Several other reports have now been issued on this subject (KARSTEN 1985; BURTON and RAO GADDE 1985; WURSTER and UPADIASHTA 1986).

UV radiation and zinc ions were thought to be of importance with respect to anthralin stability (RAAB and GMEINER 1975). It would appear that base and oxygen are of equal relevance (COLIN et al. 1981) and as a result stable formulations are best prepared under oxygen-free conditions.

C. Structure-Activity: New Derivatives

KREBS and SCHALTEGGER (1969) published an important article in this area which essentially concluded that the minimal structure for anti-psoriatic activity is the 1-hydroxyanthr-9-one. WIEGREBE and GERBER (1979) studied 1,8,9-triacetoxyanthralin which is devoid of side effects and postulated that non-specific esterases liberate mixtures containing substances which correspond to varying degrees of hydroxyl acylation, including free anthralin. MUSTAKALLIO (1979, 1980) and BRANDT and MUSTAKALLIO (1983) presented a series of analogues bearing an acyl group at C-10. They studied single and repeated application of these derivatives to psoriatic patients and found the 10-butyryl derivative to be four times less irritating than anthralin following daily application of 8 mmol/kg for 3 weeks. SCHALTEGGER et al. (1982) demonstrated that 10-acyl derivatives on mild acid hydrolysis are potential sources of anthralin. This is summarized in Fig. 4.

D. Mode of Action

In view of the instability of anthralin in aqueous solution at pH 7 (CAVEY et al. 1982), it is dangerous to speculate on the mode of action of this drug unless concomitant stability studies are carried out. Early ideas on the inhibitory potency of anthralin on enzymes regulating glucose utilization, such as glucose-6-phosphate dehydrogenase (G6-PDH), the key enzyme in the hexose monophosphate shunt, which is elevated in psoriasis (RASSNER 1972), have to be reconsidered in the light of work with irradiated solutions of anthralin. RAAB and GMEINER (1975) suggested that the potent inhibitor of this enzyme was a decomposition product. It has been established by CAVEY et al. (1982) that anthralin, or the only two known decomposition products, the dimer and the quinone, are weak inhibitors of G6-PDH. This suggests that in vivo it is an unstable breakdown product, or the redox process itself, which is responsible for this effect. It is generally believed that anthralin acts by decreasing cell proliferation. This has been demonstrated in cell culture with human fibroblasts (JACQUES and REICHERT 1981); in the hairless mouse after topical application (FISHER and MAIBACH 1975) and in the mouse by suppressing tetradecanoyl phorbol acetate-induced increase in DNA synthesis and ornithine decarboxylase induction (DE YOUNG et al. 1981). From experiments with T98G cells CLARK and HANAWALT (1982) conclude that anthralin and the quinone may intercalate into DNA, but neither compound affects UV-stimulated excision repair. A direct interaction of anthralin with DNA, as proposed at first

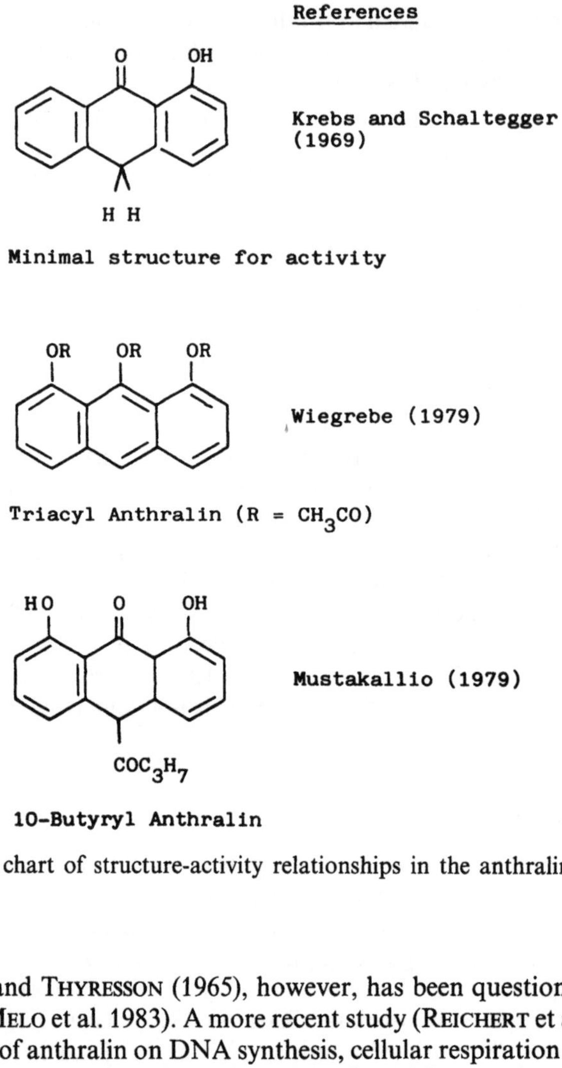

Fig. 4. Summary chart of structure-activity relationships in the anthralin field (SHROOT 1987)

by SWANBECK and THYRESSON (1965), however, has been questioned (CARON et al. 1982; SA E MELO et al. 1983). A more recent study (REICHERT et al. 1985) compares the effect of anthralin on DNA synthesis, cellular respiration and the activity of the hexose monophosphate shunt in cultured human keratinocytes and proves that mitochondrial function is the most sensitive cellular target for the drug. This finding goes along with earlier reports describing ultrastructural changes of mitochondria in anthralin-treated psoriatic skin (SWANBECK and LUNDQUIST 1972) which are accompanied by the inhibition of cellular respiration (RAAB 1969; RAAB and PATERMANN 1966; PÄTEL 1981; PÄTEL et al. 1981; MORLIÈRE et al. 1983, 1985). Thus, it is very likely that anthralin exerts its primary action at the mitochondrial level, affecting the cellular energy supply and therewith slowing down biosynthetic processes such as DNA replication which depend on that energy. The exact nature of the interaction of anthralin with the mitochondria is not yet known, however, it has been shown (FUCHS et al. 1986) that the compound does not act as an uncoupler of the respiration chain as has been proposed (VERHAEREN 1980; MORLIÈRE et al. 1985).

E. Pharmacokinetics and Metabolism

I. In Vitro and In Vivo Studies: Human Skin

When tritium-labelled anthralin was applied to normal human skin in vitro SCHALLA et al. (1981) found that there is a dose-related influx of radioactivity estimated at 70 pmol cm^{-2} h^{-1} from petrolatum containing 0.1% anthralin. When the stratum corneum had been removed by tape-stripping a twofold increase in flux is observed. At the higher doses of 1%, anthralin flux is increased tenfold through normal skin. Thus, the barrier function of the stratum corneum is disturbed under these conditions. In earlier in vivo studies using urinary excretion data KAMMERAU et al. (1975) confirmed the reservoir function of the stratum corneum when 0.1% anthralin, containing tritiated tracer in petrolatum was applied. In addition, since the urinary excretion occurs over a long period of time (still increasing after 56 h) a depot of drug in the skin or another organ is strongly suggested. From autoradiographic studies SELIM et al. (1981) using tritiated anthralin, which had been applied to intact human skin in vitro, suggested that after a 100-min penetration period there is an accumulation of radioactivity in the upper epidermis (90 μm depth). The technique used involved the horizontal slicing of biopsy punch samples on a freeze microtome as described by SCHAEFER et al. (1981). This was confirmed by fluorescent microscopy in frozen sections biopsied following local application of anthralin. Here a brownish discoloration in the upper epidermis was noted.

In humans the quinone has been identified in the urine of patients who have been treated topically with anthralin and, conversely, when the quinone is taken orally as a cathartic (Istizin), anthralin-like irritation is observed in the perianal region (IPPEN 1981 and references therein). Thus, it is possible that the complex cycle of redox reactions may continue after the drug is absorbed and bound to serum proteins. However, under normal conditions of treatment there is no evidence of systemic toxicity or excretion of anthraquinones in the urine (NEILL et al. 1984).

II. Animal Studies

The metabolism and distribution of anthralin in the skin of the hairless rat was studied by CAVEY et al. (1985) using a plastic disc delivery system which obviated artifacts and potential problems such as decomposition in the vehicle, increasing or decreasing the quantity liberated by washing procedures to remove excess drug. These studies were carried out with both ^{3}H and ^{14}C radiolabelled material and high pressure liquid chromatographic assays for anthralin, the dimer and the quinone were performed in parallel. The accumulation of material in the skin was confirmed, but its chemical structure is unknown. Anthralin appears to be stable in the stratum corneum and reaches plateau levels very quickly. Identifiable metabolites build up slowly with time, but the major fraction which is located in the epidermis is either polymeric or bound firmly to proteins since it could not be extracted with organic solvents. In skin which had been tape-stripped prior to application of the drug more material penetrated into the skin, more extensive de-

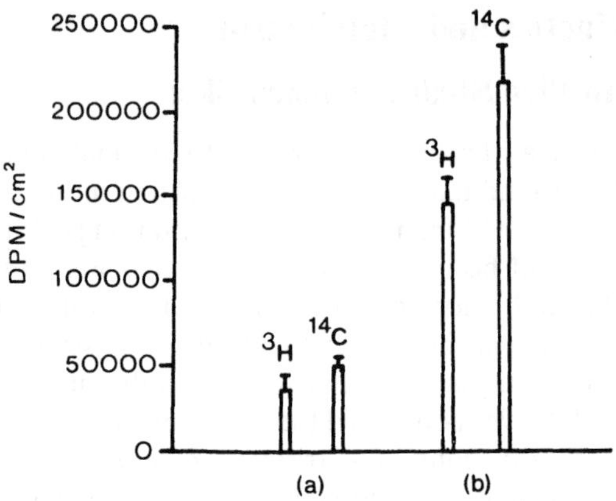

Fig. 5. Concentration expressed in radioactivity units (disintegrations per minute) after 6 h application of radiolabelled anthralin to the skin of the hairless rat: *a* intact skin; *b* prior removal of stratum corneum by tape-stripping (SHROOT 1987)

composition was evident and only traces of free anthralin could be detected, even after 15 min contact time. There was a significant difference in the levels of material recovered when ^{14}C and ^{3}H data were compared, especially after an application period of 6 h (Fig. 5).

F. Pharmacology of Anthralin Irritation

I. Quantification of the Erythematous Response

For many years it has been believed that the nature of the irritation provoked by anthralin held the key to the mode of action of the drug and its study would thus help discover more acceptable treatment regimens and also new drugs. The minimum dose required to provoke an erythematous response (MED) is 0.012% w/w chloroform (MISCH et al. 1981). The irritation is of the delayed onset type, maximizing at 72 h and subsiding between 96 and 120 h after application. There was no difference in the MED on the skin of normal subjects and uninvolved skin of psoriatics. Although the increase in skin temperature follows the time course of the irritation and the geometry of the application site, infrared thermography indicated that there is also an increased emission of infrared radiation from the area surrounding the application site (STÜTTGEN et al. 1981). LAWRENCE et al. (1984a) measured anthralin erythema by reflectance photometry and showed the advantage of using a dose-response curve rather than simple threshold (MED) measurements.

BRANDT and MUSTAKALLIO (1983) showed that when anthralin was applied daily to psoriatic patients over a 3-week period in a Finn chamber, during the first week the irritation increased, but subsequently a tachyphylaxis-like phenomenon was observed as continued application resulted in a waning of the erythema. KERSEY et al. (1981) suggested that the inflammatory response to anthralin may be partly due to a metabolite formed by aryl hydrocarbon hydroxylase (AHH) oxidation, which is induced in humans by anthralin. However, neonatal rodent epidermal AHH is not induced by anthralin (Bickers 1981). Lawrence et al. (1984 b) found that pretreatment of human skin with coal tar induces AHH activity and reduces anthralin erythema, but in further studies Lawrence and Shuster (1985 a, b) showed that both AHH activity and anthralin inflammation vary independently.

II. Mediator Studies: Indirect Methods

The effect of pharmacological antagonists and anti-inflammatory drugs on delyed erythematous response was studied in patients who received drugs either orally or by topical application. Oral aspirin, indomethacin or chlorpheniramine or topical scopolamine or betamethasone-17-valerate all failed to inhibit the erythematous response caused by the MED (Misch et al. 1981). The technique used involved the application of a measured volume of a solution of anthralin in chloroform to a glass well fixed to the skin. Kingston and Marks (1981), however, showed that prior oral administration of indomethacin can reduce the erythema caused by use-concentration of anthralin, but they only measured a threshold response. Using full dose-response curves, Lawrence and Shuster (1985 b) found that topical application of a 1% indomethacin gel had no effect on anthralin irritancy. Prior irradiation of the application site with 1–2 MED of UV-B markedly reduced the erythema produced by anthralin at 0.1% and 1% in chloroform solution (Juhlin 1981). Local application of free radical scavengers such as DL-tocopherol retinol esters and butylhydroxyanisole also inhibited the anthralin erythematous response (Finnen et al. 1984). Using KOH to inactivate anthralin in the stratum corneum Lawrence et al. (1987) showed that anthralin had to be present in the stratum corneum for 24 h to produce its full inflammatory response. Furthermore Lawrence and Shuster (1987) showed that topical arachidonic acid enhanced the inflammatory response at 6 h, reducing the subsequent response at 72 h and suggesting that cell membrane phospholipid was the normal substrate for anthralin free radical-induced inflammation.

III. Mediator Studies: Direct Methods

The levels of arachidonic acid and its metabolites are increased in the suction blister fluid of normal human volunteers who received 0.3% anthralin in yellow soft paraffin (Barr et al. 1983 a, b). Thus, since this dose is considerably higher than the MED, the results are in agreement with the observations of Kingston and Marks (1981) that the prostaglandin synthetase inhibitor indomethacin reduced anthralin-induced erythema. However, Lawrence and Shuster (1985 b) have not found anthralin-induced erythema to be inhibited by indomethacin; nor did they find it to have any effect on the augmentation of anthralin-induced erythema by topical arachidonic acid (Lawrence and Shuster 1987). Applying the same suction blister technique to the mini-pig, with either anthralin in petrolatum or plastic disc formulations, a time course of arachidonic acid release and prostaglandin production was observed which is similar to that observed in humans (Hensby et al. 1982). The capacity of anthralin to form free radicals both in solution (MARTINMAA et al. 1978) and in skin in vitro (FINNEN et al. 1984; SHROOT and BROWN 1986) may be thought to provoke lipid peroxidation in general.

G. Therapy

I. Ingram and Related Regimens

Until recently, treatment of psoriasis with anthralin was very much an art with careful attention paid to the concentration applied, the area of application, the type of psoriasis and combination therapies, all this because of the uncomfortable side effects of the drug. This large number of permutations gave rise to personalized regimes and formulations which enjoyed popularity for many years. The starting point is attributed to INGRAM (1953, 1954). In this regimen, the patient was first given a tar bath followed by exposure to UV light. A stiff Lassar's paste formulation was then carefully applied to each lesion, which was wrapped in a protective dressing after the application of talcum powder. The next morning, the paste was removed and the treatment cycle repeated. This regimen was highly successful, but could only be carried out on hospitalized patients because the formulation, in which anthralin is poorly dispersed and unstable, had to be specially prepared in the pharmacy prior to use. At high anthralin concentrations, care had to be taken to avoid contact with skin of normal appearance.

Petrolatum-based formulations are still widely used and varying concentrations of stabilizers and keratolytics such as salicylic acid and urea are incorporated. Cream formulations have been introduced which make anthralin a potential treatment for use at home (SEVILLE et al. 1979; HINDSON 1980). Treatment of the scalp is more convenient with these forms than with ointments. In general the treatment starts with a low concentration (0.05%–0.3%) which is doubled every 3–7 days up to 4%. Bathing is recommended before such application. The concentration of salicylic acid varies widely from 0.5% to 2% in petrolatum which is the common use range. The median clearance time is of the order of 14–21 days, whereas with low strength regimens, 0.03% anthralin containing 2% salicyclic acid in white petrolatum, up to 2 years treatment may be necessary to ensure clearance (BRODY 1981).

The combination of anthralin therapy with tar and UV light which was part of the Ingram regime, lost popularity because the therapeutic advantage is small (YOUNG 1970). It is even claimed to increase the tendency to relapse (SEVILLE 1975). Of potential interest is the DUVA regimen (4Hindson et al. 1983) which involves the daily application of increasing doses of anthralin in vaselin containing 2% salicylic acid. After 1 h, the ointment is washed off and the patients irradiated with 10 J cm^{-2} UV-A whose wavelength lies near the absorption maximum of anthralin. The median clearance time was 14 days in a study involving 20 patients. More extensive data are eagerly awaited since other workers do not find a clear advantage in this combination therapy (SCHAUDER and MAHRLE 1982; BRUN et al. 1984). More recently FARR et al. (1987) have shown that the disputed findings with UV radiation and anthralin depend on variation in the light source with dissociation of erythematous and therapeutic effects. In the absence of UV-C the clear beneficial effect of UV-B can be shown.

II. Short-Contact Therapy with Anthralin

DUVA therapy takes advantage of pharmacokinetic considerations discussed earlier in the chapter. Anthralin penetrates faster and in greater amounts into skin which has no stratum corneum than it does into normal skin (SCHALLA et al. 1980, 1981). This gave rise to the idea that, if the application site was washed a short time after application, the excess drug in the stratum corneum of skin of normal appearance would be decreased, thus removing the future source of irritation and staining, but the amount of drug in the lesion would still be at a sufficiently high concentration to be of therapeutic value. According to SCHAEFER et al. (1980) drug contact times of up to 1 h for 1% anthralin in petrolatum are effective in psoriasis. The reports which subsequently appeared (RUNNE and KUNZE 1982; MARSDEN et al. 1983; VERSCHOORE et al. 1983) confirmed this idea, and contact times a short as 15 min are effective. This is supported by the distribution studies in the hairless rat by CAVEY et al. (1985) which demonstrated that anthralin remains in a free form in the stratum corneum for periods of up to 1 h, after which time decomposition into insoluble substances is evident. LAWRENCE et al. (1987) showed that anthralin is normally present in the stratum corneum for 24 h, during which time it is progressively inactivated. By applying alkali at different times to destroy anthralin in the stratum corneum, they showed that with the high concentrations used in short-contact therapy, the inflammatory action required the presence of the drug for 24 h, whereas the therapeutic effect was still achieved if the anthralin was destroyed by KOH after 20–30 min. In summary, the once-daily short-contact regimen consists of the following three steps after the patient has been first titrated to establish a non-irritating initial doese:

1. Apply 0.3% or 1% anthralin in petrolatum type base for 30 min;
2. Rigorously wash out excess drug;
3. Apply an emollient to prevent drying out of the skin.

During the course of the treatment, the anthralin concentration may be increased up to 3% and the contact time up to 60 min, and with good patient compliance this effective therapy may be carried out at home. The short-contact regimen is as effective in psoriasis as other treatment schedules and is much more convenient to nurse and patient alike. The use of this regimen in day care centres, or at home and as a maintenance therapy will gain in popularity.

H. Adverse Reactions

At the clinical doses used, no evidence of systemic toxicity has been reported, even after treatments lasting for several months (FARBER and HARRIS 1970; GAY et al. 1972). DE GROOT et al. (1981) and LAWLOR and HINDSON (1982) have reported that there may be a risk of contact allergy to anthralin. There are no reports of skin cancers in humans. The staining and irritation which have been discussed at length in this chapter are reversible and dose related.

Acknowledgments. The authors would like to thank Oliver Watts, Bernard Martin, and Monika Bouiller for their help in preparing this manuscript.

References

Barr RM, Brain SD, Black AK, Camp RD, Greaves MW, Mallet AI, Wong E (1983a) Lipoxygenase products of arachidonic acid in inflamed skin. J Invest Dermatol 80:345

Barr RM, Misch KJ, Hensby C, Mallet AI, Greaves MW (1983b) Arachidonic acid and prostaglandin levels in dithranol erythema: time course. Br J Clin Pharmacol 16:715–717

Bickers D (1981) Comparative effect of anthralin and coal tar on epidermal aryl hydrocarbon hydroxylase. Br J Dermatol 105[Suppl 20]:71

Brandt H, Mustakallio KK (1983) Irritation and staining by dithranol (anthralin) and related compounds. III. Cumulative irritancy and staining during repeated chamber testing. Acta Derm Venereol (Stockh) 63:237

Brody I (1981) Treatment of psoriasis vulgaris: gentleness to the entire psoriatic skin and the use of low concentrations of anthralin. Br J Dermatol 105[Suppl 20]:109–110

Brun P, Juhlin L, Schalla W (1984) Short contact anthralin therapy of psoriasis and maintenance schedule to prevent relapses. Acta Derm Venereol (Stockh) 64:174–177

Burton FW, Rao Gadde R (1985) Analysis of anthralin in dermatological products by reverse-phase high-performance liquid chromatography. J Chromatogr 328:317–324

Caron JC; Shroot B (1981) High-pressure liquid chromatographic determination of anthralin in ointments. J Pharm Sci 70:1205–1207

Caron JC, Eustache J, Shroot B, Prota G (1982) On the interaction between anthralin and DNA: a revision. Arch Dermatol Res 274:207–214

Cavey D, Caron JC; Shroot B (1982) Anthralin: chemical instability and glucose-6-phosphate dehydrogenase inhibition. J Pharm Sci 71:980–983

Cavey D, Dickinson R, Shroot B (1985) The in vivo fate of topically applied anthralin in the skin of the hairless rat: a comparison of continuous and short contact application. Arzneimittelforschung Drug Res 35:605

Clark JM, Hanawalt PC (1982) Inhibition of DNA replication and repair by anthralin or danthron in cultured human cells. J Invest Dermatol 79:18–22

Colin M, Maignan J, Lang G, Shroot B (1981) Anthralin oxidation products: the role of C_{10} methylene and phenol groups. Br J Dermatol 105[Suppl 20]:59

Crocker HR (1888) Diseases of the skin. Lewis

De Groot AC, Nater JP, Bleumink E, de Jong MCJM (1981) Does DNCB potentiate epicutaneous sensitization to non-related contact allergens? Clin Exp Dermatol 6:139–144

Farber EM, Harris DR (1970) Hospital treatment of psoriasis. Modified anthralin program. Arch Dermatol 101:381–389

Farr PM, Diffey BL, Marks J (1987) Phototherapy and dithranol treatment of psoriasis: new lamps for old. Br Med [Clin Res] J 294:205–207

Finnen MJ, Lawrence CM, Shuster S (1984) Anthralin increases lipid peroxide formation in skin and free radical scavengers reduce anthralin irritancy. Br J Dermatol 111:717

Fischer LB, Maibach H (1975) The effect of anthralin and its derivatives on epidermal cell kinetics. J Invest Dermatol 64:338–341

Fuchs J, Zimmer G, Wölbing RH, Milbradt R (1986) On the interaction between anthralin and mitochondria: a revision. Arch Dermatol Res 279:59–65

Gay MW, Moore WJ, Morgan JM, Montes LF (1972) Anthralin toxicity. Arch Dermatol 105:213–215

Hensby C, Juhlin L, Chatelus A, Civier A, Schaefer A, Greaves MW, Black AK, Barr R, Fourtanier A (1982) Mini-pig skin as a potential alternative for predicting human skin pharmacological reactions. Int J Immunopharmacol 4:352

Hindson C (1980) Treatment of psoriasis of the scalp. An open assessment of 0.1% dithranol in a 17% urea base. Clin Trials J 17:131–136
Hindson C, Diffey B, Lawlor F, Downey A (1983) Dithranol-U.V.A. phototherapy (PUVA) for psoriasis: a treatment without dressings. Br J Dermatol 108:457–460
Ingram JT (1953) Approaches to psoriasis. Br Med J 2:591
Ingram JT (1954) Significance and management of psoriasis. Br Med J 2:823
Ippen H (1981) Basic questions on toxicology and pharmacology of anthralin. Br J Dermatol 105[Suppl 20]:72–75
Jacques Y, Reichert U (1981) Effects of anthralin and analogues on growth and ^{3}H thymidine incorporation in human skin fibroblasts. Br J Dermatol 105[Suppl 20]:45
Juhlin L (1981) Factors influencing anthralin erythema. Br J Dermatol 105[Suppl 20]:87
Kammerau B, Zesch A, Schaefer H (1975) Absolute concentrations of dithranol and triacetyl-dithranol in the skin layers after local treatment: in vivo investigations with four different types of pharmaceutical vehicles. J Invest Dermatol 64:145–149
Karsten A (1985) Bestimmung von Dithranol in Reinsubstanz und Fertigarzneimitteln durch HPLC. Pharm Zg 130:41
Kersey P, Chapman P, Rogers S, Rawlins M, Shuster S (1981) The inflammatory response to anthralin and its relation to aryl hydrocarbon hydroxylase. Br J Dermatol 105[Suppl 20]:64–67
Kingston T, Marks R (1981) Irritant reactions to dithranol in normal subjects and psoriatic patients. Br J Dermatol 106(6):725
Krebs A, Schaltegger H (1969) Untersuchungen zur Strukturspezifität der Psoriasisheilmittel Chrysarobin und Dithranol. Hautarzt 20:204–209
Lawlor F, Hindson C (1982) Allergy to dithranol. Contact Dermatitis 8:137–138
Lawrence CM, Shuster S (1985a) Mechanism of anthralin inflammation. 2. Effect of pretreatment with glucocorticoids, anthralin and removal of stratum corneum. Br J Dermatol 113:117–122
Lawrence CM, Shuster S (1985b) Mechanism of anthralin inflammation. 1. Dissociation of response to clobetasol and indomethacin. Br J Dermatol 113:107–115
Lawrence CM, Shuster S (1987) Effect of arachidonic acid on anthralin inflammation. Br J Clin Pharmacol 24:125–131
Lawrence CM, Howel D, Shuster S (1984a) The inflammatory response to anthralin. Clin Exp Dermatol 9:336–341
Lawrence CM, Finnen MJ, Shuster S (1984b) Effect of coal tar on cutaneous aryl hydrocarbon hydroxylase induction and anthralin irritancy. Br J Dermatol 110:671–675
Lawrence CM, Shuster S, Collins M, Bruce JM (1987) Reduction of anthralin inflammation by potassium hydroxide and Teepol. Br J Dermatol 116:171–177
Marsden JR, Coburn PR, Marks JP, Shuster S (1983) Response to short term application of dithranol in psoriasis. Br J Dermatol 108(2):243
Martinmaa J, Vanhal L, Mustakallio KK (1978) Free radical intermediates produced by autoxidation of 1,8-dihydroxy-9-anthrone (dithranol) in pyridine. Experientia 34:872–873
Misch K, Davies M, Greaves M, Coutts A (1981) Pharmacological studies of anthralin erythema. Br J Dermatol 105[Suppl 20]:82–86
Morlière P, Dubertret L, Sae Melo T, Salet T, Santus R (1983) Modifications of mitochondria by dithranol. A morphological and biochemical study. J Invest Dermatol 80:350
Morlière P, Dubertret L, Sa E Melo T, Salet C, Fosse M, Santus R (1985) The effects of anthralin (dithranol) on mitochondria. Br J Dermatol 112:509–515
Mustakallio KK (1979) Irritation and staining by dithranol (anthralin) related compounds. 1. Estimation with chamber testing and contact thermography. Acta Derm Venereol (Stockh) 59[Suppl 85]:125
Mustakallio KK (1980) Irritation and staining by dithranol (anthralin) and related compounds. 2. Structure-activity relationships among 10-*meso*-substituted acyl analogues. Acta Derm Venereol (Stockh) 60:169
Neill SM, Bugrein A, Coulson IH, Greaves MW (1984) Toxicologic study of anthralin in an aqueous cream formulation. Therap Clin 34:563–566
Pätel M (1981) Calorimetric screening test for dermatologically active drugs on human skin fibroblast-cultures. Thermochim Acta 49:123–129

Pätel M, Schaarschmidt B, Reichert U (1981) Calorimetric and manometric measurements on human skin fibroblasts in culture. Br J Dermatol 105[Suppl 20]:60–61
Raab W (1969) Die Wirkung externer Antipsoriatica auf die Gewebsatmung menschlicher und tierischer Haut. Ach Klin Exp Dermatol 234:44–51
Raab W, Gmeiner B (1975) Influence of ultraviolet light, various temperatures and zinc ions on anthralin (dithranol). Biochemical and chemical investigations. Dermatologica 150:267–276
Raab W, Patermann F (1966) Die Wirkung externer Antipsoriatika auf die Zellatmung. Arch Klin Exp Dermatol 226:144–152
Rassner G (1972) Enzyme der Epidermis. Spezielle Aspekte. Arch Dermatol Forsch 244:48–52
Reichert U, Jacques Y, Grangeret M, Schmidt R (1985) Antirespiratory and antiproliferative activity of anthralin in cultured human keratinocytes. J Invest Dermatol 84:130–134
Runne U, Kunze J (1982) Short duration (minutes) therapy with dithranol for psoriasis. Br J Dermatol 106:135–140
Sa E Melo MT, Dubertret L, Prognon P, Gond A, Mahuzier G, Santus R (1983) Physicochemical properties and stability of anthralin in model systems and human skin. J Invest Dermatol 80:1–6
Schaefer H, Farber EM, Goldberg L, Schalla W (1980) Limited application period for dithranol in psoriasis. Br J Dermatol 102:571–573
Schaefer H, Zesch A, Stüttgen G (1981) skin permeability. In: Marchionini A (ed) Handbuch der Haut- und Geschlechtskrankheiten, Ergänzungswerk, vol 1/4B. Springer, Berlin Heidelberg New York
Schalla W, Bauer E, Wesendahl C, Goldberg L, Farber EM, Schaefer H (1980) Penetration studies in short-term therapy with dithranol. Arch Dermatol Res 267:203
Schalla W, Bauer E, Schaefer H (1981) Skin permeability of anthralin. Br J Dermatol 105[Suppl 20]:104–108
Schaltegger A, Bloch U, Krebs A (1982) On the stability of some antipsoriatic active anthralin (dithranol) derivatives. A qualitative high performance liquid chromatography investigation. Dermatologica 165:363–368
Schauder S, Mahrle G (1982) kombinierte Einstundentherapie der Psoriasis mit Anthralin und UV-Licht. Hautarzt 33:206–209
Selim MM, Goldberg LH, Schaefer H, Bishop SC, Farber EM (1981) Penetration studies on topical anthralin. Br J Dermatol 105[Suppl 20]:101–103
Seville RH (1975) Simplified dithranol treatment for psoriasis. Br J Dermatol 93:205–208
Seville RH (1981) Advances in the use of anthralin. J Am Acad Dermatol 5:319–321
Seville RH, Walker SB, Whitefield M (1979) Dithranol cream. Br J Dermatol 100:475
Shroot B, Brown C (1986) Free radicals in skin exposed to anthralin and its derivatives. Arzneimittelforschung 36:1253–1255
Shrot B (1987) Anthraline. Ann Dermatol Venereol 114:1605–1615
Swanbeck G, Lundquist P (1972) Ultrastructural changes of mitochondria in dithranol-treated psoriatic epiderm. Acta Derm Venereol (Stockh) 52:94–98
Swanbeck G, Thyresson N (1965) Interaction between dithranol and nucleic acid. A possible mechanism for the effect of dithranol on psoriasis. Acta Derm Venereol (Stockh) 45:344–348
Verhaeren E (1980) Mitochondrial uncoupling activity as a possible base for a laxative and antipsoriatic effect. Pharmacology 20[Suppl I]:43–49
Wurster DE, Upadrashta SM (1986) Simultaneous quantitation of 1,8,9-anthracenetriol, 1,8-dihydroxy-9,10-anthraquinone and 1,8,1′,8′-tetrahydroxy-10,10′-dianthrone by reversed-phase high performance liquid chromatography. J Chromatogr 362:71–78

CHAPTER 35

The Treatment of Acne

J. R. MARSDEN and S. SHUSTER

A. Introduction

There is a direct relationship between the severity of acne and the rate of sebum excretion (CUNLIFFE and SHUSTER 1969a; BURTON and SHUSTER 1971) and increased sebum production is the primary causal factor (SHUSTER 1985). Secondary bacterial colonisation of the pilo-sebaceous follicle with mainly *Propionibacterium acnes* (MARPLES 1974) results in the characteristic inflammatory lesions. Consequently acne responds best to drugs which reduce sebum production, followed in effectivness by anti-microbials. There is little evidence for the opinion that obstruction of the pilo-sebaceous ducts is causally related to acne (see SHUSTER 1985) although several treatments are alleged to work by unblocking them. Despite its simplicity, failure of treatment is common; and this is as often due to inadequate explanation by the doctor as to misunderstanding by the patient. The slow onset of action of most acne treatments and the persistence with which they need to be used (LANCET 1982) needs emphasising from the outset. There is no evidence that dietary, sexual, hygienic or cosmetic practices affect the severity of acne.

Because of toxicity the most potent sebostatic drugs are used only in severe acne, whereas anti-microbials are used in all grades of severity. There is a major placebo effect in acne (FRY and RAMSAY 1966; POCHI and STRAUSS 1973; CHRISTIANSEN et al. 1974) which probably explains the alleged effects of many available treatments.

B. Sebostatic Drugs

Although sebum production in humans is mainly due to endocrine modulation (SHUSTER and THODY 1974; THODY and SHUSTER 1989 and Chap. 14 Volume I) because many hormones are involved complete control of acne by endocrine mechanisms is not feasible. The alternative of non-endocrine inhibition of sebaceous cells is proving a more satisfactory approach.

I. Endocrine Inhibitors

1. Anti-androgens

Although it is clear that many hormones affect sebaceous gland function, the largest contribution is that of androgens (SHUSTER 1982). Hence anti-androgens are the most effective endocrine sebaceous inhibitors and they work either by direct competition for the cytosolic androgen receptor (e.g. cyproterone acetate, cimetidine, flutamide) or by inhibiting the intracellular interconversion of androgens. Although the conversion of testosterone to dihydrotestosterone by 5α-reductase is generally believed to be critical for the sebaceous response to androgen (see Chaps. 14, 15 Volume I) dihydrotestosterone is only a weak androgen agonist in the human sebaceous gland (COOPER et al. 1979) where it role is dubious (SHUSTER 1982). Thus, the reason patients with 5α-reductase deficiency do not develop acne (PETERSON et al. 1977) is likely to be a developmental effect of dihydrotestosterone in early life and not a modulatory effect on sebum production in the adult (SHUSTER 1982).

Cyproterone acetate (CPA) is the most potent anti-androgen and reduces sebum excretion rate (SER) by a maximum of 65–70% by 4 weeks of treatment with 100 mg/day (BURTON et al. 1973) with commensurate improvement in acne apparent by 8 weeks (MARSDEN et al. 1984a). The response is dose related with a negligible effect with doses less than 10 mg/day. Thus, it is probable that any therapeutic action of commercially available low dose CPA preparations is due to their oestrogen content. Its action is to reduce sebaceous lipogenesis with little qualitative change in lipid composition (MARSDEN et al. 1984a). The unwanted effects of CPA include reduction in libido and potency (which restricts its use in males to short periods only) and breast tenderness, weight gain and mild adrenal suppression (CHAPMAN 1982). In females it is combined with ethinyloestradiol both to prevent conception, as CPA feminises male foetuses, and to reduce its progestogenic effects. A suitable regimen CPA 50–100 mg/day from day 5 to day 14 of the menstrual cycle, and 50 μg ethinyloestradiol from day 5 to day 25 (HAMMERSTEIN and CUPCEANCU 1969). The long-term side effects of this treatment are unclear and it is progestogenic (LANCET 1983). Other anti-androgens, e.g. cimetidine (LYONS et al. 1979), spironolactone (KEAHEY et al. 1983) or 17α-methylnortestosterone (STRAUSS and POCHI 1970) have only a small effect and topical anti-androgens which could minimise the unwanted effects of systemic treatment are surprisingly ineffective (STRAUSS and POCHI 1970; SIMPSON et al. 1979; LYONS and SHUSTER 1982). In the case of topical CPA this is presumably because its action is due to a metabolite (LYONS and SHUSTER 1983).

2. Oestrogens and Oral Contraceptives

Ethinyloestradiol 250 μg given cyclically reduces SER by about 40% (POCHI and STRAUSS 1973), and has a modest therapeutic effect; 50 μg, the dose used in several oral contraceptive pills, would be expected to reduce SER by about 20%, with little therapeutic effect. However, the greater the dose or potency of progestogen in the contraceptive pill, the smaller the decrease in SER and the most suitable combination in this respect if 50 μg ethinyloestradiol and 1 mg norethisterone acetate (PYE et al. 1977).

3. Corticosteroids

Prednisone can reduce SER in females because of suppression of adrenal androgen production, but the effect is too small to be useful on its own. Prednisone combined with an oral contraceptive (SAIHAN and BURTON 1980) seems to have no advantage over the CPA-ethinyloestradiol combination. Topical preparations of corticosteroids with either neomycin or chloramphenicol are available, but there is little objective evidence of efficacy; furthermore, corticosteroids produce an acne-like perioral dermatitis. Injection of small amounts of triamcinolone into inflamed acne cysts produces rapid improvement.

II. Inhibitors of Sebaceous Lipogenesis

Drugs which directly interfere with the metabolic pathways involved in sebaceous lipid production would be expected to reduce SER. The unsaturated fatty acid eicosatetraynoic acid inhibits cholesterol synthesis; systemic treatment reduced SER by over 50% within a few weeks (STRAUSS et al. 1967) and improved acne. It has a small effect topically (BURTON and SHUSTER 1970). Despite these promising findings this compound or its analogues have not been developed (CUNLIFFE and SHUSTER 1969b). There are not other therapeutically useful inhibitors of lipogenesis although few further studies have been done.

III. Direct Action on Sebaceous Cell: Isotretinoin

This synthetic derivative of vitamin A was found by chance to have a beneficial effect in acne. It causes profound atrophy of sebaceous cells within 3 weeks of systemic treatment (LANDTHALER et al. 1980) with major inhibitory changes in sebaceous lipogenesis (MARSDEN et al. 1984a). The decrease in SER is dose related (CUNLIFFE et al. 1985) with maximum inhibition of about 90% by 4 weeks with doses above 0.8 mg kg^{-1} day^{-1} (FARRELL et al. 1980; MARSDEN et al. 1984a) and a decrease of about 45% with 0.05 mg kg^{-1} day^{-1} (MARSDEN et al. 1984b). Treatment is given for 12–16 weeks and minor side effects such as cheilitis and facial dermatitis are very common (GOLDSTEIN et al. 1982). Dose-related increases in serum cholesterol and triglyceride and decreases in high density lipoprotein cholesterol occur during treatment (MARSDEN et al. 1983) and there are small increases in indices of liver function and decreases in circulating thyroid hormone (MARSDEN et al. 1984c). These changes revert to normal within 4 weeks of stopping the drug. As isotretinoin is teratogenic (ROSA 1983) an effective contraceptive method must be used until at least 1 month after treatment has stopped. As the unwanted effects of isotretinoin are not separable from the therapeutic effect, and as their risk is not yet clear, use of the drug is usually limited to severe and moderately severe acne. Hence, the development of an effective topical preparation or a systemic analogue with a better therapeutic ratio is required.

C. Anti-microbials

A variety of anti-microbials are effective by virtue of their action on *P. acnes* which is the predominant pilo-sebaceous organism (MARPLES 1974). None reduce SER (FRY and RAMSAY 1966; EADY et al. 1982a). Numbers of *P. acnes* do not correlate with SER or acne severity (COVE et al. 1980a, b). This lack of correlation may be technical (SHUSTER 1985) and the effect of anti-microbials may be to impair bacterial function more than bacterial numbers. There is probably little difference in the efficacy of the various agents used and both inflamed and non-inflamed lesions improve.

I. Tetracycline

Although this has been the mainstay of treatment for moderate and severe acne for over 30 years, the number of trials which adequately assess its effect are small (AKERS et al. 1975). Its advantages are that it is safe (AKERS et al. 1975; SAUER 1976), cheap and can be taken just once daily. Its disadvantages are that for adequate absorption it *must* be taken on an empty stomach with 100 ml or so of water 30 min before food, and like other antibiotics in acne, the onset of its effect is slow. For convenience the best method is to take a single dose first thing in the morning. Treatment must be continued for 4–6 months at least, although the initial dose of 1.0–1.5 g/day can usually then be reduced. Dose-response data are not available for tetracycline (or other antibiotics) in acne, but severe disease may require up to 2 g/day (BAER et al. 1976). It is likely that combined tratment with topical benzoyl peroixde produces a greater effect than tetracycline alone. Failure to respond is usually because of incorrect use, inadequate dosage, or duration of treatment (MARSDEN 1985). Other forms of tetracycline have little if any therapeutic advantage and are much more expensive (HUBBELL et al. 1982). Unwanted effects from tetracycline are uncommon (AKERS et al. 1975) considering the extent of its use; they include dysphagia and oesophagitis, nausea, diarrhoea and candidiasis; benign intracranial hypertension is rare. Tetracyline should be avoided in children and pregnant women because it permanently discolours developing teeth. Gram-negative folliculitis can occur after long-term treatment with tetracycline and other antibiotics: it consists of pustules of inflamed nodules and may respond to ampicillin or co-trimoxazole. Resistance of *P. acnes* to tetracyclines is uncommon, but is probably increasing (LEYDEN et al. 1983) as is resistnce of other organisms (CHOPRA et al. 1981). Topical formulations of tetracycline have only a small effect on acne (EADY et al. 1982b; MILLS and KLIGMAN 1983) and have the unwanted effects of staining the skin yellow and making it fluoresce.

II. Erythromycin

This antibiotic is effective in acne in doses of 0.5–1.5 g/day given for several months (BLEEKER et al. 1981), but because its use is associated with increased bacterial resistance (LILLY and LOWBURY 1978) it should only be used in the few patients who do not respond adequately to tetracycline.

III. Co-trimoxazole

Systemic treatment, usually 960 mg/day, is effective (NORDIN et al. 1978). The frequency of mild unwanted effects is similar to tetracycline, but more severe adverse reactions can occur, e.g. severe erythema multiforme. Resistance to co-trimoxazole is increasing and its use in acne cannot now be recommended.

IV. Clindamycin

Because of the occasional development of pseudomembranous colitis due to *Clostridium difficile*, the drug is now rarely used systemically in acne although it is effective (AKERS et al. 1975). The emergence of resistant organisms is also a problem (MARPLES and KLIGMAN 1971). When it is used topically as a 1% solution the response is similar to that obtained with tetracycline 0.5 g/day (GRATTON et al. 1982), but increased resistance amongst staphylococci and *P. acnes* is likely (EADY et al. 1982a). Small amounts of clindamycin are absorbed during topical use (BARZA et al. 1982), but nonetheless colitis is rare.

V. Systemic versus Topical Antibiotics

Adequate studies comparing the efficacy of systemic and topical antibiotics are not available; nor is there sufficient comparison of topical antibiotics with other topical treatments although it is clear that topical erythromycin is no more effective than 5% benzoyl peroxide lotion (BURKE et al. 1983). Analogous to the situation following the increased topical use of gentamicin (LANCET 1981), it is likely that bacterial resistance will result from the widespread topical use of antibiotics in acne. Until the risks and benefits of topical antibiotics in acne are clear, their use is difficult to justify.

VI. Benzoyl Peroxide

This compound has been used as a skin antiseptic for over 40 years and is a popular topical treatment for acne. It is a powerful oxidising agent and has a rapid and potent broad-spectrum anti-bacterial effect (KLIGMAN et al. 1977). *P. acnes* numbers are reduced to a similar extent after 1 month of treatment with either 2.5% or 10% concentrations (LEYDEN et al. 1980). The drug is not sebostatic (CUNLIFFE et al. 1983). The preparation (lotion or gel are equally effective; CUNLIFFE and HOLLAND 1981) should be applied to *all* of the affected skin, left on initially for 30–60 min and then washed off. The duration of application is increased to 8–10 h over 10–20 days according to tolerance and this approach minimises the unwanted soreness and erythema which is otherwise a frequent reason for stopping treatment. There is no dose-response data, but a concentration of 5% is effective and well tolerated. Although some improvement may be apparent by 7–14 days (SCHUTTE et al. 1982), treatment must be continued for at least 2–3 months (CUNLIFFE and HOLLAND 1981) to obtain the maximum response and thereafter to maintain it. In mild acne benzoyl peroxide is effective alone, but is combined with antibiotics in more severe grades. Although irritant dermatitis is common, allergic contact dermatitis is infrequent, affecting at most 1%–2% of patients (CUNLIFFE et al. 1981). Although systemic absorption and elimination occur after topical application of benzoyl peroxide (NACHT et al. 1981) no adverse effects from this have been reported. It is important to warn patients that the drug can bleach clothing.

VII. Other Anti-microbials

There is no good evidence that other topical antiseptics such as chlorhexidine or povidone-iodine have a therapeutic effect in acne, although they are commonly used as detergent lotions or in medicated soaps. Preparations containing and abrasive are no more effective than the vehicle alone (FULGHUM et al. 1982).

D. Miscellaneous Treatments

I. Tretinoin (Retinoic Acid)

Although much used, evidence of its effect is conflicting. Unlike isotretinoin, it appears not to be sebostatic (KLIGMAN et al. 1975) and does not reduce bacterial numbers. It is claimed to reduce numbers of comedones (KLIGMAN et al. 1975), but in some studies the results are little different from placebo (NORDIN et al. 1981) and there is similar disagreement about the respone of inflamed lesions (CHRISTIANSEN et al. 1977; NORDIN et al. 1981). Irritant dermatitis is very common and compliance with treatment is often poor. Its place intreatment has not been established.

II. Ultraviolet Radiation

Some patients improve considerably after exposure to natural sunlight, but some are made worse (MILLS and KLIGMAN 1978). Although exposure to artifical UV is widely used in the treatment of acne there have been few studies of effect or mode of action. It is often used with topical benzoyl peroxide and systemic tetracycline for mild and moderately severe acne.

III. Superficial X-ray Therapy

The dose of X-radiation required to reduce the size of sebaceous glands and SER may carry a risk of radiation damage (EPSTEIN 1978) and this treatment is no longer used. Less penetrating radiation (Grenz rays) has no effect on sebaceous glands or acne (DE GROOT et al. 1963).

E. Conclusion

The treatment of acne follows locially from an understanding of its cause. However, until sebostatic drugs are available which are effective topically or are less toxic systemically, sebostatic drugs such as isotretinoin and CPA are used only for severe and moderately severe acne. For the remainder, treatment is anti-microbial usually with systemic tetracycline and topical benzoyl peroxide and is reasonably if not totally effective. The large number of other preparations available have no proven value and are therapeutically irrelevant.

References

Akers WA et al. (1975) Review article: systemic antibiotics for treatment of acne vulgaris. Arch Dermatol 111:1630–1636
Baer RL, Leshan SM, Shalita AR (1976) High dose tetracycline therapy in severe acne. Arch Dermatol 112:479–481
Barza M, Goldstein JA, Kane A et al. (1982) Systemic absorption of clindamycin hydrochloride after topical application. J Am Acad Dermatol 7:208–214
Bleeker J, Hellgren L, Vincent J (1981) Effect of systemic erythromycin stearate on inflammatory lesions and skin surface fatty acids in acne vulgaris. Dermatologica 162:342–349
Burke B, Eady EA, Cunliffe WJ (1983) Benzoyl peroxide versus topical erythromycin in the treatment of acne vulgaris. Br J Dermatol 108:199–204
Burton JL, Shuster S (1970) Topical tetraynoic acid and sebum excretion. Br J Dermatol 82:626–627
Burton JL, Shuster S (1971) The relationship between seborrhoea and acne vulgaris. Br J Dermatol 54:600–601
Burton JL, Laschet U, Shuster S (1973) Reduction of sebum excretion in man by the anti-androgen cyproterone acetate. Br J Dermatol 89:487–490
Chapman M (1982) Side effects of anti-androgen therapy. In: Jeffcoate SL (ed) Androgens and anti-androgen therapy. Wiley Chichester, pp 169–178
Chopra I, Howe TG, Linton AH et al. (1981) The tetracyclines: prospects at the beginning of the 1980s. J Antimicrob Chemother 8:5–21
Christiansen JV, Gadborg E, Ludvigsen K (1974) Topical tretinoin, vitamin A acid (Airol) in acne vulgaris. Dermatologica 148:82–89
Christiansen JV, Holm P, Reymann F (1977) The retinoic acid derivative Roll-1430 in acne vulgaris. Dermatologica 154:219–227
Cooper MF, McGibbon D, Wilson PD, Shuster S (1979) Androgenic control of the human sebaceous gland. J Invest Dermatol 72:267
Cove JH, Cunliffe WJ, Holland KT (1980a) Acne vulgaris: is the bacterial population size significant? Br J Dermatol 102:277–280
Cove JH, Holland KT, Cunliffe WJ (1980b) An analysis of sebum excretion rate, bacterial population and the production rate of free fatty acids in human skin. Br J Dermatol 103:383–386
Cunliffe WJ, Holland KT (1981) The effect of benzoyl peroxide on acne. Acta Derm Venereol (Stockh) 61:267–269
Cunliffe WJ, Shuster S (1969a) Pathogenesis of acne. Lancet 1:685–687
Cunliffe WJ, Shuster S (1969b) The effect of inhibitors of cholesterol synthesis on sebum secretion in patients with acne. Br J Dermatol 81:280–282
Cunliffe WJ, Clayden AD, Gould D, Simpson NB (1981) Acne vulgaris – its aetiology and tratment. Clin Exp Dermatol 6:461–469
Cunliffe WJ, Stanton C, Forster RA (1983) Topical benzoyl peroxide increases the sebum excretion rate in patients with acne. Br J Dermatol 109:577–579
Cunliffe WJ, Jones DH, Pritlove J, Parkin E (1985) Long term benefits of isotretinoin in acne. In: Saurat J (ed) Retinoids: new trends in research therapy. Karger, Basel, pp 242–251
De Groot WP, Verbeck AM, Woerdman MJ (1963) Value of Grenz ray therapy for acne vulgaris. Dermatologica 126:319–325
Eady EA, Holland KT, Cunliffe WJ (1982a) The use of antibiotics in acne therapy: oral or topical administration. J Antimicrob Chemother 10:89–115
Eady EA, Holland KT, Cunliffe WJ (1982b) Should topical antibiotics be used for the treatment of acne vulgaris? Br J Dermatol 107:235–246
Epstein E (1978) Thyroid neoplasm after radiation therapy for acne. Arch Dermatol 114:53–55
Farrell LN, Strauss JS, Stranieri AM (1980) The treatment of severe cystic acne with 13-*cis*-retinoic acid. J Am Acad Dermatol 3:602–611
Fry L, Ramsay CA (1966) Tetracycline in acne vulgaris. Br J Dermatol 78:653–660

Fulghum BD, Catalano PM; Childers RC (1982) Abrasive cleansing in the management of acne vulgaris. Arch Dermatol 118:658–659
Goldstein JA, Socha-Szott A, Thomsen RJ et al. (1982) Comparative effect of isotretinoin and etretinate on acne and sebaceous gland secretion. J Am Acad Dermatol 6:760–765
Gratton D, Raymond GP, Guertin-Larochelle S et al. (1982) Topical clindamycin versus systemic tetracycline in the treatment of acne. J Am Acad Dermatol 7:50–53
Hammerstein J, Cupceancu B (1969) Behandlung des Hirsutisums mit cyproteronacetat. Dtsch Med Wochenschr 94:829–834
Hubbell CG, Hobbs ER, Rist T, White JW (1982) Efficacy of minocycline compared with tetracycline in treatment of acne vulgaris. Arch Dermatol 118:989–992
Keahey TM, Martinez N, Blasco L et al. (1983) Suppression of sebum excretion following treatment of acne and hirsutism with spironolactone. J Invest Dermatol 80:359
Kligman AM, Mills OH, McGinley KJ (1975) Acne therapy with tretinoin in combination with antibiotics. Acta Derm Venerol 74[Suppl] (Stockh) 111–115
Kligman AM, Leyden JJ, Stewart R (1977) New uses for benzoyl peroxide: a broad spectrum antimicrobial agent. Int J Dermatol 16:413–417
Lancet (1981) Leading article. Of gentamicin and staphylococci. 2:127–128
Lancet (1982) Leading article. Topical dilemmas in acne treatment. 2:1138–1139
Lancet (1983) Leading article. Oral contraceptives and neoplasia. 2:947
Landthaler M, Kummermehr J, Wagner A et al. (1980) Inhibitory effects of 13-*cis*-retinoic acid on human sebaceous glands. Arch Dermatol Res 269:297–309
Leyden JJ, McGinley K, Mills OH et al. (1980) Topical antibiotics and topical antimicrobial agents in acne therapy. In: Juhlin L, Rorsman H, Strauss JS (eds) Acne: dermatology symposia in Lund. Upplands, Uppsala
Leyden JJ, McGinley KJ, Cavalieri S et al. (1983) *Propionibacterium acnes* resistance to antibiotics in acne patients. J Am Acad Dermatol 8:41–45
Lilly HA, Lowbury EJL (1978) Antibiotic resistance of *Staphylococcus aureus* in burns unit after stopping routine prophylaxis with erythromycin. J Antimicrob Chemother 4:545–550
Lyons F, Shuster S (1982) Sex difference in response of the human sebaceous gland to topical flutamide. Br J Dermatol 107:697–699
Lyons F, Shuster S (1983) Indirect evidence that the action of cyproterone acetate is due to a metabolite. Clin Endocrinol (Oxf) 19:53–55
Lyons F, Cook LJ, Shuster S (1979) Inhibition of sebum excretion by an H1 blocker. Lancet 1:1376–1377
Marples RR (1974) The microflora of the face and acne lesions. J Invest Dermatol 62:326–331
Marples RR, Kligman AM (1971) Ecological effects of oral antibiotics on the microflora of human skin. Arch Dermatol 103:148–153
Marsden JR (1985) Evidence that method of use, dose and duration of treatment with benzoyl peroxide and tetracyline determines response of acne. J R Soc Med 78[Suppl 10]:25–28
Marsden JR, Laker MF, Shuster S (1983) Isotretinoin and serum lipids. Lancet 1:134
Marsden JR, Shuster S, Lyons F (1984 a) Effect of 13-*cis*-retinoic acid and cyproterone acetate on acne severity, sebum excretion rate, dermal and epidermal lipogenesis, serum lipids and liver function tests. In: Retinoid therapy – a review of clinical and laboratory research. Cunliffe WJ, Miller AJ (eds) MTP, Boston, pp 267–276
Marsden JR, Laker MF, Ford GP, Shuster S (1984 b) Effect of cyproterone acetate on the response of acne to isotretinoin. Br J Dermatol 110:697–702
Marsden JR, Trinnick T, Laker MF, Shuster S (1984 c) Effects of isotretinoin on serum lipids and lipoproteins, liver and thyroid function. Clin Chim Acta 143:243–251
Mills OH, Kligman AM (1978) Ultraviolet phototherapy and photochemotherapy of acne vulgaris. Arch Dermatol 114:221–223
Mills OH, Kligman AM (1983) Drugs that are ineffective in the treatment of acne vulgaris. Br J Dermatol 108:371–374
Nacht S, Yeung D, Beasley JN et al. (1981) Benzoyl peroxide: percutaneous penetration and metabolic disposition. J Am Acad Dermatol 4:31–34

Nordin K, Hallander H, Fredriksson T et al. (1978) A clinical and bacteriological evaluation of the effect of sulphamethoxazole trimethoprim in acne vulgaris resistant to prior therapy with tetracyclines. Dermatologica 157:245–253
Nordin K, Fredriksson T, Rylander C (1981) Ro-11-1430 a new retinoic acid derivative for the topical treatment of acne. Dermatologica 162:104–111
Peterson RE, Imperato-McGuinley J, Gautier T, Sturla E (1977) Male pseudohermaphroditism due to steroid 5-α-reductase deficiency. Am J Med 62:170–190
Pochi PE, Strauss JS (1973) Sebaceous gland suppression with ethinyloestradiol and diethylstilboestrol. Arch Dermatol 108:210–214
Pye RJ, Meyrick G, Pye MJ, Burton JL (1977) Effect of oral contraceptives on sebum excretion rate. Br Med J 2:1581–1582
Rosa FW (1983) Teratogenicity of isotretinoin. Lancet 2:513
Saihan EM, Burton JL (1980) Sebaecous gland suppression in female acne patients by combined glucocorticoid-oestrogen therapy. Br J Dermatol 103:139–142
Sauer G (1976) Safety of long-term tetracycline therapy for acne. Arch Dermatol 112:1603–1605
Schutte H, Cunliffe WJ, Forster RA (1982) The short term effects of benzoyl peroxide lotion on the resolution of inflamed acne lesions. Br J Dermatol 106:91–94
Shuster S (1982) The sebaceous glands and primary cutaneous virilism. In: Jeffcoate SL (ed) Androgens and anti-androgen therapy. Wiley, Chichester, pp 1–21
Shuster S (1985) Acne: the ashes of a burnt out controversy. Acta Derm Venereol [Suppl] (Stockh) 120:43–46
Shuster S, Thody AJ (1974) The control and measurement of sebum secretion. J Invest Dermatol 62:172–190
Simpson NB, Bowden PE, Forster RA, Cunliffe WJ (1979) The effect of topically applied progesterone on sebum excretion rate. Br J Dermatol 100:687
Strauss JS, Pochi PE (1970) Assay of anti-androgens in man by the sebaceous gland response. Br J Dermatol 6[Suppl]:33–42
Strauss JS, Pochi PE, Whitman EN (1967) Suppression of sebaceous gland activity with eicosa-5:8:11:14 – tetraynoic acid. J Invest Dermatol 48:492–493
Thody AJ, Shuster S (1989) Control and function of sebaceous glands. Physiol Rev 69:383–416

CHAPTER 36

Pharmacology of Anti-androgens in the Skin

D. S. THOMSON

A. Introduction

Anti-androgens are compounds which prevent androgens from expressing their activity at target sites (DORFMAN 1970), and are therefore considered of potential therapeutic value in diseases where androgenic stimulation appears to play a pathogenic role. Thus, dermatological conditions involving hair follicles and sebaceous glands, such as androgenic alopecia, idiopathic hirsutism, acne and seborrhoea, where androgen involvement is recognised, should benefit from such therapy. Clinical experience in recent years, using systemic anti-androgens to treat females with these conditions, has been encouraging. In males, however, anti-androgen treatment has been restricted to male hypersexuality and tumours of the prostate (NEUMANN and JACOBI 1982) for obvious toxicological reasons.

B. Mechanism of Action of Androgens

It is generally believed that the mechanism of action of androgens in skin appendages (sebaceous glands, hair follicles) is similar to that of other androgen-sensitive tissues such as the prostate, seminal vesicle and epididymis (PRICE 1975; TAKAYASU 1979). In these tissues the response to testosterone is dependent on its conversion to dihydrotestosterone (DHT) by the enzyme 5α-reductase (BRUCHOVSKY and WILSON 1968; EWING et al. 1975; FARNWORTH and BROWN 1963). Both male and female skin is also capable of forming DHT from testosterone (HAY 1977; WILSON and WALKER 1969; VOIGT et al. 1970), as are human sebaceous glands (TAKAYASU et al. 1980). It may not, however, be the only active metabolite. Androgen metabolism in skin is complex and androstenedione and 5α-androstanediols have also been identified and may be important (MAUVAIS-JARVIS 1977). In addition the skin is capable of metabolising dehydroepiandrosterone (DHA) to testosterone, DHT and 3α-androstanediol (THOMAS and OAKE 1974), suggesting that testosterone may not be the principal intracellular androgen.

Current concepts of testosterone metabolism in skin suggest that DHT plays a major role in the pathophysiology of androgenic alopecia, acne and idiopathic hirsutism (PRICE 1975). Increased DHT formation has been found in acne-bearing skin (SANSONE and REISNER 1971) and in women with idiopathic hirsutism (JENKINS 1973; THOMAS and OAKE 1974). The most convincing evidence, however, has come from the study of male pseudohermaphroditism (type 2) (IMPERATO-MCGINLEY et al. 1974; PETERSON et al. 1977; WALSH et al. 1974). These patients have a deficiency of the enzyme 5α-reductase (SAENGER et al. 1978) which prevents the conversion of testosterone to DHT. At birth the defect causes an incomplete differentiation of the male external genitalia. At puberty the voice deepens, there is normal muscle development and psychosexual orientation is unequivocally male. The patients do not, however, develop normal beard growth, nor do they develop temporal recession of the hairline or acne. These observations demonstrate the selective roles of testosterone and DHT during both embryogenesis and puberty, and in particular the role of DHT in the development of acne and androgenic alopecia.

Nevertheless, the importance of 5α-reductase activity has been questioned in acne (COOPER et al. 1978) in that enzyme activity does not appear to correlate with neutral lipid synthesis. A correlation, however, does exist between 3 β-hydroxysteroid dehydrogenase activity and lipid synthesis, suggesting a more important role for DHA than testosterone and DHT in the maintenance of seborrhoea (COOPER et al. 1979). It therefore still remains to be determined which metabolic process constitutes the critical event in the activation of the cell, leading subsequently to the stimulation of sebum synthesis. The view that 5-hydroxytryptamine and related 5α-reductase activity of human skin have little part to play in other than sexual skin is discussed by SHUSTER (1982) and in Chap. 4.

The specific androgen metabolite formed in the target cell cytoplasm binds to specific cytosol receptors and is translocated into the nucleus, where the androgen-receptor complex binds to an acceptor that is mainly chromatin (LIAO et al. 1972). This nuclear binding initiates the subsequent transcription or translational events in the cell (MAINWARING and JONES 1975).

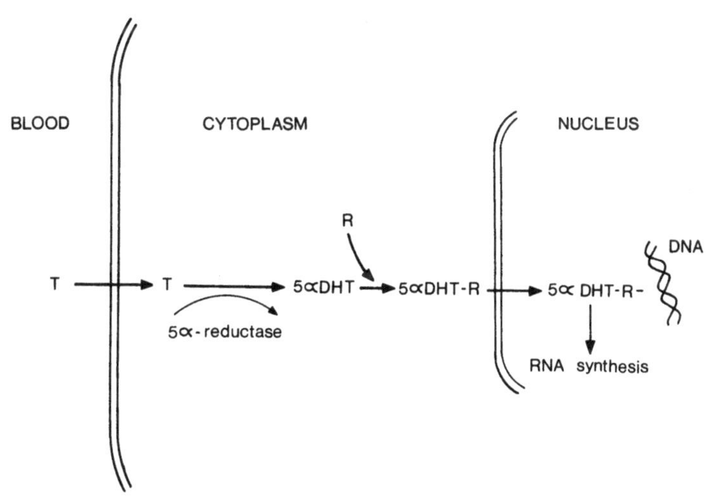

Fig. 1. The interaction of testosterone with a dihydrotestosterone-dependent androgen target cell. (Abbreviations: *T* testosterone; *5αDHT* dihydrotestosterone; *R* cytosol receptors; *5αDHT-R* dihydrotestosterone-receptor complex)

C. Anti-androgens: Mode of Action and Chemistry

Anti-androgens, by definition, prevent androgen expression at target sites. The inhibitory effect of these substances should, therefore, be differentiated from compounds which have central hypothalamic action or act directly on the gonads to inhibit androgen secretion (NEUMANN and JACOBI 1982). On the basis of the mechanisms of cellular action of androgens, two major processes are considered for anti-androgen attack: (a) the inhibition of 5α-reductase, thereby preventing the conversion of testosterone to DHT; and (b) competitive inhibition of the binding of androgen to the cytosol receptor. It has been suggested that 5α-reductase inhibitors should not be classified as anti-androgens (NEUMANN and SCHLEUSENER 1981). However, on biological and clinical grounds, 5α-reductase inhibitors are still regarded as having anti-androgenic activity (BURTON 1979; PRICE 1975; STEWART and POCHI 1978), and shall therefore be discussed as such.

I. Inhibitors of 5α-Reductase

The structural features required for steroidal 5α-reductase inhibitors are a 3-keto-Δ^4 structure in the A ring and a 17β-substituent (Fig. 2) (VOIGT and HSIA 1973a). Naturally occurring steroids having a structural resemblance to testosterone such as corticosterone, androstenedione and progesterone, therefore, are effective inhibitors of 5α-reductase, by competing with testosterone for the binding site on the enzyme. Androstenedione and corticosterone, however, are unsuitable for use because of their adverse effects. Progesterone is one of the most potent inhibitors of 5α-reductase (HSIA and VOIGT 1974; VOIGT et al. 1970; VOIGT and HSIA 1973a). In male pubic skin in vivo, progesterone produces a $>70\%$ inhibition of 5α-reductase activity in skin biopsies taken from the treated area (MAUVAIS-JARVIS et al. 1974).

An alternative, hormonally inactive compound, 4-androsten-3-one 17β-carboxylic acid (Fig. 2), is also a potent inhibitor of testosterone 5α-reduction, as is its methyl ester (HSIA and VOIGT 1974; VOIGT et al. 1970; VOIGT and HSIA 1973a, b). When they are applied topically to hamster costovertebral organs (see Sect. D) along with testosterone, the androgenic effect on the gland is inhibited. These compounds should in theory have considerable potential as safe topical anti-androgens.

II. Cytosol Receptor Blockers

The anti-androgen which has sustained clinical interest in recent years is cyproterone acetate (HAMADA et al. 1963). It was originally developed as a progestogen, but in animal experiments was found to prevent the masculinisation of male foetuses. Cyproterone acetate is therefore a powerful anti-androgen which is also a potent progestogen. It is a hydroxyprogesterone derivative (Fig. 2) and other steroids with a close chemical similarity have also been developed (e.g. chlormadinone acetate). These compounds act by competitive inhibition of the binding of androgen to the cytosol receptor (ADACHI and KANO 1972; BAULIEU and JUNG 1970, FANG et al. 1969; FANG and LIAO 1969). It has been suggested that receptor

5α-reductase inhibitors

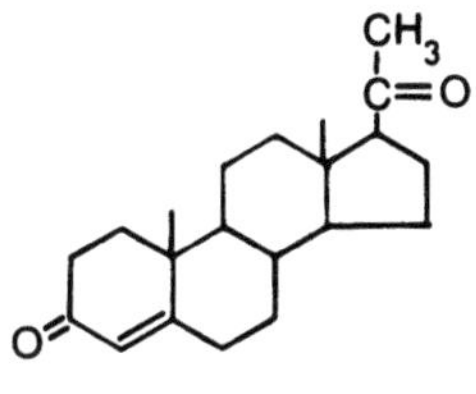

PROGESTERONE

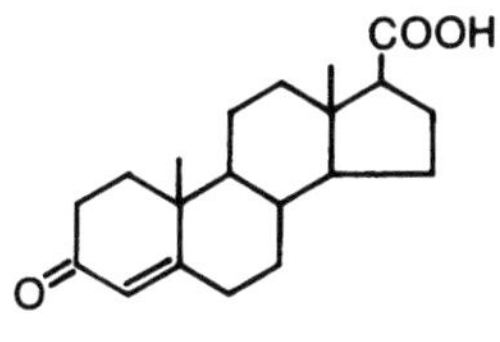

4-ANDROSTEN-3-ONE-17β-CARBOXYLIC ACID

Cytosol receptor blockers

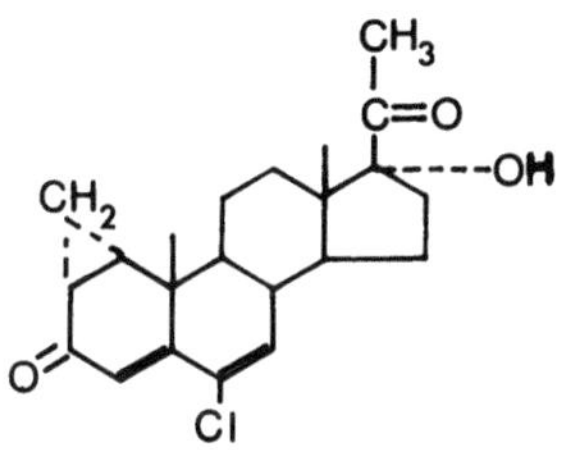

CYPROTERONE

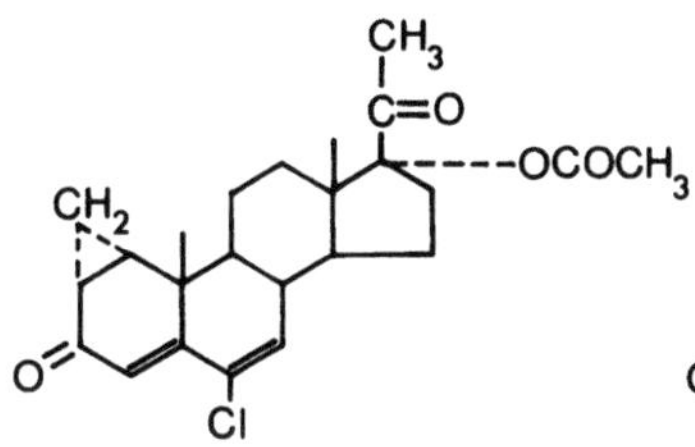

CYPROTERONE ACETATE

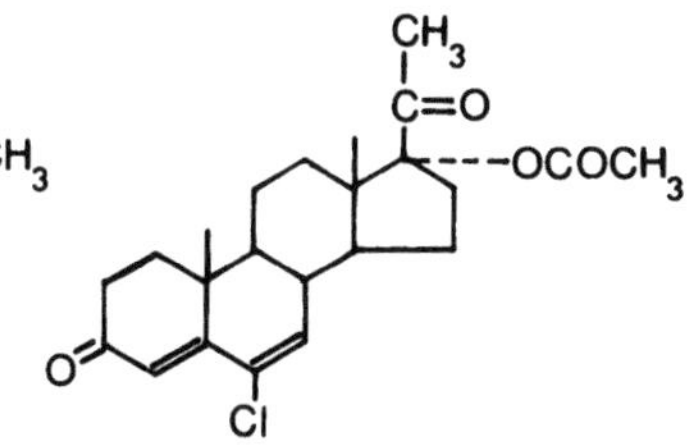

CHLORMADINONE ACETATE

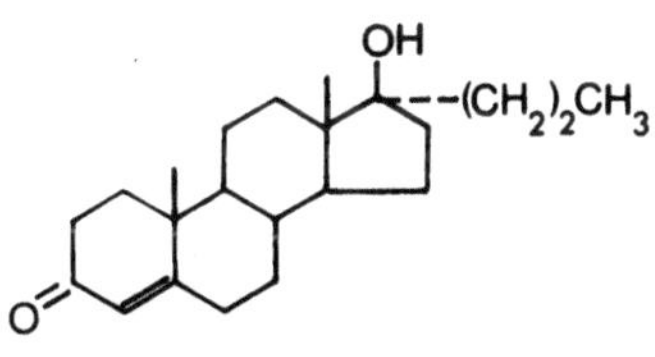

17α PROPYLTESTOSTERONE

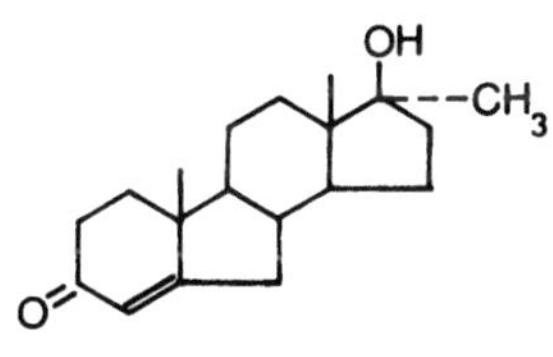

17α-METHYL-B-NORTESTOSTERONE

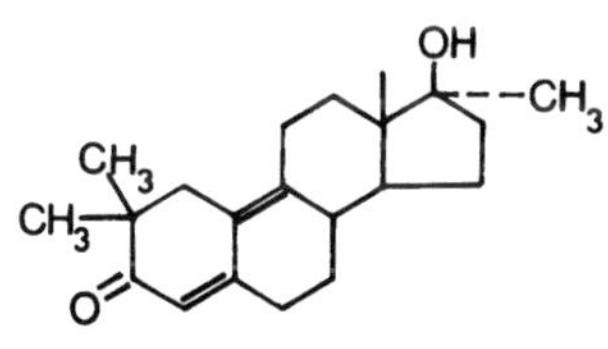

R 2956

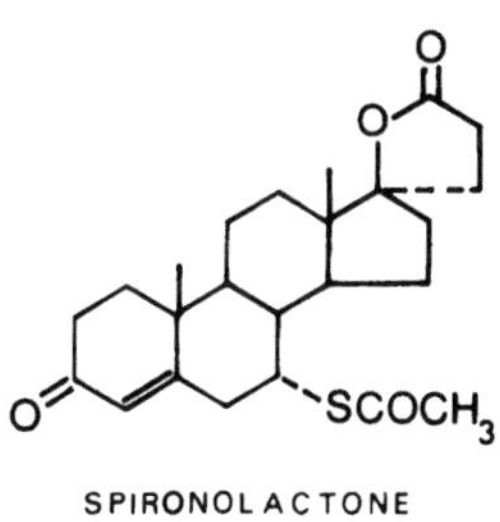

SPIRONOLACTONE

Fig. 2. Steroidal anti-androgens

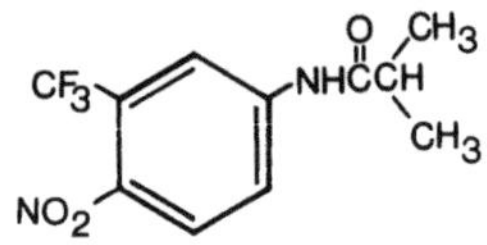

FLUTAMIDE

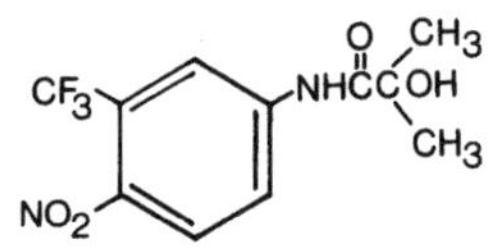

HYDROXYFLUTAMIDE

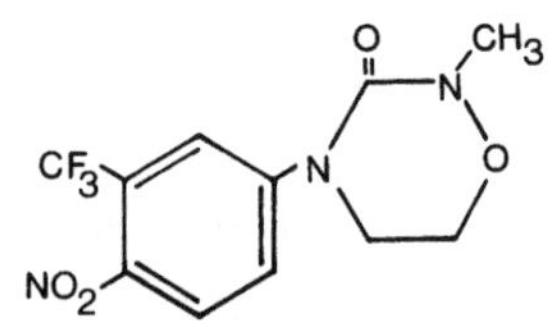

RU 22930

Fig. 3. Non-steroidal anti-androgens

translocation is prevented and the nuclear concentration of DHT reduced (BRUCHOVSKY 1980), but it is not known if the androgen receptor-cyproterone acetate complex enters the nucleus. The 5α-reduction of testosterone to DHT is not suppressed by cyproterone acetate (BELHAM and NEAL 1971; SZALAY et al. 1975). Other steroidal anti-androgens include 17α-propyltestosterone (FERRARI et al. 1980), which is also a progestogen, but appears suitable for topical use because large topical doses failed to elicit systemic hormone activities, 17α-methyl-*B*-nortestosterone (SAUNDERS et al. 1964) and R 2956 (RAYNAUD et al. 1977).

Anti-androgens such as cyproterone acetate have additional progestational and anti-gonadotropic properties, while their unesterified free alcohols (e.g. cyproterone) have no such effects and are therefore classified as "pure" anti-androgens (NEUMANN et al. 1980). A similar "pure" anti-androgen, which has received considerable attention is the non-steroidal compound flutamide (Fig. 3) (NERI et al. 1972; PEETS et al. 1974). This compound, however, has been shown to be more effective in vivo than in vitro and it is now considered that the hydroxylated metabolite, hydroxyflutamide, is probably the active molecule (KATCHEN and BUXBAUM 1975). An additional non-steroidal anti-androgen, structurally related to flutamide and of potential interest in the skin, is RU 22930. Studies in animals have shown that this compound, while having anti-androgenic properties, only demonstrates activity when applied topically (RAYNAUD et al. 1977).

A possible drawback to the systemic use of "pure" anti-androgens, such as cyproterone or flutamide, is their lack of additional central action. Their anti-androgenic effect extends to the hypothalamus where they inhibit the negative feedback action of testosterone. This in turn stimulates the release of gonadotropin and consequently stimulates further testosterone secretion. The increase in testosterone levels, therefore, gradually overcomes the anti-androgenic action of the drug (NEUMANN et al. 1980). The anti-gonadotropic effects of cyproterone acetate, on the other hand, prevent these stimulatory feedback effects and thus produce a much more effective androgen suppression.

III. Spironolactone

Spironolactone (Fig. 2) is an anti-hypertensive drug which competes with aldosterone for mineralocorticoid receptors in the kidney. In use, however, side effects such as gynaecomastia and decreased libido have been shown to be associated with anti-androgenic activity (CORVOL et al. 1975). Clinical investigation, however, failed to demonstrate anti-sebotrophic activity (STEWART and POCHI 1978).

D. Animal Models: Anti-androgens and Sebaceous Gland Function

The realisation of the causative importance of androgens in the pathogenesis of certain skin conditions has prompted the search for drugs which might be effective, non-toxic antagonists of androgen action. To this end numerous animal models have been investigated which exploit the androgen responsiveness of either sebaceous glands or of specialised pheromone-producing organs. While most of these model systems have been extremely useful in helping to elucidate the influence of both androgens and anti-androgens on sebaceous gland function, none are free from drawbacks, because of differences in hormone responsiveness, route of drug administration or difficulty in the measurement of responses. Nevertheless, some of the tissues investigated show close similarities in sebaceous gland structure and function to those of human glands and offer relative ease of quantitation of effects.

The animal models which have received most attention include the hamster costovertebral organ (BURDICK and HILL 1970; FROST et al. 1973; TAKAYASU and ADACHI 1970) the rat preputial (EBLING et al. 1975; KRÄHENBÜHL and DESAULLES 1969) and sebaceous glands (ARCHIBALD and SHUSTER 1970; EBLING and SKINNER 1967; NIKKARI 1965), the gerbil abdominal gland (GLENN and GRAY 1965) and the hamster ventral ear glands (PLEWIG and LUDERSCHMIDT 1977). Various methods have been developed to measure sebaceous gland function. The simplest approaches determine gland size or weight from biopsied tissues. Sebum production has been measured directly by either extraction of surface lipid by immersing the animal in a suitable solvent (ARCHIBALD and SHUSTER 1970) or by estimating the changes in hair fat content following detergent washing (EBLING and SKINNER 1967). Attempts have also been made to determine secretion rates by gravimetric measurement (EBLING et al. 1981) in a similar fashion to the method developed originally for the estimation of human sebum secretion rates (STRAUSS and POCHI 1961). Histology has been widely used to assess changes in sebaceous glands; cell proliferation can be determined by mitotic counts or autoradiography following tritiated thymidine pulsing, and gland area quantified by planimetry.

Using these measuring techniques to study the effects of drugs on sebaceous gland function, however, is dependent on allowing sufficient time for the gland parameters to change and therefore the experiments tend to be of fairly long duration. The great variety of models and measuring techniques suggest that the ideal system does not exist, and further refinements will continue to be developed for some time. The animal model attracting most interest at present utilises the ventral ear glands of the hamster. It is claimed that because the glands possess an infundibulum, lobules and a pilary unit, and are under androgen control, they are much closer to the ideal than the more specialised glands studied in the past. A recent study of the hormonal control of hamster ear sebaceous gland lipogenesis (HALL et al. 1983), utilising [^{14}C]acetate incorporation into sebum lipids, offers a simple and reliable method of investigating the action of anti-androgens, or any other potential anti-sebotrophic agents, on sebaceous gland function.

E. Clinical Evaluation of Anti-androgens

Cyproterone acetate has been used widely to treat women with acne and hirsutism, and has been found to inhibit both sebum secretion (BURTON et al. 1973) (this study was undertaken in male sexual offenders) and terminal hair growth (ISMAIL et al. 1974). The need to prevent menstrual problems, particularly withdrawal bleedings, and pregnancy during therapy necessitated the use of the so-called reversed sequential therapy (HAMMERSTEIN et al. 1975). This involves treatment with 100 mg cyproterone acetate from days 5 to 14, and 0.05 mg ethinyloestradiol from days 5 to 25 of each menstrual cycle. A marked improvement in more than 90% of acne patients was seen within 3 months, but the response in hirsute women was less impressive although 70% had improved by 6 months. Androgenic alopecia did not respond well, with only about half the patients showing improvement by the end of 1 year. Side effects were mild and resembled those produced by other oral contraceptives. Cyproterone acetate has been claimed to have some glucocorticoid activity and, while depression of adrenocorticotropin has not been seen as a clinical problem in women, high dose therapy to girls with idiopathic precocious puberty has resulted in abnormally low plasma cortisol levels (JEFFCOATES et al. 1976; VONMUHLENDAHL et al. 1977).

Because of the potent progestational partial effect of cyproterone acetate it also appears to be suitable for use at lower doses as an oral contraceptive. Cyproterone acetate 2 mg in combination with 0.05 mg ethinyloestradiol has been introduced as a single pill to be taken on 21 consecutive days of each menstrual cycle. A number of studies have shown that this regime gives effective contraception. Good results are claimed in severe acne and mild hirsutism (AYDINLIK and LACHNIT-FIXSON 1977; SCHMIDT-ELMENDORF and STEYER 1977) but others have attributed this to the oestrogen content since the dose of cyproterone used is too small (MARSDEN et al. 1983, 1984).

F. Topical Anti-androgens

At the present time there is no satisfactory topical anti-androgen therapy. It is difficult to reconcile this with the fact that anti-androgens are known to act peripherally and would be expected to have a topical action on pilo-sebaceous units. A number of anti-androgens have been evaluated topically in the clinic, including progesterone (SIMPSON et al. 1979), cyproterone acetate (CUNLIFFE et al. 1969; PYE et al. 1976), LYONS and SHUSTER (1983), 17-propyl testosterone (LYONS and SHUSTER 1981) and flutamide (LYONS and SHUSTER 1982). The studies showed that anti-androgens given by this route were either singularly ineffective or at best produced marginal improvement in seborrhoea which was sex related in one study (LYONS and SHUSTER 1982); clinical improvement was not demonstrated. It is interesting to speculate whether topical anti-androgens will ever play an important role in the treatment of acne, hirsutism or androgenic alopecia.

References

Adachi K, Kano M (1972) The role of receptor proteins in controlling androgen action in the sebaceous glands of hamsters. Steroids 19:567–574
Archibald A, Shuster S (1970) The measurement of sebum secretion in the rat. Br J Dermatol 82:146–151
Aydinlik S, Lachnit-Fixson U (1977) Diane – eine Gestagen-Östrogenkombination mit Antiandrogenwirkung. Med Monatsschr 31:425–429
Baulieu EE, Jung I (1970) A prostatic cytosol receptor. Biochem Biophys Res Commun 38:599–606
Belham JE, Neal GE (1971) Testosterone action in rat ventral prostate. The effects of diethylstilboestrol and cyproterone acetate on the metabolism of [^{3}H] testosterone and the retention of labelled metabolites by rat ventral prostate in vivo and in vitro. Biochem J 125:81–91
Bruchovsky N (1980) Molecular action of androgens and antiandrogens. In: Hammerstein J, Lachnit-Fixson U, Neumann F, Plewig G (eds) Androgenization in women. Acne, seborrhoea, androgenic alopecia and hirsutism. Symposium, Berlin. Excerpta Medica, Amsterdam, pp 7–20
Bruchovsky N, Wilson JC (1968) The conversion of testosterone to 5α-androstan-17β-ol-3-one by rat prostate in vivo and in vitro. J Biol Chem 243:2012–2021
Burdick KH, Hill R (1970) The topical effect of the antiandrogen chlormadinone acetate and some of its chemical modifications on the hamster costovertebral organ. Br J Dermatol 82 [Suppl 6]:19–25
Burton JL (1979) Antiandrogen therapy in dermatology: a review. Clin Exp Dermatol 4:501–507
Burton JL, Laschet U, Shuster S (1973) Reduction of sebum excretion in man by the antiandrogen cyproterone acetate. Br J Dermatol 89:487–490
Cooper MF, Hay JB, McGibbon D, Shuster S (1978) Correlations between androgen metabolism and sebaceous gland activity. Trans Biochem Soc 6:970–972
Cooper MF, McGibbon D, Wilson PD, Shuster S (1979) Androgenic control of the human sebaceous gland (abstr). J Invest Dermatol 72:267
Corvol P, Michaud A, Menard J, Freifeld M, Mahoudeau J (1975) Antiandrogenic effects of spirolactones: mechanism of action. Endocrinology 97:52–58
Cunliffe WJ, Shuster S, Cassels Smith AJ (1969) The effect of topical cyproterone acetate on sebum secretion in patients with acne. Br J Dermatol 81:200–201
Dorfman RI (1970) Biological activity of antiandrogens. Br J Dermatol 82 [Suppl 6]:3–8
Ebling FJ, Skinner J (1967) The measurement of sebum production in rats treated with testosterone and oestradiol. Br J Dermatol 79:386–392

Ebling FJ, Ebling E, Randall V, Skinner J (1975) The effects of hypophysectomy and of bovine growth hormone on the responses to testosterone of prostate, preputial, Harderian and lachrymal glands and of brown adipose tissue in the rat. J Endocrinol 66:401–406
Ebling FJ, Randall VA, Skinner J (1981) Local suppression of sebum secretion in rats by topical cyproterone acetate in ethanol. J Invest Dermatol 77:458–463
Ewing L, Brown B, Irby DC, Jardine I (1975) Testosterone and 5α-reduced androgen secretion by rabbit testes-epididymides in vitro. Endocrinology 96:610–617
Fang S, Liao S (1969) Antagonistic action of antiandrogens on the formation of a specific dihydrotestosterone-receptor protein complex in rat ventral prostate. Mol Pharmacol 5:428–431
Fang S, Anderson KM, Liao S (1969) Receptor proteins for androgens: on the role of specific proteins in selective retention of 17βhydroxy-5α-androstan-3-one by rat ventral prostate in vivo and in vitro. J Biol Chem 244:6584–6595
Farnworth WE, Brown JR (1963) Metabolism of testosterone by the human prostate. JAMA 183:436–439
Ferrari RA, Chakrabarty K, Creange JE, Beyler AL, Potts GO, Schane HP (1980) Endocrine profile of Topterone, a topical antiandrogen, in three species of laboratory animals. Methods Find Exp Clin Pharmacol 2:65–69
Frost P, Giegel JL, Weinstein GD, Gomez EC (1973) Biodynamic studies of hamster flank organ growth: hormonal influences. J Invest Dermatol 61:159–167
Glenn EM, Gray J (1965) Effect of various hormones on the growth and histology of the gerbil (*Meriones unguiculatus*) abdominal sebaceous gland pad. Endocrinology 76:1115–1123
Hall DWR, Van der Hoven WE, Noordzij-Kamermans NJ, Jaitly KD (1983) Hormonal control of hamster ear sebaceous gland lipogenesis. Arch Dermatol Res 275:1–7
Hamada H, Neumann F, Junkmann K (1963) Intrauterine antimaskuline Beeinflussung von Rattenfeten durch ein stark gestagenwirksames Steroid. Acta Endocrinol (Copenh) 44:380–388
Hammerstein J, Meckies J, Leo-Rossberg I, Moltz L, Zielske F (1975) Use of cyproterone acetate (CPA) in the treatment of acne, hirsutism and virilism. J Steroid Biochem 6:827–836
Hay JB (1977) A study of the in vitro metabolism of androgens by human scalp and pubic skin. Br J Dermatol 97:237–246
Hsia SL, Voigt W (1974) Inhibition of dihydrotestosterone formation: an effective means of blocking androgen action in hamster sebaceous glands. J Invest Dermatol 62:224–227
Imperato-McGinley J, Guerrero L, Gautier T, Peterson RE (1974) Steroid 5α-reductase deficiency in man; an inherited form of male pseudohermaphroditism. Science 186:1213–1215
Ismail AAA, Davidson DW, Souka AR, Barnes EW, Irvine WJ, Kilimnik H, Vanderbeeken Y (1974) The evaluation of the role of androgens in hirsutism and the use of a new antiandrogen "cyproterone acetate" for therapy. J Clin Endocrinol Metab 39:81–95
Jeffcoates WJ, Edwards CRW, Rees LH, Besser GM (1976) Cyproterone acetate. Lancet 2:1140
Jenkins JS, Ash S (1973) The metabolism of testosterone by human skin in disorders of hair growth. J Endocrinol 59:345–351
Katchen B, Buxbaum S (1975) Disposition of a new, non-steroid, antiandrogen, α, α, α-trifluoro-2-methyl-4′-nitro-*m*-propionotoluidide (flutamide) in men following a single oral 200 mg dose. J Clin Endocrinol Metab 41:373–379
Krähenbühl C, Desaulles PA (1969) Interactions between α-MSH and sex steroids on the preputial glands of female rats. Experientia 25:1193–1195
Liao S, Liang T, Tymoczko JL (1972) Structural recognitions in the interactions of androgens and receptor proteins in their association with nuclear acceptor components. J Steroid Biochem 3:401–408
Lyons F, Shuster S (1981) Effect of topical 17α-propyltestosterone on sebum excretion in man. Br J Dermatol 104:685–686

Lyons F, Shuster S (1982) Sex difference in response of the human sebaceous gland to topical flutamide. Br J Dermatol 107:697–699

Lyons F, Shuster S (1983) Indirect evidence that the action of cyproterone acetate on the skin is due to a metabolite. Clin Endocrinol 19i:53–55

Mainwaring WIP, Jones DM (1975) Influence of receptor complexes on the properties of prostate chromatin, including its transcription by RNA polymerase. J Steroid Biochem 6:475–481

Marsden JR, Shuster S, Lyons F (1983) Is low dose cyproterone acetate and ethinyl oestradiol ("Diane") effective in acne? (letter) Lancet 2:215

Marsden JR, Shuster S, Lyons F (1984) Effect of 13-*cis*-retinoic acid and cyproterone acetate on acne severity, sebum excretion rate, dermal and epidermal lipogenesis, serum lipids and liver function tests. In: Cunliffe WJ, Miller AJ (eds) Retinoid therapy – a review of clinical and laboratory research. MTP, Boston, pp 267–276

Mauvais-Jarvis P (1977) Androgen metabolism in human skin: mechanisms of control. In: Martini L, Motta M (eds) Androgens and antiandrogens. Raven, New York, pp 229–245

Mauvais-Jarvis P, Kuttenn F, Baudot N (1974) Inhibition of testosterone conversion to dihydrotestosterone in men treated percutaneously by progesterone. J Clin Endocrinol Metab 38:142–147

Neri R, Florance K, Koziol P, Van Cleave S (1972) A biological profile of a nonsteroidal antiandrogen SCH 13521 (4′-nitro-3′-trifluoromethylisobutyranilide). Endocrinology 91:427–437

Neumann F, Jacobi GH (1982) antiandrogens in tumour therapy. In: Furr BJA (ed) Clinics in oncology, Ed. by vol 1. Saunders, Philadelphia, pp 41–64

Neumann F, Schleusener A (1981) Pharmacology of cyproterone acetate with special reference to the skin. In: Vokaer R, Fanta D (eds) Combined antiandrogen-estrogen therapy in dermatology, proc of the DIANE symposium. Excerpta Medica, Amsterdam, pp 19–51

Neumann F, Schleusener A, Albring M (1980) Pharmacology of antiandrogens. In: Hammerstein J, Lachnit-Fixson U, Neumann F, Plewig G (eds) Androgenization in women. Acne, seborrhoea, androgenic alopecia and hirsutism. Symposium, Berlin. Excerpta Medica, Amsterdam, pp 147–192

Nikkari T (1965) Composition and secretion of the skin surface lipids of the rat; effects of dietary lipids and hormones. Scand J Clin Lab Invest 17 [Suppl 85]:1–140

Peets EA, Henson MF, Neri R (1974) On the mechanism of the antiandrogenic action of flutamide (α-α-α-trifluoro-2-methyl-4′-nitro-*m*-propionotoluidide) in the rat. Endocrinology 94:532–540

Peterson RE, Imperato-McGinley J, Gautier T, Sturla E (1977) Male pseudohermaphroditism due to steroid 5α-reductase deficiency. Am J Med 62:170–191

Plewig G, Luderschmidt C (1977) Hamster ear model for sebaceous glands. J Invest Dermatol 68:171–176

Price VH (1975) Testosterone metabolism in the skin. a review of its function in androgenetic alopecia, acne vulgaris and idiopathic hirsutism including recent studies with antiandrogens. Arch Dermatol 111:1496–1502

Pye RJ, Burton JL, Harris JI (1976) Effect of 1% cyproterone acetate in cetomacrogol cream BPC (Formula A) on sebum excretion rate in patients with acne. Br J Dermatol 95:427–428

Raynaud JP, Azadian-Boulanger G, Bonne C, Perronet J, Sakiz E (1977) Present trends in antiandrogen research. In: Martini L, Motta M (eds) Androgens and antiandrogens. Raven, New York, pp 281–293

Saenger P, Goldman AS, Levine LS, Korthschutz S, Muecke EC, Katsumata M, Doberne Y, New MI (1978) Prepubertal diagnosis of steroid 5α-reductase deficiency. J Clin Endocrinol Metab 46:627–634

Sansone G, Reisner RM (1971) Differential rates of conversion of testosterone to dihydrotesterone in acne and in normal human skin. A possible pathogenic factor in acne. J Invest Dermatol 56:366–372

Saunders HL, Holden K, Kerwin JF (1964) The antiandrogenic activity of 17α-methyl-β-nortestosterone (SKF-7690). Steroids 3:687–698

Schmidt-Elmendorff H, Steyer M (1977) Klinische Erfahrungen mit einem niedrigdosierten Antiandrogen (Cyproteron-azetat) und Äthinylöstradiol bei Frauen mit Virilisierungserscheinungen. Geburtshilfe Frauenheilkd 37:297–299

Shuster S (1982) The sebaceous glands and primary cutaneous virilism. In: Jeffcoate SL (ed) Androgen sand anti-androgen therapy. Wiley, Chichester, pp 1–21

Simpson NB, Bowden PE, Forster FA, Cunliffe WJ (1979) The effect of topically applied progesterone on sebum excretion rate. Br J Dermatol 100:687–692

Stewart ME, Pochi PE (1978) Antiandrogens and the skin. Int J Dermatol 17:167–179

Strauss JS, Pochi PE (1961) The quantitative gravimetric determination of sebum production. J Invest Dermatol 36:293–298

Szalay R, Krieg M, Schmidt H, Voigt KD (1975) Metabolism and mode of action of androgens in target tissues of male rats. V. Uptake and metabolism of cyproterone acetate and its influence on the uptake and metabolism of testosterone in target organs and peripheral tissues. Acta Endocrinol (Copenh) 80:592–602

Takayasu S (1979) Metabolism and action of androgens in the skin. Int J Dermatol 18:681–692

Takayasu S, Adachi K (1970) Hormonal control of metabolism in hamster costovertebral glands. J Invest Dermatol 55:13–19

Takayasu S, Wakimoto H, Itami S, Sano S (1980) Activity of testosterone 5α-reductase in various tissues of human skin. J Invest Dermatol 74:187–191

Thomas JP, Oake RJ (1974) Androgen metabolism in the skin of hirsute women. J Clin Endocrinol Metab 38:19–22

Voigt W, Hsia SL (1973 a) Further studies on testosterone 5α-reductase of human skin: structural features of steroid inhibitors. J Biol Chem 248:4280–4285

Voigt W, Hsia SL (1973 b) The antiandrogenic action of 4-androsten-3-one-17β-carboxylic acid and its methyl ester on hamster flank organ. Endocrinology 92:1216–1222

Voigt W, Fernandez EP, Hsia SL (1970) Transformation of testosterone into 17β-hydroxy-5α-androstan-3-one by microsomal preparations of human skin. J Biol Chem 245:5594–5599

Von Muhlendahl KE, Korth-Schultz S, Muller-Hess R, Helge H, Webb B (1977) Cyproterone acetate and adrenocortical function. Lancet 1:1160–1161

Walsh PC, Madden JD, Harrod MJ, Goldstein JL, MacDonald PC, Wilson JD (1974) Familial incomplete male pseudohermaproditism type 2: decreased dihydrotestosterone formation in pseudovaginal perineoscrotal hypospadias. N Engl J Med 291:944–949

Wilson JD, Walker JD (1969) The conversion of testosterone to 5α-androstan-17β-ol-3-one (dihydrotestosterone) by skin slices of man. J Clin Invest 48:371–379

CHAPTER 37

The Effect of Drugs on Hair

N. B. SIMPSON

A. Introduction

Hair has lost most of its physiological importance in humans. However, alteration of hair growth, whether natural or drug-induced, causes great psycho-social upset. The majority of the published literature concerning the effect of drugs on hair has therefore come from clinicians, with smaller contributions from biochemists and physiologists. Owing to the large inter-species differences in hair growth and moulting patterns, results from animals have been cited only where they contribute to the understanding of the situation that occurs in humans.

B. The Hair Growth Cycle

Foetal hair starts to develop during the 12th week of intra-uterine life and from the time of first formation hair follicles undergo repeated cycles of growth and rest. Cyclical activity is a charcteristic of hair and is influenced by age, sex and body site, but can be affected by physiological, pathological and pharmacological factors. The sequence (Fig. 1) (KLIGMAN 1959) can be most easily understood by starting at the end of the active growth phase. This phase is called metanagen or anagen VI. The cycle follows the descriptive phases:

Metanagen ⟶	Catagen ⟶	Telogen
↑		↓
Anagen V		Anagen I
↑		↓
Anagen IV	⟵ Anagen III	⟵ Anagen II

I. Catagen

The mechanism underlying the initiation of catagen and the follicular cycle is unknown. Catagen only lasts a few days. Cells of the matrix cease mitotic activity, keratinisation of the inner root sheath continues, but melanocytic activity stops, leading to a non-pigmented, club-shaped terminal portion. The lower third of the follicle shortens, the connective tissue sheath appears thickened and corrugated and the inner root sheath disappears. Dermal papilla cells follow the follicle in its shortened form and remain in contact with the external root sheath where it forms a sac enclosing the follicular germ cells. The close spatial relationship be-

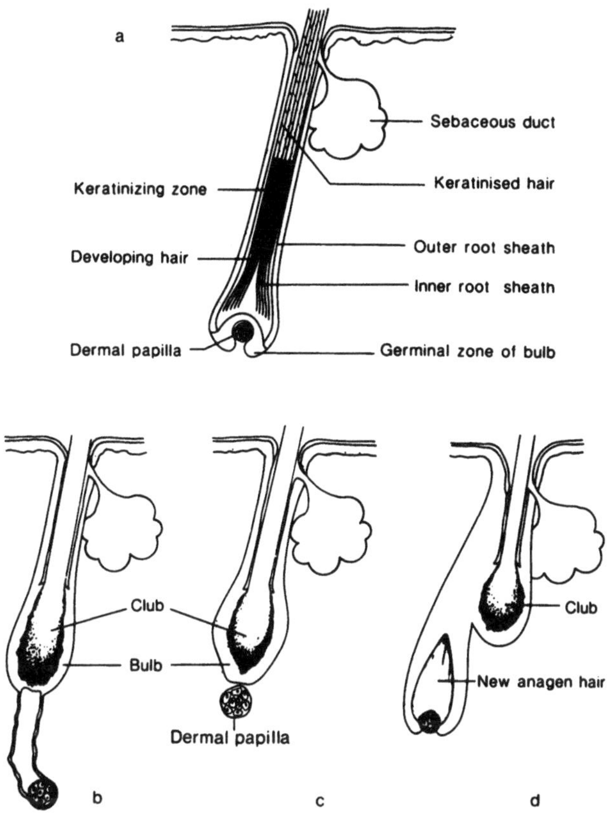

Fig. 1 a–d. A diagrammatic representation of the hair growth cycle. **a** metanagen (anagen stage VI), **b** catagen, **c** telogen, **d** anagen stage II

tween dermal papilla cells and the follicle is very important because initiation of the next cycle cannot occur in the absence of these cells (OLIVER 1966; HORNE et al. 1986).

II. Telogen

Telogen is the resting phase of the cycle. The club-shaped hair is held in the external root sheath sac by intercellular connections and retained in the follicle until metanagen of the next cycle is well established. Sometimes a telogen hair may be retained for several cycles.

III. Anagen

Phases anagen I–anagen V are collectively known as proanagen. Plucking of the old club hair can initiate anagen prematurely. In stage I the dermal papilla cells increase in size and show increased RNA synthesis; germinal cells at the base of the sac show increased mitotic activity. The follicle lengthens and engulfs the dermal papilla cells during stage II. A new internal root sheath is produced from the matrix cells in stage III and melanocytic activity occurs in stage IV. Also in stage IV the keratinising zone forms just below the sebaceous duct. The hair emerges from the inner root sheath in stage V and, finally, stage VI or metanagen starts when the hair emerges at the skin surface and lasts until the onset of catagen. Vellus hair is the fine non-pigmented body hair seen in prepubertal children. Terminal hair is coarse, often pigmented and best chracterised by adult sexual hair.

C. Physiological Changes in the Hair Cycle

Humans are not susceptible to the effects of seasonal or photoperiod-related moulting although some physiological conditions such as pregnancy may affect the hair growth cycle (Sect. H). Follicular dynamics have been studied on the human scalp and very little information is available for other body sites. Consequently the figures quoted below are for scalp hair follicles. Anagen lasts approximately 1000 days and telogen approximately 100 days (ORENTREICH 1969). Therefore, the anagen : telogen (A : T) ratio is about 9 : 1 because the percentage of catagen hairs is very small. There are approximately 10000 hair follicles at birth with a density of 1135 cm^{-2}. Changes in surface area cause a fall in density to 795 cm^{-2} by the end of the first year. Further decline to 615 cm^{-2} by age 30 years 485 cm^{-2} by 50 years is due to the natural destruction of hair follicles. Average figures for bald scalp were 330 cm^{-2} (age 45–70 years) and 280 cm^{-2} (70–85 years) (GIACOMETTI 1965).

D. Assessment of Drug Effects on Hair

For methods of clinical assessment and cell culture see Chap. 5.

E. Hormones and Hair Growth

Oestrogens slow the rate of hair growth and prolong the duration of anagen. Telogen is shortened by thyroxine and prolonged by glucocorticoids. Androgens are essential, in both sexes, for the development of sexual hair on the body (conversion of vellus to terminal hair) and androgenetic alopecia on the scalp (conversion of terminal to vellus hair). Therefore the same hormones appear to have an opposite effect at different body sites. The response to androgens is determined genetically, but may be affected by previous exposure to sex steroids. HAMILTON (1958) showed that beard development was abolished by castration of males before the age of 16 years; while castration between 16 and 20 years led to a reduction in the extent of beard growth and the same operation after 21 years had no effect. Thus, the magnitude of the response of hair to androgens may be set permanently by modification of gene expression following puberty (SHUSTER 1982). The same factors may therefore limit the potential for reversal by anti-androgen therapy (Sect. G. I. 1).

There is widespread acceptance of a model for androgen metabolism in target organs (see Chap. 36) in which testosterone is converted to dihydrotestosterone (DHT) by the enzyme 5α-reductase; DHT then binds to a cytoplasmic receptor and the DHT-receptor complex is translocated to the nucleus where it interacts with DNA. Androgen receptors have been demonstrated in plucked hair roots (FAZEKAS and SANDOR 1972). Both hair roots (SCHWEIKERT and WILSON 1974) and hair follicles obtained by microdissection (HAY and HODGINS 1978) show 5α-reductase activity. Although the model was derived from experiments on human prostate, the importance of 5α-reductase and the cytoplasmic receptor can be seen from patients with deficiency syndromes (GRIFFIN et al. 1982). There is therefore potential for therapeutic attack by reducing plasma androgen levels and competing for 5α-reductase activity and/or DHT receptor binding sites within the target cells.

F. Hirsutism

Hirsutism is the growth of coarse terminal hair by a female in all or part of the male pattern. There is a clinical spectrum from gross masculinisation with menstrual disturbance to mild cosmetic embarrassment. The association with androgens is easily made in the presence of virilisation resulting from tumours of the ovary or adrenal cortex. The association between hirsutism and polycystic ovarian disease has been reviewed by YEN (1980) and is the commonest cause of familial hirsutism. Patients with normal urinary 17-oxosteroid excretion have been labelled as "idiopathic". However, most patients have plasma testosterone slightly above the normal range (GIVENS et al. 1975; BARDIN and LIPSETT 1976) and sex hormone binding globulin below normal (ROSENFELD 1971). SHUSTER (1972) showed increased sebum production, sweat rate and skin collagen content of women with marked hirsutism and coined the descriptive phrase "primary cutaneous virilism". Claims for increased androgen metabolism in suprapubic skin (MAUVAIS-JARVIS 1977; KUTTEN et al. 1980) and the hairy areas (THOMAS and OAKE 1974) of patients with hirsutism probably reflect the increased mass of androgen target tissue in whole skin biopsies (HAY and HODGINS 1978; SIMPSON et al. 1983) and accord with Shuster's hypothesis. These structures may provide a greater soruce of enzymes for a local (and possibly systemic) supply of potent androgens (GOMPEL et al. 1986; GREEP et al. 1986), but there is no evidence whether this is a primary or secondary effect.

I. Hormonal Therapy for Hirsutism

1. Reduction of Circulating Androgen Levels

Adrenal suppression by dexamethasone (CASEY et al. 1966; ABRAHAM et al. 1976) and ovarian suppression have produced limited success. Systemic medroxyprogesterone acetate, which blocks gonadotropin release, had a short-term beneficial effect (ETTINGER and GOLDITCH 1977). Nafarelin, a gonadotropin-releasing hormone agonist, reduced hirsutism in a small number of subjects (ANDREYKO et al.

1986). Long-term treatment with this type of drug could provide important evidence as to the primary or secondary nature of cutaneous virilism if hirsutism or seborrhoea were re-estblished in the presence of reduced levels of circulating androgens. Oestrogens raise plasma sex hormone binding globulin levels and thereby lower the free testosterone level. However, while suppressing sebum production (see Chap. 4), oestrogen therapy alone is ineffective against hirsutism.

2. Anti-androgens (see Chap. 36)

HAMMERSTEIN and CUPCEANCU (1969) reported successful treatment with cyproterone acetate (100 mg) in combination with ethinyloestradiol as an ovulatory suppressant. Cyproterone acetate is a potent progestogen and works as an anti-androgen by competitive binding to the cytosol receptor (see Chap. 36). DEWHURST et al. (1977) and EBLING et al. (1977) have confirmed the effectiveness of Hammerstein's regime by objective criteria. Lower doses of cyproterone acetate in the oral contraceptive pill Diane (2 mg cyproterone acetate with 50 μg ethinyloestradiol daily), while moderately effective in acne, do not work in hirsutism, but may be useful as a maintenance therapy after improvement on the higher dose (HOLDAWAY et al. 1985). Spironolactone is a potent competitive inhibitor for DHT receptor binding sites and inhibits testosterone biosynthesis (MENARD et al. 1974). SHAPIRO and EVRAN (1980) reported therapeutic promise in hirsutism, but subsequent studies have shown that spironolactone alone is much less effective than Hammerstein's regime (DORRINGTON-WARD et al. 1985; CHAPMAN et al. 1986). One small study of spironolactone in combination with an oral contraceptive (CHAPMAN et al. 1985) has demonstrated more encouraging results. CULLBERG et al. (1985) and DEWIS et al. (1985) showed a slow reduction in acne and hirsutism with the potent progestogen desogestrel (150 μg) combined with ethinyloestradiol (30 μg).

Despite the clinical effectiveness of systemic anti-androgen therapy and limited reports of the efficacy of intradermally injected medroxyprogesterone acetate (SCHMIDT et al. 1985), topical application of anti-androgens has been without clinical benefit (see Chap. 36). The rapid relapse of hirsutism following anti-androgen therapy (UNDERHILL and DEWHURST 1979) suggests that the cutaneous virilisation experienced by adult women is due to modified gene expression and further persuance of this therapeutic approach seems doomed to failure.

G. Male-Pattern Baldness (Androgenetic Alopecia)

HAMILTON (1942) established a link between androgens and baldness in genetically predisposed males and showed that eunuchs did not become bald until some time after testosterone replacement therapy. Inerhitance is probably autosomal dominant with incomplete penetrance; males and females are equally affected, but the rate and extent of baldness are affected by exposure to androgenic hormones. The acceleration of hair loss due to some relatively androgenic contraceptive pills is good evidence for this effect. Hair roots from the scalps of balding males show no in vitro differences in androgen metabolism or in the conversion of androgens to oestrogens regardless of whether they are plucked from areas predestined to lose or retain hair (SCHWEIKERT and WILSON 1974; SCHWEIKERT et al. 1975). However, HODGINS et al. (1985) have shown an increase in 17β-hydroxysteroid dehydrogenase activity in hairs plucked from the vault of the scalps of children and young adults with a strong family history of androgenetic alopecia, but not yet expressing baldness themselves. This enzyme converts testosterone to the weaker androgen androstenedione and thereby reduces the androgen drive within the target cell.

I. Treatments for Androgenetic Alopecia

1. Anti-androgens

Progression of baldness in men may be halted by cyproterone acetate, but the loss of libido makes therapy impractical. Hair loss stops, the telogen count falls and some new terminal hairs form in women (EKOE et al. 1980), but the effect is temporary and lost on withdrawal of therapy. The situation is therefore similar to hirsutism in which competitive inhibition is only of temporary benefit, owing to modified gene expression. An approach based on the findings of HODGINS et al. (1985) to alter the androgenic environment of genetically susceptible individuals before puberty may affect the onset and extent of baldness in childhood.

2. Non-hormonal Therapy

After centuries of unsuccessful patent remedies for baldness GROVEMAN et al. (1985) showed a definite placebo response. The only drug to demonstrate reversal of baldness in controlled clinical trials is minoxidil. This drug is a piperidinopyrimidine derivative and a potent vasodilator which is effective orally for severe hypertension and frequently causes unwanted hypertrichosis with occasional reversal of androgenetic alopecia (ZAPPACOSTA 1980). When dissolved in ethanol for topical application, minoxidil has shown conversion of vellus to terminal hair in up to 30% of individuals (VANDERVEEN et al. 1984; DE VILLEZ 1985; OLSEN et al. 1985). Regrowth occurred from the periphery, but complete covering of the bald area was seen in less than 10% of responders after 6–9 months of treatment. There was a dose-response correlation up to 2% (OLSEN et al. 1986). Recent onset of baldness with small size of bald area predisposed to a good result. The mode of action is unknown, but a shortening of the vellus hair cycle with histological thickening of growing hairs has been seen in scalps of stump-tailed macaques (*Macaca arctoides*) (UNO et al. 1985). WESTER et al. (1984) found an increase in cutaneous blood flow of the scalp in a small study, but did not include a positive control and failed to demonstrate a dose-response correlation. Minoxidil sulphate may be an active metabolite following oral administration (JOHNSON et al. 1982); skin demonstrates sulphatase and sulphotransferase activity, but this aspect has yet to be explored. Minoxidil has no effect on human hair bulb papilla cells and hair root sheath-fibroblasts in culture (KATSUOKA et al. 1987). Topical minoxidil appears to be a safe therapy with only local side effects of irritation and low incidence of contact dermatitis (TOSTI et al. 1985) receiving much attention. There are no published data on the outcome following treatment, but the indications are that clinical regression occurs after 3 months to the state of baldness that would have existed if no treatment had been applied. Combination therapies may be beneficial in the future; BAZZANO et al. (1986) found in a 1-year study that topical tretinoin and 0.5% minoxidil was equally efficient as 2% minoxidil alone, but VERMORKEN (1983) failed to show synergism with cyproterone acetate.

H. Pregnancy and the Contraceptive Pill

Following parturition the A:T ratio may reach parity and many women experience rapid hair loss. This so-called post-partum telogen effluvium (KLIGMAN 1961) is usually reversed in 3–12 months. A similar situation has been noted 3–4 months after stopping the combined contraceptive pill (DAWBER and CONNOR 1971; GRIFFITHS 1973). Anecdotal reports of diffuse hair loss as a result of the contraceptive pill are difficult to evaluate scientifically. GRIFFITHS (1973) failed to identify hair loss in most women on the contraceptive pill. However, in women genetically predisposed to acne, hirsutism or androgenetic alopecia the use of contraceptive pills with a relatively "androgenic" progestogen may precipitate or aggravate these conditions. Desogestrel (150 μg) with ethinyloestradiol (30 μg) or the 2 mg cyproterone acetate pill could therefore be used for contraception in these women.

J. Treatments for Alopecia Areata and Alopecia Totalis

Alopecia areata is a common, distressing disorder of unknown aetiology. It is thought to have an immune basis and is associated with several autoimmune diseases. The variable and unpredictable natural history of alopecia areata has given rise to many therapeutic claims. Traditional remedies using irritant substances are still employed, but lack scientific validation. Systemic glucocorticoids promote rapid regrowth and pigmentation, but relapse is common on withdrawal. Intralesional steroids can promote regrowth which is often confined to the site of injection, giving the appearance of a "doll's head" and localised dermal atrophy is a frequent consequence. Topical glucocorticoids are without effect.

First details of regrowth following induction of contact dermatitis of the scalp with dinitrochlorobenzene (DNCB) (ROSENBERG and DRAKE 1976), squaric acid dibutylester (SABE) (HAPPLE et al. 1980) and diphencyprone (HAPPLE et al. 1983) have been followed by disappointing reports when resistant cases were included [DNCB – DAMAN and ROSENBERG (1978); WARIN et al. (1979); SABE – BARTH et al. (1985); TOSTI et al. (1986); diphencyprone – ORRECHIA and RABBIOSI (1985)]. PUVA may be successful (WEISSMANN et al. 1978), but proof is lacking and maintenance therapy is difficult. Immunomodulators, cyclosporin A (GEBHART et al. 1986) and methisoprinol (GALBRATH et al. 1984) look promising in view of the likely immune basis for alopecia areata, but only solitary case reports have been published. Early reports of succesful treatment with topical minoxidil (WEISS et al. 1981; FENTON and WILKINSON 1983; WEISS et al. 1984) have not been repeated. The author's experience of 3% topical minoxidil in a multi-centre trial for alopecia totalis and universalis showed no better effect than placebo (FENTON et al. 1985).

K. Drugs Causing Hair Loss or Hair Gain

I. Drug-induced Hypertrichosis

Hypertrichosis is an increase in hair growth over the trunk, hand or face without any predilection for the male sexual pattern. Drugs which cause hypertrichosis as a side effect are:

Systemic

Acetazolamide	Methoin
Benoxaprofen	Minoxidil
Cortisone	Penicillamine
Cyclosporin A	Phenytoin sodium
Diazoxide	Psoralens
Diphenylhydantoin	Streptomycin
Ethotoin	Tienilic acid

Local

Minoxidil	Tretinoin
Sodium tetradecylsulphate	Vaccination

The mode of action is unknown in every case although minoxidil has been studied extensively (see Sect. G.I.2). Increased hair from diazoxide therapy affects children, but is uncommon in adults (KOBLENZER and BAKER 1968); diphenylhydantoin affects the extensor aspects of the limbs predominantly. Psoralen with natural sunlight (SINGH and LAL 1967) and as part of PUVA chemotherapy may cause hypertrichosis in up to 65% of patients (RAMPEN 1985). Generalised hypertrichosis with regrowth of male pattern baldness has been reported with benoxaprofen (FENTON et al. 1982) and minoxidil (ZAPPACOSTA 1980). Cyclosporin A causes increased hair in up to 80% of patients, affecting the face and eyebrows with reversal after stopping treatment (HARPER et al. 1984). Topical cyclosporin A is difficult to apply, owing to its lipid nature, but PENDRY and ALEXANDER (1982) have shown stimulation of hair growth on nude mice. Localised hair growth has followed compression sclerotherapy with sodium tetradecylsulphate (MARKS 1974).

II. Drug-induced Hair Loss

The following drugs have been reported to have caused hair loss as a side effect.

Allopurinol	Chloramphenicol
Arsenic	Cimetidine
Aspirin	Clofibrate
L-Asparaginase	Clomiphene citrate
Bismuth	Colchicine
Bleomycin	Coumarin anticoagulant
Boric acid	Cyclophosphamide
Bromocriptine	Cyproterone acetate
Carbamazepine	Dactinomycin
Carbon monoxide	Danazol

Diethylcarbamazine
Dipyridamole
Doxorubicin
Ethionamide
Etoposide
Etretinate
Fenofibrate
Gentamicin sulphate
Guanethidine
Heparinoids
Ibuprofen
Idoxuridine
Inanedione
Indomethacin
Interferon
Iodine
Isophosphamide
Isotretinoin
Levamisole
Levodopa
Lithium
Mepacrine
Mercury
Mesalazine
Methisazone
Methotre xate
Methyl-CCNU
Methyserguide
Metoprolol
Mitomycin
Mitozantrone
Morphine
Nadolol
Nafoxidine
Nicotinic acid
Nicotinyl alcohol
Nitrofurantoin sodium
Norethisterone
Oestrogens
Oral contraceptives
Para-aminosalicylate
Phenindione
Potassium thicyanate
Procainamide
Propranolol
Selenium sulphide
Sodium aurothiomalate
Sodium valproate
Spironolactone
Sulphasalazine
Tamoxifen
Terfenadine
Thalliumacetate
Thiampenicol
Trimethadione
Troxidone
Vincristine
Vindesine
Vitamin A

Many are case reports with little documentation. The original references can be obtained from *Martindale The Extra Pharmacopoeia* (REYNOLDS 1982).

1. Cytostatic Agents

Drugs used to treat malignancy may, because they affect dividing cells, lead to hair loss. These drugs arrest mitosis and the type of response observed is a function of the dose of drug administered. The alkylating agents: methotrexate and actinomycin D (CROUNSE and VAN SCOTT 1960), cyclophosphamide BRAUN-FALCO 1961) and colchicine (MALKINSON and LYNFIELD 1959) produce mitotic arrest of the germinal matrix, causing so-called anagen effluvium. Hair growth continues, but because the major affect is on the lower matrix cells a constriction zone of the hair shaft is formed, resulting in a weakness which may fracture 3–6 days later. Higher doses of these drugs may induce atrophy of the matrix and permanent hair loss. Thus, high dosage "pulse" regimes may be expected to have a greater effect than continuous low dosage (STOLL 1974). Reduction of scalp blood flow by ice, cooling packs and tourniquets may limit the extent of hair loss. The temporary nature of this type of hair loss following single exposure has led to its use in defleecing sheep (FEIL and LAMOUREUX 1974). There have been anecdotal reports of qualitative microscopic changes in hair following immunosuppression (KORANDA 1974) and the author has seen one patient who developed curly hair following prednisolone and another following methotrexate therapy.

2. Epidermal Growth Factor

Epidermal growth factor (EGF) was originally isolated from mouse salivary glands (COHEN 1962) and is identical to human urogastrone (GREGORY 1975). In cell culture systems EGF is a potent mitogen with powerful effects on keratinocytes (RHEINWALD and GREEN 1977). Receptors for EGF have been demonstrated on basal cells of the hair shaft and the outer root sheath (GREEN and COUCHMAN 1985) which has led to speculation that EGF might be a chemical messenger between the dermal papilla and the hair matrix. EGF increased proliferation of human hair bulb papilla cells and hair root sheath fibroblasts in culture (KATSUOKA et al. 1987). Infusion of EGF to castrated sheep stimulated mitotic activity in the basal cells of the epidermis and sebaceous glands (MOORE et al. 1984), but inhibited growth of wool follicles (HOLLIS et al. 1983), leading to shedding of the fleece. Cytostatic drugs produce the same morphological pattern of hair change and alterations in structural proteins that are seen in sheep after EGF treatment (GILLESPIE et al. 1982). The evident discrepancy between in vitro stimulation and in vivo retardation of hair growth may be explained by the many systemic effects of EGF which include a dramatic, but temporary fall in serum calcium (MOORE et al. 1986).

L. Hair Colour and Shape

Chloroquine interferes with the phaeomelanin pathway (SAUNDERS et al. 1959) and therefore affects redheads and blondes. Mephenesin, fluorobutyrophenone and triparanol cause lightening of dark hair, but the mechanism is unknown. Diazoxide may induce red discoloration. Topical medicaments may stain hair dithranol-brown; clioquinol – yellow; resorcin – yellow; benzyl peroxide – white). Curly hair may be induced by retinoids (GRAHAM et al. 1985), immunosuppressants and cytostatic drugs, but the mechanism is unknown.

References

Abraham G, Maroulis G, Buster J, Chang J, Marshall J (1976) Effect of dexamethasone on serum cortisol and androgen levels in hirsute patients. Obstet Gynecol 47:395–402
Andreyko JL, Monroe SE, Jaffe RB (1986) Treatment of hirsutism with a gonadotrophin-releasing hormone agonist. J Clin Endocrinol Metab 63:854–859
Bardin WC, Lipsett MB (1976) Testosterone and androstenedione blood production rates in normal women and women with idiopathic hirsutism and polycystic ovaries. J Clin Invest 46:891–902
Barth JH, Darley CR, Gibson JR (1985) Squaric acid dibutyl ester in the treatment of alopecia areata. Dermatologica 170:40–42

Bazzano GS, Terezakas N, Galen W (1986) Topical tretinoin for hair growth promotion. J Am Acad Dermatol 15:880–883
Braun-Falco O (1961) Klinik und Pathomechanismus der Endoxan-Alopecia als Beitrag zum Wesen cytostatischer Alopecia. Arch Klin Exp Dermatol 212:194–216
Casey JH, Burger HG, Kent JR, Kellie AE, Moxham A, Nabarro J, Nabarro JDN (1966) Treatment of hirsutism by adrenal and ovarian suppression. J Clin Endocrinol Metab 26:1370–1374
Chapman MG, Dowsett M, Dewhurst CJ, Jeffcoate SL (1985) Spironolactone in combination with an oral contraceptive: an alternative treatment for hirsutism. Br J Obstet Gynaecol 92:983–985
Chapman MG, Katz M, Dowsett M, Hague W, Jeffcoate SL, Dewhurst CJ (1986) Spironolactone in the treatment of hirsutism. Acta Obstet Gynecol Scand 65:349–350
Cohen S (1962) Isolation of a mouse submaxillary gland protein accelerating incisor eruption and eyelid opening in the newborn animal. J Biol Chem 237:1555–1562
Crounse RG, Van Scott EJ (1960) Changes in scalp hair roots as a measure of toxicity from cancer chemotherapeutic drugs. J Invest Dermatol 35:83–90
Cullberg G, Hamberger L, Mattsson L-A, Mobacken H, Samsioe G (1985) Effects of a low-dose desogestrel-ethinylestradiol combination on hirsutism, androgens and sex hormone binding globulin in women with a polycystic ovary syndrome. Acta Obstet Gynecol Scand 64:195–202
Daman LA, Rosenberg EW (1978) Treatment of alopecia areata with DNCB. Arch Dermatol 114:1036–1038
Dawber RPR, Connor BL (1971) Pregnancy, hair loss and the pill. Br Med J iv:234
De Villez RL (1985) Topical minoxidil therapy in hereditary androgenetic alopecia. Arch Dermatol 121:197–202
Dewhurst CJ, Underhill R, Goldman S, Mansfield M (1977) The treatment of hirsutism with cyproterone acetate (an antiandrogen). Br J Obstet Gynaecol 84:119–123
Dewis P, Petsos P, Newman M, Anderson DC (1985) Thre treatment of hirsutism with a combination of desogestrel and ethinyl oestradiol. Clin Endocrinol (Oxf) 22:29–36
Dorrington-Ward P, McCartney ACE, Holland S, Scully J, Carter G, Alaghband-Zadeh J, Wise P (1985) The effect of spironolactone on hirsutism and female androgen metabolism. Clin Endocrinol (Oxf) 23:161–167
Ebling FJ, Thomas AK, Cooke ID, Randall VA, Skinner J, Cawood M (1977) Effect of cyproterone acetate on hair growth, sebaceous secretion and endocrine parameters in a hirsute subject. Br J Dermatol 97:371–381
Ekoe TM, Burchardt P, Revedi B (1980) Treatment of hirsutism, acne and alopecia with cyproterone acetate. Dermatologica 160:398–404
Ettinger B, Golditch IM (1977) Medroxyprogesterone acetate for the evaluation of hypertestosteronism in hirsute women. Fertil Steril 28:1285–1288
Fazekas AG, Sandor T (1972) Metabolism of androgens by isolated human hair follicles. J Steroid Biochem 3:485–491
Feil VJ, Lamoureux C-JH (1974) Alopecia activity of cyclophosphamide metabolites and related compounds in sheep. Cancer Res 34:2596–2598
Fenton DA, Wilkinson JD (1983) Topical minoxidil in the treatment of alopecia areata. Br Med J [Clin Res] 287:1015–1017
Fenton DA, English JS, Wilkinson JD (1982) Reversal of male pattern baldness, hypertrichosis and accelerated hair and nail growth in patients receiving benoxaprofen. Br Med J [Clin Res] 284:1228–1229
Fenton DA, Wilkinson JD, Young E, Dodd H, Simpson NB, Desselberger U, Friedman P, White S, Hinchley H (1985) Topical minoxidil in the treatment of alopecia totalis (poster). Br J Dermatol 113 [Suppl 29]:31
Galbrath GMP, Thiers BH, Fudenberg HH (1984) An open-label trial of immunomodulation therapy with inosiplex (Isoprinosine) in patients with alopecia totalis and cell-mediated immunodeficiency. J Am Acad Dermatol 11:224–230
Gebhart W, Schmidt JB, Schemper M, Spona J, Kopsa H, Zazgornik J (1986) Cyclosporin-A induced hair growth in human renal allograft recipients and alopecia areata. Arch Dermatol Res 278:238–240

Giacometti L (1965) The anatomy of the human scalp. In: Montagna W, Dobson RL (eds) Hair growth. Pergamon, Oxford (Advances in biology of skin, vol IX)
Gillespie JM, Marshall RC, Moore GPM, Panaretto BA, Robertson DM (1982) Changes in the proteins of wool following treatment of sheep with EGF. J Invest Dermatol 79:197–200
Givens JR, Anderson RN, Ragland JB, Wiser WL, Umstot ES (1975) Adrenal function in hirsutism I. Diurnal change and response of plasma androstenedione, testosterone, 17 hydroxyprogesterone, cortisol, LH and FSH to dexamethasone and ½ unit of ACTH. J Clin Endocrinol Metab 40:988–1000
Gompel A, Wright F, Kuttenn F, Mauvais-Jarvais P (1986) Contribution of plasma androstenedione to 5 alpha-androstanediol glucuronide in women with idiopathic hirsutism. J Clin Endocrinol Metab 62:441–444
Graham RM, James MP, Ferguson DJP, Guerrier CW (1985) Acquired kinking of the hair associated with etretinate therapy. Clin Exp Dermatol 10:426–431
Green MR, Couchman JR (1985) Differences in human skin between the epidermal growth factor receptor distribution detected by EGF binding and monoclonal antibody recognition. J Invest Dermatol 85:239–245
Greep N, Hoopes M, Horton R (1986) Androstanediol glucuronide plasma clearance and production rates in normal and hirsute women. J Clin Endocrinol Metab 62:22–27
Gregory H (1975) Isolation and structure of urogastrone and its relationship to epidermal growth factor. Nature 257:325–327
Griffin JE, Leshin M, Wilson JD (1982) Androgen resistance syndromes. Am J Physiol 243:81–87
Griffiths WAD (1973) Diffuse hair loss and oral contraceptives. Br J Dermatol 88:31–36
Groveman HD, Ganiats T, Klauber MR (1985) Lack of efficacy of polysorbate 60 in the treatment of male pattern baldness. Arch Intern Med 145:1454–1458
Hamilton JB (1942) Male hormone stimulation is prerequisite as an incitant in common baldness. Am J Anat 71:451–480
Hamilton JB (1958) Age, sex and genetic factors in the regulation of hair growth in man: a comparison of Caucasian and Japanese populations. In: Montagna W, Ellis RA (eds) The biology of hair growth. Academic, New York, pp 399–433
Hammerstein J, Cupuceancu B (1969) Behandlung des Hirsutismus mit Cyproteronacetat. Dtsch Med Wochenschr 94:829–834
Happle R, Kalvaran KJ, Buchner U, Echternecht-Happle K, Goggleman W, Summer JH (1980) Contact allergy as a therapeutic tool for alopecia areata: application of squaric acid dibutylester. Dermatologica 161:289–297
Happle R, Hausen BM, Wiesner-Menzel L (1983) Diphencyprone in the treatment of alopecia areata. Acta Derm Venereol (Stockh) 63:49–52
Harper JI, Kendra JR, Desai S, Staughton RCD, Barrett AJ, Hobbs JR (1984) Dermatological aspects of the use of cyclosporin A for prophylaxis of graft versus host disease. Br J Dermatol 110:469–474
Hay JB, Hodgins MB (1978) Distribution of androgen metabolising enzymes in isolated tissues of human forehead and axillary skin. J Endocrinol 79:29–39
Hodgins MB; Murad S, Simpson NB (1985) A search for variation in hair follicle androgen metabolism which might be linked to male pattern baldness. Br J Dermatol 113:794 (abstr)
Holdaway IM; Croxson MS, Ibbertson HK, Sheehan A, Knox B, France J (1985) Cyproterone acetate as initial treatment and maintenance therapy for hirsutism. Acta Endocrinol (Copenh) 109:522–529
Hollis DE, Chapman RE, Panaretto BA, Moore GPM (1983) Morphological changes in the skin and wool fibres of merino sheep infused with mouse epidermal growth factor. Aust J Biol Sci 36:419–434
Horne KA, Jahoda CA, Oliver RF (1986) Whisker growth induced by implantation of cultured virbrissa dermal papilla cells in the adult rat. J Embryol Exp Morphol 97:111–124
Johnson GA, Barsuhn KJ, McCall JM (1982) Sulfation of minoxidil by liver sulfotransferase. Biochem Pharmacol 31:2949–2954

Katsuoka K, Schell H, Wessel B, Hornstein OP (1987) Effects of epidermal growth factor, fibroblast growth factor, minoxidil and hydrocortisone on growth kinetics of human hair bulb papilla cells and root sheath fibroblasts cultured in vitro. Arch Dermatol Res 279:247–250
Kligman AM (1959) The human hair cycle. J Invest Dermatol 33:307–316
Kligman AM (1961) Pathologic dynamics of human hair loss I. Telogen effluvium. Arch Dermatol Syphilol 83:175–198
Koblenzer PJ; Baker L (1968) Hypertrichosis lanuginosa associated with diazoxide therapy in prepubertal children: a clinicopathologic study. Ann NY Acad Sci 150:373–383
Koranda FC (1974) Hair changes in immunosuppressed patients. In: Brown AC (ed) First human hair symposium. Medcom, New York, p 91
Kutten F, Mowszowicz I, Mauvais-Jarvis P (1980) Androgen metabolism on human skin. In: Mauvais-Jarvis P, Vickers CFH, Wepierre J (eds) Percutaneous absorption of steroids. Academic,London, pp 99–121
Malkinson FD, Lynfield YL (1959) Colchicine alopecia. J Invest Dermatol 33:371–384
Marks CG (1974) Localised hirsuties following compression sclerotherapy with sodium tetradecyl sulphate. Br J Surg 61:127–128
Mauvais-Jarvis P (1977) Androgen metabolism in human skin: mechanisms of control. In: Martini L, Motta M (eds) Androgens and antiandrogens. Raven, New York, pp 229–245
Menard RH, Stripp B, Gillette JR (1974) Studies on the destruction of adrenal and testicular cytochrome P-450 by spironolactone. J Biol Chem 254:1726–1733
Moore GPM, Panaretto BA, Wallace ALC (1984) Treatment of ewes at different stages of pregnancy with EGF: effects on wool growth and plasma concentrations of growth hormone, prolactin, placental lactogen and thyroxine and on foetal development. Acta Endocrinol (Copenh) 105:558–566
Moore GPM, Wilkinson M, Panaretto BA, Delbridge LW, Posen S (1986) Epidermal growth factor causes hypocalcaemia in sheep. Endocrinology 118:1525–1529
Oliver RF (1966) Whisker growth after removal of the dermal papilla and lengths of the follicle in the hooded rat. J Embryol Exp Morphol 15:331–347
Olsen EA, Weiner MS, De Long ER, Pinnell SR (1985) Topical minoxidil in early male pattern baldness. J Am Acad Dermatol 13:185–192
Olsen EA, De Long ER, Weiner MS (1986) Dose-response study of topical minoxidil in male pattern baldness. J Am Acad Dermatol 15:30–37
Orentreich N (1969) Scalp hair regeneration in man. In: Montagna W, Dobson RL (eds) Hair growth. Pergamon, Oxford (Advances in biology of skin, vol IX)
Orrechia G, Rabbiosi G (1985) Treatment of alopecia areata with diphencyprone. Dermatologica 171:193–196
Pendry A, Alexander P (1982) Stimulation of hair growth on nude mice by cyclosporin A. In: White DJG (ed) Proceedings of an international conference on cyclosporin A. Cambridge. Elsevier Biomedical, Amsterdam
Rampen FHJ (1985) Hypertrichosis: a side-effect of PUVA therapy. Arch Dermatol Res 278:82–83
Reynolds JEF (1982) Martindale the extra pharmacopoeia, 28th edn. The Pharmaceutical Press, London
Rheinwald JG, Green H (1977) Epidermal growth factor and the mulitplication of cultured human epidermal keratinocytes.Nature 265:461–424
Rosenberg EW, Drake L (1976) Alopecia areata: discussion. Arch Dermatol 112:256
Rosenfield RL (1971) Plasma testosterone binding globulin and indexes of the concentration of unbound plasma androgens in normal and hirsute subjects. J Clin Endocrinol Metab 32:717–728
Saunders TS, Fitzpatrick TB, Seiji M, Brunet P, Rosenbaum EE (1959) Decrease in human hair colour and feather pigment of fowl following chloroquine diphosphate. J Invest Dermatol 33:87–90
Schmidt JB, Huber J, Spona J (1985) Medroxyprogesterone acetate therapy in hirsutism. Br J Dermatol 113:161–165
Schweikert HU, Wilson JD (1974) Regulation of human hair growth by steroid hormones, i. Testosterone metabolism in isolated hairs. J Clin Endocrinol Metab 38:811–819

Schweikert HU, Milewich L, Wilson JD (1975) Aromatisation of androstenedione by isolated human hairs. J Clin Endocrinol Metab 40:313–417
Shapiro G, Evran S (1980) A novel use of spironolactone: treatment of hirsutism. J Clin Endocrinol Metab 51:429–432
Shuster S (1972) Primary cutaneous virilism or idiopathic hirsuties? Br Med J 2:285–286
Shuster S (1982) The sebaceous glands and primary cutaneous virilism. In: Jeffcoate SL (ed) Androgens and antiandrogen therapy. Wiley, Chichester
Simpson NB, Cunliffe WJ, Hodgins MB (1983) The relationship between the in vitro activity of 3β-hydroxysteroid dehydrogenase Δ^{4-5} isomerase in human sebaceous glands and their secretory activity in vitro. J Invest Dermatol 81:139–144
Singh G, Lal S (1967) Hypertrichosis and hyperpigmentation with systemic psoralen treatment. Br J Dermatol 79:501–503
Stoll BA (1974) Evaluation of cyclophosphamide dosage schedules in breast cancer. Br J Cancer 24:475–483
Thomas JP, Oake RJ (1974) Androgen metabolism in the skin of hirsute women. J Clin Endocrinol Metab 38:19–22
Tosti A, Bardazzi F, De Padova MP, Caporieri GM, Melino M, Veronesi S (1985) Contact dermatitis to minoxidil. Contact Dermatitis 13:275–276
Tosti A, De Padora MP, Minghetti G, Veronesi S (1986) Therapies versus placebo in the treatment of patchy alopecia areata. J Am Acad Dermatol 15:209–210
Underhill R, Dewhurst J (1979) Further clinical experience in the treatment of hirsutism with cyproterone acetate. Br J Obstet Gynaecol 86:139–141
Uno H, Cappas A, Schlagel C (1985) Cyclic dynamics of hair follicles and the effect of minoxidil on the bald scalps of stumptailed macaques. Am J Dermatopathol7:283–297
Vanderveen EC, Ellis CN, Kang S, Case P, Headington JT, Voorhees JJ, Swanson NA (1984) Topical minoxidil for hair regrowth. J Am Acad Dermatol 11:416–421
Vermorken AJM (1983) Reversal of androgenetic alopecia by minoxidil: lack of effect of simultaneously administered intermediate doses of cyproterone acetate. Acta Derm Venereol (Stockh) 63:268–269
Warin AP, Hehir ME, Du Vivier A (1979) Alopecia areata treated with DNCB. Clin Exp Dermatol 4:385–387
Weiss VC, West DP, Mueller CE (1981) Topical minoxidil in alopecia areata. J Am Acad Dermatol 5:224–226
Weiss VC, West DP, Fu TS, Robinson LA, Cook B, Cohen RL, Chambers DA (1984) Alopecia areata treated with topical minoxidil. Arch Dermatol 120:457–463
Weissmann I, Hoffmann C, Wagner G, Plewig G, Braun-Falco O (1978) PUVA therapy for alopecia areata. Arch Dermatol Res 262:333–336
Wester RC, Maibach HI, Guy RH, Novak E (1984) Minoxidil stimulates cutaneous blood flow in human balding scalps: pharmacodynamics measured by laser doppler velocimetry and photopulse plethysmography. J Invest Dermatol 82:515–517
Yen SSC (1980) The polycystic ovary syndrome. Clin Endocrinol (Oxf) 12:177–208
Zappacosta AR (1980) Reversal of baldness in a patient receiving minoxidil for hypertension. N Engl J Med 303:1480–1481

CHAPTER 38

Photochemotherapy

R. S. STERN, J. A. PARRISH, and B. JOHNSON

A. Definition and History

Photochemotherapy is the use of a chemical plus exposure to nonionising electromagnetic radiation to achieve a biological effect which neither has by itself. Whether given topically or parenterally, the chromophore or photosensitiser must reach viable cells which subsequently absorb photons of radiation. Psoralen-UV-A (PUVA) therapy was discovered thousands of years ago and remains the best example of successful photochemotherapy (PARRISH et al. 1974); it will be discussed later on. A further example of photochemotherapy is the use of phototoxic chemicals to treat cancer. The term photoradiation therapy (PRT) is used by DOUGHERTY et al. (1983) to describe the use of a photodynamic haematoporphyrin derivative (Hpd) which localises in malignant tumour and inflamed tissues and produces necrosis of tumour tissues upon subsequent red light exposures at or near 630 nm. Other phototoxic dyes localise in tumour tissues. TANENIER and JESIONEK (1903) used eosin and sunlight to cause regression of skin tumours and GRANELLI et al. (1975) used intravenous haematoporphyrin to treat transplanted gliomas in rats. Acridine orange, a phototoxic dye, also causes tumour regression in mice exposed to argon laser radiation after oral ingestion of the compound (TOMSON et al. 1974). The first report of complete regression of experimental tumours owing to selective phototoxicity was with intravenous Hpd and red light from a filtered xenon arc lamp (DOUGHERTY et al. 1978).

Hpd seemed suitable for clinical use because it is preferentially retained in tumour tissue for days, and because its photodynamic effect is elicited with the more penetrating long wavelength visible radiation. Hpd PRT appears to involve production of singlet oxygen with alteration in the plasma membrane, leading to cell death (DOUGHERTY et al. 1982). As Hpd is widely distributed intracellularly, singlet oxygen may cause damage by a variety of radical intermediates. The intracellular targets and repair mechanisms are not known. The explanation of the preferential retention of Hpd by tumour tissue is unknown, as is the chemical identity of its photochemically active or fluorescent components. The fluorescence of tumour tissue is not well correlated with uptake of either ^{3}H- or ^{14}C-labelled Hpd (GOMER and DOUGHERTY 1979), suggesting that different components or products of Hpd may be responsible for fluorescence, binding and/or phototoxic effects. One major side effect of Hpd PRT is that normal skin remains photosensitive to visible light for weeks or months and severe phototoxicity has been observed after accidental light exposure.

B. Photobiology

Electromagnetic radiation comprises a continuous spectrum of wavelengths varying from fractions of an Ångstrom (1 Å $=10^{-10}$ m) to millions of metres (Fig. 1). The portion of the spectrum which is of direct concern to human photobiology extends from about 200 to 800 nm. This radiation penetrates skin, causing electronic excitation when absorbed by molecules, a process dependent upon the configuration of the molecule and photon energy.

Ultraviolet-induced photochemistry within skin cells can lead to a host of chemical, metabolic and structural changes, most of which are injurious. If enough cell injury occurs in the skin, an inflammatory reaction follows after a latent period of several hours. The erythema component has received most attention because it is easily observed by non-invasive techniques, but many changes including epidermal cell death, DNA damage, melanogenesis, and decrease in threshold to subsequent exposure occur after suberythemogenic exposures to UV. The action spectrum for delayed erythema is obtained by plotting the reciprocal of the erythemal threshold dose at each wavelength. The exact shape of the curve from 250 to 290 nm varies depending upon experimental method. At wavelengths longer than 290 nm erythemal effectiveness falls abruptly so that radiation at 320–400 nm is over 1000 times less erythemogenic (Fig. 2).

Arbitrary divisions of the UV range (Figs. 1 and 2) are used in medical photobiology because terrestrial sunlight does not usually include wavelengths shorter than 290 nm and because wavelengths between 290 and 320 nm are the most strongly erythemogenic. Wavelengths shorter than 180 nm are absorbed by air and are therefore called the *vacuum UV*. Because UV (180–290 nm) is efficient at killing one-celled organisms, it is often called *germicidal radiation*. Because of its physical and photobiological properties, UV-B (290–320 nm) is often referred to as the *sunburn (or erythema) spectrum* or middle wavelength UV. Because many substances emit visible fluorescence when irradiated with long wavelength UV, UV-A (320–400 nm) is sometimes referred to as *black light*.

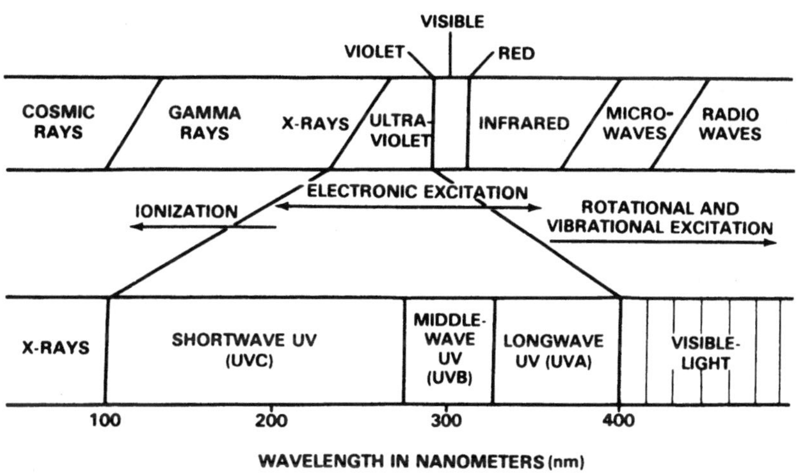

Fig. 1. Electromagnetic spectrum with expanded scale for ultraviolet radiation

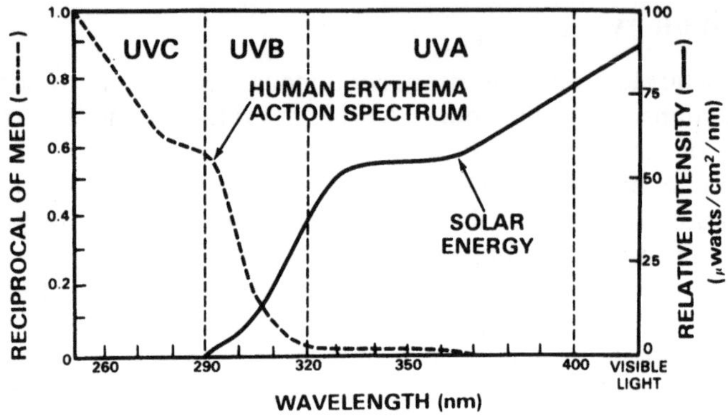

Fig. 2. Diagrammatic representation of human erythema action spectrum and terrestrial solar spectrum. These curves can roughly define the UV wavebands described in the text

Phototoxicity or cell injury by photons is central to many forms of photochemotherapy of skin disease. It is assumed that the radiation-induced cellular alterations and cascade of tissue reactors in normal skin also occur in diseased skin and that changes in the latter may lead to improvement of diseased skin without unacceptable damage to normal skin. Ultraviolet radiation may decrease abnormal hyperproliferation, interfere with function of abnormal cells or kill cells intimately involved in pathophysiological expression of disease. These effects may be augmented by photosensitisation chemicals. Abnormal cells may be more sensitive to certain phototoxic effects.

C. Photochemistry and Photobiology in Relation to Photochemotherapy

Only absorbed energy can produce change (the Grotthus-Draper law) although absorbed energy may be dissipated harmlessly as heat or it may be re-emitted as fluorescence or phosphorescence. Nonetheless, where the conditions are appropriate, a photochemical reaction will occur.

The electrons of an atom may be depicted as being distributed around the nucleus in successive shells, the orbitals of which represent discrete energy levels. Normally, the electrons occupy the lowest possible energy levels giving the "ground state" of the atom. If an external energy source is applied, as seen with an electric discharge in a low pressure mercury arc lamp, the electrons in the mercury atoms are raised to the characteristic higher levels of the "excited states" which are inherently unstable with a half-life around 10^{-9} s. The electrons return to the ground state and the absorbed energy is emitted as radiation in the form of particles of energy, photons or quanta, the magnitude of which is directly related to the difference in energy levels of the excited and ground states. The quantum energy is inversely related to the wavelength of the emitted radiation and each atomic species, with its unique electron content and distribution, will emit radiation in a specific wavelength pattern. Similarly, the absorption of UV or vis-

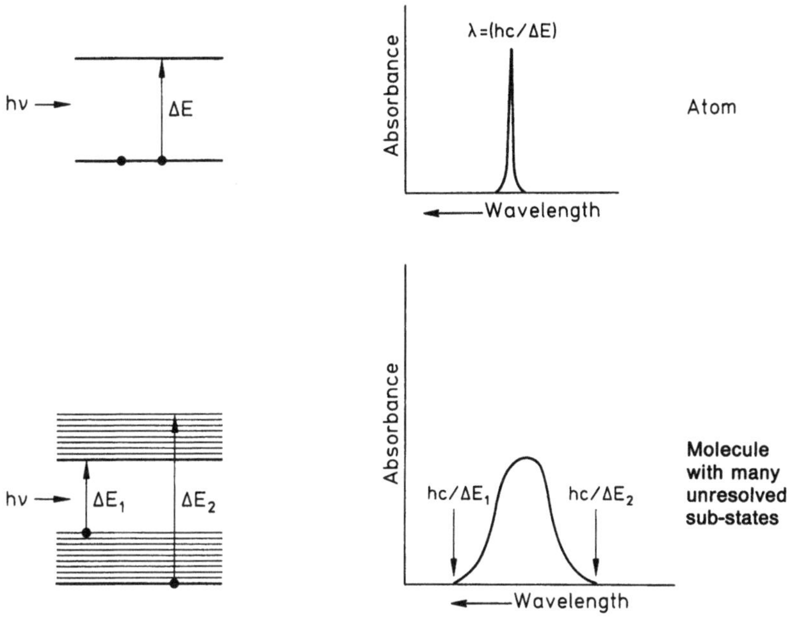

Fig. 3. The derivation of an absorption spectrum. (Adapted from CLAYTON 1970)

ible radiation depends on the energy level differences available so that each atomic species has its own characteristic absorption pattern. This is the basis of the wavelength dependence for both absorption spectroscopy, by which a chemical may be partially identified, and action spectroscopy by which the wavelengths involved in photochemical and photobiological reactions may be defined. With atoms, single-step transitions between discrete energy levels result in absorption spectra comprised of single lines at wavelengths corresponding to these energy differences (Fig. 3). With the more complex molecular interactions with radiation, the basic principles still apply. However, whereas the atom has clearly defined ground and excited states, molecules have numbers of sub-states, derived from the different modes of vibration and rotation of atomic nuclei and groups of nuclei in the molecule. Therefore, the single-step transition of the atom is replaced by a range of possible steps and the close spacing of the spectral lines resulting from these steps, plus the effects of interaction between the molecule and its surroundings, leads to the formation of a broad absorption band (Fig. 3). In this way, organic chemicals and complex biomolecules in solution have characteristic absorption spectra (Fig. 4).

In the ground state, electrons exist in pairs, with their spins in opposing directions to give a net electron spin of zero in what is called a singlet state. With appropriate irradiation, one electron of a pair is raised to a higher energy level. The spin direction is retained in this initial step and the molecule is at this stage in an "excited singlet state". However, in intersystem crossing, which requires no further irradiation, the electron spin may be reversed to give rise to an excited triplet state with a net electron spin of 1. The reversal of spin direction cannot be induced

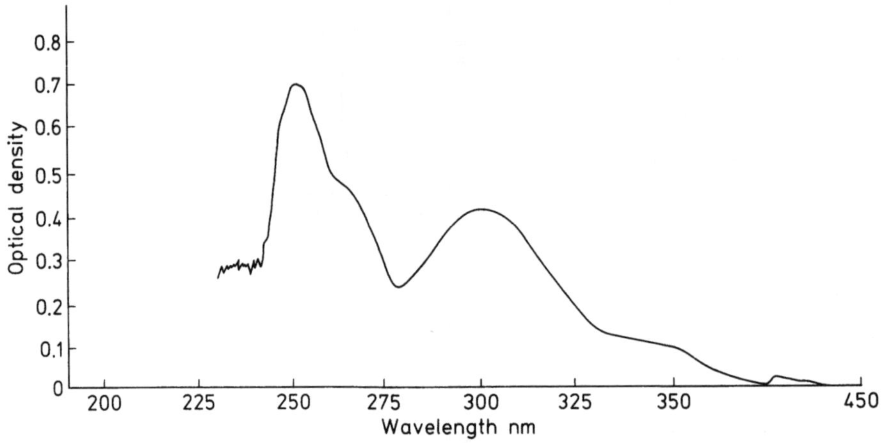

Fig. 4. The absorption spectrum of 8-methoxypsoralen in chloroform. While the major absorption peaks are in the UV-C and UV-B, there is significant absorption in the UV-A wavelengths above 325 nm

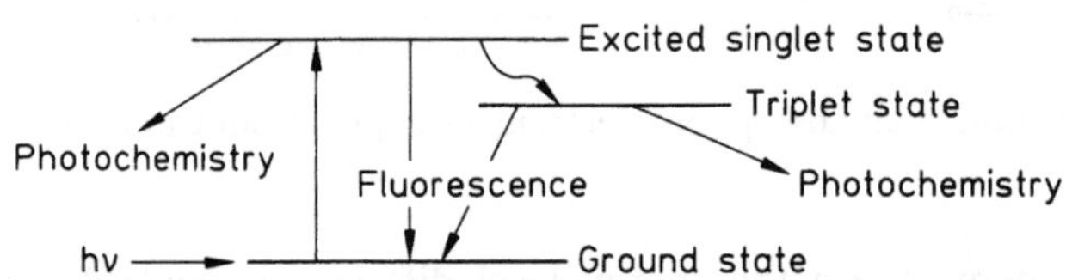

Fig. 5. Pathways for electron excitation and relaxation giving rise to fluorescence or photochemistry

directly by absorbed energy and the direct transition from the ground state to excited triplet does not occur. The excited singlet state is short lived, existing for around 10^{-9} s, but the triplet state is "metastable" having a lifetime between 10^{-3} s and, in rare instances, 100 s or more. This longer lived excited state obviously has a greater potential for chemical change. In certain instances, the absorbed energy is sufficient to remove an electron completely from an atom, giving rise to an unbalanced, often highly reactive molecule in a free radical state (MASON and CHIGNELL 1982).

The absorption of energy by a chemical may result only in fluorescence or phosphorescence and the analysis of this light emission may be useful in identifying the chemical involved and may also give some information concerning the possible mechanisms for photochemical change. However, photochemical reactions may result from either excited singlet or triplet states (Fig. 5), the absorbed energy leading to the rearrangement of molecular structure, the dissociation of existing chemical bonds or the formation of new bonds and the production of free radicals with subsequent chemical change. When chemical change takes place it may take various forms (TURRO and LAMOLA 1977). A simple break of molecular bonding, as occurs in the photolysis of hydrogen iodide, may also occur in more complex molecules such as phenothiazines and halogenated salicylanilides with the release of halogen atoms. Molecular bond rearrangement is seen in the UV-induced breakage of disulphide bonds leading to enzyme inactivation and in the UV-induced opening of the second benzene ring of 7-dehydrocholesterol to form pre-vitamin D_3. A third reaction category is photochemical isomerisation as in the *cis* to *trans* change in visual pigment or the *trans* to *cis* change in urocanic acid. These are essentially unimolecular reactions. Bimolecular reactions are more complex, but have greater relevance for photochemotherapy. Typical of such reactions are the oxygen-independent UV-induced thymine dimer formation in DNA and the UV-A-induced covalent binding of psoralens to DNA. Where oxygen is involved, as in the protoporphyrin photosensitised oxidation of cholesterol, it is the oxygen itself which is part of the bimolecular reaction, being transformed into the highly reactive excited singlet state.

D. Photosensitisation

Photosensitisation may be defined as the action of one component (the photosensitiser) in a system which causes another component (the substrate) to react to radiation (LAMOLA 1974). It is a process in which abnormal reactions are induced in a system by the introduction of a specific, radiation absorbing substance (BLUM 1964; PATHAK 1969; SPIKES 1977). Within the context of photochemotherapy, the process is usually considered in terms of supposedly harmless radiation (UV-A and visible) being absorbed by a "foreign" molecular species, the cell and tissue reactions involved in the therapeutic effects resulting from the dissipation of the absorbed energy. The pathways available for this energy dissipation vary, particularly in relation to the molecular structure of the photosensitiser, the nature of the biomolecule with which it becomes associated and whether or not molecular oxygen is involved.

Because the lifetime of the photosensitiser excited singlet state (designated ^1P) is very short, between 10^{-9} and 10^{-6} s, the possibility of interaction with neighbouring molecules to produce a photosensitised reaction is low. Such interactions are more probable with the longer lifetime of the excited triplet and it is a general rule that photosensitisation proceeds via the triplet state (^3P).

$$P_0 \longrightarrow {}^1P \longrightarrow {}^3P \longrightarrow \text{Photosensitisation}$$

Fluorescence (from ^1P and ^3P) $\downarrow P_0$ $\quad$ $\downarrow P_0$

Although photosensitising chemicals generally exhibit fluorescence, there is no correlation between the intensity of fluorescence and photosensitising potential.

There are various pathways by which the triplet state photosensitiser may change substrate (S) molecules. First, in what are classified as type I reactions (FOOTE 1974) electron or hydrogen transfer gives rise to reactive free radical forms of both substrate and photosensitiser. Alternatively, the substrate may be oxidised directly by the highly reactive superoxide-hydrogen peroxide products of interactions between free radical photosensitiser and oxygen. Type II, energy transfer, reactions involve the transfer of excitation energy from the triplet state photosensitiser directly to ground state oxygen. Attempts to differentiate between type I and type II mechanisms necessarily simplify a complex situation. Both may operate for the same overall photosensitised reaction; the pathway may change, depending on pH, nature of the solvent or concentration of photosensitiser. However, in general, photosensitisers such as anthraquinones and benzophenone, in which the major energy transition is relatively low, react through a type I mechanism. Acridine, xanthenes such as eosin Y and rose bengal, thiophenes such as methylene blue, and the photosensitising porphyrins such as protoporphyrin and haematoporphyrin and its derivatives produce excited singlet oxygen (Spikes 1982).

Most photosensitised reactions are oxygen dependent and the term "photodynamic action" is reserved for such reactions. They may cause damage to lipid membranes, mainly through lipid peroxidation, enzyme denaturation and damage to nucleic acids, mainly through effects on guanine bases. A close spatial relationship between the photosensitiser and substrate molecules is required, even when excited singlet oxygen is involved. For photosensitisation by the furocoumarins and the psoralens in particular, these relationships are of major importance.

E. The Psoralens

The psoralens derive their name from the leguminous plant, *Psoralea corylifolia*, seeds of which were used for centuries in the ayurvedic treatment for leucoderma (Soine 1964). Psoralen (Fig. 6), the photoactive constituent of these seeds, was characterised as a furocoumarin compound (Jois et al. 1933). The furocoumarins generally are heterocyclic molecules formed through the condensation of a furan ring onto the benz-α-pyrone, coumarin. The complex nomenclature (Scott et al. 1976) is simplified by using the conventional ring numbering system for the coumarin and furan moieties, with primes on the numbers for the furan ring (Fig. 6) (Spath 1937). Although there are numerous ways in which the molecular condensation might occur, there appear to be only two forms in nature. The psoralens are specifically linear furocoumarins in which the furan ring is joined at its 3′,2′ bond to the 6,7 bond of the coumarin. The second form, seen in the angular furocoumarins related to angelicin (isopsoralen) isolated from the umbelliferous

Psoralen

Angelicin

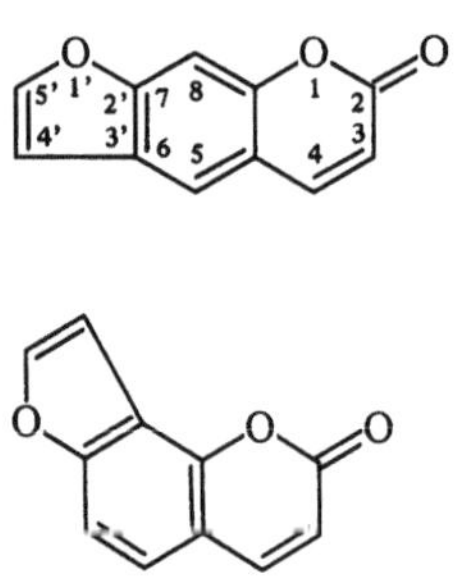

Fig. 6. The molecular structures of psoralen and angelicin

plant, *Angelica archangelica*, has the condensation at the 7,8 bond of the coumarin and the furan ring is reversed (Fig. 6).

While many different forms of both psoralens and angelicins have been synthesised (MUSAJO and RODIGHIERO 1962; PATHAK et al. 1967) they occur naturally in limited numbers, confined mainly to plants of the Umbelliferae, Rutaceae, Moraceae and Leguminosae and are listed by PATHAK et al. (1962). SOINE (1964) lists 19 psoralens, 4 dihydropsoralens, 5 angelicins and 6 dihydroangelicins in a concise table of these naturally occurring coumarins (see also MITCHELL and ROOK 1979 for furocoumarin-containing plants causing photosensitivity).

The first furocoumarin derived from plant material was isolated by Kallbruner in 1834 from the perfume constituent bergamot oil (bergapten, FOWLKS 1959) and later characterised as 5-methoxypsoralen (5-MOP); xanthotoxin (THOMS 1911) was characterised as 8-methoxypsoralen (8-MOP). These compounds were shown to cause abnormal skin reactions to sunlight after contact with extracts from figs (*Ficus carica*), common rue (*Ruta graveolens*) and various Umbelliferae (KUSKE 1938). The coumarins have a number of pharmacological effects including anti-coagulant, vasodilator, oestrogenic, anti-bacterial and sedative actions (SOINE 1964) whereas the furocoumarins are relatively inactive apart from cutaneous photosensitisation.

Of the naturally occurring furocoumarins, 8-MOP, 5-MOP and psoralen of the linear compounds and angelicin of the angular type have been subject to most study.

Recent studies suggest 5-MOP may have equal therapeutic efficacy to that of 8-MOP, but a lower risk of the photosensitivity (TANEW et al. 1988).

Of the many synthesised furocoumarins, trimethylpsoralen has been the major compound of interest, but increasingly, new and varied compounds such as 3-carbethoxypsoralen (AVERBECK et al. 1978) different methylangelicins (AVERBECK et al. 1984) and pyridopsoralens (DUBERTRET et al. 1984) are being produced and their effects investigated.

I. Pharmacology

The first studies of pharmacological aspects of the psoralens were concerned with the hepatotoxic potential of oral 8-MOP in humans. No evidence was found for any change in liver function after daily doses of 30 mg over 3 months (LABBY et al. 1959) or 20 mg over 3 weeks (TUCKER 1959). Subsequently, this finding was confirmed in over 1000 of the many subjects undergoing PUVA therapy for psoriasis (MELSKI et al. 1977) and any liver damage associated with 8-MOP ingestion is apparently restricted to those with predisposing factors such as previously damaging treatment with methotrexate or excessive alcohol intake (FREEMAN and WARIN 1984).

Early studies of psoralen pharmacokinetics and metabolism were concentrated on 4,5′,8-trimethylpsoralen (TMP) and the treatment of vitiligo, but included investigations of 8-MOP, and demonstrated a maximum plasma concentration 2–3 h after ingestion in experimental animals and humans. This coincided with the peak of cutaneous photosensitivity to UV-A. Examination of urine and faecal specimens showed that 90% was excreted as metabolites, mainly in the urine, in the first 12 h and the remainder during the next 12 h. Repeated doses failed to result in tissue accumulation of the drugs (PATHAK et al. 1974) although more recent studies with mice indicate that some accumulation may occur in liver, gall bladder, kidney and retina (WULF 1984). It is not clear whether this is 8-MOP or a metabolite, but it is probably the latter. Since no parent drug was detected in the urine, the psoralens are obviously not only absorbed rapidly from the gut, but also rapidly metabolised, a desirable feature of a drug used in a treatment such as photochemotherapy. The early studies associated with clinical trials of PUVA therapy were mainly concerned with plasma concentrations of 8-MOP 2 h after ingestion (THUNE and VOLDEN 1977; STEINER et al. 1978; HENSBY 1978; HERFST et al. 1978) the time of supposed maximal photosensitivity (KLIGMAN and GOLDSTEIN 1973) and therefore of maximal therapeutic efficacy. After standard doses of 0.5–0.6 mg/kg body weight with the commercial preparations of 8-MOP available, considerable inter-individual variation was found, from around 50 to greater than 300 ng/ml. This is a greater variation than may be explained through inconsistency in an individual drug preparation (HENSBY 1978) and is only partially explained by differences in drug preparation (LJUNGGREN et al. 1980; MENNE et al. 1981) or the effects of dietary variation (EHRSSON et al. 1979; ROELANDTS et al. 1981; HERFST and WOLFF 1982). A more detailed pharmacokinetic study to determine the plasma 8-MOP profile over a 24-h period (Fig. 7) shows a typical rapid build-up to a peak plasma concentration (123 ng/ml) at 2 h after ingestion followed by a similar rapid decay ($t_{1/2}=1.9$ h) until little 8-MOP is detected at 12 h and none at 24 h (STEVENSON et al. 1981). This appears to be the normal pharmacokinetic behaviour of the drug (STEINER et al. 1977; THUNE and VOLDEN 1977; BUSCH et al. 1978) although with different drug preparations, the peak plasma level may be reached earlier (GAZITH et al. 1978; THUNE 1978). This earlier peaking may also be observed in fasting, as opposed to non-fasting subjects (HERFST and WOLFF 1982). However, especially in psoriasis patients who do

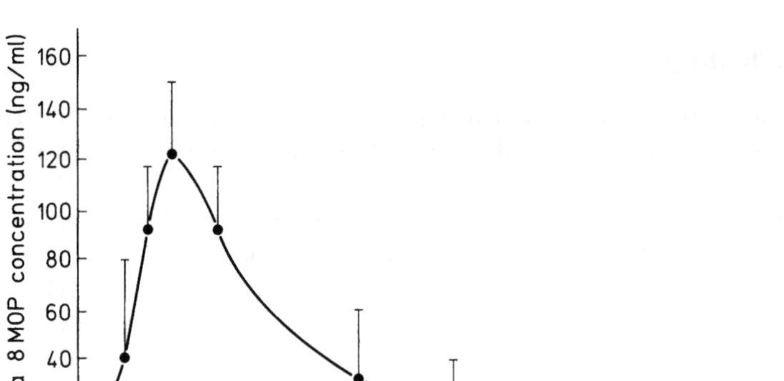

Fig. 7. Plasma 8-MOP concentrations following a single oral dose of 0.5–0.6 mg/kg body weight in four subjects with psoriasis who responded well to standard PUVA therapy. Means + standard errors of the means (STEVENSON et al. 1981)

not respond well to oral PUVA therapy, an abnormal plasma profile may be obtained with either consistently low levels or a prolonged latent period for peak level attainment (WAGNER et al. 1979; STEVENSON et al. 1981). The pathophysiology underlying these differences is not understood.

Newer psoralen formulations have greater bioavailability. Peak levels are higher and more rapidly achieved (SULLIVAN et al. 1986). Unfortunately, gastrointestinal and central nervous system toxicity may be more frequent with these preparations (DOVER and STERN 1986).

It is now established that the bioavailability of 8-MOP, expressed in terms of plasma concentration, depends on a metabolic "first-pass" effect which is readily saturable (SCHMID et al. 1980a) and that a 4-fold increase in dose may produce a 25-fold increase in plasma level. It is clear that, especially with "quick release" preparations, minor variations around the level of the first-pass saturation may produce considerable variations in final peak plasma levels (BRICKL et al. 1984). However, any consideration of variations in blood levels of 8-MOP should take into account the finding that serum concentrations are higher than those in plasma (WULF and ANDREASEN 1981). In reviewing the possible contribution of drug formulation differences to these variations, BRICKL et al. (1984) suggest that 8-MOP in solution, as described by STOLK et al. (1980) and HONIGSMANN et al. (1982) produce high and reproducible plasma levels in a short time, but that crystalline formulations produce widely varying plasma levels and even the best is inferior to a soft gelatin capsule preparation containing 8-MOP in a dissolved state.

The typical 8-MOP plasma profile obtained is indicative of a simple, one-compartment, first-order process. However, studies of metabolite excretion in the urine suggest a two-compartment model (SCHALLA et al. 1976) although some confusion in the interpretation of these results may arise because the radiolabelling required for such studies may lead to the formation of tritiated water which has a longer biological life than the psoralen metabolites (WULF and ANDREASEN 1981). Nonetheless, close examination of the relationship between cutaneous photosensitivity and plasma levels of 8-MOP do reveal that peak photosensitivity occurs later than the peak plasma level and remains high for up to 2 h after the plasma level has begun to fall (BRICKL et al. 1984). Determinations of 8-MOP in suction blister fluid shows lower absolute levels than in plasma, but a close correlation in kinetics (HERFST et al. 1980; KORTING et al. 1982; LAUHARANTA et al. 1982). It is clear that suction blister fluid is no better an indication of "active" psoralen in the skin than is plasma and the prolonged retention of 8-MOP has been demonstrated in rats where an overall correlation between plasma levels and skin content is apparent, but where the skin concentration remains at a high level when the plasma level has begun to fall (ROELANDTS et al. 1983).

Rectal administration of 8-MOP provides rapid and consistent peak plasma levels of the same order as the best oral preparations and the absorption and excretion may occur more rapidly (STOLK et al. 1981). Topical applications, as 1% 8-MOP in various vehicles, if applied to one-half of the body surface, should produce plasma levels similar to those obtained with routine oral 8-MOP therapy (KAMMERAU et al. 1976). However, topical administration of 8-MOP through bath treatments in which subjects are immersed for 10 min in 150 l warm water containing 60 mg 8-MOP, produces serum levels of only 3–15 ng/ml although the treatment is highly successful (HUUKSONEN et al. 1984).

Plasma levels obtained with 5-MOP solutions given orally are around the same as those of 8-MOP (SCHMID et al. 1981), but with other preparations, 5-MOP levels are consistently lower, about 30–100 ng/ml (STOLK et al. 1981; NITSCHE and MASCHER 1982) possibly because 5-MOP is less water soluble than 8-MOP by a factor of 6.

The metabolic pattern for 8-MOP was compared in humans, rats and dogs by BUSCH et al. (1978). Following oral administration of ^{14}C-labelled 8-MOP, no unchanged parent compound was detected in the urine, but several polar metabolites were present, some of which were conjugates, and some polar compounds were also detected in faeces. This result is similar to that obtained by MANDULA et al. (1976) who found at least four fluorescent metabolites of 8-MOP in the urine of treated mice and humans. PATHAK et al. (1975) cited unpublished studies indicating that the biotransformation of 8-MOP in mice involved hydroxylation, glucuronide formation, epoxide formation and opening of the pyrone ring. KOLIS et al. (1979) studied the metabolism of 8-MOP in the dog, finding that the opening of the furan ring or the pyrone ring, followed by conjugation, is a major pathway in this species. Three of the metabolites obtained resulted from opening of the furan ring and were designated as A (7-hydroxy-8-methoxy-2-oxo-2*H*-1-benzopyran-6-acetic acid), B (α-7-dihydroxy-8-methoxy-2-oxo-2*H*-1-benzopyran-6-acetic acid), and D, an unknown conjugate of A at the 7-hydroxy position. The fourth, formed by opening the pyrone ring, was also an unidentified conjugate, designated C.

This general pattern has been confirmed for 8-MOP metabolism in rats and humans and it is clear that opening of the pyrone ring at the 1,2 position and demethylation to form 8-hydroxypsoralen are two minor components of the process (SCHMID et al. 1980b; EHRSSON et al. 1978). The major pathways for the metabolism of 8-MOP are shown in Fig. 8, as presented by SCHMID et al. (1980b) and it is interesting that BLAIS et al. (1982) detected a psoralen metabolite with fluorescence characteristics similar to those of M_2 in this figure, in suction blister fluid from subjects on PUVA treatment. This is, in turn, the metabolite designated A by KOLIS et al. (1979). The metabolism of 5-MOP is essentially the same as that of 8-MOP and takes place, apparently exclusively, in the liver. The microsomal fraction is primarily involved with a major activity being ascribed to the cytochrome P-450 system (BRICKL et al. 1984). Some indication that 8-MOP is metabolised by this system in vitro is presented by MANDULA and PATHAK (1979), but metabolism is only minimal. However, if optimal conditions are used, in particular an 8-MOP concentration of 10^4 *M*, the in vitro metabolism of 8-MOP is unequivocally demonstrated (MOHAMED 1984). No similar results were obtained with skin. The metabolism of TMP in vitro is entirely different from that of the other psoralens, consisting of oxidation at the methyl groups, yielding alcohols and carboxylic acids (MANDULA and PATHAK 1979). The major metabolite, derived from the action of mixed function oxidases, is 4,8-dimethyl,5′-carboxypsoralen, previously identified in the urine of mice and humans after oral administration of TMP (MANDULA et al. 1976). A second metabolite, identified in the in vitro study was 4,8-dimethyl,5′-hydroxymethylpsoralen and it is apparent that

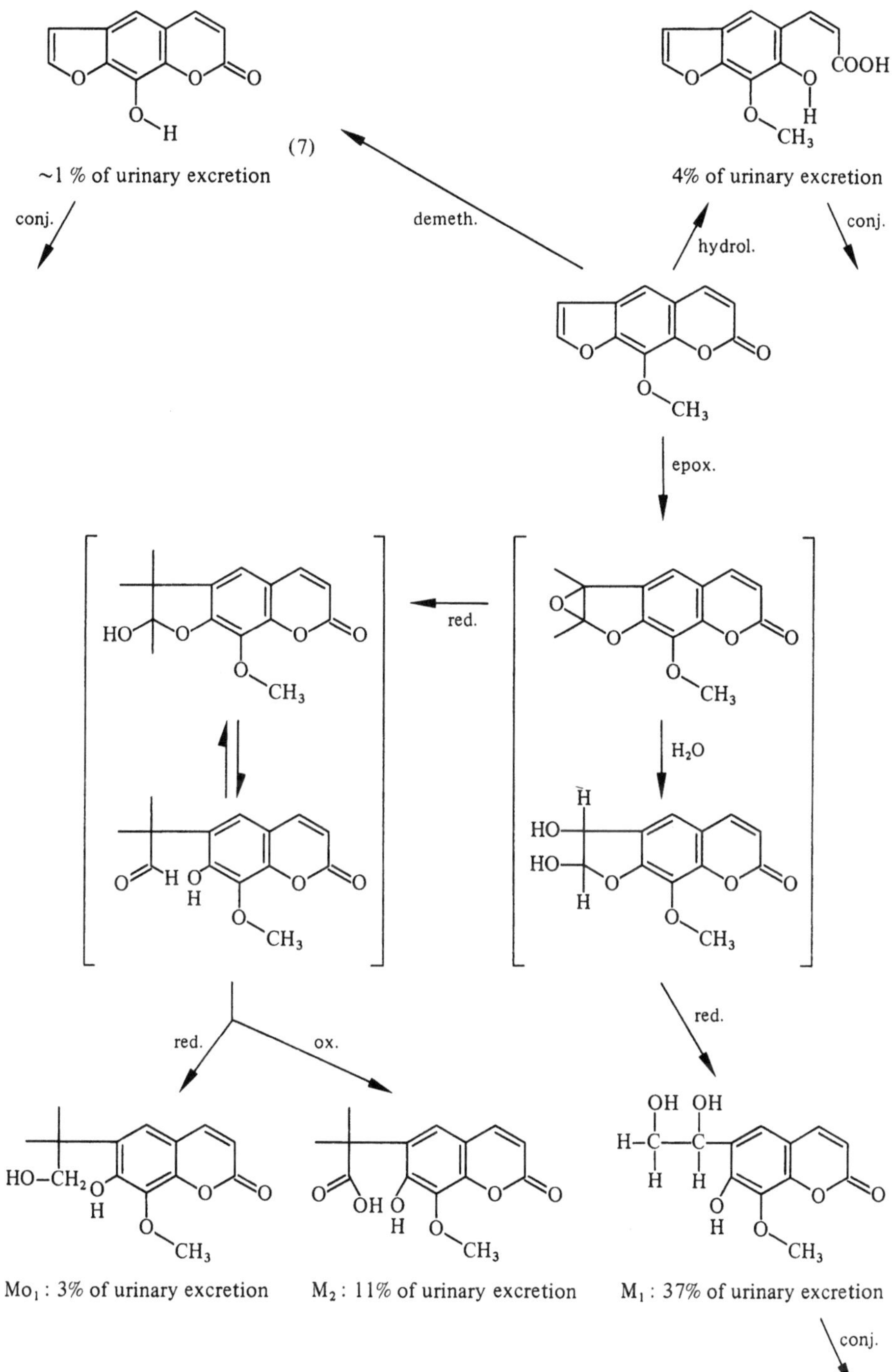

Fig. 8. Proposed metabolic pathway for 8-MOP (SCHMID et al. 1980b)

other, less easily identifiable metabolites are formed. The phototoxic potential of the dimethylcarboxypsoralen is significantly less than that of TMP or 8-MOP (PATHAK et al. 1983) although its methyl ester appears to be quite active. Unpublished observations (cited by PATHAK et al. 1983) purport to indicate that the metabolites of 8-MOP are photoreactive. Overall, the indications are that the difference in photosensitising efficacy of TMP and 8-MOP when given orally is made up from the relatively low solubility of TMP, its more rapid metabolism and the formation of non-photosensitising metabolites. The varying reports in the literature make this matter somewhat contentious still (CHAKRABARTI et al. 1983) and more work is required. While orally administered 8-MOP in mice and rats was shown to be capable of inducing liver aryl hydrocarbon hydroxylase (AHH), ethylmorphine *N*-demethylase and cytochrome P-450 (although to a lesser extent than do barbiturates) TMP had no effect. No effect of topical 8-MOP on cutaneous AHH activity was detected (BICKERS et al. 1982).

II. Photochemistry

Although the first studies of psoralen photochemistry clearly established the possibility of free radical production by exposure to UV-A (PATHAK et al. 1961) the photosensitised killing of bacteria with 8-MOP is independent of oxygen (OGINSKY et al. 1959) unlike that of toluidine blue (MATTHEWS 1963). Moreover, other tests for oxygen-dependent photosensitisation such as photodynamic haemolysis, and the photo-oxidation of α-terpinen and serum proteins were unrewarding (MUSAJO and RODIGHIERO 1962). It appeared that the mechanisms involved in psoralen photosensitisation were different from those for the majority of known photosensitisers. Evidence for the photosensitised interaction between psoralens and DNA was first presentd by MUSAJO et al. (1965). It is now clear that the major photobiological effects of the psoralens are mediated through the UV-A-induced covalent binding of psoralen molecules into chromosomal DNA to produce monofunctional photoadducts and bifunctional, interstrand cross-links with pyrimidine bases, thymines in particular, in the DNA double helix (MUSAJO and RODIGHIERO 1972; PATHAK et al. 1976; SONG and TAPLEY 1979).

Where psoralens are taken into cells, there is a preferential association with DNA. The psoralen molecule is held within the DNA double helix by weak binding forces so that an intercalation complex is formed with the planar furocoumarin molecule between two base pairs of the DNA. Unless the system is saturated, this dark reaction is transient and the psoralen diffuses quite rapidly out of the cell. However, if the system is exposed to UV-A, cycloaddition photoproducts are formed by covalent binding between the psoralens and thymines with which they are associated. Further exposure to UV-A leads to the formation of cross-links. When angular furocoumarins such as angelicin are examined, it is seen that the molecular configuration within the DNA results in a spatial arrangement which makes it impossible for cross-links to occur, and only monoadducts are formed. The relative photosensitising potential of various common furocoumarins has been summarised by PATHAK et al. (1974). Psoralen photoadducts have been isolated from DNA, microorganisms, mammalian cells in culture and mammalian epidermis irradiated in vivo (MUSAJO and RODIGHIERO 1972; PATHAK et al. 1974; PATHAK et al. 1976). At the cellular level, psoralen photosensitisation results in an early inhibition of DNA synthesis and cell replication, induction of mutation, and cell death. Although the photochemical lesions may be repaired (RODIGHIERO and DALL'ACQUA 1976) no evidence for unscheduled DNA synthesis, an indication for excision repair, has been obtained in human skin undergoing photochemotherapy (BISHOP 1979). A more complex, excision-recombination type of repair may operate for psoralen photoadducts.

Attempts to differentiate the relative contributions of monoadducts and cross-links to the lethal and mutation effects of psoralen photosensitisation in bacteria (SEKI et al. 1978; BRIDGES et al. 1979) and mammalian cells (COPPEY et al. 1979) have produced equivocal results. Using the monofunctional angelicin or 3-carbethoxypsoralen, or bifunctional 8-MOP but with a two-stage irradiation, the second exposure, given after unbound psoralen has been washed out of the cells, allows the demonstration of any differential effects which might exist. Both types of lesion contribute to the inhibitory, mutagenic and lethal effects. However, cross-linking, which constitutes only some 10% of the total photoadducts produced with 8-MOP, may be more efficient by an order of magnitude. Increased levels of chromosome aberrations result from both forms of psoralen photoadduct (ASHWOOD-SMITH et al. 1977). Increased numbers of sister chromatid exchanges are detected in human lymphocytes exposed in vitro to relatively low concentrations of 8-MOP (0.004–4 μg/ml) and exposure doses of UV-A in the range of 1 J/cm^2 (MOURELATOS et al. 1977). Concentrations of 8-MOP above 0.1 μg/ml are required to produce a significant photosensitised inhibition of mitogen-stimulated DNA synthesis in lymphocytes (KRUGER et al. 1978) although there is an inverse relationship between 8-MOP concentration down to 0.02 μg/ml and UV-A dose for the inhibition obtained in growing cultures of 3T3 cells (POHL and CHRISTOPHERS 1979). Similar results have been obtained with Ehrlich ascites cells (MUSAJO et al. 1967), Chinese hamster cells (BEN-HUR and ELKIND 1973) and human skin fibroblasts in culture (BADEN et al. 1972; POHL and CHRISTOPHERS 1978).

TMP at a final concentration of 2 μM and UV-A at approximately 0.2 J/cm^2 produced an average 70% inhibition of DNA synthesis in mouse epidermis within 4 h of irradiation in vitro (WALTER et al. 1973). Topical 8-MOP at concentrations around 1% produces a photosensitised inhibition of DNA synthesis in mouse (EPSTEIN and FUKUYAMA 1975) and guinea pig (PULLMAN et al. 1980) epidermis. However, when hairless mice are treated with oral 8-MOP sufficient to elicit a UV-A-induced oedematous skin reaction, no inhibition of DNA synthesis is observed unless the epidermis was previously made hyperproliferative (FRITSCH et al. 1979). Although the dividing cells of psoriatic epidermis may be exposed to 8-MOP at concentrations around 0.1 μg/ml and therefore the 8-MOP concentration and UV-A exposure dose required to inhibit DNA synthesis in vitro appear to be attained in the PUVA treatment of patients with psoriasis, no early inhibition has been observed (HELL et al. 1979; GOLDBERG et al. 1980; HOFFMANN et al. 1980). Finally, it has proved difficult to detect psoralen DNA photoadducts in the epidermis of guinea pig (CECH et al. 1979) or human (ZAREBSKA et al. 1984) skin after therapeutic doses of PUVA. The possibility remains, therefore, that although the photosensitised damage to skin as obtained in phytophotodermatitis may be due to lesions in DNA, the molecular changes involved in photochemotherapy remain to be established.

III. Acute Effect of PUVA on Normal Skin

Because serum levels of psoralen and skin photosensitivity peak approximately 1–3 h after oral administration, UV-A exposures are usually given 2 h after ingestion of the crystalline drug or 1.5 h after the gelatin capsule. The duration of photosensitivity after topical application depends on local psoralen concentration and there is some evidence that the stratum corneum acts as a reservoir. Topical application of very low concentrations (<0.003%, aqueous) may lead to photosensitivity which lasts less than 1 h. With higher concentrations, the skin may remain photosensitive for more than 24 h.

Exposure of psoralen-treated skin to UV-A causes injury to skin cells and a subsequent erythematous inflammation. Melanogenesis is evident several days later. Large doses cause swelling, blistering, pruritus, pain and desquamation. UV-B-induced sunburn appears in 4–6 h and peaks 12–24 h after exposure, whereas PUVA erythema appears about 24 h after exposure and peaks at 48–72 h, occasionally even later.

The mediators of PUVA erythema have not been identified (see GANGE and PARRISH 1984). UV-B causes increased levels of PGD_2, PGE_2 and 6-oxo-PGF_1; in contrast no increase in PGE_2, PGE_2 and arachidonic acid are detectable in suction blister fluid 24–72 h after PUVA (PLUMMER et al. 1978) and no alteration in prostaglandin metabolites are found in the serum or urine (KORNHAUSER 1980). In contrast PUVA enhances oxygenation of arachidonic acid and prostaglandin synthesis by skin homogenates in vitro (LORD et al. 1976).

Neither corticosteroids (GREENWALD et al. 1981) nor indomethacin modify PUVA-induced erythema in humans (MORISON et al. 1977), although the latter inhibits UV-B erythema. This is consistent with the observations on prostaglandin levels in the skin of the mouse. PUVA phototoxicity is characterised more by oedema, the development of which is inhibited by indomethacin and 5,8,11,14-eicosatetraynoic acid (ETYA), an inhibitor of both the cyclooxygenase and the lipoxygenase pathways (CLEAVER et al. 1981).

Anti-oxidants such as α-tocopherol inhibit the phototoxic response to PUVA. These agents are only effective if present in the skin during irradiation (POTAPENKO et al. 1984). This suggests that the creation of active oxygen species may be important for PUVA's phototoxic effect. Guinea pigs made neutropenic with cyclophosphamide show normal erythema responses to PUVA, but not to UV-B, suggesting that a neutrophil product may be important only for the latter (EAGELSTEIN et al. 1979). PUVA degranulates mast cells (KONRAD et al. 1976), but the kallikrein inhibitor aprotinin (Trasylol) does not alter PUVA erythema.

A single PUVA exposure sufficient to cause erythema results in histological changes in both dermis and epidermis different from those seen with comparable erythema induced by UV-A, UV-B or UV-C alone (ROSARIO et al. 1979). Epidermal changes occur later than with UV-B and UV-C; fewer dyskeratotic cells are seen in the first 2–3 days and dermal changes of endothelial swelling, nuclear dust and extravasation of blood are more marked after UV-A than after shorter wavelength radiation. These findings persist for 7–21 days after exposure.

The formation of DNA cross-links appears to be of prime importance in the development of delayed erythema following PUVA exposure (GANGE et al. 1984). The relative proportion of monofunctional and cross-links formed after application of 8-methoxypsoralen depends on the wavelength of UV-A. For example, exposure with wavelengths greater than 380 nm favours formation of monofunctional adducts, which are only slowly repaired, and which can be converted to DNA cross-links with subsequent exposure to broad-band (320–380 nm) UV-A (GANGE et al. 1984).

PUVA pigmentation appears similar to normal melanogenesis. Ultrastructural alterations within melanocytes are evident within minutes; there is enhanced activity of tyrosinase and an increase in the rate of transfer of melanosomes to keratinocytes. Repeated PUVA increases the number and size of functional melanocytes. Pigmentation is maximal 5–7 days after exposure and lasts weeks to months. As with UV-B, ability to tan with PUVA is genetically determined.

IV. Clinical Uses of PUVA

Psoralens from plants and sunlight have been used for more than 3000 years for the treatment of depigmentation of skin (FITZPATRICK and PATHAK 1959). In the last decade improved dosimetry from artificial sources has led to a variety of new applications for vitiligo, psoriasis, mycosis fungoides, lichen planus, alopecia areata and photosensitivity disorders.

1. Vitiligo

Vitiligo is an acquired disorder characterised by loss of melanocytes and pigmentation in circumscribed areas of skin. Psoralens used to treat vitiligo are topical and oral TMP and methoxsalen, and unsubstituted psoralen followed by exposure to sunlight or UV-A. In areas of repigmentation following PUVA therapy, hypertrophic and hyperactive melanocytes have been noted, but melanocyte mitoses are not evident. The melanocytes in newly repigmented areas produce melanosomes which are longer than those found in surrounding skin (ORTONNE et al. 1979). Success rates vary substantially and complete repigmentation is infrequent. In a recent study of 73 patients with vitiligo treated with topical PUVA, only 7 repigmented completely and less than half repigmented at least half of the affected area (GRIMES et al. 1982). Repigmentation and melanocyte proliferation occurs first around hair follicles (ORTONNE et al. 1980) and is usually long-standing; central lesions repigment more quickly than acral lesions. Any improvement that might occur begins after 50–100 PUVA treatments with either oral TMP or 8-MOP (PATHAK et al. 1984). Oral TMP is safer than topical application when sunlight is used.

In one in three patients little or no repigmentation occurs and in the remainder it is unpredictable and variable. Patients should be informed about the necessity of prolonged treatment, the apparent worsening by contrast as normal skin tans from treatment, and the likelihood of incomplete repigmentation. The time, cost, potential risk of UV and partial response limits the use of PUVA to patients who find the disease socially disabling. More carefully controlled clinical trials are necessary to establish the therapeutic use of PUVA in vitiligo.

2. Psoriasis

Psoriasis is a hyperproliferative disorder of unknown aetiology. The therapeutic effect of sunlight has long been appreciated. Early studies with topical psoralens and UV-A demonstrated that photochemotherapy could alleviate the symptoms of psoriasis, but unfortunately, the phototoxic response to topical psoralen proved too variable. The availability of fluorescent lamps with sufficient UV-A output to achieve phototoxicity following oral administration of methoxsalen allowed out-patient photochemotherapy for psoriasis. The initial report of PUVA's effectiveness (PARRISH et al. 1974) has been confirmed by studies of thousands of patients in many parts of the world. Possible mechanisms include selective cell killing, decreased recruitment, suppression of DNA synthesis and cellular proliferation, and correction of an immunological derangement or its effect on dermal blood vessels. In spite of PUVA's substantial efficacy in clearing psoriasis, it certainly does not cure this chronic disease and relapses occur. Some, but not all, patients can be kept free of disease with PUVA maintenance programmes. Because long-term PUVA toxicity will be dose related, the current emphasis is on the develpment of therapeutic techniques that will reduce total long-term PUVA exposure. Efforts to reduce cumulative dose during the clearing phase continues. The "more is less" technique, widely used in Europe (WOLFF and HONIGSMANN 1983), is designed to administer maximum doses and frequency of treatment early in the course of therapy before PUVA-induced melanisation and thickened stratum corneum and other unknown factors reduce the effect of treatment. The advantage appears to be that clearing is achieved in fewer treatments and therefore at lower total cumulative dose. The "plateau" approach has also been suggested. After patients begin to improve significantly, the UV-A dose is no longer increased, but is held at a constant dose throughout the clearing phase. While this may make it necessary to use a greater number of treatments, the UV-A dose per treatment is decreased.

In addition, two, often complementary, approaches to reducing cumulative PUVA exposure during long-term therapy have been advanced, combination therapy and cyclic therapy. The basis for these approaches is that the simultaneous or sequential administration of treatments with substantially different mechanisms of action or potential toxicities may give an additive or synergistic therapeutic effect with less than additive toxicity.

The first combination therapy to receive wide acceptance combined oral aromatic retinoid administration and PUVA (FRITSCH et al. 1978; K. MOMTAZ and J. A. PARRISH 1982, unpublished work). In patients treated with the aromatic retinoid and PUVA, some, but not complete, normalisation of keratinocytic structure was noted (LAUHARANTA et al. 1981 a). Patients receiving PUVA and retinoids exhibit decreases in their polyamine levels (KANERVA et al. 1982). Recent reports suggest that 13-*cis*-retinoic acid, which unlike the aromatic retinoid is available in the United States, may also be valuable as adjuvant therapy with

PUVA (LAUHARANTA et al. 1981 b). The possible anti-carcinogenic effect of the oral retinoids remains controversial (HONIGSMANN and WOLFF 1983). Other combination therapies that have been successfully used for psoriasis include UV-B-PUVA, methotrexate-PUVA and hydroxyurea-PUVA. Because of the difference in the time from exposure to peak erythema of UV-A and PUVA, combined therapy with UV-B and PUVA permits the simultaneous administration of nearly phototoxic doses of each radiation. Patients treated with this combination clear in half the number of treatments and one-quarter the cumulative dose compared with patients treated with PUVA or UV-B alone (EPSTEIN 1981). The extent to which combination therapies lower the cutaneous toxicity of treatment or the duration of remission is still to be determined.

Therapies combining systemic anti-proliferative agents such as methotrexate or hydroxyurea also rely on the principle of combined therapeutic effect and non-additive toxicity of different mechanisms. Clearing with methotrexate-PUVA occurs in an average of 12 treatments. In the combined methotrexate-PUVA protocol, methotrexate is begun 4 weeks before PUVA and is gradually reduced in dose as PUVA dose is increased (MORISON et al. 1981). Phototoxicity from the simultaneous administration of PUVA and methotrexate withdrawal appears infrequently. Experience with hydroxyurea and PUVA used together is more limited, but selected patients unable to tolerate methotrexate and/or high doses of PUVA have been successfully maintained on this regimen. Combined treatments have recently been reviewed (PARRISH 1981).

Combination therapy utilising topical agents and PUVA has been less uniformly successful. Both dithranol and topical steroids appear to accelerate the initial clearing of psoriasis when combined with PUVA (GOULD and WILSON 1978; SCHMOLL et al. 1978; MORISON et al. 1978 a; HANKE et al. 1979), but more rapid relapse has been reported following clearing with these combined therapies (FRITSCH et al. 1978). Periodic treatment with PUVA after initial clearing with dithranol has been shown to extend the length of remission significantly (MARKS et al. 1984). Topical and oral methoxsalen photochemotherapy have been used together and this seems most useful for clearing recalcitrant plaques.

Psoriasis variants including palmo-plantar psoriasis, pustular psoriasis and erythrodermic psoriasis have all been treated successfully with PUVA. Special radiation sources are used to treat palmar and plantar psoriasis (MORISON et al. 1978 b; MURRAY and WARIN 1979). Combination of PUVA and the aromatic retinoid, etretinate, appears to be especially effective in treating palmo-plantar pustular psoriasis (LAWRENCE et al. 1984; TANEW et al. 1988). Topical PUVA therapy has been less successful for palmo-plantar pustulosis than oral therapy (JANSEN and MALMIHARJU 1981). Pustular and erythrodermic psoriasis is more difficult to treat, but some groups have had substantial success (HONIGSMANN et al. 1977). While one group reports nail improvement with PUVA, this observation has not been repeated by others (MARX and SCHUR 1980).

Methoxsalen has not been the only oral psoralen used for PUVA therapy. Limited reports indicate that 5-MOP and 3-carbethoxypsoralen may be useful in the treatment of psoriasis. Whether the ratio of toxicity to efficacy is different for these agents compared with methoxsalen remains speculative. One potential advantage of 5-MOP is its apparently lower phototoxicity. The angelicins, including

6-methylangelicin, have received recent attention as possible alternatives to 8-MOP in the treatment of psoriasis. 6-Methyangelicin appears to form monoadducts, and a preliminary study suggests this compound may be less erythemogenic and mutagenic than 8-MOP and yet still helpful in the treatment of psoriasis (GUIOTTO et al. 1984), but this work remains unconfirmed by larger trials.

That PUVA therapy used alone or in combination with other agents is effective in clearing psoriasis is unquestioned. Continued maintenance or periodic retreatment is required. In a 16-centre study patients who relied on PUVA alone required an average of 30 treatments per year for maintenance. The frequency of maintenance treatments increased with increasing severity of psoriasis (MELSKI and STERN 1982). Therefore, over a period of years, patients continuing to rely on PUVA therapy may receive hundreds of treatments. After 5 years, patietns enrolled in the 16-centre study who were actively using PUVA when interviewed averaged a total of 190 treatments or a total exposure of more than 2500 J/cm^2. Since any long-term toxicity is likely to be dose related, and may be increased by continuous therapy, in addition to the combination therapies which lower the total dose of PUVA to clearing and the initial maintenance dose, cyclic therapy for long-term control of psoriasis is receiving increasing attention.

Cyclic therapy is based on the principle that different psoriasis treatments have different long-term risks and that a particular treatment's risks may be reduced if it is used intermittently rather than continuously. Cyclic therapy including PUVA may use PUVA alone or as part of a combination clearing therapy and relies on alternative topical agents for maintenance. Patients cleared with UV-B or dithranol may also have longer remissions when maintenance with PUVA is employed following initial clearing (VAN DER KERKHOF and MALI 1981) although the duration has not been well determined. The duration of remission of PUVA and dithranol therapy are comparable (MARKS et al. 1981). It has been suggested that varying therapies may maintain therapeutic response as well as reducing overall toxicity.

3. Mycosis Fungoides

Mycosis fungoides is a cutaneous T-cell lymphoma. The Harvard group demonstrated substantial improvement in plaque and tumour stage mycosis fungoides (GILCHREST et al. 1976) and subsequent studies have confirmed this effect. In one study 58% of 51 pre-tumour patients achieved complete remission and 84% of 25 tumour stage cases achieved a complete or partial remission of skin lesions (MOLIN et al. 1981), but relapses were frequent and internal dissemination has been observed despite clearing of skin lesions by PUVA. The Austrian group showed excellent results in patients with eczematous and plaque lesions of mycosis fungoides. After nearly 4 years, almost half of such patients were in remission and many had only required one course of photochemotherapy. Patients with early tumours did poorly; a majority died within 6 years of first treatment with PUVA (HONIGSMANN et al. 1984). The Mayo Clinic also reports high success rates in parapsoriasis and early (stage I or II) mycosis fungoides, but poor responses in more advanced cases (POWELL et al. 1984).

The mechanisms by which PUVA improves mycosis fungoides are not well understood. Lymphocyte infiltrate in the skin is greatly reduced, but lymph node involvement and the atypical circulating lymphocytes of Sezary's syndrome are unaltered by PUVA. Disseminated cutaneous T-cell lymphoma has also been treated with extracorporeal photochemotherapy (EDELSON et al. 1987). This may permit more intense treatment of lymphocytes than traditional PUVA. Disappearance of abnormal photosensitivity and improvement of immunological function in mycosis fungoides has also been observed after PUVA. Controlled trials are required to establish whether PUVA or photopheresis has an effect on the course of the disease or is merely a palliative therapy albeit more acceptable to many than alternative treatments. Likewise trials of combination or cyclic therapy are required, e.g. the combination of topical nitrogen mustard with PUVA.

4. Urticaria Pigmentosa

PUVA diminishes the mast cell infiltrate and symptoms of itching and dermatographism (CHRISTOPHERS et al. 1978) with a diminished urinary excretion of histamine metabolites. Responses in common chronic urticaria are no greater than to placebo and UV-A irradiation (OLAFSSON et al. 1986).

5. Light-related Dermatoses

Surprisingly, PUVA has been proved to be useful in the treatment of a number of photosensitivity disorders, including polymorphous light eruption, photosensitive eczema, solar urticaria and persistent light reaction (GSCHNAIT et al. 1978). The mechanism of the therapeutic benefit is unknown although there is an increased barrier to UV penetration by an increase in melanisation and thickness of stratum corneum (JANSEN et al. 1982). PUVA also appears useful in the treatment of chronic actinic dermatitis (HINDSON et al. 1985).

6. Eczema

Although the primary aetiological events in eczema are incompletely understood, the condition resembles a type IV immune reaction. Since such reactions are inhibited by PUVA both by local and systemic effects on afferent and efferent components of the immune response (see Sect. E. V. 2) eczema might be expected to respond to PUVA. Unfortunately, and unlike the situation with psoriasis, there have been few clinical trials of response and relapse. More treatments appear to be required both to clear and maintain clearing of severe eczema than psoriasis and this requirement limits the therapy's usefulness. Recalcitrant eczema of the palms and soles is said to respond to PUVA, but there have been too few studies to define the use of PUVA in this and other eczematous dermatoses.

7. Other Disorders

A T-cell lymphocytic infiltrate is characteristic of lichen planus and is cleared by PUVA with clinical regression (ORTONNE et al. 1978). Keratosis lichenoids chronica and pityriasis lichenoides are said to respond to PUVA with long-term remissions (BOELEN et al. 1982). Other diseases which have been reported as responding to treatment with PUVA include chronic cutaneous graft-versus-host reaction (HYMES et al. 1985), pityriasis alba (ZAYNOUN et al. 1986) and granuloma annulare (HINDSON et al. 1987). Somewhat paradoxically, transient acantholytic disease, a disorder often exacerbated by sunlight, can be suppressed with PUVA.

V. Treatment Principles

After nearly a decade, oral methoxsalen and UV-A from fluorescent bulbs remain the most frequently used type of photochemotherapy.

1. Dosimetry

Since it is easier to control the dose of UV-A radiation, drug dose is kept constant. Essential features for a UV-A source (LEVIN and PARRISH 1975) include providing a uniform dose on the patient's entire skin surface, to deliver a minimal phototoxic dose (MPD) in 5 min or less. The output of fluorescent bulbs currently in use in the UV-A range generally ranges from 5 to 15 mW/cm^2, but because bulbs employing different phosphors may have very different emission spectra, an apparently equal dose in J/cm^2 from different bulbs will not necessarily give equal biological responses.

The usual dose of crystalline 8-MOP is 0.6 mg/kg: higher doses have a risk of nausea. Because the drug is not standardised methoxsalen preparations that differ in manufacturer or even batch may have very different activities (ANDERSON et al. 1980). Serum levels may vary more than six-fold between individuals (STEVENSON et al. 1981). Differences in absorption kinetics may account for some treatment problems with prolongation of the time to peak plasma methoxsalen concentrations (WAGNER et al. 1979). The newer gelatin capsule is more rapidly, more uniformly, and more completely absorbed. Peak sensitisation occurs 60–90 min after oral administration. Standard dosage is 0.4 mg/kg. In spite of lower dosage, CNS and gastrointestinal side effects appear to be more frequent (DOVER and STERN 1986).

Topical psoralen photochemotherapy produces persistent photosensitivity (because of psoralen in the stratum corneum), a less predictable phototoxic response and uneven pigmentation. Furthermore, serum levels following whole body topical application of methoxsalen are comparable to those achieved with oral administration (NIELD and SCOTT 1982).

With a constant 8-MOP dose, the dose of UV-A that can be tolerated is estimated either from skin type (i.e. history of burning and pigmentation after sun exposure) or by doses of UVA to 1 cm^2 areas of normal skin and assessment of erythema and pigmentary responses (WOLFF et al. 1977). In practice, the latter has no advantage over schedules based on skin type.

Continuing exposure to PUVA induces pigmentation and thickening of the stratum corneum which act as UV-A filters. Therefore, maintaining erythema requires gradually increasing doses of UV-A with subsequent treatments. Because of the delayed time course of PUVA erythema, patients are most often treated at 2- to 3-day intervals.

2. Immunological Effects

In vitro acute exposures to PUVA result in decreased viability and function of lymphocytes. In vivo studies of PUVA's acute effect on lymphocytes in humans have given conflicting results, but recent evidence suggests that long-term PUVA therapy reduces helper/inducer T-lymphocytes (MOSCICKI et al. 1982). In addition to its apparent effects on lymphocytes, PUVA appears to impair both induction and elicitation of delayed hypersensitivity responses in the skin, both in animals and in the course of treatment of psoriasis (MOSS et al. 1985). Depressed responses to dinitrochlorobenzene (DNCB) (STRAUSS et al. 1980; MOSS et al. 1981) and suppression of pre-existing nitrogen mustard sensitivity (VOLDEN et al. 1978) have been observed with PUVA. These changes in delayed hypersensitivity may reflect alterations in lymphocyte function or result from the PUVA-related decrease in Langerhans' cell function as the antigen-presenting cell in the skin (FRIEDMANN 1981) or other mechanisms (KALIMO et al. 1983).

Immunological changes may be a risk factor in the development of skin tumours although MOSS et al. (1981) doubt this because they found similar changes with a dithranol regimen which has been used for many years. There is no evidence that the immunosuppression produced by PUVA increases the risk of skin infections. PUVA-induced alterations in DNA may be antigenic and increase the risk of lupus erythematosus (LE) in predisposed individuals. While case reports of LE in PUVA-treated patients exist, there has been no systematic study of a possible association between PUVA exposure and collagen vascular disease and no significant relationship has been found between PUVA exposure and the risk of a positive anti-nuclear antibody test (STERN et al. 1979a).

3. Other Skin Changes

Photosensitivity reactions to drugs and polymorphous light eruption can be exacerbated by PUVA as well as UV-B radiation. The exacerbation of pemphigoid may be related to the reduced dermo-epidermal adhesion provoked by PUVA (ROBINSON et al. 1978; FRIEDMANN et al. 1987). Furthermore, the response of atopic eczema suggests that PUVA may induce beneficial as well as adverse effects in the immune system.

4. Skin Cancer

PUVA is a mutagen in bacteria, yeast and human fibroblasts; it leads to monofunctional and bifunctional adducts with DNA pyrimidines; it causes chromosomal damage in cultured mammalian cells and sister chromatid exchanges in human lymphocytes; it leads to rapid induction of tumours in animals (GRIFFIN 1959) and dysplastic cells in human epidermis. Atypical nuclear features have been noted in keratinocytes and melanocytes from patients treated with PUVA. A 2.1-year, 16-centre PUVA follow-up study of 1380 patients in the United States documented a nearly three-fold increase in the risk of cutaneous carcinoma in patients treated with PUVA (STERN et al. 1979b), with a higher than expected proportion of squamous cell carcinomas in areas not exposed to sun. Similar conclusions were reached from a 5-year prospective study which also suggested that the risk of squamous cell carcinoma after adjustment for age, geographic residence and skin type is higher in patients with greater exposure to PUVA, and in patients with prior exposure to other carcinogens including UVB- and ionising radiation, and tar therapy (STERN et al. 1980). The 16-centre follow-up study now includes more than 10 years of follow-up on its original cohort. When skin tumours occurring at least 58 months after first treatment are considered separately, there is a clear relationship between increasing exposure to PUVA and increasing risk of squamous cell carcinoma. Compared with population expected values, the PUVA cohorts' overall risk of squamous cell cancer is more than ten times greater. Patients who have received high doses of PUVA have risk of squamous cell cancer more than 60-fold greater than that observed in the general population (STERN et al. 1988). These studies have been criticised for lack of controls and others have not found an increase in the prevalence of non-melanoma skin tumours in PUVA-treated patients (ROENIGK et al. 1981; LASSUS et al. 1981) whilst others (HONIGSMANN et al. 1980) found a risk only in those individuals who had prior exposure to arsenic, ionising radiation or methotrexate. These differences may be due to dose, duration, skin type and the small numbers of patients and short periods of investigation in studies with negative findings.

Although psoralens increase melanocytic activity (ZAYNOUN et al. 1977; ORTONNE et al. 1980) and despite the relationship of UV exposure and melanoma an increased risk has not been reported in any of the larger PUVA studies. PUVA-induced freckles which may be stellate and persistent have larger and more numerous melanocytes than normal skin or solar lentigines with large nuclei, nucleoli and increased euchromatin (KANERVA et al. 1981). Aberrant morphology has also been noted (ROS et al. 1983). Similar changes are seen in dysplastic naevi, which are associated with an increased risk of melanoma. After 10 years of follow-up, there is no significant increase in the risk of melanoma in the PUVA cohort.

5. Skin Ageing

Actinic changes resembling those seen with high intensity UV-B have been reported after PUVA treatment. The severity and frequency of these changes appear to be dose dependent (GRIFFIN 1959; PIERARD and PIERARO-FRANCHIMONT 1984; STERN et al. 1985a). So far changes reported include colloid bodies and amyloid-like deposits at the dermo-epidermal junction, alterations in blood vessels, and elastic tissue changes that are dose dependent and persist for at least 1 year after discontinuing therapy (HASHIMOTO and KUMAKIRI 1979; ZELICKSON et al. 1980). The long-term significance of these findings and their reversibility is not yet clear.

6. Miscellaneous Cutaneous Toxicities Associated with PUVA Therapy

Pain, acute phototoxicity, onycholysis, acneiform eruptions, a seborrhoeic dermatitis-like facial rash and a variety of other cutaneous side effects have been reported. Itch is commonest and can be severe; in one study its severity after treatment correlated best with its severity before treatment (ROGERS et al. 1981 a).

7. Ophthalmological Risks

8-MOP binds to retinal tissue (LAFOND et al. 1984). The lens is a good absorber of UV-A and is susceptible to permanent damage by photosensitisers because it is completely encapsulated, never sheds its cells and has poor repair mechanisms (see LERMAN 1979). Systemically administered methoxsalen enters the lens of human and other species where it may remain for up to 12 h whereas UV-A photoproducts may persist indefinitely.

Whereas large doses of PUVA cause cataracts in laboratory animals, therapeutic doses in humans do not (PARRISH et al. 1982). Reports of changes in unprotected eyes of PUVA patients (WAGNER et al. 1978; LERMAN et al. 1983) are suggestive, but require confirmation. Although there is no evidence of an increased incidence of cataracts in patients with adequate eye protection, after 5 years a dose-dependent increase in some types of lenticular opacities was noted. Only further study can determine the clinical significance of these changes (STERN et al. 1985 b). Eye protection remains an integral part of therapy.

8. Systemic Toxicity of PUVA

There is remarkably little systemic effect of PUVA. Immunological changes have already been discussed. Inevitably various hypersensitivity reactions have been described, but a causal relationship has not been established. Vitamin D synthesis increases, but serum changes remain in the physiological range (MOSS et al. 1985) and there is no change in major endocrine function (ROGERS et al. 1981 b).

References

Anderson KE, Menne T, Gammeltoft M, Hjorth N, Larsen E, Solgaard P (1980) Pharmacokinetic and clinical comparison of two 8-methoxypsoralen brands. Arch Dermatol Res 268:23–29

Ashwood-Smith MJ, Grant EL, Heddle JA, Friedman GB (1977) Chromosome damage in Chinese hamster cells sensitized to near ultraviolet light by psoralen and angelicin. Mutat Res 43:377–381

Averbeck D, Moustacchi E, Bisagni E (1978) Biological effects and repair of damage photo induced by a derivative of psoralen substituted at the 3,4 reaction site. Photoreactivity of this compound and lethal effect in yeast. Biochim Biophys Acta 518:464–481

Averbeck D, Dubertret L, Craw M, Truscott TG, Dall'Acqua T, Rodighiero P et al. (1984) Photophysical photochemical and photobiological studies of 4'-methylangelicins, potential agents for photochemotherapy. Farmaco [Sci] 39:57–69

Baden HP, Parrington JM, Delhanty JDA, Pathak MA (1972) DNA synthesis in normal and xeroderma pigmentosum fibroblasts following treatment with 8-methoxypsoralen and long-wave UV light. Biochim Biophys Acta 262:247–252

Ben-Hur E, Elkind MM (1973) Psoralen plus near ultraviolet light inactivation of cultured Chinese hamster cells and its relation to DNA cross links. Mutat Res 18:315–322

Bickers DR, Mukhtar H, Molica SJ, Pathak MA (1982) The effect of psoralens on hepatic and cutaneous drug metabolizing enzymes and cytochrome P-450. J Invest Dermatol 79:201–205
Bishop SC (1979) DNA repair synthesis in human skin exposed to ultraviolet radiation used in PUVA (psoralen and UV-A) therapy for psoriasis. Br J Dermatol 101:399–405
Blais J, Dubertret L, Gaboriau F, Vigny P (1982) Fluorescence detection of an 8-methoxypsoralen metabolite in human interstitial fluid. Photochem Photobiol 35:423–426
Blum HF (1964) Photodynamic action and diseases caused by light. Hafner, New York
Boelen RE, Faber WR, Lambers JCCA, Cormane RH (1982) Long-term follow-up of photochemotherapy in pityriasis lichenoides. Acta Derm Venereol (Stockh) 62:442
Brickl R, Schmid J, Koss FW (1984) Clinical pharmacology of oral psoralen drugs. Photodermatol Clin Exp 1:174–185
Bridges BA, Mottershead RP, Knowles A (1979) Mutation induction and killing of *Escherichia coli* by DNA adducts and cross links: a photobiological study with 8-methoxypsoralen. Chem Biol Interact 27:221–235
Busch U, Schmid J, Koss PF, Zipp H, Zimmer A (1978) Pharmacokinetics and metabolite pattern of 8-methoxypsoralen in man following oral administration as compared to the pharmacokinetics in rat and dog. Arch Dermatol Res 262:255–265
Cech T, Pathak MA, Biswas RK (1979) An electron microscopic study of the photosensitized cross-linking of DNA in guinea pig epidermis by psoralen derivatives. Biochim Biophys Acta 562:342–360
Chakrabarti SG, Grimes PE, Minus HR, Kenney JA, Pradhan TK (1983) Serum concentrations of trimethylpsoralen after oral administration: reply. J Invest Dermatol 81:289–290
Christophers E, Honigsmann H, Wolff K, Langner A (1978) PUVA-treatment of urticaria pigmentosa. Br J Dermatol 98:701
Clayton RK (1970) Light and living matter, vol 1. The physical part. McGraw Hill, New York
Cleaver L, Gange RW, Folsom KT (1981) Skin edema due to PUVA in the hairless mouse and effects of antihistamine on indomethacin. J Invest Dermatol 76:322
Coppey J, Averbeck D, Moreno G (1979) Herpes virus production in monkey kidney and human skin cells treated with angelicin or 8-methoxypsoralen plus 365 nm light. Photochem Photobiol 29:797–801
Diette KM, Momtaz K, Stern RS, Arndt KA, Parrish JA (1984) Role of ultraviolet A in phototherapy for psoriasis. J Am Acad Dermatol 11:441–447
Dougherty TJ, Kaufman JE, Goldfarb A, Weishaupt KR, Boyle DG, Mittelman A (1978) Photoradiation therapy for the treatment of malignant tumours. Cancer Res 38:2628–2635
Dougherty TJ, Boyle DG, Weishaupt KR (1982) Photoradiation therapy of human tumors. In: Regan JD, Parrish JA (eds) The science of photomedicine. Plenum, New York, pp 625–638
Dougherty TJ, Weishaupt KR, Boyle DG (1983) Photoradiation therapy of malignant tumors. In: de Vita U, Hellman S, Rosenberg S (eds) Principles and practice of oncology. Lippincott, Philadelphia
Dover JS, Stern RS (1986) Psoralens. New formulation! Better formulation? Arch Dermatol 122(7):763–765
Dubertret L, Averbeck D, Bisagni E, Moron J, Moustacchi E, Billardon C et al. (1984) Photobiological and phototherapeutic properties of new monofunctional pyrido psoralens. 9th International Photobiology Congress, Philadelphia
Eagelstein WH, Sakai M, Mizuno N (1979) Ultraviolet radiation induced inflammation and leukocytes. J Invest Dermatol 72:59–63
Edelson R et al. (1987) Treatment of cutaneous T-cell lymphoma by extracorporeal photochemotherapy. N Engl J Med 316:297–303
Ehrsson H, Eksborg S, Wallin I (1978) Metabolism of 8-methoxypsoralen in man: identification and quantification of 8-hydroxypsoralen. Eur J Drug Metab Pharmacokinet 2:125–128

Ehrsson H, Nilsson SO, Ehrnebo M, Wallin I, Wennersten G (1979) Effect of food on kinetics of 8-methoxsalen. Clin Pharmacol Ther 25:161–171
Epstein JH (1981) Effects of retinoids on ultraviolet-induced carcinogenesis. J Invest Dermatol 76:144–146
Epstein JH, Fukuyama K (1975) Effects of 8-methoxypsoralen induced phototoxic effects on mammalian epidermal macromolecule synthesis in vivo. Photochem Photobiol 21:325–330
Fitzpatrick TB, Pathak MA (1959) Historical aspects of methoxsalen and furocoumarins. J Invest Dermatol 32:225–228
Foote CS (1974) Photoreactions with molecular oxygen. In: Schenck GO (ed) Progress in photobiology. Proceedings of the VI International Photobiology Congress
Fowlks WL (1959) The chemistry of the psoralens. J Invest Dermatol 32:249–254
Freeman K, Warin AP (1984) Deterioration of liver function during PUVA therapy. Photodermatol Clin Exp 1:147
Friedmann PS (1981) Disappearance of epidermal Langerhans cells during PUVA therapy. Br J Dermatol 105:219–221
Friedman PS, Coburn P, Dahl MG, Duffey BL, Ross J, Ford GP, Parker SC, Bird P (1987) UVA-induced blisters, complement deposition, and damage to the dermoepidermal junction. Arch Dermatol 123:1471–1477
Fritsch PO, Honigsmann H, Jaschke E et al. (1978) Augmentation of oral methoxsalen-photochemotherapy with an oral retinoic acid derivative. J Invest Dermatol 70:178–182
Fritsch PO, Gschnait F, Kaaserer G, Brenner W, Chaikittisilpa S, Honigsmann H et al. (1979) PUVA suppresses the proliferative stimulus produced by stripping on hairless mice. J Invest Dermatol 73:188–190
Gange RW, Parrish JA (1984) Cutaneous phototoxicity due to psoralens. In: Pathak MA, Dunnick JK (eds) Photobiologic, toxicologic and pharmacologic aspects of psoralens. Natl Cancer Inst Monogr 66:117–126
Gange RW, Levins PC, Anderson RR, Parrish JA (1984) Cutaneous photosensitization by 8-methoxypsoralen: order dependent synergism between radiation less than 380 nm and broadband UVA. J Invest Dermatol 82:594–597
Gazith J, Schalla W, Schaefer H (1978) 8-Methoxypsoralen – Gas chromatographic determinations and serum kinetics. Arch Dermatol Res 263:215–222
Gilchrest BA, Parrish JA, Tanenbaum L et al. (1976) Oral methoxsalen photochemotherapy of mycosis fungoides. Cancer 38:683–689
Goldberg LH, Cox AJ, Abel EA (1980) The mitotic index in psoriatic plaques and their response to PUVA therapy. Br J Dermatol 102:401–405
Gomer CJ, Dougherty TJ (1979) Determination of [3H] and [^{14}C]hemoatoporphyrin derivative distribution in malignant and normal tissue. Cancer Res 39:146–151
Gould PW, Wilson L (1978) Psoriasis treated with clobetasol propionate and photochemotherapy. Br J Dermatol 98:133–136
Granelli S, Diamond I, McDonagh AF, Wilson CB, Nielson SL (1975) Photochemotherapy of glioma cells by visible light and hematoporphyrin. Cancer Res 35:2567–2570
Greenwald JS, Parrish JA, Jaenicke KF, Anderson RR (1981) Failure of systemically administered corticosteroids to suppress UVB induced erythema. J Am Acad Dermatol 5:197–202
Griffin AC (1959) Methoxsalen in ultraviolet carcinogenesis in the mouse. J Invest Dermatol 32:367–372
Grimes PE, Minus HR, Chakrabarti SG, Enterline J, Halder R, Gough JE, Kenney JA (1982) Determination of optimal topical photochemotherapy for vitiligo. J Am Acad Dermatol 7:771–778
Gschnait F, Honigsmann H, Brenner W, Fritsch P, Wolff K (1978) Induction of UV light tolerance by PUVA in patients with polymorphous light eruption. Br J Dermatol 99:293
Guiotto A, Rodighiero P, Manzini P, Pastorini G, Bordin F, Buccichetti F et al. (1984) 6-Methylangelicins: a new series of potential photochemotherapeutic agents for the treatment of psoriasis. J Med Chem 27:959–967

Hanke CW, Steck WD, Roenigk HH (1979) Combination therapy for psoriasis. Psoralen plus longwave ultraviolet radiation with betamethasone valerate. Arch Dermatol 115:1074–1077

Hashimoto K, Kumakiri M (1979) Colloid-amyloid bodies in PUVA-treated human psoriatic patients. J Invest Dermatol 72:70–80

Hell E, Hodgson C, Hanna V (1979) Psoralen photochemotherapy of psoriasis. Br J Dermatol 101:293–298

Hensby CN (1978) The qualitative and quantitative analysis of 8-methoxypsoralen by HPLC-UV and GLC-MS. Clin Exp Dermatol 3:355–361

Herfst MJ, de Wolff FA (1982) Influence of food on the kinetics of 8-methoxypsoralen in serum and suction blister fluid in psoriatic patients. Eur J Clin Pharmacol 23:75–80

Herfst MJ, Koot-Gronsveld EAM, Wolff FA (1978) Serum levels of 8-methoxypsoralen in psoriasis patients using a new fluorodensitometric method. Arch Dermatol Res 262:1–6

Herfst MJ, Edelbroek PM, de Wolff FA (1980) Determination of 8-MOP in suction blister fluid and serum by liquid chromatography. Clin Chem 26:1825–1828

Hindson C, Spiro J, Downey A (1985) PUVA therapy of chronic actinic dermatitis. Br J Dermatol 113:157–160

Hindson TC, Spiro JG, Cochrane H (1987) PUVA therapy of diffuse granuloma annulare. Chem Exp Dermatol 13:26–27

Hoffman C, Landthaler M, Plewig G, Braun-Falco O (1980) In vivo autoradiography and planimetry of epidermis in psoriatics under PUVA therapy. Arch Dermatol Res 267:61–70

Honigsmann H, Wolff K (1983) Isotretinoin-PUVA for psoriasis. Lancet i:236

Honigsmann H, Gschnait F, Konrad K et al. (1977) Photochemotherapy for pustular psoriasis (von Zumbusch). Br J Dermatol 97:119–126

Honigsmann H, Wolff K, Gschnait F, Brenner W, Jaschke E (1980) Keratoses and non-melanoma skin tumors in long-term photochemotherapy (PUVA) J Am Acad Dermatol 3:406–414

Honigsmann H, Jaschke E, Nitsche V, Brenner W, Waltraut R, Wolff K (1982) Serum levels of 8-methoxypsoralen in two different drug preparations: correlation with photosensitivity and UV-A dose requirements for photochemotherapy. J Invest Dermatol 79:233–236

Honigsmann H, Brenner W, Rauschmeier W, Konrad K, Wolff K (1984) Photochemotherapy for cutaneous T cell lymphoma. A follow-up study. J Am Acad Dermatol 10:238–245

Huuskonen H, Koulu L, Wilen G (1984) Quantitative determination of methoxsalen in human serum, suction blister fluid and epidermis by gas chromatography mass spectrometry. Photodermatol Clin Exp 1:137–140

Hymes SR, Morison WL, Farmer ER, Walters LL, Tutschka J, Santos GW (1985) J Am Acad Dermatol 12:30–37

Jansen CT, Malmiharju T (1981) Inefficacy of topical methoxsalen plus UVA for palmoplantar pustulosis. Acta Derm Venereol (Stockh) 61:354

Jansen CT, Karvonen J, Malmiharju T (1982) PUVA therapy for polymorphous light eruptions: comparison of systemic methoxsalen and topical trioxsalen regimens and evaluations of local protective mechanisms. Acta Derm Venereol (Stockh) 62:317–329

Jois HS, Manjunath BL, Rao SV (1933) Chemical examination of the seeds of *Psoralea corylifolia* (Linn). J Indian Chem Soc 10:41–46

Kalimo K, Koulu L, Jansen CT (1983) Effect of a single UVB or PUVA exposure on immediate and delayed skin hypersensitivity in humans. Correlation to erythema response and Langerhans cell depletion. Arch Dermatol Res 275:374–378

Kammerau B, Klebe U, Zesch A, Schaefer H (1976) Penetration, permeation and resorption of 8-methoxypsoralen. Arch Dermatol Res 255:31–42

Kanerva L, Niemi KM; Lassus A (1981) Hyperpigmentation and hypopigmentation of the skin after longterm PUVA therapy. Light electron microscopic observations on three patients. J Cutan Pathol 8:199–213

Kanerva L, Lauharanta J, Niemi KM, Lassus A (1982) Light and electron microscopy of psoriatic skin before and during retinoid (Ro 10-9359) and retinoid-PUVA treatment. J Cutan Pathol 9:175–188
Kligman AM, Goldstein FP (1973) Ineffectiveness of trioxsalen as an oral photosensitizer. Arch Dermatol 107:413–414
Kolis SJ, Williams TH, Postma EJ, Sasso GJ, Confalone PN, Schwartz MA (1979) The metabolism of [^{14}C]methoxysalen by the dog. Drug Metab Dispos 7:220–225
Konrad K, Gschnait F, Honigsmann M, Wolff K (1976) UVA mediated psoralen tissue interactions: subcellular effects. J Invest Dermatol 66:258
Kornhauser A (1980) Molecular aspects of phototoxicity. Ann NY Acad Sci 346:398–418
Korting HC, Schafer-Korting M, Roser-Maas E, Mutschler E (1982) Determination of 8-methoxypsoralen levels in plasma and skin suction blister fluid by a new sensitive fluorodensitometric method. Arch Dermatol Res 272:9–20
Kruger JO, Christophers E, Sclaak M (1978) Dose effects of 8-methoxypsoralen and UV-A in cultured human lymphocytes. Br J Dermatol 98:141–147
Kuske H (1938) Experimentelle Untersuchungen zur Photosensibilisierung der Haut durch pflanzliche Wirkstoffe; Lichtsensibilisierung durch Furocumarine als Ursache veschiedener phytogener Dermatosen. Arch Dermatol Syphilol 178:112–123
Labby DH, Imbrie JD, Fitzpatrick TB (1959) Studies of liver function in subjects receiving methoxsalen. J Invest Dermatol 32:273–276
Lafond G, Roy PE, Grenier R (1984) Lens opacities appearing during therapy with methoxsalen and longwave ultraviolet radiation. Can J Ophthalmol 19:173–175
Lamola AA (1974) Fundamental aspects of the spectroscopy and photochemistry of organic compounds; electronic energy transfer in biologic systems; and photosensitization. In: Fitzpatrick TB (ed) Sunlight and man. University of Tokyo Press, Tokyo, pp 17–55
Lassus A, Reunala T, Idanpaa-Heikkila J, Juvakoski T, Salo O (1981) PUVA treatment and skin cancer: a follow-up study. Acta Derm Venereol (Stockh) 61:141–145
Lauharanta J, Juvakoski T, Lassus A (1981 a) A clinical evaluation of the effects of an aromatic retinoid (Tigason), combination of retinoid and PUVA, and PUVA alone in severe psoriasis. Br J Dermatol 104:325
Lauharanta J, Kousa M, Kapyaho K, Linnamaa K, Mustakallio K (1981 b) Reduction of increased polyamine levels in psoriatic lesions by retinoid and PUVA treatments. Br J Dermatol 105:267
Lauharanta J, Juvakoski T, Kanerva L, Lassus A (1982) Pharmacokinetics of 8-methoxypsoralen in serum and suction blister fluid. Arch Dermatol Res 273:111–114
Lawrence CM, Marks J, Parker S, Shuster S (1984) A comparison of PUVA-etretinate and PUVA-placebo for palmoplantar pustular psoriasis. Br J Dermatol 110:221–226
Lerman S (1979) Radiant energy and the eye. MacMillan, New York
Lerman S, Megaw JM, Gardner KM, Drake L (1983) Ocular and cutaneous manifestations of PUVA therapy: a review. J Toxicol Cutan Ocular Toxicol 1:257–266
Levin RE, Parrish JA (1975) Phototherapy of vitiligo. Lighting design and application. 5:35–43
Ljunggren B, Carter DM, Albert J, Reid T (1980) Plasma levels of 8-methoxypsoralen determined by high pressure liquid chromatography in psoriatic patients ingesting drugs from two manufacturers. J Invest Dermatol 74:59–62
Lord JT, Ziboh V, Poitier J, Leggett G, Penneys NS (1976) The effects of photosensitizers and ultraviolet radiation on the biosynthesis and metabolism of prostaglandins. Br J Dermatol 95:397–406
Mandula BB, Pathak MA (1979) Metabolic reactions in vitro of psoralens with liver and epidermis. Biochem Pharmacol 28:127–132
Mandula BB, Pathak MA, Dudek G (1976) Photochemotherapy: identification of a metabolite of 4,5′,8-trimethylpsoralen. Science 193:1131–1135
Marks J, Rogers S, Chadkirk B, Shuster S (1981) Clearance of chronic plaque psoriasis by anthralin-subjective and objective assessment and comparison with photochemotherapy. Br J Dermatol 105:96

Marks JM, Lawrence CM, Corbett M, Coburn P, Parker S, Shuster S (1984) Influence of prophylactic photochemotherapy on incidence of relapse of psoriasis cleared initially with dithranol. Br Med J [Clin Res] 288:95–96
Marx JL, Schur RK (1980) Response of psoriatic nails to oral photochemotherapy. Arch Dermatol 116:1023
Mason RP, Chignell CF (1982) Free radicals in pharmacology and toxicology-selected topics. Pharmacol Rev 33:189–211
Matthews MM (1963) Comparative studies of lethal photosensitization of *Sarcina lutea* by 8-methoxypsoralen and by toluidine blue. J Bacteriol 85:322–331
Melski JW, Stern RS (1982) Annual rate of psoralen and ultraviolet-A treatment of psoriasis after initial clearing. Arch Dermatol 118:404–408
Melski JW, Tanenbaum L, Parrish JA, Fitzpatrick TB, Bleich HL and 28 participating investigators (1977) Oral methoxsalen photochemotherapy for the treatment of psoriasis: a co-operative clinical trial. J Invest Dermatol 68:328–335
Menne T, Andersen KE, Larsen E, Solgaard P (1981) Pharmacokinetic comparison of seven 8-methoxypsoralen brands. Acta Derm Venereol (Stockh) 61:137–140
Mitchell J, Rook A (1979) Botanical dermatology. Greengrass, Vancouver
Mohamed Z (1984) Some aspects of the pharmacology of 8-methoxypsoralen. Thesis, University of Dundee
Molin L, Thomsen K, Volden G, Groth O (1981) Photochemotherapy (PUVA) in the tumor stage of mycosis fungoides: a report from the Scandinavian mycosis fungoides study group. Acta Derm Venereol (Stockh) 61:52–54
Morison WL, Paul BS, Parrish JA (1977) The effects of indomethacin on longwave ultraviolet induced delayed erythema. J Invest Dermatol 68:130–133
Morison WL, Parrish JA, Fitzpatrick TB (1978a) Controlled study of PUVA and adjunctive topical therapy in the management of psoriasis. Br J Dermatol 98:125–132
Morison WL, Parrish JA, Fitzpatrick TB (1978b) Oral methoxsalen photochemotherapy of recalcitrant dermatoses of palms and soles. Br J Dermatol 99:297–302
Morison WL, Momtaz K, Parrish JA, Fitzpatrick TB (1981) Combined methotrexate-PUVA therapy in the treatment of psoriasis. J Am Acad Dermatol 6:46–51
Moscicki RA, Morison WL, Parrish JA et al. (1982) Reduction of the fraction of circulating helper-inducer T cells identified by monoclonal antibodies in psoriatic patients treated with long-term psoralen/ultraviolet-A radiation (PUVA). J Invest Dermatol 79:205–208
Moss C, Friedmann PS, Shuster S (1981) Photochemotherapy, anthralin and DNCB responsiveness in psoriasis. J Invest Dermatol 76:432
Moss C, Friedmann PS, Shuster S (1985) Susceptibility and amplification of sensitivity in contact dermatitis. Clin Exp Immunol 61:223–241
Mourelatos D, Faed MJW, Johnson BE (1977) Sister chromatid exchanges in human lymphocytes exposed to 8-methoxypsoralen and longwave UV radiation prior to incorporation of bromo-deoxyuridine. Experientia 33:1091–1092
Murray D, Warin AP (1979) Phototherapy for persistent palmoplantar pustulosis. Br J Dermatol 101(17):13–14
Musajo L, Rodighiero G (1962) The skin-photosensitizing furocoumarins. Experientia 18:153–161
Musajo L, Rodighiero G (1972) Mode of photosensitizing action of furocoumarins. In: Giese A (ed) Photophysiolgy, vol 7. Academic, London, pp 115–147
Musajo L, Rodighiero G, Colombo G, Torlone V, Dall'Acqua F (1965) Photosensitizing furocoumarins: interaction with DNA and photo-inactivation of DNA containing viruses. Experientia 23:22–24
Musajo L, Visentini P, Baccichetti F, Razzi MA (1967) Photo-inactivation of Ehrlich ascites tumour cells in vitro obtained with skin-photosensitizing furocoumarins. Experientia 23:335–336
Nield VS, Scott LV (1982) Plasma levels of 8-methoxypsoralen in psoriatic patients receiving topical 8-methoxypsoralen. Br J Dermatol 106:199–203
Nitsche V, Mascher H (1982) 5-MOP: Bioverfügbarkeit und Pharmakokinetik. Arzneimittelforschung 32:1338–1341

Oginsky EL, Green GS, Griffith DG, Fowlks WL (1959) Lethal photosensitization of bacteria with 8-methoxypsoralen to long wavelength ultraviolet radiation. J Bacteriol 78:821–833
Olafsson J, Larkö O, Roupe G, Granerus G, Bengtsson U (1986) Treatment of chronic urticaria with PUVA or UVA plus placebo: a double-blind study. Arch Dermatol Res 278:228–231
Ortonne JP, Thivolet J, Sannwad C (1978) Oral photochemotherapy in the treatment of lichen planus (LP). Br J Dermatol 99:77–78
Ortonne JP, MacDonald DM, Micoud A, Thivolet J (1979) PUVA-induced repigmentation of vitiligo: a histochemical and ultrastructural study. Br J Dermatol 101:1–12
Ortonne JP, Schnitt D, Thivolet J (1980) PUVA-induced repigmentation of vitiligo: scanning electron microscopy of hair follicles. J Invest Dermatol 74:40–42
Parrish JA (1981) Phototherapy and photochemotherapy of skin diseases. J Invest Dermatol 77:167–171
Parrish JA, Fitzpatrick TB, Tanenbaum L, Pathak MB (1974) Photochemotherapy of psoriasis with oral methoxsalen and long wavelength ultraviolet light. N Engl J Med 291:1207–1212
Parrish JA, Stern RS, Fitzpatrick TB (1982) Evaluation of PUVA-1980: its basic nature and toxicity. In: Moschella SL (ed) Dermatology update: review for physicians. Elsevier-North Holland, New York, pp 313–338
Pathak MA (1969) Basic aspects of cutaneous photosensitization. In: Urbach F (ed) The biologic effects of ultraviolet radiation (with emphasis on the skin). Pergamon, Oxford, pp 489–511
Pathak MA, Allen B, Ingram DIE, Fellman JH (1961) Photosensitization and the effect of ultraviolet radiation on the production of unpaired electrons in the presence of furocoumarins (psoralens). Biochim Biophys Acta 54:506–515
Pathak MA, Daniels F Jr, Fitzpatrick TB (1962) Presently known distribution of furocoumarins (psoralens) in plants. J Invest Dermatol 39:225–239
Pathak MA, Worden LR, Kaufman KD (1967) Effect of structural alterations on the photosensitizing potency of furocoumarins (psoralens) and related compounds. J Invest Dermatol 48:103–118
Pathak MA, Kramer DM, Fitzpatrick TB (1974) Photobiology and photochemistry of furocoumarins (psoralens). In: Fitzpatrick TB (ed) Sunlight and man. University of Tokyo Press, Tokyo, pp 335–368
Pathak MA, Mandula B, Nakayama Y, Parrish JA, Fitzpatrick TB (1975) Cutaneous photosensitization and in vivo metabolism of psoralens. J Invest Dermatol 64:279
Pathak MA, Fitzpatrick TB, Parrish JA, Biswas R (1976) Photochemotherapeutic, photobiological and photochemical properties of psoralens. In: Castellani A (ed) Research in photobiology. Plenum, New York, pp 267–281
Pathak MA, Marciani MS, Guitto A, Rodighiero G (1983) A study of the relationship between photosensitizing and therapeutic activity of 4,5′,8-trimethylpsoralen and its major metabolite 4,8-dimethyl, 5′carboxypsoralen. J Invest Dermatol 81:533–539
Pathak MA, Mosher DB, Fitzpatrick TB (1984) Safety and therapeutic effectiveness of 8-methoxypsoralen, 4,5′,8-trimethylpsoralen, and psoralen in vitiligo. Natl Cancer Inst Monogr 66:165–173
Pierard GE, Pieraro-Franchimont C (1984) Xerodermoid induced by photochemotherapy. Am J Dermatopathol 6:397–410
Plummer NA, Hensby CN, Warin AP, Camp RD, Greaves MW (1978) Prostaglandins E_2, F_2 and arachidonic acid levels in irradiated and unirradiated skin of psoriatic patients receiving PUVA treatment. Clin Exp Dermatol 3:367–369
Pohl J, Christophers E (1978) Photoinactivation of cultured skin fibroblasts by sublethal doses of 8-methoxypsoralen and longwave ultraviolet light. J Invest Dermatol 71:316–319
Pohl J, Christophers E (1979) Dose effects of 8-methoxypsoralen and long wave UV-light in 3T3 cells: evaluation of a phototoxic index. Experientia 35:247–249
Potapenko A, Abijev GA, Pistsov M, Roshchupkin D, Vladimisov Y, Pliguett F et al. (1984) PUVA-induced erythema and changes in mechanoelectrical properties of skin inhibition by tocopherols. Arch Dermatol Res 276:12–16

Powell FC, Spiegel GT, Muller SA (1984) Treatment of parapsoriasis and mycosis fungoides: the role of psoralen and longwave ultraviolet light A (PUVA). Mayo Clin Proc 59:538–546
Pullman H, Galosi A, Jakobeit C, Steigleder GK (1980) Effects of selective ultraviolet phototherapy (SUP) and local PUVA treatment on DNA synthesis in guinea pig skin. Arch Dermatol Res 267:37–42
Robinson JK, Baughman RD, Provost TT (1978) Bullous pemphigoid induced by PUVA therapy. Br J Dermatol 99:709–713
Rodighiero G, Dall'Acqua F (1976) Biochemical and medical aspects of psoralens. Photochem Photobiol 24:647–653
Roelandts R, van Boven M, Deheyn T, van der Stichele G, de Greef H, Daenens P (1981) Dietary influences on 8-MOP plasma levels in PUVA patients with psoriasis. Br J Dermatol 105:569–572
Roelandts R, van Boven M, Adriaens P, de Schryver F, de Greef H (1983) The relationship between 8-methoxypsoralen skin and blood levels. J Invest Dermatol 81:331–333
Roenigk HH Jr, Caro WA and 12 participating investigators (1981) Skin cancer in the PUVA-48 co-operative study. J Am Acad Dermatol 4:319–324
Rogers SCF, Marks JM, Shuster S (1981 a) Itch following photochemotherapy for psoriasis. Acta Dermatovenereol 61:178–180
Rogers SCF, Shuster S, Marks JM, Penny RJ, Thody AJ (1981 b) The effects of photochemotherapy on endocrine secretion in patients with psoriasis. Acta Derm Venereol (Stockh) 61:350–352
Ros AM, Wennersten G, Lagerholm B (1983) Long-term photochemotherapy for psoriasis: a histopathological and clinical follow-up study with special emphasis on tumour incidence and behaviour of pigmented lesions. Acta Derm Venereol (Stockh) 63:215–221
Rosario R, Mark GJ, Parrish JA, Mihm MC Jr (1979) Histological changes produced in the skin by equally erythemogenic doses of UVA, UVB, UVC and UVA with psoralens. Br J Dermatol 101:299–308
Schalla W, Schaefer H, Kammerau B, Zesch A (1976) Pharmacokinetics of 8-methoxypsoralen (8-MOP) after oral and local application. J Invest Dermatol 66:258–259
Schmid J, Prox A, Zipp H, Foss FW (1980 a) The use of stable isotopes to prove the saturable first-pass effect of methoxssalen. Biochem Mass Spectrometry 7:560–564
Schmid J, Prox A, Reuter A, Zipp H, Koss FW (1980 b) The metabolism of 8-methoxypsoralen in man. Eur J Drug Metab Pharmacokin et 5:81–92
Schmid J, Brickl R, Busch U, Koss FW (1981) Comparison of pharmacokinetics, pharmacodynamics and metabolism of 8-MOP, 5-MOP and TMP. In: Cahn J, Forlot P, Grupper C, Meybeck A, Urbach F (eds) Psoralens in comsetics and dermatology. Pergamon, Oxford, pp 109–116
Schmoll M, Henseler T, Christophers E (1978) Evaluation of PUVA, topical corticosteroids, and the combination of both in the treatment of psoriasis. Br J Dermatol 99:693–702
Scott BR, Pathak MA, Mohn GR (1976) Molecular and genetic basis of furocoumarin reactions. Mutat Res 39:29–74
Seki T, Nozu K, Kondo S (1978) Differential causes of mutation and killing in *Escherichia coli*: monoadducts and cross links. Photochem Photobiol 27:19–23
Serrano G, Aliaga A, Lorente M (1988) Reactive perforating collagenosis responsive to PUVA. Int J Dermatol 27:118–119
Soine TO (1964) Naturally occurring coumarins and related physiological activities. J Pharm Sci 53:231–264
Song PA, Tapley JJ Jr (1979) Photochemistry and photobiology of psoralens. Photochem Photobiol 29:1177–1197
Spath E (1937) Die natürlichen Cumarine. Ber Deutsch Chem Ges 70A:83–117
Spikes JD (1977) Photosensitization. In: Smith KC (ed) The science of photobiology. Plenum, New York, pp 87–110
Spikes JD (1982) Photodynamic reactions in photomedicine. In: Regan JD, Parrish JA (eds) The science of photomedicine. Plenum, New York, pp 113–144

Steiner I, Prey T, Gschnait F, Washuttl J, Greiter F (1977) Serum level profiles of 8-methoxypsoralen after oral administration. Arch Dermatol Res 259:299–301
Steiner I, Prey T, Gschnait F, Washuttl J, Greiter F (1978) Serum levels of 8-methoxypsoralen 2 hours after oral administration. Acta Derm Venereol (Stockh) 58:185–188
Stern RS (1986) Long-term use of psoralens and ultraviolet A for psoriasis: evidence for efficacy and cost savings. J Am Acad Dermatol 14:520–526
Stern RS, Melski JW (1982) Long-term continuation of psoralen and ultraviolet-A treatment of psoriasis. Arch Dermatol 118:400–403
Stern RS, Thibodeau LA, Kleinerman RA et al. (1979a) Antinuclear antibodies and oral methoxsalen photochemotherapy (PUVA) for psoriasis. Arch Dermatol 115:1320–1324
Stern RS, Thibodeau LA, Kleinerman RA, Parrish JA, Fitzpatrick TB (1979b) Risk factors and increased incidence of cutaneous carcinoma in patients treated with oral methoxsalen photochemotherapy for psoriasis. N Engl J Med 300:809–813
Stern RS, Kleinerman RA, Parrish JA, Fitzpatrick TB, Bleich HL (1980) Phototoxic reactions to photoactive drugs in patients treated with PUVA. Arch Dermatol 116:1269–1271
Stern RS, Parrish JA, Fitzpatrick TB (1985a) Ocular findings in patients treated with PUVA. J Invest Dermatol 85:269–73
Stern RS, Parrish JA, Fitzpatrick TB, Bleich HL (1985b) Actinic degeneration in association with long-term use of PUVA. J Invest Dermatol 84:135–138
Stern RS, Lange R et al. (1988) Non-melanoma skin cancer occurring in patients treated with PUVA five to ten years after first treatment. J Invest Dermatol 91:120–124
Stevenson IH, Kenicer KJA, Johnson BE, Frain-Bell W (1981) Plasma 8-methoxypsoralen concentrations in photochemotherapy of psoriasis. Br J Dermatol 104:47–51
Stolk L, Kammayer A, Cormane RH, van Zweiten PA (1980) Serum levels of 8-MOP: difference between two oral methods of administration. Br J Dermatol 103:417–420
Stolk L, Siddiqui AH, Kammeyer A, Cormane RH, van Zwieten PA (1981) Serum and saliva levels of 8-methoxypsoralen after rectal administration as a micro-enema. Br J Dermatol 104:447
Strauss GH, Greaves M, Price M et al. (1980) Inhibition of delayed hypersensitivity reaction in skin (DNCB test) by 8-methoxypsoralen photochemotherapy. Lancet 2:556–559
Sullivan TJ, Walter JL, Kouba RF, Maiwald DC (1986) Bioavailability of a new oral methoxsalen formulation. A serum concentration and photosensitivity response study. Arch Dermatol 122(7):768–771
Tanenier H, Jesionek A (1903) Therapeutische Versuche mit fluoreszierenden Stoffen. Muench Med Wochenschr 1:2042–2044
Tanew A, Ortel B, Rappersberger K, Honigsmann H (1988) 5-Methoxypsoralen (Bergapten) for photochemotherapy. Bioavailability, phototoxicity, and clinical efficacy in psoriasis of a new drug preparation. J Am Acad Dermatol 18:333–338
Thoms H (1911) Über die Konstitution des Xanthotoxins und seine Beziehungen zu Bergapten. Ber Deutsch Chem Ges 44:3325–3332
Thune P (1978) Plasma levels of 8-methoxypsoralen and phototoxicity studies during PUVA treatment of psoriasis with Meladinin tablets. Acta Derm Venereol (Stockh) 58:149–151
Thune P, Volden G (1977) Photochemotherapy of psoriasis with relevance to 8-methoxypsoralen plasma level and low intensity irradiation. Acta Derm Venereol (Stockh) 57:351–355
Tomson SH, Emmett EA, Fox SH (1974) Photodestruction of mouse epithelial tumors after acridine orange and argon laser. Cancer Res 34:3124–3127
Tucker HA (1959) Clinical and laboratory tolerance studies in volunteers given oral methoxsalen. J Invest Dermatol 32:277–279
Turro NJ, Lamola AA (1977) Photochemistry. In: Smith KC (ed) The science of photomedicine. Plenum, New York, p 63
Van der Kerkhof PLM, Mali JWH (1981) Low dose PUVA maintenance in psoriasis following Ingram therapy. Br J Dermatol 104:681–684

Volden G, Molin L, Thomsen K (1978) PUVA-induced suppression of contact sensitivity to mustine hydrochloride in mycosis fungoides. Br Med J 2:865–866
Wagner G, Hofmann C, Busch U, Schmid J, Plewig G (1979) 8-MOP plasma level in PUVA problem cases with psoriasis. Br J Dermatol 101:285–292
Wagner J, Manthrope R, Philip P, Frost F (1978) Pre-leukemia development in patients with psoriasis treated with 8-methoxypsoralen and ultraviolet light (PUVA treatment). Scand J Haematol 21:299–304
Walter JR, Voorhees JJ, Kelsey WH, Duell EA (1973) Psoralen plus black light inhibits epidermal DNA synthesis. Arch Dermatol 107:861–865
Wolff K, Honigsmann H (1983) Clinical aspects of photochemotherapy. In: Baden HP (ed) Pharmacology and therapeutics. Chemotherapy of psoriasis. Pergamon, New York
Wolff K, Gschnait F, Honigsmann H, Konrad K, Parrish JA, Fitzpatrick TB (1977) Phototesting and dosimetry for photochemotherapy. Br J Dermatol 96:1
Wulf HC (1984) Distribution and accumulation af radioactivity in mice following administration of [^{14}C]8-MOP and [^{3}H]8-MOP: an autoradiographic study. Photodermatol Clin Exp 1:293–297
Wulf HC, Andreasen MP (1981) Changes in radioactivity in rat organs after [^{3}H]8-MOP medication followed by UVA and "sunlight". Arch Dermatol Res 271:157–163
Zarebska Z, Jarzabek-Chorzelska M, Rzesa G, Glinski W, Pawinska M, Chorzelski T et al. (1984) Detection of DNA-psoralen photoadducts in situ. Photochem Photobiol 39:307–312
Zaynoun S, Konrad K, Gschnait F et al. (1977) The pigmentary response to photochemotherapy. Arch Derm Venereol (Stockh) 57:431–440
Zaynoun S, Jaber LAA, Kurban AK (1986) Oral methoxsalen photochemotherapy of extensive pityriasis alba. J Am Acad Dermatol 15:61–65
Zelickson AS, Mottaz JH, Zelickson BD, Muller SA (1980) Elastic tissue changes in skin following PUVA therapy. J Am Acad Dermatol 5:186

CHAPTER 39

Dapsone and Sulphapyridine

T. M. TWOSE

> What has fascinated dermatologists for over 30 years is why sulphapyridine and dapsone have such a remarkable effect on dermatitis herpetiformis
> (ALEXANDER 1975)

A. Introduction

Dermatitis herpetiformis (DH) is a chronic vesiculobullous disorder, involving an intensely itchy papulovesicular eruption of the skin. Both dapsone (4,4′-diaminodiphenylsulphone, DDS) and sulphapyridine (SP), two chemically related drugs, quickly relieve the itching and dramatically bring the skin lesions under control.

Neither drug was developed for this purpose, and despite 40 years or more of use in DH, their mechanisms of action in this disorder are still not known. Both drugs are anti-bacterial by virtue of their *p*-aminobenzoic acid antagonistic properties. However, no bacterial aetiology has been established for DH,and several lines of evidence rule out anti-bacterial modes of action in DH. DH is uncommon and the effects of these drugs might be regarded as a convenient curiosity were it not for evidence of their therapeutic effects in a wider range of dermatoses.

This chapter focuses on this intriguing and elusive property of DDS and SP. Information will be drawn together which suggests that they have actions which differ from those of other anti-inflammatory drugs. Possible modes of action will also be discussed. A better understanding of these drugs could help unravel the pathogenetic mechanisms of the diseases which respond to them and might provide a better rationale for their clinical use. Furthermore, if they are prototypes of a new class of drug, it could aid the development of better and safer compounds possessing this dapsone-like action. Historical aspects will not be considered since development of DDS and SP as anti-microbials has been reviewed by LONG and BLISS (1939), HAWKING and LAWRENCE (1950) and ROSE (1964).

B. Chemical Aspects

I. Structures of DDS, SP and Their Circulating Metabolites

DDS (1) and SP (2) are both anilines (Fig. 1) which determines some of their common metabolic routes and toxic actions, and could contribute to their shared therapeutic effects. Both compounds undergo *N*-acetylation in vivo (Fig. 1; structures 1 a, 1 c, 2 a and 2 c). At least in the case of DDS, *N*-oxidation also occurs to give *N*-hydroxy DDS derivatives (Fig. 1; structures 1 b and 1 c). These hydroxylamines readily undergo redox reactions, and are implicated in the development of adverse effects (see Sect. H). *N*-Hydroxy metabolites of SP have not been reported, but ring hydroxylation is a major route of metabolism, yielding compounds 2 b

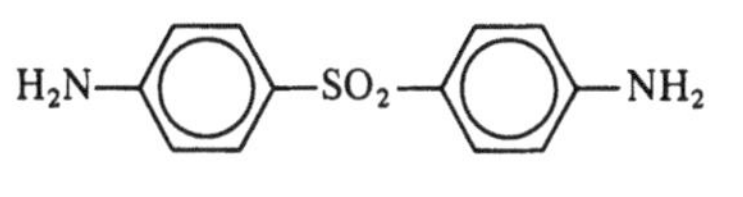

(1)

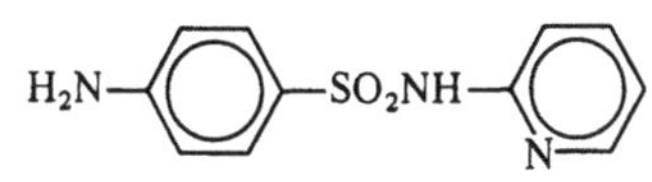

(2)

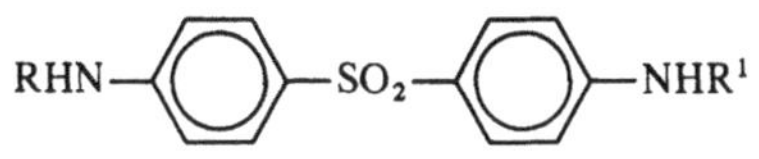

	R	R^1
1a	CH_3CO	H
1b	HO	H
1c	HO	CH_3CO

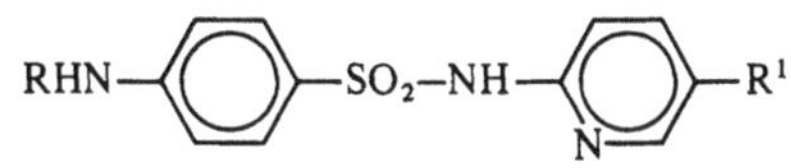

	R	R^1
2a	CH_3CO	H
2b	H	HO
2c	CH_3CO	HO

Fig. 1. Dapsone (*1*), sulphapyridine (*2*) and circulating metabolites

and 2c (Fig. 1). Other more polar, rapidly cleared metabolites (e.g. glucuronides, sulphates) are also formed, but are not further discussed here since they do not appear to contribute to the biological activity of the parent drug.

II. Assay Methods

DDS and SP can be measured by the Bratton and Marshall (1937) method based on chemical reaction at free arylamine groups. To measure certain metabolites (e.g. 1a, 2a) prior acid hydrolysis is required. A sensitive fluorimetric method is also available for DDS and when combined with selective extraction (Peters et al. 1970) or high pressure liquid chromatography (HPLC) separation (Murray et al. 1978) it can be used for quantitating metabolites. Rapid HPLC methods (using UV absorption detection) have been described for DDS and its metabolites (Carr et al. 1978) and for SP and its metabolites (Fischer and Klotz 1978). The earlier chemical methods were sensitive to 1 µg/ml, but the newer HPLC techniques can detect 20 ng/ml, depending on the detector system. The measurement of hydroxy metabolites in serum is more difficult, and investigation of these has generally required measurements of their conjugates or derivatives after excretion in the urine (Schröder and Campbell 1972; Israili et al. 1973; Uehleke and Tabarelli 1973).

C. Pharmacokinetics and Metabolism in Humans

Studies in laboratory animals, and in vitro, are important for understanding the actions of these drugs in humans. However, both qualitative and quantitative differences exist in the ways laboratory animals and humans handle these drugs and care must be taken in the design and interpretation of in vivo experiments. It cannot be assumed that the parent drugs are the active moieties; knowledge of metabolites and their concentrations in humans forms an essential basis for in vitro work.

For both DDS and SP, *N*-acetylation is a major route of metabolism. This is subject to the same genetic polymorphism as is the acetylation of isoniazid and other drugs. A single oral 200 mg dose of DDS is rapidly absorbed in fast acetylators to give mean maximum serum concentrations, after 4 h, of 1.8 μg/ml for DDS, and 1.8 μg/ml for acetyl-DDS. In slow acetylators DDS levels are higher (2.3 μg/ml) and, as expected, acetyl-DDS levels are lower (0.4 μg/ml). The serum half-life does not depend on acetylation status and is approximately 20 h (LAMMINTAUSTA et al. 1979). DH patients show similar results (ALEXANDER et al. 1970) and, predictably, blood levels are higher with repeated daily dosing (BEIGUELMAN et al. 1974).

Acetylation phenotype also influences the kinetics of SP. In fast acetylators the half-life is 4.6 h compared with 13 h in slow acetylators (FISCHER and KLOTZ 1979). Acetyl-SP is cleared three times faster than SP (SCHRÖDER and CAMPBELL 1972) and cannot be deacetylated in vivo. However, SP is more erratically absorbed after oral doses than is DH. Mean peak blood levels from a 1.25 g oral dose are 16.7 μg/ml (fast acetylators) and 21.6 μg/ml (slow acetylators) with 63% and 25%, respectively, of the total plasma SP in the acetylated form (FISCHER and KLOTZ 1979).

The second major route of metabolism of DDS is *N*-hydroxylation. *N*-hydroxy-DDS and its derivatives (e.g. an azoxy derivative and glucuronides) may account for one-third of the excreted drug. *N*-Hydroxy-DDS is readily oxidised to a nitroxide. Both compounds are implicated in causing methaemoglobinaemia and haemolysis, which occur in the majority of patients treated with DDS. Nothing is known about blood or tissue concentrations of these metabolites, but the high incidence of these adverse effects implies that the concentrations are significant.

5-Hydroxylation of the pyridine ring is the other major route of SP metabolism, yielding compounds 2b and 2c (Fig. 1). Between 7% and 12% of total SP circulates as the glucuronides of these metabolite (DAS and DUBIN 1976) and 48%–63% of total SP is excreted in these forms. Little unconjugated hydroxy metabolite was found in the serum (SCHRÖDER and CAMPBELL 1872).

D. Therapeutic Effects in Dermatoses

Over 75% of DH patients benefit from DDS or SP treatment. Different approaches are used by different prescribers in finding suitable dosages for each patient, but effective doses of DDS are generally below 300 mg/day, and of SP below 4 g/day (see LANG 1979; BERNSTEIN and LORINCZ 1981).

Itching is relieved within hours of starting therapy with either drug and skin lesions improve within 2–7 days. Drug withdrawal, even after years of therapy, invariably leads to a rapid relapse. The rapidity of the onset of their effects and of relapse suggests they are acting by a direct pharmacological mechanism. These and other considerations (LORINCZ and PEARSON 1962) also largely exclude an anti-microbial mode of action.

It is therefore an enigma that clinical response does not correlate with blood concentration. Response is dosage dependent, but varies widely from patient to patient. Successfully controlled patients had DDS blood levels ranging from 0.6 to 7.3 μg/ml (Ellard et al. 1974). DDS and SP are aromatic amines, with sulphone/sulphonamide moieties in the *para* position. These features determine their shared anti-bacterial actions, and are also present in many other sulphanilamide drugs which are effective as anti-bacterials in humans. However, the therapeutic activity in DH of DDS and SP does not extend to these other sulphanilamides (Alexander 1975). This is an important observation which might provide clues to the mechanism of action of DDS and SP in DH, and might also be useful for testing hypotheses about the mechanism of action. Clues to the modes of action of these drugs can be sought in what is known about the pathogenesis of DH (see Katz and Strober 1978) and of the other sulpha-responsive dermatoses.

In DH, skin lesions begin with dermal papillary microabscesses containing neutrophil and eosinophil leucocytes which cause damage at the dermal-epidermal junction, leading eventually to vesiculation and blister formation. In the dermal papillae there are generally deposits of IgA, often accompanied by C3. Patients also suffer from a gluten-sensitive enteropathy (GSE) (Shuster et al. 1968), which some believe to have causal significance, since they find gluten-free diets to be an effective treatment (Fry et al. 1973; Katz et al. 1980). DDS and SP therapy, even continued for years, fail to affect GSE, IgA or C3 deposits (Dahl 1977; van der Meer et al. 1979). Assuming that these do play a role in pathogenesis, the drugs' site of action lies elsewhere.

There are several reasons for suspecting that these drugs may in some way affect the functions of the polymorphonuclear leucocytes which are so prominent in the lesions of DH, and which may be the ultimate effectors of skin damage in this disease. Two related dermatoses, pemphigus and bullous pemphigoid, do not generally respond to DDS or SP. Polymorphonuclear leucocytes are not prominent in these conditions. Those patients who do respond may have an overlap syndrome with greater polymorph involvement (Person and Rogers 1977).

Though response to DDS has been considered by some to confirm the diagnosis of DH, it is now accepted that other dermatoses may respond to these drugs. Some of these conditions are listed (after McDougall 1979) in Table 1, but diseases with known infectious aetiologies have been excluded. The doses of DDS and SP used in these disorders are similar to those used in DH, but relatively few

Table 1. Some conditions which have been reported to respond to DDS or SP

Acne conglobata	Herpes gestationis
Acne (cystic)	Pemphigus vulgaris
Acrodermatitis continua (Hallopeau)	Pustular bacterid
Allergic vasculitis	Pyoderma gangrenosum
Australia antigen polyarteritis	Pyoderma vegetans
Bullous permphigoid	Relapsing polychondritis
Dermatitis repens	Subcorneal pustular dermatosis
Discoid (nummular) eczema	Rheumatoid arthritis
Erythema elevatum diutinum	Vasculitic urticaria
Follicular mucinosis	

studies have been reported and efficacy cannot be said to have been established conclusively in all cases. However, continuous therapy has been required in most studies and interruption of therapy has led to rapid recrudescence of disease activity. Comprehensive reviews are provided by LORINCZ and PEARSON (1962), LINN (1962), ALEXANDER (1975), PLEWIG and KLIGMAN (1975), MCDOUGALL (1979), LANG (1979), and BERNSTEIN and LORINCZ (1981).

Among the disorders listed in Table 1 are those involving microabscesses or chronic pustulation, various forms of vasculitis, and those involving a more diffuse inflammatory attack on connective tissues. The polymorphonuclear leucocyte is dominant in many of these lesions and as many authors have suggested, efficacy in these apparently diverse disorders may reflect an action on a common pathogenetic mechanism involving this cell. It would be valuable to explore this idea further in clinical studies, for instance by searching for exceptions to the rule or testing its power in predicting diseases which would respond to DDS or SP.

E. Therapeutic Effects in Rheumatoid Arthritis

MCCONKEY et al. (1976) and SWINSON et al. (1981) reported the successful use of DDS to treat rheumatoid arthritis. Erythrocyte sedimentation rates and C-reactive protein levels fell during treatment; this may also represent evidence of a basic action against chronic inflammatory disease mechanisms.

F. Anti-inflammatory Actions in Animals

Several inflammation models in animals respond to DDS (Table 2) at doses ranging from 10 to 200 mg/kg. The extensive studies of LEWIS et al. (1978) and GEMMEL et al. (1979) show that the anti-inflammatory action of DDS differs from that of aspirin-like anti-inflammatory drugs. It has pronounced activity against reverse passive Arthus-type reactions in rats, which was evident even at low doses. These findings are especially interesting in view of the singular importance of polymorphonuclear leucocytes (PMN) in the mediation of these reactions (COCHRANE 1974). By contrast, the relevance of the activity shown in adjuvant arthirits and cotton wool pellet granulomas must be questioned since the doses required for an effect (50–150 mg/kg^{-1} day^{-1}) approach toxic levels. BARRANCO (1974) observed that DDS prevented cartilage destruction in rabbits treated with toxic doses of vitamin A, he attributed this to inhibition of lysosomal enzymes. Such an action could explain many of the therapeutic properties of DDS.

G. Comparative Pharmacokinetics and Metabolims

There are major interspecific differences in the disposition of DDS and SP which clearly complicate studies in animal models. The serum half-life ranges from 1.2 to 14.5 h in six species, compared with approximately 20 h in humans. Dogs and mice proved to be poor acetylators while rats, some rabbits, and humans are good acetylators (PETERS et al. 1971). *N*-Oxidation is a major metabolic route in hu-

Table 2. Animal models of inflammation which respond to DDS

Model	Dose range	References
A. Single-dose studies[a]		
Carrageenan oedema (rat)	36–160	Capstick and Lewis (1977)
Kaolin-induced oedema (rat)	11–100	Lewis et al. (1978); Gemmel et al. (1979)
Zymosan-induced oedema (rat)	10–100	Lewis et al. (1978); Gemmel et al. (1979)
Reverse passive Arthus-type reactions (rat)	10–100	Lewis et al. (1978); Gemmel et al. (1979)
UV erythema (guinea pig)	ED_{50}–160	Lewis et al. (1978)
B. Multiple dose studies[b]		
Reverse passive Arthus reaction (guinea pig)	2– 5	Ruzicka et al. (1981)
Active Arthus reaction (guinea pig)	5– 50	Thompson and Souhami (1975)
Cotton wool pellet granuloma (rat)	200	Capstick and Lewis (1977)
Carrageenan-impregnated cotton wool pellet granuloma (rat)	50–200	Lewis et al. (1978)
Adjuvant arthritis (rat)	100–200	Capstick and Lewis (1977), Lewis et al. (1978)
Hypervitaminosis A (rabbit)	25	Barranco (1974)

[a] Doses mg/kg.
[b] Doses mg kg^{-1} day^{-1}.

mans, but is minor in cats and dogs, and negligible in guinea pigs (Israili et al. 1973; Uehleke and Tabarelli 1973). Such differences could lead to large differences in the dose required to produce equivalent effects in different species, especially if the active moiety is a metabolite. It is of interest that in the dog (in which DDS undergoes minimal acetylation or *N*-oxidation), the drug is effective against DH and subcorneal pustular dermatosis at doses similar to those which are effective in humans (Halliwell et al. 1977; McKever and Dahl 1977).

Less information is available on the comparative disposition of SP. There are large differences between half-lives in humans (13 h in fast acetylators, 4–6 h in slow acetylators) and laboratory animals such as rats (1.8 h) and rabbits (0.79 h) (Yamazaki et al. 1968). There are also large species differences in the degree of acetylation – none in dogs, a little in mice, extensive in rabbits (Marshall and Litchfield 1937).

H. In Vitro Actions and Possible Mechanisms of Action

There is no evidence that DDS or SP inhibit lysosomal enzymes directly, other than at high concentrations (Fraki and Hopsu-Havu 1977; Mier and van den Hark 1975) and this cannot therefore explain the results in the rabbit hypervitaminosis A model. DDS and SP do not inhibit chemotaxis or phagocytosis of PMN at concentrations within the therapeutic range. However, at relatively low concentrations, DDS and SP both inhibit PMN-mediated cytocidal activity

(STENDAHL et al. 1978; MOLIN and STENDAHL 1980). DDS has also been shown to inhibit phagocytosis-stimulated chemiluminescence (WEBSTER et al. 1984) and PMN-mediated oxidative inactivation of α_1-antiprotease (THERON et al. 1985). These specialised functions of PMN all involve reduced oxygen species such as hydrogen peroxide, superoxide anion and hydroxyl ion generated in response to phagocytic stimuli, and also involve the PMN enzyme myeloperoxidase. STENDAHL et al. (1978) report that DDS competitively inhibits myeloperoxidase, but not the generation of hydrogen peroxide, while others find a concentration-dependent reduction in hydrogen peroxide and hydroxyl ion generation, but without effect on superoxide anion (NIWA et al. 1984). At therapeutic doses partial inhibition of the PMN-oxygen metabolite-myeloperoxidase system by DDS would be expected and this might explain the therapeutic effects of the drug in DH (WOZEL 1987). It is known that phagocytosing PMN damage bystander cells in a myeloperoxidase-mediated reaction (CLARK and KLEBANOFF 1975), and DDS has been shown to partially inhibit an in vitro model of this process (MARTIN and KACHEL 1985) albeit at relatively high concentrations. Bystander cell damage may occur in the microabscesses in DH, and it has been suggested that reversible inhibition of this process might explain the therapeutic effects of DDS and SP (STENDAHL et al. 1978).

All patients taking DDS have some degree of haemolysis and methaemoglobinaemia (MANFREDI et al. 1979). These adverse effects are dose related and are thought to be caused by the *N*-hydroxy metabolite (KRAMER et al. 1972) which is required only in catalytic amounts to put a severe oxidant stress on red cell metabolism. Glutathione, NADPH and the hexose monophosphate shunt are especially affected. This is normally well tolerated because of increased production of NADPH. Because these processes are invariably occurring in DDS-treated patients we must ask whether they could also explain the therapeutic effects. Certain PMN functions (e.g. reduction of oxygen to superoxide anion and hydrogen peroxide) do depend on NADPH, and decreased availability of NADPH might have an anti-inflammatory effect. The *N*-hydroxy metabolites of DDS (see Fig. 1; structures 1 b and 1 c) are more redox active than the parent compound and would be expected to be more active on the PMN-oxygen metabolite-myeloperoxidase pathway. Compound 1 b is more potent than DDS as a myeloperoxidase inhibitor (T. M. TWOSE 1984, unpublished work).

In vitro investigations suggest strongly that the site of action of DDS and SP is in the specialised PMN oxidative pathways, but further evidence is still required. Studies need to be extended to the metabolites of DDS and SP. Drug-metabolising liver microsomal systems such as those used to investigate methaemoglobin-forming metabolites (THAUER et al. 1968) could be used. Studies are also required on chemically related sulphanilamides which are not active in DH.

J. Concluding Remarks

DDS and SP are effective in several skin disorders. The simplest explanation for their therapeutic effects is that these apparently diverse diseases all share an underlying disease mechanism which is modified by these drugs. We have outlined evidence which suggests that the cellular target is the polymorphonuclear leucocyte, and that the drugs possess an anti-inflammatory action of a fundamentally novel type, directed at these cells. Evidence from studies in animal models of inflammation appears to support this hypothesis, but as yet, the actual mechanism of action has not been elucidated.

References

Alexander JO'D (1975) Dermatitis herpetiformis, vol 4. Major problems in dermatology. Saunders, Philadelphia

Alexander JO'D, Young E, McFadyen T, Fraser NG, Duguid WP, Meredith EM (1970) Absorption and excretion of ^{35}S-dapsone in dermatitis herpetiformis. Br J Dermatol 83:620–631

Barranco VP (1974) Inhibition of lysosomal enzymes by dapsone. Arch Dermatol 110:563–566

Beiguelman B, Pinto W, El-Guindy MM, Krieger H (1974) Factors affecting the level of dapsone in blood. Bull WHO 51:467–471

Bernstein JE, Lorincz AL (1981) Sulphonamides and sulphones in dermatologic therapy. Int J Dermatol 20:81–88

Bratton AC, Marshall EK (1937) A new coupling component for sulphanilamide determination. J Biol Chem 128:537

Capstick RB, Lewis DA (1977) Aspects of the biochemistry and toxicology of dapsone and azapropazone. Drugs Exp Clin Res 2:79–87

Carr K, Oates JA, Nies AS, Woosley RL (1978) Simultaneous analysis of dapsone and monoacetyldapsone employing high performance liquid chromatography. Br J Clin Pharmacol 6:421–427

Clark RA, Klebanoff SJ (1975) Neutrophil mediated tumour cell cytotoxicity: role of the peroxidase system. J Exp Med 141:1442–1447

Cochrane CG (1974) The Arthus reaction: a model of neutrophil and complement mediated injury. In: Zweifach BW (ed) The inflammatory process, 2nd edn, vol 3. Academic, New York

Dahl MGC (1977) Dapsone in dermatitis herpetiformis and other immunological diseases. Lancet i:429

Das KM, Dubin R (1976) Clinical pharmacokinetics of sulphasalazine. Clin Pharmacokinet 1:406–425

Ellard GA, Gammon PT, Savin JA, Tan RS-H (1974) Daspone acetylation in dermatitis herpetiformis. Br J Dermatol 90:441–444

Fischer C, Klotz U (1978) Determination of sulphapyridine and its major metabolites in plasma by high pressure liquid chromatography. J Chromatogr 146:157–162

Fischer C, Klotz U (1979) High performance liquid chromatographic determination of aminosalicylate, sulphapyridine and their metabolites: its application to pharmacokinetic studies in man. J Chromatogr 162:237–243

Fraki JE, Hopsu-Havu VK (1977) Inhibition of human skin proteinases by chloroquine, dapsone and sulphapyridine. Arch Dermatol Res 259:113–115

Fry L, Seah PP, Riches DJ, Hoffbrand AV (1973) Clearance of skin lesions in dermatitis herpetiformis after gluten withdrawal. Lancet 1:288–291

Gemmel DK, Cottney J, Lewis AJ (1979) Comparative effects of drugs on four paw oedema models in the rat. Agents Actions 9:107–116

Halliwell REW, Schwartzman RM, Ihrke PJ, Goldschmidt MH, Wood MG (1977) Dapsone for the treatment of pruritic dermatitis (dermatitis herpetiformis and subcorneal pustular dermatosis) in dogs. J Am Vet Med Assoc 170:697–703

Hawking F, Lawrence JS (1950) The sulphonamides. Lewis, London

Israili ZH, Cucinelli SA, Vaught J, Davis E, Lesser JM, Dayton PG (1973) Studies of the metbolism of dapsone in man and experimental animals: formation of *N*-hydroxy metabolites. J Pharmacol Exp Ther 187:138–151

Katz SI, Strober W (1978) The pathogenesis of dermatitis herpetiformis. J Invest Dermatol 70:63–75
Katz SI, Hall RP, Lowley TJ, Strober W (1980) Dermatitis herpetiformis: the skin and the gut. Ann Intern Med 93:857–874
Kramer PA, Glader BE, Li T-K (1972) Mechanism of methemoglobin formation by diphenylsulphones. Effect of 4-amino-4′-hydroxyaminodiphenylsulphone and other *p*-substituted derivatives. Biochem Pharmacol 21:1265–1274
Lammintausta K, Kangas L, Lammintausta R (1979) The pharmacokinetics of dapsone and acetylated dapsone in serum and saliva. Int J Clin Pharmacol Bipharm 17:159–163
Lang PG (1979) Sulfones and sulfonamides in dermatology today. J Am Acad Dermatol 1:479–492
Lewis AJ, Gemmel DK, Stimson WH (1978) The anti-inflammatory profile of dapsone in animal models of inflammation. Agents Actions 8:578–586
Linn HW (1962) The use of dapsone in dermatology. Aust J Dermatol 6:203–207
Long PH, Bliss EA (1939) The clinical and experimental use of sulphanilamide, sulphapyridine and allied compounds. McMillan, New York
Lorincz AL, Pearson RW (1962) Sulphapyridine and sulphone-type drugs in dermatology. Arch Dermatol 85:42–56
Manfredi G, de Panfilis G, Zampetti M, Allegra F (1979) Studies on dapsone induced hemolytic anemia 1. Methemoglobin production and G-6-PD activity in correlation with dapsone dosage. Br J Dermatol 100:427–432
Marshall EK, Litchfield JT (1937) Some aspects of the pharmacology of sulphapyridine. J Pharmacol 67:454–475
Martin WJ, Kachel DL (1985) Reduction of neutrophil mediated injury to pulmonary endothelial cells by dapsone. Am Rev Resp Dis 131:544–547
McConkey B, Davies P, Crockson RA, Crockson AP, Butler M, Constable TJ (1976) Dapsone in rheumatoid arthritis. Rheumatol Rehabil 15:230–234
McDougal AC (1979) Dapsone. Clin Exp Dermatol 4:139–142
McKever PJ, Dahl MV (1977) A disease in dogs resembling human subcorneal pustular dermatosis. J Am Vet Med Assoc 170:704–708
Mier PD, van den Hark JJMA (1975) Inhibition of lysosomal enzymes by dapsone. Br J Dermatol 93:471–472
Molin L, Stendahl O (1980) The effects of salazopyrine and its active components on human polymorphonuclear leukocyte function in relation to ulcerative colitis. Acta Med Scand 206:451–457
Murray JF, Gordon GR, Galledge CC, Peters JH (1978) Chromatographic-fluorimetric analysis of antileprotic sulphones. J Chromatogr 107:67–72
Niwa Y, Sakane T, Miyachi Y (1984) Dissociation of inhibitory effects of dapsone in the generation of oxygen intermediates – in comparison with colchicine and other scavengers. Biochem Pharm 33:2355–2360
Person JR, Rogers RS (1977) Bullous pemphigoid responding to sulphapyridine and the sulphones. Arch Dermatol 113:610–616
Peters JH, Gordon GR, Colwell WT (1970) The fluorimetric measurement of 4,4′-diaminodiphenylsulphone and its acetylated derivatives in plasma and urine. J Lab Clin Med 76:338–348
Peters JH, Gordon GR, Biggs JT, Levy L (1971) The disposition of dapsone and monoacetyl dapsone in laboratory animals. Int J Lepr 39:101–102
Plewig G, Kligman AM (1975) Acne morphogenesis and treatment. Springer, Berlin Heidelberg New York
Rose FL (1964) Medicine's debt to the sulphonamide group. Chem Ind (Lond), 23 May: 858–865
Ruzicka T, Bower A, Gluck S, Born M (1981) Effects of dapsone on passive Arthus rection and chemotaxis and phagocytosis of polymorphonuclear leukocytes. Arch Dermatol Res 270:347–351
Schröder H, Campbell DES (1972) Absorption, metabolism and excretion of salicylazosulphapyridine in man. Clin Pharmacol Ther 13:539–551

Schröder H, Evans DAP (1972) The polymorphic acetylation of sulphapyridine in man. J Med Genet 9:168–171
Shuster S, Watson AJ, Marks J (1968) Coeliac syndrome in dermatitis herpetiformis. Lancet 1:1101–1106
Stendahl O, Molin L, Dahlgren C (1978) The inhibition of polymorphonuclear leukocyte cytotoxicity by dapsone. A possible mechanism in the treatment of dermatitis herpetiformis. J Clin Invest 62:214–220
Swinson DR, Zlosnick J, Jackson L (1981) Double-blind trial of dapsone against placebo in the treatment of rheumatoid arthritis. Ann Rheum Dis 40:235–239
Thauer RK, Meiforth A, Uehleke H (1968) Methämoglobinbildung durch Sulphonamide im System Leberhomogenat, Erythrocyten, NADPH und Sauerstoff. Naunyn Schmiedebergs Arch Pharmacol 252:291–296
Theron A, Anderson R (1985) Investigation of the protective effects of the antioxidants ascorbate, cysteine and dapsone on the phagocyte-mediated oxidative inactivation of human α_1-protease inhibitor in vitro. Am Rev Resp Dis 132:1049–1054
Thompson DM, Souhami R (1975) Suppression of Arthus reaction in the guinea pig by dapsone. Proc R Soc Med 68:213–274
Uehleke H, Tabarelli S (1973) *N*-Hydroxylation of 4,4′-diaminodiphenylsulphone (dapsone) by liver microsomes, and in dogs and humans. Naunyn Schmiedebergs Arch Pharmacol 278:55–68
Van der Meer JB, Poey H, van der Putte SCJ, Nefkens MJ, Baart de la Faille-Knyper EH (1979) Effects of dapsone and gluten free diet on the immune complexes in the skin of patients with dermatitis herpetiformis. Br J Dermatol 101:232–233
Webster GF, Alexander JC, McArthur WP, Leyden TJ (1984) Inhibition of chemiluminescence in human neutrophils by dapsone. Brit J Dermatol 110:657–663
Wozel von G (1987) Zum Wirkungsmechanismus von Dapsone bei chronisch entzündlichen Dermatosen. Teil 1: Beeinflussung der Funktionen polymorphkerniger Leukozyten. Dermatol Mon Schr 173:569–576
Yamazaki M, Aoki M, Kamada A (1968) Biological activities of drugs in physicochemical factors affecting the excretion of sulphonamides in rats. Chem Pharm Bull (Tokyo) 16:721–727

CHAPTER 40

Zinc Deficiency

D. J. Atherton

A. Introduction

The importance of zinc deficiency was not fully appreciated until Moynahan (1974), associated it with acrodermatitis enteropathica, a potentially fatal disorder of infants. Devotion of a chapter of this book to zinc reflects its particular importance for the skin and I will particularly review those situations where zinc may be of therapeutic value; excellent reviews of the metabolic role of zinc are Aggett and Harries (1979), Chesters (1980) and Weismann (1980).

B. Zinc Deficiency

That the link between zinc deficiency and the clinical features of acrodermatitis enteropathica was discovered by a dermatologist (Moynahan 1974) illustrates the way in which the skin reflects tissue zinc. Systemic manifestations precede the skin changes, but are non-specific and therefore easily overlooked. The clinical features are shown in their approximate order of appearance in Table 10.

Subclinical zinc deficiency is difficult to detect as plasma zinc concentrations are not a good guide to tissue availability. Zinc in the plasma is almost entirely bound to proteins, particularly albumin and α_2-macroglobulin, and to amino acids and changes in the plasma zinc concentration reported in a wide variety of diseases more often reflect changes in the levels of these carriers than changes in the amount of zinc available to cells. Measurements of zinc in hair, saliva and urine are not helpful and interest has recently turned to measurement of intracellular zinc although it is unclear that this will provide any better guide to zinc status than plasma levels. The value of skin content remains to be established (Michaelsson et al. 1980). In view of these problems clinical response to zinc administration is still the best evidence of zinc deficiency.

Table 1. Clinical features of zinc deficiency

Acute	Photobia; peri-orificial eczematous eruptions; bullae around nails and in flexor creases of fingers and toes; paronychia and nail dystrophy; diarrhoea
Acute and chronic	Anorexia and impairment of taste; depression, apathy and irritability; delayed wound healing
Chronic	Growth retardation; diffuse alopecia; acneiform facial eruption; erythrokeratodermal plaques on extensor surfaces; loss of libido

I. Acrodermatitis Enteropathica

This acute zinc deficiency syndrome is due to an abnormality of zinc absorption by the gut (AGGETT and HARRIES 1979; ATHERTON et al. 1979) which is an autosomal recessive trait usually presenting in infancy, with weeping eczematous lesions appearing at the mucocutaneous junctions when breast-feeding is stopped. There is irritability, photophobia, and later diarrhoea and failure to thrive. There may be paronychia, periungual bullae and nail shedding and the hair is progressively lost. Plasma zinc concentrations are almost always low and the serum alkaline phosphatase is frequently reduced. The response to oral zinc is dramatic. Individual requirements vary greatly and change with time, being greatest during periods of rapid growth, puberty, and during intercurrent illness. A subtle change in mood may signal early deficiency consequent upon increased demand and the parents should be aware of this. Treatment is given in divided doses about 30 min before meals, as absorption is decreased when given with food. Zinc sulphate provides 22.5 mg elemental zinc per 100 mg and is given to babies in an elixir form, in a daily dose of approximately 10 mg/kg. In older children and adults, zinc sulphate may be given in the same dose in capsules, though its absorption is less efficient in this form.

Gastrointestinal irritation not uncommonly causes nausea and may be overcome by administration during meals. There is a risk of toxicity both with acute overdosage (MURPHY 1970), and with chronic administration, which may induce hypocupraemia (PRASAD et al. 1978) and decrease high density lipoproteins (HOOPER et al. 1980). Treatment therefore needs to be regularly monitored, both clinically and by plasma zinc and copper determination.

II. Prematurity and Bottle Feeding

About 60% of the zinc present in the foetus at term is transferred during the last 10 weeks of pregnancy (WIDDOWSON et al. 1974). Premature infants therefore have a considerable requirement for zinc. They may also have difficulty absorbing zinc, particularly when bottle-fed (WIDDOWSON et al. 1974; DAUNCEY et al. 1977). Although human milk contains less zinc than cow's milk, some of the zinc is associated with low molecular weight compounds picolinic acid (EVANS and JOHNSON 1980) or citrate, whereas it is entirely associated with high molecular weight compounds in cow's milk (COUSINS and SMITH 1980) from which zinc is less well absorbed by the neonatal gastrointestinal tract. There is a risk of zinc deficiency in infants born at term who are fed cow's milk without zinc supplement (WALRAVENS and HAMBIDGE 1976) and in infants born pre-term however they are fed (AGGETT et al. 1980). Some women produce breast milk deficient in zinc (ZIMMERMAN et al. 1982) and their offspring are likely to develop symptomatic zinc deficiency if born prematurely and exclusively breast-fed. It seems probable that this defect is genetically determined and an analogous disorder is known in mice (PILETZ and GANSCHOW 1978). It remains to be established whether zinc deficiency in breast-fed premature infants occurs exclusively in the offspring of such mothers. Zinc sulphate in doses of 3 mg kg^{-1} day^{-1} is is usually adequate for the treatment of acquird zinc deficiency.

III. Availability of Zinc in the Diet

The best sources of dietary zinc are meat and fish. Cereals and vegetables have a similar content of zinc, but gastrointestinal absorption is less because of complexing with phytate and fibre (OBERLEAS and HARLAND 1977). Therefore a largely vegetable diet may not provide sufficient available zinc, particularly during those periods of high zinc requirement. Synthetic diets used in the treatment of inborn errors of metabolism and dietary intolerance in children may require supplementation with zinc (THORN et al. 1978). Self-selected diets are often deficient in total zinc content (SANDSTEAD 1973) though precise zinc requirements have not yet been established for humans. During a survey of hair zinc in apparently normal American children, HAMBIDGE et al. (1978) found that a number had very low levels of hair zinc in association with anorexia, hypogeusia and retarded growth. Small zinc supplements corrected taste acuity and hair zinc content. In low-income families in an American city they subsequently found nearly 40% had heights below the tenth centile. When compared with a group of children from middle-income families these children had significantly lower hair and plasma zinc levels. Small oral zinc supplements led to growth acceleration (HAMBIDGE et al. 1978). Thus, dietary zinc deficiency may be a common cause of poor growth in children of low-income families.

IV. Intravenous Feeding

Acute zinc deficiency may develop in patients nourished intravenously for long periods (see WEISMANN 1980) owing to their low zinc content, excessive urinary loss where amino acid solutions are used and an increased requirement for zinc. Trace element mixtures are now available for incorporation into intravenous feeding regimens.

V. Malabsorption and Inflammatory Bowel Disease

Low plasma zinc levels and impaired taste occur in patients with gluten enteropathy and may influence the response to a gluten-free diet (LOVE et al. 1978). Malabsorption of zinc and symptomatic deficiency occur in Crohn's disease (MCCLAIN et al. 1980) and growth retardation in children with chronic inflammatory bowel disease may be partly due to zinc deficiency (NISHI et al. 1980).

VI. Treatment with Chelating Agents and Diuretics

Chelating agents, especially D-penicillamine and edetates, have complex effects on zinc handling and both increases and decreases in gastrointestinal absorption as well as increased urinary zinc loss are reported. Clinical features suggestive of zinc deficiency are reported in patients receiving chelating agents (KLINGBERG et al. 1976) and increased urinary loss of zinc has been reported from diuretics (WEBSTER 1980), but it is not known whether this can lead to zinc deficiency.

VII. Burns and Skin Diseases

Low plasma zinc levels are frequently found in patients with burns and various skin diseases (GREAVES and BOYDE 1967; MORGAN et al. 1980). Total body zinc appears unchanged (HAWKINS et al. 1976) and it is therefore unlikely that these patients become symptomatically zinc deficient. However, the matter is not totally settled, especially in serious chronic skin disease in childhood with associated retardation of growth.

VIII. Dialysis

A low plasma zinc is common in patients on regular haemodialysis for chronic renal failure and impairment of taste and appetite may contribute to their persisting malnourishment (MAHAJAN et al. 1980). Its relationship to impaired sexual function is disputed (ANTONIOU et al. 1977; BROOK et al. 1980).

IX. Alcoholism and Cirrhosis

Alcoholics have an increased urinary zinc loss (RUSSELL 1980) and an inadequate dietary intake. Symptomatic zinc deficiency has been described in them (WEISMANN et al. 1980).

C. Zinc and Wound Healing

Evidence that zinc accelerates wound healing is conflicting (PORIES et al. 1967; Barcia 1970; Husain 1969; Greaves and IVE 1972) as is the relationship of response to serum zinc concentration ((HALLBOOK and LANNER 1972; PHILLIPS et al. 1977). A major problem is the unreliability of the serum zinc value as an index of tissue zinc. The consensus view is that tissue zinc deficiency delays healing, and that zinc therapy is helpful only in patients deficient in zinc. Treatment is justified when there are clinical grounds for suspecting that a patient with delayed healing may be zinc deficient and the serum zinc concentration is low.

D. Zinc and Acne Vulgaris

Eruptions resembling acne vulgaris and exacerbations of pre-existing lesions have been reported during a zinc-deficient diet (BAER et al. 1978). However, serum zinc has mostly been found to be normal in acne and therapeutic trials of oral zinc in acne vulgaris have not yielded convincing results (see CUNLIFFE et al. 1979; VERMA et al. 1980).

E. Zinc and Herpes Simplex

Zinc ions irreversibly inhibit replication of herpes viruses by a selective effect on their DNA polymerase (see FRIDLENDER et al. 1978). Topical zinc sulphate reduced the degree of vaginal inflammation, the incidence of encephalitis and mortality following vaginal inoculationof immature virgin mice with herpes virus (TENNICAN et al. 1980). Encouraging results were reported from an open study of topical 0.5% zinc sulphate for recurrent labial and genital herpes simplex in humans (WAHBA 1980) and in the prevention of recurrent cutaneous herpes simplex by application of 0.025–0.05% zinc sulphate solution during and after an acute attack (BRODY 1981). Controlled studies of this safe and simple treatment are now required.

F. Possible Mechanisms of the Manifestations of Zinc Deficiency

Remarkably little insight has been gained into the mechanisms of the cutaneous manifestations of zinc deficiency although it has many effects.

I. Immunological Responsiveness

Zinc has a role in immune responsiveness (see GOOD et al. 1979) and the eczematous eruption of acute zinc deficiency at mucocutaneous junctions is suggestive of impaired local resistance to microorganisms originating in the gastrointestinal and respiratory tracts and is similar to *Candida albicans* infections in severely immunocompromised patients.

II. Essential Fatty Acid Metabolism

There are similarities between the manifestations of zinc deficiency and those of essential fatty acid (EFA) deficiency and parenteral EFA supplementation is said to reverse the effects of zinc deficiency (CUNNANE and HORROBIN 1980). Some of the effects of zinc deficiency may result from the structural role of EFA in cell membranes or as precursors for prostaglandin synthesis.

III. Cell Membrane Stability

Zinc has an effect on the functional integrity of biological membranes (CHVAPIL 1976) and may protect them from free radical oxidation (LANCET 1978).

IV. Vitamin A Metabolism

Zinc is important for vitamin A metabolism (SOLOMONS and RUSSELL 1980) and vitamin A is important for keratinocyte function.

V. Buccal Epithelial Proliferation

The proliferation of buccal epithelium and alteration in number and distribution of epithelial membrane-coating granules (ASHRAFI et al. 1980) in zinc-deficient experimental animals has some analogies with that observed in psoriatic epidermis.

VI. Collagen

Zinc is required for collagen synthesis and for the activity of human skin collagenase (Seltzer et al. 1977) and may in this way be related to wound healing.

References

Aggett PJ, Harris JT (1979) Current status of zinc in health and disease states. Arch Dis Child 54:909–918

Aggett PJ, Atherton DJ, More J, Davey J, Delves HT, Harries JT (1980) Symptomatic zinc deficiency in a breast-fed pre-term infant. Arch Dis Child 55:547–550

Antoniou LD, Shalhoub RJ, Sudhaker T, Smith JC Jr (1977) Reversal of uraemic impotence by zinc. Lancet 2:895–898

Ashrafi SH, Meyer J, Squier CA (1980) Effects of zinc deficiency on the distribution of membrane-coating granules in rat buccal epithelium. J Invest Dermatol 74:425–432

Atherton DJ, Miller DPR, Aggett PJ, Harries JT (1979) A defect in zinc uptake by jejunal biopsies in acrodermatitis enteropathica. Clin Sci 56:505–507

Baer MT, King JC, Tamura T, Margen S (1978) Acne in zinc deficiency. Arch Dermatol 114:1093

Barcia PJ (1970) Lack of acceleration of healing with zinc sulphate. Ann Surg 172:1048–1050

Brody I (1981) Topical treatment of recurrent herpes simplex and post-herpetic erythema multiforme with low concentrations of zinc sulphate solution. Br J Dermatol 104:191–194

Brook AC, Johnston DG, Ward MK, Watson MJ, Cook DB, Kerr DNS (1980) Absence of a therapeutic effect of zinc in the sexual dysfunction of haemodialysed patients. Lancet 2:618–620

Chesters JK (1980) Biochemical functions of zinc in animals. World Rev Nutr Diet 32:135–164

Chvapil M (1976) Effects of zinc on cells and biomembranes. Med Clin North Am 60:799–812

Cousins RJ, Smith KT (1980) Zinc binding properties of bovine and human milk in vitro: influence of changes in zinc content. Am J Clin Nutr 33:1083–1087

Cunliffe WJ, Burke B, Dodman B, Gould DJ (1979) A double-blind trial of zinc sulphate/citrate complex and tetracycline in the treatment of acne vulgaris. Br J Dermatol 101:321–325

Cunnane SC, Horrobin DF (1980) Parenteral linoleic and γ-linoleic acids ameliorate the gross effects of zinc deficiency. Proc Soc Exp Biol Med 164:583–588

Dauncey J, Shaw JCL, Urman J (1977) The absorption and retention of magnesium, zinc and copper by low birth weight infants fed pasteurised human milk. Pediatr Res 11:991–997

Evans GW, Johnson PE (1980) Characterization and quantitation of zinc-binding ligand in human milk. Pediatr Res 14:876–880

Fridlender B, Chejanovsky N, Becker Y (1978) Selective inhibition of herpes simplex virus type I DNA polymerase by zinc ions. Virology 84:551–554

Good RA, Fernandez G, West A (1979) Nutrition, immunity and cancer – a review. Part I: influence of protein or protein-calorie malnutrition and zinc deficiency on immunity. Clin Bull 9:3–12

Greaves MW, Boyde TRC (1967) Plasma zinc concentrations in patients with psoriasis, dermatoses and venous leg ulceration. Lancet 2:1019–1020

Greaves MW, Ive FA (1972) Double-blind trial of zinc sulphate in the treatment ochronic venous leg ulceration. Br J Dermatol 87:632–634

Hallbook T, Lanner E (1972) Serum zinc and healing of venous leg ulcers. Lancet 2:780–782
Hambidge KM, Chaves MN, Brown RM, Walravens PA (1978) Zinc supplementation of low-income pre-school children. In: Kirchgessner M (ed) Trace element metabolism in man and animals, vol 3. Technische Universität, München, Freising-Weihenstephan, pp 296–299
Hawkins T, Marks JM, Plummer VM, Greaves MW (1976) Whole body monitoring and other studies of zinc 65 metabolism in patients with dermatological diseases. Clin Exp Dermatol 1:243–252
Hooper PL, Visconti L, Garry PH, Johnson GE (1980) Zinc lowers high density protein-cholesterol levels. J A M A 244:1960–1961
Husain SL (1969) Oral zinc sulphate in leg ulcers. Lancet 1:1069–1071
Klingberg WG, Prasad AS, Oberleas D (1976) Zinc deficiency following penicillamine therapy. In: Prasad AS, Oberleas D (eds) Trace elements in human helath and diseae, vol 1. Academic, New York, pp 51–65
Lancet (1978) Editorial: A radical approach to zinc. Lancet i:191–192
Love AHG, Elmes M, Golden MJ, McMaster D (1978) Zinc deficiency and coeliac disease. In: Kirchgessner (ed) Trace element metabolism in man and animals. Technische Universität München, Freising-Weihenstephan, pp 357–362
Mahajan SK, Prasad AS, Lambujon J, Abbasi AA, Briggs WA, McDonald FD (1980) Improvement of uremic hypogeusia by zinc: a double-blind study. Am J Clin Nutr 33:1517–1521
McClain C, Soutor C, Zieve L (1980) Zinc deficiency: a complication of Crohn's disease. Gastroenterology 78:272–279
Michaelsson G, Ljunghall K, Danielson BG (1980) Zinc in epidermis and dermis in healthy subjects. Acta Dermato Venereol (Stockh) 60:295–299
Morgan MEI, Hughes MA, McMillan EM, King I, Mackie RM (1980) Plasma zinc in psoriasis in patients treated with local zinc applications. Br J Dermatol 102:579–583
Moynahan EJ (1974) Acrodermatitis enteropathica: a lethal inherited human zinc deficiency disorder. Lancet 1:399–400
Murphy JV (1970) Intoxication following ingestion of elemental zinc. J A M A 212:2119–2120
Nishi Y, Lifshitz F, Bayne MA, Daum F, Silverberg M, Aiges H (1980) Zinc status and its relation to growth retardation in children with chronic inflammatory bowel disease. Am J Clin Nutr 33:2613–2621
Oberleas D, Harland BF (1977) Nutritional agents which affect metabolic zinc status. In: Brewer GJ, Prasad AS (eds) Zinc metabolism: current aspects in health and disease. Liss, New York, pp 11–24
Phillips A, Davidson M, Greaves MW (1977) Venous leg ulceration: evaluation of zinc treatment, serum zinc and rate of healing. Clin Exp Dermatol 2:395–399
Piletz JE, Ganschow RW (1978) Lethal milk mutation results in dietary zinc deficiency in nursing mice. Am J Clin Nutr 31:560–562
Pories WJ, Henzel JH, Rob CG, Strain WH (1967) Acceleration of wound healing in man with zinc sulphate given by mouth. Lancet 1:121–124
Prasad AS, Brewer GJ, Schoomaker EB, Rebbani P (1978) Hypocupraemia induced by zinc therapy in adults. J A M A 240:2166–2168
Russell RM (1980) Vitamin A zinc metabolism in alcoholism. Am J Clin Nutr 33:2741–2749
Sandstead HH (1973) Zinc nutrition in the United States. Am J Clin Nutr 26:1251–1260
Seltzer JL, Jeffrey JJ, Eisen AZ (1977) Evidence for mammalian collagenases as zinc ion metalloenzymes. Biochim Biophys Acta 485:179–187
Solomons NW, Russel RM (1980) The interaction of vitamin A and zinc: implications for nutrition. Am J Clin Nutr 33:2031–2040
Tennican P,, Carl G, Frey J, Thies L, Chvapil M (1980) Topical zinc in the treatment of mice infected intravaginally with herpes genitalis virus. Proc Soc Exp Biol Med 164:593–597

Thorn JM, Aggett PJ, Delves HT, Clayton BE (1978) Mineral and trace element supplement for use with synthetic diets based on comminuted chicken. Arch Dis Child 53:931–938
Verma KC, Saini AS, Dhamija SK (1980) Oral zinc sulphate therapy in acne vulgaris: a double-blind trial. Acta Derm Venereol (Stockh) 60:337–340
Wahba A (1980) Topical application of zinc solutions: a new treatment for herpes simplex infections in the skin. Acta Derma Venereol (Stockh) 60:175–177
Walravens PA, Hambidge KM (1976) Growth of infants fed a zinc supplemented formula. Am J Clin Nutr 29:1114–1121
Webster PO (1980) Urinary zinc excretion during treatment with different diuretics. Acta Med Scand 208:209–212
Weismann K (1980) Zinc metabolism and the skin. In: Rook, Savin (eds) Recent advances in dermatology, vol 5. Churchill Livingstone, Edinburgh, pp 109–129
Weismann K, Hoyer H, Christensen E (1980) Acquired zinc deficiency in alcoholic cirrhosis: report of 2 cases. Acta Derm Venereol (Stockh) 60:447–449
Widdowson EM, Dauncey J, Shaw JCL (1974) Trace elements in foetal and early postnatal development. Proc Nutr Soc 33:275–284
Zimmermann AW, Hambidge KM, Lepow ML, Greenberg RD, Stover ML, Casey CE (1982) Acrodermatitis in breast fed premature infants: evidence for a defect of mammary zinc secretion. Pediatrics 69:176–183

CHAPTER 41

The Ichthyoses

R. MARKS

A. Introduction

The ichthyoses are a heterogeneous group of skin disorders characterised by persistent and generalised scaling without inflammation (MARKS and DYKES 1978). They are either primarily genetic disorders or acquired as a response to a variety of diseases, particularly malabsorption and neoplasia. The metabolic faults underlying most of the diseases in this group are still obscure. So far none has been shown to be a primary disorder of keratin synthesis. A defect of lipid synthesis appears to be common in many of the ichthyoses (SUMMERLY and YARDLEY 1967; COOPER et al. 1980; DYKES et al. 1978) and this defect is primary in essential fatty acid (EFA) deficiency (Chap. 4, Vol. 87/I) (PROTTEY 1976), Refsum's syndrome (DAVIES et al. 1977; REYNOLDS et al. 1978), sex-linked ichthyoses in which all cells are deficient in a steroid sulphatase (SHAPIRO et al. 1978; WILLIAMS and ELIAS 1981) and after drugs given to reduce lipid synthesis such as nicotinic acid (PARSONS and FLYNN 1959), butyrophenones (SIMPSON et al. 1964) and triparanol (WINKELMANN et al. 1963). Increased intracorneal cohesion has been found in ichthyosis and all scaling disorders (MARKS et al. 1981) and the increase parallels the severity of the ichthyotic state. The increased binding forces between corneocytes explains the scaling as cells desquamate in clumps rather than individually.

B. Treatment of the Ichthyoses

I. Treatment Directed Towards an Underlying Cause

This is possible only in the acquired ichthyoses and Refsum's syndrome. Ablative or adequately suppressive therapy for a neoplastic disorder responsible for acquired ichthyosis will improve the skin as will stopping the particular lipid lowering drug or correcting EFA deficiency by surgical or medical treatment of the underlying bowel condition or by topical EFA (see Chap. 4, Vol. 87/I).

In *Refsum's syndrome* decreased α-hydroxylation (STEINBERG et al. 1967) allows phytanic acid from green vegetables to accumulate in the blood and lipid fractions of most tissues (REYNOLDS et al. 1978) whilst the concentrations of arachidonic, linoleic and linolenic acids fall. Presumably this derangement of lipid metabolism results in the neural degeneration and ichthyosis which may be analogous to EFA deficiency. Improvement follows dietary exclusion of green vegetables, a major source of materials broken down to phytanates and which cannot be further metabolised in this condition (SAHGAL and OLSEN 1975).

The ichthyosis of *EFA deficiency* is corrected by treating its cause, by dietary supplements or by topical application of EFA (see Chap. 4, Vol. 87/I). The essential fatty acids are long chain unsaturated compounds that are needed in the diet because they are not synthesised in the body. EFA deficiency occurs in infants who have received diets containing low levels of linoleic acid (HANSEN et al. 1958) and in adults with small bowel disease or intestinal bypass operations, e.g. for obesity (PRESS et al. 1974). In this deficiency state 5,8,11-eicosatrienoic acid, an abnormal fatty acid metabolite, accumulates in the blood. The condition can be reproduced in animals (PROTTEY 1976) where it is characterised by epidermal hyperplasia (LOWE and STOUGHTON 1977). Its relationship to prostaglandins (see Chap. 4, Vol. 87/I) and to zinc deficiency (see Chap. 40) have not been resolved.

II. Symptomatic Treatments

1. General Management

Environmental conditions which dry the stratum corneum with cracking from inflexibility lead to chapping and fissuring and should be avoided in ichthyotic patients. The most important situations are a low ambient humidity, most commonly from central heating without humidification, and a low ambient temperature and high air flow. Frequent wetting of the skin, particularly with hot water containing detergents or soaps, and subsequent drying by evaporative loss will likewise make the stratum corneum brittle so that it cracks during movement. In patients with ichthyosis this sequence can lead to considerable disability and apart from discomfort the damaged horny layer predisposes the skin to infection, dermatitis and itch (see Chap. 7).

2. **Retinoids** (see Chap. 25)

Despite their disadvantages retinoids are uniquely effective for some severe ichthyotic disorders and long-term treatment with etretinate should be considered in all patients who are severely disabled by ichthyosis.

3. Topical Therapies

a) Emollients

These agents improve the appearance, decrease the discomfort and improve the function of skin affected by ichthyosis. Their application provides an occlusive oily film which increases hydration of the abnormal stratum corneum. Since hydration is the main determinant of the brittleness it allows a greater range of movement and diminishes painful cracking and fissuring. The appearance of the skin also improves because the application reduces light refraction by the scales (FARR et al. 1984). Many emollient preparations are available, based on mineral oils of the paraffin hydrocarbon type, lanolin or vegetable oils. There is surprisingly little evidence on comparative effect and since patients have unpredictable and clear. but as yet unexplained, preferences, in practice emollients are chosen by the patient's trial and error of preparations they find comfortable and are prepared to use. Because they are soon rubbed off, the action of emollients is short lived, and to obtain the optimum effect they should be used as frequently as necessary to maintain a cover of severely affected areas. A convenient way of applying emollients is as a bath oil. These are discussed in Chap. 32.

b) Osmotic Agents

Water retention by the stratum corneum can be achieved osmotically (MIDDLETON 1969) and 15% urea has been used for this (SWANBECK 1968) together with 1% hydrocortisone where there is associated eczema (see Chap. 32). More controlled stuides of both the physical and therapeutic effects of this treatment are required.

c) Keratolytic (Desquamatory) Agents

The term keratolytic is misleading as there is no evidence that any of the materials in this class of compounds lyse keratin. Their action is to enhance desquamation although how they do this is uncertain. Their main use in the ichthyotic disorders is to reduce the thick hyperkeratotic scales and plaques thereby optimising the effect of emollients in smoothing and softening the skin surface. Salicylic acid is the most useful and most frequently employed desquamatory agent. Concentrations of 1%–6% are usual, but up to 20% may be used on thick warty areas. It is formulated in a variety of vehicles, mostly ointments, but seems particularly effective when 70% propylene glycol is included in the preparation although critical trials of formulations have not been done. Other desquamatory agents used topically in ichthyosis include organic hydroxy acids such as lactic, pyruvic and tartaric acids as well as the retinoids, the best known of which is all-*trans*-retinoic acid (tretinoin). These agents are discussed in more detail.

d) Essential Fatty Acids

Sunflower seed oil has been found to be effective topically (see Chap. 4, Vol. 87/I) in EFA deficiency (PROTTEY 1976), but the few clinical trials done with sunflower and evening primrose seed oils topically have shown no effect in various other ichthyoses.

References

Cooper MF, Wilson PD, Hartop PJ, Shuster S (1980) Acquired ichthyosis and impaired dermal lipogenesis in Hodgkin's disease. Br J Dermatol 102:689–693
Davies MG, Marks R, Dykes PJ, Reynolds DJ (1977) Epidermal abnormalities in Refsum's disease. Br J Dermatol 97:401–406
Dykes PJ, Marks R, Smith P (1978) Profile of epidermal metabolic activity in autosomal dominant ichthyosis and small bowel disorders. Br J Dermatol 98:611–618
Farr PM, Diffey BL, Steele MC (1984) A preliminary study on the in vivo transmission of light through psoriatic plaques. Photodermatol 1:87–90
Hansen AE, Haggard ME, Boelsche AN, Adam DJD, Wiese MF (1958) Essential fatty acids in infant nutrition. III. Clinical manifestations of linoleic acid deficiency. J Nutr 66:565–576
Lowe NJ, Stoughton RB (1977) Essential fatty acid deficient hairless mouse: a model of chronic epidermal hyperproliferation. Br J Dermatol 96:155–162

Marks R, Dykes PJ (1978) The ichthyoses. MTP, Lancaster
Marks R, Finlay AY, Holt PJA (1981) Severe disorders of keratinization: effects of treatment with Tigason (etretinate). Br J Dermatol 104:667–674
Middleton JD (1969) The effect of temperature on extensibility of isolated stratum corneum and its relation to skin chapping. Br J Dermatol 81:717
Parsons WB, Flinn JH (1959) Reduction in serum cholesterol levels and β-lipoprotein cholesterol levels by nicotinic acid. Arch Intern Med 103:783–790
Press M, Kikuchi H, Shimoyama T, Thompson GR (1974) Diagnosis and treatment of essential fatty acid deficiency in man. Br Med J 2:247–250
Prottey C (1976) Essential fatty acids and the skin. Br J Dermatol 94:579–587
Reynolds DJ, Marks R, Davies MG, Dykes PJ (1978) The fatty acid composition of skin and plasma lipids in Refsum's disease. Clin Chim Acta 90:171–177
Sahgal V, Olsen WO (1975) Heredopathia atactica polyneuritiformis (phytanic acid storage disease): a new case with special reference to dietary treatment. Arch Intern Med 135:585–587
Shapiro LJ, Weiss R, Webster D, France JT (1978) X-linked ichthyosis due to steroid sulphatase deficiency. Lancet 1:70–72
Simpson GM, Blair JH, Cranswick EH (1964) Cutaneous effects of a new butyrophenone drug. Clin Pharmacol Ther 5:310–321
Steinberg D, Herndon JH, Uhlendorf BW, Mize CE, Avigan J, Milne GWA (1967) Refsum's disease: nature of the enzyme defect. Science 156:1740–1742
Summerly R, Yardley HJ (1967) Cholesterol synthesis in ichthyosis vulgaris. Br J Dermatol 79:378–385
Swanbeck G (1968) A new treatment of ichthyosis and other hyperkeratotic conditions. Acta Derma Venereol (Stockh) 48:123
Williams ML, Elias PM (1981) Stratum corneum lipids in disorders of cornification. J Clin Invest 68:1404–1410
Winkelmann RK, Perry HO, Achor RWP, Kirby TJ (1963) Cutaneous syndromes produced as side effects of triparanol therapy. Arch Dermatol 87:372–377

CHAPTER 42

Tropical Skin Diseases

G. H. RÉE

A. Introduction

Chronic infections – parasitic, bacterial or fungal – dominate tropical dermatology. All the diseases to be considered in this chapter are difficult to treat and their eventual control is likely to be difficult.

B. Onchocerciasis (River Blindness)

Onchocerciasis is a chronic filarial infection of skin, eye and subcutaneous tissues, transmitted by blackflies (*Simulium* spp.) and found in tropical Africa and in small foci in Central and South America. Though biological and biochemical variants of *Onchocerca volvulus* exist, these are of epidemiological rather than individual importance. The adult worm lives in and migrates through subcutaneous tissues. The microfilariae which emerge from the adult female worm are found especially in the skind and eyes. Itching is the commonest symptom and appears to be related to death of microfilariae. In chronic infections fibrosis, loss of elasticity, oedema and pigmentary changes occur in the skin. Diethylcarbamazine (DEC, Banocide, Hetrazan) and suramin (Antrypol, Germanin, Moranyl) are the drugs most widely used in the management of this disease.

DEC has no microfilaricidal action in vitro, but in vivo stimulates rapid mobilization and redistribution of microfilariae and permits host killing of these by histiocytes and eosinophils. The mechanism of this process is uncertain, but may be through an action on the microfilarial cuticle and the subsequent unmasking of antigenic determinants with which antibody and phagocytes can react. Treatment with DEC is usually accompanied by a severe exacerbation of symptoms known as the Mazzotti reaction. Within a few hours of the first dose of DEC, swelling and oedema of the skin, intense pruritus, fever, tachycardia and headache occur. The blood pressure may fall and respiratory distress may be precipitated. The immunological mechanisms that underlie the reaction are uncertain though activation of complement and the release of vasoactive agents seem likely. DEC is rapidly absorbed from the gastrointestinal tract and peak serum concentrations are found after 1–2 h. Some DEC is metabolized though none of the metabolites are microfilaricidal. In an attempt to reduce the severity of the Mazzotti reaction DEC has been given transepidermally as a 1% or 2% lotion. The results of trials have shown that DEC can be absorbed through skin, that killing of microfilariae occurs and that severe reactions may still be precipitated. At present therefore DEC lotion should not be used in the management of patients with onchocerciasis. The major limitation of DEC is its inability to kill the adult female worm. Recurrence of symptoms is therefore likely.

Suramin, a complex naphthalene derivative, appears to be macrofilaricidal and offers a chance for a radical cure. It is given intravenously which limits its usefulness in control schemes. Furthermore, the drug is strongly bound to plasma proteins and a low concentration is maintained for several weeks. Immediate and cumulative toxic reactions can occur. After an intravenous injection vomiting and pruritus may occur and within 24 h photophobia, blindness and paraesthesia, especially of the hands and feet. Damage to the kidneys leads initially to a reversible albuminuria and aminoaciduria, but if treatment is continued, irreversible renal damage results. It is claimed that small doses of suramin have equal parasitological results, but reduced toxicity (Hawking 1981).

Ivermectin is a semi-synthetic macrocyclic lactone with broad anti-parasitic activity which was originally introduced for veterinary use. A number of trials, both open and blind, have shown that ivermectin is an effective microfilaricide in onchocerciasis which induces significantly less reaction than DEC and does not aggravate any of the eye lesions. Furthermore, the parasite count (microfilaria per milligram skin) remain at a significantly lower level 6–12 months later than following treatment with DEC.

Ivermectin is not lethal to adult worms, but appears to affect embryogenesis in the female adult *O. volvulus*. A major advantage of ivermectin over DEC and suramin is that the drug is given in a single dose, in contrast to the multiple doses of the other drugs. It has been shown to be safe in mass chemotherapeutic trials in West Africa and is being used as an adjunct to conventional control schemes. Drugs of the benzimidazole class show some promise as anti-onchocercal agents, though none is yet entirely satisfactory.

C. Cutaneous Leishmaniasis

The cutaneous leishmaniases (CL) are a heterogeneous group of cutaneous sores caused by diverse species of *Leishmania*. All are transmitted by species of sandfly. Most cutaneous leishmaniases are self-limiting and followed by resistance to reinfection. Mucosal extension after apparent self-cure is characteristic of the Brazilian form of the disease. Diffuse cutaneous leishmaniasis (DCL) is a form of CL in which cell-mediated immune responses to infection do not occur, a clear pathological analogy with lepromatous leprosy. In endemic areas of CL most patients are never treated. In temperate climates patients with CL are usually visitors who are not prepared to wait for natural cure. Treatment does prevent the development of scarring and is usually given with sodium antimony gluconate (Pentostam) a pentavalent antimony compound. This drug is given parenterally, may rarely cause anaphylaxis and is less toxic than the trivalent antimony compounds used in the treatment of schistosomiasis. The usual adult dose is 20 mg antimony per kilogram body weight (maximum 850 mg) i.v. daily for 10–30 days. The drug is rapidly cleared by the kidneys and has virtually no accumulation in any organ. Doses of Pentostam in excess of 20 mg antimony kg may be toxic. Old World

DCL does not respond to Pentostam. Treatment with a diamidine such as pentamidine is required and may need to be repeated for several courses. The usual adult dose of pentamidine is 3–4 mg base per kilogram body weight i.m. once or twice a week for several months. It binds to liver and kidney and is only slowly excreted, mainly in the urine. Pentamidine may lead to disturbances of carbohydrate metabolism and an insulin-sensitive diabetes has been described in up to 10% of patients treated with it. Accidental intravenous injection may lead to hypotensive collapse. Cycloguanil pamoate, rifampicin, metronidazole and other drugs have been used for CL, but there is little evidence of benefit. Amphotericin B has been used in the treatment of mucocutaneous CL which has not responded to other treatments.

D. Leprosy

Leprosy is a chronic infection of the skin and nerves caused by *Mycobacterium leprae*. The slow evolution of the disease is attributable to the slow generation time of the organism which has been estimated to be 8–15 days during the logarithmic multiplication phase in mouse foot-pads. The clinical spectrum reflects the immune response of the host. Most of those infected produce an excellent response and never develop a recognized clinical illness. The major therapeutic problem in leprosy relates to the development of dapsone resistance. Secondary dapsone resistance occurs in patients with multibacillary leprosy treated with dapsone for at least 5 years. Monotherapy, with low and irregular dosage all encourage secondary resistance. Primary dapsone resistance results from the transmission of *M. leprae* from patients with secondary resistance whose relapse was not recognized and adequately treated. The only way to prevent the spread of dapsone resistance is to use multiple drug therapy in order to eliminate any dapsone-resistant organisms present and to prevent the emergence of such organisms. The four major drugs in use for leprosy today are: clofazimine, rifampicin, dapsone, and prothionamide.

Clofazimine is a reddish-brown riminophenazine derivative having both anti-bacterial and anti-inflammatory effects: its anti-bacterial effects are not affected by sulphone resistance. After oral administration, clofazimine is well absorbed and is particularly taken up by fat cells and macrophages. A brownish discoloration of skin affecting particularly diseased skin and skin exposed to direct sunlight occurs in most patients treated with the drug. Though accumulation and crystallization of clofazimine occurs in many organs, including the heart, liver, kidneys, pancreas, lymph glands and gastrointestinal tract, the main side effects have been on the gastrointestinal tract. Dyspepsia, anorexia, nausea and diarrhoea have all been described. Clofazimine crosses the placenta and gives the characteristic brownish discoloration to babies and also to maternal milk – the latter often a source of maternal anxiety (HASTINGS et al. 1976).

Rifampicin is well absorbed after oral administration and has a rapid bactericidal action. Blood levels tend to be lower after the patient has received the drug for several weeks; there is therefore a pharmacological basis for intermittent therapy.

Intermittent therapy may however lead to a number of rare side effects of which the commonest is the "flu" syndrome. Fever, chills and aches and pains occur shortly after a dose of rifampicin and symptoms last for several hours. This syndrome occurs most commonly when rifampicin is given once weekly and has not been reported with once-monthly schedules. Other complications of intermittent therapy include shock, haemolysis and renal failure. Mycobacterial resistance to rifampicin has been reported.

Ethionamide and prothionamide have bacteriostatic and bactericidal effects and may have special value in destroying dormant, intracellular organisms.

Dapsone is a well-tolerated drug, well absorbed after oral administration and largely excreted via the urine. It is bacteriostatic rather than bactericidal. Some tissue accumulation occurs in experimental animals and tissue levels are usually slightly higher than concurrently determined plasma levels. There is an impressively long list of side effects, some of which (e.g. haemolysis) are dose dependent, some idiosyncratic (e.g. Stevens-Johnson syndrome, agranulocytosis, liver damage). *Mycobacterium leprae* appears to find, in the Schwann cells of peripheral nerves, a protected environment in which the organism can multiply, safe from immunological attack. Dapsone-sensitive organisms can be found in peripheral nerves even several years after adequate dapsone therapy, suggesting that the drug may not penetrate into the endoneurium of peripheral nerves (Committee on Experimental Chemotherapy 1976).

The course of treatment of leprosy is punctuated and complicated by the occurrence of reactions (lepra reactions). Type I (reversal) reactions occur particularly in tuberculoid and borderline leprosy and reflect changes in cell-mediated immune responses. Type II reactions occur particularly in multibacillary leprosy and are the result of immune complex vasculitis. The clinical manifestations of type II reactions are protean, including erythema nodosum leprosum, iritis, neuritis, arthritis, fever, albuminuria and others. A variety of drugs have been used in the management of these distressing features, but thalidomide offers a unique treatment. This drug appears to shorten the duration of the reaction, allows the continued use of anti-leprosy drugs in maximum doses and shortens the period of hospitalization. The mechanism of action of thalidomide in type II reactions is unknown, but may be related to its teratogenicity as only teratogenic derivatives are active against reactions. The usual management is to start treatment with 100 mg three times a day, then to reduce the dose gradually to a maintenance dose of 50 or 100 mg daily as symptoms improve. The dangers of this drug in women of reproductive years do not need to be emphasized. Steroids should be reserved in leprosy for the treatment of acute neuritis only.

E. Yaws

Yaws is a chronic disease of skin and bone caused by *Treponema pertenue*, an organism serologically and morphologically identical to, but biologically distinct from *T. pallidum*. The hope that yaws had been eradicated by the extensive penicillin campaigns of the 1950s has been frustrated by the reappearance of this disease, particularly in West Africa. As before, yaws occurs particularly in sparsely clad children living in areas of high ambient temperatures and humidity. Yaws remains highly responsive to penicillin. Benzathine penicillin 1.2 or 2.4×10^6 units as a single i.m. dose will give blood levels of about 0.015 units/ml on day 1 to 0.03 units/ml on day 14. As the treponeme is sensitive to blood levels of 0.01 units/ml this single injection is generally curative. Benzathine penicillin has all the complications of crystalline penicillin. HACKETT and GUTHE (1956) stated that "the ultimate eradication [of yaws] must depend upon the removal of factors that favour the transmission of the disease. While treatment with penicillin ... rapidly reduces its public health and economic importance, further and continued action is needed to consolidate that initial achievement." These points were generally ignored and yaws inevitably returns. While clinicians in temperate climates are unlikely to see any cases of early yaws, the serology of the disease may cause confusion with syphilis. If a symptomatic individual from a yaws endemic region is found to be serologically positive it is wise to assume that latent syphilis is the cause and to treat accordingly.

References

Committee on Experimental Chemotherapy (1976) Experimental chemotherapy in leprosy. Bull WHO 53:425–433

Hackett CJ, Guthe T (1956) Some aspects of yaws eradication. Bull WHO 15:869–896

Hastings RC, Jacobson RR, Trautman JR (1976) Longterm clinical toxicity studies with clofazimine (B663) in leprosy. Int J Lepr 44:287–293

Hawking F (1981) Chemotherapy of filariasis. Antibiot Chemother 30:135–162

Subject Index

Handbook of Experimental Pharmacology

Volume 51
Uric Acid

Volume 52
Snake Venoms

Volume 53
Pharmacology of Ganglionic Transmission

Volume 54: Part 1
Adrenergic Activators and Inhibitors I

Part 2
Adrenergic Activators and Inhibitors II

Volume 55
Psychotropic Agents

Part 1
Antipsychotics and Antidepressants

Part 2
Anxiolytics, Gerontopsychopharmacological Agents and Psychomotor Stimulants

Part 3
Alcohol and Psychotomimetics, Psychotropic Effects of Central Acting Drugs

Volume 56, Parts 1 + 2
Cardiac Glycosides

Volume 57
Tissue Growth Factors

Volume 58
Cyclic Nucleotides

Part 1
Biochemistry

Part 2
Physiology and Pharmacology

Volume 59
Mediators and Drugs in Gastrointestinal Motility

Part 1
Morphological Basis and Neurophysiological Control

Part 2
Endogenous and Exogenous Agents

Volume 60
Pyretics and Antipyretics

Volume 61
Chemotherapy of Viral Infections

Volume 62
Aminoglycoside Antibiotics

Volume 63
Allergic Reactions to Drugs

Volume 64
Inhibition of Folate Metabolism in Chemotherapy

Volume 65
Teratogenesis and Reproductive Toxicology

Volume 66
Part 1: **Glucagon I**

Part 2: **Glucagon II**

Springer-Verlag
Berlin Heidelberg
New York London
Paris Tokyo Hong Kong

Springer

Handbook of Experimental Pharmacology

Springer-Verlag
Berlin Heidelberg
New York London
Paris Tokyo Hong Kong

Volume 67
Part 1
Antiobiotics Containing the Beta-Lactam Structure I

Part 2
Antibiotics Containing the Beta-Lactam Structure II

Volume 68, Parts 1 + 2
Antimalarial Drugs

Volume 69
Pharmacology of the Eye

Volume 70
Part 1
Pharmacology of Intestinal Permeation I

Part 2
Pharmacology of Intestinal Permeation II

Volume 71
Interferons and Their Applications

Volume 72
Antitumor Drug Resistance

Volume 73
Radiocontrast Agents

Volume 74
Antieliptic Drugs

Volume 75
Toxicology of Inhaled Materials

Volume 76
Clinical Pharmacology of Antiangial Drugs

Volume 77
Chemotherapy of Gastrointestinal Helminths

Volume 78
The Tetracycline

Volume 79
New Neuromuscular Blocking Agents

Volume 80
Cadmium

Volume 81
Local Anesthetics

Volume 82
Radioimmunoassay in Basic and Clinical Pharmacology

Volume 83
Calcium in Drug Actions

Volume 84
Antituberculosis Drugs

Volume 85
The Pharmacology of Lymphocytes

Volume 86
The Cholinergic Synapse

Volume 87
Part 1
Pharmacology of the Skin I

Part 2
Pharmacology of the Skin II

Volume 88
Drugs for the Treatment of Parkinson's Disease

Volume 89
Antiarrhythmic Drugs

Volume 90
Part 1
Catecholamines I

Part 2
Catecholamines II

Springer

MIX
Papier aus verantwortungsvollen Quellen
Paper from responsible sources
FSC® C105338

If you have any concerns about our products,
you can contact us on
ProductSafety@springernature.com

In case Publisher is established outside the EU,
the EU authorized representative is:
Springer Nature Customer Service Center GmbH
Europaplatz 3, 69115 Heidelberg, Germany

Printed by Libri Plureos GmbH
in Hamburg, Germany